Biofuel and Biorefinery Technologies

Volume 12

This book series provides detailed information on recent developments in biofuels & bioenergy and related research. The individual volumes highlight all relevant biofuel production technologies and integrated biorefinery methods, describing the merits and shortcomings of each, including cost-efficiency. All volumes are written and edited by international experts, academics and researchers in the respective research areas.

Biofuel and Biorefinery Technologies will appeal to researchers and post-graduates in the fields of biofuels & bioenergy technology and applications, offering not only an overview of these specific fields of research, but also a wealth of detailed information.The ideas and technologies presented in this book series contribute to the UN Sustainable Development Goal 7: *Affordable and Clean Energy*.

Carlos Ricardo Soccol •
Gonçalo Amarante Guimarães Pereira •
Claude-Gilles Dussap •
Luciana Porto de Souza Vandenberghe
Editors

Liquid Biofuels: Bioethanol

Springer

Editors
Carlos Ricardo Soccol
Department of Bioprocess Engineering and Biotechnology
Federal University of Paraná
Curitiba, Brazil

Gonçalo Amarante Guimarães Pereira
Laboratory of Genomics and BioEnergy, Institute of Biology, Department of Genetics, Evolution and Bioagents
UNICAMP
Campinas-SP, Brazil

Claude-Gilles Dussap
Institut Pascal, Université Clermont Auvergne, CNRS, Clermont Auvergne INP, UMR6602
Clermont-Ferrand, France

Luciana Porto de Souza Vandenberghe
Department of Bioprocess Engineering and Biotechnology
Federal University of Paraná
Curitiba, Brazil

ISSN 2363-7609 ISSN 2363-7617 (electronic)
Biofuel and Biorefinery Technologies
ISBN 978-3-031-01243-3 ISBN 978-3-031-01241-9 (eBook)
https://doi.org/10.1007/978-3-031-01241-9

This Springer imprint is published by the registered company Springer Nature Switzerland AG
The registered company address is: Gewerbestrasse 11, 6330 Cham, Switzerland

Preface

Clean and renewable alternatives have been searched to reduce the dependence on petroleum fossil fuels. In this context, biofuels appear as a sustainable way to reach these goals. Since the 1970s, bioethanol has been produced from different feedstocks including mainly sugarcane and corn. The bioethanol industry has boosted not only the development of new bioprocesses technologies, but also the automotive industry with the fabrication of flex fuels vehicles. The bioethanol-processing chain appears as one of the most important, sustainable, and low-cost solutions for biofuel, bioenergy, and biochemicals production, together with the establishment of new biorefinery concepts and circular bioeconomy. The exact impact of this contribution is difficult to determine, as calculations depend on their entire life cycle, which depends on several direct (raw material processing, transportation, operational units employed, among others) and indirect (land use change, for example) factors. However, estimates based on Renewable Fuels Association (RFA) data have shown that worldwide emissions of 500 million tonnes of CO_2 equivalent have been avoided due to the replacement of gasoline by ethanol in vehicle combustion engines (Sydney et al. 2018). Of this amount, 56% corresponds to ethanol produced from sugarcane, about 26% to ethanol from corn and 18% to other feedstocks.

Therefore, the introduction of new low-carbon technologies, with the concomitant conversion of sugars from lignocellulosic material for 2G bioethanol production, and the development of high-biomass (energy), has created a new agro-industrial way. Furthermore, the 3G bioethanol production and the involved microalgae technology are also reported. This book presents all aspects about bioethanol production, including history, fundamentals, research involved, government incentives, contributions for CO_2 mitigation, recent advances, and innovations. It is divided into four sections: (I) ***First-generation ethanol***; (II) ***Second-generation ethanol***; (III) ***Third-generation ethanol***; and (IV) ***Economy, sustainability, and the future of bioethanol production***. The main employed feedstocks and environmental sustainability of bioethanol production, as long as the existent governmental incentives guarantee the viability and continuity of biofuels' programs, will be discussed. The present and future of bioethanol biorefinery in terms of the exploitation of feedstocks, with the integrated production of other biofuels, bioelectricity, biopolymers, organic acids, enzymes, and other biomolecules, as well as the use of process-generated liquid and solid by-products and/or wastes, will be described. Finally, as a differential, this book brings the very important aspects of bioethanol biorefineries—life cycle and sustainability, carbon credits and the bioethanol industry, governmental programs and incentives (Renovabio), ethanol motorization, development of hybrid motors and biofuels electrification, new technologies, patents, and innovation.

Curitiba, Brazil — Carlos Ricardo Soccol
Campinas-SP, Brazil — Gonçalo Amarante Guimarães Pereira
Clermont-Ferrand, France — Claude-Gilles Dussap
Curitiba, Brazil — Luciana Porto de Souza Vandenberghe

Reference

Sydney EB, Novak AC, Rosa D, ABP M, Brar SK, Larroche C, Soccol CR (2018) Screening and bioprospecting of anaerobic consortia for biohydrogen and volatile fatty acid production in a vinasse based medium through dark fermentation. Process Biochem 67:1–7. https://doi.org/10.1016/j.procbio.2018.01.012

Contents

Chapter 1
First Generation Bioethanol: Fundamentals—Definition, History, Global Production, Evolution

Emmanuel Bertrand and Claude-Gilles Dussap

Abstract Bioethanol is a fuel of plant origin obtained by the fermentation of simple sugars into alcohol. It can be used blended with fuels of fossil origin or almost pure in adapted vehicles (E-85, flex-fuel technology) particularly developed in Brazil. From a technological point of view, 3 generations coexist depending on the origin (from edible plants, lignocellulosic residues, or algae) of the sugars used for the fermentation. However, to date, the first generation still accounts for more than 95% of the 109 billion liters of ethanol available on the global market in 2019 despite many controversies about competition with foodstuffs and indirect changes in land use associated with the first generation. The obligation to reduce greenhouse gases emissions (GHG) to stay as close as possible to the 1.5 °C temperature raise objective by the end of the century that has been set in the 2015 Paris agreement, associated with the many changes in governments energy and mobility policies favoring electric vehicles in some parts of the world have led to a great uncertainty to foresee the future of this first-generation bioethanol technology.

1.1 Definition

Biofuels are alternative fuels produced from biomass, and almost exclusively used in the transportation sector in 2020. A few applications in the marine and aviation sectors are under development. Biofuels are blended up to 10–20% with fossil fuels to be used without any modification of the conventional thermal engines. When the

E. Bertrand (✉)
Biodiversité et Biotechnologie Fongiques, Aix-Marseille Université, INRAE, UMR1163, Marseille, France
e-mail: emmanuel.bertrand@univ-amu.fr

C.-G. Dussap
Institut Pascal, Université Clermont Auvergne, CNRS, Clermont Auvergne INP, UMR6602, Clermont-Ferrand, France

LabEx IMobS3, Université Clermont Auvergne, Clermont-Ferrand, France

C. R. Soccol et al. (eds.), *Liquid Biofuels: Bioethanol*, Biofuel and Biorefinery Technologies 12, https://doi.org/10.1007/978-3-031-01241-9_1

Table 1.1 Comparative assessment of the third generations bioethanol concerning their impact on the three pillars of sustainable development adapted from (DGRIS 2021) potential risks appear in bold font

Bioethanol	Environment	Society	Economy
First-generation	Increased risks of deforestation Increased use of pesticides and fertilizers Mitigation of greenhouse gases emissions	New outcomes possibilities for farmers Competition with food and feed	Very intensive mass production with possible scale economies Possibilities to valorize other co-products
Second generation	Lower risk of land-use changes	Diversification of raw materials Lower risk of competition with food and feed	Technology not fully deployed in the market because of prices volatility
Third generation	Very low risk of land-use change Increased need for phosphorus	Very low risk of competition with food and feed	The technology us not mature enough yet

volume of incorporation is very high, technical adaptations of the engines are necessary. Like conventional fuels, biofuels can be divided into two main categories: the gasoline sector includes bioethanol and *tert*-butyl ethyl ether (ETBE). They are manufactured from sugars to produce ethanol. The diesel sector includes biodiesels and hydrotreated vegetable oils (HVH). These fuels are manufactured from oils to form fatty acid esters. The production of biofuel has evolved into different generations, depending on the type of biomass used. The first generation bioethanol is derived from traditional agricultural products, such as corn starch in the United States, saccharose from sugarcane in Brazil, or beetroot in Europe. Therefore, the edible part of the plant is used to produce ethanol. Although their arrival on the market is later than expected, demonstrators are currently being industrialized to produce a second-generation using agro-industrial side products such as wheat straw, sugarcane bagasse, or other woody materials specially grown for this emerging industry (poplar, beechwood, pine, miscanthus). Finally, research and development work are currently underway to use emerging microorganisms such as algae that are possibly grown in non-arable land or even offshore. Following the 2008 food protests, a distinction between conventional and advanced biofuels was established by European and US regulations. The advanced biofuels, defined as the second and third generation ones, must be derived from agricultural feedstocks that do not threaten food security or pose risks in terms of land-use change (such as deforestation) (Table 1.1).

1.2 History

The rise of biofuels and bioethanol is tightly related to the industrial revolution in the nineteenth century. In the 1850s, ethanol was already being used as a fuel for street lighting and in 1876, Nicolaus Otto, the inventor of the four-stroke internal combustion engine, used ethanol to power his invention. In 1908, Henri Ford's Model T was the first industrially large-scale powered by ethanol vehicle. In the 1920s and 1930s, ethanol was combined with gasoline for the first time as an octane enhancer, and it was in high demand during World War II owing to fuel shortages. However, due to technical improvement of drilling methods to extract fossil oils and the discovery of huge petroleum resources at the beginning of the twentieth century (especially in Iran and the middle east), ethanol was considered economically less competitive than its fossil counterparts as a fuel for modern public and individual transportation and generally only used in a situation of shortages (Bajpai 2021). The revival of bioethanol occurred during the 1970s due to the two oil crises (Ayadi et al. 2016). In 1973, The Organization of Petroleum Exporting Countries (OPEC) caused gasoline shortages by suddenly raising its selling prices and delaying crude oil exportation to the United States of America and most of the Western countries. The OPEC move brought attention to the fact that most of the industrially developed countries were heavily dependent on foreign energy. The spotlight has turned once more to alternative fuels such as ethanol. Since the modest beginnings of a significant ethanol business in the 1970s, technology has progressed steadily, resulting in the development of lower-cost techniques and generating larger amounts of fuel ethanol while also being more efficient in their use of fossil fuel inputs. Brazil began a government-mandated ethanol production program in 1975 and has subsequently pushed the manufacture of flex-fuel vehicles (FFVs) and ethanol-only automobiles.

Today, more than 80% of Brazil's automobile production has flex-fuel capability and Brazil is the second world's largest producers of ethanol, thanks to its early start in ethanol production and its geographic advantage in cultivating sugar cane. Brazil is so efficient that a gallon of ethanol costs roughly a dollar to manufacture. The Brazilian ethanol market has evolved into a self-sustaining system and is no longer reliant on government regulation and subsidies (Timilsina and Shrestha 2014). In the United States and Europe, the sector was also encouraged to revitalize the farming sector during a period of agricultural surplus and bioethanol was seen as an environmentally friendly opportunity to diversify the sector outcomes (Gnansounou and Dauriat 2011).

Interestingly, the Clean Air Act mandates the Environmental Protection Agency of the United States (EPA) to develop national ambient air quality guidelines for certain common and pervasive contaminants based on the most recent research to safeguard human health and welfare across the country. Particulate matter (also known as particle pollution), ozone, sulfur dioxide, nitrogen dioxide, carbon monoxide, and lead are six major "criteria pollutants" for which the EPA has established air quality guidelines. During the 1990s, Methyl Tertiary Butyl Ether (MTBE, later abandoned due to possible leakages and groundwater contamination) and ETBE

(Ethyl Tertiary Butyl Ether, a compound obtained by the equimolar addition of ethanol on isobutene) were successively used as a gasoline additive to increases octane rating and has been utilized to satisfy air pollution reduction targets set by the 1990 Clean Air Act Amendments in the U.S. EPA (2017). In the United States, the usage of ether oxygenates like ETBE in reformulated gasoline was successfully phased out in 2006; nevertheless, the use and manufacturing of these oxygenates has persisted in Europe and Asia.

The awareness of the negative impact of fossil fuels, such as the persistence of heavy pollutants in major cities in the 1990s has been gradually transformed into a deeper understanding of the risks associated with global negative impacts such as greenhouse gas emissions and global warming, and the role that alternative fuels such as bioethanol could play in its mitigation.

In 2005, the first Renewable Fuels Standard became law as part of the energy policy of the United States. The Energy Independence and Security Act of 2007 signed by the United States President set a target of 36 billion gallons of renewable fuel per year by 2022. However, the current production in the United States is closer to 16 billion gallons per year. "The new Renewable Fuels Standard, which currently guides national ethanol policy, states that only 15 billion gallons of production should be produced from corn grain (starch), while the remaining 22 billion should be produced from other advanced and cellulosic feedstock sources. Similarly, after its first directive on renewable energies was issued in 2009, the European Union proposed a revised version in 2018. This new version does not set an increase in the blending proportion of ethanol and other renewable fuels in the energy mix used for transportation but promotes the replacement of the first-generation produced fuels by more advanced ones (Directive REDII EU 2018/2001). The recent inclusion of the environmental impacts of fuels in the American and European legislation has for the first time given biofuels a clear competitive advantage over conventional fuels.

1.3 Global Production

After taking the lead for the first time in 2006, the United States was the undisputed leader in 2019 with 54% of world production corresponding to 61,000 billion liters and more than half of the world's exports, as presented in Fig. 1.1. Latin America, with Brazil, was the world's second largest ethanol producer with 29% (30,000 billion liters) but its weight on the world markets was more limited because a large part of its production was destined for its domestic needs. The European Union with only 5% (5500 billion liters) comes third but is the world leader for biodiesel production. Two-thirds of the world's ethanol production was made from corn and more than a quarter from sugarcane. In 2020, bioethanol is still largely produced from first-generation agricultural materials. Among the 284 ethanol facilities listed by the Renewable Fuel Association operating in the United States of America, only 8 are of second-generation technology. International trade in biofuels remained relatively small in volume (excluding intra-regional trade). Exports accounted for

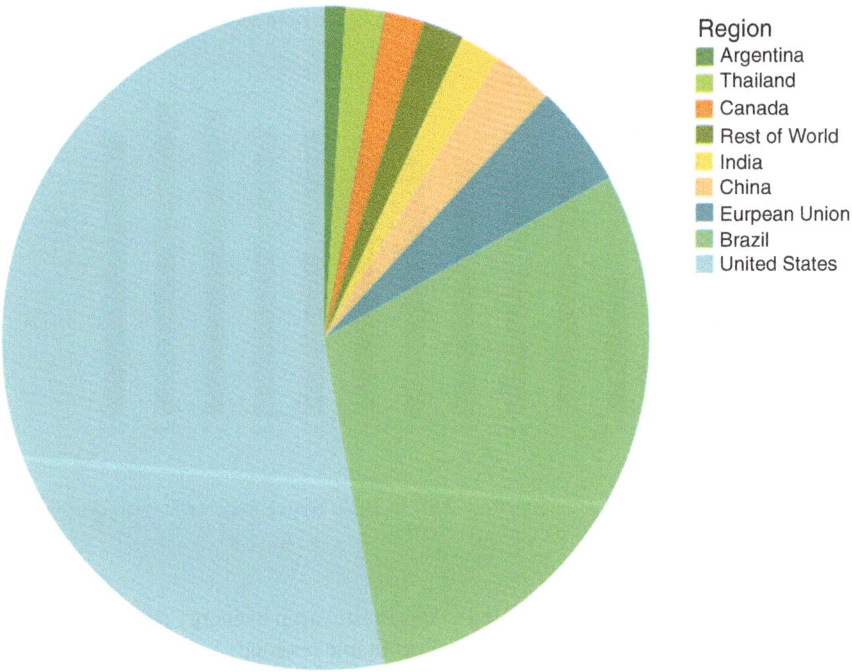

Fig. 1.1 Share of the bioethanol market in 2019. Source Renewable Fuels Association, Ethanol Industry Outlook 2020 (AFDC 2021)

less than 9% of world global ethanol production, and 14% for biodiesel over the period 2017–2019. Most of the support measures for the biofuel industry around the world are aimed at supporting domestic producers. Most countries created tariff barriers to limit imports and protect their biofuel sector. The United States has a positive export trade balance in 2019 of 3080 million liters of ethanol, Brazil comes second with only 125 million liters, while Europe and Asia are net importers with 1135 and 2745 million liters respectively.

After a strong increase mainly driven by the emergence of the United-States as a key player in the ethanol industry, the growth in biofuel production has slowed down considerably over the 2010–2019 decade, as it is presented in Fig. 1.2, and market prices have fallen. Global biofuel production has increased ninefold in 20 years, but the growth of biofuel production has slowed down significantly since 2010. While the average annual growth rate was 18.1% from 2000 to 2009, this rate has dropped to only 3.9% between 2010 and 2019. This high volatility in the energy market can be linked to the financial crisis of 2008 as well as a negative image of biofuels associated with the consumers with an unprecedented rise in the price of agricultural commodities causing hunger-related riots in various parts of the world. It has since been shown, however, that these increases were due more to speculation in the economic markets and lack of regulation than to the increased use of biofuels themselves. Besides the introduction of new unconventional oil sources such as

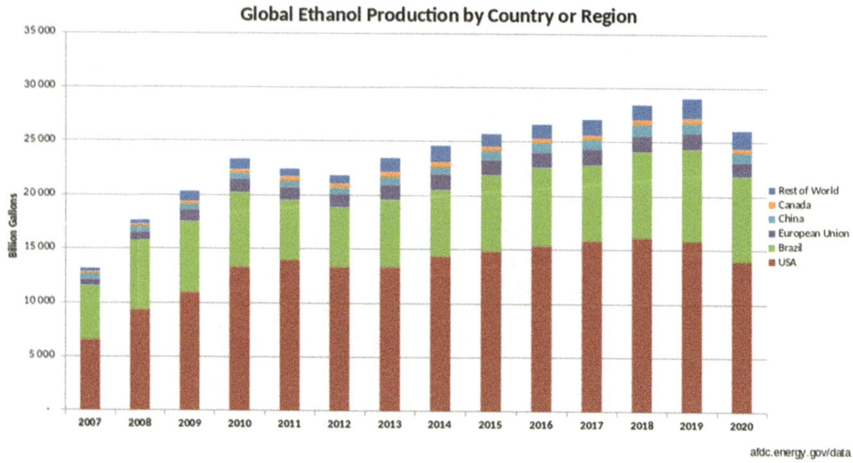

Fig. 1.2 Global bioethanol production from 2007 to 2020. Source Renewable Fuels Association, Ethanol Industry Outlook 2007–2020 (AFDC 2021)

schist gas or oil sands also contributed to this drastic drop in energy prices. It is likely that this uncertainty about the prospect of profitable revenues, coupled with the lack of diversity of products from the second generation biorefinery, is the reason for its relative delay in reaching the market.

The biofuel industry has been hard hit by the Covid-19 crisis. Global production fell by 11.6% in 2020, compared to 2019. This is the first time that a decline in global biofuel production is observed in the last 20 years. The containment measures imposed around the world, the shutdown of many economies, and the closure of borders have drastically reduced the demand for fuels for all types of transportation. The decline in production of biofuels has been even more severe than that of fossil fuels probably because of its relatively high need for human manpower concerning fieldwork. The gradual recovery of the economy in 2021 indicates that biofuel production will start to rise again with the International Energy Agency forecasting +11–13%. The growth will be based on the return of oil consumption and the maintenance or reinforcement of public policies in favor of biomass-based fuels in the framework of the Green Deal. But the volume of global biofuel production in absolute terms will remain slightly below its 2019 level for the year 2021. Although the pandemic's income losses and local supply chain disruptions have undoubtedly increased food insecurity in many developing countries, global food consumption has been largely unaffected due to the inelastic demand for most agricultural commodities in the short term (Elleby et al. 2020). This is not the case for biofuel with a decrease of about 10% of its production. With the recovery of economies, the coronavirus crisis has added new challenges on the supply chains to ensure food and food workers' safety (Vargas-Ramella et al. 2021), and conjectural production losses in agricultural sectors associated with the lack of labor during critical periods such as the 2020 harvests due to the lockdown. It should be noted that in 2021, agricultural commodity prices are at a historically high level close to that of 2008 (Porc 2021), raising fears of a new episode of competition between food and biofuel uses. This event may be only temporary, but the increase in population, the growing artificialization of agricultural land on the urban periphery of major cities, and the competition for land use make new episodes in the future a realistic scenario. Strategic studies find that high agricultural commodity prices combined with government support for biofuels may encourage farmers to switch from food to energy crops (Brown et al. 2021).

In the context of drastic changes in the technologies associated with the mobility of goods and people, it is difficult to predict the global evolution of biofuel consumption. However, some real threats could reduce the consumption of biofuels significantly by 2029. The Covid-19 crisis could mark a lasting decline in mobility as people have experienced other ways of living such as distance-working. The electrification of the automotive industry in many countries will reduce the global demand for fuel. Public policy support will continue to be one of the pillars of biofuels development. But this support is increasingly fragile, especially within the EU. In addition to limiting support for the first-generation biofuels, EU directives are giving preference to other technologies such as electromobility in the automotive sector and green hydrogen. First-generation biofuels, made from agricultural feedstocks will continue to dominate global production. Corn and sugarcane will remain the heart of biofuel geopolitics as the share of advanced biofuels is not expected to exceed 10% of global production by 2029. Furthermore, the biofuels industry will be particularly vulnerable to the threats posed by climate change to the stability of agricultural supplies. Population growth and the reduction of arable land could rekindle the risk of land-use conflicts. Highly dependent on the agricultural sector, biofuels should remain a commodity produced and consumed locally. The volume of international trade in biofuels is expected to decline by 2029, as will the share of the global production internationally sold.

Brazilian biofuel production is expected to grow by 2029. One of the main drivers for the growth of the Brazilian ethanol market is the RenovaBio plan. It was signed in January 2018 and went into effect in June 2020, this program aims to reduce the intensity of CO_2 emissions from the transport sector while supporting the Brazilian biofuel industry. Similarly, the expected growth of the Brazilian car fleet, particularly in the flex-fuel segment, will support domestic demand for ethanol. Ethanol production growth in China is expected to slow down by 2029. The authorities could remove the 10% ethanol incorporation requirement for the entire country, which was introduced in 2017 to dispose of surplus corn stocks as the corn production has jumped from 82 Mt. in 2009 to 209 Mt. in 2016. But the sharp reduction in stocks in 2018 has removed the main motivation for increasing the use of ethanol in transportation. China may prefer other alternative solutions to achieve carbon neutrality in transport, especially in the transportation, particularly in the electric vehicle sector, where its industry is already one of the best performing in the world. According to the IEA (Africa Energy Outlook 2019), biofuels accounted for less than 0.1% of transport energy consumption in Africa. Biofuel production is primarily used in decentralized rural electrification programs in West African countries to power village-level electricity grids. However, a few countries are beginning to show a willingness to promote the biofuel industry and have recently passed legislation. South Africa enacted in March 2019 a new legislative framework to promote biofuels through incorporation obligations, motivated by the desire to reduce its dependence on oil imports and reduce its trade deficit (International Energy Agency 2021).

1.4 Contributions to Climate Change Mitigation and the Environment

Biofuel generation has attracted significant attention considering environmental pollution, growing demand for energy linked to the demographic transition, and possible fossil fuel scarcity. Because of their long-term viability and ecological friendliness, hydrogen and ethanol are seen as auspicious energy carriers. Ethanol synthesis by photosynthetic plants utilizing CO_2 and solar energy offers an appealing solution for long-term green fuel production. However, biocontamination of raw materials, which restricted biomass or causes aggregations of feedstocks, typically weakened or even crushed ethanol manufacturing scaling up operations and required that seasonal crops are converted in a short time. Developing reliable biocontamination management techniques and favoring raw materials that can be stored over the long term to compensate for seasonal production variations (integrated first and second generation biorefineries) is critical to continue improving ethanol production. On an industrial scale, high gravity (VHG) technology was used to make ethanol from molasses, as well as by-product formation estimation. VHG with continuous airflow is a sustainable approach for reducing the cost of ethanol production from molasses on a commercial scale. Different source materials containing simple sugars that may be immediately treated to fermentation are used in a variety of methods. The cheapest production of ethanol employing the current technology is using sugarcane, such as in Brazil, and starch crops, such as in Europe. The energy output of ethanol production varies from 1.7 to 3.8 depending on the materials used and is shown to be inversely related to the processing time of the materials fermented (Sharif et al. 2021).

In terms of the mitigation of greenhouse gases, a huge variability of results has been observed depending on the scope and boundaries of the life cycle analyses studies (Bertrand et al. 2016). However, according to a detailed comprehensive meta-analysis sustainability study, first-generation bioethanol is just as beneficial as second-generation bioethanol for a viable climate strategy provided good agricultural and manufacturing practices are implemented (Dammer et al. 2017). The findings show that the European Union Renewable Energy Directive REDII systematic discrimination against first-generation biofuels is based on no scientific grounds. Further reducing the use of first-generation fuels in the EU's energy mix would be detrimental to the greenhouse gases mitigation saving. According to the

Commission's REDII, a specific transportation objective should be scrapped, and first-generation fuels should be phased out and replaced by second-generation fuels. These steps are intended to guarantee that Europe meets its ambitious climate goals without jeopardizing food security. All the investigated bioethanol feedstocks, particularly first-generation (sugar, starch) and second-generation (lignocellulosic, waste-based), provide considerable positive effects as well as limitations for a realistic climate strategy (Dammer et al. 2017). When it comes to the often-debated detrimental impact of first-generation biofuels on food security, the data indicate that the great land efficiency of first-generation crops (particularly sugar beet) and protein-rich co-products offsets the competition for arable land (especially wheat and corn). In this sense, the use of short rotation crops needed for biofuels creates a lot more competition for arable land because they take up a lot more area. Greenhouse gas emissions are reduced significantly in all feedstocks. While second-generation fuels outperform first-generation fuels in this aspect, the advantage is heavily diluted when abatement costs are included in the calculation. Reducing greenhouse gases emissions using second-generation biofuels is costly, and it precludes far more cost-effective climate action from being taken elsewhere. In this regard, sugar beet and sugar cane's major advantage are their great land efficiency. No other biomass produces more bioethanol per hectare. Additional positive characteristics are high GHG reductions and cheap GHG abatement costs. Co-products are utilized as animal feed, and the infrastructure and logistics are well-developed. The primary drawbacks are the effects of intensive agriculture on biodiversity, water, air, and soil—although these effects are restricted to small regions due to the high land efficiency and could be further improved using good farming management practices. The protein-rich co-products of starch crops, which are good animal feed, are their major strength. The land efficiency of starch crops is lower than that of wood crops. The GHG reductions are expected to be smaller than those achieved by the other choices.

However, the infrastructures and logistics are well developed and should be taken into consideration when an urgent decrease in GHG is needed.

A significant nitrogen fertilizer application may have detrimental implications in intensive farming. Ammonia is produced by the Haber process at 200 to 400 pressures and 450 °C in the reaction of nitrogen from the air and natural gas. This is a high-energy and high-cost process that contributes to the connection between fossil fuel prices and food crop prices. As a result, the nitrogen cycle at the agricultural parcel needs to be closed naturally. Even though bioelectricity created in cogeneration from sugarcane bagasse in Brazil helps to make the entire process economically viable even without subsidies, it also leads to less nitrogen being fixed in the soils and worries about sustainability. Alternative agricultural practices, such as co-cultivation of annual cereal-legume intercrops, have shown to increase feedstock yields without the need for additional fertilizers. Its environmental toxicity has been demonstrated in numerous studies. Short-term profits, the percentage of plants utilized for electric and heat generation and those who return to the soil, must be balanced to sustain long-term resource management. Vinasses, a final by-product of biomass distillation from sugar crops and one of the primary wastes of ethanol production, has also garnered a lot of attention due to environmental concerns over its improper and indiscriminate disposal in soils and water. Vinasses also add to greenhouse gas emissions during storage, transportation, and application to soils. Vinasses, rather than being a waste product, might be a valuable resource for creating biogas via anaerobic fermentation (Bertrand et al. 2016). This biogas generation may be used to generate the necessary N-fertilizer to divorce agricultural practices from the heavy use of fossil fuels. Biohydrogen and volatile fatty acids can also be generated by selecting suitable anaerobic consortiums (Sydney et al. 2014).

The water requirements of first-generation biorefineries is another natural concern. Water consumption for dry mill ethanol production is typically 4–4.7 gallons of water per gallon ethanol. These figures have decreased with time, although they were formerly as high as 15 liters/gallon. The importance of water reuse in reducing total water use cannot be overstated because of the extremely concentrated salts that result from recycling water, it is impossible to significantly reduce current typical water usage. To keep the cycle going, the evaporated water must be recovered. Despite falling per gallon ethanol requirements, the amount of water required by the ethanol sector continues to rise due to an increase in the number of plants and their capacity (Bajpai 2021). A recent study, however, suggests that certain species of marine yeast might be utilized to make bioethanol from saltwater. On an experimental scale, the researchers demonstrated that this sort of fermentation procedure may be done without sacrificing productivity when compared to existing manufacturing methods. This technique is especially important in nations that already have water shortages, such as in the Middle East, Australia, portions of India, and sections of the United States. According to Zaky's research, in addition to requiring less energy and being more cost-effective, the new process has the potential to create as much as 7 liters of excellent quality freshwater as a byproduct every 1 liter of bioethanol (Zaky et al. 2018).

1.5 Conclusion and Perspectives

The economic feasibility of first-generation bioethanol has been demonstrated on a large industrial scale. The United States and Brazil are the two biggest producers, with corn and sugarcane being the most common crops. Despite the numerous concerns about the first-generation bioethanol's long-term viability, such as the effects on land-use change, water use, potential contamination of soils with distillation residues, and competition for food and feed production, many potential routes for making this production greener are emerging. In this context, integrated biorefineries are a promising way to diversify the feedstocks available, resulting in smaller facilities and more efficient supply chains, to more efficiently valorize bagasse from sugarcane and corn stover, or to exploit the potential of microalgae to capture carbon dioxide produced during fermentation steps. This is a chance to take advantage of this large-scale successful deployment to gain expertise in developing the most promising processing schemes for next-generation facilities that are currently experiencing economic viability concerns that are slowing down their coming to the market (Susmozas et al. 2020).

The technology used in today's vehicles is undergoing a major transformation. Several techniques, such as electric energy, various types of batteries, and hydrogen fuel cell cars, are getting a lot of attraction. However, the use of liquid fuels in internal combustion engines will likely continue to be one of the best solutions in some transportation sectors, such as aviation, marine shipping, or heavy vehicles, where modern engine integration is still a difficulty (Ortar and Ryghaug 2019). Furthermore, the pressing requirement to reduce greenhouse gases emissions means that this energy shift must consider the engines used in actual cars and the readily available technologies, among them bioethanol, rather than expecting new technologies that may experience delays in their development. In this regard, the development of biofuels from renewable raw materials is required to meet the timeframe set in the fuel decarbonization policy, while taking advantage that biofuels are simple to integrate into logistic transportation networks (Zailani et al. 2020).

Acknowledgments The authors acknowledge the support from the French government research program "Investissements d'avenir" through the IMObS3 Laboratory of Excellence (ANR-10-LABX-16-01).

References

Africa Energy Outlook (2019) International Energy Agency SpecialReport available online at ww.iea.org/africa2019 (last time accessed 13/06/2022)

AFDC (2021) Renewable Fuels Association, Ethanol Industry Outlook 2008-2021 reports. https://ethanolrfa.org/resources/annual-industry-outlook. Accessed 8 Oct 2021

Ayadi M, Sarma SJ, Pachapur VL, Brar SK, Cheikh RB (2016) History and global policy of biofuels. In: Soccol CR, Brar SK, Faulds C, Ramos LP (eds) Green fuels technology. Springer, Cham, pp 1–14

Bajpai P (2021) Historical perspectives. In: Developments in bioethanol. Springer Singapore, Singapore, pp 15–20

Bertrand E, Pradel M, Dussap C-G (2016) Economic and environmental aspects of biofuels. In: Soccol CR, Brar SK, Faulds C, Ramos LP (eds) Green fuels technology. Springer, Cham, pp 525–555. https://doi.org/10.1007/978-981-15-8779-5_2

Brown B, Schoney R, Nolan J (2021) Assessing the food vs. fuel issue: an agent-based simulation. Energy Policy 159(112553). https://doi.org/10.1016/j.enpol.2021.112553

Dammer L, Carus M, Piotrowski S, Puente, A, Breimayer E, de Beus CL (2017) Sustainable first- and second-generation bioethanol for Europe. https://renewable-carbon.eu/publications/product/sustainable-first-and-second-generation-bioethanol-for-europe-%e2%88%92-full-version/. Accessed October 2021

DGRIS (2021) Observatoire de la sécurité des flux et des matières énergétiques, Direction Générale des relations internationales et de la stratégie, Perspectives d'évolution des biocarburants: jeux des acteurs et enjeux financiers, C4 du séminaire n°7 avril 2021. https://www.defense.gouv.fr/content/download/626092/10405149/file/202104-biocarburants_Energie-CR-s%C3%A9minaire-7.pdf. Accessed October 2021

Directive REDII (EU) 2018/2001 of the European Parliament and of the Council of 11 December 2018 on the promotion of the use of energy from renewable sources available online at https://

eur-lex.europa.eu/legal-content/EN/TXT/?uri=uriserv:OJ.L_.2018.328.01.0082.01.ENG&toc=OJ:L:2018:328:TOC. Accessed September 2021

Elleby C, Domínguez IP, Adenauer M, Genovese G (2020) Impacts of the COVID-19 pandemic on the global agricultural markets. Environ Resour Econ 76(4):1067–1079. https://doi.org/10.1007/s10640-020-00473-6

Gnansounou E, Dauriat A (2011) Life-cycle assessment of biofuels. In: Pandey A, Larroche C, Ricke SC et al (eds) Biofuels, alternative feedstocks, and conversion processes. Academic, Oxford, pp 25–50

International Energy Agency (2021) World Energy Outlook 2021. https://www.iea.org/reports/world-energy-outlook-2021. Accessed September 2021

Ortar N, Ryghaug M (2019) Should all cars be electric by 2025? The electric car debate in Europe. Sustainability 11(7):1868. https://doi.org/10.3390/su11071868

Porc O, Nova Price indices June 2021. https://renewable-carbon.eu/publications/product/nova-price-indices-june-2021-png/. Accessed October 2021

Sharif N, Munir N, Hasnain M, Naz S, Arshad M (2021) Environmental impacts of ethanol production system. In: Arshad M (ed) Sustainable ethanol and climate change. Springer, Cham, pp 205–223. https://doi.org/10.1007/978-3-030-59280-6_10

Susmozas A, Martín-Sampedro R, Ibarra D, Eugenio ME, Iglesias R, Manzanares P, Moreno AD (2020) Process strategies for the transition of 1G to advanced bioethanol production. PRO 8(10):1310. https://doi.org/10.3390/pr8101310

Sydney EB, Larroche C, Novak AC, Nouaille R, Sarma SJ, Brar SK, Letti LAJ, Soccol VT, Soccol CR (2014) Economic process to produce biohydrogen and volatile fatty acids by a mixed culture using vinasse from sugarcane ethanol industry as nutrient source. Bioresour Technol 159:380–386. https://doi.org/10.1016/j.biortech.2014.02.042

Timilsina GR, Shrestha A (2014) An overview of global markets and policies. In: Timilsina GR, Zilberman D (eds) The impacts of biofuels on the economy, environment, and poverty. Springer, New York, pp 1–14

U.S. EPA (2017) IRIS Toxicological review of Ethyl Tertiary Butyl Ether (ETBE) (External Review Draft, 2017). U.S. Environmental Protection Agency, Washington, DC, EPA/635/R-17/016

Vargas-Ramella M, Lorenzo JM, Bohrer BM, Pateiro M, Cantalapiedra JJ, Franco D (2021) A year following the onset of the COVID-19 pandemic: existing challenges and ways the food industry has been impacted. Foods 10(10):2389. https://doi.org/10.3390/foods10102389

Zailani S, Iranmanesh M, Foroughi B, Kim K, Hyun SS (2020) Effects of supply chain practices, integration and closed-loop supply chain activities on cost-containment of biodiesel. Rev Manag Sci 14(6):1299–1319. https://doi.org/10.1007/s11846-019-00332-9

Zaky AS, Greetham D, Tucker GA, Du C (2018) The establishment of a marine focused biorefinery for bioethanol production using seawater and a novel marine yeast strain. Sci Rep 8(1):12127. https://doi.org/10.1038/s41598-018-30660-x

Chapter 2
Feedstocks for First-Generation Bioethanol Production

Arion Zandoná Filho, Adenise Lorenci Woiciechowski, Luis Alberto Junior Letti, Luis Alberto Zevallos Torres, Kim Kley Valladares-Diestra, and Carlos Ricardo Soccol

Abstract Ethanol from biomass was the first fuel used by man in Otto cycle engines. Its substitution by petroleum derivatives followed naturally due to the logarithmic growth of the world's demand for energy and limitations in agricultural growth. The so-called first generation (1G) ethanol produced from biomass with important levels of easily fermentable sugars or lignocellulosic material that will be hydrolyzed and fermented. The first economically viable materials from saccharine fermentations were sugar cane and sugar beets, starchy fermentations were corn and cassava. The ability to ferment C5 and C6 sugars using classical or GMO yeast strains such as *Saccharomyces cerevisiae* makes it possible nowadays to use almost any type of biomass.

The definition of the use of one of these raw materials depends on factors such as its availability in quantity and frequency, storage organization, tax incentives offered by countries or regions, non-competition with food markets, the price of petroleum derivatives, and the culture of using clean energy that minimizes greenhouse gas (GHG) emissions. The technology for producing ethanol from 1G cereals is well known and there are many plants operating in the world, based on cereals. The interest subsidy for the creation of new ethanol plants is in common use in several countries in Asia, Africa, and the Americas, which justifies the preference for its mixture in gasoline.

A. Zandoná Filho (✉)
Department of Chemical Engineering, Federal University of Parana, Curitiba, Brazil
e-mail: a.zandona@ufpr.br

A. Lorenci Woiciechowski · L. A. J. Letti · L. A. Zevallos Torres · K. K. Valladares-Diestra · C. R. Soccol
Department of Bioprocesses Engineering and Biotechnology, Federal University of Paraná, Curitiba, Brazil
e-mail: adenise@ufpr.br; letti@ufpr.br; kim.valladares@ufpr.br; soccol@ufpr.br

C. R. Soccol et al. (eds.), *Liquid Biofuels: Bioethanol*, Biofuel and Biorefinery Technologies 12, https://doi.org/10.1007/978-3-031-01241-9_2

2.1 Introduction

The natural raw material of greatest human interest, since the beginning time, is lignocellulosic biomass due to its versatile and renewable character. This source, which is around 450 million years old, is an indirect way of harnessing solar electromagnetic radiation that chemically converts and stores energy through photosynthesis.

Scientists aiming to replace energy from fossil sources with a clean and renewable energy source are incessantly looking for biomass with high productivity, resistance to climatic variations, easy-to-use technologies, and processing without significant environmental impacts. One of these alternatives uses residues from agroindustry and forestry to produce biofuels. The apex of this vision focuses on technologies known as first and second-generation fuels (e.g., bioethanol).

Engineering and bioengineering studies were needed to define the way to use carbohydrates from biomass, whether by saccharine or starchy processes leading to sucrose and starch fermentation, respectively. After judging your potential of biomass and technology is possible to implement the biological and thermochemical conversion for utilizing biomass to produce bioethanol. Biorefineries in various regions of the world, whether raw sugar mills or those that exclusively produce bioethanol, do not operate year-round, and must adopt other raw materials as a complement. The classic example is sugarcane and sweet sorghum, which have different harvest times, but can be processed into bioethanol using the same industrial equipment.

Over the past 35 years, governments and human society have been following with enthusiasm and concern the growing demand for natural and energy resources. This induced by the consumption of goods and services. Technological development in the production of economically sustainable and environmentally friendly biofuels has been stimulated by the stock and investment market and by governments that have established specific programs for their non-food agricultural matrices. The biggest link in this problem is the population growth that induces a disproportionate consumption in the entire production chain of inputs, following an infinitely worrying spiral.

The agronomic issues throw the proper characterization of a given plant biomass within critical for the optimization of its use for ethanol because this is provided valuable information about properties, storage and handling criteria, theoretical yields and potential environmental problems that are related to large scale industrial operations.

2.2 Production Feedstocks for 1G Ethanol Around Policy and Programs

To start with the correct definition of First-Generation Bioethanol, there is a direct relationship with biomass (raw material) which is a market commodity and used as a food source. Ethanol is produced from the fermentation of sugars (polysaccharides,

C6-glucose sugars) using classical or Genetically Modified Organisms (GMO) yeast strains such as *Saccharomyces cerevisiae, Zymomonas mobilis,* and *Escherichia coli* for improved ethanol fermentation (Rosales-Calderon and Arantes 2019).

The production of 1G ethanol occurs through the biological fermentation of hexoses such as glucose. 'Sugars' is the name for all types of monosaccharides and disaccharides found in nature. The sugars used in the production of first-generation ethanol are mainly extracted from crops rich in sugars or polymers such as starch. In Brazil, sugarcane juice is used to produce first-generation ethanol, while in other countries other raw materials are used, such as: Sweet potatoes (5.5% sugar), beetroot (8.0% sugar) and sweet corn (6.3% sugar). Other sugars ingredients include molasses, honey, maple syrup, and corn sweeteners, which are composed of varying levels of glucose, fructose, and sucrose (Canadian Sugar Institute 2021).

Therefore, the main raw material used in the 1G ethanol production process are feedstocks such as sugar cane, corn, wheat, among others. Sugar crops, such as sugar cane, sugar beet, and sweet sorghum, can be used as feedstocks for both conventional biofuels process (ethanol via fermentation of sugar) and/or advanced biofuels.

The main differences between corn and sugarcane ethanol are in the form of production. Sugarcane ethanol is simpler to be obtained, requiring fewer unit operations, followed by approximately 10.0 h of fermentation for conversion to bioethanol. However, sugarcane has less sugar than corn, and a one ton of sugarcane produces 89.5 liters of ethanol while corn can produce up to 407 liters of ethanol. The difference is in the chemical conversion of corn starch, leading to the fermentation process which can accept 70 h or more (Barros and Woody 2020).

The world leaders in the production of 1G ethanol are the United States and Brazil, which together produce more than 80% of the world's ethanol. The 1G ethanol industry in the United States is quite strong and with large technological bases already developed for the efficient production of ethanol from corn. The United States total 1G ethanol production in 2020 was 13,926 million gallons (~63 million L) representing 53% of world production (RFA 2021). Due to the high volume of 1G ethanol production, normally 30% of corn production in the United States is used in ethanol production process (Mohanty and Swain 2019). Corn production in the United States was 347 million tons in 2019, which represents a total of 104.1 million tons of corn directed to the production of 1G ethanol to the production of 1G ethanol (FAO 2020). However, a small percentage (3%) of 1G ethanol could be produced from wheat and since 2021 sorghum is also being used in small quantities as raw material (RFA 2021).

The great production capacity of 1G ethanol in the United States allows it to export large volumes of this biofuel, registering the export of 1463 million gallons (6.6 million L) worldwide in 2019 (U.S. GRAINS 2019). On the other hand, Brazil's 1G ethanol production in 2019 was 7.93 million gallons (~36 million L), being the second largest producer worldwide with 30% of 1G ethanol produced from sugarcane. According to Brazilian Association of the Sugar Cane Industry (UNICA 2020), in the 2020/2021 season, 657 million tons of sugar cane were produced in Brazil, which allowed a production of 32.0 thousand million liters of 1G ethanol and the production of 41.5 million tons of edible sugar. In addition, new plants to

Table 2.1 Source used in the ethanol 1G production cost by countries and global productivity

Country	Main sources	Ethanol cost (USD/L)	Production (millions of L)	World percentage (%)
Unite State	Corn	0.41	63,308.8	53
Brazil	Sugarcane	0.44	36,050.5	30
European Union	Sugar beet, wheat, and corn	0.47	5682.6	5
China	Corn	0.49	4000.6	3
India	Molasses	0.37	2341.2	2
Canada	Corn and wheat	–	1945.7	2
Thailand	Molasses	–	1818.4	2

produce ethanol from corn are being implemented and the by-products of sugar refining (molasses) are also used as raw material to produce 1G ethanol. Due to its large domestic market, Brazil exports only a part of its ethanol produced, which represents 7.18% of its domestic production (2.3 million liters of 1G ethanol).

The advantage of sugarcane is its productivity, since in one hectare it is possible to obtain up to 90.0 tons of sugarcane, capable of producing up to 8000 liters of ethanol. Corn yields a maximum of 20.0 tons per hectare, which can be transformed into 3500 liters of ethanol.

Although the United States and Brazil concentrate the largest bioethanol production worldwide, the European Union, China, India, Canada, Thailand, and Argentina are emerging as major producers of 1G ethanol from raw materials rich in sugar (FAO 2020).

For example, in the US the cost of raw materials is a major concern in relation to current 1G ethanol production methods, because the raw material cost of corn ethanol is on average 58% of total production, this cost is based on data from the last 15 years (2007–2021) in a typical production plant in the US using the dry milling process.

The production of 1G ethanol in 2020 by the European Union, China, and India was 5.7, 4.0 and 2.2 million liters. Together these three regions/countries account for 10% of world production. The source of raw materials turns out to be truly diverse according to the country and environmental conditions. Especially within the European Union, where corn, wheat, and sugar beet it used for ethanol production. In contrast, China and India produce their alcohol from corn and molasses derived from sugar production respectively (Table 2.1). However, due to their still low productivity, the European Union and Canada are two regions where 1G ethanol is imported, with an approximate of 1.5 billion liters (U.S. GRAINS 2019).

The role of raw materials rich in sugars and starch in the production of 1G ethanol is fundamental and defines the final cost of ethanol, representing 70% of the total process costs. For this reason, the need to increase the productivity of crops reducing their costs is particularly important.

The highest production crop worldwide is corn (1148.5 Mtonne), wheat (765.8 Mtonne), sugar cane (1949.3 Mtonne), rice (755.5 Mtonne) and palm fruits (410.7 Mtonne) (FAO 2020) (Fig. 2.2).

Table 2.1 was assembled using official websites: Canadian Sugar Institute (2021), CONAB (2021), MAPA (2021), FAO (2020), Renewable Fuels Association (2020a, b), and United States Department of Agriculture - Foreign Agricultural Service (2021).

As it might be seen four of the highest ethanol producers' crops are rich in sugars or starch that are used or can serve as raw material to produce 1G ethanol. There is precisely a direct relationship between the largest ethanol producers and the predominant type of crop in each country. The production of ethanol from corn is still slightly more expensive.

In 2020, the green fuel ethanol production was around the world 98.6 million m^3. The U.S. and Brazil were the biggest players and producers in the world in our century for green fuel ethanol were with a participation of 53.0 and 30.0% of world production, respectively. The European Union accounted for 5.0% of the world production followed by India (2.0%), Canada (2.0%), Thailand (2.0%) and Argentina (1.0%). The rest of the world reached only 2.0% of the fuel ethanol world production Renewable Fuels Association (2020a, b) (Fig. 2.1).

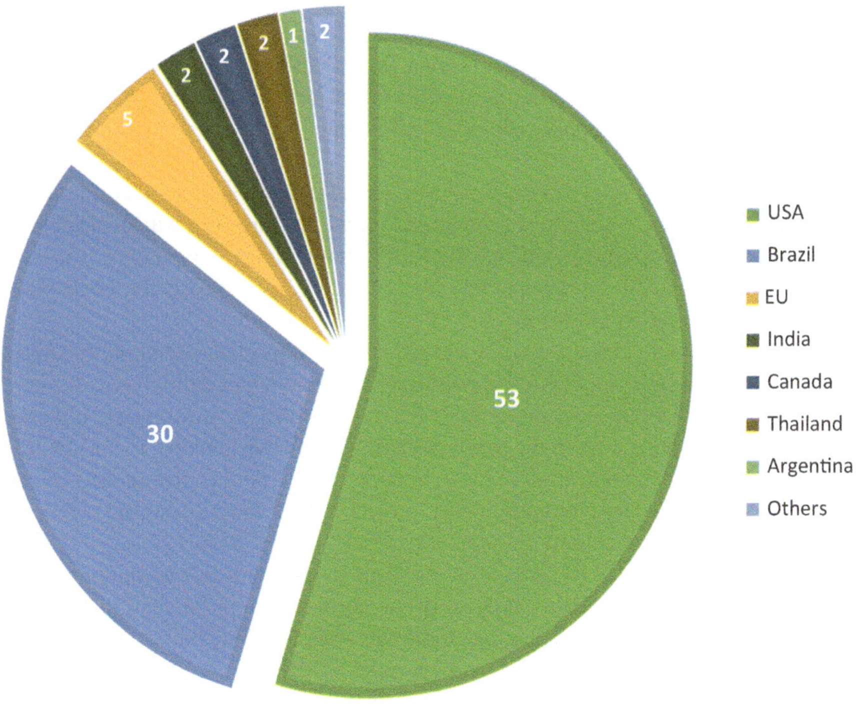

Fig. 2.1 Global ethanol 1G production, by regions/countries (in % of total production)—adapted from Renewable Fuels Associating (2020a, b)

According to Renewable Fuels Association (RFA), the main feedstock type for ethanol production in the U.S. are corn starch (94.0%), corn/sorghum/cellulosic biomass/waste (3.4%), corn/sorghum (2.1%), cellulosic biomass (0.5%) and waste sugars/alcohol/starch (0.1%) (Renewable Fuels Association 2020a, b).

In Brazil, the two major feedstocks employed in bioethanol production are sugarcane and corn. In 2019, 96% of the ethanol produced came from sugarcane (USDA 2020). However, the ethanol produced using corn as feedstock has been growing fast during 2020/21 (being sugar cane responsible for 90.8% of ethanol production and corn for the remaining 9.2%) (https://www.conab.gov.br/info-agro/safras/cana).

In 2020, 49.5% of the ethanol produced in the European Union was from corn, followed by wheat (18.5%) and sugar (17.8%). The ethanol produced using other cereals and starch-rich crops accounted for only 6.3% (European Renewable Ethanol 2020).

Now after understanding this projection is possible to represent the data's such as in Fig. 2.2, Data were selected and worked from FAOSTAT with the aim of illustrating the world agricultural production based on green energy sources, either of oilseed or lignocellulosic origin. It is evident that in the main producing countries in the Southern hemisphere and below the Tropic of Cancer (South America, Central America and Mexico) sugarcane production is in the majority (green color) while in the Northern hemisphere wheat and maize production stand out (blue and yellow). Also noteworthy is the growth of rice production (purple) in Asia (FAO 2020).

Businesses and government institutions around the world are migrating their sugar production plants to produce bioethanol as well. Following the treadmill, gasoline should have a higher percentage of ethanol in the mixture, as is already the case in Brazil.

Brazil is moving, in some regions, to replace sugarcane with corn, taking advantage of the infrastructure of sugarcane mills that are idle between harvests and expands the opportunity to purchase raw materials at competitive prices in regions with production surplus and that present high logistical costs for the disposal of a product with low added value.

The use of regional biomass makes bioethanol an important substitute for fossil fuels, with advantages such as sustainability, and good adaptability. Over the past 30 years, the development of green fuels has been driven by government policies. Many countries and areas have authorized laws and regulations to ensure the minimization of environmental impacts and reduction of the greenhouse effect. With the support of governments, many projects were commercialized, and raw materials boosted the economy (Soccol et al. 2010; Barros 2020).

Governments provided privileges and financed investments and production costs, reducing subsidies for the entire process of the agro- and sugar-alcohol industry, to reduce the investment risk. As a result of this support, bioethanol started to be marketed in several states, districts, and countries.

Political debate and market reserves continue to be discussed among partner countries about the impact of green fuels on climate change, the extension of plantation areas, and food security.

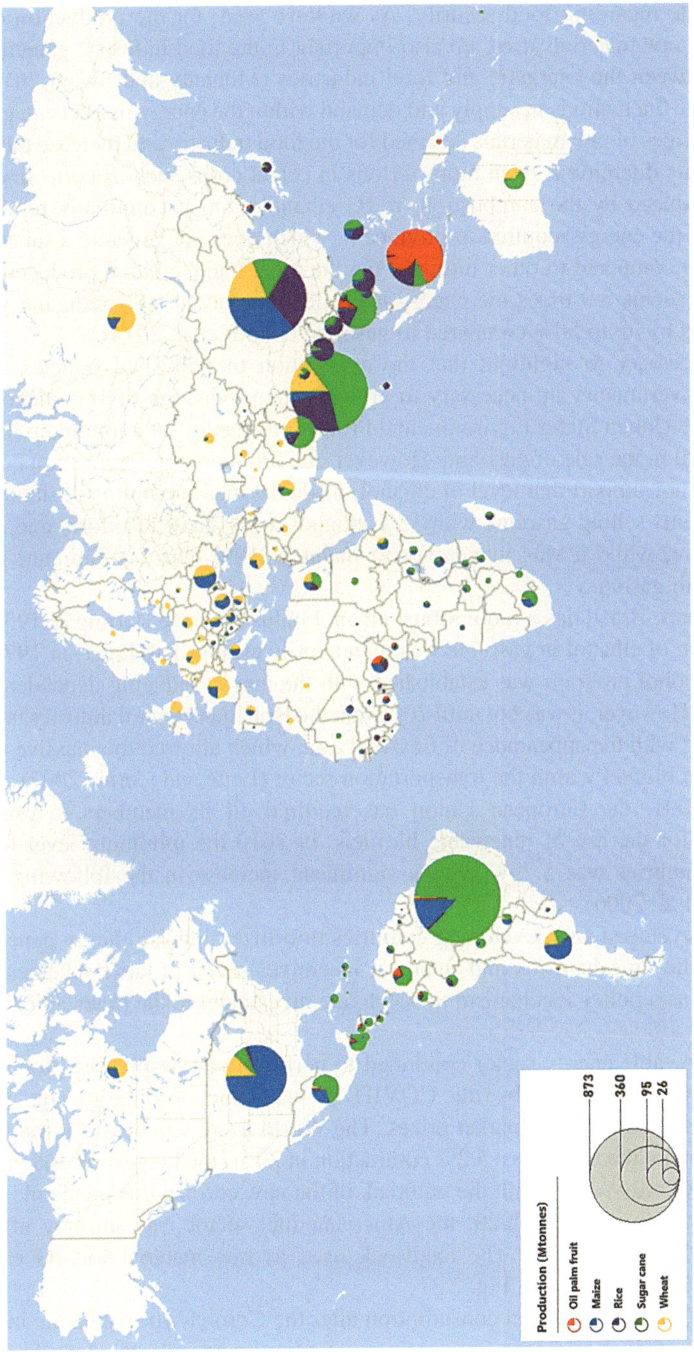

Fig. 2.2 Mains crops distribution by country around of the world (FAO 2020)

Although the production of biofuels is a positive alternative to fossil fuels, there is still a debate regarding food security. As we have seen, for the production of 1G ethanol the raw materials used are also important in the food industry, generating a conflict between the energetic and food industries (Mohanty and Swain 2019). In addition, the fluctuations in supply and demand within the energy market can greatly affect the prices of raw materials destined for the food industry and increase the price of food. This dilemma is even more serious in cereal crops such as corn, since the energy produced by the combustion of 1G ethanol from it commonly represents three times the energy required to produce it, which does not generate a substantial energy gain compared to other biofuels. On the other hand, ethanol produced from sugar cane yields 8.0 times the energy needed to produce it, also reducing greenhouse gases by up to 50% compared to gasoline (Chum et al. 2014).

It is necessary to highlight that the application of published policies of the different governments are necessary to promote the production of 1G ethanol as a biofuel. The United States Environmental Protection Agency set a minimum limit of 15% ethanol in the sale of gasoline. However, the limitation of combustion engines that would not support that level of ethanol, made the measure optional, stating that establishments selling gasoline with 15% ethanol would have to inform and signal this measure, while it was mandatory to maintain 10% ethanol minimum in the mixture with gasoline.

For its part, Brazil has a more robust ethanol industry, since starting in 1933, the combination of ethanol in gasoline was mandatorily established. Later, in 1975, the national alcohol program was established with the aim of reducing dependence on petroleum. However, it was not until 2003 that the Brazilian ethanol industry became more robust with the appearance of flex-fuel cars, which allowed the massive use of ethanol as a biofuel within the transportation sector (Leite and Cortez 2007).

Since 2001, the European Union has required all its members to establish legislation for the use of renewable biofuels. In 2010 the minimum level for the member countries was 5.75% with a significant increase in the following years (Mojović et al. 2006).

Many developed and developing countries prioritize biomass energy generation through policy mechanisms and financial incentives. Feed-in tariff schemes were introduced as a policy mechanism to accelerate investment in the renewable energy sector.

The renewable energy sector continued to perform well despite the global economic slowdown, caused by the COVID-19 pandemic, leading to cuts in tax incentives and commodity market prices. The World Bank's forecast for the global economy is that there will be a 5.2% contraction in 2021, the biggest recession since the end of World War II. With the outbreak of the new coronavirus pandemic in the United States in mid-March/20, the American fuel sector was strongly affected (Barros 2020; EPA 2020). The feedstock used as raw material for 1G ethanol production is synthesized in Fig. 2.3.

News of growth in market consumption after the Coronavirus pandemic in 2021 points to a gradual reopening of world economies, especially the US. The corn futures market is a clear response to the increase in demand for this cereal, also

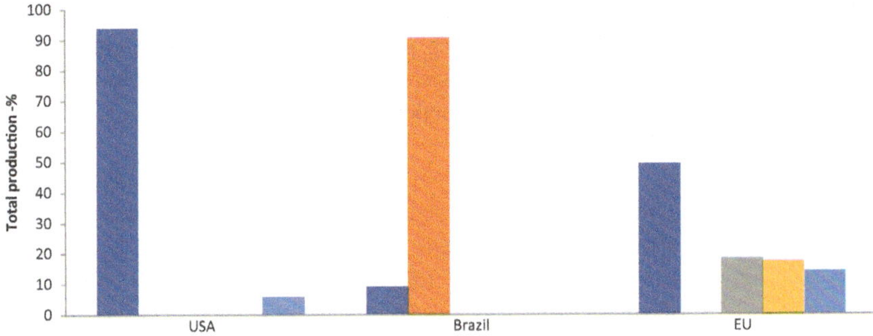

	USA	Brazil	EU
Corn	94	9.2	49.5
Sugarcane	0	90.8	0
Wheat	0	0	18.5
Sugarbeat	0	0	17.8
Others	6	0	14.2

Fig. 2.3 Feedstock used as raw material for 1G ethanol production—distribution in the main countries/regions (data given in % for total production for each country/region)

aiming at the fuel market. According to the USDA greater domestic availability of corn and the uncertainties surrounding the market, at this time after COVID-19, the scenarios for the demand for ethanol production stood out because the sector is responsible for the consumption of more than one-third of the US corn crop (FAO 2020; U.S. GRAINS 2019; Barros 2020).

2.3 Sugar-Based Feedstocks

2.3.1 Sugarcane

Sugarcane processing for 1G ethanol production consists in milling and extracting the sugarcane juice. In general, ethanol can be produced from cane saccharides such as glucans (cellulose and β-glucans), hemicelluloses (xyloglucans and heteroxylans) and pectin after hydrolysis and fermentation. However, some monosaccharides (mostly pentoses) are more difficult to ferment to fuel ethanol than hexoses such as glucose. The juice is sent to a juice treatment system to remove impurities (minerals, salts, organic acids, dirt and fine particles) prior to fermentation (Oliveira et al. 2015). Juice treatment consists of separation of fibers and sand in screens, heating of juice from 30 °C to 70 °C, lime addition with a second heating (up to 105 °C), air removal (flash), and flocculant polymer addition and final removal of impurities through clarification process. Clarified juice is then concentrated to achieve adequate sugar concentration for fermentation. Besides ethanol, a sub-product from distillation process is potassium-rich vinasse, which can be employed for ferti-irrigation of the fields, reducing costs of chemical fertilizers (Lopes et al. 2016). Currently, some sugarcane ethanol plants have implemented a process for anaerobic production of biogas from the vinasse. Raizen company recently implemented a biogas producing plant in 2020. The biogas plant, located in Sao Paulo (Brazil), can make use of vinasse and filtration cake from sugarcane juice extraction as raw material. The plant possesses an energy producing capacity of 21 MW (Raizen 2020).

In Brazil, from the entire production of sugarcane, around 50% is destined for ethanol production, while the remaining sugarcane is employed for sugar production (this ratio can vary from year to year tough, according to market offer and demand). Approximately, each ton of sugarcane yields 85 liters of ethanol and Brazilian plants produce 10–15 L of vinasse for each liter of ethanol generated. The volumes may vary depending on the technology used in the fermentation process (Pazuch et al. 2017).

2.3.2 Sugar Beet

Sugar beets require more energy to produce sugar from than sugarcane, because unlike sugarcane sugar beet does not have a by-product like bagasse that can be burned to produce energy. On the hand, the tops of sugar beets and the pulp left after extraction of sugar are by-products used as animal feed for cows and sheep.

Ethanol production process from sugar beets starts shredding the beets into thin chips, called cossettes, to facilitate sugar extraction. The cossettes are washed in a counter-current continuously agitated tank (for 1 h) with high temperature water to draw the sugar into a solution called juice. Then, the washed cossettes are pressed to remove the remaining juice. The residual pressed beets, known as pulp, sent to a drying plant to use it as animal feed. Impurities are removed from the crude juice adding lime and bubbling CO_2 before the spent material is filtrated out. After carbonation, SO_2 is pumped through the juice to neutralize the solution. The juice may require to be concentrated to be an adequate substrate for the yeast that would use it during fermentation. The recovery of ethanol is done by distillation (Bowen et al. 2010; NNFCC 2019).

2.4 Starch-Based Feedstocks

2.4.1 Corn

In the U.S., the two major processes for 1G ethanol production from corn are: dry mill (90.9%) and wet mill process (9.1%). Dry milling is preferred over wet mill process due to the low capital and operating costs (Bušić et al. 2018). Dry milling grounds the corn to fine particles to ease subsequent liquefaction step. In the liquefaction step (85 °C, 1–2 h), the milled substrate is mixed with water and α-amylase enzymes. Then, the mixture cooled down to 30–35 °C and supplemented with glucoamylase enzymes and yeast for simultaneous saccharification and fermentation (SSF) process for 40–50 h. After distillation, ethanol is obtained as the main product. On the other hand, the protein-rich stillage is dried to a 27% protein product known as distillers dried grains with soluble (DDGS), commercialized as animal feeding (Susmozas et al. 2020). In wet milling, the grain is first separated into its basic components through soaking. After steeping, the slurry is processed through grinders to separate the corn germ. The remaining fiber, gluten and starch components are further segregated. The gluten component (protein) is filtered and dried to produce animal feed. The remaining starch can then be fermented into ethanol, using a process like the dry mill process (Renewable Fuels Association 2020a, b).

On average, 100 kg of corn processed by a dry mill ethanol plant produces: 43.2 L of denatured fuel ethanol, 28.2 kg of distillers' grain animal feed (10% moisture), 1.4 kg of corn distillers' oil and 29.5 kg of captured biogenic carbon dioxide (further employed in bottling, food processing, dry ice production, among others). Additionally, ethanol biorefineries produce distiller's grains, gluten feed and gluten meal (Renewable Fuels Association 2020a, b).

2.4.2 Wheat

Wheat processing for ethanol production starts with milling as in traditional flour mills. Once the wheat has been milled into flour, water is added to it forming a slurry. Then, the slurry is cooked using steam and enzymes are incorporated to thin the mixture and convert starch into sugar. The fermentation process is similar to conventional brewery but on a larger scale. After fermentation, distillation takes place to separate ethanol from the protein and fiber. The proteins and fiber obtained from distillation step are pressed to remove most of the water. Some of the solid can be commercialized in a moist form or as a syrup feed. Other portion can be dried and pelletized to form an animal feed product (Vivergo Fuels 2017).

2.4.3 Cassava

Cassava is the third source of carbohydrates for human consumption in the world. Cassava is cultivated in countries with warm and moist tropical climate. The tubers grow well on soils of relatively low fertility where the cultivation of other crops would be difficult or uneconomical.

After harvesting cassava, the roots are chopped into chips and dried usually in the sun. Dried chips can be stored for months. However, depending on the storage temperature, approximately a 5% reduction of starch yield is obtained in 8-month storage.

The ethanol production process from cassava is very similar to those of corn and wheat. As in corn processing, on an industrial scale, cassava processing can be

Table 2.2 Main co-products generated during the processing of feedstock for 1G ethanol production, their co-products and main destination

Feedstock	Main co-products	Main destination of co-products	References
Sugarcane	Bagasse and vinasse	Ethanol 2G, biogas, fertirrigation	Pazuch et al. (2017)
Sugarbeet	Sugar beet tops and pulp	Animal feed	Bowen et al. (2010), NNFCC (2019)
Corn	Distiller dried grain	Animal feed, biogenic CO_2	RFA (2021)
Wheat	Protein and fiber rich residues	Syrup feed, animal feed	Vivergo Fuels (2017)
Cassava	Distiller dried grains, cassava pulp	Biogas	Kuiper et al. (2007)

carried out with two different technologies: dry-grind process and wet mill process. The main differences between the two processes are the feedstocks preparation steps and the number and type of by-products obtained. As in corn processing, once the starch has been recovered, the fermentation and distillation processes are similar in both dry-grind and wet mill facilities. Table 2.2 shows the main feedstock used for 1G ethanol production and their respective main co-products and some common destination.

Wet milling begins soaking the chips in and acid to soften the material and separate the fibers. Then, starch and proteins are separated. On the other hand, dry grinding process starts grinding the chips. Then, water added to the ground material, and the mixture is cooked and mixed with enzymes. This process obtains distiller dried grains with soluble (DDGS) as only by-product after fermentation step. However, the use of this by-product is limited due to its high fiber content. Additionally, cassava pulp, also known as root cake, is the residue remaining after extraction of starch from the grinded root. This residue can be further employed for biogas production (Kuiper et al. 2007).

There is a consensus in the literature regarding the processes that use sugarcane to produce G1 bioethanol as being the best raw material in all aspects (energy balance, GHG emission savings and production cost).

Production of ethanol from sugarcane is in close competition with the sugar market, leading to a reduction in biofuel production in countries such as Brazil. Ethanol from corn is limited by a similar paradox with the increasing value of food on the world's market. Although very advantageous for the producers, increases in the sugar price are a problem for the bioethanol business.

2.5 Conclusion and Future Prospective

The fuels obtained from biomass have reached a crucial step in the substitution of petroleum derivatives by representing an alternative with low greenhouse gas emissions and an almost inexhaustible production capacity. The other great advantage is associated with the processes, technologies used, and infrastructure already installed that allows it to be adapted to different raw materials (whether saccharinic ethanol, starch ethanol or cellulosic ethanol). A wide variety of renewable energy sources are being studied, leading to the belief that new biomasses or better cultivars using new fermentation strains can lead to more environmentally sustainable and economically balanced processes. Promising results, even at pilot scale, identify different feedstocks as promising sources for biorefineries. Ethanol made of sustainable cellulosic feedstock is standard fuel for future, because this biomass is the biggest and renewable in biosphere.

Acknowledgments The financial support provided by Federal University of Paraná is acknowledged. The support of Carlos Ricardo Soccol and Adenise Lorenci Woiciechowski in Biotechnological and Bioprocesses Engineering Department appreciated as well.

Conflict of Interest The authors declare no conflict of interest.

References

Barros S (2020, September) Biofuels annual. USDA-United States Department of Agriculture. Global Agricultural Information Network, Report no. BR2020-0032

Barros S, Woody K (2020, October) Corn ethanol production booms in Brazil. USDA-United States Department of Agriculture, Global Agricultural Information Network, Report no. BR2020-0041

Bowen E, Kennedy SC, Clark WM (2010) Ethanol from sugar beets: a process and economic analysis. Worcester Polytechnic Institute, Worcester

Bušić A, Mardetko N, Kundas S, Morzak G, Belskaya H, Šantek MI, Komes D, Novak S (2018) Bioethanol production from renewable raw materials and its separation and purification: a review. Food Technol Biotechnol 56:289–311. https://doi.org/10.17113/ftb.56.03.18.5546

Canadian Sugar Institute (2021). https://sugar.ca/sugar-basics/sources-of-sugar. Accessed 21 Sep 2021

Chum HL, Warner E, Seabra JEA, Macedo IC (2014) A comparison of commercial ethanol production systems from Brazilian sugarcane and US corn. Biofuels Bioprod Biorefining 8: 205–223. https://doi.org/10.1002/bbb.1448

Companhia Nacional de Abastecimento – CONAB (2021) Série Histórica das Safras. Brasília. https://www.conab.gov.br/info-agro/safras/serie-historica-das-safras. Accessed 21 July 2021

European Renewable Ethanol, Key figures (2020). https://www.epure.org/wp-content/uploads/2021/09/210823-DEF-PR-European-renewable-ethanol-Key-figures-2020-web.pdf. Accessed 20 July 2021

FAO (2020) FAOSTAT [WWW Document]. http://www.fao.org/faostat/en/#data/QC. Accessed 20 July 2021

Kuiper K, Ekmekci L, Hamelinck B, Hettinga C, Meyer KWS (2007) Bio-ethanol from Cassava. http://www.probos.net/biomassa-upstream/pdf/FinalmeetingEcofys.pdf

Leite RC, Cortez LAB (2007) O Etanol Combustível no Brasil. In: Biocombustíveis No Brasil: Realidades e Perspectivas

Lopes ML, de Paulillo SC, Godoy LAR, Cherubin AMS, Lorenzi FHC, Giometti CD, Bernardino AN, de Amorim Neto HB, de Amorim HV (2016) Ethanol production in Brazil: a bridge between science and industry. Braz J Microbiol 47:64–76. https://doi.org/10.1016/j.bjm.2016.10.003

Ministério da Agricultura, Pecuária e Abastecimento. Exportações Brasileiras de Etanol - Comércio Exterior Brasileiro 2019 (2021) MAPA, Brasília. http://www.agricultura.gov.br/assuntos/sustentabilidade/agroenergia/etanol-comercio-exterior-brasileiro. Accessed 15 July 2021

Mohanty SK, Swain MR (2019) Chapter 3 - Bioethanol production from corn and wheat: food, fuel, and future. In: Bioethanol production from food crops. Elsevier, New York. https://doi.org/10.1016/B978-0-12-813766-6/00003-5

Mojović L, Nikolić S, Rakin M, Vukasinović M (2006) Production of bioethanol from corn meal hydrolyzates. Fuel 85:1750–1755. https://doi.org/10.1016/j.fuel.2006.01.018

NNFCC (2019) An assessment of the opportunities for sugar beet production and processing in the Scotland. http://www.nnfcc.co.uk/tools/assessment-of-the-opportunities-for-sugar-beet-production-and-processing-in-the-uk-nnfcc-project-nnfcc-07-017/at_download/file. Accessed 28 July 2021

Oliveira M, Dias DS, Maciel R, Eduardo P, Cavalett O, Eduardo C, Rossell V, Bonomi A (2015) Sugarcane processing for ethanol and sugar in Brazil. Environ Dev 15:35–51. https://doi.org/10.1016/j.envdev.2015.03.004

Pazuch FA, Nogueira CEC, Souza SNM, Micuanski VC, Friedrich L, Lenz AM (2017) Economic evaluation of the replacement of sugar cane bagasse by vinasse, as a source of energy in a power plant in the state of Paraná, Brazil. Renew Sust Energ Rev 76:34–42. https://doi.org/10.1016/j.rser.2017.03.047

Raizen, Raízen inaugura planta de biogás e consolida portfólio de energias renováveis, Raizen (2020). https://www.raizen.com.br/sala-de-imprensa/raizen-inaugura-planta-de-biogas-e-consolida-portfolio-de-energias-renovaveis. Accessed 25 July 2021

Renewable Fuels Association (2020a) Annual fuel ethanol production, (2020). https://ethanolrfa.org/statistics/annual-ethanol-production/. Accessed 22 Sep 2021

Renewable Fuels Association (2020b) Ethanol Industry Outlook. https://ethanolrfa.org/library/rfa-publications. Accessed 22 Sep 2021

RFA (2021) Annual U.S. & World Fuel Ethanol Production-Renewable Fuel Association [WWW Document]. https://ethanolrfa.org/statistics/annual-ethanol-production/. Accessed 27 Sep 21

Rosales-Calderon O, Arantes V (2019) A review on commercial-scale high-value products that can be produced alongside cellulosic ethanol. Biotechnol Biofuels 12:240

Soccol CR, Vandenberghe LPDS, Medeiros ABP, Karp SG, Buckeridge M, Ramos LP, Pitarelo AP, Ferreira-Leitão V, Gottschalk LMF, Ferrara MA, Bon EPS, Moraes LMP, Araújo A, Torres FAG (2010) Bioethanol from lignocelluloses: status and perspectives in Brazil. Bioresour Technol 101(13):4820–4825

Susmozas A, Martín-Sampedro R, Ibarra D, Eugenio ME, Iglesias R, Manzanares P, Moreno AD (2020) Process strategies for the transition of 1G to advanced bioethanol production. Processes 8:1–45. https://doi.org/10.3390/pr8101310

U.S. GRAINS (2019) Global fuel ethanol production by country [WWW document]. https://grains.org/wp-content/uploads/2020/09/Global_ethanol_consumption_gallons.pdf. Accessed 27 Sep 21

UNICA - The Brazilian Sugarcane Industry Association (2020) Observatório da cana [WWW Document]. https://observatoriodacana.com.br/historico-de-producao-e-moagem.php?idMn=32&tipoHistorico=4&acao=visualizar&idTabela=2493&safra=2020%2F2021&estado=RS%2CSC%2CPR%2CSP%2CRJ%2CMG%2CES%2CMS%2CMT%2CGO%2CDF%2CBA%2CSE%2CAL%2CPE%2CPB%2CRN%2CCE%2CPI%2CMA%2CTO%2CPA. Accessed Sep 21

United States Department of Agriculture (2021) Foreign Agricultural Service. Sugar: World Markets and Trade. USDA, Washington, maio. Disponível em. https://usda.library.cornell.edu/concern/publications/z029p472x?locale=en. Accessed 20 July 2021

United States Department of Agriculture-USDA (2020) Foreign Agricultural Service, Corn Ethanol Production Booms in Brazil

US Environmental Protection Agency – EPA (2020) Renewable Fuel Standard Program (RFS) regulatory impact analysis. Accessed 15 Sep 2021

Vivergo Fuels, Our Process - How we do it? (2017). https://vivergofuels.com/process/ Accessed 22 Sep 2021

Chapter 3
Microorganisms and Genetic Improvement for First and Second Generation Bioethanol Production

Gilberto Vinícius de Melo Pereira, Bruna Leal Maske, Dão Pedro de Carvalho Neto, Alexander da Silva Vale, Elisângela Muynarsk, Maria Giovanna Binder Pagnoncelli, Susan Grace Karp, Vanessa Bassi Pregolini, and Carlos Ricardo Soccol

Abstract This chapter explores the main microbial breeding techniques for 1G and 2G bioethanol production, including classical genetic strategies (induced mutations, clonal selection, sexual hybridization, artificial hybridization, and evolutionary engineering) and those based on genetic transformation (deletion and regulation of genes, pooled-segregant whole genome sequence analysis, and CRISPR/CAS9 system). *Saccharomyces cerevisiae* has long been a popular model organism for breeding biological research; however, genetic manipulation of non-model micro-organisms (e.g., *Z. mobilis* and *Escherichia coli*) has also been explored mainly for 2G ethanol production. Tools based on classical genetics are generally random and, therefore, less efficient. However, these techniques have the advantage of not needing prior knowledge about the gene of interest (facilitated procedure), and the microorganisms generated are not considered genetically modified. On the other hand, modifications based on genetic transformation result in more targeted improvements and overproduction of metabolites, although they are more expensive techniques and require extensive knowledge of intracellular biochemical pathways and regulatory mechanisms.

G. V. de Melo Pereira (✉) · B. Leal Maske · D. P. de Carvalho Neto · A. da Silva Vale · E. Muynarsk · S. G. Karp · C. R. Soccol
Department of Bioprocess Engineering and Biotechnology, Federal University of Paraná, Curitiba, Paraná, Brazil

M. G. Binder Pagnoncelli · V. Bassi Pregolini
Department of Chemistry and Biology, Federal University of Technology – Paraná (UTFPR), Curitiba, Paraná, Brazil

C. R. Soccol et al. (eds.), *Liquid Biofuels: Bioethanol*, Biofuel and Biorefinery Technologies 12, https://doi.org/10.1007/978-3-031-01241-9_3

3.1 Introduction

First generation (1G) ethanol is produced from traditional agricultural crops such as maize, sugarcane, sugarbeet, and sweet sorghum. Except in the case of amylaceous biomass (maize) that must be hydrolysed to glucose units, the sugars are readily available to be fermented by microorganisms and converted to ethanol. Second generation (2G) ethanol is produced from crop residues usually composed of lignocellulosic biomass, from woody biomass or from dedicated crops such as switchgrass and, therefore, requires additional pretreatment steps to make sugars available for fermentation.

Yeasts, particularly *Saccharomyces cerevisiae*, are the most widely used microorganisms for industrial ethanol production. However, this species is unable to ferment pentoses. For this purpose, other yeast species have been explored, such as *Candida shehatae*, *Pichia stipitis* and *Pachysolen tannophilus*, or even bacterial strains of *Zymomonas mobilis*, *Thermoanaerobacterium saccharolyticum*, and *Thermoanaerobacter ethanolicus*. Engineered strains of *S. cerevisiae*, *Z. mobilis* and *Escherichia coli* were also reported in the literature for ethanol production from pentoses (Kuhad et al. 2011).

Despite the immense unexplored microbial diversity, high strengths and specific characteristics, required for industrial fermentation processes, may not be found in nature. While in nature microbial physiological traits are primarily intended for reproduction, many fermentative processes require optimization of processes and traits that may not be relevant for survival. The techniques used for the improvement of microbial strains can be classified as classical genetics (induced mutations, clonal selection, sexual hybridization, artificial hybridization and evolutionary engineering) and those based on genetic transformation (deletion or regulation of genes, pooled-segregant whole genome sequence analysis, and CRISPR/CAS9 system). In this chapter, the main microbial breeding techniques used for 1G and 2G ethanol production will be addressed. Specifically, we discuss what strain improvement programs, based classical and genetic-directioned transformation, have helped to increase fermentation productivity and the resulting decreases in costs in the bioethanol production industry.

3.2 Concept of Fermentation Process for 1G and 2G Ethanol Production

The industrial fermentation process configuration for ethanol production varies according to the raw material, the fermenting microorganism, and the technological inputs (e.g., for the recovery of fermentation gas). Ethanol production from sugarcane is the simplest and most cost-effective process because the sugars extracted from the raw material (sucrose, glucose, and fructose) are readily fermentable by *S. cerevisiae*. Sugarcane stalks are crushed, releasing the juice and the bagasse, which is directed to an efficient energy co-generation system. The juice is treated to remove impurities and contaminants and fermented. The Melle-Boinot process with cell recirculation is a recommended alternative, used in 75% of Brazilian facilities. This process can also be adapted to 2G ethanol production. Fermentation is conducted in (usually) open tanks starting with a yeast inoculum that occupies approximately 1/3 of the volume of the tank. The fermentation is operated as a fed-batch to avoid excessive foam formation. When the tank is full and the consumption of the carbon source is complete, the fermented broth is centrifuged to separate the yeast biomass, with is directed to acid treatment (H_2SO_4, pH 2.5–3.0, 1 h) to inactivate weak cells and contaminants. This yeast cream is then directed to a new fermentation cycle, while the clarified wine is distilled, yielding ethanol and a liquid residue named vinasse. Through this process, more than 90% of the yeast cells are reused, as the viability is maintained between 90–80% (Karp et al. 2021).

Ethanol production from maize has some additional steps, necessary to convert starch into fermentable sugars. The raw material is usually dry-grinded, water is added to produce a slurry which is then gelatinized at high temperature and liquefied by the action of α-amylase enzymes. The broth is then saccharified by the action of glucoamylases, yielding a glucose-rich broth which is then fermented by microorganisms, usually *S. cerevisiae*. Simultaneous saccharification and fermentation (SSF), where glucoamylase and yeast are applied at the same time, is also possible. Ethanol is then recovered by distillation and the yeast biomass, together with other residual solids originated from corn, are separated from the vinasse by centrifugation, originating a co-product named wet distiller's grains (WDG). After drying, this material can be commercialized as distiller's dried grains (DDG). When the remaining vinasse is concentrated through evaporation yielding a syrup, this can be incorporated to the WDG and dried, yielding the so-called DDG with solubles (DDGS) (Kumar and Singh 2019).

Finally, 2G ethanol production from lignocellulosic biomass is in great part dependent on pretreatment and saccharification steps to release fermentable sugars. The aim of the pretreatment is to separate lignin and reduce the crystallinity of cellulose. Then, saccharification takes place by the action of cellulase and hemicellulase complexes, releasing hexoses (mostly glucose) and pentoses (xylose and arabinose). The next challenge is the conversion of both hexoses and pentoses to ethanol. This step often requires the use of modified yeasts that cannot be recirculated in the process. Further information on pretreatment and saccharification of lignocellulosic biomass and 2G ethanol fermentation can be found elsewhere (Santos et al. 2020).

3.3 Prospection and Isolation Strategies of Microorganisms for Ethanol Production

Microbial bioethanol production route emerged in the last decades. Demand for higher productivity, lower costs and the use of sustainable alternatives resulted in massive exploration of consolidated bioprocessing (CBP) microbes including bacteria, yeast and fungi. Biofuel production depends on finding efficient strains for entire large scale fermentation process (Parisutham et al. 2014). Ideal microorganisms for biofuel production should perform enzyme hydrolysis of lignocellulosic biomass and be able to utilize multiple complex sugar substrates (Kim et al. 2010). Complete utilization of pentose-rich and hexose sugars from residues (e.g. agricultural waste) and tolerance to inhibitory compounds generated during the pretreatment step are key points in strain selection (Adegboye et al. 2021). Inhibitory compounds during fermentation include aromatic molecules derived from lignin, acetate from hemicellulose and aldehydes from sugar metabolism. Besides resistance to these end products and high ethanol production rates, the strains must present high growth rates, withstand high temperatures and low pH, and have high metabolic fluxes to biosynthesize single fermentation products. Microbial physiology must be chosen in accordance to technical operation issues of large scale bioreactors. For example, ability to survive at high temperatures is desired as it makes temperature control during the process easier and more cost-effective. Greater temperatures also improve reaction rates, decrease the viscosity of the fermentation medium, and reduce contamination risk during the process. In the same way, the ability to tolerate lower pH can be useful to reduce bacterial contaminants (McMillan and Beckham 2017).

Taking into consideration the aforementioned characteristics, the selection of microorganisms can be made by isolation and identification of native strains from the environment, or through genetic manipulation of model organisms (Table 3.1). The discovery of native strains can be extremely advantageous for industrial bioprocess. Ethanologenic extremophile bacteria, for example, have a physiological structure and enzymes that are highly resistant to industrial processes, such as high temperature, pressure and salinity (Adegboye et al. 2021). In this sense, the isolation of extremophiles, thermophilic and thermotolerant is promising for bioethanol production. They are ubiquitous in nature and can be retrieved from geothermal areas (hot springs), soil, deep sea vents, river bank sediments, artificial habitats (self-heated compost piles, solid waste or sewage sludge), and thermally treated foods. Most of thermophilic species are obligate or facultative anaerobes due to the low oxygen concentrations of their native geothermal environments (Scully and Orlygsson 2015).

Other microbial features, such as stress tolerance and production of essential enzymes to bioethanol generation (e.g. xylanolytic and cellulolytic), are searched among the strains. Sources for the recovery of ethanol-producing microorganisms include diverse environmental samples, ranging from agroindustry residues, soil, organic waste, animal rumen and insect gut (Table 3.1). The choice of environment and isolation strategy depends on the necessity of the fermentation process. Field et al. (2015) isolated *S. cerevisiae* NCYC 2826 from grape must in Norwich located in England, seeking a furfural resistant strain. Lignocellulose must be pretreated with steam explosion and enzymatic hydrolysis to release glucose for bioethanol production. This process can lead to glucose and xylose dehydration to furfural and hydroxymethylfurfural (HMF), respectively, which are toxic to yeast growth and ethanol fermentation. Furfural resistant strains are, then, highly appreciated. After

Table 3.1 Sources of isolation of ethanologenic microbial strains

Microbial group/species	Environmental source	Characteristic	References
S. cerevisiae NCYC 2826	Grape must	Furfural resistance	Field et al. (2015)
S. cerevisiae NCYC 3451	Wort (beer spoilage yeast)	Furfural resistance	Field et al. (2015)
S. cerevisiae strains NCYC 3284 and NCYC 3312	Soil	Furfural resistance	Field et al. (2015)
S. paradoxus NCYC 3277	Oak bark	Furfural resistance	Field et al. (2015)
Candida tropicalis, Pichia kudriavazevii, Candida tropicalis, P. kudriavazevii, S. cerevisiae, and	Rotten fruits	Thermophilic	Choudhary et al. (2017)
S. cerevisiae Wickerhamomyces anomalus JRC7	Distillery waste	Thermophilic	Choudhary et al. (2017)
Herbivorax saccincola	Bovine manure compost	Xylanolytic enzymes	Aikawa et al. (2018)
Chryseobacterium	Cattle dung	Cellulase–xylanase producer	Tan et al. (2018)
Acidothermus cellulolyticus 11B	Acidic hot springs	Tri-functional enzyme having endo-xylanase, arabinofuranosidase and acetyl xylan esterase activities	Shahid et al. (2018)
Anoxybacillus kamchatkensis NASTPD13	Hot Springs	Thermostable alkaline xylanase	Yadav et al. (2018)
Caldicoprobacter sp. CL-2	Bovine manure compost	Xylanase activity	Widyasti et al. (2018)
Thermoanaerobacter sp. DBT-IOC-X2	Hot spring	Thermophilic	Singh et al. (2018)
S. cerevisiae Fm17	Grape marc	Thermotolerant	Favaro et al. (2013)
Bacillus pumilis, B. licheniformis, Paenibacillus dendritiformis, B. cereus and *pseudomonas aeruginosa*	Soil samples cultivated on sugarcane bagasse	Cellulolytic and ethanologenic	Chaudhary et al. (2017)
Candida sp. *S. cerevisiae*	Wastewater-grown microalgae	Ferment saccharified microalgae sugars	Romero-Frasca et al. (2021)
Pseudozyma hubeiensis STAG 1.7, *Hannaella pagnoccae* STAG, and *C. tropicalis* TS32	Insect gut	Xylanolytic activity	Tiwari et al. (2020)
Klebsiella sp. *PRW-1*	Agricultural waste	Cellulolytic enzymes	Waghmare et al. (2014)

(continued)

Table 3.1 (continued)

Microbial group/species	Environmental source	Characteristic	References
Bacillus licheniformis	Sugarcane bagasse	Alkali-thermophilic	Ahmad et al. (2017)
Pichia kudriavzevii KVMP10	Soil	Thermotolerant	Koutinas et al. (2016)
Thermoanaerobacter pentosaceus	Rapeseed straw	Extreme thermophilic	Tomás et al. (2013)
Saccharomycetaceae *Candida* sp. *Kluyveromyces marxianus* *Pichia kudriavzevii* *S. cerevisiae* *Pichia kudriavzevii D1C*	Mango pulp–peel compost	Thermo- and osmo-tolerant	Dandi et al. (2013)
S. cerevisiae	Grape wine and medicinal herbs	Tolerant to stresses associated with second-generation biofuel production	Dubey et al. (2016)
S. paradoxus NBRC 0259	Natural and fermentative habitats	–	Ota et al. (2013)
Trichoderma harzianum WF5	Decaying wood	Glycosyl hydrolases	Kaushal et al. (2016)
Spathaspora passalidarum CMUWF1–2	Soil	Thermotolerant	Rodrussamee et al. (2018)
Saccharomyces cerevisiae F111 and a *Kluyveromyces marxianus* WR12	Distillery	Thermotolerant	Abdel-Fattah et al. (2000)
Candida bombi, Wickerhamomyces anomalus and Torulaspora delbrueckii	Soil and sugar-rich habitats (floral nectar or sugar beet thick juice)	–	Ruyters et al. (2015)
Clostridium thermocellum	Compost sample	Thermophilic	Lv and Yu (2013)
Klebsiella oxytoca THLC0409	Lignocelluloses-degrading microflora	–	Tran et al. (2011)
S. cerevisiae *Kluyveromyces marxianus*	Bio-ethanol production plants Beer plant "Cocoa" fermentations	–	Pereira et al. (2014)
Geobacillus thermoglucosidasius	Waste compost	Thermophilic	Fong et al. (2006)
Thermoanaerobacter mathranii sp.	Hot spring	Extremely thermophilic	Larsen et al. (1997)
Thermoanaerobacter J1	Hot spring	Thermophilic	Jessen and Orlygsson (2012)

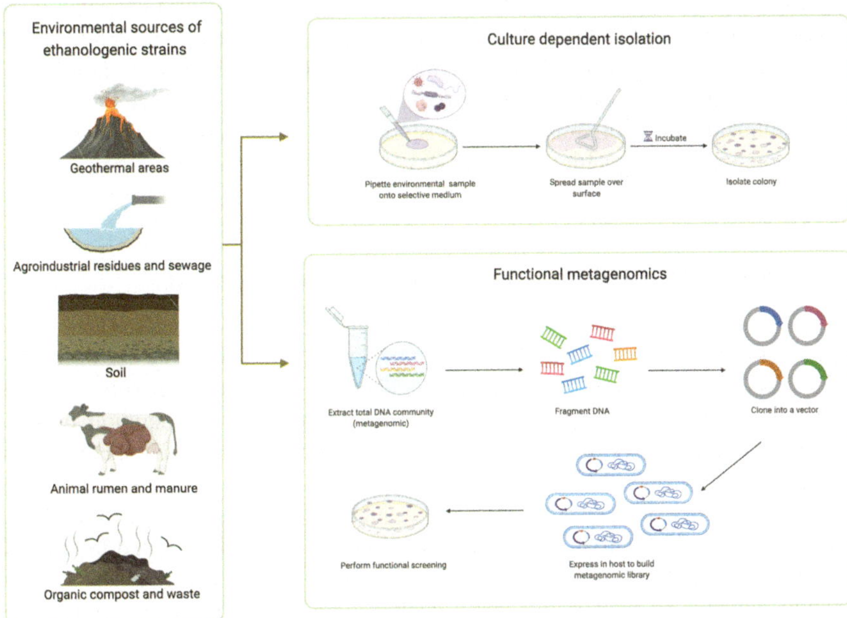

Fig. 3.1 Schematic workflow for bioprospecting ethanol-producing microorganisms

isolation of microorganism with desired characteristics, identification is performed through extraction of genetic material and sequencing (Liti et al. 2009). Identification is usually performed through ITS region sequencing for eukaryotes and 16S rDNA for prokaryotes. Successful use of wild strains to produce biofuel demand deep understanding of their physiology and metabolism. Despite time consuming efforts for bioprospecting novel strains, it is a promising alternative.

Another strategy to access novel ethanologenic strains from environmental samples is through metagenomics. The screening is based on the use of probes and primers specific to genes of interest (Fig. 3.1). Metagenomic analysis estimates whole microbial community of the sample, including uncultivable microorganisms, sub-dominant populations, and late-growing species, facilitating prospecting potential strains (de Melo Pereira et al. 2020). Total community DNA can be sequenced or fragmented, cloned into a vector and introduced to a host to construct a metagenomic library. This method is called 'functional metagenomics' and focus on searching for phenotypes or novel genes of interest that are expressed in the host cells (Loaces et al. 2016). Several functional enzymes are identified from natural sources, such as cow rumen, termite gut, soil, and plant- or algae-associated.

Multifunctional glucanase and xylanase were identified by functional metagenomics in a bovine microbiota library for ethanol production on cane biomass (Loaces et al. 2017). Besides searching novel microbial strains, metagenome can improve bioethanol production by tracking and understanding microbial dynamics responses through fermentation process. Commercial scale silage inoculated with ethanologenic yeasts had their ethanol production yield boosted by the presence of indigenous lactic acid bacteria (Gallagher et al. 2018). Large-scale bioethanol production using complex residues for fermentation generally is affected by action of environmental microorganisms. In this sense, the microbiota of these systems still needs to be unveiled to achieve even more cost-effective processes.

3.4 Diversity of Ethanol-Producing Microorganisms

Ample range of microorganisms performs bioethanol production. The use of microbial strains to achieve a high yield process depends on the type of substrate and fermentation process physical conditions. Bioethanol of first generation utilizes homogeneous substrates for fermentation process, such as starch and sucrose. The mesophilic ethanologenic *Saccharomyces cerevisae* and *Zymomonas mobilis* are the pioneers in history of large-scale ethanol production (Scully and Orlygsson 2015). *S. cerevisae* is a facultative anaerobic yeast, and it is naturally adapted to ethanol fermentation with alcohol yield up to 90%. In addition, *S. cerevisae* have high ethanol tolerance as well as for some chemical inhibitors (Limayem and Ricke 2012). *Z. mobilis* is an ethanologenic gram-negative bacterium with greater alcohol yield up to 97%, which does not require additional oxygen to growth. However, *Z. mobilis* is highly sensitive to osmotic pressure, oxidative stresses and ethanol presence (Tan et al. 2016). In addition, it uses Entner Doudoroff pathway leading to unwanted end-products formation during catabolism of sucrose and fructose (He et al. 2014).

In the 2G bioethanol fermentation, lignin needs to be removed and the sugar polymer needs to be released to turn them accessible for the hydrolytic and enzymatic steps. This additional pretreatment step is the main difference between the utilization of simple biomass from 1G (sugars and starch) compared to complex biomass. Also, the sugar fraction released in the medium is more heterogeneous as compared with first generation substrates (Scully and Orlygsson 2015). In this sense, the interest on strains capable of using complex substrates has increased. Mesophilic natural ethanologenic yeast species, such as *Pichia stipilis*, *Kluyveromyces marxianus*, *Candida tropicalis* and *Candida shehatate,* are promising in replace *S. cerevisiae* for ethanol production from lignocellulosic biomass. These microbial species are able to ferment xylose (one of the most abundant pentoses sugars) to ethanol (Selim et al. 2018). *P. stiplis* is a facultative anaerobic yeast with the best ethanol yield (82%) from xylose. Also, *P. stiplis* can ferment cellobiose, glucose, and galactose. However, this yeast has some drawbacks, such low fermentation efficiency at low pH, sensitivity to chemical inhibitors, and require micro-aerophilic conditions to reach peak performance (Limayem and Ricke 2012). Studies had tested the co-culture of *Zymomonas mobilis*, *Pichia stipites*, and *Pachysolen tannophilus*. *P. tannophilius* is also a mesophilic, ethanologenic yeast. It is aerobic and also ferment xylose. Results proved potential synergistic utilization in mixtures containing glucose and xylose to ethanol production (Fu and Peiris 2008; Fu et al. 2009).

The mixing operations of second generation bioethanol are facilitated at elevated temperatures due to reduced viscosity and greater substrate addition. The mass transfer rates are higher and the risk of mesophilic contamination is lower (Turner et al. 2007). Thermophilic microorganisms are, then, suitable for second generation ethanol production as they have high operating temperatures and diverse enzyme apparatus to utilize wide range of substrates compared to *S. cerevisae* and *Z. mobilis*. Examples of highly ethanologenic thermophilic includes *Clostridium, Caldanaerobacter, Thermoanaerobacter, Thermoanaerobacterium, Bacillus, Geobacillus, Paenibacillus*, and Caloramator. The cultivation of thermophilic microorganisms usually does not require extensive mixing, cooling, or heating of the fermentation vessel, and it is possible to recovering ethanol from the fermentation broth by in situ vacuum distillation (Scully and Orlygsson 2015). Also, thermophiles frequently tolerate extremes pH and salt values during fermentation, have low nutritional necessity, and are generally recognized as safe (GRAS).

Notable bacteria species are *Thermoanaerobacterium saccharolyticum, T. ethanolicus*, and *Clostridium thermocellum*. They are extreme anaerobic bacteria and resistant to high temperatures (above 70 °C). They ferment wide variety of sugars, display cellulolytic activity, and amenability to genetic modification (Limayem and Ricke 2012). Thermoanaerobacterium are commonly amylolytic and xylanolytic with optimal temperature between 55 °C–65 °C. Thermoanaerobacterium strains have been isolated from geothermal areas and from thermally treated foods (Scully and Orlygsson 2015). Thermoanaerobacter species have similar characteristics. They have saccharolytic enzymes and produces ethanol, lactate, and acetate. Other thermophilic ethanologenic bacteria genera include *Caldicellulosiruptor, Caloramator* (e.g. *C. boliviensis*), *Geobacillus* (e.g. G. *thermoglucosidasius*) and *Paenibacillus*. Promising thermophilic yeast also produce ethanol. Yeasts have advantages over bacteria due to the thickness of their cell walls, less nutritional requirements, large sizes, resistance to contamination, and growth at low pH (Selim et al. 2018). Classic example is *Kluveromyces marxianus*, which is able to grow above 52 °C, reducing cooling cost in the process as well as contamination rates and ferments broad spectrum of sugars. However, some drawbacks, such as low ethanol tolerance and excess of sugars, have been shown to affect the yield of ethanol in *K. marxianus* (Limayem and Ricke 2012).

3.5 Process-Driven Selection Strategy for Recovery of Ethanol-Producing Microorganisms

Some relevant characteristics must be taken into account for bacteria and yeasts to be considered ethanologenic, such as (1) good ethanol production in short fermentation time; (2) high ethanol tolerance; (3) stress resistance (e.g. pH, osmolarity, temperature); (4) tolerance to repeated fermentation cycles; (5) low glycerol production; (6) non-flocculant; and (7) genetic stability (Brexó and Sant'Ana 2017; Jacobus et al. 2021). These characteristics are commonly desired in strains used for first-generation bioethanol production from sugarcane juice. However, the stress generated in microorganisms used for bioethanol production from corn starch is slightly different, mainly because these fermentations generate high levels of ethanol (13–23% v/v) compared to sugarcane (7–13% v/v). In addition, the fermentation of corn starch does not allow the yeasts used in the process to be recycled, because the high concentration of solids present in the fermented material makes the cell recycling process unfeasible (Favaro et al. 2019).

In recent years, different studies have demonstrated the microbiological profile of ethanol fermentation, which is initiated by a high diversity comprising yeasts of the genera *Saccharomyces*, *Schizosaccharamyes*, *Kluyveromyces*, *Pichia*, and some bacteria such as *Lactobacillus*, *Leuconostoc*, *Acetobacter*, *Bacillus Escherichia*, *Klebsiella*, and *Zymomonas* (Basso et al. 2008; Dhaliwal et al. 2011; Muthaiyan et al. 2011; Madeira-Jr and Gombert 2018). Molecular studies using the PCR-fingerprinting method based on microsatellite primer $(GTG)_5$ were used to characterize the yeast population dynamics along the fermentation period in six distilleries from bioethanol and revealed the marked presence of *S. cerevisiae*. Furthermore, the results showed that the baker's yeast strains used as starter cultures were rapidly replaced by the indigenous yeasts that were present in the substrate (da Silva-Filho et al. 2005).

Subsequently, a study by Basso et al. (2008) evaluated the microbial ecology and dynamics for 12 years (1993–2005) in several Brazilian distilleries. During this period, 1160 samples were collected and plated on a YPD culture medium. After yeast growth, about 11 colonies were submitted to a karyotyping analysis using pulsed-field gel electrophoresis (PFGE) technique. From this, 300 indigenous strains were selected. Initially, the isolates were subjected to laboratory fermentations using industrial substrates (sugarcane juice and molasses) as a carbon source, mimicking the industrial process with the cell recycling step. Only six strains (PE-2; SA-1; CAT-1; VR-1; BG-1; JP-1) were selected because they had excellent fermentation performance and lack of undesirable characteristics such as foaming and flocculation. These results suggest that, although some strains dominate the fermentation and remain viable until the end of the process, it is not a guarantee that they exhibit all the desirable fermentation characteristics.

It is known that the high presence of wild yeasts and bacteria cancause significant losses, because these contaminating microorganisms compete with the starter culture for carbon sources, macronutrients, and micronutrients. The 1% reduction in ethanol production can have a great impact on the financial health of alcohol industries since some of them work with narrow profit margins (Muthaiyan et al. 2011). Monitoring these contaminants is not straightforward and analyses can often take weeks. Therefore, molecular techniques have been successfully employed in some distilleries as reported by De Souza Liberal et al. (2005). The procedure consists of DNA extraction from pure or mixed cultures and use for amplified ribosomal DNA (rDNA-PCR) fragment sizes to identify the presence of non-*Saccharomyces* yeasts in the fermentation. The amplified DNA fragments of the contaminants ranged from 400 to 700 bp, while those of *Saccharomyces* were greater than 800 bp (De Souza Liberal et al. 2005, 2007). These procedures, together with other methods such as PCR-fingerprinting with single-primed amplification reactions using the $(GTG)^5$ primer and DNA sequencing, have obtained satisfactory results. These methodologies, however, are still expensive and do not fit the routine of some distilleries (De Souza Liberal et al. 2007).

Bioethanol production is one of the greatest examples of a bioprocess ever undertaken. Although the process has a high yield (>90.0% theoretical yield), these values have remained unchanged for about 20 years. Therefore, innovative technologies are needed as conventional solutions have not presented great impacts (Brexó and Sant'Ana 2017). Among them, performing the fermentation at high temperatures (≥45 °C) seems to be an interesting alternative, because this condition has the potential to (1) decrease the microbial contamination rate; (2) decrease the amount of energy and water needed to cool the fermenters; (3) decrease the energy used in the distillation step (separation of ethanol from the fermented broth) (Costa et al. 2001; Madeira-Jr and Gombert 2018). However, all of the commercially S. cerevisiae starter cultures are metabolically inefficient at temperatures above 40 °C. Therefore, for this bioprocess to be viable, new thermotolerant yeast strains need to be isolated and characterized (Della-Bianca and Gombert 2013; Jacobus et al. 2021).

The process of isolation and selection of thermotolerant microorganisms is similar to the methodologies used by Basso et al. (2008), mentioned above. Nonetheless, it is recommended that isolation of thermotolerant strains be performed at temperatures ≥35 °C (Banat et al. 1992). Recently, a work conducted by Madeira-Jr and Gombert (2018) evaluated the ethanol production capacity of 20 *Kluyveromyces marxianus* strains. Interestingly, *K. marxianus* strains NCYC 3396 and UFV-3 showed promising results, as they were able to produce ethanol at 48 °C using glucose as a carbon source, and yields were similar to those exhibited by *S. cerevisiae* at 37 °C. However, when the authors mimicked the actual conditions of an alcohol industry - i.e. addition of molasses and ethanol to the culture medium and five cycles of cell recycling, they observed that cell viability was severely compromised after the first cycle at 45 °C and 48 °C. Taking into consideration the above-mentioned characteristics, metabolic and evolutionary engineering techniques can be used to improve NCYC 3396 and UFV-3 strains and introduce them as starter cultures in the industry.

Therefore, the genetic improvement of strains belonging to the genus *Kluyveromyces* presents great potential because, besides growing at temperatures ≥40 °C, they can ferment other sugars such as xylose (Signori et al. 2014). These metabolic peculiarities enable the use of these microorganisms in a more efficient way for second-generation ethanol production. During the saccharification process of lignocellulosic biomass, it is necessary to use high temperature due to the hydrolytic enzymes used in the process Thus, thermotolerant yeast can be used without the need to cool the reactors after hydrolysis, decreasing the process time leading to increased productivity (Moreno et al. 2013; Gao et al. 2014).

3.6 Classical Genetic Tools

3.6.1 Adaptive *Evolution*

Several strategies were developed to improve yeast strains for industrial application based on multitolerance performance. Methods such as adaptive evolution, protoplast fusion, direct mating and mass mating could be used to develop 1G and 2G-ethanol production strains (Fig. 3.2).

Mutations promote adaptation in a changing environment—an event assumed to occur by chance and drive evolution. However, rates of spontaneous mutation are approximately 10^6 per generation (Pretorius 2000). Stressful conditions are imposed on yeast in the industrial production of bioethanol, exerting a selective pressure on the population of wild strains. The industrial process can select strains with multiple stress tolerance, but not necessarily result in variants with desirable attributes (high ethanol yield, low foam formation, no flocculation, low glycerol formation, amongst others) (Basso et al. 2008).

Adaptive evolution mimics natural evolution through random mutation of microorganisms' own genes, which create strains with specific enhanced characteristics (Sandberg et al. 2019). Directed evolution can be used towards the desired phenotypes present in the original population to improve the chosen trait. Yeast strains can be exposed to increasing selective pressure, forcing the desired phenotype to be obtained, relying on the occurrence of genetic variations, mutations, and natural selection (Mans et al. 2018). This process takes place over subsequent generations, where mutation events select the strains best adapted to the stress environment used (Zheng et al. 2021). This technique was used to select microbial strains for very high-gravity bioethanol fermentation by a freeze–thaw method combined with stress shock selection, since the increase in ethanol concentration during fermentation leads to a decrease in viability and, consequently, a reduction in yield (Zhang et al. 2019). In another example, it was possible enhancing acetic acid tolerance ($3\ g\ l^{-1}$ up to $13\ g\ l^{-1}$) in batch cultures of *Saccharomyces cerevisiae* through adaptive laboratory evolution for bioethanol 2G production (Gurdo et al. 2018).

3.6.2 Protoplast Fusion

Studies have been carried out to assess the effect of ploidy on evolution adaptive, demonstrating a difference between haploids and diploids (Selmecki et al. 2015; Voordeckers et al. 2015; Gurdo et al. 2018). Diploid yeasts showed improved xylose fermentation phenotypes when compared to haploid strains (Chen et al. 2012). The experimental evolution for adaptation to high ethanol was carried out in populations of haploid, diploid and tetraploid S. cerevisiae for approximately 2 years in a turbidostatic, showing changes in ploidy and phenotypes (Voordeckers et al. 2015).

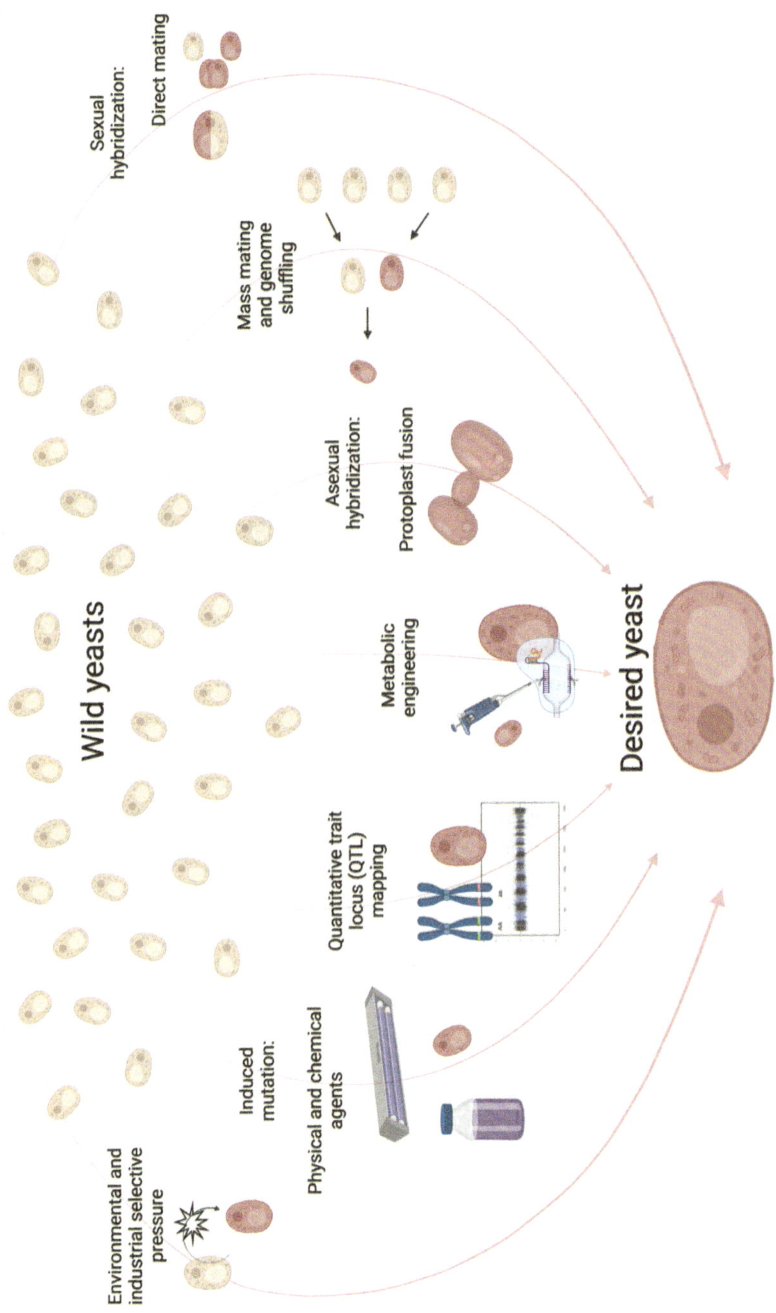

Fig. 3.2 Schematic representation of the different microbial breeding methods used for IG and 2G ethanol production

Another viable approach to developing new yeast strains for 1G and 2G bioethanol is sexual hybridization, such as direct mating, mass mating and genome shuffling. This is possible because industrial strains can sporulate when exposed to conditions of nutritional deficiency and in the presence of a non-fermentable carbon source, such as potassium acetate (Gerke et al. 2006).

Yeasts can be classified according to the reproductive cycle in homothallic or heterothallic. Homothallism is the ability of haploid cells to change the mating type, forming cells with the opposite mating type from that of the cell of origin. This is because homothallic species are capable of expressing HO gene, which encodes an endonuclease responsible for converting mating type "a" to mating type "α", or vice versa. In this way, haploid cells are able to cross each other and become diploid (Katz Ezov et al. 2010; Haber 2012). The heterothallic strains have a stable mating type, and the crossing can occur through haploid spores originating from different tetrads or originating from the same tetrad, but they are not able to self-diploidized (Hicks and Herskowitz 1977).

Hybridization is a technique to obtain new species with distinct advantages for the originating organism (Rainieri and Pretorius 2000). To obtain hybrids with better performances, the degree of reproductive isolation of the parent species and the hybrids generated are the main limiting factors, since the hybrids must be self-fertile and sufficiently capable of reproducing in isolation and maintaining a distinct strain (Greig et al. 1997). Heterosis, a phenomenon that results in strains with characteristics superior to the parental strains, will depend on the strains used and the effectiveness of the technique (Lippman and Zamir 2007). Corroborating these ideas, Shapira et al. (2014) used genetic models to explain heterosis and concluded that complementation of the recessive allele domain, within-locus interactions, and epistasis, collaborate with hybrid vigor. From the gene expression, the generated hybrid presents resources in the function of metabolic pathways and cellular processes (Schnable and Springer 2013).

3.6.3 Direct Mating

Direct mating, although time-consuming to perform when compared to other protocols of crossing, is an effective way of obtaining improved strains and crossing can be done inter- or intra- species (Sipiczki 2008). After the sporulation period of the selected strains, ascus containing four spores are formed are exposed to enzymatic digestion, and the haploids are released. With the aid of a micromanipulator, the tetrads are dissected, and the haploids obtained are typified according to the mating type for subsequent crossing (Hicks and Herskowitz 1977). The main advantage of targeted cross is that the parental haploids can be fully phenotyped before performing the cross, which increases the chance of producing a hybrid with the desired characteristics.

The direct crossing associated with adaptive evolution technique was utilized between engineered xylose-fermenting strains of *S. cerevisiae* with wild strains from various *Saccharomyces* species for the production of cellulosic ethanol (2G). Genome sequencing showed hybrids underwent rearrangements, duplications, and deletions, resulting in improved efficiency fermentative and, in some cases, heterosis (Peris et al. 2017a).

3.6.4 Mass Mating

Mass mating technique allows producing a large number of haploid cells coupled to the advantage of genetic diversity exploitation. This process induces the combination of useful mutations from several different individuals (Horinouchi et al. 2020). It is a strain improvement technique which allows cultivating an expressive number of haploid cells from different parental strains, allowing the generation of quick and random crosses; however, the cross is random, and the diploid formed can be originated from the cross from two strains or between the same strain. This type of crossing allows the crossing of heterothallic and homothallic strains and strains that have low reproductive efficiency (Tao et al. 2012). In this method, the steps of sporulation and digestion of the wall of the asci are similar to directional crossing; however, it is not necessary to dissect the tetrads and identify the sexual type, which makes the process faster to be performed. Successful mass mating was realized to improve high acetic acid tolerance present in 2G bioethanol (Meijnen et al. 2016).

Genome shuffling is a method to improve biotechnological processes that is related to mass crossing. However, repeated cycles of sporulation and mating are carried out with the advantage of exploiting genetic diversity through the recombination of genomic information from various parental strains (Giudici et al. 2005). It has been used in experiments related to metabolism (the production of metabolites can be recombined during several rounds of the genome) and stress tolerance (Giudici et al. 2005; Jin et al. 2009; Meijnen et al. 2016). This method can be used individually or associated with other technologies.

In order to obtain a strain of *S. cerevisae* with better performance for high gravity fermentation, the metabolic engineering technique was associated with three cycles of the genome shuffling process and, in the end, a hybrid with lower glycerol production and higher ethanol yield was selected (Tao et al. 2012). Recombinants of *S. cerevisiae* were produced to tolerate inhibitory present in hydrolysates of lignocellulosic biomass by genome shuffling, and mutants were identified via whole-genome resequencing (Chen et al. 2015).

Protoplast fusion can be used for hybridization with interspecific or intergeneric species, yeasts with sporulation deficient or unstable mating type. First step is the generation of protoplasts, *i.e.,* yeast without cell wall. It is achieved with enzymes in stable osmotic medium. After, protoplast fusion and cell wall regeneration are performed. However, chromosomal loss or increase the ploidy can occur, and the characteristics obtained from both parents are difficult to estimate. Interesting results have been obtained using protoplast fusion for high-ethanol-producing (Ge et al.

2014) and bioethanol production by a flocculent hybrid obtained by the crossing between *S. cerevisiae* and *S. bayanus* (Choi et al. 2010).

3.6.5 Classical Genetic Tools

Mutations correspond to an inheritable change in the nucleic acid base sequence in the organism of a genome, resulting in small genetic changes. These mutations can promote changes, which can be either beneficial or harmful, and they can be spontaneous or induced. From advances in recombinant DNA technology and the use of synthetic DNA, it has become possible the induction of specific mutations, in specific genes. Induced mutations are deliberately carried out at specific sites in the genome—this procedure has been called mutagenesis (Steensels et al. 2014).

Many microorganism' isolation and selection strategies have been employed to find new improved strains; however, the enormous biodiversity makes this task quite complex. It is reported that the stress tolerance in microorganisms is regulated by complex gene interaction. Thus, mutagenesis has been demonstrated as a powerful strategy to enhance tolerances in yeast strains. Within the knowledge of genetic engineering, regarding to induced mutations, it has been possible to obtain more versatile microorganisms which are capable to improve the productivity and the yield of the processes. A combination of mutagenesis, selection and cross-stress protection methods can be used to improve and optimize bio-ethanol production from new feedstocks. Some recent studies have demonstrated the possibility of mutation in ethanol-producing microorganisms, using chemical and physical mutagens, such as *N*-methyl-*N′*-nitro-*N*-nitrosoguanidine (MNNG), ethyl methane sulfonate (EMS), ultravioleta (UV) radiation, X-ray, and atmospheric and room temperature (ARTP) (Kumari and Pramanik 2012).

Ultraviolet mutagenesis of *S. cerevisiae* is one the most effective methods to improve a strain with tolerance ability against glucose and ethanol. A UV mutagenic *S. cerevisiae* strain produced 122 g/L of ethanol before temperature optimization (Yi et al. 2018). Mutants of *Pichia stipitis* NBRC1687 were obtained after UV mutagenesis having higher ethanol yield from xylose (Watanabe et al. 2011).

EMS mutagenesis stimulates the activation of specific molecular stress response mechanisms with results in higher levels of resistance to stress conditions. The mutants have better ethanol yields and fermentation efficiency than the parent strain, after that they have been exposed to high concentrations of ethanol (Hemmati et al. 2012). A commercial *S. cerevisiae* subjected to mutagenesis using EMS produced bioethanol 17.3% more than the wild type (Mobini-Dehkordi et al. 2008).

The ethanol production by yeast strain can also be improved by sequential mutagenesis using EMS, MNNG, and UV radiation. Kumari and Pramanik (2012) obtained a improved *S. cerevisiae* strain by using multiple stresses during fermentation of glucose-xylose mixture, producing around 48 g/L ethanol and ethanol yield of .0,295 g/g. *S. cerevisiae* KF-7 mutants with improved ethanol or heat tolerance were obtained by combining ARTP mutagenesis and several rounds of genome shuffling (Wang et al. 2021). The mutant strain had a better development under various stress conditions compared with the parental strain. The ethanol production was increased 10.14%–81.02% under high ethanol, high temperature, and high osmosis stress conditions. ARTP is a mutagenesis based on radio-frequency atmospheric-pressure glow discharge plasma, which features higher mutation rates than UV radiation or chemical mutagens while maintaining low treatment temperatures (Wang et al. 2021).

3.7 Pooled-Segregant Whole Genome Sequence Analysis

The classical development of bioethanol yeast strains is based on the selection of desirable discrete phenotypes or through the manipulation of previously established genetic diversity using random or directed mutations, hybridizations, or adaptative evolution. Although satisfactory results have been achieved in high ethanol productivity/yield and sugar consumption (Peris et al. 2017b; Yi et al. 2018), the high selective pressure imposed by these methods can result in the perpetuation of disadvantageous mutations that may affect the expression of polygenic phenotypes of interest (Swinnen et al. 2012a). Flocculation, metabolization of non-fermentable carbon sources, and tolerances to stressful conditions are some of the heritable traits that may be affected, once its expression results from the summative and regulatory effects of multiple genetic positions, referred as quantitative trait loci (QTL) (Nieduszynski and Liti 2011). Based on this premise, QTL mapping relies on the linkage between the identification of molecular markers and the variations of mean values observed for the quantitative phenotype in the individuals (MacKay et al. 2009). Single nucleotide polymorphisms (SNPs), simple repeated sequences (microsatellites), and polymorphic insertion/deletion (indel) are the molecular markers most widely used (Fig. 3.2). Due to this complex genetic architecture, allied with epistatic effects and gene-environment interactions, methods capable of identifying collectively the QTL are necessary.

A straightforward approach for the QTL mapping associated to quantitative traits of industrial interest is the pooled-segregant model, also known as reciprocal hemizygosity. In this approach, the natural distribution of QTL is observed in segregants generated from the cross between two parentals distinguishable by the trait of interest. A haploid parental with superior phenotype (trait$^+$) is mated with a reference strain lacking the phenotype of interest (trait$^-$). The resulting diploid hybrid (F1) strain is sporulated and a tetrad dissection is performed in order to generate meiotic segregants genetically distinct. This genetic diversity is supported by the principle of meiotic recombination events that occurs at chromosomal and inter-chromosomal levels (Swinnen et al. 2012a). Successive *n* generations can be performed through intercrosses between F1 segregants in order to minimize sampling effects over meiotic recombination across the genome (Darvasi and Soller 1995). Subsequently, the segregant tetrads from F1-*n* are screened through microfermentations for the presence of the trait of interest, which are then pooled and the genomic DNA of each individual is extracted. The sample size of segregants and markers can be determined statistically, but there is not a consensus and it will depend on the intended QTL mapping power and resource availability. A whole genome sequencing is performed and compared with parental strains, where statistical frameworks are developed for the identification of QTL based on the relative allele frequency of both parentals (trait$^+$ and trait$^-$) in the segregant pool (Magwene et al. 2011a). It is important to highlight that the potential loss of energetically cost traits due to human influence during domestication of yeast strains (*i.e.,* specialty *S. cerevisiae* in wine, bread, and ethanol fermentation) did not result in a significant standing variation, which turns them suitable for the investigation of quantitative traits (Liti and Louis 2012).

Over the past two decades, sequencing technology suffered an expressive revolution with the development of the massive parallel sequencing provided by next-generation sequencing (NGS) platforms, which allowed a significant reduction of the sequencing costs per kilobase and increased the coverage accuracy of whole genome sequencing (Zhou et al. 2010). Briefly, these techniques constitute on the fragmentation of genomic DNA into short sequences (250 up to 700 bp), where each DNA molecule is physically separated individually into supports or beads, allowing millions of polymerase-chain reactions (PCR) to occur simultaneously. The generated data (reads) can be: (*i*) aligned to reference sequences of previously sequenced genome, a process known as re-sequencing; or (*ii*) assembled de novo, where overlapped sequences are compared to create continuous stretches of sequences (*contigs* and *scaffolds*). In order to overcome assembling errors, the latter technique requires the integration of different NGS platforms and a more intensive bioinformatics analysis to increase total coverage (Ekblom and Wolf 2014). A discussion regarding whole-genome sequencing and the technology involved will be later presented in this chapter.

In practice, the pooled-segregant model using NGS techniques aligns the whole genome sequences from the screened meiotic segregants against the sequences of a parental. Statistical analysis is then performed to discriminate the differences between each SNP, indel, or microsatellites present in the selected chromosomal positions of screened and control (random unselected segregants) pools (Swinnen et al. 2012a). This recent, high-resolution QTL mapping has been used as an efficient tool for elucidating the genetic basis necessary for the development of industrial-grade yeast strains, a process known as reverse metabolic engineering. Hubmann et al. (2013) used the pooled-segregant whole genome sequencing approach for the identification of SNP involved on the non-selectable trait of glycerol production, a byproduct generated under hyperosmotic pressure that reduces the ethanol yields. After an initial screen for low glycerol yield in 52 *Saccharomyces cerevisiae* potential parental, a cross between the strains CBS6412 (trait$^+$; used in sake production) and Ethanol Red (trait$^-$; used in corn bioethanol production) resulted in selection of 20 trait$^+$ meiotic segregants. Whole-genome sequencing revealed the presence of a major QTL at the gene *SSK1* located in the chromosome XII, which is responsible for the multi-step cascade signaling of the transmembrane osmosensor Sln1 associated to osmoadaptation (Udom et al. 2019). The construction of a plasmid vector containing the allele identified in the meiotic segregants (*ssk1*$^{E330N...K356N}$) and the introduction on the trait^{-} parental revealed a better effect regarding low glycerol production in comparison to the deletion of gene *SSK1*.

A recent study performed by Wang et al. (2019) evaluated the causative QTL associated with thermotolerance in meiotic segregants, which resulted from the cross between the *S. cerevisiae* industrial strains ScY01 (trait^{+}; generated through adaptative evolution from Ethanol Red) and W65 (trait^{-}; wild-type). The authors identified three SNP in the alleles *VPS34, VID24*, and *DAP1* associated with the superior phenotype, presenting protective effects over the cell membrane through increase of trehalose accumulation and reduction of membrane fluidity. Lignocellulosic hydrolysates generated from alkaline or acid pretreatments for 2G ethanol production generates high concentration of growth inhibitors, mainly furfural, 5-hydroxymethylfurfural (5-HMF), and acetic acid, which can result in the prolongation of the latency phase and inhibitory effects against alcohol dehydrogenase, aldehyde dehydrogenase, and pyruvate dehydrogenase (Modig et al. 2002; Swinnen et al. 2014). Adopting a successive intercrosses approach between F1 segregants, resultant from the cross between S. cerevisiae Ethanol Red (trait^{-}) and JT22689 (trait^{+}; fermenting must of sturm), Meijnen et al. (2016) identified a SNP that increases the translation of allele *HAA1*, which is responsible for encoding transcriptional factors that regulates the transcription of plasma membrane proteins associated to weak acid tolerance (Takabatake et al. 2015). These studies reveal that pooled-segregant whole genome sequencing model is a promising tool for the development of metabolic engineered yeasts with desirable quantitative traits for bioethanol production. However, the requirement to perform several small-fermentation batches for each segregant, allied with costly chemical analysis, can be stated as major bottlenecks for more scientific investigations (Swinnen et al. 2012a).

3.8 Deletion or Regulation of Genes

Besides the identification, modification, addition, or overexpression of genes of interest, another possible avenue for the metabolic engineering of bioethanol producing yeasts is the deletion or regulation of undesirable genes (Fig. 3.2). Deletion or knockout (represented with a △) refers to the partial or total removal of a given gene, which blocks the encoding enzymes/transcriptional factors/proteins and, consequently, increases ethanol yield through elimination of byproduct formation pathways (Garrigues et al. 2021). A traditional practice is the restriction of the unwanted gene followed by ligation of the generated blunt-ends in the double-strand DNA. Even though the method is efficient and relatively simple, it becomes laborious and costly when there is a need to silence several genes simultaneously. In this sense, numerous strategies proposed the construction of cassettes for a one-step procedure. An example is the Cre-*loxP* mediated system, where a marker cassette is flanked by short-length sequences (*lox, loxP, lox2272,* and *loxLE/RE*) that acts as recognition sites for the excision by the Cre-recombinase (Sauer 1987). The advantage of this method is that it allows multiple deletions depending on the orientation and position of the *lox* fragments. Other systems, such as *Delitto perfetto* (Storici et al. 2001), DelsGate (García-Pedrajas et al. 2010), and CRISPR/Cas9 system (Min et al. 2016), has also been proposed. For a more in-depth study of the cassette mechanisms developed for gene deletion in yeast, the authors suggest the work of Fraczek et al. (2018).

These silencing mechanisms have been recently used in the modification of yeasts for the bioethanol producing optimization. González-Siso et al. (2015) transformed a *Kluyveromyces lactis* strain with a deletion cassette for the excision of the nucleotides +21 to +1523 from the *KlNDI1* gene, which is responsible for encoding an internal, mitochondrial NADH dehydrogenase (ubiquinone oxidoreductase). The absence of this enzyme results in a redox imbalance in the cell due to the accumulation of NADH in the interior of the cell, forcing it to shift from a respiratory to a fermentative metabolism. Although this study demonstrated the application of the *K. lactis* Δ*klndi1* mutant in the production of bioethanol from cheese whey, it could be applied to the generation of 1G ethanol due to the optimized production from glucose, high ethanol tolerance, and low ethanol consumption as a secondary carbon source. Another study proposed the deletion of genes involved in the down-regulation of galactose metabolism (*GLK1, MIG1,* and *MIG2*) in *S. cerevisiae* during bioethanol production from hydrolysates of red seaweed (*Gracilaria verrucose*) rich in galactose (Sukwong et al. 2020). The simultaneous deletion of the three genes revealed an incremental effect over the galactose uptake and consumption rate, and ethanol yield, showing superior values when compared to the control strain and mutants with only one gene deleted (Δ*glk1,* Δ*mig1,* and Δ*mig2* mutants).

Besides the improvement of ethanol production, the deletion of genes that can increase the tolerance to growth inhibitors present in lignocellulosic hydrolysates has also been an effective strategy. Fujitomi et al. (2012) revealed the enhancement of ethanol production in presence of acid acetic, acid formic, and furfural in rice hydrolysate after deleting the gene *PHO13* in the xylose-fermenting *S. cerevisiae* strain BY4741. Despite the mechanisms associated to this phenotype has not been determined, previous studies demonstrated that the deletion of this particular gene promotes the upregulation of the pentose pathway through changes in the transcriptional profiles of genes involved in redox balance, mainly NADPH restoration, during xylose metabolism (Kim et al. 2015; Xu et al. 2016). A recent study adopted a direct approach: deletion of the plasma membrane acetate transporter gene (*ADY2*) (Zhang et al. 2017). The authors evidenced that *S. cerevisiae* Δ*ady2* mutant showed an enhanced ethanol production in the presence of acetic acid, ethanol, and H_2O_2 stressors for allowing a reduction in the intracellular acetic acid and reactive oxygen species (ROS), and improvement of membrane integrity. These studies demonstrate the plethora of existing possibilities for the development of new bioethanol-producing microorganisms, allowing to overcome several bottlenecks associated to modifications in specific metabolic pathways.

3.9 Whole-Genome Sequencing

Due to the millennial relationship between *Saccharomyces* strains and humans, the first eukaryotic genome sequenced was the one of the yeast *Saccharomyces cerevisiae* S288c in 1996 (Goffeau et al. 1996). This data was essential to provide the first insight into the genetics and physiology of this microbial group, and also allowed the development of various “omics” techniques such as transcriptome, metabolome, proteome, and others. However, the sequencing methodologies and equipment available in the 1990s had a high cost and low performance. Therefore, the need to obtain genetic information drove the development of next-generation sequencing (NGS) technologies. These new methodologies (e.g., Pyrosequencing, Illumina, Ion-torrent, PacBio, and Nanopore) feature reduced costs, high capacity to generate data in a short time. Recently, Pereira et al. (2020) and Reuter et al. (2015) published a detailed review of these sequencing platforms and the main advantages and disadvantages of each method.

Due to these technological advances, over 1000 genomes of *S. cerevisiae* have been sequenced and deposited National Center for Biotechnology Information (NCBI) (Peter et al. 2018). Some examples are reported in Table 3.2. The gene sequences identified in these studies were of paramount importance as they served as basis for a deeper understanding of the cellular mechanisms related to carbohydrate metabolism and stress tolerance for industrial conditions (Goffeau et al. 1996; Kvitek et al. 2008).

Alcoholic fermentations have been conducted for centuries mainly for the production of foods and beverages (McGovern et al. 2004). Although several bacterial and fungal genera are observed to be involved in these biotechnological processes, the yeasts belonging to the genus *Saccharomyces*, also called *Saccharomyces* “sensu

Table 3.2 Bioethanol-producing *Saccharomyces cerevisiae* strains with published genome

Strain	Platform	Assembly size (Mb)	Predicted number of genes	References
AY291 (PE-2)	Illumina MiSeq/ pyrosequencing	11.6	5.880	Argueso et al. (2009)
CAT-1	Pyrosequencing	~12	6.652	Babrzadeh et al. (2012)
VR1-5B (VR-1)	Illumina HiSeq	11.1	NA	Swinnen et al. (2012b)
YJS329	Pyrosequencing	NA	5.602	Zheng et al. (2012)
M3707	Illumina HiSeq	~11.5	~6.000	Brown et al. (2013)
ISO12	Illumina MiSeq	11.4	NA	Wallace-Salinas et al. (2015)
BG-1	Illumina HiSeq	11.7	5.607	Coutouné et al. (2017)
NY1308	Pyrosequencing	ND	ND	Zhang et al. 2018)

NA not available

stricto" (i.e., yeasts specialized for growth in sugar-rich environments) that dominate these fermentations (Fay and Benavides 2005; Stambuk 2019).

These observations led to the discussion if *Saccharomyces* sensu stricto arose from a domestication process carried out by man or if it appeared in nature and was later selected by human populations. The most accepted hypothesis today is that the *S. cerevisiae* species originated in natural environments and then underwent successive processes of domestication. This hypothesis has been supported mainly by comparative genomics analyses of various *S. cerevisiae* strains, as environmental strains showed higher genetic variability compared to domesticated strains (Borneman et al. 2011; Sicard and Legras 2011; Peter et al. 2018). The higher genetic diversity observed in wild strains may be associated with their efficiency in changing asexual reproduction to sexual reproduction under nutritional stress. Moreover, this microbial group shows resistance to several abiotic factors, while the domesticated strains lost some of these characteristics over time and ended up specializing to their new niche (Magwene et al. 2011b; Peter et al. 2018). Interestingly, the dispersal of these microorganisms by man has also allowed different species to crossbreed. The yeasts from these crosses have a mosaic genome, that is, strains that inherit characteristics from both parental lines. However, these genomic characteristics are only found in domesticated yeasts (Legras et al. 2007; Liti et al. 2009).

However, bioethanol production is a relatively new bioprocess, especially when compared to wine, bread, and beer fermentations. Moreover, the fermentative challenges encountered by microorganisms in distilleries differ significantly from the conditions encountered by yeasts in fermented foods and beverages (Mukherjee et al. 2014). For example, the work performed by da Silva-Filho et al. (2005) and Basso et al. (2008) demonstrated that the baker's yeast strains used as starter cultures were rapidly replaced by wild yeasts in a cell recycling system, demonstrating the genetic potential of the non-domesticated strains. Therefore, identifying and characterizing the genetic variability of these microorganisms is the first step to understand the metabolism and propose metabolic engineering approaches, aiming to improve the fermentative performance of these yeasts for bioethanol production.

The genome of the strain (JAY29) derived from industrial line PE-2 was the first *S. cerevisiae* genome isolated from the Brazilian bioethanol industry to be sequenced (Argueso et al. 2009). After this study, the genome of other industrial strains used for bioethanol production began to be determined and some characteristics of these genomes are shown in Table 3.2 Interestingly, the results showed that the Brazilian strains are diploid, while other industrial strains used for wine, beer and bread fermentation may have polyploid or aneuploid genomes (Borneman et al. 2011). The strains from the bioethanol industry also showed a highly heterozygous genome, in addition to having fewer transposable elements and several structural polymorphisms between homologous chromosomes when compared to the reference genome, i.e., strain S288c (Argueso et al. 2009; Borneman et al. 2011; Babrzadeh et al. 2012).

According to Argueso et al. (2009), the high incidence of polymorphism, especially in telomeres, may be associated with the efficiency of these strains in producing ethanol and with their ability to adapt to the industrial environment. In addition, the increased copy number of telomeric genes *SNZ* and *SNO*, responsible for the production of thiamine (vitamin B1) and pyridoxine (vitamin B6), were observed in five industrial strains (PE-2, CAT-1, BG-1, SA-1, and VR-1). These genes may be associated with the fermentative success of these microorganisms, as these vitamins are essential for yeast metabolism in high sugar environments, and also protect the cells from oxidative stress (Argueso et al. 2009). Interestingly, the expression of the *SUC2* gene encoding the extracellular enzyme β-fructosidase (invertase) responsible for sucrose hydrolysis was not observed in any of the five strains, suggesting that the conversion of this sugar is not a limiting factor for the fermentation of these strains. It is also speculated that these genetic changes were adaptive and selected in the industrial environment (Stambuk et al. 2009).

Recently, genome sequencing of the *S. cerevisiae* strain NY1308 used to produce second-generation ethanol identified 43 unique genes, and phylogenetic analyses indicated that a large part of these genes was obtained by horizontal transfer. However, during the fermentation process, the NY1308 strain tends to flocculate due to the high presence of some inhibitors that are produced from the pretreatment of the cellulosic feedstock. The flocculation ability was linked to the presence of the NYn43 gene, and deletion of this gene reduced the flocculation rate and increased the fermentative capacity of strain NY1308. Unfortunately, much of the genomes showed in Table 3.1 have not been analyzed in detail, being only deposited/announced. Therefore, this genomic information can still be explored and used for industrial applications (Zhang et al. 2018).

3.10 CRISPR/Cas9

The CRISPR structure is generally located close to the cas genes (CRISPR-associated genes), considered as a large, polymorphic gene family. These genes encode proteins that carry nucleotide domains, such as nucleases, helicases, and ligases. Four of the cas genes have already been identified. The Cas1 protein is the only one found in all species that contains the CRISPR locus and, like the Cas2 protein, it acts as a nuclease. The Cas3 and Cas4 proteins are considered helicase and exonucleases, respectively. CRISPR–Cas9 system has been used for genome editing in *S. cerevisiae* and other yeasts (Fig. 3.2). In *S. cerevisiae*, the first report using CRISPR/Cas9 provides a foundation for a simple and powerful genome engineering tool (DiCarlo et al. 2013). Xue et al. (2018) used the CRISPR/Cas9 technology to disrupt the alcohol dehydrogenase (ADH) 2 gene via complete deletion of the gene and introduction of a frameshift mutation in the ADH2 locus in *Saccharomyces cerevisiae*. The authors observed an ethanol yield improving by up to 74.7% compared with the yield obtained using the native strain.

3.11 Concluding Remarks

The ethanol fuel production process is a well-established commercial technology, but still needs to be improved. Among these factors inherent to the process and which can still be improved, it has been highlighted the replacement of feedstock used as source of nutrients for the microorganism evolved in the fermentation process. Nowadays a large amount of biomass waste has been generated worldwide and this waste is rich in nutrients; moreover, it has potential to be used as a renewable energy source. Due to the origin of the raw material, the generate biomass waste is constituted by a lignocellulosic structure, which must be disrupted before being used in fermentative process. The microbial cells cannot metabolize all the complex polysaccharides available. Recent research has focused on strategies for the decomposition of this complex biomass into its primary constituents, through chemical or enzymatic hydrolysis. In addition, strains used in ethanol production must be able to metabolize sugars which are presented in the lignocellulosic biomass.

As the ethanol production process results in high concentrations of ethanol in the culture medium, the yield is reduced as yeast growth is inhibited due to toxigenic effects. Furthermore, the productivity can be influenced by the reduced performance of yeasts at high initial concentrations of glucose. Therefore, it is desirable that the strains are tolerant to these high concentrations of ethanol and glucose at any stage of the process. Microorganisms which are resistant to temperature and pH variations are also good options for the ethanol production process, considering that commonly used bioreactors devoid strict temperature and pH control. Other challenge is to conduce a simultaneous saccharification and fermentation, the optimal temperature for ethanol fermentation is generally below 37 °C and for enzymatic saccharification is above 50 °C. Ethanol producing strains must be tolerant to high saccharification temperatures. The increases in fermentation productivity and the resulting decreases in costs have come about mainly by using breeding strategies. The techniques reported in this chapter are useful for obtaining new strains and for the identification of new genetic targets to be used in strain improvement programs.

Acknowledgements The authors acknowledge the financial support given by CNPq and CAPES.

References

Abdel-Fattah WR et al (2000) Isolation of thermotolerant ethanologenic yeasts and use of selected strains in industrial scale fermentation in an Egyptian distillery. Biotechnol Bioeng 68(5): 531–535. https://doi.org/10.1002/(SICI)1097-0290(20000605)68:5<531::AID-BIT7>3.0.CO;2-Y

Adegboye MF et al (2021) Bioprospecting of microbial strains for biofuel production: metabolic engineering, applications, and challenges. Biotechnol Biofuels 14(1):1–21. https://doi.org/10.1186/s13068-020-01853-2

Ahmad QUA et al (2017) Moderate alkali-thermophilic ethanologenesis by locally isolated Bacillus licheniformis from Pakistan employing sugarcane bagasse: a comparative aspect of aseptic and

non-aseptic fermentations. Biotechnol Biofuels 10(1):1–20. https://doi.org/10.1186/s13068-017-0785-1
Aikawa S et al (2018) Characterization and high-quality draft genome sequence of Herbivorax saccincola A7, an anaerobic, alkaliphilic, thermophilic, cellulolytic, and xylanolytic bacterium. Syst Appl Microbiol 41(4):261–269. https://doi.org/10.1016/j.syapm.2018.01.010
Argueso JL et al (2009) Genome structure of a Saccharomyces cerevisiae strain widely used in bioethanol production. Genome Res 19(12):2258–2270. https://doi.org/10.1101/gr.091777.109
Babrzadeh F et al (2012) Whole-genome sequencing of the efficient industrial fuel-ethanol fermentative Saccharomyces cerevisiae strain CAT-1. Mol Gen Genomics 287(6):485–494. https://doi.org/10.1007/s00438-012-0695-7
Banat IM, Nigam P, Marchant R (1992) Isolation of thermotolerant, fermentative yeasts growing at 52°C and producing ethanol at 45°C and 50°C. World J Microbiol Biotechnol 8(3):259–263. https://doi.org/10.1007/BF01201874
Basso LC et al (2008) Yeast selection for fuel ethanol production in Brazil. FEMS Yeast Res 8(7): 1155–1163. https://doi.org/10.1111/j.1567-1364.2008.00428.x
Borneman AR et al (2011) Whole-genome comparison reveals novel genetic elements that characterize the genome of industrial strains of Saccharomyces cerevisiae. PLoS Genet 7(2). https://doi.org/10.1371/journal.pgen.1001287
Brexó RP, Sant'Ana AS (2017) Impact and significance of microbial contamination during fermentation for bioethanol production. Renew Sust Energ Rev 73(February 2016):423–434. https://doi.org/10.1016/j.rser.2017.01.151
Brown SD et al (2013) Genome sequences of industrially relevant Saccharomyces cerevisiae strain M3707, isolated from a sample of distillers yeast and four haploid derivatives. Genome Announc 1(3). https://doi.org/10.1128/genomeA.00323-13
Chaudhary N, Qazi JI, Irfan M (2017) Isolation and identification of cellulolytic and ethanologenic bacteria from soil. Iran J Sci Technol Trans A Sci 41(3):551–555. https://doi.org/10.1007/s40995-017-0282-1
Chen MT et al (2012) Generation of diploid Pichia pastoris strains by mating and their application for recombinant protein production. Microb Cell Factories 11:1–18. https://doi.org/10.1186/1475-2859-11-91
Chen Z et al (2015) Characteristics and kinetic study on pyrolysis of five lignocellulosic biomass via thermogravimetric analysis. Bioresour Technol 192:441–450. https://doi.org/10.1016/j.biortech.2015.05.062
Choi GW et al (2010) Bioethanol production by a flocculent hybrid, CHFY0321 obtained by protoplast fusion between Saccharomyces cerevisiae and Saccharomyces bayanus. Biomass Bioenergy 34(8):1232–1242. https://doi.org/10.1016/j.biombioe.2010.03.018
Choudhary J, Singh S, Nain L (2017) Bioprospecting thermotolerant ethanologenic yeasts for simultaneous saccharification and fermentation from diverse environments. J Biosci Bioeng 123(3):342–346. https://doi.org/10.1016/j.jbiosc.2016.10.007
Costa AC et al (2001) Factorial design and simulation for the optimization and determination of control structures for an extractive alcoholic fermentation. Process Biochem 37(2):125–137. https://doi.org/10.1016/S0032-9592(01)00188-1
Coutouné N et al (2017) Draft genome sequence of Saccharomyces cerevisiae Barra Grande (BG-1), a Brazilian industrial bioethanol- producing strain. Genome 5(September):1–2. https://doi.org/10.1128/genomeA.00111-17
da Silva-Filho EA et al (2005) Yeast population dynamics of industrial fuel-ethanol fermentation process assessed by PCR-fingerprinting. Antonie Van Leeuwenhoek 88(1):13–23. https://doi.org/10.1007/s10482-004-7283-8
Dandi ND, Dandi BN, Chaudhari AB (2013) Bioprospecting of thermo- and osmo-tolerant fungi from mango pulp-peel compost for bioethanol production. Anton Leeuw Int J Gen Mol Microbiol 103(4):723–736. https://doi.org/10.1007/s10482-012-9854-4
Darvasi A, Soller M (1995) Advanced intercross lines, an experimental population for fine genetic mapping. In: Genetics. Elsevier, New York, pp 1199–1207

de Melo Pereira GV et al (2020) An updated review on bacterial community composition of traditional fermented milk products: what next-generation sequencing has revealed so far? Crit Rev Food Sci Nutr 0(0):1–20. https://doi.org/10.1080/10408398.2020.1848787

De Souza Liberal AT et al (2005) Contaminant yeast detection in industrial ethanol fermentation must by rDNA-PCR. Lett Appl Microbiol 40(1):19–23. https://doi.org/10.1111/j.1472-765X.2004.01618.x

De Souza Liberal AT et al (2007) Identification of Dekkera bruxellensis as a major contaminant yeast in continuous fuel ethanol fermentation. J Appl Microbiol 102(2):538–547. https://doi.org/10.1111/j.1365-2672.2006.03082.x

Della-Bianca BE, Gombert AK (2013) Stress tolerance and growth physiology of yeast strains from the Brazilian fuel ethanol industry. Anton Leeuw Int J Gen Mol Microbiol 104(6):1083–1095. https://doi.org/10.1007/s10482-013-0030-2

Dhaliwal SS et al (2011) Enhanced ethanol production from sugarcane juice by galactose adaptation of a newly isolated thermotolerant strain of Pichia kudriavzevii. Bioresour Technol 102(10): 5968–5975. https://doi.org/10.1016/j.biortech.2011.02.015

DiCarlo JE et al (2013) Genome engineering in Saccharomyces cerevisiae using CRISPR-Cas systems. Nucleic Acids Res 41(7):4336–4343. https://doi.org/10.1093/nar/gkt135

Dubey R, Jakeer S, Gaur NA (2016) Screening of natural yeast isolates under the effects of stresses associated with second-generation biofuel production. J Biosci Bioeng 121(5):509–516. https://doi.org/10.1016/j.jbiosc.2015.09.006

Ekblom R, Wolf JBW (2014) A field guide to whole-genome sequencing, assembly and annotation. Evol Appl 7(9):1026–1042. https://doi.org/10.1111/eva.12178

Favaro L et al (2013) Exploring grape marc as trove for new thermotolerant and inhibitor-tolerant Saccharomyces cerevisiae strains for second-generation bioethanol production. Biotechnol Biofuels 6(1):1–14. https://doi.org/10.1186/1754-6834-6-168

Favaro L, Jansen T, van Zyl WH (2019) Exploring industrial and natural Saccharomyces cerevisiae strains for the bio-based economy from biomass: the case of bioethanol. Crit Rev Biotechnol 39(6):800–816. https://doi.org/10.1080/07388551.2019.1619157

Fay JC, Benavides JA (2005) Evidence for domesticated and wild populations of saccharomyces cerevisiae. PLoS Genet 1(1):0066–0071. https://doi.org/10.1371/journal.pgen.0010005

Field SJ et al (2015) Identification of furfural resistant strains of Saccharomyces cerevisiae and Saccharomyces paradoxus from a collection of environmental and industrial isolates. Biotechnol Biofuels 8(1):1–8. https://doi.org/10.1186/s13068-015-0217-z

Fong JCN et al (2006) Isolation and characterization of two novel ethanol-tolerant facultative-anaerobic thermophilic bacteria strains from waste compost. Extremophiles 10(5):363–372. https://doi.org/10.1007/s00792-006-0507-2

Fraczek MG, Naseeb S, Delneri D (2018) History of genome editing in yeast. Yeast 35(5):361–368. https://doi.org/10.1002/yea.3308

Fu N, Peiris P (2008) Co-fermentation of a mixture of glucose and xylose to ethanol by Zymomonas mobilis and Pachysolen tannophilus. World J Microbiol Biotechnol 24(7):1091–1097. https://doi.org/10.1007/s11274-007-9613-2

Fu N et al (2009) A novel co-culture process with Zymomonas mobilis and Pichia stipitis for efficient ethanol production on glucose/xylose mixtures. Enzym Microb Technol 45(3): 210–217. https://doi.org/10.1016/j.enzmictec.2009.04.006

Fujitomi K et al (2012) Deletion of the *PHO13* gene in Saccharomyces cerevisiae improves ethanol production from lignocellulosic hydrolysate in the presence of acetic and formic acids, and furfural. Bioresour Technol 111:161–166. https://doi.org/10.1016/j.biortech.2012.01.161

Gallagher D et al (2018) Dynamic bacterial and fungal microbiomes during sweet sorghum ensiling impact bioethanol production. Bioresour Technol 264(March):163–173. https://doi.org/10.1016/j.biortech.2018.05.053

Gao Y et al (2014) Ethanol production from high solids loading of alkali-pretreated sugarcane bagasse with an SSF process. Bioresources 9(2):3466–3479. https://doi.org/10.15376/biores.9.2.3466-3479

García-Pedrajas MD et al (2010) Molecular and cell biology methods for fungi. Springer Protocols 638:55–76. https://doi.org/10.1007/978-1-60761-611-5

Garrigues S, Martínez-Reyes N, de Vries RP (2021) Genetic engineering for strain improvement in filamentous fungi. In: Zaragoza Ó, Casadevall A (eds) Encyclopedia of mycology. Elsevier, Kidlington, pp 489–504. https://doi.org/10.1016/b978-0-12-819990-9.00006-8

Ge J et al (2014) Construction and analysis of high-ethanol-producing fusants with co-fermentation ability through protoplast fusion and double labeling technology. PLoS One 9(9). https://doi.org/10.1371/journal.pone.0108311

Gerke JP, Chen CTL, Cohen BA (2006) Natural isolates of Saccharomyces cerevisiae display complex genetic variation in sporulation efficiency. Genetics 174(2):985–997. https://doi.org/10.1534/genetics.106.058453

Giudici P et al (2005) Strategies and perspectives for genetic improvement of wine yeasts. Appl Microbiol Biotechnol 66(6):622–628. https://doi.org/10.1007/s00253-004-1784-2

Goffeau A et al (1996) Life with 6000 genes. Science 274(October):546–567

González-Siso MI et al (2015) Improved bioethanol production in an engineered *Kluyveromyces lactis* strain shifted from respiratory to fermentative metabolism by deletion of NDI1. Microb Biotechnol 8(2):319–330. https://doi.org/10.1111/1751-7915.12160

Greig D et al (1997) Greig, Plays 1. Methuen, London, pp 2–5

Gurdo N et al (2018) Improved robustness of an ethanologenic yeast strain through adaptive evolution in acetic acid is associated with its enzymatic antioxidant ability. J Appl Microbiol. https://doi.org/10.1111/jam.13917

Haber JE (2012) Mating-type genes and MAT switching in Saccharomyces cerevisiae. Genetics 191(1):33–64. https://doi.org/10.1534/genetics.111.134577

He MX et al (2014) Zymomonas mobilis: a novel platform for future biorefineries. Biotechnol Biofuels 7(1):1–15. https://doi.org/10.1186/1754-6834-7-101

Hemmati N, Lightfoot DA, Fakhoury A (2012) A mutated yeast strain with enhanced ethanol production efficiency and stress tolerance. Atlas J Biol 2(2):100–115. https://doi.org/10.5147/ajb.2012.0092

Hicks JB, Herskowitz I (1977) Interconversion of yeast mating types. II. Restoration of mating ability to sterile mutants in homothallic and heterothallic strains. Genetics 85(3):373–393. https://doi.org/10.1093/genetics/85.3.373

Horinouchi T, Maeda T, Kotani H, Furusawa C (2020) Suppression of antibiotic resistance evolution by single-gene deletion. Sci Rep 10(1):1–9

Hubmann G et al (2013) Quantitative trait analysis of yeast biodiversity yields novel gene tools for metabolic engineering. Metab Eng 17(1):68–81. https://doi.org/10.1016/j.ymben.2013.02.006

Jacobus AP et al (2021) Saccharomyces cerevisiae strains used industrially for bioethanol production. Essays Biochem 65(2):147–161. https://doi.org/10.1042/ebc20200160

Jessen JE, Orlygsson J (2012) Production of ethanol from sugars and lignocellulosic biomass by thermoanaerobacter J1 isolated from a hot spring in Iceland. J Biomed Biotechnol 2012. https://doi.org/10.1155/2012/186982

Jin ZH et al (2009) Enhanced production of spinosad in saccharopolyspora spinosa by genome shuffling. Appl Biochem Biotechnol 159(3):655–663. https://doi.org/10.1007/s12010-008-8500-0

Karp SG et al (2021) Bioeconomy and biofuels: the case of sugarcane ethanol in Brazil. Biofuels Bioprod Biorefin 15(3):899–912. https://doi.org/10.1002/BBB.2195

Katz Ezov T et al (2010) Heterothallism in Saccharomyces cerevisiae isolates from nature: effect of HO locus on the mode of reproduction. Mol Ecol 19(1):121–131. https://doi.org/10.1111/j.1365-294X.2009.04436.x

Kaushal R, Sharma N, Dogra V (2016) Molecular characterization of glycosyl hydrolases of Trichoderma harzianum WF5 - a potential strain isolated from decaying wood and their application in bioconversion of poplar wood to ethanol under separate hydrolysis and fermentation. Biomass Bioenergy 85:243–251. https://doi.org/10.1016/j.biombioe.2015.12.010

Kim JH, Block DE, Mills DA (2010) Simultaneous consumption of pentose and hexose sugars: an optimal microbial phenotype for efficient fermentation of lignocellulosic biomass. Appl Microbiol Biotechnol 88(5):1077–1085. https://doi.org/10.1007/s00253-010-2839-1

Kim SR et al (2015) Deletion of PHO13, encoding haloacid dehalogenase type IIA phosphatase, results in upregulation of the pentose phosphate pathway in *Saccharomyces cerevisiae*. Appl Environ Microbiol 81(5):1601–1609. https://doi.org/10.1128/AEM.03474-14

Koutinas M et al (2016) High temperature alcoholic fermentation of orange peel by the newly isolated thermotolerant Pichia kudriavzevii KVMP10. Lett Appl Microbiol 62(1):75–83. https://doi.org/10.1111/lam.12514

Kuhad RC et al (2011) Bioethanol production from pentose sugars: current status and future prospects. Renew Sust Energ Rev 15(9):4950–4962. https://doi.org/10.1016/J.RSER.2011.07.058

Kumar D, Singh V (2019) Bioethanol production from corn. In: Serna-Saldivar SO (ed) Corn, 3rd edn. AACC International Press, St Paul, pp 615–631

Kumari R, Pramanik K (2012) Improvement of multiple stress tolerance in yeast strain by sequential mutagenesis for enhanced bioethanol production. J Biosci Bioeng 114(6):622–629. https://doi.org/10.1016/j.jbiosc.2012.07.007

Kvitek DJ, Will JL, Gasch AP (2008) Variations in stress sensitivity and genomic expression in diverse S. cerevisiae isolates. PLoS Genet 4(10):31–35. https://doi.org/10.1371/journal.pgen.1000223

Larsen L, Nielsen P, Ahring BK (1997) Thermoanaerobacter mathranii sp. nov., an ethanol-producing, extremely thermophilic anaerobic bacterium from a hot spring in Iceland. Arch Microbiol 168(2):114–119. https://doi.org/10.1007/s002030050476

Legras JL et al (2007) Bread, beer and wine: Saccharomyces cerevisiae diversity reflects human history. Mol Ecol 16(10):2091–2102. https://doi.org/10.1111/j.1365-294X.2007.03266.x

Limayem A, Ricke SC (2012) Lignocellulosic biomass for bioethanol production: current perspectives, potential issues and future prospects. Prog Energy Combust Sci 38(4):449–467. https://doi.org/10.1016/j.pecs.2012.03.002

Lippman ZB, Zamir D (2007) Heterosis: revisiting the magic. Trends Genet 23(2):60–66. https://doi.org/10.1016/j.tig.2006.12.006

Liti G, Louis EJ (2012) Advances in quantitative trait analysis in yeast. PLoS Genet 8(8):e1002912. https://doi.org/10.1371/journal.pgen.1002912

Liti G et al (2009) Population genomics of domestic and wild yeasts. Nature 458(7236):337–341. https://doi.org/10.1038/nature07743

Loaces I et al (2016) Improved glycerol to ethanol conversion by E. coli using a metagenomic fragment isolated from an anaerobic reactor. J Ind Microbiol Biotechnol 43(10):1405–1416. https://doi.org/10.1007/s10295-016-1818-7

Loaces I, Schein S, Noya F (2017) Ethanol production by Escherichia coli from Arundo donax biomass under SSF, SHF or CBP process configurations and in situ production of a multifunctional glucanase and xylanase. Bioresour Technol 224:307–313. https://doi.org/10.1016/j.biortech.2016.10.075

Lv W, Yu Z (2013) Isolation and characterization of two thermophilic cellulolytic strains of Clostridium thermocellum from a compost sample. J Appl Microbiol 114(4):1001–1007. https://doi.org/10.1111/jam.12112

MacKay TFC, Stone EA, Ayroles JF (2009) The genetics of quantitative traits: challenges and prospects. Nat Rev Genet 10(8):565–577. https://doi.org/10.1038/nrg2612

Madeira-Jr JV, Gombert AK (2018) Towards high-temperature fuel ethanol production using Kluyveromyces marxianus: on the search for plug-in strains for the Brazilian sugarcane-based biorefinery. Biomass Bioenergy 119(March):217–228. https://doi.org/10.1016/j.biombioe.2018.09.010

Magwene PM, Willis JH, Kelly JK (2011a) The statistics of bulk segregant analysis using next generation sequencing. PLoS Comput Biol 7(11):e1002255. https://doi.org/10.1371/journal.pcbi.1002255

Magwene PM et al (2011b) Outcrossing, mitotic recombination, and life-history trade-offs shape genome evolution in Saccharomyces cerevisiae. Proc Natl Acad Sci U S A 108(5):1987–1992. https://doi.org/10.1073/pnas.1012544108

Mans R, Daran JMG, Pronk JT (2018) Under pressure: evolutionary engineering of yeast strains for improved performance in fuels and chemicals production. Curr Opin Biotechnol 50:47–56. https://doi.org/10.1016/j.copbio.2017.10.011

McGovern PE et al (2004) Fermented beverages of pre- and proto-historic China. Proc Natl Acad Sci U S A 101(51):17593–17598. https://doi.org/10.1073/pnas.0407921102

McMillan JD, Beckham GT (2017) Thinking big: towards ideal strains and processes for large-scale aerobic biofuels production. Microb Biotechnol 10(1):40–42. https://doi.org/10.1111/1751-7915.12471

Meijnen JP et al (2016) Polygenic analysis and targeted improvement of the complex trait of high acetic acid tolerance in the yeast *Saccharomyces cerevisiae*. Biotechnol Biofuels 9(1):5. https://doi.org/10.1186/s13068-015-0421-x

Min K et al (2016) *Candida albicans* gene deletion with a transient CRISPR-Cas9 system. mSphere 1(3):e00130–e00116. https://doi.org/10.1128/msphere.00130-16

Mobini-Dehkordi M et al (2008) Isolation of a novel mutant strain of Saccharomyces cerevisiae by an ethyl methane sulfonate-induced mutagenesis approach as a high producer of bioethanol. J Biosci Bioeng 105(4):403–408. https://doi.org/10.1263/jbb.105.403

Modig T, Lidén G, Taherzadeh MJ (2002) Inhibition effects of furfural on alcohol dehydrogenase, aldehyde dehydrogenase and pyruvate dehydrogenase. Biochem J 363(3):769–776. https://doi.org/10.1042/0264-6021:3630769

Moreno AD et al (2013) Comparing cell viability and ethanol fermentation of the thermotolerant yeast Kluyveromyces marxianus and Saccharomyces cerevisiae on steam-exploded biomass treated with laccase. Bioresour Technol 135:239–245. https://doi.org/10.1016/j.biortech.2012.11.095

Mukherjee V et al (2014) Phenotypic evaluation of natural and industrial Saccharomyces yeasts for different traits desirable in industrial bioethanol production. Appl Microbiol Biotechnol 98(22): 9483–9498. https://doi.org/10.1007/s00253-014-6090-z

Muthaiyan A, Limayem A, Ricke SC (2011) Antimicrobial strategies for limiting bacterial contaminants in fuel bioethanol fermentations. Prog Energy Combust Sci 37(3):351–370. https://doi.org/10.1016/j.pecs.2010.06.005

Nieduszynski CA, Liti G (2011) From sequence to function: insights from natural variation in budding yeasts. Biochim Biophys Acta Gen Subj 1810(10):959–966. https://doi.org/10.1016/j.bbagen.2011.02.004

Ota A et al (2013) Production of ethanol from mannitol by the yeast strain Saccharomyces paradoxus NBRC 0259. J Biosci Bioeng 116(3):327–332. https://doi.org/10.1016/j.jbiosc.2013.03.018

Parisutham V, Kim TH, Lee SK (2014) Feasibilities of consolidated bioprocessing microbes: from pretreatment to biofuel production. Bioresour Technol 161:431–440. https://doi.org/10.1016/j.biortech.2014.03.114

Pereira FB et al (2014) Industrial robust yeast isolates with great potential for fermentation of lignocellulosic biomass. Bioresour Technol 161:192–199. https://doi.org/10.1016/j.biortech.2014.03.043

Pereira GVM et al (2020) Lactic acid bacteria: what coffee industry should know? Curr Opin Food Sci 31:1–8. https://doi.org/10.1016/j.cofs.2019.07.004

Peris D et al (2017a) Hybridization and adaptive evolution of diverse Saccharomyces species for cellulosic biofuel production. Biotechnol Biofuels 10(1):1–19. https://doi.org/10.1186/s13068-017-0763-7

Peris D et al (2017b) Hybridization and adaptive evolution of diverse *Saccharomyces* species for cellulosic biofuel production. Biotechnol Biofuels 10(1):78. https://doi.org/10.1186/s13068-017-0763-7

Peter J et al (2018) Genome evolution across 1,011 Saccharomyces cerevisiae isolates species-wide genetic and phenotypic diversity. Nature

Pretorius IS (2000) Tailoring wine yeast for the new millennium: novel approaches to the ancient art of winemaking. Yeast 16(8):675–729. https://doi.org/10.1002/1097-0061(20000615)16:8<675::aid-yea585>3.3.co;2-2

Rainieri S, Pretorius IS (2000) Selection and improvement of wine yeasts. Ann Microbiol 50(1):15–31

Reuter JA, Spacek DV, Snyder MP (2015) High-throughput sequencing technologies. Mol Cell 58(4):586–597. https://doi.org/10.1016/j.molcel.2015.05.004

Rodrussamee N, Sattayawat P, Yamada M (2018) Highly efficient conversion of xylose to ethanol without glucose repression by newly isolated thermotolerant Spathaspora passalidarum CMUWF1-2. BMC Microbiol 18(1):1–11. https://doi.org/10.1186/s12866-018-1218-4

Romero-Frasca E et al (2021) Bioprospecting of wild type ethanologenic yeast for ethanol fuel production from wastewater-grown microalgae. Biotechnol Biofuels 14(1):1–10. https://doi.org/10.1186/s13068-021-01925-x

Ruyters S et al (2015) Assessing the potential of wild yeasts for bioethanol production. J Ind Microbiol Biotechnol 42(1):39–48. https://doi.org/10.1007/s10295-014-1544-y

Sandberg TE et al (2019) The emergence of adaptive laboratory evolution as an efficient tool for biological discovery and industrial biotechnology. Metab Eng 56(August):1–16. https://doi.org/10.1016/j.ymben.2019.08.004

Santos F et al (2020) Production of second-generation ethanol from sugarcane. In: Sugarcane biorefinery, technology and perspectives. Elsevier, New York, pp 195–228. https://doi.org/10.1016/B978-0-12-814236-3.00011-1

Sauer B (1987) Functional expression of the *cre-lox* site-specific recombination system in the yeast *Saccharomyces cerevisiae*. Mol Cell Biol 7(6):2087–2096. https://doi.org/10.1128/mcb.7.6.2087-2096.1987

Schnable PS, Springer NM (2013) Progress toward understanding heterosis in crop plants. Annu Rev Plant Biol 64:71–88. https://doi.org/10.1146/annurev-arplant-042110-103827

Scully SM, Orlygsson J (2015) Recent advances in second generation ethanol production by thermophilic bacteria. Energies 8(1):1–30. https://doi.org/10.3390/en8010001

Selim KA et al (2018) Bioethanol a microbial biofuel metabolite; new insights of yeasts metabolic engineering. Fermentation 4(1). https://doi.org/10.3390/fermentation4010016

Selmecki AM et al (2015) Polyploidy can drive rapid adaptation in yeast. Nature 519(7543):349–351. https://doi.org/10.1038/nature14187

Shahid S, Tajwar R, Akhtar MW (2018) A novel trifunctional, family GH10 enzyme from Acidothermus cellulolyticus 11B, exhibiting endo-xylanase, arabinofuranosidase and acetyl xylan esterase activities. Extremophiles 22(1):109–119. https://doi.org/10.1007/s00792-017-0981-8

Shapira R et al (2014) Extensive heterosis in growth of yeast hybrids is explained by a combination of genetic models. Heredity 113(4):316–326. https://doi.org/10.1038/hdy.2014.33

Sicard D, Legras JL (2011) Bread, beer and wine: yeast domestication in the Saccharomyces sensu stricto complex. C R Biol 334(3):229–236. https://doi.org/10.1016/j.crvi.2010.12.016

Signori L et al (2014) Effect of oxygenation and temperature on glucose-xylose fermentation in Kluyveromyces marxianus CBS712 strain. Microb Cell Factories 13(1):1–13. https://doi.org/10.1186/1475-2859-13-51

Singh N et al (2018) Bioethanol production potential of a novel thermophilic isolate Thermoanaerobacter sp. DBT-IOC-X2 isolated from Chumathang hot spring. Biomass Bioenergy 116(January):122–130. https://doi.org/10.1016/j.biombioe.2018.05.009

Sipiczki M (2008) Interspecies hybridization and recombination in Saccharomyces wine yeasts. FEMS Yeast Res 8(7):996–1007. https://doi.org/10.1111/j.1567-1364.2008.00369.x

Stambuk BU (2019) Yeasts: the leading figures on bioethanol production. In: Ethanol as a green alternative fuel: insight and perspectives. Nova Science Publishers, New York, pp 57–92

Stambuk BU et al (2009) Industrial fuel ethanol yeasts contain adaptive copy number changes in genes involved in vitamin B1 and B6 biosynthesis. Genome Res 19(12):2271–2278. https://doi.org/10.1101/gr.094276.109

Steensels J et al (2014) Improving industrial yeast strains: exploiting natural and artificial diversity. FEMS Microbiol Rev 38(5):947–995. https://doi.org/10.1111/1574-6976.12073

Storici F, Lewis LK, Resnick MA (2001) In vivo site-directed mutagenesis using oligonucleotides. Nat Biotechnol 19(8):773–776. https://doi.org/10.1038/90837

Sukwong P et al (2020) Improvement of bioethanol production by Saccharomyces cerevisiae through the deletion of GLK1, MIG1 and MIG2 and overexpression of PGM2 using the red seaweed Gracilaria verrucosa. Process Biochem 89:134–145. https://doi.org/10.1016/j.procbio.2019.10.030

Swinnen S, Thevelein JM, Nevoigt E (2012a) Genetic mapping of quantitative phenotypic traits in *Saccharomyces cerevisiae*. FEMS Yeast Res 12(2):215–227. https://doi.org/10.1111/j.1567-1364.2011.00777.x

Swinnen S et al (2012b) Identification of novel causative genes determining the complex trait of high ethanol tolerance in yeast using pooled-segregant whole-genome sequence analysis. Genome Res 22(5):975–984. https://doi.org/10.1101/gr.131698.111

Swinnen S et al (2014) The fraction of cells that resume growth after acetic acid addition is a strain-dependent parameter of acetic acid tolerance in *Saccharomyces cerevisiae*. FEMS Yeast Res 14(4):642–653. https://doi.org/10.1111/1567-1364.12151

Takabatake A, Kawazoe N, Izawa S (2015) Plasma membrane proteins Yro2 and Mrh1 are required for acetic acid tolerance in *Saccharomyces cerevisiae*. Appl Microbiol Biotechnol 99(6): 2805–2814. https://doi.org/10.1007/s00253-014-6278-2

Tan F et al (2016) Using global transcription machinery engineering (gTME) to improve ethanol tolerance of Zymomonas mobilis. Microb Cell Factories 15(1):1–9. https://doi.org/10.1186/s12934-015-0398-y

Tan H et al (2018) A bifunctional cellulase–xylanase of a new Chryseobacterium strain isolated from the dung of a straw-fed cattle. Microb Biotechnol 11(2):381–398. https://doi.org/10.1111/1751-7915.13034

Tao X et al (2012) A novel strategy to construct yeast saccharomyces cerevisiae strains for very high gravity fermentation. PLoS One 7(2). https://doi.org/10.1371/journal.pone.0031235

Tiwari S et al (2020) Xylanolytic and Ethanologenic potential of gut associated yeasts from different species of termites from India. Mycobiology 48(6):501–511. https://doi.org/10.1080/12298093.2020.1830742

Tomás AF et al (2013) Extreme thermophilic ethanol production from rapeseed straw: using the newly isolated Thermoanaerobacter pentosaceus and combining it with Saccharomyces cerevisiae in a two-step process. Biotechnol Bioeng 110(6):1574–1582. https://doi.org/10.1002/bit.24813

Tran DT et al (2011) Ethanol production from lignocelluloses by native strain Klebsiella oxytoca THLC0409. Waste Biomass Valorization 2(4):389–396. https://doi.org/10.1007/s12649-011-9082-6

Turner P, Mamo G, Karlsson EN (2007) Potential and utilization of thermophiles and thermostable enzymes in biorefining. Microb Cell Factories 6:1–23. https://doi.org/10.1186/1475-2859-6-9

Udom N et al (2019) Coordination of the cell wall integrity and high-osmolarity glycerol pathways in response to ethanol stress in *Saccharomyces cerevisiae*. Appl Environ Microbiol 85(15): e00551–e00519. https://doi.org/10.1128/AEM.00551-19

Voordeckers K et al (2015) Adaptation to high ethanol reveals complex evolutionary pathways. PLoS Genet 11(11):1–31. https://doi.org/10.1371/journal.pgen.1005635

Waghmare PR et al (2014) Production and characterization of cellulolytic enzymes by isolated klebsiella sp. PRW-1 using agricultural waste biomass. Emir J Food Agric 26(1):44–59. https://doi.org/10.9755/ejfa.v26i1.15296

Wallace-Salinas V et al (2015) Cell periphery-related proteins as major genomic targets behind the adaptive evolution of an industrial Saccharomyces cerevisiae strain to combined heat and hydrolysate stress. BMC Genomics 16(1):1–16. https://doi.org/10.1186/s12864-015-1737-4

Wang Z et al (2019) QTL analysis reveals genomic variants linked to high-temperature fermentation performance in the industrial yeast. Biotechnol Biofuels 12(1):59. https://doi.org/10.1186/s13068-019-1398-7

Wang L et al (2021) Improving multiple stress-tolerance of a flocculating industrial Saccharomyces cerevisiae strain by random mutagenesis and hybridization. Process Biochem 102:275–285. https://doi.org/10.1016/j.procbio.2020.12.022

Watanabe T et al (2011) A UV-induced mutant of Pichia stipitis with increased ethanol production from xylose and selection of a spontaneous mutant with increased ethanol tolerance. Bioresour Technol 102(2):1844–1848. https://doi.org/10.1016/j.biortech.2010.09.087

Widyasti E et al (2018) Biodegradation of fibrillated oil palm trunk fiber by a novel thermophilic, anaerobic, xylanolytic bacterium Caldicoprobacter sp. CL-2 isolated from compost. Enzym Microb Technol 111:21–28. https://doi.org/10.1016/j.enzmictec.2017.12.009

Xu H et al (2016) *PHO13* deletion-induced transcriptional activation prevents sedoheptulose accumulation during xylose metabolism in engineered *Saccharomyces cerevisiae*. Metab Eng 34:88–96. https://doi.org/10.1016/j.ymben.2015.12.007

Xue T et al (2018) Improved bioethanol production using CRISPR/Cas9 to disrupt the ADH2 gene in Saccharomyces cerevisiae. World J Microbiol Biotechnol 34(154):1–12. https://doi.org/10.1007/s11274-018-2518-4

Yadav P et al (2018) Production, purification, and characterization of thermostable alkaline xylanase from Anoxybacillus kamchatkensis NASTPD13. Front Bioeng Biotechnol 6(May). https://doi.org/10.3389/fbioe.2018.00065

Yi S et al (2018) Screening and mutation of *Saccharomyces cerevisiae* UV-20 with a high yield of second generation bioethanol and high tolerance of temperature, glucose and ethanol. Indian J Microbiol 58(4):440–447. https://doi.org/10.1007/s12088-018-0741-1

Zhang M et al (2017) Deletion of acetate transporter gene *ADY2* improved tolerance of *Saccharomyces cerevisiae* against multiple stresses and enhanced ethanol production in the presence of acetic acid. Bioresour Technol 245:1461–1468. https://doi.org/10.1016/j.biortech.2017.05.191

Zhang K et al (2018) Genetic characterization and modification of a bioethanol-producing yeast strain. Appl Microbiol Biotechnol 102(5):2213–2223. https://doi.org/10.1007/s00253-017-8727-1

Zhang Q et al (2019) Adaptive evolution and selection of stress-resistant Saccharomyces cerevisiae for very high-gravity bioethanol fermentation. Electron J Biotechnol 41:88–94. https://doi.org/10.1016/j.ejbt.2019.06.003

Zheng DQ et al (2012) Genome sequencing and genetic breeding of a bioethanol Saccharomyces cerevisiae strain YJS329. BMC Genomics 13(1). https://doi.org/10.1186/1471-2164-13-479

Zheng Y et al (2021) Genetic diversity for accelerating microbial adaptive laboratory evolution. ACS Synth Biol 10(7):1574–1586. https://doi.org/10.1021/acssynbio.0c00589

Zhou XG et al (2010) The next-generation sequencing technology: a technology review and future perspective. Sci China Life Sci 53(1):44–57. https://doi.org/10.1007/s11427-010-0023-6

Chapter 4
Enzymatic Hydrolysis of Feedstocks for 1G Bioethanol Production

Adenise Lorenci Woiciechowski, Luiz Alberto Junior Letti, Susan Grace Karp, Arion Zandoná Filho, Luis Alberto Zevallos Torres, Walter José Martinez Burgos, Carolina Mene Savian, and Carlos Ricardo Soccol

Abstract The search for new alternatives of fuels to compose an innovative and diversified energetic matrix available for countries is being the goal of a lot of research done worldwide. There are some essential steps to be well successful in this research. Especially for the biofuels development technology, involving the biofuels production through fermentative processes, the knowledge of the substrate composition and the alternatives of microorganisms and enzymes used at the process is fundamental. This chapter offers an overview of the most important agro industrial biomasses available to compose the substrates to produce 1G bioethanol, including data about the composition, world production, and others, discussing aspects important for each one. The enzymatic pretreatment is an essential step in converting amylolytic feedstocks into direct fermented sugar used as carbon source for alcoholic fermentation. This chapter brings informations about amylolytic enzymes, their activities in different conditions, industrial producers of amylolytic enzymes, producer microorganisms and the most recent published data about the use of amylolytic enzymes at the bioethanol production.

4.1 Introduction

For many reasons, including environmental, economic, social, and technical aspects, the search for new alternatives for energetic sources is the objective of a significant number of researches done nowadays for many researchers' groups all around the world either in public or private institutions, researcher's laboratories, industries and

A. Lorenci Woiciechowski (✉) · L. A. J. Letti · S. G. Karp · L. A. Zevallos Torres · W. J. Martinez Burgos · C. Mene Savian · C. R. Soccol
Department of Bioprocess Engineering and Biotechnology, Federal University of Paraná, Curitiba, Brazil
e-mail: adenise@ufpr.br

A. Zandoná Filho
Department of Chemical Engineering, Federal University of Parana, Curitiba, Brazil

C. R. Soccol et al. (eds.), *Liquid Biofuels: Bioethanol*, Biofuel and Biorefinery Technologies 12, https://doi.org/10.1007/978-3-031-01241-9_4

others. Among the energy sources developed can be cited, the photovoltaic energy using the solar light available mainly in tropical countries, wind energy, chemical energy, and bioenergy with biofuels, bioethanol, biodiesel, biogas, biohydrogen, for example.

When it comes to biofuels, significant amount of money is being employed in the diverse areas of knowledge, since the research on the development of new raw materials, including new vegetable cultivars, agro-industry residues, industrial processing residues, liquid or solid, to work as substrate to produce biofuels through biotechnological processes, the development of new technologies to pretreat these raw material do make them an available substrate for the fermentative processes, until the search of new biofuels alternatives, like biohydrogen.

In this sense, it is important to know very deeply, that the available material to compose the substrate for a fermentative process, precede a careful characterization of the raw material, to specify any necessary pretreatment to suit the substrate composition to the target microorganism.

This chapter gives an overview of the most important feedstock available for 1G bioethanol production through fermentative processes, the enzymes microorganisms producer and the enzymes necessary to prepare substrates for 1G bioethanol production.

4.2 Starch Feedstock for 1G Bioethanol

Starch is found mainly in plant tissues and constitutes an important energy reserve. Chemically, it is a polymer of glucose, and hence classified as a carbohydrate. When the glucose chain is "linear" (α-1,4-linked glucose residues), it is called amylose, and the branched-chain starch is called amylopectin (α-1,4-linked glucose residues plus much α-1,6-linked glucose branched points) (Zhang et al. 2021a). The branched are around 5–6% of total links in amylopectin (Rolland-Sabaté et al. 2012).

Starch-containing feedstocks are those raw materials that contain starch in large quantities. These include grain crops (e.g., corn, barley, wheat or grain sorghum) and root/tubular crops (e.g. cassava, potato, sweet potato, Jerusalem artichoke, cactus, or arrowroot) (Bušić et al. 2018). The production of fuel ethanol from starch was first introduced in the United States at the beginning of the twentieth century, and today most of the 1G ethanol production comes from corn (Krajang et al. 2021).

In 2020, United States was reported as the largest bioethanol producer globally (48.2%), using maize/corn as the primary feedstock. Brazil was the second bioethanol producer (26.2%), where sugarcane and maize were the principal feedstocks. China and the European Union were the third and fourth bioethanol producers accounting for 8.1% and 4.9% of the global ethanol produced, using maize, cassava, sugar beet, and wheat as the main feedstocks (OECD Library 2020).

For bioethanol production from starch-containing feedstocks, it is necessary to perform the starch hydrolysis (primarily by α-amylase and glucoamylase) into glucose syrup, which can be converted into ethanol by yeast Saccharomyces cerevisiae. This step is an additional cost compared to the bioethanol production from sugar-containing feedstocks (Bušić et al. 2018).

However, some of the feedstocks currently used for 1G ethanol production present lower ethanol yield per unit area of cultivation when compared to that of other crops. Average yields of cassava, sweet potato, wheat, rice, sorghum, and starchy corn are approximately 31.25, 30.00, 9.00, 7.31, 6.25, and 6.00 MT/ha/year (metric ton/ha/year), respectively (Krajang et al. 2021).

4.2.1 Maize/Corn

The world corn production was 1148 million tons in 2019 where the United States was reported as the major ethanol producer contributing with 30.2% of the world production followed by China (22.7%), Brazil (8.8%), and Argentina (5.0%) (Food and Agriculture Organization of the United Nations 2019).

Corn is one of the most important sources of starch worldwide, commonly used for water retention and as gelling bulking agent, despite constituting one of the essential sources of simple sugars for ethanol production worldwide. Its average composition in terms of amylose and amylopectin can be seen in Table 4.1 (Ibrahim et al. 2019).

4.2.2 Wheat

Wheat is the most widely grown cereal crop in the world. Wheat is sensitive to various adverse weather conditions and abiotic stresses, significantly reducing yields (Harkness et al. 2020).

Table 4.1 Average composition of starch, amylose and amylopectin for the most important starchy crops

Crop	Starch content (g/100 g of dry material)	Amylose content (g/100 g of dry material)	Amylopectin content (g/100 g of dry material)	References
Corn	>70	24.6	75.4	Ibrahim et al. (2019)
Wheat	Around 70	Average of 25	Average of 75	Shevkani et al. (2017)
Cassava	84.5	19.5	Around 80	Bertoft Eric et al. (2010), Oladunmoye et al. (2014)
Rice	Around 90	9.0–29.5	Variable	Aoki et al. (2012)
Sorghum	54.7	16.5	83.5	Gerrano et al. (2014)
Sweet potato	66.6	17.0	83.0	Tortoe et al. (2017)

In 2019, the world wheat production reached 765 million tons. China and India were the two major wheat producers accounting for 17.4% and 13.5% of the world production, respectively. The United States was the fourth wheat producer after Russia, contributing with 6.8% and 9.7% of the global production, respectively (Food and Agriculture Organization of the United Nations 2019).

Wheat starch is widely variable according to wheat variety and also depends on environmental and agronomical conditions. For example, starch from soft wheat presents higher lipids and protein contents, which confer them lower paste viscosity, while starch from hard wheat presents lower gelatinization temperatures and enthalpies. These properties of wheat starch are strongly dependent on the amylose and amylopectin contents. In general, the average percentage of amylose and amylopectin are 25 and 75, respectively. However, common varieties may present from 18% to 29% of amylose, while durum varieties on the 17–28% range, and spelled varieties between 30% and 33% (Shevkani et al. 2017).

4.2.3 Cassava

Cassava is cultivated in countries with warm and moist tropical climate. The tubers grow well on soils of relatively low fertility where the cultivation of other crops would be difficult or uneconomical (Kuiper et al. 2007). Regarding to the economic production, cassava is harvested, preferably, with one cycle (10–12 months).In contrast, industrial cassava is harvested with one or two cycles (18–24 months) (de Oliveira Aparecido et al. 2020).

The global production of cassava achieved 303 million tons in 2019, where Nigeria produced 19.5% of the total production, followed by Congo (13.2%) and Thailand (10.2%) (Food and Agriculture Organization of the United Nations 2019). Ghana and Brazil were the fourth (7.4%) and fifth (5.8%) cassava producers.

The economic viability of bioethanol production from cassava in different countries is currently under investigation. Pretreatment of cassava tubers for bioethanol production includes the following operations: cleaning, peeling, chipping and drying. After that, the dried cassava chips are used for bioethanol production (Bušić et al. 2018).

Regarding to cassava starch, it can be used for food industry like other starches, but the main concern is the reduction of cyanide levels. When this aspect is carefully taken to account, cassava starch can repair high-quality flours, usually mixed with wheat flour (Oladunmoye et al. 2014).

4.2.4 Rice

Rice is grown in almost all continents except in low-temperature regions where plant growth is impossible. Rice is a C3 agricultural crop like wheat (Hussain et al. 2020).

Approximately 755 million tons of rice were produced in the world in 2019. China and India were the two major producers accounting for 28.0% and 23.5% of the global production, respectively (Food and Agriculture Organization of the United Nations 2019).

Rice starch composition depends on several factors, such as variety, extraction methodology (isolation), and environmental conditions (Wani et al. 2012). Compared to other starches, rice starch has unique characteristics like creamy, smooth and bland taste (Wani et al. 2012). Its average composition is shown in Table 4.1.

4.2.5 Sorghum

Sorghum is a short-cycle C4 plant that produces high amounts of biomass and fermentable sugars demanding low agricultural inputs and water, adaptable to marginal and drought-affected lands (López-Sandin et al. 2021).

In 2019, United States produced 8.6 million tons of sorghum, representing 15.0% of the world's production. Nigeria and Ethiopia were the following major sorghum producers, accounting for 11.5% and 9.1% of the global production, respectively (Food and Agriculture Organization of the United Nations 2019). By 2020, in the United States, there were nine industrial plants producing ethanol from sorghum (López-Sandin et al. 2021).

Like other starchy crops, sorghum starch can widely vary in composition, depending on several factors, but mainly due to genotypic differences. A study with 22 varieties has shown that starch corresponds to around 54.7% of total dry matter, while amylose equals about 16.5% of total dry weight and amylopectin 83.5% (Gerrano et al. 2014).

4.2.6 Sweet Potato

Sweet potato presents wide adaptability to marginal lands in areas that range from the tropics to temperate zones. At high latitudes, sweet potato plants require less chemical pesticide and fertilizer treatment (Kwak 2019). Another competitive advantage of the sweet potato is its short life cycle, varying from 5 to 6 months, allowing two harvests per year. The minimum requirement for sweet potato cultivation is a frost-free period lasting at least 4 months (Kwak 2019; Rizzolo et al. 2021).

China is the largest sweet potato producer globally (103 million tons) in 2019, accounting for 56.6% of the world's production. Other major sweet potato producers are African countries like Malawi (6.4%), Nigeria (4.5%), Tanzania (4.3%), Uganda (2.1%), Angola (1.8%) (Food and Agriculture Organization of the United Nations 2019).

A research has studied 12 varieties of sweet potatoes and has concluded they have non-uniform compositions (Tortoe et al. 2017). The total content of starch varies between 48.9 to 77.0 grams of starch per gram of dry weight (average of 66.6 g/ 100 g), while numbers for amylose have ranged from 10.1 to 20 g/100 g (average of 17.0 g/100 g) and for amylopectin from 79.8 to 89.9 g/100 g (average of 83.0 g/ 100 g).

4.3 Role of Enzymes in Starch Saccharification

For most industrial applications of starch (including its use as raw material for the1G ethanol production), it needs to be hydrolyzed into smaller fractions to release oligomers or even its glucose monomers depending on the case. The enzymes which act in the prior steps of starch hydrolysis cleave the α-1,4 glycosidic bonds of the linear chains, namely: α-amylase, β-amylase, glucoamylase and α-glucosidase. They act synergistically with the debranching chain enzymes, responsible for breaking the starch side chains (α-1,6 glycosidic bonds) (Xia et al. 2021).

In general means, the amylases can be classified as endoamylases, when capable of hydrolyzing internal α-1,4 glycosidic bonds, thus releasing oligomers of different types and lengths. The exoamylases are just able to hydrolyze terminal α-1,4 glycosidic bonds, releasing glucose monomers.

The classification of debranching chain enzymes depends on their substrate specificity and also in their catalytic mode. The main four groups are amylo, α-1,6-glucosidases, glucoamylases, pullulanases, and isoamylases. The glucoamylases (also called amylo-glucosidases) act on α-1,6 and α-1,6 bonds, and are responsible for removing glucose units from the polysaccharide chain ends. Pullulanases and isoamylases act more specifically debranching at α-1,6 links, being preferred industrially for starch processing, especially when amylopectin is present in higher amounts (Xia et al. 2021).

The following sections will deal with the most important industrial enzymes used for starch hydrolysis, and the recent developments and advances in this.

4.4 Amylolytic Enzymes

4.4.1 Currently Used Amylases

Amylases are enzymes capable of hydrolysate glycosidic bonds of the starch molecules, acting in amylose and amylopectin chains, releasing fermentable sugar for bioethanol production (Fig. 4.1) (de Souza et al. 2019). The main amylases commonly used in industries are α-amylases, β-amylases, isoamylases, pullulanases and glucoamylases (de Souza et al. 2019; Singh et al. 2014). α and β-amylases are *endo*-acting enzymes capable of cleaving α-1,4 glycosidic linkages present inside the

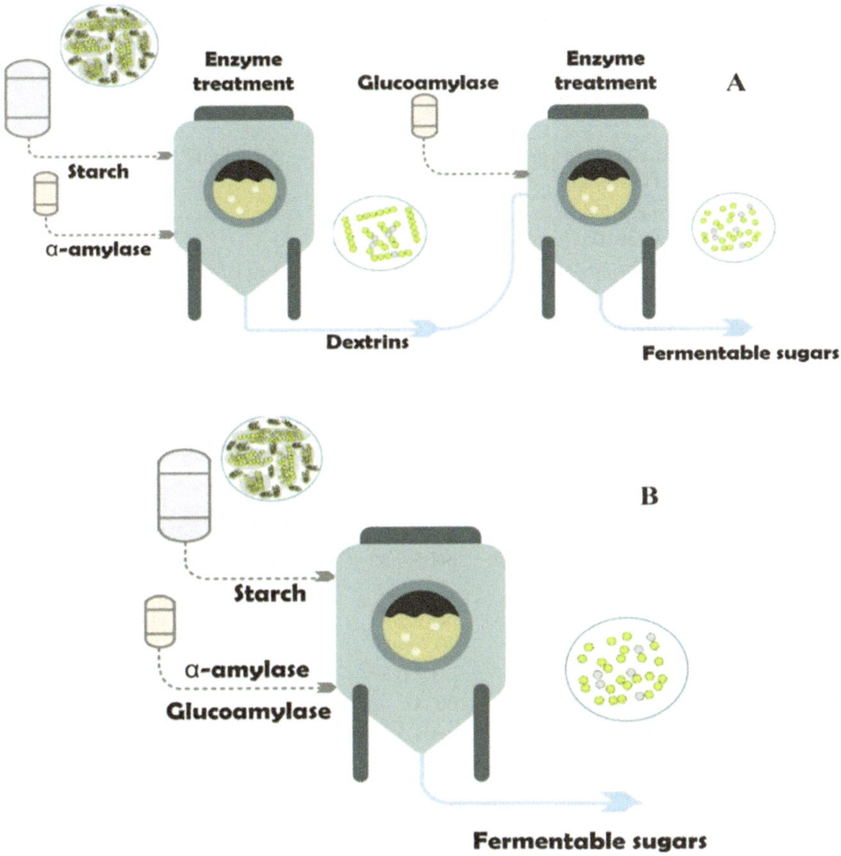

Fig. 4.1 (**a**) Enzymatic starch hydrolysis in two steps. (**b**) Simultaneous liquefaction and saccharification of starch

starch molecule structure, releasing non-reducing ends where glucoamylases can act, subsequently. These glucoamylases are *exo*-acting that can cleave α-1,4 glycosidic linkages from those non-reducing ends (Cripwell et al. 2020); however, its action can be interrupted by α-1,6 bonds, leaving branch points, which are debranching by isoamylases and pullulanases. Therefore, for starch-completely hydrolyzation, amylases must act synergically (Hii et al. 2012; Xia et al. 2021).

These enzymes are produced by various organisms, such as plants, animals, bacteria, and fungi (Gopinath et al. 2017). Usually, only a limited number of natural enzymes tolerates the harsh physicochemical conditions in large-scale production; therefore, several researches aim to isolate enzymes from microorganisms from adverse environment, such as mangroves, heavy metal-rich soil, hot springs, and marsh. However, bacteria and fungi are the suitable source for enzyme production in large-scale, as *Bacillus subtilis*, capable of producing α-amylase, *Aspergillus niger*,

Table 4.2 Amylases producers microorganisms

Enzyme	Production source	Enzymatic assay conditions	Enzymatic activity	References
α-Amylase (EC 3.2.1.1)	*Bacillus subtilis*	pH 7 35–55 °C	–	Shiau et al. (2003)
	Bacillus licheniformis ATCC 9945a	pH 6.5 70 °C	5.2 IU/mL[a]	Božić et al. (2011)
	Aspergillus oryzae IFO-30103	pH 5.0 50 °C	–	Ramachandran et al. (2004)
Glucoamylase (EC 3.2.1.3)	*Aspergillus awamori* MTCC 6652	pH 5.0 70 °C	140 U/mL[b]	Negi and Banerjee (2009)
	Aspergillus niger	pH 4.0–5.0 55–65 °C	–	Ono et al. (1988)
	Corticiumrolfsii AHU 9627	pH 4.5 60 °C	–	Nagasaka et al. (1998)
	Humicola sp.	pH 4.7 55 °C	56.63 U/mg	Riaz et al. (2007)
Isoamylase (EC 3.2.1.68)	*Pseudomonas amyloderamosa*	pH 3.5 52 °C	–	Harada et al. (1972)
	Rhizopus oryzae PR7 MTCC 9642	pH 5.5 55 °C	52.4 U/mg	Ghosh et al. (2020)
	Bacillus lentus JNU3	pH 6.5 70 °C	318 U/mL	Li et al. (2013)
Pullulanase (EC 3.2.1.41)	*Bacillus acidopullulyticus*	pH 5.0 60 °C	220 U/mg	Kusano et al. (1988)
	Bacillus deramificans	pH 4.5 55 °C	1567.9 U/mL	Zou et al. (2014)
	Paenibacillus lautus DSM 3035	pH 7.0 40 °C	11.1 U/mL	Chen et al. (2017)

[a]IU: International Units
[b]U: Units

which produces glucoamylases, and many others are shown in Table 4.2, that are easily cultivated, and are optimal amylase producers, compared to the other organism mentioned (Saleem and Ebrahim 2014; Zaferanloo et al. 2014).

One of the methods used to search for industrial applicable enzymes consists of strains isolation, followed by a genome sequencing to identify the enzymes (Fasim et al. 2021). Other methods, including genetic engineering and recombinant DNA technology, are applied to improve enzyme stability, increase production and substrate specificity (Corbin et al. 2016; Gopinath et al. 2017), leading to nearly 90% of commercial enzymes being engineered (Adrio and Demain 2014).

The level of enzyme production and its stability in a specific range of pH and temperature, varies from one microorganism to another, as well as hydrolysis yield in each starch feedstock (Fasim et al. 2021), this was verified by Szambelan et al. (2020), when four different blends of enzymes were applied in pairs in sorghum starch. One experiment was carried out by adding a *Bacillus licheniformis*α-amylase for simultaneous gelatinization and liquefaction, at 100 °C for 1 h, followed by the addition of *Aspergillus niger* glucoamylase for saccharification at 60 °C for 100 min; in contrast, the second experiment consisted of separate gelatinization at 100 °C for 1 h, and subsequential liquefication at 80 °C for 20 min using *Bacillus stearothermophilus* α-Amylase; then, the addition of *Aspergillus niger* glucoamylase and protease for saccharification at 60 °C for 100 min. The hydrolysis yields obtained in the first and the second experiment reached 55.4% and 72.8%, respectively. de Farias et al. (2020), applied the same liquefaction and saccharification processes in two other feedstocks (Cassava starch and Taioba starch) using the same enzyme blends and resulting in slightly different hydrolysis yields (52% and 50%, respectively);this shows that enzymes from different microorganisms have different hydrolytic action. Other experiment using different feedstock and enzymes are shown in Table 4.3.

Starch from many sources can be hydrolyzed generating simple sugar used to produce bioethanol in industrial plants with different process configuration. Also, there is a great variety of enzymes available obtained from many microorganisms with different properties to be used at the starch hydrolysis. These facts allow a lot of possible process configurations to set this process. The conventional "hot" process consists of a feedstock cooked for a few minutes in high temperatures, then a stage of liquefication at 80–90 °C with a thermostable α-amylase followed by saccharification with a glucoamylase to hydrolysate starch into fermentable sugar (Bothast and Schlicher 2005). Alternatively, the "cold" process converts raw/granular starch feedstock directly into glucose using modified raw starch-degrading enzymes (van Zyl et al. 2012), so 40–50% of operational and total costs can be save, only by exclusion of cooking and liquefaction stages at high temperatures (Brown et al. 2020).

Many companies have developed specific products for bioethanol platforms that adopt both hot and cold processes, adapting the enzymes content according to the type of treatment (van Zyl et al. 2012). Anmylex® BT2, from DuPont™ Genencor®, for example, has thermostable α-Amylases at 95 °C, while SAN Extra® L, from Novozymes Ltd., has Glucoamylases active at 30 to 70 °C, as shown in Table 4.4.

4.5 Advanced Amylases

4.5.1 Engineered Strains for Consolidated Bioprocess

New technologies have been developed in amylolytic strain engineering to convert starch into ethanol by a raw starch consolidated bioprocess (CBP), which consists of a single-stage process combining liquefaction, hydrolysis and fermentation of feedstock by a single organism (Zyl et al. 2007). The ideal modified strain should be capable of co-expressing α-amylase and glucoamylase to completely hydrolysate the starch feedstock (Jouzani and Taherzadeh 2015). Thus, several *Saccharomyces cerevisiae* strains were modified to express amylase genes (Cripwell et al. 2019;

Table 4.3 Enzyme hydrolysis of different biomasses

Enzyme	Enzyme source	Enzyme activity	Biomass	Process adopted	Hydrolysis condition	Hydrolysis yield	References
Amylases	*Rhizopus oligosporus* (CCT 3762)	346.8 U/g[a]	Cassava starch	Liquefaction	pH 5.5, 11 h, 55 °C	81%	Escaramboni et al. (2018)
α-Amylase Glucoamylase	Liquozyme supra 2.2X Novozymes (*Bacillus licheniformis*) AMG 300 L Novozymes (*Aspergillus niger*)	300 KNU/g[b] 260 U/mL[a]	Cassava starch	Liquefaction and saccharification	pH 5.5, 65 °C, 2 h (liquefaction) pH 4.5, 65 °C, 24 h (saccharification)	52%	de Farias et al. (2020)
α-Amylase Glucoamylase	Liquozyme supra 2.2X Novozymes (*Bacillus licheniformis*) AMG 300 L Novozymes (*Aspergillus niger*)	300 KNU/g[b] 260 U/mL[a]	Taioba starch	Liquefaction and saccharification	pH 5.5, 65 °C, 2 h (liquefication) pH 4.5, 65 °C, 24 h (saccharification)	50%	de Farias et al. (2020)
α-Amylase Glucoamylase	*Bacillus licheniformis* (Termamyl SC DS, Novozymes) *Aspergillus niger* (san extra L, Novozymes)	–	Sorghum	Simultaneous gelatinization and liquefaction	100 °C, 1 h (gelatinization and liquification) 60 °C, 100 min (saccharification)	55.4%	Szambelan et al. (2020)
α-Amylase Glucoamylase and protease	*Bacillus stearothermophilus* (Amylex BT2, Genencor international) *Aspergillus niger* (Diazyme SSF, Genencor International)	–	Sorghum	Separated gelatinization and liquefaction	100 °C, 1 h (gelatinization) 80 °C, 20 min (liquification) 60 °C, 100 min (saccharification)	72.8%	Szambelan et al. (2020)
Amylases	*Aspergillus awamori* IOC-3914	–	Babassu flour	Cold hydrolysis of raw starch with enzyme loading	pH 4.8, 72 h 50 °C–32 °C	87%	Cinelli et al. (2014)

[a]Units
[b]KNU: Kilo Novo Units

Table 4.4 Commercial enzymes for bioethanol production

Enzyme(s)	Commercial name	Enzyme activity	Source	Features	Companies	References
α-Amylases and glucoamylases	TARGEN™002	570 AGU/g[a]	*Aspergillus niger* and *A. kawachi*	pH 4.0–4.5	DuPont™ Genencor® Science	Lu et al. (2020) Nguyen et al. (2020)
α-Amylase	CG 626	–	*Trichoderma reesei*	–	DuPont™ Genencor® Science	Lu et al. (2020), Strąk-Graczyk and Balcerek (2020)
α-Amylase	Anmylex® BT2	–	*Bacillus stearothermophilus*	pH 5.5–7.0 85–95 °C	DuPont™ Genencor® Science	Szambelan et al. (2020)
α-Amylase	Termamyl® SC	133.6 KNUS/g[b]	*Bacillus lichenformis*	pH 5.0–6.0 85–95 °C	Novozymes Ltd.	Cole et al. (2019)
α-Amylase	Liquozyme®	135 KNUS/g[b]	*Bacillus lichenformis*	pH 5.5 85 °C	Novozymes Ltd.	Duvernay et al. (2013)
Isoamylase	E-ISAMY®	280 U/mL[c]	*Pseudomonas* sp.	pH 4.0–5.0	Megazyme International Ireland Ltd.	Chengyao et al. (2021))
Pullulanase	Promozyme® D2	1498 NPUN/g[d]	*Bacillus subtilis*	pH 5.0 50–65 °C	Novozymes Ltd.	Cole et al. (2019)
Pullulanase	E-PULKP®	30 U/mg[c]	*Klebsiella planticola*	pH 5.0 40 °C	Megazyme International Ireland Ltd.	Megazyme International Ireland Ltd (n.d.)
Glucoamylase	SAN Extra® L	424 AGU/g[a]	*Asperigillus niger*	pH 3.0–6.0 30–70 °C	Novozymes Ltd.	Cole et al. (2019)
Glucoamylase	Diazyme® SSF	–	*Asperigillus niger*	pH 5.0–6.0 85–110 °C	DuPont™ Genencor® Science	Szambelan et al. (2020)

[a]AGU: Amyloglucosidase units
[b]KNUS: Kilo Novo Units
[c]U: Units
[d]NPUN: New Pullulanase Unit Novozymes

Jouzani and Taherzadeh 2015), once this microorganism is easy to manipulate and allows screening of transformants without the use of antibiotic selection (Cripwell et al. 2019). Amylase genes used for engineering organisms have many possible origins and the recombinant method can vary according to each research. An *S. cerevisiae* Y294 strain modified by multi-copy vector methodology to incorporate AteAα-amylase gene from *Aspergillus terreus* and GlaA glucomalylase gene from *Aspergillus tubingensis* had a raw corn starch ethanol yield of 44% (Sakwa et al. 2018) against 46% from the same strain with AmyAα-amylase gene and GlaA glucomalylase gene from *A. tubingensis* (Viktor et al. 2013). Other research achieved better results, as shown in Table 4.5. However, no recombinant amylolytic yeast strains are capable of co-expressing α-amylase, and glucoamylase applied in any industrial CBP for ethanol production from starch (Cripwell et al. 2019).

4.5.2 Alpha-Amylase Immobilization

Since α-amylase enzymes are widely used in different industries, including ethanol production from starch (Jujjavarapu and Dhagat 2019; Pereira et al. 2017), currently, different techniques are used to increase its stability, recovery, and reuse (Atiroğlu et al. 2021). One of the strategies is the immobilization of amylases using different materials, some in the form of nanoparticles (Zhang et al. 2021b), that is, confine the enzyme in an specific space region. Among the advantages of this technique is the fact that it can reuse the enzyme, controlled product formation, adaptability, high reaction speeds, and operational flexibility, also they can be used in continuous or batch systems (Atiroğlu et al. 2021; Pereira et al. 2017). Enzymes can be immobilized by two main methods, classified as imprisonment and bond formation (Fig. 4.2). In imprisonment, the enzyme is immobilized within polymeric meshes, capsules and microcapsules are currently being evaluated. In the case of Bond formation, there is a link between the enzyme and the support. These bonds can be weak in the case of absorption, strong like covalent bonds, and can also be cross-linked (Homaei et al. 2013).

α-Amylase can be immobilized using different techniques and materials. Atiroğlu et al. (2021) used metal-organic structures for the immobilization of the α-amylase enzyme. According to the authors, the stability of the immobilized enzyme (OLB/BSA@ZIF-8)-α-amylase was significantly improved concerning the free enzyme, as it maintained 90% of the initial enzyme activity for up to 8 weeks. In fact, (OLB/BSA@ZIF-8)-α-amylase showed 81% of its initial activity after twenty cycles. (Torabizadeh and Montazeri 2020) immobilized α-amylase on magnetic nanoparticles, and according to the authors, α-amylase-immobilized maintained 84% of its initial enzymatic activity after ten cycles. Pereira et al. (2017), immobilized α-amylase on glass tubes. The immobilization yield, efficiency, and activity recovery of α-amylase were 79, 57, and 45%, respectively. Other examples of α-amylase immobilization are shown in Table 4.5. Although immobilization improves α-amylase stability, it is significantly greater when nanomaterials are

Table 4.5 *Saccharomyces cerevisiae* strains developed for raw starch consolidated bioprocess

S. cerevisiae strain	Amylase origin	Strain production method	Substrate	Maximum theoretical ethanol yield (%)	References
M2n [TLG1-SFA1]	Glucoamylase from *Thermomyces lanuginosus* and α-amylase from *Saccharomycopsis fibuligera*	δ—Sequence integration	Unmilled wheat bran	81	Cripwell et al. (2015)
MEL2 [TLG1-SFA1]	Glucoamylase from *T. lanuginosus* and α-amylase from *S. fibuligera*	δ—Sequence integration	Unmilled wheat bran	85	Cripwell et al. (2015)
M2n [TLG1-SFA1]	Glucoamylase from *T. lanuginosus* and α-amylase from *S. fibuligera*	δ—Sequence integration	Raw corn starch	55	Favaro et al. (2015)
M2n [TLG1-SFA1]	Glucoamylase from *T. lanuginosus* and α-amylase from *S. fibuligera*	δ—Sequence integration	Raw sorghum starch	62	Favaro et al. (2015)
M2n [TLG1-SFA1]	Glucoamylase from *T. lanuginosus* and α-amylase from *S. fibuligera*	δ—Sequence integration	Raw triticale starch	73	Favaro et al. (2015)
MEL2 [TLG1-SFA1]	Glucoamylase from *T. lanuginosus* and α-amylase from *S. fibuligera*	δ—Sequence integration	Raw corn starch	45	Favaro et al. (2015)
MEL2 [TLG1-SFA1]	Glucoamylase from *T. lanuginosus* and α-amylase from *S. fibuligera*	δ—Sequence integration	Raw sorghum starch	57	Favaro et al. (2015)
MEL2 [TLG1-SFA1]	Glucoamylase from *T. lanuginosus* and α-amylase from *S. fibuligera*	δ—Sequence integration	Raw triticale starch	67	Favaro et al. (2015)
M2n [TLG1-SFA1]	Glucoamylase from *T. lanuginosus* and α-amylase from *S. fibuligera*	δ—Sequence integration	Rice by-products (bran, husk, broken, discolored and unripe rice)	92	Favaro et al. (2017)
MEL2 [TLG1-SFA1	Glucoamylase from *T. lanuginosus* and α-amylase from *S. fibuligera*	δ—Sequence integration	Rice by-products (bran, husk, broken,	90	Favaro et al. (2017)

(continued)

Table 4.5 (continued)

S. cerevisiae strain	Amylase origin	Strain production method	Substrate	Maximum theoretical ethanol yield (%)	References
			discolored and unripe rice)		
Y294 [AteA-GlaA]	α-Amylase from *Aspergillus terreus* and glucoamylase from *Aspergillus tubingensis*	Multi-copy vector	Raw cornstarch	44	Sakwa et al. (2018)
ER T12	α-Amylase from *Talaromyces emersonii* and glucose from an optimized codon glucoamylase	δ—Sequence integration	Raw broken rice	100	Myburgh et al. (2019)
M2n T1	α-Amylase from *T. emersonii* and glucose from an optimized codon glucoamylase	δ—Sequence integration	Raw broken rice	100	Myburgh et al. (2019)
ER T12	α-Amylase from *T. emersonii* and glucose from an optimized codon glucoamylase	δ—Sequence integration	Raw cornstarch	86	Cripwell et al. (2019)
M2n T1	α-Amylase from *T. emersonii* and glucose from an optimized codon glucoamylase	δ—Sequence integration	Raw corn starch	94	Cripwell et al. (2019)
Y294 [TemG_Opt-TemA]	α-Amylase from *T. emersonii* and glucose from an optimized codon glucoamylase	Multi-copy vector	Raw corn starch	60	Cripwell et al. (2019)

used. Other examples of α-Amylase immobilization are shown in Table 4.6. Immobilization improves α-amylase stability, and it is considerably greater when nanomaterials are used.

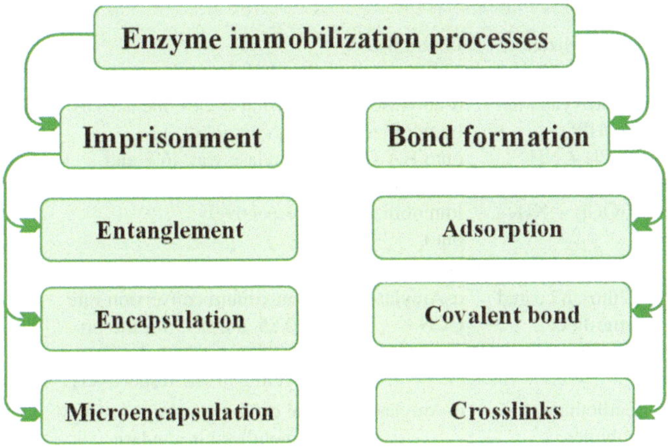

Fig. 4.2 Classification of enzymatic immobilization methods

4.6 Future Perspectives

In general, enzymes are susceptible to chemical and physical changes in the environment. Therefore, the immobilization of these catalysts with nanomaterials and/or magnetic nanomaterials promises to be one of the most effective techniques to improve the catalysts' operational stability and reusability.

4.6.1 Technological Advances

The use of enzymes to hydrolyze starch to produce bioethanol has advanced and is already implemented in several industrial facilities. To assess the most recent and relevant protected technologies, a patent search was performed in the Derwent Innovations Index Database, using the keywords [(starch*) AND (enzyme* OR amylase*)] in the field "Topic" and the code C12P-007/06 in the field "International Patent Classification" (IPC), which represents "Fermentation or enzyme-using processes to synthesize a desired chemical compound or composition, i.e., ethanol (non-beverage)". In the last 10 years, 364 patent documents were published, and most of them (63%) were filed by companies. Among these, 66 documents (18.1%) were specifically associated with the area of energy fuels. The dominant knowledge areas were chemistry (99.7%), biotechnology and applied microbiology (98.3%), food science technology (23.6%), engineering (20.6%) and polymer science (18.4%).The top assignees were Novozymes, with 24.7% of published documents, Danisco, with 14.6% and Basf Enzymes, with 3.3%. In the area of energy fuels, the top assignee was Novozymes, with 27% of patent publications, followed by Danisco with 7.6%; 54% of the patent documents were filed by companies. These data show

Table 4.6 Alpha-amylase immobilization

Amylase origin	Immobilization material	Immobilized enzyme	Results	References
α-Amylase from *Bacillus subtilis*	Hydroxyapatite: Na_2HPO_4 + $CaCl_2 + NH_3$ Zirconia: $ZrOCl_2 + NH_3$	α-amylase immobilized onto HA (1) α-Amylase immobilized onto HA/ZrO_2 (2)	After ten cycles, the activity of the immobilized α-amylase was 46% and 70%, for 1 and 2 respectively.	Almulaiky et al. (2021)
–	Chitosan coated nano-fiber (CCN)	α-Amylase-CCN	A maximum conversion rate of 0.85 and 0.99 for the process without and with recycling mode, respectively.	Gali et al. (2021)
α- amylase from *Bacillus subtilis*	Carbothane polymer dissolved in dichloromethane	α-Amylase	The catalytic activities of the immobilized α-amylase enzymes were 80% after ten cycles	El-Shishtawy et al. (2020)
α-Amylase from *Bacillus subtilis*	Sepiolite with bilayer modification	α-Amylase-BSEPA	The residual activity was 66.7% after 30 days of storage. Furthermore, the immobilized enzyme showed an activity of 44% after ten cycles.	Mortazavi and Aghaei (2020)
–	Graphene oxide (GO)—Magnetite (Fe_3O4) nanoparticles	α-Amylase-GO	The alkali tolerance of the enzyme was increased by ~20%. The immobilized amylase could be used for eleven subsequent cycles	Desai et al. (2021)
α-Amylase from *Aspergillus oryzae*	Polyvinyl alcohol (PVA)	α-Amylase-loaded	Enzymatic immobilization improves the stability of-α-amylase in a wide range of temperatures (30–70 °C).	Porto et al. (2019)
α-Amylase from *Aspergillus oryzae*	Synthesized titania/lignin	α-Amylase immobilized	The amylase showed improved thermal and chemical stability as well as its reusability.	Klapiszewski et al. (2018)
α-Amylase from *Aspergillus oryzae*	Chitosan	α-Amylase immobilized	The free and immobilized enzyme activity during 40 days of storage at 4 °C decreased 95% and 36%, respectively.	Mardani et al. (2018)

HA hydroxyapatite; *CCN* Chitosan coated nano-fiber; *BSEPA* bilayer modification of sepiolite-amylase

that the technologies related to the enzymatic saccharification of starch to produce energy fuels are mainly developed at the industrial level.

Among all patent documents, the most cited (WO2012064351-A1) (Morant et al. 2012), with 18 citations, was filed by Novozymes and described a newly isolated

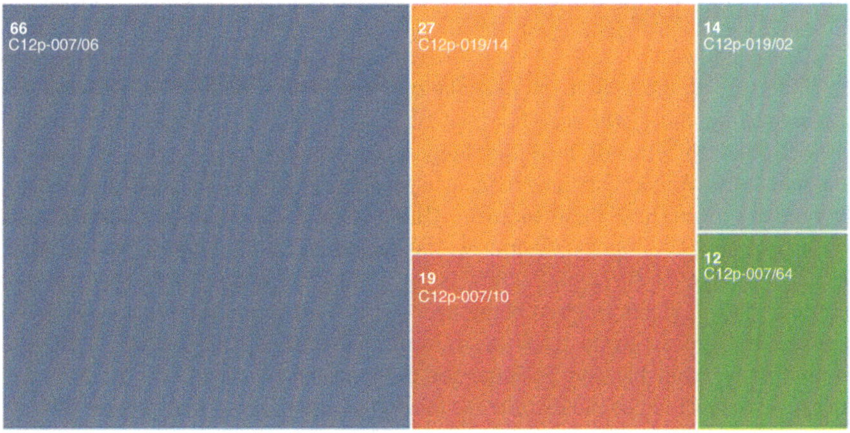

Fig. 4.3 Top-five international patent classification (IPC) codes among the 66 patent documents related to energy fuels. A search was performed on the Derwent Innovations Index Database on October 28th, 2021, using the terms [(starch* AND (enzyme* OR amylase*)) and the IPC C12P-007/06]. Note: C12P-007/06—Fermentation or enzyme-using processes to synthesize a desired chemical compound or composition, i.e., ethanol (non-beverage); C12P-019/14—Preparation of compounds containing saccharide radicals produced by the action of a carbohydrase, e.g. by alpha-amylase; C12P-007/10—Preparation of ethanol (non-beverage) from substrate containing cellulosic material; C12P-019/02—Preparation of compounds containing saccharide radicals, i.e., monosaccharides; C12P-007/64—Preparation of fats; fatty oils; ester-type waxes; higher fatty acids, i.e. having at least seven carbon atoms in an unbroken chain bound to a carboxyl group; oxidized oils or fats.

polypeptide having glucoamylase activity, useful for producing syrup and/or a fermentation product. Similarly, the document WO2013053801-A1 (Friis et al. 2013) with 16 citations, also filed by Novozymes, described anew glucoamylase variant, useful for producing syrup and/or fermentation products, and for brewing. In third position, with 15 citations, are the documents WO2012088303-A2 (Deinhammer et al. 2012) and WO2011066576-A1 (Landvik et al. 2011), both filed by Novozymes. The first described a method for producing fermentation products from starch-containing material, through liquefying the feedstock at a pH of 4.5–5 at 80–90°C with an alpha-amylase in the presence of calcium chloride (0.12 mM) and, optionally, of a thermostable protease, followed by saccharification and fermentation. The second described a new glucoamylase, useful for producing syrup and/or a fermentation product and for brewing.

The documents related to energy fuels described methods of enzymatic hydrolysis/saccharification and fermentation, novel enzymes, new strains of microorganisms, and methods to produce hydrocarbon-based fuels. The distribution of technology classifications among the documents, as represented by the IPC codes, can be seen in Fig. 4.3.

4.7 Conclusions

As important as the search for new fuels to compose a diverse and enlarge the alternatives for the world energetic matrix, is to develop innovative fuels alternatives. Biofuels, bioethanol, biodiesel, biogas, biohydrogen come to give an essential and innovative option for this. Also, it is equally important to present processes alternatives to produce these biofuels. Biofuels production through fermentative processes, the use of agro-industrial wastes and residues to compose the substrate for fermentative biofuels production is an available alternative to decrease the biofuels costs production. Even being necessary to set some additional steps at the raw material processing to reach the suitable substrate, this alternative is demonstrating to be available and essential for the environment, avoiding the area and energy demands to treat and dispose the residue of one process, that become the raw material for another process, at the concept of biorefinery and circular economy.

References

Adrio JL, Demain AL (2014) Microbial enzymes: tools for biotechnological processes. Biomol Ther 4(1):117–139. https://doi.org/10.3390/biom4010117

Almulaiky YQ, Khalil NM, El-Shishtawy RM, Altalhi T, Algamal Y, Aldhahri M, Al-Harbi SA, Allehyani ES, Bilal M, Mohammed MM (2021) Hydroxyapatite-decorated ZrO2 for α-amylase immobilization: toward the enhancement of enzyme stability and reusability. Int J Biol Macromol 167:299–308. https://doi.org/10.1016/j.ijbiomac.2020.11.150

Aoki N, Umemoto T, Hamada S, Suzuki K, Suzuki Y (2012) The amylose content and amylopectin structure affect the shape and hardness of Rice bread. J Appl Glycosci 59(2):75–82. https://doi.org/10.5458/jag.jag.jag-2011_013

Atiroğlu V, Atiroğlu A, Özacar M (2021) Immobilization of α-amylase enzyme on a protein @metal–organic framework nanocomposite: a new strategy to develop the reusability and stability of the enzyme. Food Chem 349:129127. https://doi.org/10.1016/j.foodchem.2021.129127

Bertoft Eric E, Laohaphatanalert K, Piyachomkwan K, Sriroth K (2010) The fine structure of cassava starch amylopectin. Part 2: building block structure of clusters. Int J Biol Macromol 47(3):325–335. https://doi.org/10.1016/j.ijbiomac.2010.05.018

Bothast RJ, Schlicher MA (2005) Biotechnological processes for conversion of corn into ethanol. Appl Microbiol Biotechnol 67(1):19–25. https://doi.org/10.1007/s00253-004-1819-8

Božić N, Ruiz J, López-Santín J, Vujčić Z (2011) Production and properties of the highly efficient raw starch digesting α-amylase from a Bacillus licheniformis ATCC 9945a. Biochem Eng J 53(2):203–209. https://doi.org/10.1016/j.bej.2010.10.014

Brown A, Waldheim L, Landälv I, Saddler J, Ebadian M, McMillan JD, Bonomi A, Klein B (2020) Advanced biofuels – potential for cost reduction. IEA Bionenergy 2020:1–88

Bušić A, Mardetko N, Kundas S, Morzak G, Belskaya H, Šantek MI, Komes D, Novak S, Šantek B (2018) Bioethanol production from renewable raw materials and its separation and purification: a review. Food Technol Biotechnol 56(3):289–311. https://doi.org/10.17113/ftb.56.03.18.5546

Chen SQ, Cai XH, Xie JL, Wei W, Wei DZ (2017) Structural and biochemical properties of a novel pullulanase of Paenibacillus lautus DSM 3035. Starch/Staerke 69(1–2):1–11. https://doi.org/10.1002/star.201500333

Chengyao X, Yan Q, Chaonan D, Xiaopei C, Yanxin W, Ding L, Xianfeng Y, Jian H, Yan H, Zhongli C, Zhoukun L (2021) Enzymatic properties of an efficient glucan branching enzyme and its potential application in starch modification. In: Protein expression and purification, vol 178. Elsevier, New York. https://doi.org/10.1016/j.pep.2020.105779

Cinelli BA, López JA, Castilho LR, Freire DMG, Castro AM (2014) Granular starch hydrolysis of babassu agroindustrial residue: a bioprocess within the context of biorefinery. Fuel 124:41–48. https://doi.org/10.1016/j.fuel.2014.01.076

Cole MR, Eggleston G, Gaines DK, Heckemeyer M (2019) Industrial crops & products development of an enzyme cocktail to bioconvert untapped starch in sweet sorghum processing by-products: part I. Ind Crops Prod 133(March):142–150. https://doi.org/10.1016/j.indcrop.2019.03.012

Corbin JM, Hashimoto BI, Karuppanan K, Kyser ZR, Wu L, Roberts BA, Noe AR, Rodriguez RL, McDonald KA, Nandi S (2016) Semicontinuous bioreactor production of recombinant butyrylcholinesterase in transgenic rice cell suspension cultures. Front Plant Sci 7(2016). https://doi.org/10.3389/fpls.2016.00412

Cripwell R, Favaro L, Rose SH, Basaglia M, Cagnin L, Casella S, van Zyl W (2015) Utilisation of wheat bran as a substrate for bioethanol production using recombinant cellulases and amylolytic yeast. Appl Energy 160:610–617. https://doi.org/10.1016/j.apenergy.2015.09.062

Cripwell RA, Rose SH, Favaro L, Van Zyl WH (2019) Construction of industrial Saccharomyces cerevisiae strains for the efficient consolidated bioprocessing of raw starch. Biotechnol Biofuels 12(1):1–16. https://doi.org/10.1186/s13068-019-1541-5

Cripwell RA, Favaro L, Viljoen-Bloom M, van Zyl WH (2020) Consolidated bioprocessing of raw starch to ethanol by Saccharomyces cerevisiae: achievements and challenges. Biotechnol Adv 42:107579. https://doi.org/10.1016/j.biotechadv.2020.107579

de Farias FAC, de Moretti MMS, Costa MS, BordignonJunior SE, Cavalcante KSB, Boscolo M, Gomes E, Franco CML, da Silva R (2020) Structural and physicochemical characteristics of taioba starch in comparison with cassava starch and its potential for ethanol production. Ind Crops Prod 157. https://doi.org/10.1016/j.indcrop.2020.112825

de Oliveira Aparecido LE, da Silva Cabral de Moraes JR, de Meneses KC, Lorençone PA, Lorençone JA, de Olanda Souza GH, Torsoni GB (2020) Agricultural zoning as tool for expansion of cassava in climate change scenarios. Theor Appl Climatol 142(3–4):1085–1095. https://doi.org/10.1007/s00704-020-03367-1

de Souza IA, Orsi DC, Gomes AJ, Lunardi CN (2019) Enzymatic hydrolysis of starch into sugars is influenced by microgel assembly. Biotechnol Rep 22:e00342. https://doi.org/10.1016/j.btre.2019.e00342

Deinhammer R, Clark S, Quiros M, Matthews J, Hjulmand AG, Soong C, Matsui T, Takagi S, Soong CL (2012). Producing fermentation products from starch-containing material comprises liquefying starch containing material, calcium chloride, saccharifying using carbohydrate source generating enzyme, and fermenting using a fermenting organism (Patent No. WO2012088303-A2)

Desai RP, Dave D, Suthar SA, Shah S, Ruparelia N, Kikani BA (2021) Immobilization of α-amylase on GO-magnetite nanoparticles for the production of high maltose containing syrup. Int J Biol Macromol 169:228–238. https://doi.org/10.1016/j.ijbiomac.2020.12.101

Duvernay WH, Chinn MS, Yencho GC (2013) Hydrolysis and fermentation of sweetpotatoes for production of fermentable sugars and ethanol. Ind Crops Prod 42(1):527–537. https://doi.org/10.1016/j.indcrop.2012.06.028

El-Shishtawy RM, Aldhahri M, Almulaiky YQ (2020) Dual immobilization of α-amylase and horseradish peroxidase via electrospinning: a proof of concept study. Int J Biol Macromol 163:1353–1360. https://doi.org/10.1016/j.ijbiomac.2020.07.278

Escaramboni B, Fernández Núñez EG, Carvalho AFA, de Oliva Neto P (2018) Ethanol biosynthesis by fast hydrolysis of cassava bagasse using fungal amylases produced in optimized conditions. Ind Crop Prod 112:368–377. https://doi.org/10.1016/j.indcrop.2017.12.004

Fasim A, More VS, More SS (2021) Large-scale production of enzymes for biotechnology uses. Curr Opin Biotechnol 69:68–76. https://doi.org/10.1016/j.copbio.2020.12.002

Favaro L, Viktor MJ, Rose SH, Viljoen-Bloom M, van Zyl WH, Basaglia M, Cagnin L, Casella S (2015) Consolidated bioprocessing of starchy substrates into ethanol by industrial Saccharomyces cerevisiae strains secreting fungal amylases. Biotechnol Bioeng 112(9):1751–1760. https://doi.org/10.1002/bit.25591

Favaro L, Cagnin L, Basaglia M, Pizzocchero V, van Zyl WH, Casella S (2017) Production of bioethanol from multiple waste streams of rice milling. Bioresour Technol 244:151–159. https://doi.org/10.1016/j.biortech.2017.07.108

Food and Agriculture Organization of the United Nations (2019) Crops and livestock products

Friis EP, De Maria L, Vind J, Poulsen TA, Svendsen A, Danielsen S, Lenhard RT, Friis-Madsen H, Skov LK, Friis MH (2013) New glucoamylase variant, useful for producing syrup and/or fermentation product, and for brewing (Patent No. WO2013053801-A1)

Gali KK, Soundararajan N, Katiyar V, Sivaprakasam S (2021) Electrospun chitosan coated polylactic acid nanofiber: A novel immobilization matrix for α-amylase and its application in hydrolysis of cassava fibrous waste. J Mater Res Technol 13:686–699. https://doi.org/10.1016/j.jmrt.2021.05.001

Gerrano AS, Labuschagne MT, van Biljon A, Shargie NG (2014) Genetic variability among sorghum accessions for seed starch and stalk total sugar content. Sci Agric 71(6):472–479. https://doi.org/10.1590/0103-9016-2013-0322

Ghosh B, Lahiri D, Nag M, Dash S, Ray RR (2020) Bio characterization of purified isoamylase from Rhizopus oryzae. Prep Biochem Biotechnol 50(5):453–459. https://doi.org/10.1080/10826068.2019.1706561

Gopinath SCB, Anbu P, Arshad MKM, Lakshmipriya T, Voon CH, Hashim U, Chinni SV (2017) Biotechnological processes in microbial amylase production. Biomed Res Int 2017:1–9. https://doi.org/10.1155/2017/1272193

Harada T, Misaki A, Akai H, Yokobayashi K, Sugimoto K (1972) Characterization of Pseudomonas isoamylase by its actions on amylopectin and glycogen: comparison with Aerobacter pullulanase. BBA Enzymol 268(2):497–505. https://doi.org/10.1016/0005-2744(72)90345-2

Harkness C, Semenov MA, Areal F, Senapati N, Trnka M, Balek J, Bishop J (2020) Adverse weather conditions for UK wheat production under climate change. Agric For Meteorol 282–283:107862. https://doi.org/10.1016/j.agrformet.2019.107862

Hii SL, Tan JS, Ling TC, Ariff AB (2012) Pullulanase: role in starch hydrolysis and potential industrial applications. Enzyme Res 2012. https://doi.org/10.1155/2012/921362

Homaei AA, Sariri R, Vianello F, Stevanato R (2013) Enzyme immobilization: an update. J Chem Biol 6(4):185–205. https://doi.org/10.1007/s12154-013-0102-9

Hussain S, Huang J, Huang J, Ahmad S, Nanda S, Anwar S, Shakoor A, Zhu C, Zhu L, Cao X, Jin Q, Zhang J (2020) Rice production under climate change: adaptations and mitigating strategies. In: Fahad S, Hasanuzzaman M, Alam M, Ullah H, Saeed M, Ali Khan I, Adnan M (eds) Environment, climate, plant and vegetation growth. Springer, New York, pp 659–686. https://doi.org/10.1007/978-3-030-49732-3_26

Ibrahim MIJ, Sapuan SM, Zainudin ES, Zuhri MYM (2019) Extraction, chemical composition, and characterization of potential lignocellulosic biomasses and polymers from corn plant parts. Bioresources 14(3):6485–6500. https://doi.org/10.15376/biores.14.3.6485-6500

Jouzani GS, Taherzadeh MJ (2015) Advances in consolidated bioprocessing systems for bioethanol and butanol production from biomass: a comprehensive review. Biofuel Res J 2(1):152–195. https://doi.org/10.18331/BRJ2015.2.1.4

Jujjavarapu S, Dhagat S (2019) Evolutionary trends in industrial production of α-amylase. Recent Pat Biotechnol 11(1):4–18. https://doi.org/10.2174/2211550107666180816093436

Klapiszewski Ł, Zdarta J, Jesionowski T (2018) Titania/lignin hybrid materials as a novel support for α-amylase immobilization: a comprehensive study. Colloids Surf B: Biointerfaces 162:90–97. https://doi.org/10.1016/j.colsurfb.2017.11.045

Krajang M, Malairuang K, Sukna J, Rattanapradit K, Chamsart S (2021) Single-step ethanol production from raw cassava starch using a combination of raw starch hydrolysis and fermentation, scale-up from 5-L laboratory and 200-L pilot plant to 3000-L industrial fermenters. Biotechnol Biofuels 14(1):1–15. https://doi.org/10.1186/s13068-021-01903-3

Kuiper L, Ekmekci B, Hamelinck C, Hettinga W, Meyer S, Koop K (2007) Bio-ethanol from Cassava. Ecofys

Kusano S, Nagahata N, Takahashi SI, Fujimoto D, Sakano Y (1988) Purification and properties of Bacillus acidopullulyticus Pullulanase. Agric Biol Chem 52(9):2293–2298. https://doi.org/10.1271/bbb1961.52.2293

Kwak SS (2019) Biotechnology of the sweetpotato: ensuring global food and nutrition security in the face of climate change. Plant Cell Rep 38(11):1361–1363. https://doi.org/10.1007/s00299-019-02468-0

Landvik S, Morant MD, Ayabe K, Coward-Kelly G (2011) New glucoamylase, useful for producing syrup and/or a fermentation product and for brewing (Patent No. WO2011066576-A1)

Li Y, Zhang L, Ding Z, Shi G (2013) Constitutive expression of a novel isoamylase from Bacillus lentus in Pichia pastoris for starch processing. Process Biochem 48(9):1303–1310. https://doi.org/10.1016/j.procbio.2013.07.001

López-Sandin I, Zavala-García F, Levin L, Ruiz HA, Hernández-Luna CE, Gutiérrez-Soto G (2021) Evaluation of bioethanol production from sweet sorghum variety Roger under different tillage and fertilizer treatments. Bioenergy Res. https://doi.org/10.1007/s12155-020-10215-7

Lu Y, Chae M, Vasanthan T, Bressler DC (2020) The potential of fiber-depleted starch concentrate produced through air currents assisted particle separation of barley flour in bio-ethanol production. Bioresour Technol 303. https://doi.org/10.1016/j.biortech.2020.122942

Mardani T, Sowti M, Rezaei R, Hamishehkar H (2018) Immobilization of α-amylase on chitosan-montmorillonite nanocomposite beads. Int J Biol Macromol 120:354–360. https://doi.org/10.1016/j.ijbiomac.2018.08.065

Megazyme International Ireland Ltd (n.d.) PULLULANASE M1 from Klebsiella planticola (Lot 130102b)

Morant MD, Sasa M, Ayabe K, Coward-Kelly G (2012) New isolated polypeptide having glucoamylase activity, useful for producing syrup and/or a fermentation product (Patent No. WO2012064351-A1)

Mortazavi S, Aghaei H (2020) Make proper surfaces for immobilization of enzymes: immobilization of lipase and α-amylase on modified Na-sepiolite. Int J Biol Macromol 164:1–12. https://doi.org/10.1016/j.ijbiomac.2020.07.103

Myburgh MW, Cripwell RA, Favaro L, van Zyl WH (2019) Application of industrial amylolytic yeast strains for the production of bioethanol from broken rice. Bioresour Technol 294:122222. https://doi.org/10.1016/j.biortech.2019.122222

Nagasaka Y, Kurosawa K, Yokota A, Tomita F (1998) Purification and properties of the raw-starch-digesting glucoamylases from Cotticium rolfsii. Appl Microbiol Biotechnol 50(3):323–330. https://doi.org/10.1007/s002530051299

Negi S, Banerjee R (2009) Characterization of amylase and protease produced by Aspergillus awamori in a single bioreactor. Food Res Int 42(4):443–448. https://doi.org/10.1016/j.foodres.2009.01.004

Nguyen TC, Chuky S, Luong HN, Van Nguyen H (2020) Effect of pretreatment methods on enzymatic kinetics of ungelatinized cassava flour hydrolysis. Catalysts 10(7):1–12. https://doi.org/10.3390/catal10070760

OECD Library (2020) OECD-FAO Agricultural Outlook 2020-2029

Oladunmoye OO, Aworh OC, Maziya-Dixon B, Erukainure OL, Elemo GN (2014) Chemical and functional properties of cassava starch, durum wheat semolina flour, and their blends. Food Sci Nutr 2(2):132–138. https://doi.org/10.1002/fsn3.83

Ono K, Shintani K, Shigeta S, Oka S (1988) Comparative studies of various molecular species in Aspergillus Niger Glucoamylase. Agric Biol Chem 52(7):1699–1706. https://doi.org/10.1080/00021369.1988.10868916

Pereira SE, Fernandes KF, Ulhoa CJ (2017) Immobilization of Cryptococcus flavus α-amylase on glass tubes and its application in starch hydrolysis. Starch/Staerke 69(3–4):1–8. https://doi.org/10.1002/star.201600189

Porto MDA, dos Santos JP, Hackbart H, Bruni GP, Fonseca LM, da Rosa Zavareze E, Dias ARG (2019) Immobilization of α-amylase in ultrafine polyvinyl alcohol (PVA) fibers via electrospinning and their stability on different substrates. Int J Biol Macromol 126:834–841. https://doi.org/10.1016/j.ijbiomac.2018.12.263

Ramachandran S, Patel AK, Nampoothiri KM, Chandran S, Szakacs G, Soccol CR, Pandey A (2004) Alpha amylase from a fungal culture grown on oil cakes and its properties. Braz Arch Biol Technol 47(2):309–317. https://doi.org/10.1590/S1516-89132004000200019

Riaz M, Perveen R, Javed MR, Nadeem H, Rashid MH (2007) Kinetic and thermodynamic properties of novel glucoamylase from Humicola sp. Enzym Microb Technol 41(5):558–564. https://doi.org/10.1016/j.enzmictec.2007.05.010

Rizzolo JA, Woiciechowski AL, Júnior AIM, Torres LAZ, Soccol CR (2021) The potential of sweet potato biorefinery and development of alternative uses. SN Appl Sci 3(3):1–9. https://doi.org/10.1007/s42452-021-04369-y

Rolland-Sabaté A, Sánchez T, Buléon A, Colonna P, Jaillais B, Ceballos H, Dufour D (2012) Structural characterization of novel cassava starches with low and high-amylose contents in comparison with other commercial sources. Food Hydrocoll 27(1):161–174. https://doi.org/10.1016/j.foodhyd.2011.07.008

Sakwa L, Cripwell RA, Rose SH, Viljoen-Bloom M (2018) Consolidated bioprocessing of raw starch with Saccharomyces cerevisiae strains expressing fungal alpha-amylase and glucoamylase combinations. FEMS Yeast Res 18(7):1–10. https://doi.org/10.1093/femsyr/foy085

Saleem A, Ebrahim MKH (2014) Production of amylase by fungi isolated from legume seeds collected in Almadinah Almunawwarah, Saudi Arabia. J Taibah Univ Sci 8(2):90–97. https://doi.org/10.1016/j.jtusci.2013.09.002

Shevkani K, Singh N, Bajaj R, Kaur A (2017) Wheat starch production, structure, functionality and applications—a review. Int J Food Sci Technol 52(1):38–58. https://doi.org/10.1111/ijfs.13266

Shiau RJ, Hung HC, Jeang CL (2003) Improving the thermostability of raw-starch-digesting amylase from a Cytophaga sp. by site-directed mutagenesis. Appl Environ Microbiol 69(4):2383–2385. https://doi.org/10.1128/AEM.69.4.2383-2385.2003

Singh S, Singh S, Bali V, Sharma L, Mangla J (2014) Production of fungal amylases using cheap, readily available agriresidues, for potential application in textile industry. Biomed Res Int 2014. https://doi.org/10.1155/2014/215748

Strąk-Graczyk E, Balcerek M (2020) Effect of pre-hydrolysis on simultaneous Saccharification and fermentation of native Rye starch. Food Bioprocess Technol 13(6):923–936. https://doi.org/10.1007/s11947-020-02434-9

Szambelan K, Nowak J, Szwengiel A, Jeleń H (2020) Quantitative and qualitative analysis of volatile compounds in sorghum distillates obtained under various hydrolysis and fermentation conditions. Ind Crop Prod 155:112782. https://doi.org/10.1016/j.indcrop.2020.112782

Torabizadeh H, Montazeri E (2020) Nano co-immobilization of α-amylase and maltogenic amylase by nanomagnetic combi-cross-linked enzyme aggregates method for maltose production from corn starch. Carbohydr Res 488(October 2019):107904. https://doi.org/10.1016/j.carres.2019.107904

Tortoe C, Akonor PT, Koch K, Menzel C, Adofo K (2017) Amylose and amylopectin molecular fractions and chain length distribution of amylopectin in 12 varieties of Ghanaian sweet potato (Ipomoea batatas) flours. Int J Food Prop 20(12):3225–3233. https://doi.org/10.1080/10942912.2017.1283326

van Zyl WH, Bloom M, Viktor MJ (2012) Engineering yeasts for raw starch conversion. Appl Microbiol Biotechnol 95(6):1377–1388. https://doi.org/10.1007/s00253-012-4248-0

Viktor MJ, Rose SH, Van Zyl WH, Viljoen-Bloom M (2013) Raw starch conversion by Saccharomyces cerevisiae expressing Aspergillus tubingensis amylases. Biotechnol Biofuels 6(1). https://doi.org/10.1186/1754-6834-6-167

Wani AA, Singh P, Shah MA, Schweiggert-Weisz U, Gul K, Wani IA (2012) Rice starch diversity: effects on structural, morphological, thermal, and physicochemical properties-A review. Compr Rev Food Sci Food Saf 11(5):417–436. https://doi.org/10.1111/j.1541-4337.2012.00193.x

Xia W, Zhang K, Su L, Wu J (2021) Microbial starch debranching enzymes: developments and applications. Biotechnol Adv 50:107786. https://doi.org/10.1016/j.biotechadv.2021.107786

Zaferanloo B, Bhattacharjee S, Ghorbani MM, Mahon PJ, Palombo EA (2014) Amylase production by Preussia minima, a fungus of endophytic origin: optimization of fermentation conditions and analysis of fungal secretome by LC-MS. BMC Microbiol 14(1):1–12. https://doi.org/10.1186/1471-2180-14-55

Zhang X, Guo D, Blennow A, Zörb C (2021a) Mineral nutrients and crop starch quality. Trends Food Sci Technol 114(February):148–157. https://doi.org/10.1016/j.tifs.2021.05.016

Zhang Y, Rui X, Simpson BK (2021b) Trends in nanozymes development versus traditional enzymes in food science. Curr Opin Food Sci 37:10–16. https://doi.org/10.1016/j.cofs.2020.08.001

Zou C, Duan X, Wu J (2014) Enhanced extracellular production of recombinant Bacillus deramificans pullulanase in Escherichia coli through induction mode optimization and a glycine feeding strategy. Bioresour Technol 172:174–179. https://doi.org/10.1016/j.biortech.2014.09.035

Zyl V, Willem H, Lynd LR, Den Haan R, McBride JE (2007) Consolidated bioprocessing for bioethanol production using saccharomyces cerevisiae. Adv Biochem Eng Biotechnol 108:205–235. https://doi.org/10.1007/10_2007_061

Chapter 5
Sugarcane First-Generation Bioethanol Units and Advancements in Electric Power and Biogas Production

Natália Cirqueira, Esteffany de Souza Candeo, Leonardo Barboza, Fabiana Troyner, Juliana Martins Teixeira de Abreu Pietrobelli, and Eduardo Bittencourt Sydney

Abstract Sugarcane is one of the most important industrial crops worldwide since it is remarkably rich in fermentable sugars and can be bioconverted into ethanol. This biofuel is largely used to replace traditional fossil-based fuels in tropical countries. In addition to its economic importance, ethanol plays an essential role in the biobased economy; industrial units are major examples of waste valorization. Sugarcane bagasse generated during sugarcane processing is traditionally used for heat and electric power production, whereas vinasse and filter cake are used as fertilizers in sugarcane crops. Sugarcane bagasse burning accounts for the industry's self-sufficiency in steam and electric power production, as well as for generating revenue since its excess is often sold. Vinasse and filter cake have been investigated for biogas production since biogas produces methane, which is an energy-rich molecule that can be converted into electric power or used as a renewable natural gas source; moreover, its digestate can be used in fertigation processes. The aims of the current chapter are to address sugarcane processing into ethanol, as well as to identify solid and liquid wastes generated during this process. To do so, the study investigated the current use of these wastes, as well as explored advancements observed in more efficient electric power production processes and in the development and scaling-up of anaerobic digestion technologies.

N. Cirqueira · L. Barboza · F. Troyner · J. Martins Teixeira de Abreu Pietrobelli ·
E. Bittencourt Sydney (✉)
Department of Bioprocess Engineering and Biotechnology, Federal University of Technology – Paraná, Ponta Grossa, Paraná, Brazil
e-mail: eduardosydney@utfpr.edu.br

E. de Souza Candeo
Department of Bioprocess Engineering and Biotechnology, Polytechnic Center, Federal University of Paraná, Curitiba, Paraná, Brazil

C. R. Soccol et al. (eds.), *Liquid Biofuels: Bioethanol*, Biofuel and Biorefinery Technologies 12, https://doi.org/10.1007/978-3-031-01241-9_5

5.1 Introduction

Disputes for energy sources, mainly for oil, have historically led to major political conflicts and economic crises worldwide in the twentieth century. Oil is the main nonrenewable energy source, and it moves and dictates the world's economy. However, such a dependency decreased after ethanol started to be produced and used as an alternative fuel to petroleum derivatives.

Ethanol production technology was gradually introduced in Brazil. It was initially designed to help the sugar industry; currently, the country is the second largest ethanol producer worldwide (accounting for 30% of global production), followed by the European Union (5%), China (3%) and India (2%) (RFA 2021). The United States is the largest ethanol producer in the world (53%).

Corn is the raw material used for ethanol production in the US, given its climate and territorial aspects, which are favorable to corn culture. On the other hand, ethanol production in Brazil is mostly based on sugarcane due to its tropical climate. Ethanol produced from sugar or starch is a biofuel of great environmental value since this derived fuel is cleaner and less harmful to the environment (CONAB 2008). According to Unica (2021), Brazil released 552 million tons of CO_2 in the atmosphere from the first time ethanol was used as automotive fuel in 2003 until 2021. On a global basis, the use of ethanol as biofuel in the transportation sector will avoid the emission of 160 billion ton $CO_{2\text{-eq}}$ by 2030 (Sydney et al. 2019a).

The pursuit of clean energy has become a global concern; thus, contemporary society demands constant and continuous improvements in energy production processes to help mitigate their adverse effects on the environment. Ethanol production systems have been the object of investigations focused on enabling improvements capable of reducing the polluting load of byproducts. It must be done to add greater sustainability to the ethanol production process.

The ethanol production process generates abundant byproducts, such as sugarcane bagasse, vinasse and press mud. However, despite having specific destinations, these byproducts also have high polluting potential. Therefore, different ways to use them to reduce their toxicity and to produce other byproducts that are less aggressive to the environment have been investigated. Currently, anaerobic biodigestion, which has been used to obtain two distinct products—2G ethanol and biogas—, is the main technology applied to these byproducts.

Biogas has gained prominence in recent years due to its energy generation capacity and high burning power. Biogas is generated by the decomposition of organic matter derived from waste or biomass. Its composition changes depending on the source, although it is mainly composed of methane and carbon dioxide. Such a composition, in its pure form, is harmful to the environment; however, when it is burned, its toxicity decreases due to the formation of products such as carbon dioxide and water. This process turns biogas production into an advantageous and interesting source of energy. Therefore, biogas production in ethanol production units has been encouraged to reduce impacts caused by the generation of polluting residues that cannot be reused. Such production is also encouraged based on the likelihood of generating electricity and on the prospect of having economic benefits.

The current chapter addresses the first-generation ethanol production process, and highlights the byproducts generated within this process, biogas production from waste processing, and the facilities used for such a purpose; the likelihood of using biogas to generate electric power, as well as presents the conclusions and future perspectives on biogas extraction in ethanol production units, and the new possibilities of using it.

5.2 1G Bioethanol Production Out of Sugarcane

Ethanol can be produced through many technological routes. Fermentation is a biological route and the main production method adopted worldwide. It is based on adding microorganisms capable of breaking down sugar molecules and turning them into ethanol and carbon dioxide in carbohydrate-rich liquid medium. Chemical technologies comprise ethylene hydration and acetaldehyde reduction.

Industrial ethanol production based on the fermentation of sugarcane-deriving alcohol can be described in seven stages (da Oliveira 2015): (i) *sugarcane planting and harvesting, (ii) milling, (iii) sugarcane juice treatment, (iv) pre-evaporation, (v) medium preparation, (v) fermentation, (vi) wine centrifugation, and (vii) distillation.* An additional step known as *dehydration* is required depending on the final ethanol use (Fig. 5.1).

5.2.1 Sugarcane Planting and Harvesting

Sugarcane (*Saccharum* sp.) is a semiperennial plant originating from Asia. Its culture has two ideal climate conditions, namely, a hot and humid season for its germination, tillering and vegetative development and a cold and dry season for its maturation and for sucrose accumulation in its stalks (BNDES and CGEE 2008). The chemical composition of sugarcane varies quantitatively, depending on several

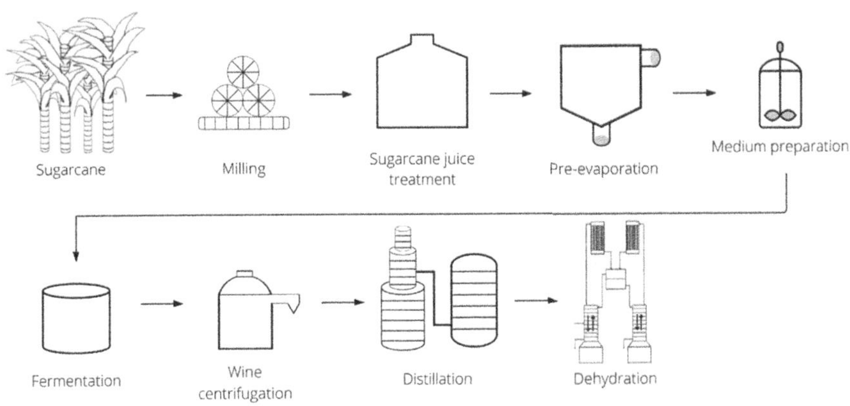

Fig. 5.1 Schematic representation of the sugarcane bioethanol production process

factors, namely, sugarcane variety, climate, plant development stage, fertilization, topping height, fertilization-irrigation with vinasse, time between cutting and processing (deterioration), and physical, chemical, and microbiological properties of the soil (de vasconcelos 2015; UFSCAR 2011). Therefore, sugarcane is composed of approximately 70% water, 13% fiber, and 17% water-soluble material. The juice obtained after sugarcane milling has 14.5% to 24% sucrose, as well as low glucose (0.2% to 1.0%) and fructose (0.0% to 0.5%) contents. The higher the sucrose content, the better the quality of the raw material. Minor components comprise amino acids, fats and waxes, dyes, acids and mineral salts (Lopes et al. 2011).

Brazil accounted for 16.9% (29,350 tons) of the global sugarcane production in 2019/2020; it was closely followed by both India (29,300 tons; 16.8%) and the European Union (17,850 tons; 10.3%) (United States Department of Agriculture, USDA). According to USDA (n.d.), all 77 producing countries have generated 174,140 tons of sugarcane worldwide. Sugarcane harvesting can be performed by manual or mechanized cutting. Based on the manual process, the whole sugarcane plant is harvested and subjected to the "cleaning" process to separate the stalks from other plant parts that have a negative effect on sugarcane juice production processing. For the mechanized harvesting process, sugarcane is chopped and transported to the processing plant, where it is cleaned through air jets.

5.2.2 *Milling*

Milling aims at sugarcane juice extraction. During this process, sugarcane bagasse is generated at a rate of 28% (m/m) (280 kg per ton of sugarcane) and used as fuel for heat and electric power generation purposes (Sun et al. 2004). Sugarcane is subjected to a defibrillator to open its cells and to make juice extraction easier before milling.

5.2.3 *Sugarcane Juice Treatment*

This step aims at reducing impurities in the juice extracted in the previous step. Eliminating unwanted material benefits the treatment process, enables obtaining a final product with better quality, and increases the lifespan of industrial equipment. Impurities found in the juice can be soluble, colloidal or insoluble; they are often sieved or removed from the juice through chemical treatments (nonsoluble impurities) (dos Santos Nunes and Finzer 2019). It is essential to correct the juice pH to avoid sucrose inversion and decomposition. The pH correction process is carried out by adding $Ca(OH)_2$ to the juice and by decanting the insolubilized components. The decanted material accounts for approximately 10% of the total juice; it is called filter cake and is used as fertilizer in crops (Lopes et al. 2011; dos Santos 2013). According to Veiga et al. (2006), each ton of processed sugarcane generates 30 kg of filter cake residue, on average.

5.2.4 Pre-evaporation

Preevaporation is carried out for sugar production purposes. The clarified sugarcane is evaporated through a heating process at 120 °C to concentrate it >80°Brix (numerical scale that measures sugar concentration in sugarcane; 1°Brix is equal to 10 g/L of dissolved solids) (da Oliveira 2015). The concentrated sugar solution, which is called molasses, is then centrifuged to recover the sugar crystals. Molasses remaining from the centrifugation process are used in the fermentation step (mixed with sugarcane juice).

5.2.5 Medium Preparation

Ethanol facilities used to produce ethanol and sugar are called annexed plants, whereas those focused on ethanol production are called autonomous plants. Autonomous plants only use sugarcane juice for fermentation purposes, whereas annexed plants use a mix comprising sugarcane juice and molasses. The standard medium used for bioethanol production with *S. cerevisiae* has 20° Brix (or approx. 200 g/L of sugar). Because fermentation is carried out in open fermenters, antibiotics are often added to eliminate competing bacteria (da Oliveira 2015).

5.2.6 Fermentation

Fermentation vats are built with cylindrical-shaped carbon steel plates; they have external and/or internal refrigeration to preferentially maintain the fermentation temperature between 32 °C and 35 °C. High cell density is ensured by yeast (*Saccharomyces cerevisiae*) recycling (the cells recovered in the centrifugation step—see next topic—are treated with acidic solution and added to the fermenter). Fermentation was carried out for 6–11 h (da Oliveira 2015), without aeration. Yeast is produced at concentration of 13% (v/v).

At the end of the fermentation process, the fermented broth, now called wine, produces 8% to 11% (v/v) ethanol. Fermentation is mostly carried out in batch mode, but few facilities in Brazil use continuous fermentation (De Vasconcelos 2015). Based on the continuous process, the fermentation medium is mixed with yeast in a first vat and continuously fed to the following fermenters (Lopes et al. 2011).

The alcoholic fermentation yield for real processes is calculated based on the stoichiometry of its reaction (Eq. 5.1).

$$C_6H_{12}O_6 \rightarrow 2\,CH_3CH_2OH + 2\,CO_2 \quad (5.1)$$

If one takes into consideration that the glucose ($C_6H_{12}O_6$) molar mass is 180 g/mol and that of ethanol (CH_3CH_2OH) is 46 g/mol, the theoretical yield of ethanol can be calculated through Eqs. (5.2)–(5.3).

$$\frac{180\text{ g of glucose}}{2\ 46\text{ g of ethanol}} = \frac{A\text{ g of glucose}}{B\text{ g of ethanol}} \tag{5.2}$$

$$B\text{ g of ethanol} = \frac{92\text{ g of ethanol} \times A\text{ g of glucose}}{180\text{ g of glucose}} \tag{5.3}$$

The calculated "B g of ethanol" corresponded to a process yield of 100%. The real process yield (R) was then calculated through Eq. (5.4) based on the real ethanol mass obtained (m) in the fermentation process.

$$R = \frac{m\text{ g of ethanol}}{B\text{ g of ethanol}} 100\% \tag{5.4}$$

On the other hand, alcoholic fermentation efficiency (E) can be expressed through Eq. (5.5).

$$E = \frac{\textit{Practical alcoholic fermentation yield}}{\textit{Theoretical alcoholic fermentation yield}} 100 \tag{5.5}$$

5.2.7 *Wine Centrifugation*

Yeast must be recovered prior to ethanol distillation. Yeasts were collected after centrifugation, diluted in water and soaked with sulfuric acid to remove contaminating bacteria prior to fermentation vat recirculation. Excess yeast can be either distilled with wine or sold as an animal feed additive. The centrifuged wine is mainly composed of ethanol, but it has low acetic acid and glycerol contents.

5.2.8 *Distillation*

According to Lopes et al. (2011), distillation systems with superimposed columns are used. Hydrated ethanol was obtained at a concentration of 96°GL after the distillation process was complete. Second grade alcohol with graduation of 92°GL is also separated; this graduation allows the alcohol to be reprocessed. The 96° GL bioethanol can be used as the final product when it is sold as automotive fuel. As an alternative, it can be subjected to a dehydration process to produce anhydrous ethanol (99.6°GL), which is used by chemical industries or as a gasoline additive.

5.2.9 Dehydration

Three technological processes are mostly used for hydrated-ethanol dehydration in Brazil: (i) azeotropic process based on the use of benzol or cyclohexane as dehydrating agent; (ii) extractive process using glycerin or mono-ethylene glycol; (iii) molecular sieve by zeolites' action.

According to CGEE (2009), anhydrous ethyl alcohol fuel (EAFC) is mostly produced through azeotropic distillation with cyclohexane at energy use ranging from 1.5 to 2.0 kg of steam per liter of ethanol. However, EAFC is the most energetically unfavorable method. The process based on glycerin or monoethylene glycol has the advantage of using a low-toxicity dehydrant; consequently, anhydrous ethanol can be applied to nobler consumption purposes, such as in pharmaceutical products or in the perfume industry (Lopes et al. 2011). Dehydration through molecular sieves is based on water adsorption by zeolites, which allows bioethanol to pass through them to generate a purer product without contaminants; such a product can be subjected to optimization and consumes approximately 0.55 kg of steam per liter of EAFC (Barreto and Coelho 2015; CGEE 2009).

5.2.10 Quantitative Process Data

The total production of the sugar-alcohol industry in Brazil, recorded for the 2020/2021 harvest, is shown in Table 5.1.

Table 5.1 Total production of the sugar-alcohol industry in Brazil, recorded for the 2020/2021 harvest. Adapted from CONAB (2021)

2020/2021 harvest	Amount
Sugarcane planted area (thousand ac)	1177.4
Sugarcane harvested area (thousand ac)	8616.1
Sugarcane production (thousand ton)	654,527.8
Sugarcane yield (ton/ac)	75,965
Sugarcane used to produce hydrated ethanol (thousand ton)	240,174.5
Sugarcane used to produce anhydrous ethanol (thousand ton)	114,293.1
Sugarcane used to produce sugar (thousand ton)	300,060.2
Hydrous ethanol production (thousand L)	20,424,611
Anhydrous ethanol production (thousand L)	9,321,812
Sugar production (thousand t)	41,254.3
Hydrous ethanol yield (L/$ton_{sugarcane}$)	85.04
Anhydrous ethanol yield (L/$ton_{sugarcane}$)	81.56
Sugar yield (t/t sugarcane)	0.14

5.3 Waste Production and Reuse in Sugarcane Bioethanol Units

The sugarcane ethanol industry generates solid, liquid and gaseous wastes with high polluting potential throughout their production process (Sydney et al. 2021) (Fig. 5.2). These wastes (Fig. 5.3) must be reused based on different recovery technologies to ensure sustainable technological development in ethanol production processes (Chen et al. 2021; Sydney et al. 2021). Waste management enables the diversification of ethanol industry byproducts; consequently, it helps increase the profit from each ton of processed sugarcane (Holanda and Ramos 2015; Karp et al. 2021).

An economically important solid waste, called bagasse, is generated after sugarcane is crushed for juice extraction; the average processing of 1000 kg of sugarcane produces 280 kg of bagasse at a moisture rate close to 50% (de Almeida and Colombo 2021). Sugarcane bagasse composition mainly comprises three fractions:

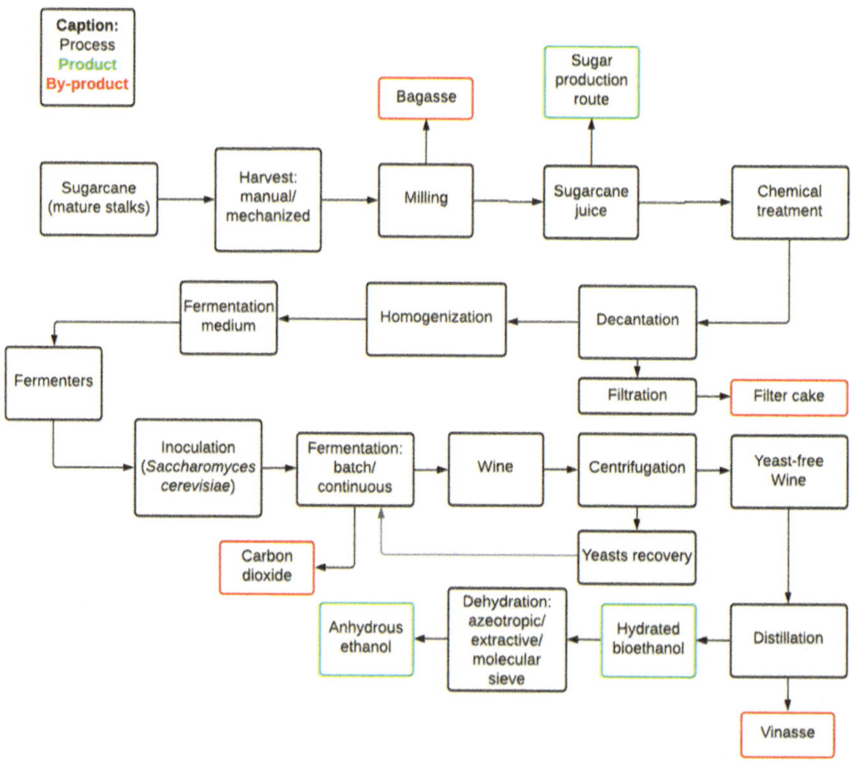

Fig. 5.2 Flowchart of the sugarcane ethanol production process depicting product (*green*) and waste (*red*) production

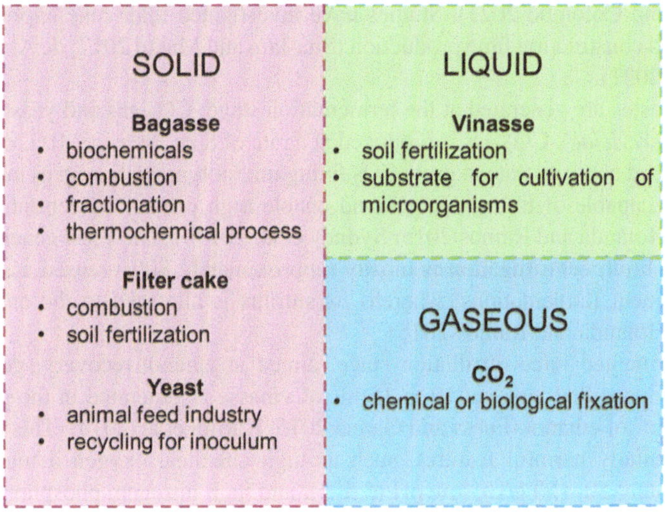

Fig. 5.3 Solid, liquid and gaseous wastes generated during sugarcane ethanol production and their applications

cellulose (46%), hemicellulose (27%) and lignin (23%), which cause high material recalcitrance (Pippo et al. 2011).

The generated sugarcane bagasse is often burned in boilers to generate the electric power necessary to keep the industry's electric motors working, whereas its surplus is exported to the energy concessionaire (Akinfalabi et al. 2020). The energy generation potential of sugarcane bagasse can also be exploited to produce biochemicals, such as second-generation ethanol and biomethane production, through aerobic and anaerobic fermentation processes, respectively (de Candeo et al. 2020). However, processes to obtain biochemicals comprise steps capable of changing the recalcitrant structure of lignocellulosic material, which often has a financial impact on these processes (Chen et al. 2021; Karp et al. 2021).

Bagasse can be fractionated into carbohydrates, such as cellulose and hemicellulose, to be used in fermentation processes and lignin to be used for combustion and electricity generation (Akinfalabi et al. 2020; Jugwanth et al. 2020). It is done to minimize bottlenecks associated with bagasse recalcitrance (Chen et al. 2021). However, another route used to harness the energy potential of bagasse lies in its destination in thermochemical processes such as pyrolysis, gasification and liquefaction, which result in low molecular weight compounds with high energy potential, or in platform chemicals (Sydney et al. 2021).

Sugarcane juice is filtered to reduce the total solid concentration before the fermentation stage; this process generates 30–100 kg of filter cake for every 1000 kg of processed sugarcane (de Almeida and Colombo 2021). Overall, the cake composition comprises 50%–79% moisture, as well as significant aluminum, manganese, zinc and iron concentrations. In addition, this waste is often used as fertilizer due to its organic quality (Holanda and Ramos 2015; Moore et al. 2017; de

Almeida and Colombo 2021). Studies have investigated filter cake applications in cement and construction lime production (Sua-Iam and Makul 2017; de Almeida and Colombo 2021).

Two wastes are generated at the fermentation stage: CO_2 gas and yeast (*Saccharomyces cerevisiae*). CO_2 gas is produced at a rate of 8 kg for every 10 L of ethanol. It can be used as a substrate to grow CO_2 fixing microorganisms, such as microalgae, which are capable of fixing this gas and enable high protein concentrations in its biomass (Holanda and Ramos 2015; Sydney et al. 2019b). The yeast generated after fermented broth centrifugation is mostly (approximately 80%) reused as inoculum for subsequent fermentations, whereas its surplus is allocated to the animal feed industry (Holanda and Ramos 2015).

The fermented juice distillation stage, aimed at ethanol recovery, generates a liquid residue called vinasse; 10 to 15 m^3 of vinasse is generated in the process to recover 1 m^3 of ethanol (Fuess and Garcia 2015; Karimi et al. 2019). This waste has environmentally harmful features, such as high chemical oxygen demand (COD approximately 50–150 g/L), low pH (3.5–5), and high nitrogen, phosphorus, potassium, calcium, magnesium and sulfur concentrations. Therefore, its reuse is a frequent object of scientific and technological studies (Moore et al. 2017; Karimi et al. 2019).

Vinasse can be used as fertilizer in hydroponic vegetable cultivation and in sugarcane plantations to minimize the use of chemical fertilizers (de Almeida and Colombo 2021). However, its application to the soil must be controlled to avoid environmental consequences such as erosion, desertification, soil salinization and water eutrophication (Moore et al. 2017; Karimi et al. 2019). An interesting way to explore the potential of this waste lies on growing microorganisms (mostly fungi and microalgae) in vinasse-based culture media to obtain high protein-rich biomass concentrations to be used in the animal feed industry (Karimi et al. 2019). Thus, vinasse can also be used to obtain biomethane, although the recalcitrance of the material can be an economic obstacle to such a process (Fuess and Garcia 2015; Moore et al. 2017).

Based on the proper management of the waste generated in the sugarcane ethanol industry, it is possible to consolidate the sustainability of the ethanol production process, promote waste recovery, reduce environmental degradation, diversify the industry's byproducts, and reduce sugarcane ethanol production costs (de Almeida and Colombo 2021).

5.4 Biogas Production Within Bioethanol Units

As presented in the previous section, the bioethanol production process generates significant amounts of waste at different levels, such as sugarcane bagasse, vinasse and filter cake, as well as water derived from washing processes. This waste can be harmful to the environment when it is poorly handled, discarded or used. However, the treatment given to environmental issues observed around the world has evolved over the years. Accordingly, environmental management has been incorporated into the routines of several companies. Industries look for opportunities to improve their processes by reusing waste to obtain value-added products that are not harmful to the environment, a fact that makes them more socially accountable and competitive in a more conscious market (ANA 2009). Anaerobic digestion is an interesting alternative to improve and add new byproducts to ethanol production, as well as to make this process more sustainable by mitigating the environmental impacts caused by waste in its raw form (Janke et al. 2019).

Anaerobic digestion has been extensively used to treat biomass or organic waste and to transform it into a potential bioenergy source through biogas production (Amon et al. 2007). Biogas can be used as a heat source and to generate electric power and different fuel gases (Ge et al. 2014). The anaerobic digestion principle lies in converting organic matter into methane and carbon dioxide based on four different reaction steps, namely, hydrolysis, acidogenesis, acetogenesis and methanogenesis (Fig. 5.4).

Overall, anaerobic digestion occurs at two basic stages. The first stage lies on using acidogenic bacteria to convert complex organic compounds (such as carbohydrates, proteins and lipids) into smaller chain compounds (such as volatile acids such as CO_2 and H_2) through extracellular enzymes. The second stage lies on using methanogenic bacteria to convert organic acids into methane and carbon dioxide gas (de Souza 1984).

Methanogenesis is considered the limiting stage of the process since, in addition to being the slowest digestion process stage, it is easily affected by environmental adversities. The operating conditions of the bioreactor, mainly pH, alkalinity and temperature, have a strong influence on the reactions taking place in it, on the effective matter-into-biogas conversion, and on the methane content. In addition, extreme cases can cause the system to collapse (Chernicharo 1997). Therefore, the anaerobic digestion of byproducts derived from bioethanol production processes and factors determining the conversion into biogas will be addressed herein.

5.4.1 Sugarcane Bagasse

Sugarcane bagasse is the byproduct derived from the sugarcane milling process; it is one of the most abundant agro-industrial wastes since it accounts for more than 540 million tons of waste production per year (Zhao et al. 2015), which corresponds to approximately 30% of the total sugarcane mass (Soccol et al. 2010). A solution found by industries to the generation of such an abundant amount of this byproduct lies in burning it in boilers and using its heat to generate electric power (Sindhu et al. 2016). In addition, the potential of using sugarcane bagasse as a substrate for 2G ethanol production has been investigated (Prajapati et al. 2020). However, this byproduct has been used as a substrate for anaerobic digestion and biogas production purposes to obtain cleaner energy (Hwu and Cai 2010; Rabelo et al. 2011; Alexandropoulou et al. 2017).

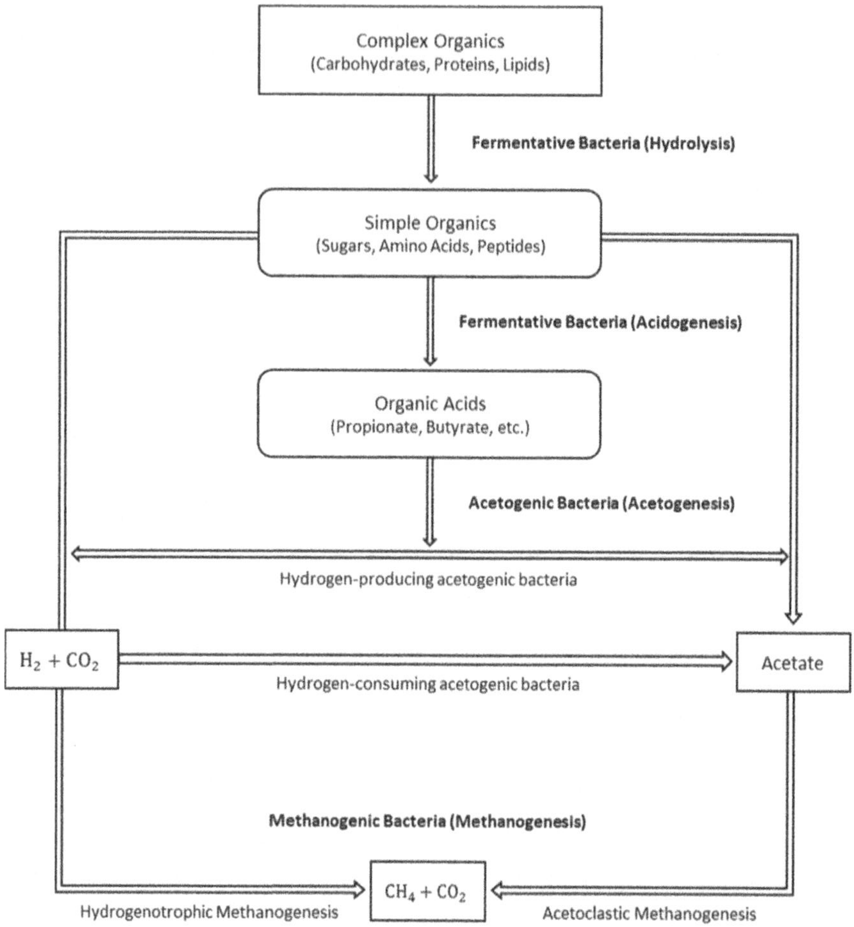

Fig. 5.4 Diagram of anaerobic digestion mechanisms. Adapted from Chernicharo (1997)

The bagasse composition of approximately 50% cellulose and 25% hemicellulose makes its application as a substrate in biodigestion processes an interesting alternative. On the other hand, the other 25% of lignin in its structure gives a recalcitrant effect to biomass, which is typical of lignocellulosic materials and hinders the biodigestion process carried out by anaerobic microorganisms. Therefore, biomass is subjected to pretreatment to make access to cellulose easier and to increase digestion performance (Pandey et al. 2000; Mustafa et al. 2018). Figure 5.5 depicts the lignocellulosic material structure before and after pretreatment application.

Currently, different technologies can be used for biomass pretreatment purposes; they are subdivided into physical, chemical and biological processes that change the lignocellulosic structure in a specific way, although with the same purpose (Zheng et al. 2014). The chemical pretreatments most often applied to sugarcane bagasse can be alkaline or acidic pretreatments. Alkaline pretreatment is applied to increase

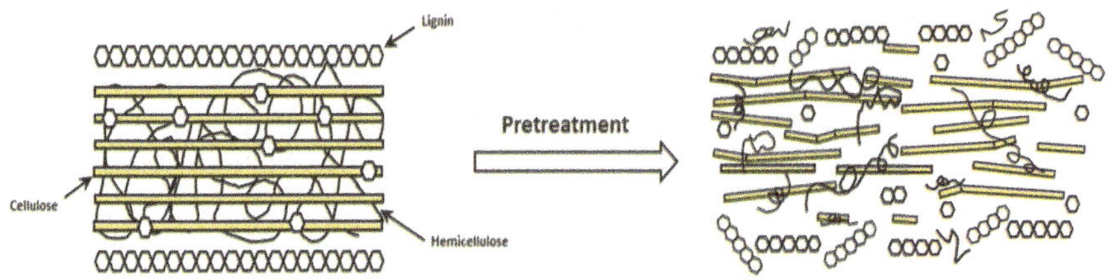

Fig. 5.5 Lignocellulosic material response to pretreatment. Adapted from Mood et al. (2013)

sugarcane bagasse porosity by distorting its fibers and breaking the bonds between lignin and other polymers (Kaur et al. 2020). On the other hand, acid pretreatment solubilizes hemicellulose to enable access to cellulose. Both acid and alkaline treatments have the disadvantage of producing reaction inhibitors, unlike the physical method, which only breaks down the sugarcane bagasse structure (Hendriks and Zeeman 2009).

In addition to pretreatment, the C:N ratio in the substrate is of paramount importance, since low C:N ratios lead to decreased biogas production (Kayhanian and Rich 1995). In addition, high nitrogen levels work as reaction inhibitors due to ammonia accumulation in the system; the ideal C:N ratio ranges from 26:1 to 30:1 (Tanimu et al. 2014).

5.4.2 *Filter Cake*

Filter cake is the waste generated in sugarcane juice clarification processes; this waste is remarkably rich in nutrients, such as nitrogen, phosphorus, potassium, sodium, iron, zinc, calcium, magnesium, and other organic compounds, in addition to water and sugarcane waste (Elsayed et al. 2008). Similar to sugarcane bagasse, filter cake is a lignocellulosic material that can be used as a great carbon source. Moreover, pretreatment is required to be used as a substrate for biodigesters. Filter cakes are the main factor accounting for increasing performance and biogas production (Janke et al. 2016a). Furthermore, filter cake codigestion with other organic materials, such as sugarcane bagasse, is recommended to balance nutrients and to improve the carbon ratio (Janke et al. 2016b). The codigestion between these two materials is interesting because it balances the C:N ratio of 26:1 of press mud, which is close to the ideal lower limit, with that of sugarcane bagasse, which presents a much higher C:N ratio ranging from 90:1 to 101:1 (López González et al. 2013).

5.4.3 Vinasse

Vinasse is the byproduct of fractional sugarcane juice distillation processes. According to estimates, 12 to 15 liters of vinasse are generated per liter of produced ethanol; thus, vinasse is the most abundant effluent derived from this process, and its main application lies in the fertigation of sugarcane plantations (Júnior et al. 2016). During the distillation process, vinasse leaves the distillation column at a temperature of approximately 90 °C and pH close to 4; this waste has a high organic load and is rich in calcium, magnesium, phosphorus and potassium. These features make this waste harmful to the environment in its raw state; however, it has great potential to be used for anaerobic digestion due to its ability to convert COD at rates higher than 50% (Wilkie et al. 2000; Romanholo Ferreira et al. 2011).

Technical Standard P4.231/2005 was regulated in São Paulo State; it establishes criteria and procedures to be followed for vinasse storage, transport and application in the soil (CETESB 2015). This application is based on soil potassium contents and is done to establish limits for soil irrigation without causing major damage to the environment, such as soil salinization, organic overload, microorganism proliferation, and soil and groundwater contamination (Fuess et al. 2017).

Vinasse application as organic matter for anaerobic digestion purposes has been gradually gaining room due to the impact of fertigation on soil and water. Thus, biodigestion is an alternative way to reduce the waste's organic load, which is associated with methane production and with its use as an energy source (Moraes et al. 2015). Pant and Adholeya (2007) pointed out a BOD removal efficiency of 80% to 90% and a biogas production efficiency of 85% to 90% based on vinasse biodigestion.

Using vinasse as a substrate proved to be interesting due to its organic and nutrient-rich features, although the system's efficiency was affected due to its low pH, a fact that required corrective measures to stabilize pH values close to neutrality. Some scholars have emphasized the use of bases as alternatives for pH recovery. España-Gamboa et al. (2012) and Kaparaju et al. (2010) used sodium bicarbonate as a corrective agent to keep the system pH at 7 in a UASB reactor. The first authors reported 69% COD removal and 84% methane content. de Barros and Duda (2016) used sodium hydroxide and pH-corrected waste in a recirculation system to stabilize the reaction.

Unlike the chemical and corrective methods of the digestion system, anaerobic codigestion stood out as a system-regulation and stabilization method, since it mixed different waste types to increase nutritional variety in the environment, a fact that helped avoid pH drops and prolonged biogas production (Dai et al. 2015). According to López González et al. (2017), vinasse codigestion with sugarcane sludge has increased methane production by 64% in comparison to the monodigestion of sugarcane sludge, a fact that evidenced its efficiency in the digestion process.

5.5 Electric Power Generation

Bioenergy is an important renewable energy source resulting from the processing of different biomasses (Cicea et al. 2019). The term "bioenergy" encompasses electric power production derived from wind, solar, marine, geothermal, biomass and waste sources, as well as the production of biomass-based fuels such as biodiesel, bioethanol and biogas (EIA 2020). Modern bioenergy production methods make great contributions to all economic sectors, with emphasis on the industrial and transport sectors (Cicea et al. 2019; EIA 2020; EPE 2021).

In addition to its global economic importance, bioenergy can be considered a sustainable technology due to its low greenhouse gas emissions (Carvalho et al. 2019; de Candeo et al. 2020). Plants absorb high carbon dioxide levels from the air during biomass growth; later on, they release carbon dioxide into the atmosphere during the combustion process for energy generation purposes. Therefore, they promote a continuous gas absorption and release cycle that, overall, enables carbon sequestration from the environment (Bayrakci Ozdingis and Kocar 2018; Cicea et al. 2019; Jiang et al. 2020).

According to updated data, these renewable energies accounted for 26% of global energy production in 2018 and 83% of Brazilian energy generation in 2019 (EIA 2021). A viable solution to help increase the internal supply of renewable energy and, consequently, reduce fossil source participation in both energy matrices lies in using sugarcane bagasse for electric power generation purposes.

Sugarcane bagasse is the fibrous waste of sugarcane plants; approximately 0.3 tons of it is generated for every 1 ton of sugarcane (Bezerra and Ragauskas 2016). This important byproduct of the sugar-alcohol industry holds approximately 50% moisture and comprises three major fractions, namely, cellulose (40%–45%), hemicellulose (30%–35%) and lignin (20%–30%) (Alves et al. 2015; Carpio and de Souza 2017). Using sugarcane bagasse for noble purposes, such as bioenergy generation, is advantageous given the abundant production of this byproduct, as well as its availability for different uses (Kapanji et al. 2019; de Candeo et al. 2020).

Large amounts of sugarcane bagasse are generated by the mechanical plant-juice extraction process adopted in the sugar and alcohol industries. These industries simultaneously produce bioethanol and sugar in an energy cogeneration system capable of converting such waste into electric power (Fig. 5.6). Cogeneration can reach energy generation up to 3.12 MWh per ton of bagasse (Carpio and de Souza 2017); thus, it can have positive economic impacts on industrial production processes (Carvalho et al. 2019).

Some steps inherent to the alcohol and sugar production process require large amounts of thermal, mechanical or electrical energy due to physical features such as temperature, pressure, solid concentration, and the amount of processed biomass. Therefore, sugarcane bagasse can be dried and stored outdoors and continuously burned in boilers to generate tons of steam that can be used as an energy source in production processes (Gongora and Villafranco 2018).

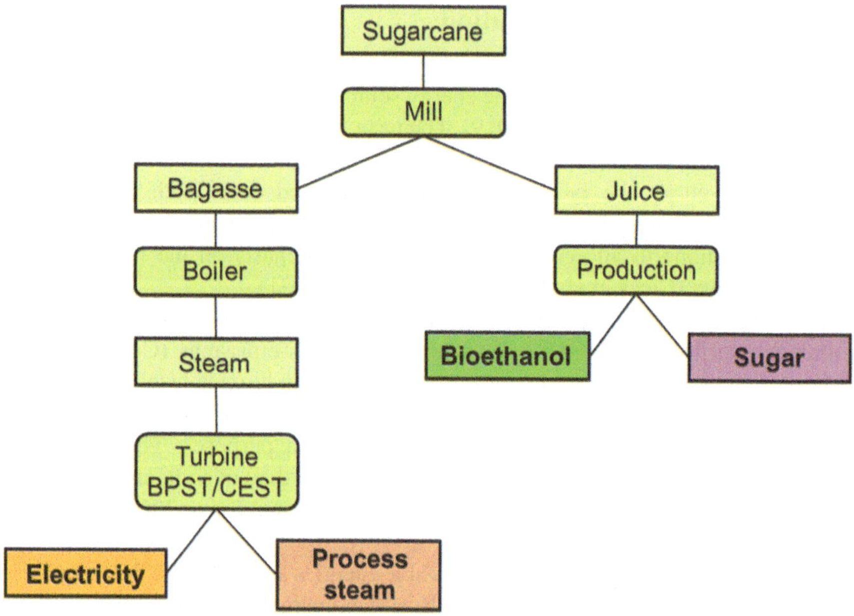

Fig. 5.6 Electric power and process steam cogeneration system in the sugar-alcohol industry

It is necessary to use high-pressure and efficient boilers (Carpio and de Souza 2017), which enable effective bagasse combustion to generate large amounts of steam, by using small amounts of fuel (Alves et al. 2015) to make energy generation from sugarcane bagasse viable. Thus, it is possible to produce larger amounts of steam than that necessary to supply production plants - approximately 0.4 tons of sugarcane bagasse for every 1 ton of processed sugarcane (Kapanji et al. 2019).

The generated steam has great potential to be used as a mechanical, thermal or electrical energy source in production plants. The steam generated in boilers (87 bar, 515 °C) is initially directed by pipes to pass through single- or multistage steam turbines. The steam's driving force propels the turbines, generates large amounts of mechanical energy and reduces both the pressure and temperature (approximately 4 bar and 144 °C) of the steam stream (Petersen et al. 2017). Part of the mechanical energy derived from the steam turbines is taken to the choppers, shredders and mills, which are used for sugarcane juice extraction, as well as to the pumps used throughout the production process (Carpio and de Souza 2017).

The exhaust steam stream derived from the turbines - with reduced temperature and pressure - is used as thermal energy in processes requiring heat exchange, such as the evaporation, drying and distillation steps. In these cases, the exhaust steam derived from the turbines is taken to the operation unit in question, wherein heat exchange between the steam stream piping and the raw material stream piping takes place.

The mechanical energy surplus generated by the steam passing through the turbines can be used as power to help generators of a cogeneration system produce electric power in production plants (Gongora and Villafranco 2018). The generated electric power is capable of making production plants self-sufficient. In addition, the electric power surplus can be exported to the energy distribution network and converted into credits for production plants (Alves et al. 2015; Bezerra and Ragauskas 2016; Cavalcanti et al. 2020).

The systems used by production plants to cogenerate steam and electric power can be of the BPST (Backpressure Steam Turbine) or CEST (Condensing Extraction Steam Turbine) type. According to the BPST-type system, the steam turbine can work at 22 bar and 300 °C and presents electrical energy production up to 99 kWh per ton of processed sugarcane (Bezerra and Ragauskas 2016). With respect to the CEST-type system, the steam turbine can work with steam streams higher than 65 bar and generate up to 121 kWh of electric power per ton of sugarcane (Bezerra and Ragauskas 2016). Overall, CEST is the system of choice to generate the electrical energy to be supplied to production plants, whereas the surplus energy is sold.

The major advances in the electric energy cogeneration sector are based on the diversification of the raw material used for combustion in boilers. Thus, combustion can be carried out by combining straw, leaf and sugarcane bagasse in boilers to generate greater amounts of steam and electric power surplus (Carvalho et al. 2019).

Another economically viable possibility of generating electric power in the sugar-alcohol industry lies in producing biogas through anaerobic digestion in ethanol production processes (see Sect. 5.4). Electrical energy can be produced from biogas or biomethane. Biomethane has calorific power similar to that of natural gas; thus, it can be used in industrial boilers for combustion and steam generation and, consequently, for electric power generation purposes (Marafon et al. 2019). This factor leads to diversification of combustion raw materials (sugarcane straw, leaf and bagasse, and biomethane) and to valorization of agro-industrial byproducts (Bechara et al. 2016; Lima et al. 2021).

However, one of the obstacles to electric power cogeneration in countries with emerging or developing economies lies in the high cost of system implementation processes (Kapanji et al. 2019). According to estimates, the cost of investing in such systems is US$ 1400 per kW, and it may suffer an annual adjustment of US$ 84.00 per kW after system installation (Gongora and Villafranco 2018).

The technological advances necessary to solidify electric power cogeneration technology in the sugar and alcohol industries comprise the use of sugarcane byproducts (straw, leaf, bagasse, vinasse), development of highly efficient steam generators and replacement of steam turbines used in grinding processes by more energy-efficient electric engines (Alves et al. 2015; Fuess and Zaiat 2018; Carvalho et al. 2019). Such technological advancements will enable generating larger amounts of steam and mechanical energy used to produce electric power; consequently, they will enable exporting greater amounts of energy to distribution networks.

The production process inherent to sugarcane byproduct conversion into electric power has advantages, such as improving the economic viability of production processes, bioethanol and sugar production sustainability, diversification of sales products to promote bioeconomy, direct generation of job positions, electric power export to energy distribution networks to minimize the use of electric power derived from fossil fuels, and industries' ability to generate electric power without depending on hydroelectric power plants or on rainfall rates observed in the region (Guo et al. 2015; Petersen et al. 2017; Carvalho et al. 2019; Kapanji et al. 2019).

Therefore, bioenergy production can be considered an important technology capable of meeting the growing demand for electric power, which is estimated at 2.1% per year by 2040 (Cavalcanti et al. 2020). This technology is also capable of reducing greenhouse gas emissions into the atmosphere (Jiang et al. 2020). Moreover, electric power derived from sugarcane biomass enables diversification of the world's energy matrix and achievement of energy generation sustainability.

5.6 Conclusion and Future Perspective

The 1G ethanol industry is acknowledged as the major example of a circular economy. All solid and liquid byproducts produced during sugarcane processing are used for some purpose. Sugarcane bagasse, vinasse and filter cake are the most relevant byproducts generated from sugarcane processing, and they have such a rich chemical composition that they have opened room for alternative technological routes for their valorization. Despite the significant efforts made to transform bagasse into biochemicals at a large scale, burning is certainly the most often adopted route because it is capable of producing electric power and steam to be used within industrial units. However, other lignocellulosic wastes, such as sugarcane straw and leaves, are mostly left in the field and have no other destination. Moreover, the low efficiency of steam as an electric power generator remains a great challenge. Filter cake and vinasse are mainly used as fertilizers to boost sugarcane growth. However, they are promising substrates for anaerobic bioconversion into methane, which can be used for energy and steam generation, and are sold as new products derived from 1G bioethanol production industries. Major challenges for commercial biogas production from these wastes are associated with process control and with sugarcane production and processing seasonality.

Acknowledgement The present study was developed based on financial support provided by the Coordination for the Improvement of Higher Education Personnel (CAPES).

Compliance with Ethical Standards The authors declare no competing financial interests or personal relationships that could have influenced the research reported in the current manuscript.

References

Akinfalabi SI, Rashid U, Ngamcharussrivichai C, Nehdi IA (2020) Synthesis of reusable biobased nano-catalyst from waste sugarcane bagasse for biodiesel production. Environ Technol Innov 18:100788. https://doi.org/10.1016/j.eti.2020.100788

Alexandropoulou M, Antonopoulou G, Fragkou E et al (2017) Fungal pretreatment of willow sawdust and its combination with alkaline treatment for enhancing biogas production. J Environ Manag 203:704–713. https://doi.org/10.1016/j.jenvman.2016.04.006

Alves M, Ponce GHSF, Silva MA, Ensinas AV (2015) Surplus electricity production in sugarcane mills using residual bagasse and straw as fuel. Energy 91:751–757. https://doi.org/10.1016/j.energy.2015.08.101

Amon T, Amon B, Kryvoruchko V et al (2007) Methane production through anaerobic digestion of various energy crops grown in sustainable crop rotations. Bioresour Technol 98:3204–3212. https://doi.org/10.1016/j.biortech.2006.07.007

ANA (2009) Manutal de conservação e reúso de água na agroindustria sucroenergética. Agencia Nacional de Águas, Federação das Indústrias do Estado de São Paulo, União da Indústrai da Cana-de-Açúcar, Centro de Tecnologia Canavieira, Brasília

Barreto TV, Coelho ACD (2015) Distillation. Sugarcane Agric Prod Bioenergy Ethanol 2015: 341–363. https://doi.org/10.1016/B978-0-12-802239-9.00016-5

Bayrakci Ozdingis AG, Kocar G (2018) Current and future aspects of bioethanol production and utilization in Turkey. Renew Sust Energ Rev 81:2196–2203. https://doi.org/10.1016/j.rser.2017.06.031

Bechara R, Gomez A, Saint-Antonin V et al (2016) Methodology for the optimal design of an integrated first and second generation ethanol production plant combined with power cogeneration. Bioresour Technol 214:441–449. https://doi.org/10.1016/j.biortech.2016.04.130

Bezerra TL, Ragauskas AJ (2016) A review of sugarcane bagasse for second-generation bioethanol and biopower production. Biofuels Bioprod Biorefin 10:634–647. https://doi.org/10.1002/bbb.1662

BNDES, CGEE (2008) Bioetanol de Cana-de-Açúcar: Energia para o Desenvolvimento Sustentável

Carpio LGT, de Souza FS (2017) Optimal allocation of sugarcane bagasse for producing bioelectricity and second generation ethanol in Brazil: scenarios of cost reductions. Renew Energy 111: 771–780. https://doi.org/10.1016/j.renene.2017.05.015

Carvalho M, Da Silva Segundo VB, De Medeiros MG et al (2019) Carbon footprint of the generation of bioelectricity from sugarcane bagasse in a sugar and ethanol industry. Int J Glob Warm 17:235–251. https://doi.org/10.1504/IJGW.2019.098495

Cavalcanti EJC, Carvalho M, da Silva DRS (2020) Energy, exergy and exergoenvironmental analyses of a sugarcane bagasse power cogeneration system. Energy Convers Manag 222: 113232. https://doi.org/10.1016/j.enconman.2020.113232

CETESB (2015) Norma P4.231: stillage - criteria and procedures for agricultural soil application. Cia Ambient do Estado São Paulo 3:1–15

CGEE (2009) Bioetanol combustível: uma oportunidade para o Brasil

Chen J, Zhang B, Luo L et al (2021) A review on recycling techniques for bioethanol production from lignocellulosic biomass. Renew Sust Energ Rev 149:111370. https://doi.org/10.1016/j.rser.2021.111370

Chernicharo CADL (1997) Reatores anaeróbios, 1st edn. Departamento de Engenharia Sanitária e Ambiental - UFMG, Belo Horizonte

Cicea C, Marinescu C, Pintilie N (2019) Smart cities using smart choices for energy: integrating modern bioenergy in consumption. Theor Empir Res Urban Manag 14:22–34

CONAB (2021) Safra Brasileira de Cana-de-Açúcar

CONAB CN de A (2008) O Etanol Como Um Novo Combustível Universal 68

da Oliveira VG (2015) Processos biotecnológicos industriais: produção de bens de consumo com o uso de fungos e bactérias, 1st edn. Érica

Dai X, Chen S, Xue Y et al (2015) Hygienic treatment and energy recovery of dead animals by high solid co-digestion with vinasse under mesophilic condition: feasibility study. J Hazard Mater 297:320–328. https://doi.org/10.1016/j.jhazmat.2015.05.027

de Almeida MA, Colombo R (2021) Production chain of first-generation sugarcane bioethanol: characterization and value-added application of wastes. Bioenergy Res. https://doi.org/10.1007/s12155-021-10301-4

de Barros VG, Duda RM, de Oliveira RA (2016) Biomethane production from vinasse in upflow anaerobic sludge blanket reactors inoculated with granular sludge. Braz J Microbiol 47:628–639. https://doi.org/10.1016/j.bjm.2016.04.021

de Candeo ES, ACN S, Hashimoto EH et al (2020) Microbial bioresources for biofuels production: fundamentals and applications. In: Biofuels production – sustainability and advances in microbial bioresources. Springer, Cham, pp 1–17

de Souza ME (1984) Fatores que influenciam a digestão anaeróbia. Rev DAE 44:7

de Vasconcelos JN (2015) Ethanol fermentation. Sugarcane Agric Prod Bioenergy Ethanol 311–340. https://doi.org/10.1016/B978-0-12-802239-9.00015-3

dos Santos MA (2013) Fontes de Energia Nova e Renovável, 1st edn. LTC

EIA (2020) Bioenergy. In: Energy Information Administration

EIA (2021) Electricity. In: Energy Information Administration

Elsayed MT, Babiker MH, Abdelmalik ME et al (2008) Impact of filter mud applications on the germination of sugarcane and small-seeded plants and on soil and sugarcane nitrogen contents. Bioresour Technol 99:4164–4168. https://doi.org/10.1016/j.biortech.2007.08.079

EPE (2021) Relatório Síntese 2021. In: Balanço Energético Nac - Empres Pesqui Energética, vol 1. EPE, Brasilia, pp 1–73

España-Gamboa EI, Mijangos-Cortés JO, Hernández-Zárate G et al (2012) Methane production by treating vinasses from hydrous ethanol using a modified UASB reactor. Biotechnol Biofuels 5: 1–9. https://doi.org/10.1186/1754-6834-5-82

Fuess LT, Garcia ML (2015) Bioenergy from stillage anaerobic digestion to enhance the energy balance ratio of ethanol production. J Environ Manag 162:102–114. https://doi.org/10.1016/j.jenvman.2015.07.046

Fuess LT, Zaiat M (2018) Economics of anaerobic digestion for processing sugarcane vinasse: applying sensitivity analysis to increase process profitability in diversified biogas applications. Process Saf Environ Prot 115:27–37. https://doi.org/10.1016/j.psep.2017.08.007

Fuess LT, Rodriges IJ, Garcia ML (2017) Fertirrigation with sugarcane vinasse: foreseeing potential impacts on soil and water resources through vinasse characterization. J Environ Sci Health A 52: 1063–1072. https://doi.org/10.1080/10934529.2017.1338892

Ge X, Matsumoto T, Keith L, Li Y (2014) Biogas energy production from tropical biomass wastes by anaerobic digestion. Bioresour Technol 169:38–44. https://doi.org/10.1016/j.biortech.2014.06.067

Gongora A, Villafranco D (2018) Sugarcane bagasse cogeneration in Belize: a review. Renew Sust Energ Rev 96:58–63. https://doi.org/10.1016/j.rser.2018.07.034

Guo M, Song W, Buhain J (2015) Bioenergy and biofuels: history, status, and perspective. Renew Sust Energ Rev 42:712–725. https://doi.org/10.1016/j.rser.2014.10.013

Hendriks ATWM, Zeeman G (2009) Pretreatments to enhance the digestibility of lignocellulosic biomass. Bioresour Technol 100:10–18. https://doi.org/10.1016/j.biortech.2008.05.027

Holanda LR, Ramos FS (2015) Reuse of waste sugarcane agribusiness and green power generation. J Clean Energy Technol 4:341–345. https://doi.org/10.18178/jocet.2016.4.5.309

Hwu C-S, Cai W-Y (2010) Enhanced biogas production of bagasse by anaerobic granular sludge under thermophilic conditions. J Biotechnol 150:21–21. https://doi.org/10.1016/j.jbiotec.2010.08.066

Janke L, Leite A, Batista K et al (2016a) Optimization of hydrolysis and volatile fatty acids production from sugarcane filter cake: effects of urea supplementation and sodium hydroxide pretreatment. Bioresour Technol 199:235–244. https://doi.org/10.1016/j.biortech.2015.07.117

Janke L, Leite AF, Nikolausz M et al (2016b) Comparison of start-up strategies and process performance during semi-continuous anaerobic digestion of sugarcane filter cake co-digested with bagasse. Waste Manag 48:199–208. https://doi.org/10.1016/j.wasman.2015.11.007

Janke L, Weinrich S, Leite AF et al (2019) Pre-treatment of filter cake for anaerobic digestion in sugarcane biorefineries: assessment of batch versus semi-continuous experiments. Renew Energy 143:1416–1426. https://doi.org/10.1016/j.renene.2019.05.029

Jiang Z, Dai Y, Du T (2020) Comparison of the energetic, environmental, and economic performances of three household-based modern bioenergy utilization systems in China. J Environ Manag 264:110481. https://doi.org/10.1016/j.jenvman.2020.110481

Jugwanth Y, Sewsynker-Sukai Y, Gueguim Kana EB (2020) Valorization of sugarcane bagasse for bioethanol production through simultaneous saccharification and fermentation: optimization and kinetic studies. Fuel 262:116552. https://doi.org/10.1016/j.fuel.2019.116552

Júnior ADNF, Koyama MH, de Júnior MM, Zaiat M (2016) Thermophilic anaerobic digestion of raw sugarcane vinasse. Renew Energy 89:245–252. https://doi.org/10.1016/j.renene.2015.11.064

Kapanji KK, Haigh KF, Görgens JF (2019) Techno-economic analysis of chemically catalysed lignocellulose biorefineries at a typical sugar mill: sorbitol or glucaric acid and electricity co-production. Bioresour Technol 289:121635. https://doi.org/10.1016/j.biortech.2019.121635

Kaparaju P, Serrano M, Angelidaki I (2010) Optimization of biogas production from wheat straw stillage in UASB reactor. Appl Energy 87:3779–3783. https://doi.org/10.1016/j.apenergy.2010.06.005

Karimi S, Soofiani NM, Lundh T et al (2019) Evaluation of filamentous fungal biomass cultivated on vinasse as an alternative nutrient source of fish feed: protein, lipid, and mineral composition. Fermentation 5:1–19. https://doi.org/10.3390/fermentation5040099

Karp SG, Bittencourt Sydney E, Lorenci Woiciechowski A et al (2021) Lignocellulosic biorefinery for value-added products: the emerging bioeconomy. In: Biomass, biofuels, biochemicals. Elsevier, New York, pp 291–321

Kaur M, Neetu VYP, Chauhan S (2020) Effect of chemical pretreatment of sugarcane bagasse on biogas production. Mater Tod Proc 21:1937–1942. https://doi.org/10.1016/j.matpr.2020.01.278

Kayhanian M, Rich D (1995) Pilot-scale high solids thermophilic anaerobic digestion of municipal solid waste with an emphasis on nutrient requirements. Biomass Bioenergy 8:433–444. https://doi.org/10.1016/0961-9534(95)00043-7

Lima DRS, de Oliveira Paranhos AG, Adarme OFH et al (2021) Integrated production of second-generation ethanol and biogas from sugarcane bagasse pretreated with ozone. Biomass Convers Biorefinery 1–17. https://doi.org/10.1007/s13399-020-01234-7

Lopes CH, Gabriel AVMD, Borges MTMR (2011) Produção de etanol a partir da cana-de-açúcar: tecnologia de produção de etanol - coleção UAB-UFSCar - Tecnologia Sucroalcooleira

López González LM, Vervaeren H, Pereda Reyes I et al (2013) Thermo-chemical pre-treatment to solubilize and improve anaerobic biodegradability of press mud. Bioresour Technol 131:250–257. https://doi.org/10.1016/j.biortech.2012.12.167

López González LM, Pereda Reyes I, Romero Romero O (2017) Anaerobic co-digestion of sugarcane press mud with vinasse on methane yield. Waste Manag 68:139–145. https://doi.org/10.1016/j.wasman.2017.07.016

Marafon AC, Salomon KR, Amorim ELC, Peiter FS (2019) Use of sugarcane vinasse to biogas, bioenergy, and biofertilizer production. Elsevier, New York

Mood SH, Golfeshan AH, Tabatabaei M et al (2013) Lignocellulosic biomass to bioethanol, a comprehensive review with a focus on pretreatment. Renew Sust Energ Rev 27:77–93. https://doi.org/10.1016/j.rser.2013.06.033

Moore CCS, Nogueira AR, Kulay L (2017) Environmental and energy assessment of the substitution of chemical fertilizers for industrial wastes of ethanol production in sugarcane cultivation in Brazil. Int J Life Cycle Assess 22:628–643. https://doi.org/10.1007/s11367-016-1074-0

Moraes BS, Zaiat M, Bonomi A (2015) Anaerobic digestion of vinasse from sugarcane ethanol production in Brazil: challenges and perspectives. Renew Sust Energ Rev 44:888–903. https://doi.org/10.1016/j.rser.2015.01.023

Mustafa AM, Li H, Radwan AA et al (2018) Effect of hydrothermal and Ca(OH)2 pretreatments on anaerobic digestion of sugarcane bagasse for biogas production. Bioresour Technol 259:54–60. https://doi.org/10.1016/j.biortech.2018.03.028

Pandey A, Soccol CR, Nigam P, Soccol VT (2000) Biotechnological potential of agro-industrial residues. I: sugarcane bagasse. Bioresour Technol 74:69–80. https://doi.org/10.1016/S0960-8524(99)00142-X

Pant D, Adholeya A (2007) Biological approaches for treatment of distillery wastewater: a review. Bioresour Technol 98:2321–2334. https://doi.org/10.1016/j.biortech.2006.09.027

Petersen AM, Van der Westhuizen WA, Mandegari MA, Görgens JF (2017) Economic analysis of bioethanol and electricity production from sugarcane in South Africa. Biofuels Bioprod Biorefin 12:224–238. https://doi.org/10.1002/bbb.1833

Pippo WA, Luengo CA, Alberteris LAM et al (2011) Energy recovery from sugarcane-trash in the light of 2nd generation biofuels. Part 1: current situation and environmental aspects. Waste Biomass Valorization 2:1–16. https://doi.org/10.1007/s12649-010-9048-0

Prajapati BP, Jana UK, Suryawanshi RK, Kango N (2020) Sugarcane bagasse saccharification using Aspergillus tubingensis enzymatic cocktail for 2G bio-ethanol production. Renew Energy 152:653–663. https://doi.org/10.1016/j.renene.2020.01.063

Rabelo SC, Carrere H, Maciel Filho R, Costa AC (2011) Production of bioethanol, methane and heat from sugarcane bagasse in a biorefinery concept. Bioresour Technol 102:7887–7895. https://doi.org/10.1016/j.biortech.2011.05.081

RFA (2021) Annual U.S. & World Fuel Ethanol Production | Renewable Fuels Association. In: RFA - Renewable Fuels Association

Romanholo Ferreira LF, Aguiar MM, Messias TG et al (2011) Evaluation of sugar-cane vinasse treated with Pleurotus sajor-caju utilizing aquatic organisms as toxicological indicators. Ecotoxicol Environ Saf 74:132–137. https://doi.org/10.1016/j.ecoenv.2010.08.042

Sindhu R, Gnansounou E, Binod P, Pandey A (2016) Bioconversion of sugarcane crop residue for value added products – an overview. Renew Energy 98:203–215. https://doi.org/10.1016/j.renene.2016.02.057

Soccol CR, de Vandenberghe LPS, Medeiros ABP et al (2010) Bioethanol from lignocelluloses: status and perspectives in Brazil. Bioresour Technol 101:4820–4825. https://doi.org/10.1016/j.biortech.2009.11.067

Sua-Iam G, Makul N (2017) Effect of incinerated sugarcane filter cake on the properties of self-compacting concrete. Constr Build Mater 130:32–40. https://doi.org/10.1016/j.conbuildmat.2016.11.033

Sun JX, Sun XF, Sun RC, Su YQ (2004) Fractional extraction and structural characterization of sugarcane bagasse hemicelluloses. Carbohydr Polym 56:195–204. https://doi.org/10.1016/j.carbpol.2004.02.002

Sydney EB, Letti LAJ, Karp SG et al (2019a) Current analysis and future perspective of reduction in worldwide greenhouse gases emissions by using first and second generation bioethanol in the transportation sector. Bioresour Technol Rep 7:100234. https://doi.org/10.1016/j.biteb.2019.100234

Sydney EB, Neto CJD, de Carvalho JC et al (2019b) Microalgal biorefineries: integrated use of liquid and gaseous effluents from bioethanol industry for efficient biomass production. Bioresour Technol 292:1–7. https://doi.org/10.1016/j.biortech.2019.121955

Sydney EB, de Carvalho JC, Letti LAJ et al (2021) Current developments and challenges of green technologies for the valorization of liquid, solid, and gaseous wastes from sugarcane ethanol production. J Hazard Mater 404:124059. https://doi.org/10.1016/j.jhazmat.2020.124059

Tanimu MI, Ghazi TIM, Harun RM, Idris A (2014) Effect of carbon to nitrogen ratio of food waste on biogas methane production in a batch mesophilic anaerobic digester. Int J Innov Manag Technol 5:116–119. https://doi.org/10.7763/IJIMT.2014.V5.497

dos Santos Nunes T, Finzer JRD (2019) A importância do tratamento do caldo de cana-de-açúcar para a produção de açúcar e etanol

Unica (2021). Brazilian sugarcane industry association. https://unica.com.br/bringbackmybluesky/. Accessed 15 Sep 2021

USDA Sugar (n.d.)

VEIGA, Carlos Frederico de Menezes, VIEIRA, Joana Rita, MORGADO IF (2006) Diagnóstico da cadeia produtiva da cana-de-açúcar do Estado do Rio de Janeiro: relatório de pesquisa, vol 1500, p 107

Wilkie AC, Riedesel KJ, Owens JM (2000) Stillage characterization and anaerobic treatment of ethanol stillage from conventional and cellulosic feedstocks. Biomass Bioenergy 19:63–102. https://doi.org/10.1016/S0961-9534(00)00017-9

Zhao Y, Chen M, Zhao Z, Yu S (2015) The antibiotic activity and mechanisms of sugarcane (Saccharum officinarum L.) bagasse extract against food-borne pathogens. Food Chem 185: 112–118. https://doi.org/10.1016/j.foodchem.2015.03.120

Zheng Y, Zhao J, Xu F, Li Y (2014) Pretreatment of lignocellulosic biomass for enhanced biogas production. Prog Energy Combust Sci 42:35–53. https://doi.org/10.1016/j.pecs.2014.01.001

Chapter 6
Corn First-Generation Bioethanol Unities with Energy and Dried Grains with Solubles (DDGS) Production

Ariane Fátima Murawski de Mello, Luciana Porto de Souza Vandenberghe, Kim Kley Valladares-Diestra, Gustavo Amaro Bittencourt, Walter José Martinez Burgos, and Carlos Ricardo Soccol

Abstract The world production of corn-bioethanol is currently led by the United States, followed by China and Brazil. As a starchy material, corn grains must pass through different pretreatment and enzymatic hydrolysis, which have an impact on final biofuel costs. In addition, a balance between the use of corn grains and their subproducts for food and feed and biofuels production has to be found. The analysis of economic and environmental impacts of bioethanol processes that affect global food prices and land use may also be conducted. During corn bioethanol production some co-products are generated, mainly dried distillers' grains with solubles (DDGS), but also other minor products. In fact, there is a great tendency to develop bioethanol processes' production integrated with the generation of other bioproducts in biorefineries so as to close the production cycle under the zero-waste point of view. With these facts in mind, this chapter presents the current situation and main advancements of the corn-bioethanol production research and innovation. Important aspects including the different steps of corn pretreatment, starch enzymatic hydrolysis, fermentation technologies, strain improvement, and valorization of different generated effluents for medium-to-high value bioproducts' production are described.

6.1 Introduction

Renewable fuels not only reduce foreign sources for energy, but contribute for the reduction of greenhouse gas (GHG) production and also promote the specific agro-industrial chain development (Mumm et al. 2014). Great attention has been given to alternative fuel sources from agriculturally produced feedstocks, where corn and sugarcane dominate as the main bioethanol sources. In 2020, the United States led

A. F. Murawski de Mello · L. Porto de Souza Vandenberghe (✉) · K. K. Valladares-Diestra · G. Amaro Bittencourt · W. J. Martinez Burgos · C. R. Soccol
Department of Bioprocess Engineering and Biotechnology, Centro Politécnico, Federal University of Paraná, Curitiba, Paraná, Brazil
e-mail: lvandenberghe@ufpr.br

C. R. Soccol et al. (eds.), *Liquid Biofuels: Bioethanol*, Biofuel and Biorefinery Technologies 12, https://doi.org/10.1007/978-3-031-01241-9_6

the production of fuel ethanol in the world followed by Brazil, with 13.9 and 7.9 billion of gallons of the biofuel, respectively (ICDA 2020).

In 2020, corn production reached around 1.1 billion metric tons where the United States alone is responsible for over one third of global production, with approximately 82.5 million acres of land devoted for corn harvesting. China and Brazil are the second and third producers with 23.4% and 9.0% of the world production, respectively (Shahbandeh 2021). In the last decades, the production and processing of corn grain as a source of biofuel was well established in the US (Mumm et al. 2014). In the last years, other countries, such as Brazil, have started to explore corn as a potential feedstock for bioethanol production, in a complementary way with sugarcane. In 2019, Brazil produced 1.33 billion liters of corn ethanol, with a projection to expand to 8 billion liters by 2028, according to the National Union of Corn Ethanol Producers (UNEM). The production expansion is expected due to Brazilian's consumption of ethanol, which is estimated to be 43 billion liters by 2029, according to the Minister of Mines and Energy/Energy Research Enterprise (MME/EPE), supported by the country's new carbon credits program, RenovaBio (Barros and Woody 2020).

However, the use of corn grain or other feedstocks for bioethanol production has to be balanced with food and feed production, with the analysis of economic and environmental impacts affecting global food prices and land use. So, co-product utilization would be a powerful tool to reduce the land usage for corn ethanol with system complementarity between fuel production and livestock nutrition, which would be an important element for biofuel lifecycle evaluation.

Over the past 35 years, a significant increase in percentage of total corn crop that is used to produce corn ethanol was observed. Approximately 45% of the annual U.S. corn crop is used by the corn ethanol industry. However, only the starch portion of the corn, which represents about 2/3 of the grain is used to produce ethanol. During corn ethanol production some co-products are generated, among them are distillers dried grains with solubles (DDGS), which are usually used as animal feed. Other minor products include wet distillers grains (WDG), corn germ meal, and corn distillers oil (CDO). About 2.8 gal of ethanol and 17 lbs. of animal feed are produced from 56 lb. bushel of corn (in a dry mill plant (Hoekman et al. 2018).

This chapter presents the advancements of corn bioethanol production including corn preparation and pretreatments, fermentation technologies, re-use of generated effluents with the description of new developments of research and innovation with the application of biorefinery concept.

6.2 Corn Preparation and Pretreatment for Bioethanol Production

Corn is a widely used crop for commercial biofuel production, with great economic importance in global scale. United States is the leader of corn ethanol production, with an estimated output of 15.8 billion gallons in 2019 (Renewable Fuels Association 2020). In corn ethanol production, its high amount of starch (61–78%) is converted to glucan by hydrolysis, and further converted to ethanol by yeast fermentation (Zhang et al. 2021). In this context, starch based ethanol is a well-established process and produces about 60% of the global ethanol, in comparison with almost 40% of sugar crops (Johnston and McAloon 2014).

Starch is the main substance used by plants to store carbohydrates, acting as energy reservoir for cereals, legumes and tubers, as well as energy source for all living organisms and an important industrial raw material (Huang et al. 2021). It consists of α-D-glucose units and it is not metabolized at commercial scale by yeast and bacteria in its original form, making necessary process steps to fractionate and convert it to glucose before fermentation. One of these polymers is amylose, which is a linear polymer of around 1000 glucose units linked by α-1-4 glyosidic bonds, and the other is amylopectin, which possess a highly branched configuration along with a branch that is linked by α-1-6 bonds in every 20 linkages (Miao et al. 2015; Huang et al. 2021). The ratio of amylose and amylopectin content in starch is associated with its physicochemical properties that affects, as well as process conditions and nature of biomass and microorganisms, the overall conversion efficiency of ethanol.

In the preparation processes of corn to ethanol production, different pretreatments and hydrolysis steps are used, where different solvents, milling and amylolytic enzymes are applied (Fig. 6.1). Firstly, dry or wet milling methods are applied to yield starch from corn grains. Dry milling is an abrasive technique responsible for 90% of milling processes applied in corn ethanol plants. It is applied in hammer mills, that break the outer coating of the seed to separate the bran and germ, obtaining a fine powder of corn flour (Erickson et al. 2005; Zhang et al. 2021). This method is a cheap technology that co-produces only fermentation residues, as CO_2 and dried distillers' grains with solubles (DDGS), which will be further discussed in this chapter. In the case of wet milling, corn kernel is fully fractionated into carbohydrates, lipids, and protein, with minimal mechanical damage and high purities potential (99.9%). The first step is corn steeping, where corn kernel is placed in sulfurous acid solution and allowed to cook, to increase its permeability and swelling, and hydrolyze disulfide bonds in proteins. Then, sequential mechanical operations of grinding and washing are applied to separate corn starch, germ, gluten and protein (Erickson et al. 2005; Zhang et al. 2021).

From corn starch to fermentable sugars, there are hydrolysis steps that employ amylolytic enzymes. The first one is the liquefaction, where thermostable α-amylase is applied, before reaching temperatures of 105–165 °C, for gelatinization of the slurry (Schwietzke et al. 2009). Thermostable α-amylases are produced by thermophilic Bacillus strains, such as *Bacillus licheniformis*, or by recombinant *Escherichia*

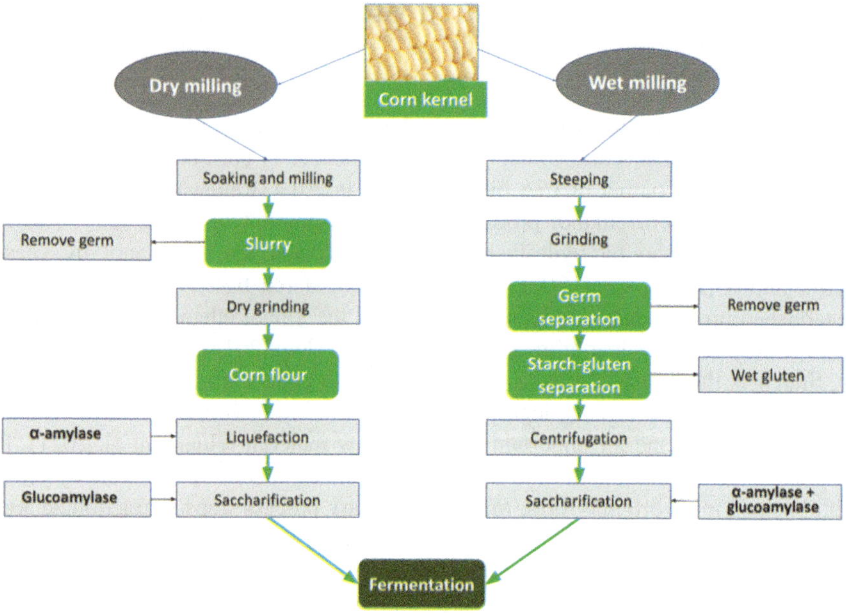

Fig. 6.1 Corn starch preparation processes before fermentation to bioethanol. Adapted from Zabed et al. (2017, 2021)

coli strains, and represents approximately 30% of the global enzyme production (Fincan et al. 2021). These hydrolases act by cleaving randomly starch molecules in α-1 → 4 glycosidic linkages, with the solubilization of almost all the amylose polymer from the corn starch granules, resulting in a swelled slurry with high viscosity. The gelatinization of starch in liquefaction step yields molecules with shorter chains, such as dextrins, maltose and maltotriose, is an important step to raise efficiency of fermentable sugars recovery in the next step (Zabed et al. 2017). The saccharification involves the application of glucoamylases in liquefied slurry to hydrolyze remaining oligosaccharides and dextrin into maltose and glucose syrups. The process conditions of pH 4.2–4.5 and 60 °C are consistent with the range over which *Aspergillus niger* or *Rhizopus* species enzymes are stable. The saccharification step can also be applied in simultaneous saccharification and fermentation (SSF) processes, that uses liquefied starches from liquefaction to be fermented by yeast and bacteria (Li et al. 2019; Zeng et al. 2020).

Pretreatment strategies can be applied in an attempt to increase starch accessibility in the enzymatic hydrolysis steps, raising overall ethanol yields. Li et al. (2018) evaluated different conditions of an ultrasound pretreatment, at a frequency of 40 kHz and starch concentration of 30% (w:v), as an acceleration step for liquefaction and saccharification of corn starch. The authors related a decrease in the relative crystallinity and changes in the surface structures of corn starch, which enhanced the liquefaction step reducing the time to achieve the optimum dextrose concentration.

The dextrose equivalent (DE) in the saccharification step, which represents the amount of reducing sugar expressed glucose per dry weight of starch in the process, raised from 44.09% for the native starch to 70.87% for the ultrasound pretreated starch. Thus, ultrasound pretreatment was suggested to enhance liquefaction and saccharification of starch-based feedstocks.

One alternative that was recently studied to enhance final sugar concentrations and improve fermentation's performance and products' yield is the optimization of substrate concentration in the starch liquefaction and saccharification steps. Li et al. (2019) evaluated solid contents from 20% to 60% in the liquefaction step using α-amylase at 90 °C, and a simultaneous saccharification and fermentation (SSF) process at 30 °C and 72 h. The authors reported that there was no significant effect of solids concentrations, in the range between 20 and 40%, on reducing sugars release after a liquefaction process during 120 min. Enzymatic hydrolysis efficiency decreased with increasing solid contents, higher than 50%. Similar results were reached by Li et al. (2015) who found that starch concentrations higher than 45% in the liquefaction step inhibited swelling and disruption of starch granules, due to incomplete gelatinization by heat pretreatment.

Due to the high energy consumption in the starch ethanol production, which is required by cooking and liquefaction steps (30–40%) (Lim et al. 2003), novel strategies have been developed to process optimization and costs reduction. The use of granular starch hydrolyzing enzymes (GSHE) introduced a non-cooking method for starch hydrolysis at sub-gelatinization temperature, with α-amylase and glucoamylase activities applied directly in starch granules. The use of GSHE in SSF of starch polymers can simplify the process and reduce the viscosity of the resulting slurry, which otherwise hinders the dispersion and mix of starch and enzymes leading to incomplete conversions (Zabed et al. 2017). Due to the sub gelatinization temperatures, higher solids contents can be employed without mixing problems between enzymes and materials in the reaction. However, to overcome sub-gelatinized process bottlenecks of incomplete hydrolysis and low rates, researchers used media supplementations, such as urea and proteases, as well as heat pretreatment steps to pre-swelling starch before SSF processes (Uthumporn et al. 2010; Li et al. 2012, 2016a; Naguleswaran et al. 2013; Pietrzak and Kawa-Rygielska 2014). Tong et al. (2019) applied a pre-heating treatment in corn starch followed by hydrolysis with 1% GSHE at 62 °C for 24 h. The authors reported glucose conversions of 61.3, 76.0, and 94.5% with 62, 65, and 70 °C, respectively, showing that heat pretreatment and pre-swelling of starch granules increased the GSHE accessibility.

6.3 Advancements in 1G Corn Bioethanol Fermentation

Fermentation is a very important stage in bioethanol production from corn, defining the efficiency degree of ethanol production according to employed conditions. The most significant factors in bioethanol fermentation are carbon source (starch, sucrose or glucose), nitrogen source (yeast extract, urea, peptone, others), mineral salts and the used microorganism with a high capacity to biotransform glucose to bioethanol (yeasts or bacteria) (Li et al. 2017). These variables are the most important in the bioethanol fermentation process, so an adequate definition of each component is always necessary to maximize process efficiency. Commonly, in corn-bioethanol production at industrial-scale, the carbon source is provided through pretreatments of corn kernels that release fermentable sugars (mainly glucose) from the starch. After the pretreatment stage and, depending on the employed method, the sugars-rich fraction may present some inhibitors that have a toxic effect on the metabolism of ethanol-producing microorganisms. Under these conditions a detoxification stage is necessary or the use of resistant strains (Greetham et al. 2019).

The raw material is responsible for 60–70% of the final cost of bioethanol, so it is necessary to minimize these costs and increase bioethanol productivity. Some studies show that the energy expenditure in the production of bioethanol from corn is still quite high, which generates a relatively small positive energy balance, which demands the optimization and improvement of the production process (Mojović et al. 2006; Mohanty and Swain 2019). In recent years, however, some improvements have been achieved, such as the selection and obtaining of ethanol hyperproducing strains like genetically modified C5/C6 yeast M11205 (Wang et al. 2019), hybrid SP2-18 with genome shuffling by fusion between *S. cerevisiae* and *P. stipites* (Jetti et al. 2019), *S. cerevisiae* MNII/cocδBEC3 transformed by an artificial zinc finger protein to improve its thermal tolerance (Khatun et al. 2017) and other strain with a high tolerance to higher concentrations of bioethanol and glucose (Favaro et al. 2019), an increase in the initial load of treated starch above 30% w/v (Puligundla et al. 2019), application of some strategies such as SSF (Szambelan et al. 2018), genetic improvements of the native strain for higher bioethanol production and tolerance (Cripwell et al. 2019; Myburgh et al. 2020) among others.

Although with these improvements, the energy balance in the production of ethanol from corn has reached positive values, there are still limiting factors in the fermentation stage that are: (a) The use of a high concentrations of glucose that inhibits the metabolism of yeasts, reducing their productivity; (b) The low tolerance of strains to high concentrations of ethanol, which generates by-products such as glycerol; (c) The energy, economic and time expenses in separate stage processes of starch hydrolysis and fermentation; (d) The high load of solids in fermentation processes that limits yeast recycle, for this reason inoculum preparation is necessary for each new stage of the process and; (e) The corn variety, with different origins and starch concentrations, also has an impact on fermentation process (Szambelan et al. 2018; Favaro et al. 2019; Mohanty and Swain 2019; Puligundla et al. 2019). For this reason, further improvements and process's optimization are still needed to make corn-ethanol sufficiently competitive compared to other biofuels and fossil fuels that are currently present in the international market.

Conventional industrial-scale bioethanol fermentation from corn is usually carried out at a concentration of 150–220 g/L of substrate (15–23% w/v dissolved solids), reaching approximately 10–15% ethanol (Mohanty and Swain 2019; Li et al.

2017). The most widely used microorganism is *Saccharomyces cerevisiae*, which produces ethanol from pyruvic acid derived from the catalysis of glucose with formation of ATP. *S. cerevisiae* is able to produce ethanol under aerobic conditions, which is advantageous at industrial scales, in batch or fed batch operation, in which the available oxygen is easily eliminated in the fermentation process. Thus, avoiding process oxidation and assimilation of final ethanol by the microorganism. In recent years, the corn ethanol production industry has optimized carbon source concentration by adopting the high gravity (25–30% w/v dissolved solids) or very high gravity ($\geq$35% w/v dissolved solids) fermentation technology. The higher solids concentration improves productivity in the bioethanol fermentation and reduce the volume of effluent due to less water requirements in the process (Puligundla et al. 2019).

Some strategies to improve ethanol productivity from corn are also focused on reducing nitrogen source costs. Yeast extract is one of the nitrogen sources used in ethanol fermentation, however, its cost is more expensive compared to other sources such as urea or ammonium sulfate, so an evaluation of nitrogen source must be conducted. Li et al. (2017) evaluated the use of different nitrogen sources derived from corn hydrolysates. Results showed the importance of yeast extract, but this nitrogen source can be partially replaced by a combination of urea (69 mM) and ammonium sulfate (26 mM). The two inorganic nitrogen sources act synergistically with yeast extract (0.6%), generating a 21% improvement in bioethanol production with a conversion yield higher than 80%. These results showed the possibility of decreasing the costs derived from the formulation of the culture medium in the production of first-generation bioethanol from corn, especially in industrial-scale productions.

The SSF method, which is widely studied, is a strategy to improve bioethanol fermentation process. This strategy allows the unification of two steps of bioethanol processes that generates energy and time savings. It consists of simultaneous enzymatic hydrolysis (liquefaction) and fermentation. This allows the progressive hydrolysis of the starch as the yeast metabolizes the released glucose, producing ethanol. In addition, this strategy allows the application of high amounts of substrate (25–40% w/v of dissolved solids) avoiding the osmotic pressure caused by the high initial amount of glucose (Szambelan et al. 2018). However, high ethanol concentrations can lead to yeast stress and, consequently, decrease cell growth and viability (Li et al. 2017). Due to the stress generated by the hyperproduction of ethanol in yeasts, different strains are evaluated, selected and genetically improved. Currently different commercial yeast strains are offered by biotechnology companies as shown in Table 6.1. These strains are specifically marketed due to their high tolerance to glucose and ethanol concentrations, with great ethanol production performances and, in some cases, tolerance to high temperatures.

Another strategy employed to optimize ethanol fermentation is the construction of microorganisms capable of producing enzymes that hydrolyze carbon sources called Consolidated Bioprocessing (CBP), facilitating and reducing production costs. The main challenge of CPB is the availability of an ideal host microorganism that can express the appropriate enzymes and with a high fermentation capacity. Genetic improvement has been performed for *S. cerevisiae* strains, because it is an

Table 6.1 Commercial yeast strains employed for industrial bioethanol production adapted from Favaro et al. (2019)

Commercial name	Characteristics	Company
Summit Ethanol Dry Yeast	High acid, temperature and ethanol tolerance; high tolerance to liberated glucose during saccharification	ABMauri Biotek [www.bioethanol.abbiotek.com]
Gen One Plus	High vigor at elevated temperatures (40 °C); ethanol tolerance; reduced glycerol production	Lesaffre Advanced Fermentations [www.lesaffreadvancedfermentations.com]
InnovaVR Lift	Glucoamylase expression, reduce enzyme addition; low nutrient addition	Novozymes [www. www.novozymes.com]
Fali Bioethanol Yeast	High acid, temperature and ethanol tolerance; high tolerance to liberated glucose during saccharification	ABMauri Biotek [www.bioethanol.abbiotek.com]
TransFerm Yieldþ	Glucoamylase expression, reduce enzyme addition; reduced glycerol production	Lallemand Biofuels & Distilled Spirits [www.lallemandbds.com]
ER-XpressTM	Glucoamylase expression, reduced enzyme addition	Lesaffre Advanced Fermentations [www.lesaffreadvancedfermentations.com]
SynerxiaVR	Glucoamylase expression, reduce enzyme addition;	Du Pont [www.dupont.com]
SafdistilTM C-70	Very robust strain for fermenting different sugar and grain substrates	Lesaffre Advanced Fermentations [www.lesaffreadvancedfermentations.com]
SafdistilTM Plus	High vigor at elevated temperatures (40 C); ethanol tolerance	Lesaffre Advanced Fermentations [www.lesaffreadvancedfermentations.com]
Ethanol RedV	High vigor at elevated temperatures (40 C); high ethanol tolerance, ideal for VHG fermentation	Lesaffre Advanced Fermentations [www.lesaffreadvancedfermentations.com]

ethanogenic strain with high yield characteristics and tolerance to ethanol. In addition, *S. cerevisiae* has a unique physiology allowing it to tolerate high concentrations of sugars (150 g/L), which is combined with an efficient passive transport of glucose to its cytosol that generates a high efficiency of glycolysis for bioconversion to ethanol (Favaro et al. 2019). Due to these characteristics, *S. cerevisiae* is used as a host for the insertion of genes capable of producing hydrolytic enzymes. In the case of ethanol produced from corn, CBP strains are genetically transformed with the insertion of genes capable of producing alpha amylases and glucoamylases, which can hydrolysate starch. Cripwell et al. (2019) evaluated the simultaneous expression of α-amylase and glucoamylase to identify the best combination of these enzymes in the hydrolysis and fermentation of raw starch. Initially, they used the strain *S cerevisiae* Y294 and genes from *Talaromyces emersonii*, the results showed a better combination of the optimized glucoamylase codon (*temG_Opt*) and the native gene of alpha amylase (*TemA*). The resulting combination of these two genes was then transferred into two δ integration gene cassettes for two commercial ethanol producing strains (Ethanol Red™ and M2n). With an initial content of 200 g/L of raw corn starch, a production of 89.4 and 98.1 g/L of ethanol was obtained for the transformed strains Ethanol Red™ and M2n, respectively, reaching a yield of 87 and 94%. Finally, a comparative experiment was carried out with the parental strains, in which they were used in the strategy of SSF with the addition of commercial enzymes. The results showed a high efficiency of the genetically modified strains, reducing the use of commercial enzymes by up to 90%, which would mean a substantial reduction in the costs of the process.

Although with the new mentioned strategies can lead to higher bioethanol production with conversions higher than 80%. This fact is due to the application of SSF and the construction of CBP that contributed for the elimination of toxicity of high concentrations of glucose. However, there is still a step that needs to be solved that is the tolerance of strains to high concentrations of ethanol. Industrial yeast strains subjected to adaptive pressure in large bioethanol factories have been selected and improved for this purpose. Another method is the simultaneous or intermittent extraction of the produced ethanol during fermentation. Kumar et al. (2018) evaluated the intermittent extraction of ethanol through evaporation with the vacuum-assisted fermentation. This new strategy allows the use of high solid contents in fermentation, extracting the ethanol produced in certain cycles by applying vacuum at fermentation temperatures of 32–34 °C, avoiding the toxicity of high ethanol concentrations and allowing the full consumption of the carbon source. The best results were obtained with the use of 40% of corn dissolved solids, application of 1 h of vacuum in determined times of 12; 24; 36 and 48 h. The production of 0.42 liters of ethanol per kg of dry corn was obtained with a conversion yield of approximately 80%. Compared with the traditional method, a total glucose consumption and an 88% increase of conversion efficiency to ethanol were achieved. In addition, the fermentation of 32% of corn (applying 1.5 h of vacuum in the 18 and 24 h times) resulted in a much faster process compared to classic strategies, reducing the fermentation time by 50%. The results undoubtedly showed the great potential of vacuum-assisted fermentation, which allows the extraction and ethanol production without significantly damaging the yeasts and decreasing the possibilities of inhibition by high concentrations of ethanol. Finally, the decrease in fermentation time allows the processing of a greater quantity of material, generating a more efficient process with higher productivities.

The different strategies used in the production of ethanol from corn can be seen in Fig. 6.2. The SHF strategy is the most used in different industries due to its easy operation and involved technology already implemented in this sector. SSF is being thoroughly evaluated and optimized at laboratory and pilot scales with excellent results and prospects for industrial implementation. In summary, new engineering technologies such as vacuum-assisted fermentation show that the ethanol production

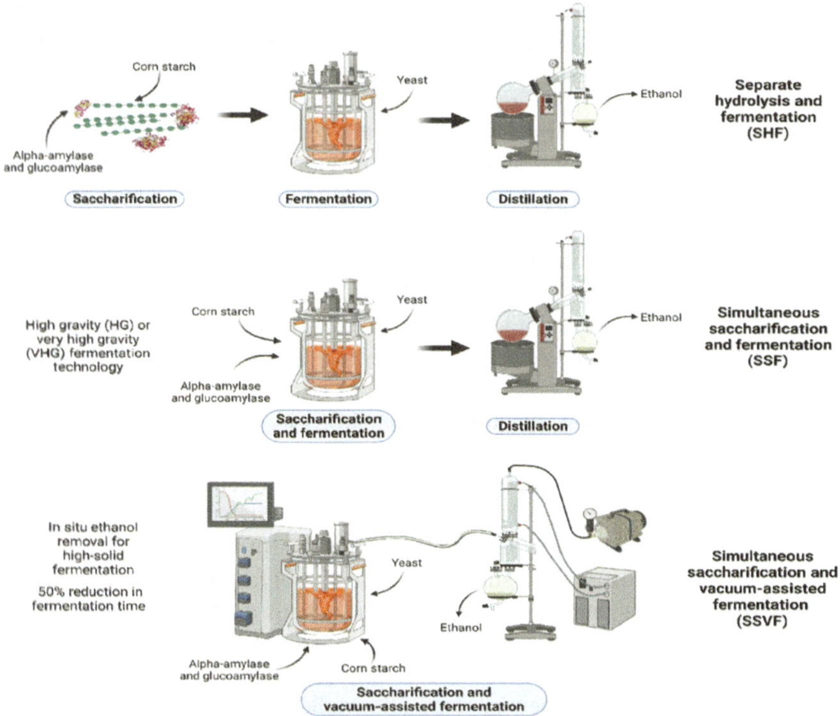

Fig. 6.2 Different strategies for bioethanol production from corn (Created with BioRender.com) showing the traditional ethanol fermentation processes, fermentation and saccharification simultaneous with increased carbon source, decreased fermentation time and increased ethanol productivity. Finally, the simultaneous saccharification and vacuum-assisted fermentation process that allows ethanol extraction in certain stages of the process, allowing greater productivity

sector is constantly developing and optimizing the process, with great prospects of generating more economically competitive biofuel.

6.4 The Corn Bioethanol Biorefinery: Energy and DDGS Production

Biorefineries can be defined as the sustainable processing of biomass into value-added products such as food, animal feed, biomolecules and energies in the form of biofuels, electricity or heat (Martinez-Burgos et al. 2021a; Sydney et al. 2021). A concern of bioethanol industries is the large amount of wastes that can be generated. Bioethanol producing industries from corn starch or sugar cane are clear examples of biorefineries, since all their residues can be used in the production of other bioproducts (Sydney et al. 2021) (Fig. 6.3).

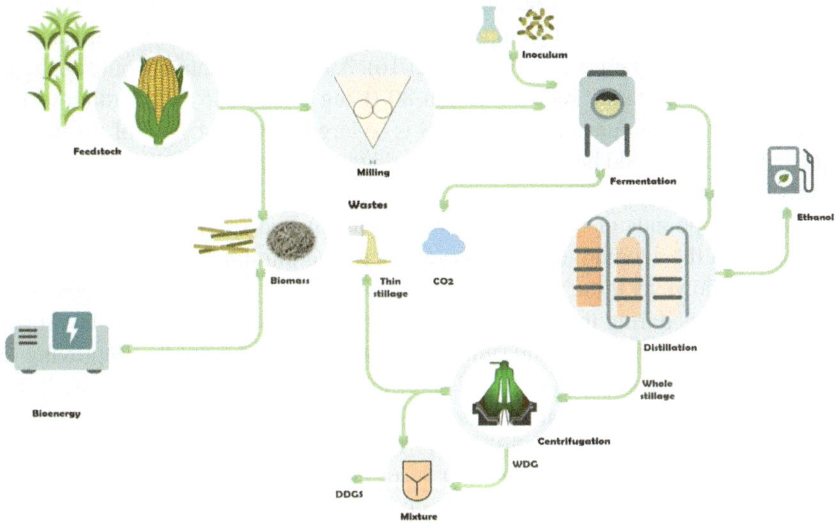

Fig. 6.3 Ethanol production from corn with the generation of solid, liquid, and gas wastes

In 2020, 98.63 million m^3 of ethanol were produced in the world (Renewable-Fuels-Association 2021). However, around 65% of ethanol came from corn (Renewable-Fuels-Association 2021; Sydney et al. 2021), which represents 64.11 million m^3 of ethanol, the main producer being the United States with 59.72 million m^3 of this value. As a result of the large amount of ethanol produced, exorbitant amounts of waste are generated, whether solid, liquid and gaseous residues.

As it was described before, the production of bioethanol from corn is performed in several stages. The last process is the separation of ethanol from other by-products through distillation processes (Reis et al. 2017). In the distillation, ethanol (the main product of interest) is separated from water, non-fermentable solids that remained from corn, and cellular biomass, a mixture known as whole stillage (WS). This by-product contains large amounts of protein, fiber, lipids and others. Subsequently, the WS is submitted to separation operations, generally centrifugation, wherein the liquid fraction is separated from the solids. The latter are known as wet distiller's grains (WDG) (Kim et al. 2008). The liquid fraction that still contains 5% of solids is known as thin stillage (TS). TS can be subjected to a drying or condensation process to form a paste with approximately 75% solids known as condensed distillers soluble (CDS). Finally, WDG and CDS can be combined and dried to produce dried distillers' grains with solubles (DDGS) in order to increase their shelf life (Reis et al. 2017). It is noteworthy that all these by-products are sources of amino acids such as arginine, histidine, isoleucine, leucine, lysine, methionine, phenylalanine, valine, alanine, aspartate, glutamate, glycine, proline and serine, which are used for the production of animal feed, except for TS (Reis et al. 2017). In fact, some compounds of the TS are separated in order to take advantage of each one of them and improve the viability of alcohol plants.

The TS contains significant amount of oils, which can be recovered through centrifugation processes or by employing precipitation aids such as precipitated and hydrophobic silica (Lewis and Shepperd 2016). Another alternative would be the use of non-ionic surfactants, such as Tween 80 (Fang et al. 2015). According to Reis et al. (2017), the amount of oil in the TS is so large, if at least 70% of the oil from bioethanol plants could be recovered, it would result in around 2 billion liters only in plants located in the United States.

Other compounds of interest that could be recovered from TS are protein fractions. One of the alternatives is the use of micro and ultrafiltration processes (Arora et al. 2010), however these processes are still highly costly. Another alternative is the use of proteases during the bioethanol production process to hydrolyze the proteins into their monomeric units and, thus, increase their use by yeasts, improving the fermentation of sugars (Reis et al. 2017). Compounds of wide industrial use, such as phytates, glycerol, lutein and zeaxanthin can also be recovered from TS. It is generated in an average proportion of 20 liters of effluent per liter of alcohol (Białas et al. 2010; Gyenge et al. 2013). In other words, in 2020, approximately 1.28 billion m^3 of effluent were generated worldwide. This residue is characterized by containing high COD and BOD_5, as well as significant concentrations of nitrogen, phosphorus, potassium, traces of sugars, such as fructose and glucose, glycerol and some organic acids, such as acetic, propionic and butyric acids (Eskicioglu et al. 2011; Choonut et al. 2015; Zangaro et al. 2018). Due to the macro and micronutrient composition of the effluent, it has been used as a biofertilizer in different types of crops, for instance in the *Brassica napus* crops, wherein great potential was observed, as it significantly promoted plant growth (Alotaibi et al. 2014).

Other wastes in ethanol production is the CO_2 are generated in the fermentation and can be estimated stoichiometrically (Eq. 6.1). It is estimated that around 48.33 million tons of CO_2 are generated in the fermentative production of ethanol with corn starch.

$$C_6H_{12}O_6- \rightarrow H_6O + 2CO_2 \tag{6.1}$$

Solid residues from lignocellulosic biomass (corn cob and corn stover) are also generated. According to Kim and Dale (2002) and Luo et al. (2009), for each m^3 of ethanol produced, around 1.6 ton of corn stover is generated; thus, in the same period, around 104.2 million tons were generated.

6.4.1 Corn Biorefinery Using Microalgae

Microalgae and cyanobacteria are photosynthetic organisms, with high rates of CO_2 fixation (approximately ten times higher than that of higher plants), high growth rates and easily applied in industry (Sydney et al. 2021). These microorganisms can use the nutrients from TS and fix the CO_2 that is generated in fermentation, which means that biorefineries based on microalgae will be able to simultaneously treat two types of waste.

Microalgae such as *Scenedesmus obliquus*, *Anabaena* sp. and the cyanobacterium *Aphanothece microscopica* have high CO_2 fixation rates around 1.4 g L^{-1} day^{-1} (Jacob-Lopes et al. 2008; Ho et al. 2012). Other microalgal species showed lower rates of CO_2 fixation. Tu et al. (2019) achieved a maximum CO_2 fixation rate of 0.209 g L^{-1} day^{-1} with *Chlorella pyrenoidosa.* In the case of *Chlorella vulgaris*, for example, the fixation rate can vary from 0.25 to 0.9 g L^{-1} day^{-1} (Sydney et al. 2010; García-Cubero et al. 2018).

Microalgae and cyanobacteria can use the nutrients from the TS effluent for their growth. These microorganisms can be employed to produce cellular protein for food supplements and food additives, such as carotenoids, antioxidants, fatty acids, polysaccharides, vitamins etc. Thus, Beigbeder et al. (2019) produced *Chlorella vulgaris* biomass with high protein content (32% w/w), carbohydrates (14% w/w), lipids (7% w/w) and significant concentrations of carotenes and chlorophylls. It is noteworthy that 85% of the total carbon was removed during the process, while the organic acids were completely consumed. Sayedin et al. (2020) removed 95.3% and 78.3% of the nitrogen and phosphorus, respectively, from the TS using *Chlorella sorokiniana.* The produced biomass contained a high content of proteins (37.8% w/w) and lipids (17.8% w/w). These organisms can also be used for the production of bioactive molecules widely employed in the cosmetic industry and for human and animal health.

Furthermore, microalgae can be used to produce bioenergy in the form of hydrogen, biodiesel and methane and, consequently remove COD and BOD_5. Soto et al. (2021) removed 45% and 75% of COD and BOD_5, respectively, from distillery stillage using the microalgae *Chlorella vulgaris.* Thus, in microalgal biorefineries different bioproducts can be produced (Fig. 6.4). However, biomass harvesting, cell disruption and metabolite extraction methods may be efficient for a efficient recovery of these medium to high-value bioproducts.

6.4.2 *Corn Biorefinery Using Bacteria and Fungi*

Bacteria and yeasts have also been used in biorefinery processes of liquid waste generated in the production of bioethanol from corn, mainly WS and TS (Nasr 2012; Fortney et al. 2021). Bacterial consortia, as well as isolated bacteria, have been used in the production of bioenergy in the form of methane and hydrogen. Westerholm et al. (2012) produced biomethane from WS and using cattle manure as inoculum. The maximum yield achieved was 0.31 L CH_4 per g volatile solids. Eskicioglu et al. (2011) produced biomethane from TS under mesophilic and thermophilic conditions, reaching yields of 49 ± 5 L $CH_4/L_{stillage}$ and 65 ± 14 L $CH_4/L_{stillage}$, respectively.

Sayedin et al. (2019) reached a maximum yield of 305 mL CH_4 per COD g and removed around 93 of COD. Nasr (2012) produced hydrogen and methane from the

TS. The latter was evaluated in two-stage and the single-stage, reaching maximum yields of 0.33 liters CH_4 per g of COD added and 0.26 liters CH_4 per g of COD added, respectively. In the case of hydrogen, the maximum yield was 19.5 L H_2/L TS and the predominant microbial species in the process were *Clostridium acetobutyricum*, *Klebsiella pneumonia*, *Clostridium butyricum* and *Clostridium pasteurianum*. In addition, some intermediate products of wide application in industry are also generated, such as the medium-chain-fatty acids and lactic acid (Martinez-Burgos et al. 2020, 2021b).

According to the United States Department of Agriculture (USDA) (2015) in the United States there are more than 2000 biogas producing plants, several of them use TS as substrate. It is estimated that the potential of biomethane production per year could reach up to 16 million metric tons per year (USDA-EPA-DOE 2015). TS has also been used in the production of other biometabolites such as lactic acid, using *Lactobacillus rhamnosus* ATCC 7469, but also in the production of probiotic biomass (Djukić-Vuković et al. 2013). West (2011) employed TS as a substrate to produce malic acid by different *Aspergillus* species, reaching a maximum yield of 0.8 g malic acid per g substrate. It is noteworthy that the fungi strains consumed around 95% of the glycerol in the effluent and 63% of the sugars. Ahn et al. (2011) used TS as a substrate for *Clostridium pasteurianum* DSM 525 to produce butanol under anaerobic conditions. Liang et al. (2012) used TS to produce *Pythium irregulare* biomass rich in protein (28% w/w) and in lipids (39% w/w) with a high content of omega-3 fatty acids. Another important application of TS is the production of polysaccharides using *Ganoderma lucidum* (Hsieh et al. 2005).

6.4.3 Corn Lignocellulosic Biomass Biorefinery

Corn biomass (corn cob and corn stover) is highly energetic. According to Luo et al. (2009) the energy content of corn biomass 4.49 to 6.41 MJ per kg of stover. Traditionally, corn biomass has been used in the production of thermal and electrical energy for the plant process itself. In fact, 90% of ethanol producing plants generate their own energy from cellulosic biomass.

Corn stover is mainly composed of three main constituents: cellulose (40%), hemicellulose (35%) and lignin (14%) (Patel and Shah 2021). Cellulose is a linear polymer with a rigid structure formed by ®- (1 $\rightarrow$ 4)-D-glycopyranose units linked by glycosidic bonds (β-1,4). Hemicellulose is composed of branched and linear heteropolymers such as: L-arabinose, D-galactose, D-glucose, D-mannose and D-xylose and lignin is a recalcitrant compound formed by a complex structure of amorphous polymers and has a hydrophobic characteristic (Li et al. 2016b; Patel and Shah 2021).

In general, the structure of lignocellulosic biomass hinders the action of microorganisms. Therefore, physical (grinding, high temperatures), chemical (acid and alkaline) and biological (microorganisms and enzymes) pre-treatments must be used to hydrolyze cellulose and hemicellulose to fermentable sugars (Li et al. 2016b), and

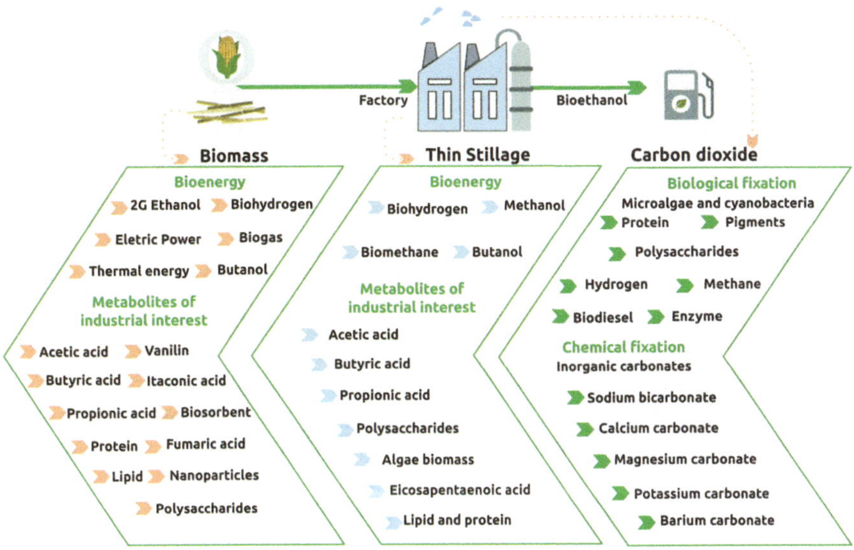

Fig. 6.4 Bioproducts generated from different waste generated in the production of ethanol from corn starch

thus produce different types of bioproducts. One of the flagship bioproducts is the production of second-generation ethanol. In fact, there are 24 industrial plants of second generation ethanol in the world (Pandiyan et al. 2019). Other bioproducts such as biomethane, biohydrogen, biodiesel, organic acids, cellular proteins, aromas, polysaccharides, among others, have also been produced using pretreated corn biomass (Fig. 6.4) (Li et al. 2016b).

6.5 Patents and Innovation in 1G Corn Ethanol

Over the last years, the demand for energy and materials has increased with a growing population, but the reliance on fossil sources has diminished due to its limited accessibility and high carbon emissions induced by its exploration (Kohli et al. 2019). Therefore, the concern for environmental sustainability and the demand for bio-based processes and products have significantly increased. Agricultural feedstocks (e.g. corn, cassava and wheat), waste streams and lignocellulosic biomass are examples of substrates that can be used for bioproducts production. The development of new processes and technologies is directly connected to innovation, which can follow an iterative cycle of idea development, invention and commercialization (Van Lancker et al. 2016). The patent system usually helps innovation since it grants protection of the developed knowledge and aids the commercialization step. Thus, in order to disclose the recent developments and innovation regarding the production of ethanol from corn, a patent search was conducted.

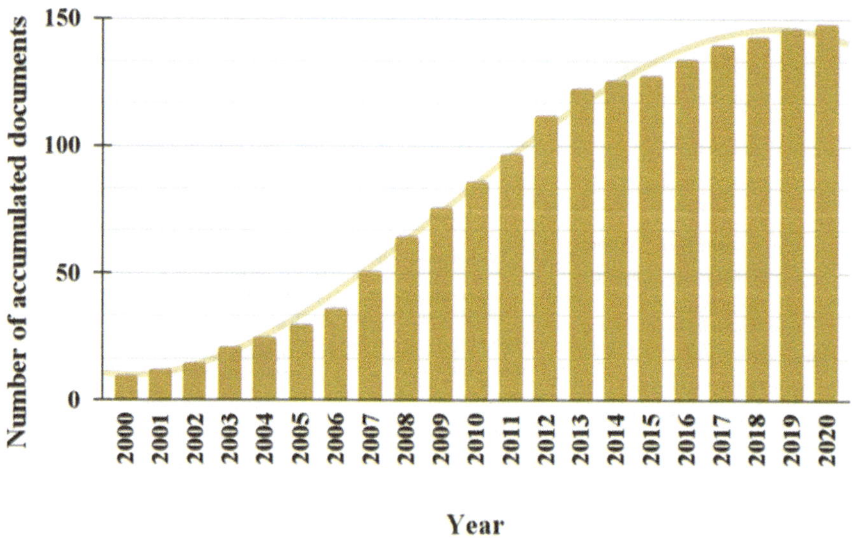

Fig. 6.5 Evolution of filed patents over the years of 2000 until 2021. Data retrieved from DII and Latipat databases

Among the available patent databases, Derwent Innovation Index (DII) and Latipat were selected for this work. The search was conducted combining the following keywords with Boolean Language: bioethanol, fuel ethanol, production, fermentation, process, fabrication, manufacturing, maize and corn. Since the topic of this study is first generation ethanol production, words that refer to lignocellulosic biomass - such as straw, stover and bran - were excluded. 230 documents were retrieved through this search and, after manual revision of titles and abstracts, 148 documents were selected and analyzed as follows.

The first records of filed documents date 1980, but until 2000, only 7 documents were found. From this year onwards, the number of filed documents has significantly increased (Fig. 6.5). In the first decade of the 2000s, the growth was more expressive since the technology was in its early development and the processes and their improvements were being constantly protected by the patent system. Besides, actions taken by the major players in the market influenced this growth. In 2001, Novozymes (a great provider of enzymes for biomass processing and bioethanol production) established a contract with the US government for the development of enzymes for production of biofuels and in 2008, inaugurated the largest unit for enzymatic fermentation focusing on bioethanol manufacture (Novozymes 2021).

However, from 2014 onwards, the number of filed documents per year hit a certain level of stability since the technology of first-generation bioethanol was well established and protected. Therefore, the focus of filed documents changed from the production of the fuel for protecting new engineered microorganisms for bioethanol fabrication (Argyros and Barret 2014; Yu et al. 2014), new enzymes that enhances starch hydrolysis (Huang et al. 2015), new transgenic crops of corn that are more

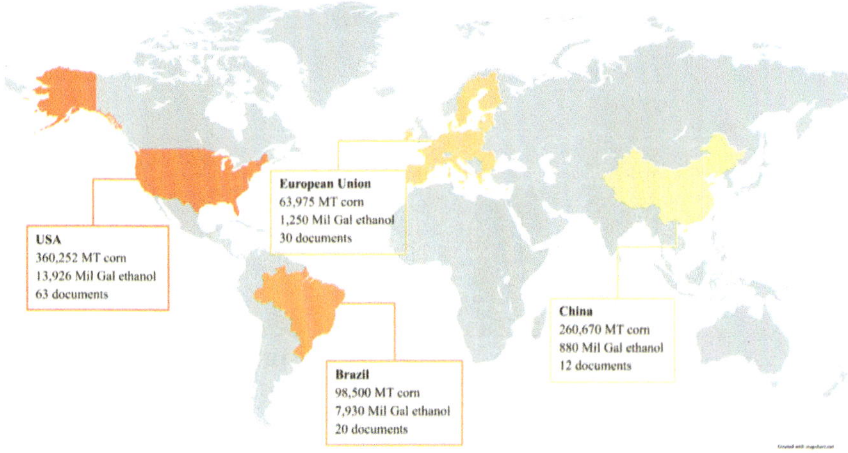

Fig. 6.6 Corn and corn-bioethanol production global producers and respective filed patents

suitable for the process (Mitchell et al. 2017) and new destinations of process byproducts, such as DDGS and yeast (Godoy 2017; Cox 2018). This shows that the interest in developing bioethanol from corn has not diminished, but the focus of technology is changing as naturally happens.

The profile of countries and regions that hold the highest number of filed documents is directly related to the profile of largest ethanol and corn producers worldwide (Fig. 6.6). The ethanol market was evaluated at 33.7 billion USD in 2020 and is projected to grow to 64.8 billion USD until 2026 (compost annual growth rate of 14%) (Markets and Markets 2021). The US was responsible for 53% of ethanol production in 2020, followed by Brazil with 30%, the European Union (EU) with 5% and China with 3% (Renewable Fuel Association 2021). On the other hand, regarding corn production, the US is the largest corn producer worldwide, followed by China, Brazil and the EU. While the bioethanol produced in the US comes mainly from maize, Brazil tends to apply sugarcane for ethanol production (Karp et al. 2021). However, in 2017 the Brazilian Corn Ethanol Union (UNEM) was founded for the promotion of first -generation bioethanol and, since then, several new plants were inaugurated and put to function, mainly in the Center-West of the country. Besides, UNEM predicted the production of 2.5 billion liters of ethanol from corn in 2020/2021, a significant increase when compared to the production of the last years (1.33 billion liters in 2019 and 720 in 2018) (Barros and Woody 2020). According to the USDA, 11 of the 19 of China's fuel ethanol licensed producers use corn as feedstock for bioethanol production (Mcgrath 2020). Regarding the EU, bioethanol is produced mainly from starchy feedstocks, including maize (Eubia 2021) and the region is also highlighted for its great enzyme production provided mainly from Novozymes, in Denmark.

In Fig. 6.7a, the profile of codes of International Patent Classification (IPC) is shown. For this analysis, all IPC codes from documents were extracted, truncated

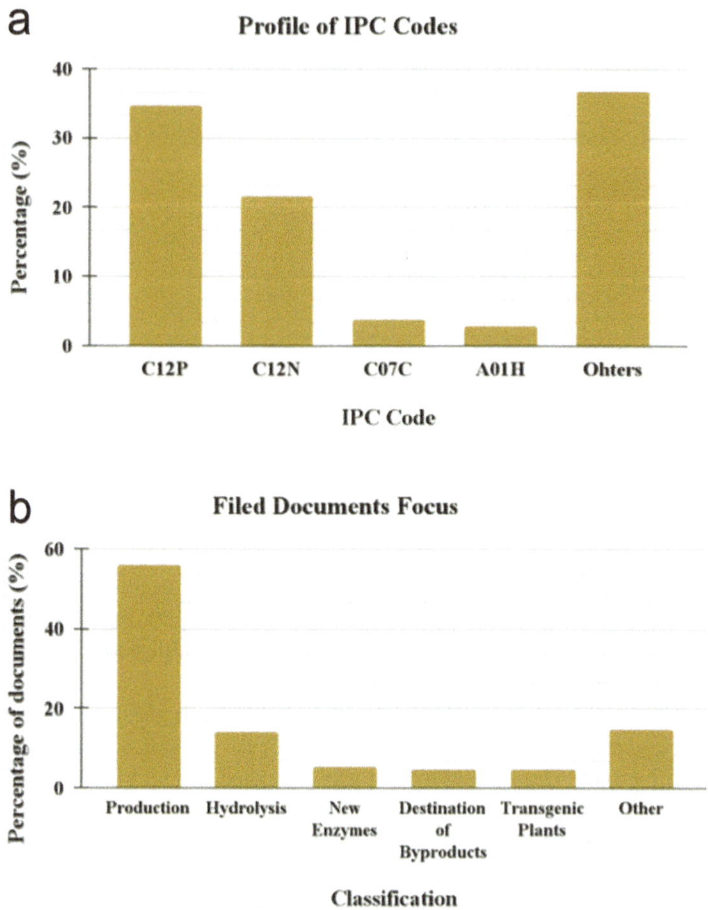

Fig. 6.7 Classification of filed patents based on (a) IPC Codes and (b) Focus

and accounted. The classification of focus of documents in Fig. 6.7b was elaborated by the authors. The most applied IPC is C12P that refers to fermentation or enzyme using process to produce a desired chemical, followed by C12N regarding microorganisms and enzymes and C07C that refers to acyclic or carbocyclic compounds. These codes are strictly related to the profile of filed documents that have the main focus in production and hydrolysis (Fig. 6.7b). Another relevant code is A01H that regards new plants (directly connected with the topic of new transgenic crops).

For the next few years, the main tendency for first generation ethanol is the integrated production of this molecule with other bioproducts in biorefineries in order to use the corn grain as a whole and, therefore, generate processes with zero waste. Besides, research and technology development in new enzymes and microorganisms for higher conversion of starch to ethanol are also in focus.

6.6 Conclusions and Perspectives

The corn-bioethanol leads the biofuel market and its production tends to increase in the world's main producing countries, United States, China and Brazil. As a biofuel, corn ethanol production chain provides some benefits such as replacement of fossil fuels, rural economic development, enhanced employment, production of valuable by-products, such as DDGS, a certain reduction of GHG and environmental impacts. However, the increase of the corn-ethanol production may also bring some problems such as water pollution, soil degradation, increased air pollutants, greater food insecurity, deforestation, monoculture and others. Different efforts must then be employed to mitigate these negative impacts with the search of new technologies and total re-use of generated solid and liquid residues in a closed and sustainable circular economic approach.

References

Ahn JH, Sang BI, Um Y (2011) Butanol production from thin stillage using Clostridium pasteurianum. Bioresour Technol 102(7):4934–4937. https://doi.org/10.1016/j.biortech.2011.01.046

Alotaibi KD, Schoenau JJ, Hao X (2014) Fertilizer potential of thin stillage from wheat-based ethanol production. Bioenergy Res 7(4):1421–1429. https://doi.org/10.1007/s12155-014-9473-1

Argyros A, Barret T (2014) Métodos Para Regular O Metabolismo De Nitrogênio Durante A Produção De Etanol A Partir De Milho Por Cepas De Leveduras Metabolicamente Modificadas Geneticamente

Arora A, Dien BS, Belyea RL, Singh V, Tumbleson ME, Rausch KD (2010) Heat transfer fouling characteristics of microfiltered thin stillage from the dry grind process. Bioresour Technol 101(16):6521–6527. https://doi.org/10.1016/j.biortech.2010.03.077

Barros S, Woody K (2020) Corn ethanol production booms in Brazil

Beigbeder J, Boboescu I, Lavoie J (2019) Thin stillage treatment and co-production of bio-commodities through finely tuned Chlorella vulgaris cultivation. J Clean Prod 216:257–267. https://doi.org/10.1016/j.jclepro.2019.01.111

Białas W, Szymanowska D, Grajek W (2010) Fuel ethanol production from granular corn starch using Saccharomyces cerevisiae in a long term repeated SSF process with full stillage recycling. Bioresour Technol 101:3126–3131. https://doi.org/10.1016/j.biortech.2009.12.090

Bioethanol Market Global Forecast to 2025 | MarketsandMarkets (2021). https://www.marketsandmarkets.com/Market-Reports/bioethanol-market-131222570.html. Accessed 26 Sep 2021

Choonut A, Yunu T, Pichid N, Sangkharak K (2015) Ethanol production from reused liquid stillage. Elsevier, New York

Cox BR (2018) Asphalt binder used in paving composition, comprises a bitumen and a post-fermentation distillers maize oil

Cripwell RA, Rose SH, Favaro L, van Zyl WH (2019) Construction of industrial Saccharomyces cerevisiae strains for the efficient consolidated bioprocessing of raw starch. Biotechnol Biofuels 12(1):201. https://doi.org/10.1186/s13068-019-1541-5

Djukić-Vuković AP, Mojović LV, Vukašinović-Sekulić MS, Nikolić SB, Pejin JD (2013) Integrated production of lactic acid and biomass on distillery stillage. Bioprocess Biosyst Eng 36(9): 1157–1164. https://doi.org/10.1007/s00449-012-0842-x

Erickson GE, Klopfenstein TJ, Adams DC, Rasby RJ (2005) Corn processing co-products manual

Eskicioglu C, Kennedy KJ, Marin J, Strehler B (2011) Anaerobic digestion of whole stillage from dry-grind corn ethanol plant under mesophilic and thermophilic conditions. Bioresour Technol 102(2):1079–1086. https://doi.org/10.1016/j.biortech.2010.08.061

Eubia (2021) Bioethanol. In: European Biomass Industry Association. https://www.eubia.org/cms/wiki-biomass/biofuels/bioethanol/. Accessed 26 Sep 2021

Fang L, Wang T, Lamsal B (2015) Synergistic effect of surfactants and silica nanoparticles on oil recovery from condensed corn distillers solubles (CCDS). Ind Crop Prod 77:553–559. https://doi.org/10.1016/j.indcrop.2015.09.031

Favaro L, Jansen T, van Zyl WH (2019) Exploring industrial and natural Saccharomyces cerevisiae strains for the bio-based economy from biomass: the case of bioethanol. Crit Rev Biotechnol 39: 800–816

Fincan SA, Özdemir S, Karakaya A, Enez B, Mustafov SD, Ulutaş MS, Şen F (2021) Purification and characterization of thermostable α-amylase produced from Bacillus licheniformis So-B3 and its potential in hydrolyzing raw starch. Life Sci 264(May 2020):1–9. doi:https://doi.org/10.1016/j.lfs.2020.118639

Fortney NW, Hanson NJ, Rosa PRF, Donohue TJ, Noguera DR (2021) Diverse profile of fermentation byproducts from thin stillage. Front Bioeng Biotechnol 9:1–14. https://doi.org/10.3389/fbioe.2021.695306

García-Cubero R, Moreno-Fernández J, García-González M (2018) Potential of Chlorella vulgaris to abate flue gas. Waste Biomass Valorization 9(11):2015–2019. https://doi.org/10.1007/s12649-017-9987-9

Godoy A (2017) Processo de reaproveitamento de biomassa de levedo, com separação de sólidos antes da destilação e recuperação de etanol do bolo úmido, na integração de fermentações alcoólicas de cana e substratos amiláceos

Greetham D, Saleh ZA, Du C (2019) Exploring the tolerance of marine yeast to inhibitory compounds for improving bioethanol production. Sustain Energy Fuels. https://doi.org/10.1039/c9se00029a

Gyenge L, Ráduly B, Barrera R, Font X, Lányi S, Ábrahám B (2013) Efficiency of biogas production from corn bioethanol by-products using different inocula. In: 4th International Youth Conference Energy (IYCE). https://doi.org/10.1109/IYCE20136604171

Ho SH, Chen CY, Chang JS (2012) Effect of light intensity and nitrogen starvation on CO2 fixation and lipid/carbohydrate production of an indigenous microalga Scenedesmus obliquus CNW-N. Bioresour Technol 113:244–252. https://doi.org/10.1016/j.biortech.2011.11.133

Hoekman SK, Broch A, Liu X (2018) Environmental implications of higher ethanol production and use in the U.S.: a literature review. Part I – Impacts on water, soil, and air quality. Renew Sust Energ Rev 81:3140–3158

Hsieh C, Hsu TH, Yang FC (2005) Production of polysaccharides of Ganoderma lucidum (CCRC36021) by reusing thin stillage. Process Biochem 40(2):909–916. https://doi.org/10.1016/j.procbio.2004.02.004

Huang H, Li X, Li G, Yu Y (2015) New mutant cellulose exoglucanases CBH I useful for producing a viscosity reducing agent for corn fuel ethanol fermentation

Huang L, Tan H, Zhang C, Li Q, Liu Q (2021) Starch biosynthesis in cereal endosperms : an updated review over the last decade. Plant Commun 2. https://doi.org/10.1016/j.xplc.2021.100237

ICDA (2020) Maps and data - global ethanol production by country or region. https://afdc.energy.gov/data/10331

Jacob-Lopes E, Scoparo CHG, Franco TT (2008) Rates of CO_2 removal by Aphanothece microscopica Nägeli in tubular photobioreactors. Chem Eng Process Process Intensif 47(8): 1365–1373. https://doi.org/10.1016/j.cep.2007.06.004

Jetti KD, Gns RR, Garlapati D, Nammi SK (2019) Improved ethanol productivity and ethanol tolerance through genome shuffling of Saccharomyces cerevisiae and Pichia stipitis. Int Microbiol 22:247–254. https://doi.org/10.1007/s10123-018-00044-2

Johnston DB, McAloon AJ (2014) Protease increases fermentation rate and ethanol yield in dry-grind ethanol production. Bioresour Technol 154:18–25. https://doi.org/10.1016/j.biortech.2013.11.043

Karp SG, Medina JDC, Letti LAJ, Woiciechowski AL, Carvalho JC, Schmitt CC, Penha RO, Kumlehn GS, Soccol CR (2021) Bioeconomy and biofuels: the case of sugarcane ethanol in Brazil. Biofuels Bioprod Biorefin 15(3):899–912. https://doi.org/10.1002/bbb.2195

Khatun MM, Yu X, Kondo A et al (2017) Improved ethanol production at high temperature by consolidated bioprocessing using Saccharomyces cerevisiae strain engineered with artificial zinc finger protein. Bioresour Technol 245:1447–1454. https://doi.org/10.1016/j.biortech.2017.05.088

Kim S, Dale BE (2002) Allocation procedure in ethanol production system from corn grain. Int J Life Cycle Assess 7(4):237–243

Kim Y, Mosier NS, Hendrickson R, Ezeji T, Blaschek H, Dien B, Cotta M, Dale B, Ladisch MR (2008) Composition of corn dry-grind ethanol by-products : DDGS, wet cake, and thin stillage. Bioresour Technol 99(2008):5165–5176. https://doi.org/10.1016/j.biortech.2007.09.028

Kohli K, Prajapati R, Sharma B (2019) Bio-based chemicals from renewable biomass for integrated biorefineries. Energies 12(2):233. https://doi.org/10.3390/en12020233

Kumar D, Juneja A, Singh V (2018) Fermentation technology to improve productivity in dry grind corn process for bioethanol production. Fuel Process Technol 173:66–74. https://doi.org/10.1016/j.fuproc.2018.01.014

Lewis S, Shepperd P (2016) Oil extraction aids in gran processing. US patent 9353332 B2, 2(12)

Li J, Vasanthan T, Bressler DC (2012) Improved cold starch hydrolysis with urea addition and heat treatment at subgelatinization temperature. Carbohydr Polym 87(2):1649–1656. https://doi.org/10.1016/j.carbpol.2011.09.061

Li Z, Liu W, Gu Z, Li C, Hong Y, Cheng L (2015) The effect of starch concentration on the gelatinization and liquefaction of corn starch. Food Hydrocoll 48:189–196. https://doi.org/10.1016/j.foodhyd.2015.02.030

Li C, Fang D, Li Z, Gu Z, Yang Q, Cheng L, Hong Y (2016a) An improved two-step saccharification of high-concentration corn starch slurries by granular starch hydrolyzing enzyme. Ind Crop Prod 94:259–265. https://doi.org/10.1016/j.indcrop.2016.08.049

Li P, Cai D, Luo Z, Qin P, Chen C, Wang Y, Zhang C, Wang Z, Tan T (2016b) Effect of acid pretreatment on different parts of corn stalk for second generation ethanol production. Bioresour Technol 206:86–92. https://doi.org/10.1016/j.biortech.2016.01.077

Li Z, Wang D, Shi Y-C (2017) Effects of nitrogen source on ethanol production in very high gravity fermentation of corn starch. J Taiwan Inst Chem Eng 70:229–235. https://doi.org/10.1016/j.jtice.2016.10.055

Li M, Li J, Zhu C (2018) Effect of ultrasound pretreatment on enzymolysis and physicochemical properties of corn starch. Int J Biol Macromol 111:848–856. https://doi.org/10.1016/j.ijbiomac.2017.12.156

Li Z, Wang D, Shi YC (2019) High-solids bio-conversion of maize starch to sugars and ethanol. Starch/Staerke 71(1–2):1–7. https://doi.org/10.1002/star.201800142

Liang Y, Zhao X, Strait M, Wen Z (2012) Use of dry-milling derived thin stillage for producing eicosapentaenoic acid (EPA) by the fungus Pythium irregulare. Bioresour Technol 111:404–409. https://doi.org/10.1016/j.biortech.2012.02.035

Lim LH, Macdonald DG, Hill GA (2003) Hydrolysis of starch particles using immobilized barley α-amylase. Biochem Eng J 13(1):53–62. https://doi.org/10.1016/S1369-703X(02)00101-8

Luo L, Van Der VE, Huppes G (2009) An energy analysis of ethanol from cellulosic feedstock – corn Stover. Renew Sust Energ Rev 13(2009):2003–2011. https://doi.org/10.1016/j.rser.2009.01.016

Martinez-Burgos WJ, Sydney EB, Paula DR, Medeiros ABP, Carvalho JC, Soccol VT, de Vandenberghe LPS, Woiciechowski AL, Soccol CR (2020) Bioresource technology biohydrogen production in cassava processing wastewater using microbial consortia : process optimization and kinetic analysis of the microbial community. Bioresour Technol 309:123331. https://doi.org/10.1016/j.biortech.2020.123331

Martinez-Burgos WJ, Bittencourt Sydney E, Bianchi Pedroni Medeiros A, Magalhães AI, de Carvalho JC, Karp SG, de Souza P, Vandenberghe L, Junior Letti LA, Thomaz Soccol V, de Melo Pereira GV, Rodrigues C, Lorenci Woiciechowski A, Soccol CR (2021a) Agro-industrial wastewater in a circular economy: characteristics, impacts and applications for bioenergy and biochemicals. Bioresour Technol 341(July):125795. https://doi.org/10.1016/j.biortech.2021.125795

Martinez-Burgos WJ, Sydney EB, de Paula DR, Medeiros ABP, De Carvalho JC, Molina D, Soccol CR (2021b) Hydrogen production by dark fermentation using a new low-cost culture medium composed of corn steep liquor and cassava processing water: process optimization and scale-up. Bioresour Technol 320. https://doi.org/10.1016/j.biortech.2020.124370

McGrath C (2020) Biofuels annual

Miao M, Li R, Huang C, Jiang B, Zhang T (2015) Impact of β-amylase degradation on properties of sugary maize soluble starch particles. Food Chem 177:1–7. https://doi.org/10.1016/j.foodchem.2014.12.101

Mitchell MC, Divi UK, Vanhercke T, Petrie JR, Singh SP, Green AG (2017) Transgenic plant or part used for obtaining seed, extract, recovered or extracted lipid or soluble protein, and for manufacturing industrial product, comprises a first exogenous polynucleotide which encodes transcription factor polypeptide

Mohanty SK, Swain MR (2019) Chapter 3 - Bioethanol production from corn and wheat: food, fuel, and future. Elsevier, New York

Mojović L, Nikolić S, Rakin M, Vukasinović M (2006) Production of bioethanol from corn meal hydrolyzates. Fuel 85(12):1750–1755. https://doi.org/10.1016/j.fuel.2006.01.018

Mumm RH, Goldsmith PD, Rausch KD, Stein HH (2014) Land usage attributed to corn ethanol production in the United States: sensitivity to technological advances in corn grain yield, ethanol conversion, and co-product utilization. Biotechnol Biofuels 7(61):1–17

Myburgh MW, Rose SH, Viljoen-Bloom M (2020) Evaluating and engineering Saccharomyces cerevisiae promoters for increased amylase expression and bioethanol production from raw starch. FEMS Yeast Res 20(6). https://doi.org/10.1093/femsyr/foaa047

Naguleswaran S, Vasanthan T, Hoover R, Bressler D (2013) The susceptibility of large and small granules of waxy, normal and high-amylose genotypes of barley and corn starches toward amylolysis at sub-gelatinization temperatures. Food Res Int 51(2):771–782. https://doi.org/10.1016/j.foodres.2013.01.057

Nasr NE (2012) Investigation of biohydrogen and biomethane production from thin stillage. Master Dissertation Civil And Environmental Engineering. The University of Western Ontario

Novozymes (2021) Novozymes' history, our heritage. https://www.novozymes.com/pt/about-us/history. Accessed 26 Sep 2021

Pandiyan K, Singh A, Singh S, Saxena AK, Nain L (2019) Technological interventions for utilization of crop residues and weedy biomass for second generation bio-ethanol production. Renew Energy 132:723–741. https://doi.org/10.1016/j.renene.2018.08.049

Patel A, Shah AR (2021) Integrated lignocellulosic biorefinery: gateway for production of second generation ethanol and value added products. J Bioresour Bioprod 6(2):108–128. https://doi.org/10.1016/j.jobab.2021.02.001

Pietrzak W, Kawa-Rygielska J (2014) Ethanol fermentation of waste bread using granular starch hydrolyzing enzyme: effect of raw material pretreatment. Fuel 134:250–256. https://doi.org/10.1016/j.fuel.2014.05.081

Puligundla P, Smogrovicova D, Mok C, Obulam VSR (2019) A review of recent advances in high gravity ethanol fermentation. Renew Energy 133:1366–1379. https://doi.org/10.1016/j.renene.2018.06.062

Reis CER, Rajendran A, Hu B (2017) New technologies in value addition to the thin stillage from corn-to-ethanol process. Rev Env Sci Biotechnol 16:175–206. https://doi.org/10.1007/s11157-017-9421-6

Renewable Fuels Association (2020) 2020 pocket guide to ethanol. In: Focus Forward

Renewable-Fuels-Association (2021) Annual World Fuel Ethanol Production. https://ethanolrfa.org/statistics/annual-ethanol-production/. Accessed 15 September 2021

Sayedin F, Kermanshahi-pour A, He QS (2019) Evaluating the potential of a novel anaerobic baffled reactor for anaerobic digestion of thin stillage: effect of organic loading rate, hydraulic retention time and recycle ratio. Renew Energy 135:975–983. https://doi.org/10.1016/j.renene.2018.12.084

Sayedin F, Kermanshahi-pour A, He QS, Tibbetts SM, Lalonde CGE, Brar SK (2020) Microalgae cultivation in thin stillage anaerobic digestate for nutrient recovery and bioproduct production. Algal Res 47:101867. https://doi.org/10.1016/j.algal.2020.101867

Schwietzke S, Kim Y, Ximenes E, Mosier N, Ladisch M (2009) Ethanol production from maize. In: Molecular genetic approaches to maize improvement. Springer, New York, pp 347–364

Shahbandeh M (2021) Corn industry worldwide - statistics & facts. In: Statistica

Soto MF, Diaz CA, Zapata AM, Higuita JC (2021) BOD and COD removal in vinasses from sugarcane alcoholic distillation by Chlorella vulgaris: environmental evaluation. Biochem Eng J. https://doi.org/10.1016/j.bej.2021.108191

Sydney EB, Sturm W, de Carvalho JC, Thomaz-Soccol V, Larroche C, Pandey A, Soccol CR (2010) Potential carbon dioxide fixation by industrially important microalgae. Bioresour Technol 101(15):5892–5896. https://doi.org/10.1016/j.biortech.2010.02.088

Sydney EB, de Carvalho JC, Letti LAJ, Magalhães AI, Karp SG, Martinez-Burgos WJ, de Candeo ES, Rodrigues C, de Vandenberghe LPS, CJD N, LAZ T, Medeiros ABP, Woiciechowski AL, Soccol CR (2021) Current developments and challenges of green technologies for the valorization of liquid, solid, and gaseous wastes from sugarcane ethanol production. J Hazard Mater 404. https://doi.org/10.1016/j.jhazmat.2020.124059

Szambelan K, Nowak J, Szwengiel A, Jeleń H, Łukaszewski G (2018) Separate hydrolysis and fermentation and simultaneous saccharification and fermentation methods in bioethanol production and formation of volatile by-products from selected corn cultivars. Ind Crop Prod 118 (March):355–361. https://doi.org/10.1016/j.indcrop.2018.03.059

Tong Z, Tong Y, Shi YC (2019) Partial swelling of granules enables high conversion of normal maize starch to glucose catalyzed by granular starch hydrolyzing enzyme. Ind Crops Prod 140 (August). https://doi.org/10.1016/j.indcrop.2019.111626

Tu R, Jin W, Han S, Fang ZX, Wang J, Wang Q, He Z, Ding W, Che L, Feng X (2019) Enhancement of microalgal lipid production in municipal wastewater: fixation of CO2 from the power plant tail gas. Biomass Bioenergy 131:105400. https://doi.org/10.1016/j.biombioe.2019.105400

USDA (2015) Biogas opportunities roadmap progress report. https://www.epa.gov/sites/default/files/2015-12/documents/biogas-roadmap-infographic.pdf. Accessed 15 September 2021

USDA-EPA-DOE (2015) Biogas opportunities roadmap progress report. https://www.energy.gov/sites/prod/files/2015/12/f27/biogas_opportunites_roadmap_progress_report_0.pdf. Accessed 15 September 2021

Uthumporn U, Zaidul ISM, Karim AA (2010) Hydrolysis of granular starch at sub-gelatinization temperature using a mixture of amylolytic enzymes. Food Bioprod Process 88(1):47–54. https://doi.org/10.1016/j.fbp.2009.10.001

Van Lancker J, Wauters E, Van Huylenbroeck G (2016) Managing innovation in the bioeconomy: an open innovation perspective. Biomass Bioenergy 90:60–69. https://doi.org/10.1016/j.biombioe.2016.03.017

Wang Z, Dien BS, Rausch KD et al (2019) Improving ethanol yields with deacetylated and two-stage pretreated corn stover and sugarcane bagasse by blending commercial xylose-fermenting and wild type Saccharomyces yeast. Bioresour Technol 282:103–109. https://doi.org/10.1016/j.biortech.2019.02.123

West T (2011) Malic acid production from thin stillage by aspergillus species. Biotechnol Lett 33(12):2463–2467. https://doi.org/10.1007/s10529-011-0720-7

Westerholm M, Hansson M, Schnürer A (2012) Improved biogas production from whole stillage by co-digestion with cattle manure. Bioresour Technol 114:314–319. https://doi.org/10.1016/j.biortech.2012.03.005

Yu L, Chen X, Li Y, Liu Y (2014) Brew yeast and application of brew yeast to manufacturing alcohol through fermentation

Zabed H, Sahu JN, Suely A, Boyce AN, Faruq G (2017) Bioethanol production from renewable sources: current perspectives and technological progress. Renew Sust Energ Rev 71:475–501. https://doi.org/10.1016/j.rser.2016.12.076

Zangaro CA, Patterson R, Gibbons WR, Woyengo TA (2018) Enhancing the nutritive value of corn whole stillage for pigs via pretreatment and predigestion. J Agric Food Chem 66(36): 9409–9417. https://doi.org/10.1021/acs.jafc.8b01943

Zeng W, Zhang B, Jiang L, Liu Y, Ding S, Chen G, Liang Z (2020) Poly(malic acid) production from liquefied corn starch by simultaneous saccharification and fermentation with a novel isolated Aureobasidium pullulans GXL-1 strain and its techno-economic analysis. Bioresour Technol 304(February):122990. https://doi.org/10.1016/j.biortech.2020.122990

Zhang R, Ma S, Li L, Zhang M, Tian S, Wang D, Liu K, Liu H, Zhu W, Wang X (2021) Comprehensive utilization of corn starch processing by-products: a review. Grain Oil Sci Technol J. https://doi.org/10.1016/j.gaost.2021.08.003

Chapter 7
Why and How: A Chronicle of Second-Generation Ethanol

Gonçalo Amarante Guimarães Pereira and Marcelo Falsarella Carazzolle

Abstract The aim of this chapter is to contextualize second-generation ethanol considering civilization's needs for fuel, jobs, and sustainability, under the logics of the Bioeconomy. We will briefly reflect on the evolution of energy use, the entry of fossil fuels into our lives and the need to move efficiently to a new energy matrix. It is in this matrix that ethanol can assume the role of a global liquid fuel. With this goal in mind, we are going to present the possibilities of producing ethanol in the necessary volume from various sources of biomass, whose use should be optimized through second-generation technologies and bioeletrification. Thus, we will explore the reasons that led to the development of the first second-generation ethanol plants, the technologies developed, and the challenges faced by the pioneers of the sector, particularly the fact that all underestimated the specific characteristics of straw and bagasse, which resulted in serious pretreatment failures. We will briefly touch on themes that will be covered in detail in other chapters of this book, such as biomass, bioelectrification and the carbon credit system, subjects that are essential to the world expansion of 2G ethanol.

7.1 A Brief History of Energy

The history of the universe is the history of energy. On our planet, the dispersion of man everywhere, in the last ten thousand years, led to the formation of our modern civilization, with significant changes in the landscape and the environment. This process becomes acute with the invention of the steam engine, which gave rise to the industrial revolution in the late eighteenth century. Machines now did the work of thousands of people, but they used the energy of millions. At that time, with the innovations taking place in a temperate climate environment, which are areas with low biomass productivity, it was no longer possible to use only the power of nature

G. Amarante Guimarães Pereira (✉) · M. Falsarella Carazzolle
Laboratory of Genomics and BioEnergy, Institute of Biology, Department of Genetics, Evolution and Bioagents, UNICAMP, Campinas-SP, Brazil
e-mail: goncalo@unicamp.br

C. R. Soccol et al. (eds.), *Liquid Biofuels: Bioethanol*, Biofuel and Biorefinery Technologies 12, https://doi.org/10.1007/978-3-031-01241-9_7

to feed the machines. It was essential to make use of fossil coal reserves, which represent a condensed photosynthesis of about 100 million years of buried forests during the Carboniferous and Permian periods (Cleal and Thomas 2005). This process progressed quickly, completely altering people's lives, who began to gather in cities and live under conditions that were often worse than those in rural areas. The industrial revolution always reminds us of London's smog, a kind of low, gray cloud, formed primarily as a result of particulates thrown into the atmosphere by factory chimneys. On the other hand, we can suggest that fossils have replaced slavery, which somewhat redeems these energy sources from the problems they cause today.

7.2 The Oil Empire

Coal reigned supreme in the generation of heat and energy, moving the most diverse machines. A relevant exception was public lighting, still made with oil distilled from whale blubber, which almost drove these animals to extinction (Adams 2020). The reason for not directly using charcoal in lighting is simple: the convenience of liquid fuel. Here we begin to see the entry of oil, also a manifestation of concentrated photosynthesis, but formed mainly because of the death and sedimentation of photosynthetic microorganisms in the primitive oceans (Schobert 2013). The distillation of this material led to lighting kerosene, a highly profitable business that generated an unbridled rush to "black gold". However, the main driver of the use of this source was its convenience as ship fuel. Churchill, still in World War I, needed ships faster than those powered by coal, a low-density fuel that occupied a lot of space and was difficult to manage (Seligmann 2018). Once again, oil solved all these problems, playing a relevant role on ending the First War, but generating a main motivation for all other wars of the twentieth century that are propagating until today.

The fact is that oil is extraordinarily convenient. It is concentrated in gigantic reserves, its exploration, refining and logistics are completely dominated and automated. One drawback is that it is an extremely capital-intensive activity, but this turned out to be a big advantage for the dominant players by creating a virtually insurmountable barrier to entry for newcomers. In addition, the financial returns are so huge that they led the big oil companies to conquer an incredible economic-political prominence in civilization, without any precedent in history. A transnational power began, beyond the borders of countries, whose specific interests generated strong geopolitical tensions and affected the fate of practically everyone (Claes 2018). In the process, we all become addicted to oil, and the energy transition can be compared to a drug detox process. Many countries, like the USA, have practically one car per inhabitant, a fact that today entirely determines our geography, cut by streets, avenues and highways. Furthermore, as mass is simply a manifestation of energy (remember $E = mc^2$), the fossil sources were indirectly converted into a population explosion, while significantly increasing the concentration of CO_2 in the atmosphere. Following this path is not a clever choice. An increase in population demands more resources and food, but the increase in CO_2 in the atmosphere changes the climate in an intense way, leading to extremes, changes in agricultural zoning and enormous risks for food production. As a result, we have the expansion of the factor that is historically the biggest reason for wars: migration (Klepp 2017).

We see the difficulties that must be faced by humanity in seeking a rational balance. Economics, in the sense of the ancient Greek word (oikos: household; nemo: distribution or allocation), should face these problems. However, in its classic definition, it is assumed that the economic system has infinite demand at one end and scarcity at the other, thus, to solve this problem, scarcity is attacked. However, with the entry of the planet's savings into the system, represented by fossil carbon and energy sources, this logic does not hold. The economy no longer must worry about scarcity, but about the perverse consequences of excess, which generates environmental destruction, overpopulation and, perhaps worst of all, inequality. As mentioned previously, capital intensive activities generate immense returns, but for few individuals. Innovations end up cutting human work, which is normally seen as a source of problems and labor conflicts, hence, the energy density of fossils ended up generating a densification of wealth in civilization, and now runs with the financial system as the main driver in economics.

If we accept this last conclusion, we invariably accept our fate of ending in the list of extinct species, keeping company with more than 99% of everything that ever lived on the planet. However, we can reverse the picture. Instead of insisting with the fossil economy, with what we could call "ThanEconomy" (referring to Thanatos, the personification of death), we need to start to improve the balance between life and environment, in what is being called Bioeconomy. For this, we will need renewable energies and, above all, the production and conversion of biomass across a whole chain of fuels and products. Perhaps as (or even more) important than energy generation is the large generation of jobs, which are permanent (unlike solar and wind sectors, which are labor intensive only during installation) and involve people from the most diverse levels of instruction. Unlike oil or coal, whose reserves are concentrated in a few places, with second-generation technologies ethanol can be produced (in greater or lesser volume) by all the countries of the world, in an activity that generates energy and, at the same time, a vast number of jobs.

7.3 Price and Value: Remunerating Externalities

The financial market is an enigma that shows the complexity of human nature. It is a powerful lever of development by mobilizing the energy (everything is energy, including money) necessary to conduct the expeditions, discoveries and inventions that gave rise to and are the basis of modern civilization. However, at some point money began to be reinvested into more money, leading to the formation of incredible fortunes, directly associated with a tremendous irregularity in the distribution of wealth. We have reached the point that much of the productive activity is intended to serve currency mining, a logic that reached its apex with the emergence of virtual currencies (for example, Bitcoin). We live in an economy whose prices are no longer defined by elements such as production cost, but rather as a perception of opportunity. This is what happens, for example, in the case of oil. Prices for this product currently bear extraordinarily little relation to its production cost, which is quite different in different regions of the globe. The price is the result of the connection of this commodity with the global nervous system that pulses with intensity and irregularity on international exchanges (Market Insights 2017). In theory, the deepening of this process can lead to an extreme and irrational optimization of "economic" activity, cutting jobs and allowing only capital-intensive activities. Thus, the financial market, if it does not connect price to value, could constitute a powerful extinction mechanism, the meteor of our civilization.

On the other hand, since the financial market exists and is powerful, a way out of the current scenario would be to use its mechanisms to financialize renewables, as is done with oil. To do this, an innovative program, called RenovaBio (Grassi and Pereira 2019), was implemented in Brazil. Fundamentally, the principle is that the use of biofuel avoids the use of fossil fuel and thus mitigates emissions. To synthesize this principle, a financial asset was created, called CBIO, which stands for one ton of fossil CO_2 mitigated by the biofuel. Therefore, the more efficient the production of biofuel (which is calculated by a specific tool—RenovaCalc), the smaller the volume needed to literally "mine" a CBIO. As a major financial innovation, the program places CBIOs as a security on the stock exchange, which can be freely acquired by any person or company. As a backing for the process, fossil fuel producers and distributors have mandatory acquisition targets for CBIOs, which must be withdrawn from the market (retired) through this operation. This system ensures minimum values for the product and avoids an eventual CBIOs inflation.

This mechanism created a form of currency to be exploited by biofuel producers. The more valued the CBIO, the greater the incentive for the biofuel to be produced with less fossil CO_2 emissions, with a net capture being even possible during the process. For example, most biofuel emissions are in the agricultural phase of the process, with the diesel used to drive the machines and with fertilization from the addition of nitrogen compounds. However, it is possible to produce large volumes of biogas in the plants, capable of moving all the machines and generating energy surpluses for the grid (both methane and electricity). In addition, several microorganisms can fix atmospheric nitrogen (Mylona et al. 1995), which would eliminate the need for external fertilization, something that is already done in soybean crops, for example. Finally, the plant's fermentation processes also generate concentrated CO_2, which is very difficult and expensive to be obtained by direct concentration from the atmosphere (Fasihi et al. 2019). In the case of alcoholic fermentation, practically pure CO_2 is obtained (Kheshgi and Prince 2005), while in methanogenesis this gas is concentrated at about 50% (Plugge 2017). As a result, this CO_2 could be sequestered underground or even used for the synthesis of other renewable fuels, such as synthetic or electro-fuels (e-fuels) which are produced by combining green hydrogen (electrolysis of water, for example) with renewable electricity and CO_2. Another alternative would be the production of biochar, a compound produced from the pyrolysis of any plant material, which is resistant to decomposition to CO_2 and dramatically improves the physicochemical properties of the soil (Ennis et al. 2012). In other words, in biorefineries it is possible to reach a counterintuitive situation in which the use of fuel leads to CO_2 sequestration, something impossible in the fossil industry. However, for this to happen, it is essential that speculative operations with CBIOs are encouraged. The valuation of this title, or currency, could be a decisive factor for investments in new 2G ethanol plants, which are, admittedly, those that lead to greater energy efficiency and lower emissions.

7.4 The Locomotive

When the oil crisis occurred in the 70s, Brazil responded by producing ethanol. However, this only made sense because an ethanol-powered car was developed, which later evolved into the flex car. Nowadays, the world is experiencing the trend of battery-powered electric cars, which became a highly desired element by society, translating these sorts of vehicles into hallmarks of modernity and sustainability. However, this is not quite the case. Metal batteries, to be produced, depend on metals, such as lithium and cobalt, that are prevalent in just a few countries in the world, and their search precedes a new geopolitical tension (Cooke 2021). In addition, these devices have low energy density and excessive cost. The production of a car with great autonomy ends up being expensive and heavy, and its level of (improbable) sustainability will depend on the nature of the electricity that was used to produce the vehicle, its operation and of the recycling efficiency (Cooke 2021). These accounts, based on Life Cycle Analysis methodologies, are essential to effectively understand the environmental impact of each technology. Methodologies that only consider the effect of vehicle output, called the Wheel to Tank, are by no means sufficient to describe the impact of vehicles.

Despite the obviousness of this reasoning, many European countries have adopted the Wheel to Tank method for assessing the environmental impact of the automobile industry (Friedl et al. 2021). This was probably not a decision taken by rationality, but by lobby. Once converted in regulation, all the auto industry is investing heavily in battery electrification, something that is likely to lead to great regrets in the near future. Having said that, the electric motor is a desirable alternative. Its energy efficiency is far superior to that of the combustion engine, and this rationalizes and expands energy use. Therefore, the problem is not electrification, but the use of metal batteries as fuel reserve.

An approach to this scenario is given by Fuel Cells, which are based on principles known for a long time and which are now being refined for use on a large scale (İnci et al. 2021). Some commercial vehicles, such as Hyundai Tucson/ix35 FCEV, Toyota Mirai FCEV, and Honda Clarity FCEV are for sale in North America, Europe, and Asia since 2014 (Hardman and Tal 2018). In all these, the vehicles are equipped with Proton Exchange Membrane technology, PEM, which uses super pure hydrogen (For details, see Wang et al. 2020). Briefly explaining the principle, molecular hydrogen is cleaved at the anode and separated into its protons and electrons. The anode is separated from the cathode by a membrane impermeable to electrons but permeable to protons. These protons will then cross the membrane and react with oxygen at the cathode, where the hydrogen electrons will also migrate. As a result, water is formed, and an electrical current is generated that will power a traction battery or the electric motor itself. These cells, which have been commercially used for some time, have the advantage of being dynamic, delivering power as hydrogen is fed. However, they use expensive metals to conduct the reactions and work at a low temperature (40–120 °C), which facilitates the poisoning of the system. In addition, they need super pure hydrogen for their operation, and with hydrogen being a highly flammable gas, presents a series of logistical challenges, with high storage and transport costs. Finally, although there are many plans to produce green hydrogen, coming from the hydrolysis of water from renewable electricity (wind and solar), currently it is obtained mainly from the reform of natural gas (Martin and Saikawa 2017), which is a fossil source of carbon. As a result, this is an arrangement that does not generate effective results for the decarbonization of the atmosphere, although only water is emitted in its exhaust.

An alternative would be solid oxide cells, SOFC (for details, see Peng et al. 2021), which can be fueled by biofuels, in particular liquid ethanol. In this case, a reformer, which may or may not be coupled to the cell, catalyzes the ethanol into hydrogen and CO_2, passing through the synthesis gas. This hydrogen, which has a relatively low degree of purity, goes to an anode that separates the electrons from the protons. In this case, protons are trapped in the anode while the electrons migrate along a wire and form an electrical current. At the cathode, molecular oxygen (O_2) is converted to oxygen anion (O^{-2}), an ion that can cross the solid oxide "membrane" (usually a metal-supported ceramic material), which is done as if the atom was jumping between the holes in a Swiss cheese (referring to the gaps within the crystalline structures) (Chroneos et al. 2011). Upon reaching the anode, these anions combine with the hydrogen protons, generating water. Unlike the case of PEM cells, there is no commercial vehicle available with this technology. In 2016, Nissan presented in Brazil a prototype plugin, in which the SOFC cell functioned as a range extender (Nissan 2016). In this case, the vehicle was equipped with a 20-liter ethanol tank and had a range of 600 km, being refueled at regular gas stations, using hydrous ethanol. Nissan claims to be continuing the development of this technology for small vehicles, but initial developments should perhaps focus on utility vehicles such as buses and trucks, which have more room for such equipment. In addition, the purchase of commercial vehicles is based on more rational criteria, such as cost per kilometer driven and depreciation. In this case, visual attractiveness and appeal play a minimal role, if any. Finally, customers are starting to demand that companies tend to lower emissions during the transport of their goods, which is not achievable with the use of diesel. Thus, SOFC cells would bring the ideal solution, supplying high autonomy, low (or even no) emission and rapid refueling in ordinary fuel stations.

There is enormous interest in making this technology viable in the shortest possible time, which would be a great achievement (Velandia Vargas and Seabra 2021). With 2G ethanol technologies and adequate public policies, it is possible to produce ethanol for the world, even considering the lower efficiency of combustion engines (Milanez et al. 2015). However, it would be much more suitable if this biofuel could be used in a more noble way, as a CO_2 battery for electrification. In this model, we could develop a sophisticated circular economy for mobility: the atmospheric CO_2 would be fixed through photosynthesis into biomass, which would then be hydrolyzed and fermented into ethanol and biomethane to be used in SOFC-type fuel cells. With this strategy, all the world would be able to produce ethanol. All countries, to a greater or lesser extent, have biomass of some kind available, whether as a dedicated energy crop, such as sugarcane or corn, or waste harvesting of different cultures. This would bring a brighter future for humanity, who could then keep fossil sources as safety reserves for moments of crisis or as a failsafe.

7.5 Planting Oil

The beauty of oil lies in its energy density, abundance and ease of production. If oil were inexhaustible and did not increase the concentration of CO_2 in the atmosphere, it would be a perfect source. Unfortunately, this is not the case. Therefore, the challenge is to plant agricultural crops that can achieve high productivity per hectare in order to compete with oil. The higher the energy density of the crop, the easier and cheaper it will be to process. Brazil showed that this would be possible with sugarcane, a crop that produces around 100–120 t/ha of wet biomass, generating approximately 80 t/ha of stalk (where the sugar cane juice is stored), which guarantees an ethanol production of around 6000 l/ha (Goldemberg 2008). However, when we look at sugarcane in detail, we realize that its productive potential is much greater. Due to the inexistence of a technological alternative to 1G ethanol, the improvement of sugarcane took place in the direction of maximizing the production of sucrose, a reserve sugar whose accumulation signals the interruption of photosynthesis (Paul and Driscoll 1997). If instead of soluble sugar, sugarcane accumulates insoluble sugar, such as cellulose and hemicellulose, the biomass productivity can be much greater. This is the principle behind the energy cane (de Abreu et al. 2020), whose first varieties were recently developed and are already on the market (Carvalho-Netto et al. 2014). In these plants, lignin metabolism seems to be significantly different, leading to the formation of a structure capable of sustaining a greater amount of biomass (de Abreu et al. 2020). In addition, the growth strategy is different from that of sugarcane, with early leaf development capable of producing enough sugar to sustain robust root growth. It is in these roots that perhaps the greatest difference between the varieties can be seen. While sugarcane has small and shallow roots, energy cane shows an intense ramification of these structures, which reach great depth and present strong resistance to trampling by machines (Matsuoka et al. 2014). In addition, energy cane does not require frequent replanting, and may keep high productivity for more than a decade in a stable manner, which greatly reduces its production cost. Therefore, with second-generation technologies capable of solubilizing insoluble sugar, energy cane has the potential to double or even triple sugarcane productivity, while also allowing planting in areas with less rainfall and soil fertility. Considering the acceleration of climate change, these new varieties could become a real revolution in the sector (Dos Santos et al. 2016b).

To put this productivity increase in perspective, an article in published 2015 by the Brazilian Development Bank (BNDES) showed that Brazil has around 190 million hectares of pasture, at least half of it with low productivity, but with enormous potential for energy cane. The use of this cane in this area, combined with 2G ethanol technologies, could increase ethanol productivity per hectare from 6800 to 24,800 l (Milanez et al. 2015). It is estimated that the use of 75 million hectares would be enough to supply the planet with a volume of ethanol (1.86 trillion liters) capable of replacing the annual consumption of gasoline (1.3 trillion liters). Although this was a theoretical and strategic exercise, it shows the potential for producing ethanol to meet world needs only considering the Brazilian area. When we look at other nations, several countries also have potential to produce ethanol on a large scale. As an example, a recent study by the WWF showed that Africa, a continent that is expected to have 5 of the 10 most populous countries in the world by 2100 (Ezeh et al. 2020), has around 540 million hectares available for the production of biofuels (Fischer et al. 2019). In other words, instead of the intense movement of people migrating from tropical countries to other continents, with the production of ethanol, in particular 2G-ethanol, we will be able to generate jobs in abundance and a fuel that can be used everywhere, decarbonizing the planet, and improving quality of life.

In addition to sugarcane and energy cane, another plant worthy of note is Agave (Raya et al. 2021). Species of this genus can have extremely high productivity, coming close to the numbers reached by sugarcane, but in semi-arid regions (Nobel 1991). From these plants, Tequila and Mezcal are produced in Mexico (Gschaedler Mathis et al. 2017), and sisal fiber in Brazil, China and several countries in Africa (Townsend and Sette 2016). In Brazil alone, we have a large semi-arid region called "Sertão" (a word derived from big desert, written as "desertão" in Portuguese), which covers almost 90 million hectares and where Agave can achieve great productivity. Interestingly, this plant has not yet been used for fuel production, although many studies present the obviousness of this alternative (Holtum et al. 2011; Nazir et al. 2019). A plausible reason for this oversight is the difficulty in using the reserve sugar of these plants, which is a fructose polymer (Lopez et al. 2003; Ávila-Fernández et al. 2009) and is not readily fermented by ordinary yeasts (unlike sugarcane juice). Thus, processing costs would justify the production of beverages but not fuel. However, the strategy changes completely if the plant can have all its biomass used for the production of 2G-ethanol (Cushman et al. 2015). This is a terrific opportunity, and the Agave value chain must be urgently developed, especially in view of the acceleration of climate change. Under this new scenario, semi-arid regions are expected to be expanded across the entire planet and the development of Agave could be an important anticipation for this scenario. It is, without a doubt, a relevant choice for the planet and one of the "oil-like" sources that can be planted. We need pioneers in this field, and companies like Ausagave (Holtum et al. 2011) in Australia are taking the first steps in this direction. In Brazil, we are bringing new varieties and encouraging companies and entrepreneurs to develop this opportunity.

7.6 Ode to Ethanol

Ethanol is an incredible molecule, bringing all the advantages of an efficient fuel, like gasoline, but being produced from biomass. To better understand the advantages, let's make some comparisons. Let's start with methane, an extremely important molecule that can also be efficiently produced from biomass (in this case, biomethane). This molecule has only one carbon and four hydrogens, an excellent energy ratio, but the disadvantage of being a gas. Therefore, its use requires energy for compression and has a whole logistical difficulty for storage and transportation (Svensson 2013). To solve this problem, one alternative is to produce methanol, a molecule that incorporates one oxygen, thereby converting it into a liquid (Zakaria and Kamarudin 2016). However, the density of this fuel is much lower than that of ethanol (15.8 MJ/l) and it is toxic.

If we continue our reasoning by adding carbons to the carbon chain, the next molecule we get is ethanol. As we know, it is a liquid molecule, with high energy density (21.1 MJ/l), which represents about 65% of that of gasoline (https://www.energy.gov/eere/fuelcells/hydrogen-storage). Furthermore, it is non-toxic, being widely used by humanity as a "soul additive" since the dawn of civilization. As a fuel, the presence of an oxygen in the molecule brings the great advantage of near-complete burning. In the case of hydrocarbons, a considerable number of molecules end up presenting folds and the formation of compact nuclei, which are inaccessible to oxygen. In this way, small "coals" are generated, which are the basis of particulates (Hergueta et al. 2018), today perceived as the biggest public health problem caused by the combustion engine in large cities (Valavanidis et al. 2008). With the combustion of ethanol, this problem is strongly reduced. Furthermore, ethanol is an excellent antiknock, significantly increasing the octane of the fuel (Park et al. 2010). Simply put, when gasoline is compressed, it detonates at a certain pressure. The better the end-product, the greater its resistance to this compression, which is based on top of gasoline, comprised of 8-carbon molecules (octane). To stabilize this characteristic and increase compressive strength, specific molecules are added to gasoline, such as toluene or tetraethyl lead. However, in addition to being expensive, these substances are highly toxic, with great carcinogenic potential (Mehlman 1990). In Brazil, ethanol, with all its qualities, including its lower price (in comparison to additives), plays this role. The benefits of these actions can be easily verified in the city of São Paulo, the fourth most populous city in the world, which meanwhile ranks only 54th in the ranking of most polluted cities. Its air quality is rated as good, with an US AQI (Air Quality Index) of 34 against New Delhi's 165, 107 from

Mexico City and 51 from Los Angeles (https://www.iqair.com/world-air-quality-ranking).

But ethanol is not limited to a "drop in" fuel and can be easily converted into other molecules. For example, with hydrogenation, it is possible to generate ethylene, which is the basis for polyethylene, one of the most used plastics in our daily lives (Zhang and Yu 2013). Similarly, with dehydration, oligomerization, and hydrogenation ethanol can be converted into Jet Fuel (Brooks et al. 2016), perhaps the most desired renewable fuel today. In principle, biomass (including ethanol) can be converted into any other molecule, turning it into a petrochemical-like industry (Galembeck 2018). Certainly, molecules from petroleum, which do not have the hanging oxygen, are more easily convertible. However, the side effect on the atmosphere justifies the increased challenge generated by this oxygen. It is in the realm of Bioeconomy to find the financial adjustments, based on carbon credit policies, a topic that will be dealt with in another chapter.

In summary, ethanol is a very suitable molecule for the future of civilization, capable of replacing oil with great advantages. Consequently, the point of attention is its availability, how to produce it in large volumes, safely and sustainably.

7.7 Ethanol Is Just the Tip of the Iceberg of a Long Value Chain

Brazil has a long tradition in the production and use of ethanol as fuel, a factor directly associated with the production of Sugarcane, which began during the country's colonization. Since 1925, ethanol was mixed with gasoline in small proportions, being produced from molasses which resulted from sugar manufacturing (Leite and Cortez 2007). However, it was in the 70s, with the oil crisis, when its use soared. In a few years, between 1975 and 1985, Brazilian production jumped from less than 1 billion liters to over 10 billion, with several positive side effects (Goldemberg et al. 2004). The first was the development of the Alcohol Car, a vehicle equipped with an engine calibrated for the use of hydrated ethanol, having 8% of water. Over time, this vehicle has evolved in its performance, having been fundamental for the development of Brazilian automotive engineering. In 1985, about 90% of Brazil's car fleet was alcohol powered. However, in the late 1980s, the drop in oil prices and the concentration of the ethanol industry led to the dissatisfaction of the owners of vehicles using ethanol, a fuel that was then more expensive than gasoline (de Moraes et al. 2017). To solve this, the auto industry acted again. In 2003 the first Flexfuel car was launched, a Brazilian innovation that allowed the engine to run with any relative amount of ethanol-gasoline, something that continues to exist only in Brazil (Brito et al. 2019). For a brief time, technology dominated the market, with most vehicles being produced with this type of engine.

A second "side effect" was leftover bagasse. With increased milling due to increased ethanol production, bagasse became a major nuisance for mills, which had no use for it. However, in the 1987s it was understood that it would be possible to generate excess energy and use it in the system (Olivério and Ferreira 2010). This has been done and in 2020 the supply of bioelectricity for the national energy grid by the sugar-alcohol sector reached a volume of 22,604 GWh, with 83% of this total offered in the dry season—between May and November—, a period of reduced water in the reservoirs of hydroelectric plants. As it is low carbon, it is estimated that the generation of bioelectricity from sugarcane in 2020 has additionally avoided the emission of 6.3 million tons of CO_2, a mark that would only be achieved with the cultivation of 44 million native trees throughout 20 years. Despite this impact, currently, only 15% of the energy generated from sugarcane waste is used in the national energy grid, which, instead of using this source, has used fossil fuel thermoelectric plants to meet the demand (De Souza 2020; UNICA 2021).

The third major effect was vinasse, which had been traditionally a tragic source of pollution in rivers (Marinho et al. 2014), when it was realized that it could be used in fertigation, as an excellent quality fertilizer (Prado et al. 2013). However, it has a large load of carbon that is not kept in the soil, being converted into CO_2 without conducting any work. To face this and take advantage of the opportunity represented by this residual biomass (which we could call co-product), Brazilian mills are starting to produce bioelectricity on a large-scale using vinasse and filter cake (Coelho and Goldemberg 2019). In 2021, the Raízen-Geo Biogas Plant began commercial operation, with an installed capacity of 21 MW (Rural 2021), one of the greatest biogas installation in the world.

7.8 Ethanol Around the World

Brazil's success in developing integrated biorefineries has not gone unnoticed by the world, which has shown waves of interest in ethanol. Remarkably, the movement did not develop with intensity in countries with similar climatic conditions, capable of producing sugarcane. For example, in South America only Colombia produces ethanol on any scale to blend with gasoline in proportion that range from 4% to 10% (Voegele 2021). In Africa, only Zimbabwe, Malawi, Kenya, Ethiopia and South Africa, with a production bellow 5 million liters (Amigun et al. 2011; Deenanath et al. 2012). In Australia, production reaches only about 450 million liters (Voegele 2020a), while in Asia, only more recently some countries decided to increase production. China was expected to have 18 first-generation ethanol plants in operation by the end of 2020, up from 14 in 2019, with a combined operation capacity of 6.578 billion liters (Voegele 2020b). India, the country where sugarcane come from, has kept its production limited to around 2 billion liters in the last 10 years, a volume less than necessary to comply with its gasoline blending policy (Dey et al. 2021). This situation changed in 2021, with the government's decision to increase the blending rate to 10%, which should encourage its own production and that of other Asian countries.

Surprisingly, the Brazilian initiative was perceived by the United States, a country with a predominantly temperate climate, with a very limited capacity to produce sugarcane, but an enormous capacity to produce corn. Motivated by the possibility of reducing dependence on oil and thereby easing geopolitical tensions, the US started a broad program of corn ethanol production, which in a short time reached more than double the Brazilian production (Chum et al. 2014). However, the process in the USA is quite different from the Brazilian one and from the sugarcane system. Initially, the production of ethanol from corn is not self-sustainable in energy. For the plants to work, it is necessary to import energy, which is done using fossil sources, particularly natural gas. Considering the Renewable Energy Rate (RER), measured by life cycle data such as total renewable energy produced per unit of fossil energy consumed, sugarcane systems showed RER values of 7.0 in 2002 to 9.4 in 2009, while the average US RER ranged from 1.1 to 1.7 from 2000 to 2010 (Chum et al. 2014). These relations were irrelevant at the beginning, when the motivation was only to have cheaper and safer energy sources, but they became somewhat limiting when the underlying issue to solve is environmental. In addition, corn is a food of foremost importance to society and its use for biofuels ended up evoking a huge worldwide concern, which was translated by the "Food X Fuel" dilemma (Pimentel et al. 2009). This concept was strongly propagated from the 1990's onwards, casting doubts even on the sustainability of Brazilian ethanol, whose production began to be accused of taking agriculture and livestock towards the Amazon. In other words, the production of ethanol (currently, about 10 million ha of sugarcane plantations that represent 1% of the national territory) would be, indirectly, taking part in the devastation of the tropical forest (Solomon 2010). This discussion, which has been conducted in a very emotional way, was one of the main reasons for the lack of interest in 1G ethanol, mainly in Europe, which has an enormous influence on world public opinion. The second reason, and perhaps the most important, would be the concentration of about 85% of ethanol production in just two countries, Brazil and the USA (RFA 2021), which would make the distribution problem even worse than with oil. However, the production of ethanol in the USA, despite the problems pointed out, brought light to the world to think about producing ethanol through different strategies, as it was from corn processing that emerged the industrial inspiration for the use of cellulose.

7.9 The Second-Generation Ethanol Race

A "boom" of interest in cellulosic ethanol appeared around 2007 caused, again, by the increase in the oil price and by the perception that, "this time", prices would have reached a new equilibrium level, above the magic number of U$ 100 (Macrotrends 2021). This perspective has led to real hysteria in the sector, with a large number of companies announcing the development of miraculous technologies and their determination to invest in 2G-ethanol production (Waltz 2008). There were big companies, such as DuPont, DSM, Poet, BP and Abengoa, closely followed by startups and technology companies that presented ready-made solutions for the systems to work. These solutions were mainly concentrated on enzymes and second-generation yeasts, areas where technological bottlenecks were believed to be. Effectively, the rapid hydrolysis of lignocellulosic material is not an easy task, nor is it simple to "convince" yeasts to make ethanol from pentoses. In both cases there is a need to subvert the logic of nature, which always demands intelligence and energy. Efforts paid off and efficient systems were developed, which gave rise to a race to find out who would be the first to produce 2G ethanol industrially. After all, this would mean opening a real pot of gold for the licensing of the technology by all countries of the world. Investments were frantic, as the payoff seemed too great.

However, although the most difficult challenges were overcome, Pre-treatment was completely underestimated. The main reason for this was the fact that the industry has a long and wide experience with biomass processing of the pulp and paper industry (Lamberg et al. 2012). Unlike enzymes and yeasts, largely developed by startups and technology companies specialized in other areas (for example, Novozymes and DSM), in the case of biomass, there were large technology companies, willing to enter this new sector and presenting all their training in the form of contracts with performance guarantees (Phillips et al. 2013). However, despite the self-confidence and good intentions, there was a new factor, something that seemed like a small detail, but that had a devastating effect: the nature of the biomass.

Below, we'll briefly address the "rocket technology", the development of enzymes and yeasts that encouraged the somewhat premature departure of the 2G journey. However, it is from these pendular and painful movements that civilization advances.

7.10 Need to Hit Before Cutting

The fragmentation of "food", which can be understood as an external source of concentrated energy, was one of the main evolutionary drives that allowed for the diversity of life as we know it today. In the beginning, there were only microorganisms, capable of synthesizing everything that was needed just by absorbing basic nutrients. As populations increased, it has become extremely attractive to take advantage of the energy that some other would have already concentrated (the "food"). Thus begins the choice of processes, physical and enzymatic, that lead to the degradation of biomass of others for the reconstruction of the biomass from himself. This is basically the principle of any ecological interaction including biofuels production. Then, if we consider the evolution of lignocellulosic biomass as "food", it tried to protect itself from this fate through an intricate web of polymers that today constitute cellulose, hemicellulose and lignin. Typically, most of the energy is in cellulose, which makes up about 40 to 50% of the different biomasses (Chen 2014). Considering that it was reasonably protected, the lignocellulose was able to host an incredible amount of glucose in its structure. Simply put, getting your hands on this potentially sweet energy source, through the controlled digestion of cellulose, is the "holy grail" of 2G ethanol.

The path for this began during World War II, in the Japan campaign, when American soldiers saw the degradation of their uniforms and tents by the action of the *Trichoderma reesei* fungus (Paloheimo et al. 2016). This problem was quickly converted into an opportunity, assuming that this phenomenon was due to the production of cellulolytic enzymes by these fungi. With this, a large body of research was developed, several enzymes were found, isolated, and characterized, as well as many other promising species of microorganisms were, and are, being developed as platforms for the production of enzymes. In a simple way, three main classes of cellulases were initially recognized (for a review, please see Bhati and Shreya 2021): (i) Exocellulases, or cellobiohydrolases, which cleave polymers from the ends, starting from the reducing end (type I) or non-reducing end (type II), with cleavages that normally lead to the release of cellobiose disaccharide; (ii) Endocellulases, which cleave amorphous cellulose internally, generating tips for the work of exocellulases; and (iii) Beta-glycosidases, which cleave cellobiose to glucose. This enzymatic arrangement was quite efficient in ruining tents and uniforms, produced from purified cellulose. However, in biomass the situation is much more complex. Cellulose bundles organize themselves into a crystalline structure that prevents water penetration and is recalcitrant to access by enzymes. As a result, even after pre-treatment, where the aim is to expose cellulose, the efficiency of the enzymes was insufficient to turn second-generation ethanol into a good business. There was an unacceptable amount of unconverted material, which would destroy value and make any attractive return on investment unachievable.

This scenario began to change with the discovery and characterization of oxidative cellulases, whose activity in cellulose deconstruction was suggestively called auxiliary (Vaaje-Kolstad et al. 2010). Enzymes now called AA9 and AA10 (Auxiliary Activity) play a fundamental role in the loosening of cellulosic fibers, allowing a significant increase in the efficiency of "true cellulases". Its mechanism of action is not based on a traditional hydrolysis, but on an oxidation, which leads these enzymes to be also known as LPMO, Lytic Polysaccharide Monooxygenase (Rani Singhania et al. 2021).

AA9 and AA10 do something particularly unique when they interact with the crystalline part of cellulose: by reducing the copper present in its active center by an enzyme (e.g., cellobiose dehydrogenase—CDH) or a small reducing molecule (e.g., ascorbic acid), it extracts a hydrogen from carbons from the substrate, simultaneously adding an oxygen after successive electronic rearrangements. This process leads to the formation of an aldonic acid, that is, a carbonic acid coming from the oxidation of an aldehyde group in the sugar, which destabilizes the glycosidic reaction and generates points and amorphous cellulose for the action of "true cellulases" (Corrêa et al. 2016). This mechanism significantly increases the efficiency of the enzyme cocktail, and it would not be an exaggeration to state that it was a "game changer". In practice, the hydrolysis of pretreated biomass using cocktails with and without these enzymes leads to efficiency changes in the order of 20–30%, which is probably the range that makes the process feasible (data not published). It is especially important to realize that we are dealing with the production of commodities, that is, products that are not very differentiated and whose economic attractiveness depends almost exclusively on their production efficiency. Any hiccup in the process, any increase in cost, can simply make the activity unfeasible. On the other hand, any small gain, multiplied by millions of liters to be produced, can make the process extremely attractive. This was the case with LPMOs.

It is interesting to note that the mechanism of action of these enzymes involves the transfer of electrons from one molecule to another. In this scenario, several compounds can be donors, such as lignin derivatives, different organic acids or even substances that inhibit fermentation, such as furfurals, which here can become substrate for a more efficient enzymatic hydrolysis (Kracher et al. 2016). Another point is the aeration of hydrolysis reactors, which was not considered in the original design of most 2G plants. Like what happens in fermentation, it is possible that the addition of oxygen helps the action of LPMOs and improves the efficiency of hydrolysis. Thus, research on this class of enzyme should bring many advances to 2G ethanol processes, integrating hydrolysis and fermentation.

Finally, a crucial point for the hydrolysis step is the location of enzyme production. Currently, given the frustration of most 2G ethanol production initiatives, the enzyme market has not yet expanded. In practice, there is only one active manufacturer, Novozymes, which produces the cellulolytic cocktails in its European industrial unit. This initiative started in 2000, with the first enzyme cocktail launched in 2010, initiating the Cellic line (Cellic CTec2/HTec2), which was later enhanced with the addition of LPMOs (Cellic CTec3/HTec3) (Paloheimo et al. 2016). In principle, the production of highly efficient industrial enzymatic cocktails will require a blend of enzymes secreted by different strains or species of microorganisms. Today, with genetic engineering techniques, it is much easier and simpler to modify the strains to complement the missing enzymes in them (Bhati and Shreya 2021). However, if enzymes are sold for use elsewhere, these cocktails must be optimized, purified and stabilized prior to shipment. If this shipment is made to countries that have restrictions with genetically modified organisms, there will be more costs associated with the regulation of the product, which will possibly have to receive additional treatments to avoid the presence of transgenic remains or potentially transforming material. In summary, all these steps, which are extraordinary technological achievements, represent a significant and inconvenient added cost that can collide with the barrier of economic attractiveness. Thus, we are possibly still living in a moment of transition, in which the lack of consolidation in the sector is limiting the options and generating the production of these enzymes in conditions that increase their cost to the limit of economic viability. A magic number for enzyme cost, which would be the biggest wish of the industry, would be the up to US$ 100.00 per ton of ethanol, which still doesn't seem to be feasible. Today, Raízen and Granbio purchase enzymes from Novozymes, which have excellent quality. However, Clariant, in the Sunliquid process, makes the local production of enzymes, directly integrated into the process (Rarbach and Söltl 2013).

On site production of enzymes is expected to be the future for 2G ethanol and Clariant's practical results are eagerly awaited. In principle, it makes perfect sense.

Initially, in the plants there is the substrate necessary for the growth of fungi and the secretion of enzymes. Furthermore, since the proportion of different enzymes must be adapted to the specific biomass, there is nothing better than feeding the fungi with this biomass, which will induce them to produce the necessary arrangement. However, no strain of natural fungus is currently capable of secreting the necessary enzymes in the correct proportion (Bhati and Shreya 2021). Therefore, this is a field that requires great attention and investment for the development of genetically modified microorganisms, which are robust and efficient, capable of growing and producing within the plant environment, just as yeasts do during fermentation. It is a challenge and a fantastic opportunity, which will possibly be the object of the development of new biotech startups.

7.11 Ethanol Machines

Even before dogs, yeasts were perhaps the first organisms domesticated by man, although they didn't know it. It is not difficult to imagine the interesting human reactions to eating a fermented fruit or extract, actions that brought them closer to the gods. When we think about the production of 1G ethanol from sugarcane, we see that the process is basically a formatting of nature, with interventions mainly of a logistical nature. In the sugarcane plant, sucrose already diluted in water is stored in small bags (the storage parenchyma cells), from where it is easily extracted by the physical (and brutal) process of grinding. Once placed in tanks, the sugarcane juice, which has the soluble sugar sucrose (a dimer of glucose and fructose) as its main component, is accessed by microorganisms present in the plant itself or in the environment (de Souza et al. 2016), eager to use this abundant and easy energy. If nothing is intentionally added to this wort, there will be a microbiological war in the middle, which will normally be won by natural yeasts, able to produce ethanol. Of course, the purpose of yeast is not to fuel cars. The strategy behind this action is to fight bacteria based on the microbicidal properties of ethanol. Therefore, the greater the amount of yeast, the more efficient the conversion will be, and the less sugar will be lost in this war. Thus, one of the most important industrial interventions for sugarcane ethanol is the addition of yeast to the must to increase efficiency and speed of conversion.

In this industrial process, yeasts are the true production reactors, with all their metabolism prepared for this activity (for details, please see Zamora 2009). Initially, yeasts secrete invertase into the periplasmic space of their cells, where sucrose is cleaved into glucose and fructose and quickly internalized from the activity of sugar transporters (Özcan and Johnston 1999). Once inside the cells, these sugars follow the glycolysis route to pyruvate—a 3-carbon molecule already prepared for decarboxylation—which, in principle, could be metabolized by two ways: (i) the respiration pathway, in which the pyruvate is decarboxylated and connected with Coenzyme A, generating Acetyl-CoA and reducing a NAD+ to NADH; this step, performed by the enzyme complex Pyruvate Dehydrogenase, prepares the acetate for the Krebs Cycle and by passing it through the respiratory chain, which generates about 36 ATPs for each hexose molecule; (ii) in the second, much simpler way, pyruvate is simply decarboxylated by the enzyme Pyruvate Decarboxylase (Flikweertt et al. 1996) and converted into acetaldehyde, a compound with two carbons. If we consider the glycolysis reactions like a river flowing to the sea, the decarboxylation of pyruvate would be a great waterfall. It is an exothermic, highly efficient reaction, responsible for the incredible speed of fermentation. From there, a second enzyme, Alcohol Dehydrogenase (ADHI), reduces acetaldehyde to ethanol using hydrogens from an NADH that was charged during earlier stages of glycolysis. This second pathway produces only 2 ATP molecules per hexose, two ethanol molecules and two CO_2 molecules, which recycle in the atmosphere, as they were captured by sugarcane photosynthesis. However, in the presence of high concentrations of glucose, this is the preferred route for yeasts, even if they are in contact with oxygen and can breathe. The name of this phenomenon is the "Crabtree Effect" (Zamora 2009), a brilliant adaptive process that made the yeasts, faced with the "fight or flight" dilemma, to choose to fight the bacteria and produce an efficient chemical weapon capable of containing them. Therefore, it appears that the first goal of the yeast, when consuming the hexoses, is not so much to produce energy (ATP), but to protect itself from bacteria through the production of ethanol. If the fermentation process is not interrupted, after some time a second stage of metabolism starts with the consumption of ethanol. In this case, another Alcohol Dehydrogenase (ADHII) is produced, and this converts the ethanol into acetaldehyde preparing it to enter the airway (GANCEDO 1992). In other words, after the fermentation explosion, yeast does not enter a stationary phase, but a slow growth phase, literally "tasting" the ethanol produced in the earlier phase.

However, when we think about second-generation ethanol, this logic must be adapted. The lignocellulosic hydrolysate, unlike sugarcane juice or starch hydrolysate, holds a significant amount of C5 sugars, normally represented by xylose. In large numbers, about 40–45% of soluble sugars are hexoses, while approximately 25% are pentoses (Canilha et al. 2012). That is, it's too much sugar to simply be ignored. Or, to put it in other words, if pentose cannot be converted to ethanol, there is no possible economic feasibility for 2G ethanol (the commodity dilemma). Thus, the existence of 2G ethanol is directly dependent not only on efficient enzymes, but on yeasts capable of converting xylose into ethanol quickly. As we saw before, alcoholic fermentation, to be cheap, cannot be a sterile process and is the result of microbial warfare. Therefore, any delay leads to colonization of the hydrolysate by bacteria, which in turn produce organic acids that are toxic to yeasts and stop fermentation. It is not enough just to be able to ferment xylose. Yeast needs to rush.

There is in nature a large variety of yeasts capable of consuming xylose, but only a few are reported to have the ability to ferment this sugar (Jeffries 2006). The evolutionary reason is relatively simple. Free xylose usually appears during the degradation of lignocellulosic material in the soil or in the digestive system of herbivores (animals or insects). Thus, being in relatively small quantity and being slowly produced, it does not seem to make much sense to rapidly ferment this material. It is, therefore, a quite different situation from alcoholic fermentation, in which yeasts colonize material that has excess free glucose, such as in a fallen fruit or a decaying sugar cane stalk. Thus, when we look closely at the pentose metabolism (please refer to Horecker 2002; Jeffries 2006 for details), we see the existence of an intricate set of reactions, with a lot of isomerization and production of intermediate compounds for the general metabolism. When starting from glucose, the first step is decarboxylation, which aims to generate a reducing potential from the NADP+ loading. This cofactor is fundamental in the synthesis of membranes and many other molecules in the cell, having as its main mission the metabolic construction. The pentose generated, Ribulose-5P, is then ready to undergo isomerization and degradation, generating in the end glyceraldehyde-3P and fructose-6P, which connect directly to glycolysis and are thereby able to enter the ethanol production pathway. Some yeast species, like *Spathaspora passalidarum* and *Scheffersomyces stipitis*, do this naturally and there is an intense field of research to make them commercial organisms producing 2G ethanol (Nakanishi et al. 2017). Even some bacteria species have been considered for this purpose (Dien et al. 2003). However, the great machine for producing ethanol is effectively *Saccharomyces cerevisiae*. Thus, from the beginning it was clear that yeasts should be taught how to consume pentoses to produce ethanol.

From a metabolic engineering standpoint, the challenge is to make these organisms produce xylulose, a 5-carbon ketosis that yeasts are already able to metabolize. For that there are two paths, a reductive way, or an isomerization way. The reductive pathway was the first to be developed, in pioneering work by Nancy Ho (Purdue University) that used three xylose-metabolizing genes: a xylose reductase, a xylitol dehydrogenase (both from *Pichia stipitis*), and a xylulokinase gene (from *Saccharomyces cerevisiae*) (Ho et al. 1998). Basically, D-xylose is initially reduced to xylitol from xylose reductase, which uses NADPH as a cofactor. In turn, xylitol is oxidized to d-xylulose with which it uses NAD+ as a cofactor. Thereafter, xylulose is phosphorylated by xylulose kinase and enters the pentose phosphate pathway. While brilliant, this strategy has some problems. Initially, it needs reducing potential for the first reaction, a work that is undone in the next reaction. In principle, this appears to be just a cofactor unloading and reloading. However, the cofactors are different: the first reaction uses NADP while the second uses NAD. Therefore, the chance of imbalance and physiological impasse is real. In addition, it needs energy for conversion.

The second way is direct and energy efficient. It was known that some bacteria had the enzyme xylose isomerase which transformed xylose directly into xylulose without the need for energy input or the use of a cofactor (Walfridsson et al. 1996). The first attempts in this direction were made without success. The enzyme simply did not work in the yeast environment or was not efficient. This situation changed in 2003, when the xylose isomerase from the anaerobic cellulolytic fungus *Piromyces* sp. was proved to conduct the reaction satisfactorily (Kuyper et al. 2003). However, just inserting this gene is not enough for yeast to efficiently produce ethanol. A number of other genes in the pathway must have their function optimized, such as the xylulose kinase, which is normally done by the overexpression of these genes (Jeffries 2006). Furthermore, other less obvious systems, such as the cell's iron metabolism (Kwak and Jin 2017), needs to be altered in ways that metabolic engineering is unable to predict. To achieve this, the process of guided evolution has been needed, which is an educated version of trial and error.

As a principle, the transformed cells (modified with the genes that are known as important for the xylose metabolism) are grown in medium having a low concentration of glucose and a higher concentration of xylose. Cells that don't use xylose efficiently will grow rapidly with the use of glucose but will then slow down when this source is exhausted. However, cells with occasional mutations, capable of growing well in xylose, will continue to multiply and increase their relative frequency in the medium. Thus, after about 24–48 h, an aliquot of the culture is transferred to another flask with the same fresh medium, a process that is repeated for several generations. If evolution is possible, the expectation is that after some transfers the flask density will increase significantly, which will mean that xylose consuming cells have evolved. In some cases, the resulting cells can grow, but not ferment xylose, which is an undesirable effect.

This evolution process, with greater or lesser changes, was used by several laboratories in the search for these optimized strains, which were successfully achieved (Kwak and Jin 2017). In our group, we also work on the development of these strains, however we have adopted a slightly different strategy. Instead of using laboratory strains, which are easier to manipulate but less robust, we started to domesticate Brazilian industrial strains, naturally selected over decades in the environment of 1G ethanol plants, and that's where we made the modifications. In an earlier work, considered by the yeast scientific community as one of the seminal articles in the area, we have made the first genome of an industrial strain (SGD 2013). In this work, we noticed the great variability present in diploids and the agility of these strains in suffering (probably induced by stress) alterations in their genomes, mainly from gene conversion (Argueso et al. 2009; Argueso and Pereira 2010). Thus, these strains, unlike those normally used in the laboratory, which had their genomes stabilized over time, seemed to us the ideal organisms to undergo rapid evolution. That's exactly what we did and, surprisingly, we have achieved high-performance strains in a few weeks, identifying the mutations responsible for the phenotypes (Dos Santos et al. 2016a). This strain, named Celere 2 L, is now responsible for the fermentation at Granbio's Bioflex 1 plant, carrying out the 2G fermentation step in about 24 h.

7.12 Ready to Go?

The pulp and paper industry, which began in the early days of the Industrial Revolution, developed mainly based on wood. In plants, this material is the result of a metabolic process that creates a practically inert, rigid arrangement of molecules that can be compared to the wall of a house (for details, see Chen 2014). In this analogy, the building blocks are cellulose molecules, formed by glucose polymers joined by β-1,4 bounds. This kind of connection, which seems to be a small detail (glucose in starch are connected by α-1,4 bounds), put these monomers in an arrangement that favors the formation of a crystalline structure. These "bricks" are then joined by the "cement" of hemicellulose, a branched heteropolymer made up of several diverse types of sugars, including many pentoses such as xylose (usually the most abundant in grasses). Finally, "iron" is represented by lignin, a complex chemical network formed by the polymerization of cyclic alcohols (coumaryl, coniferyl and sinapyl) in branched bonds, which generates great rigidity to the structure. These three components, cellulose, hemicellulose and lignin, which correspond to approximately 45%, 25% and 26%, respectively, are firmly intertwined from various chemical cross-links, produced at the time of their synthesis and secretion by plant cells (McFarlane et al. 2014).

In pulp and paper industry, the aim is to separate the cellulose from the other constituents and the entire process is conducted based on physicochemical principles. Considering the Kraft method (Kleppe 1970), the wood logs are mechanically fragmented into chips and placed in a digester. In this reactor, the fragments are subjected to cooking in the presence of sodium hydroxide and sodium sulfide, which leads to the dissolution of lignin and hemicellulose, and release of cellulose fibers. These fibers then undergo a purification and bleaching treatment, while the lignin, hydrolyzed hemicellulose and added chemicals go to burning and recycling systems. It is important to mention that the process has an energy surplus.

It is seen that the entire process is done, obviously, without any biological concern. The goal here is to obtain the cellulose fibers themselves and not monomerize them into glucose or xylose. However, the pulp and paper industry saw the opportunity in the second generation and understood that it would be able to make the necessary adaptations without much effort. Basically, the use of chemical substances would have to be limited and the reaction conditions controlled to avoid the production of toxic substances for fermentation. To achieve this goal, the main method chosen was hydrothermal cooking, with or without the addition of small concentrations of acids or bases (Phillips et al. 2013). Around this principle, technologies were developed, some of them quite creative, such as Proesa (Michele Rubino 2013), a continuous process that used reactors with different conditions to obtain a fractional extraction of cellulose and hemicellulose without generating fermentation inhibitors (for details, see Bezerra et al. 2020). Basically, the chopped biomass entered a first vertical reactor, where it was treated with water vapor and low pressure for a certain retention time. Under these conditions, most of the hemicellulose was liquefied and would have to be removed by the action of a dewatering screw, which would also have the role of transporting the processed biomass to a second reactor. This second, vertical reactor would run at a higher pressure, with a short retention time, and would be expelled from the system in small explosions (Steam Explosion) caused by the opening of a rotary valve. This energy would make the final loosening of the fibers, which would be transported by pressure to hydrolysis tanks. These tanks would also receive the cooled, liquefied hemicellulose, a stream that would serve to reduce the temperature of the hydrolysis reactor to conditions suitable for the enzymes. It looked perfect.

Alternative systems, but following similar principles, were also developed seeking a separation of currents in the initial stage of the process (Chandel et al. 2019). In some cases, horizontal reactors equipped with conveyor screws. In these reactors, water vapor and diluted acid or base are added, all under a certain pressure. The material outlet can then be coupled to explosion valves and, depending on the intensity of the chemical treatment, be fractionated by centrifugation. In this case, the liquefied hemicellulose would be separated from the loosened cellulose fibers, which go on to enzymatic hydrolysis. It is important to mention that this hydrothermal treatment is normally sufficient to hydrolyze most of the hemicellulose. Therefore, in the processes where this kind of separation takes place, the hemicellulose stream is practically ready to be fermented by 2G yeasts.

When around 2009 technology companies reported that enzymes and yeast were ready, there was a real uproar in the industry (Waltz 2008). Nothing else seemed to be missing, as the pulp industry giants had enough ability and tradition to supply the technology needed to put the 2G ethanol industry on its feet. The final push was given by the behavior of oil price, which surpassed the US$ 150.00 in the euphoria before the 2008 global financial crisis, collapsed with the crisis, but quickly returned to an intense growth, surpassing the US$ 100.00 between 2009 and 2010 and staying around that level until mid-2014 (Macrotrends 2021). This time it didn't feel like a brief rally anymore, but a movement that would have come to stay, promoted mainly by the political movements of the Middle East, which became known as the Arab Spring. It seemed then that $100.00 would be the new level for oil. Thus, many companies and entrepreneurs decided to build their plants based on the experiences obtained in pilot plants, accepting the risk of scale up (Waltz 2008). Examples of this systematic can be seen in Beta Renewables, Poet-DSM, DuPont, Abengoa, Granbio, Raízen, CTC, among others (Fletcher 2014). In all these cases, the fundamental stage of the Demonstration Plants was missing, which were replaced by the desire to be the first producer of second-generation ethanol.

Unfortunately, none of these initiatives worked at first. From the second half of 2014, the price of a barrel of oil began to fall sharply, reaching values well below U$ 50.00 in 2015 (Macrotrends 2021). When that happened, many of the pioneers gave up on the initiatives, including big companies like DuPont and DSM. As a result, the impression remained on the market that it would not yet be possible to produce 2G ethanol, that there would be no mature technology for this and that, perhaps, this was an impossible idea (Marques 2018). Thus, a perfect storm was formed from the following tripod: the technology did not work at first, the part that did not work was exactly the one that is capital intensive (represented by the pre-treatment and movement of biomass), and the price of oil plummeted, making the main thrust for second-generation development no longer exist.

7.13 Why Didn't the Second-Generation Work at First?

But after all, why didn't it work? The reason is bewilderingly simple: the structure of biomass. Amazingly, the companies totally underestimated the fact that leaf and bagasse biomass was completely different from wood chip biomass, the material that was widely used in pilot plant trials and that is the basic raw material for the pulp and paper industry.

To understand this point, we need to consider factors related to the evolution of plants (in a botanic sense) and their structures. When we look at a tree, it is clear that the trunk is a support structure, which has evolved to resist water. It is dry and hydrophobic, with a high concentration of crystalline cellulose and lignin. When we look at a cane leaf or stalk, we see the reverse situation. Its structure is irregular, with many vesicles for liquid storage and cellulose bundles largely tied together by hemicellulose and other branched sugar polymers. In this case, the lignin is distributed in such a way as to support mainly the conducting vessels, but in a significantly different arrangement from that found in wood, which is explained by the smaller size of these plants (Anderson and Akin 2008). That is, there is no need to support structures as big and heavy as a tree trunk and it is essential that they keep water. Furthermore, these herbaceous plants developed during evolution a series of mechanisms to avoid, or at least mitigate, herbivory, to make them less attractive to insects and animals. Among these mechanisms is the deposition of silica (Currie and Perry 2007), which can constitute 2–6% of the dry weight of the leaves of grasses and can be easily noticed by the cuts we receive when we enter a sugarcane field. The edges of the cane leaves hurt as if they were sharp blades.

Thus, these differences in the property of these types of biomasses, which did not play a relevant role in pilot plants—in which the processes were conducted separately and with little biomass movement—, proved to be decisive in industrial installations. It was clear that some devices designed for wood could not perform the same job with non-woody biomass. Some points stood for significant challenges, common to practically all initiatives, although there is not yet available literature with individualized reports. In view of this, let's consider just a few issues that are widely known to industry pioneers.

7.14 The Long Journey of Biomass

To start up the process, biomass must be transported on conveyors to feed the digesters. When this is done with wood chips, this transport is more homogeneous, with a relatively regular density, which facilitates the entry of material into the pressurized reactor. There is a technological challenge in feeding a pressure reactor, which is like trying to put more food into a pressure cooker that is already cooking. In this case, if we open the pan abruptly, the pressure will naturally be relieved by throwing the food out. In the pulp industry, the solution to this problem came through a brilliant mechanism for the formation of a continuous stopper with the biomass itself, which manages to keep pressure while the material enters the reactor (Plug Screw Feeder Mechanism). In simple terms, the biomass is fed with water vapor into a horizontal reactor, inside which a screw turns. The biomass enters and is compressed by the screw, forming a plug (a kind of stopper). This plug presses a plunger at the end, which closes the reactor inlet. At this end, the pressure of the screw causes the plug to displace the plunger when its pressure is greater than the closing force, which causes the part of the plug that reached this point to disintegrate and enter the reactor, feeding the system (for details, see Dai et al. 2012). This process is commonly used in the cellulose industry and the woody material can generate highly consistent plugs, capable of sealing the system and avoiding pressure return. However, the same does not occur with non-woody biomass. In this case, air pockets can be formed during the compression of the material, which often results in biomass return and feed stoppage. Thus, it was realized belatedly, already during the operation of the plants, that the biomass feeding system could not simply be copied from the traditional industry but needed to be designed specifically for the characteristics of straw and bagasse.

Once fed, the behavior of biomass in the cooking reactor was also different. When wood is chopped into chips, cooked at elevated temperature and pressure, and exploded, what occurs is the fragmentation of the material, with the loosening of the cellulose fibers and the liquefaction of the hemicellulose. In this state, the liquid fraction can be easily drained by dewatering screws, allowing the solid fraction to go ahead to enzymatic hydrolysis. However, the behavior of bagasse and straw is quite different. With high pressure heating, the material hydrates strongly, gaining a porridge-like consistency. The water penetrates like a sponge, and it is not possible to make the later dewatering with screws. In this case, attempts to transport it by this means lead to the material sticking to the screw and causing it to lock. Hence, the system cannot operate with continuity, making the process unfeasible (Marques 2018).

Finally, silica took its toll. A phenomenon that became clear, for all industrial plants, was the rapid wear of the screws and valves, which seemed to be undergoing an erosion process. Even when manufactured with stainless steel, with a quality and thickness superior to that normally used in the industry, the equipment had an exceptionally low durability. This was a particularly challenging occurrence in the pipelines that received material under pressure after the steam explosion. In this case, the silica acted like a projectile, traveling at high speeds, and damaging the inner wall of the pipe. Something similar happens in the screw of horizontal reactors, which suffer strong wear on the edges of the fins and in a brief time lose the capacity to transport the biomass. When this problem appeared, the first impression was that the silica would come from the field, because of the drag of dirt during the harvesting of the material. Thus, a series of biomass washing protocols were integrated into the systems, but without effectiveness. Of course, biomass brings a lot of soil and sand on its surface, but the main problem is the silica that is part of the plant's constitution (see Saini et al. 2014).

Thus, these factors caused large industrial plants, theoretically ready to go, to suffer great frustration in their plans, not being able to reach the projected productions or even to run continuously. As a result, most of the initiatives were interrupted, with great losses for its shareholders and leading to disbelief in relation to the technology's viability (Marques 2018). However, after the first shock, this scenario could be investigated more calmly, perceiving in it the ordinary pattern of technological disruptions. The desire to be first causes pioneers to underestimate the difficulties in view of the potential prize to be achieved. Thus, the entrepreneurial rush in the search for this "Holy Grail" meant that the characteristics of the raw material were not considered, which was vital for the functioning of the pre-treatment and movement of biomass, precisely the stages that are capital intensive. Therefore, the correct conclusion is not that the technology is unfeasible, but only that the pre-treatment system ordinarily used in the pulp and paper industry is not suitable for the processing of bagasse and straw, raw materials that are fundamental for the economic return of the 2G process. In some mills, such as the one in Crescentino, Italy, it was proved that the machinery would work well with wood chips. However, the final cost of ethanol would make the initiative completely unfeasible in any other country in the world. Consequently, other solutions would have to be looked for.

7.15 The Chronicle of Second-Generation Pioneers

Although the difficulties of processing cheap biomass have generated a real shock wave in the sector, three pioneers have resisted and history should recognize them as the heroes of 2G, those who found the problems and had the courage to double the bet. Three companies are noteworthy: (i) Granbio, a startup that developed the first 2G ethanol plant in the southern hemisphere, located in the state of Alagoas, northeastern Brazil (http://www.granbio.com.br/); (ii) Raízen, a company resulting from the Joint Venture between the giants Cosan and Shell, which are respectively the largest producers of ethanol from sugarcane and oil in the world; Raízen installed its 2G ethanol plant attached to the Costa Pinto 1G ethanol plant, in the municipality of Piracicaba, São Paulo (https://www.raizen.com.br/); and (iii) Clariant, with the Sunliquid process (Rarbach and Söltl 2013), which has a demonstration plant in Straubing—Munich, Germany, and announced its first commercial plant in the municipality of Podari, Romania. Let's talk a little about each of them.

Granbio was founded in 2011 within an idea of integrating raw materials to 2G ethanol, with the understanding that the success of 2G ethanol would depend on cheap biomass, which is Brazil's great differential (Dos Santos et al. 2016b). To achieve this, the company started the development of a new biomass, energy cane, with productivity far superior to that of sugar cane and containing much larger amounts of fiber (Carvalho-Netto et al. 2014). In addition, it started a pioneering operation to collect sugarcane straw, having been responsible, in 2014–2015, for the formation of the largest straw deposit in the world (with around 250.000 t). Its original process was licensed from Beta Renewables, a spin-off from the Mossi Guisolfi group, which had developed the Proesa technology (Rubino 2013). This process was based on two stages of hydrothermal reaction, without the addition of chemicals, with communication between reactors made by a dewatering screw. After cooking, the biomass is subjected to a steam explosion that transported the material to the hydrolysis process in two-stage reactors, followed by batch fermentation. The enzyme solution was supplied by Novozymes and the yeasts by DSM. For the reasons described above, the system did not perform due to difficulties in the pre-treatment and movement of biomass. Hydrolysis and fermentation worked successfully. To solve this, Granbio went into its own development, with the acquisition of the USA's company American Process (Kennedy 2019), which included the possession of a pilot plant located in the city of Thomaston, Georgia. From this initiative, several changes were made to the process (see Bezerra et al. 2020), which eliminated the pre-treatment in two stages, solved the problems related to material wear and biomass feeding, developed systems to move the biomass by pressure, not screws, and added refiners for homogenization of pre-treated biomass. In this way, it was possible to stabilize the system and be able to run continuously and reliably. The enzymes used are from Novozymes and the yeast, Celere 2, was developed by the company itself based on the genotype of Brazilian industrial yeasts (Dos Santos et al. 2016a). It is a high-performance strain, capable of fermenting the must in about 24 h. It is important to mention that the process is completely independent, with no interaction with first generation plants for the fermentation of sugars. The must is the result of the integral treatment of the biomass, without current separation. Its consistency is like that of a dense, dark porridge, resulting from the direct hydrolysis of the pre-treated material. Yeasts conduct fermentation in this medium without any clear difficulty and keeping bacterial contamination under control. After fermentation, ethanol is distilled, the vinasse is filtered and the cake, holding the remains of the process and rich in lignin, is used to generate electricity in a fluidized bed boiler. Boiler sharing is the only point of cooperation between Bioflex and the 1G Caeté plant, together with which the 2G unit was installed. In other words, Granbio's process produces 2G ethanol directly from biomass, allowing direct tracking between the biomass fed and the ethanol produced. Bioflex's technology has now been proven and Granbio has signed a partnership with NextChem, Maire Tecnimont's subsidiary for energy transition, for licensing the technology worldwide (Maire Tecnimont 2020).

In the case of Raízen, the company works with another system, intensively taking advantage of the synergy with a 1G plant, including the use of bagasse. The original technology was developed by Iogen (Tolan 2002), undergoing updates and adaptations throughout the industrial operation, which pointed out the bottlenecks and difficulties not identified during the piloting of the process, which was carried out in Ottawa, Canada. Basically, the system uses a continuous horizontal reactor, fed with diluted sulfuric acid and with the biomass being moved from a screw conveyor. After the residence time, the liquid fraction is separated from the solid by centrifugation, going to a C5 fermentation unit, which is mixed with sugarcane molasses using technology developed in partnership with Novozymes. Molasses have to has the role of adding nutrient and facilitating C5 fermentation. The solid part, consisting mainly of cellulose and lignin, goes to enzymatic hydrolysis, with enzymes also from Novozymes. The C6 thus released is filtered and conveyed to the plant's fermentation vats, where it is co-processed with the 1G fermentation wort. The filter cake, rich in lignin, is then used in boilers to produce electricity and steam. Thus, the strong interaction between the 1G and 2G plants can be seen, with the volume of 2G ethanol produced being calculated gravimetrically from the biomass fed into the system and subtracting the volume of 1G ethanol produced. Despite the first problems, common to all the pioneers, Raízen managed to overcome the bottlenecks and today exports its 2G ethanol to Europe and the USA, markets that attribute significant prizes to this product due to its exceptionally low carbon intensity. About 80 million liters have been exported since the start of the operation (Raízen 2021) and, most importantly, the company announced the construction of its second 2G ethanol plant (Bossle 2021) and raised U$ 1,15 billion, in the biggest Brazil IPO of 2021, to expand capacity (Kennedy 2021).

Finally, there is Clariant, with the Sun Liquid process (Rarbach and Söltl 2013). In this case, we have a hydrothermal treatment without chemical addition, biomass movement with limited use of screws and energy self-sufficiency. In this technology, there are two points that differentiate it from others (Mitchell 2017). Initially, the enzymes are produced locally using the pre-treated biomass as substrate. This system brings a great advantage in this aspect, as it avoids all the logistic inefficiency associated with enzyme importation. Furthermore, as the enzymes are produced from the same biomass that will be hydrolyzed, there is a greater chance of secretion of a specialized cocktail for that biomass. The amount of material used for this purpose should be around 10% of the biomass fed, which probably offsets the cost of external purchase of enzymes. A second point of distinction is the separation of ethanol, which was originally supposed to be done by adsorbing the molecule on special "sponges", avoiding the energy waste of distillation. Currently the process seems to combine adsorption with distillation. This system, with the integration of processes, was demonstrated in a pilot plant with a nominal capacity of 1000 t/year (Mitchell 2017). The first commercial plant, with a capacity of 50,000 tons/year, is scheduled to start operating at end 2021 in Podari, Romania (Clariant 2021).

In summary, 2G ethanol is back. Now, it is no longer in its infancy, but after the turnaround of a painful adolescence. That doesn't mean there won't be more problems. Only that the naivety of the early days no longer exists, and the teams are prepared for the new challenges. One lesson learned is that oil price fluctuations are not a good adviser. The main driver, which really makes sense, is climate change. Without it, there would be no need for so much effort to make ethanol from cellulose. However, with the current scenario, 2G ethanol has become urgent. After all, the pioneers were right, and their efforts will pay off.

7.16 Conclusions and Future Perspectives

The second-generation ethanol process is more alive than ever. The first initiatives underestimated the difficulties inherent to short-cycle biomass, such as bagasse and straw, which showed the need for customized solutions for its pre-treatment and industrial logistics. This phase, which left profound scars, is being overcome and pioneers have already announced consolidation of their technologies, construction of second unities and inauguration of new plants. In parallel to the industrial phase, there is a great concentration of efforts for the development of more productive and robust biomass, such as energy cane and agave, which prepare the environment for a scenario of climate change and changes in agricultural zoning. With these components it will be possible to produce ethanol in all countries of the world, particularly in tropical zones, where this activity can be a great driver of employment and income, including carbon credits. Finally, electrification is a path of no return, but the metallic battery does not seem to be a practical universal alternative. However, this expensive and inefficient device may be replaced, with great advantage, by the CO_2 batteries represented by biofuels, whose energy can be converted into electricity by the fuel cells. This can be a great model for the future of mobility.

References

Adams S (2020) American Lucifers: the dark history of artificial light, 1750–1865 by Jeremy Zallen (review). J South Hist 86:695–696

Amigun B, Musango JK, Stafford W (2011) Biofuels and sustainability in Africa. Renew Sust Energ Rev 15:1360–1372

Anderson WF, Akin DE (2008) Structural and chemical properties of grass lignocelluloses related to conversion for biofuels. J Ind Microbiol Biotechnol 35:355–366

Argueso JL, Pereira GAG (2010) Perspective: indigenous sugarcane yeast strains as ideal biological platforms for the delivery of next generation biorefining technologies. Int Sugar J 112:86–89

Argueso JL, Carazzolle MF, Mieczkowski PA et al (2009) Genome structure of a Saccharomyces cerevisiae strain widely used in bioethanol production. Genome Res 19:2258–2270

Ávila-Fernández Á, Rendón-Poujol X, Olvera C et al (2009) Enzymatic hydrolysis of fructans in the tequila production process. J Agric Food Chem 57:5578–5585

Bezerra PXO, De Farias Silva CE, Soletti JI, de Carvalho SHV (2020) Cellulosic ethanol from sugarcane straw: a discussion based on industrial experience in the northeast of Brazil. Bioenergy Res 143(14):761–773

Bhati N, Shreya SAK (2021) Cost-effective cellulase production, improvement strategies, and future challenges. J Food Process Eng 44:1–11

Bossle R (2021) Raízen irá investir em nova planta de etanol de segunda geração. Novacana. novacana.com/n/etanol/2-geracao-celulose/raizen-investir-nova-planta-etanol-segunda-geracao-250621

Brito TLF, Islam T, Stettler M et al (2019) Transitions between technological generations of alternative fuel vehicles in Brazil. Energy Policy 134:1–10

Brooks KP, Snowden-Swan LJ, Jones SB et al (2016) Low-carbon aviation fuel through the alcohol to jet pathway. Biofuels Aviation 2016:109–150

Canilha L, Chandel AK, Dos Santos S, Milessi T et al (2012) Bioconversion of sugarcane biomass into ethanol: an overview about composition, pretreatment methods, detoxification of hydrolysates, enzymatic saccharification, and ethanol fermentation. J Biomed Biotechnol 2012:989572

Carvalho-Netto OV, Bressiani JA, Soriano HL et al (2014) The potential of the energy cane as the main biomass crop for the cellulosic industry. Chem Biol Technol Agric 11:1–8

Chandel AK, Albarelli JQ, Santos DT et al (2019) Comparative analysis of key technologies for cellulosic ethanol production from Brazilian sugarcane bagasse at a commercial scale. Biofuels Bioprod Biorefin 13:994–1014

Chen H (2014) Chemical composition and structure of natural lignocellulose. In: Biotechnology of lignocellulose. Springer, Dordrecht, pp 25–71

Chroneos A, Yildiz B, Tarancón A et al (2011) Oxygen diffusion in solid oxide fuel cell cathode and electrolyte materials: mechanistic insights from atomistic simulations. Energy Environ Sci 4:2774–2789

Chum HL, Warner E, Seabra JEA, Macedo IC (2014) A comparison of commercial ethanol production systems from Brazilian sugarcane and US corn. Biofuels Bioprod Biorefin 8:205–223

Claes DH (2018) The politics of oil-producer cooperation. Routledge, New York. https://doi.org/10.4324/9780429495977

Clariant (2021) Sunliquid® cellulosic ethanol plant in Romania. In: Clariant.com. https://www.clariant.com/pt/Corporate/Events/2021/10/Global-Media-Call-sunliquid-Cellulosic-Ethanol-Plant

Cleal CJ, Thomas BA (2005) Palaeozoic tropical rainforests and their effect on global climates: is the past the key to the present? Geobiology 3:13–31

Coelho ST, Goldemberg J (2019) Sustainability and environmental impacts of sugarcane biofuels. In: Sugarcane biofuels. Springer, Dordrecht, pp 409–444

Cooke P (2021) The lithium wars: from Kokkola to the Congo for the 500 mile battery. Sustainability 13(8):4215

Corrêa TLR, dos Santos LV, Pereira GAG (2016) AA9 and AA10: from enigmatic to essential enzymes. Appl Microbiol Biotechnol 100:9–16

Currie HA, Perry CC (2007) Silica in plants: biological, biochemical and chemical studies. Ann Bot 100:1383–1389

Cushman JC, Davis SC, Yang X, Borland AM (2015) Development and use of bioenergy feedstocks for semi-arid and arid lands. J Exp Bot 66:4177–4193

Dai J, Cui H, Grace JR (2012) Biomass feeding for thermochemical reactors. Prog Energy Combust Sci 38:716–736

de Abreu LGF, Grassi MCB, de Carvalho LM et al (2020) Energy cane vs sugarcane: watching the race in plant development. Ind Crop Prod 156:112868

de Moraes MAFD, Rodrigues L, Kaplan S (2017) The sugarcane industry and the use of fuel ethanol in Brazil: history, challenges, and opportunities. In: Handbook of bioenergy economics and policy, vol II. Springer, Cham, pp 39–63

De Souza ZJ (2020) Bioelectricity of sugarcane: a case study from Brazil and perspectives. In: Sugarcane biorefinery, technology and perspectives. Elsevier, New York, pp 255–279

de Souza RSC, Okura VK, Armanhi JSL et al (2016) Unlocking the bacterial and fungal communities assemblages of sugarcane microbiome. Sci Rep 61(6):1–15

Deenanath ED, Iyuke S, Rumbold K (2012) The bioethanol industry in sub-Saharan Africa: history, challenges, and prospects. J Biomed Biotechnol 2012:416491

Dey B, Roy B, Datta S, Singh KG (2021) Comprehensive overview and proposal of strategies for the ethanol sector in India. Biomass Convers Biorefinery. https://doi.org/10.1007/s13399-021-01546-2

Dien BS, Cotta MA, Jeffries TW (2003) Bacteria engineered for fuel ethanol production: current status. Appl Microbiol Biotechnol 63:258–266

Dos Santos LV, Carazzolle MF, Nagamatsu ST et al (2016a) Unraveling the genetic basis of xylose consumption in engineered Saccharomyces cerevisiae strains. Sci Rep 6:38676

Dos Santos LV, De Barros Grassi MC, Gallardo JCM et al (2016b) Second-generation ethanol: the need is becoming a reality. Ind Biotechnol 12:40–57
Ennis CJ, Evans AG, Islam M et al (2012) Biochar: carbon sequestration, land remediation, and impacts on soil. Microbiology 42:2311–2364
Ezeh A, Kissling F, Singer P (2020) Why sub-Saharan Africa might exceed its projected population size by 2100. Lancet 396:1131–1133
Fasihi M, Efimova O, Breyer C (2019) Techno-economic assessment of CO_2 direct air capture plants. J Clean Prod 224:957–980
Fischer G, Tramberend S, van Velthuizen H, et al (2019) Sustainable aviation biofuel potential in sub-Saharan Africa. A systems analysis investigation into the current and future potential for biofuel feedstock production. World Wildlife Fund for Nature – South Africa
Fletcher K (2014) POET-DSM, DuPont, Abengoa begin commissioning cellulosic plants. Biomass Mag. http://biomassmagazine.com/articles/10511/poet-dsm-dupont-abengoa-begin-commissioning-cellulosic-plants
Flikweertt MT, Van Der Zandens L, Janssent WM, Steensma HY, Van Dijken JP, Pronk JT et al (1996) Pyruvate decarboxylase: an indispensable enzyme for growth of Saccharomyces cerevisiae on glucose. Yeast 12:247–257
Friedl M, Bach C, Bollien A et al (2021) Effects of European CO_2-regulations for vehicles on the European energy system. European Research Institute for Gas and Energy Innovation, Brussels
Galembeck F (2018) Synergy in food, energy and advanced materials production from biomass. Pure Appl Chem 90:109–119
Gancedo JM (1992) Carbon catabolite repression in yeast. Eur J Biochem 206:297–313
Goldemberg J (2008) The Brazilian biofuels industry. Biotechnol Biofuels 1:6. https://doi.org/10.1186/1754-6834-1-6
Goldemberg J, Coelho ST, Nastari PM, Lucon O (2004) Ethanol learning curve - the Brazilian experience. Biomass Bioenergy 26:301–304
Grassi MCB, Pereira GAG (2019) Energy-cane and RenovaBio: Brazilian vectors to boost the development of biofuels. Ind Crop Prod 129:201–205
Gschaedler Mathis AC, Acevedo F, Aroca G (2017) Tequila and Pisco. In: Current developments in biotechnology and bioengineering: food and beverages industry. Elsevier, New York, pp 469–486
Hardman S, Tal G (2018) Who are the early adopters of fuel cell vehicles? Int J Hydrog Energy 43: 17857–17866
Hergueta C, Tsolakis A, Herreros JM et al (2018) Impact of bio-alcohol fuels combustion on particulate matter morphology from efficient gasoline direct injection engines. Appl Energy 230:794–802
Ho NWY, Chen Z, Brainard AP (1998) Genetically engineered Saccharomyces yeast capable of effective cofermentation of glucose and xylose. Appl Environ Microbiol 64:1852–1859
Holtum JAM, Chambers D, Morgan T, Tan DKY (2011) Agave as a biofuel feedstock in Australia. GCB Bioenergy 3:58–67
Horecker BL (2002) The pentose phosphate pathway. J Biol Chem 277:47965–47971
İnci M, Büyük M, Demir MH, İlbey G (2021) A review and research on fuel cell electric vehicles: topologies, power electronic converters, energy management methods, technical challenges, marketing and future aspects. Renew Sust Energ Rev 137:110648
Jeffries TW (2006) Engineering yeasts for xylose metabolism. Curr Opin Biotechnol 17:320–326
Kennedy HT (2019) Brazil's GranBio acquires USA's American process. Biofuels Digest. https://www.biofuelsdigest.com/bdigest/2019/03/17/brazils-granbio-acquires-usas-american-process/?utm_campaign=shareaholic&utm_medium=whatsapp&utm_source=im
Kennedy HT (2021) Raízen raises $1.15B in biggest Brazil IPO of the year. Biofuels Digest. https://www.biofuelsdigest.com/bdigest/2021/08/15/raizen-raises-1-15b-in-biggest-brazil-ipo-of-the-year/
Kheshgi HS, Prince RC (2005) Sequestration of fermentation CO_2 from ethanol production. Energy 30:1865–1871

Klepp S (2017) Climate change and migration. Oxford Res Encycl Clim Sci. https://doi.org/10.1093/ACREFORE/9780190228620.013.42
Kleppe P (1970) Kraft pulping. Tappi 53:35–47
Kracher D, Scheiblbrandner S, Felice AKG et al (2016) Extracellular electron transfer systems fuel cellulose oxidative degradation. Science 352:1098–1101
Kuyper M, Harhangi HR, Stave AK et al (2003) High-level functional expression of a fungal xylose isomerase: the key to efficient ethanolic fermentation of xylose by Saccharomyces cerevisiae? FEMS Yeast Res 4:69–78
Kwak S, Jin YS (2017) Production of fuels and chemicals from xylose by engineered Saccharomyces cerevisiae: a review and perspective. Microb Cell Factories 16:82. https://doi.org/10.1186/s12934-017-0694-9
Lamberg J-A, Ojala J, Peltoniemi M, Särkkä T (2012) Research on evolution and the global history of pulp and paper industry. O Papel 74:51–54
Leite RC, Cortez LABE (2007) O Etanol Combustível no Brasil. In: Biocombustíveis no Brasil: realidades e perspectivas, pp 60–75. https://www.agencia.cnptia.embrapa.br/Repositorio/etanol3_000g7gq2cz702wx5ok0wtedt3xdrmftk.pdf
Lopez MG, Mancilla-Margalli NA, Mendoza-Diaz G (2003) Molecular structures of Fructans from Agave tequilana Weber var. azul. J Agric Food Chem 51:7835–7840
Macrotrends (2021) Crude oil prices - 70 year historical chart. In: Macrotrends https://www.macrotrends.net/1369/crude-oil-price-history-chart. Accessed 11 Oct 2021
Maire Tecnimont (2020) GranBio and NextChem sign partnership to develop cellulosic ethanol market. Maire Tecnimont. https://www.mairetecnimont.com/en/media/press-releases/granbio-and-nextchem-sign-partnership-develop-cellulosic-ethanol-market-non-price-sensitive
Marinho JFU, Correia JE, de Marcato AC et al (2014) Sugar cane vinasse in water bodies: impact assessed by liver histopathology in tilapia. Ecotoxicol Environ Saf 110:239–245
Market Insights (2017) Oil prices explained: putting a dollar value on a barrel of crude. Oil Sand Magazine. https://www.oilsandsmagazine.com/market-insights/oil-prices-explained-how-to-value-a-barrel-of-crude
Marques F (2018) Obstacles in the way. Revista Pesquisa Fapesp 268:59–63. https://revistapesquisa.fapesp.br/en/obstacles-in-the-way/
Martin G, Saikawa E (2017) Effectiveness of state climate and energy policies in reducing power-sector CO_2 emissions. Nat Clim Chang 7:912–919
Matsuoka S, Kennedy AJ, dos Santos EGD et al (2014) Energy cane: its concept, development, characteristics, and prospects. Adv Bot 2014:1–13
McFarlane HE, Döring A, Persson S (2014) The cell biology of cellulose synthesis. Annu Rev Plant Biol 65:69–94
Mehlman MA (1990) Dangerous properties of petroleum-refining products: carcinogenicity of motor fuels (gasoline). Teratog Carcinog Mutagen 10:399–408
Milanez AY, Nyko D, Valente MS et al (2015) De promessa a realidade: como o etanol celulósico pode revolucionar a indústria da cana-de-açúcar – uma avaliação do potencial competitivo e sugestões de política pública. BNDES Setorial 41:237–294
Mitchell M (2017) Sunliquid® - delivering a proven technology solution for commercial cellulosic ethanol production. In: CTBE - Brazilian National Laboratory of Ethanol (ed) CTBE cellulosic ethanol conference. http://pages.cnpem.br/2gbioethanol/wp-content/uploads/sites/85/2017/06/Martin_Mitchell_Clariant.pdf
Mylona P, Pawlowski K, Bisseling T (1995) Symbiotic nitrogen fixation. Plant Cell 7:869–885
Nakanishi SC, Soares LB, Biazi LE et al (2017) Fermentation strategy for second generation ethanol production from sugarcane bagasse hydrolyzate by Spathaspora passalidarum and Scheffersomyces stipitis. Biotechnol Bioeng 114:2211–2221
Nazir M, Sadat S, Soltani Howyzeh M (2019) The effect of different hormone combinations on direct and indirect somatic embryogenesis in Agave Americana. Plant Physiol 9:2739–2747
Nissan (2016) Nissan unveils world's first FCEV with SOFC running on bioethanol. Fuel Cells Bull 2016:3. https://doi.org/10.1016/S1464-2859(16)30231-0

Nobel PS (1991) Achievable productivities of certain CAM plants: basis for high values compared with C3 and C4 plants. New Phytol 119:183–205
Olivério JL, Ferreira FM (2010) Cogeneration – a new source of income for sugar and ethanol mills or bioelectricity—a new business. Proc Int Soc Sugar Cane Technol 27:1–13
Özcan S, Johnston M (1999) Function and regulation of yeast hexose transporters. Microbiol Mol Biol Rev 63:554–569
Paloheimo M, Haarmann T, Mäkinen S, Vehmaanperä J (2016) Production of industrial enzymes in Trichoderma reesei. In: Gene expression systems in fungi: advancements and applications. Springer, Cham, pp 23–57
Park C, Choi Y, Kim C et al (2010) Performance and exhaust emission characteristics of a spark ignition engine using ethanol and ethanol-reformed gas. Fuel 89:2118–2125
Paul MJ, Driscoll SP (1997) Sugar repression of photosynthesis: the role of carbohydrates in signalling nitrogen deficiency through source: sink imbalance. Plant Cell Environ 20:110–116
Peng J, Huang J, Wu X, Xu Y et al (2021) Solid oxide fuel cell (SOFC) performance evaluation, fault diagnosis and health control: a review. J Power Sources 505:230058. https://doi.org/10.1016/J.JPOWSOUR.2021.230058
Phillips RB, Jameel H, Chang HM (2013) Integration of pulp and paper technology with bioethanol production. Biotechnol Biofuels 61(6):1–12
Pimentel D, Marklein A, Toth MA et al (2009) Food versus biofuels: environmental and economic costs. Hum Ecol 37:1–12
Plugge CM (2017) Biogas. Microb Biotechnol 10:1128–1130
Prado RDM, Caione G, Campos CNS (2013) Filter cake and vinasse as fertilizers contributing to conservation agriculture. Appl Environ Soil Sci 2013:581984. https://doi.org/10.1155/2013/581984
Raízen (2021) Relatório Anual 2021. https://www.raizen.com.br/relatorioanual/2021/pdf/raizen-rs2021-pt.pdf
Rani Singhania R, Dixit P, Kumar Patel A et al (2021) Role and significance of lytic polysaccharide monooxygenases (LPMOs) in lignocellulose deconstruction. Bioresour Technol 335:125261. https://doi.org/10.1016/J.BIORTECH.2021.125261
Rarbach M, Söltl Y (2013) Sunliquid®: sustainable and competitive cellulosic ethanol from agricultural residues. Chimia (Aarau) 67:732–734
Raya FT, Marone MP, Carvalho LM et al (2021) Extreme physiology: biomass and transcriptional profiling of three abandoned Agave cultivars. Ind Crop Prod 172:114043. https://doi.org/10.1016/J.INDCROP.2021.114043
RFA (2021) Annual world fuel ethanol production. In: Renewable Fuels Association. https://ethanolrfa.org/markets-and-statistics/annual-ethanol-production
Rubino M (2013) Proesa technology. In: BIO. Montreal, pp 1–25. https://www.bio.org/sites/default/files/legacy/bioorg/docs/beta%20renewables%20proesa%20technology%20june%202013_bio_michele_rubino.pdf
Rural G (2021) Conheça a 1a usina do Brasil a gerar energia elétrica em escala comercial com resíduos da cana. Globo Rural. https://g1.globo.com/economia/agronegocios/globo-rural/noticia/2021/07/04/conheca-a-1a-usina-do-brasil-a-gerar-energia-eletrica-em-escala-comercial-com-residuos-da-cana.ghtml
Saini JK, Saini R, Tewari L (2014) Lignocellulosic agriculture wastes as biomass feedstocks for second-generation bioethanol production: concepts and recent developments. 3 Biotech 54: 337–353. https://doi.org/10.1007/S13205-014-0246-5
Schobert H (2013) Formation of fossil fuels. In: Chemistry of fossil fuels and biofuels. Springer, New York, pp 103–128
Seligmann MS (2018) War lord in training: Churchill and the royal navy during the first world war. In: Finest Hour 182, p 20. https://winstonchurchill.org/publications/finest-hour/finest-hour-182/war-lord-in-training-churchill-and-the-royal-navy-during-the-first-world-war/
SGD (2013) Seminal yeast literature. In: Sacharomyces genome database. https://wiki.yeastgenome.org/index.php/Seminal_Yeast_Literature

Solomon BD (2010) Biofuels and sustainability. Ann N Y Acad Sci 1185:119–134
Svensson M (2013) Biomethane for transport applications. In: The biogas handbook: science, production and applications. Elsevier, New York, pp 428–443. https://doi.org/10.1533/9780857097415.3.428
Tolan JS (2002) Iogen's process for producing ethanol from cellulosic biomass. Clean Technol Environ Policy 3:339–345
Townsend T, Sette J (2016) Natural Fibres and the World Economy. RILEM Bookseries 12:381–390. https://doi.org/10.1007/978-94-017-7515-1_30
UNICA (2021) Bioeletricidade pode ajudar a salvar o Brasil da falta de energia. In: UNICA Notícias. https://unica.com.br/noticias/bioeletricidade-pode-ajudar-a-salvar-o-brasil-da-falta-de-energia/
Vaaje-Kolstad G, Westereng B, Horn SJ et al (2010) An oxidative enzyme boosting the enzymatic conversion of recalcitrant polysaccharides. Science 330:219–222
Valavanidis A, Fiotakis K, Vlachogianni T (2008) Airborne particulate matter and human health: toxicological assessment and importance of size and composition of particles for oxidative damage and carcinogenic mechanisms. J Environ Sci Health C 26:339–362
Velandia Vargas JE, Seabra JEA (2021) Fuel-cell technologies for private vehicles in Brazil: environmental mirage or prospective romance? A comparative life cycle assessment of PEMFC and SOFC light-duty vehicles. Sci Total Environ 798:149265. https://doi.org/10.1016/J.SCITOTENV.2021.149265
Voegele E (2020a) Report: Australia's biofuel consumption to remain minimal in 2020. Ethanol Prod Mag. http://ethanolproducer.com/articles/17609/report-australiaundefineds-biofuel-consumption-to-remain-minimal-in-2020
Voegele E (2020b) Report: China may lower E10 mandate to E5. Ethanol Prod Mag. http://ethanolproducer.com/articles/17618/report-china-may-lower-e10-mandate-to-e5
Voegele E (2021) Colombia temporarily reduces ethanol blend mandate. Ethanol Prod Mag. http://www.ethanolproducer.com/articles/18235/colombia-temporarily-reduces-ethanol-blend-mandate
Walfridsson M, Bao X, Anderlund M et al (1996) Ethanolic fermentation of xylose with Saccharomyces cerevisiae harboring the Thermus thermophilus xylA gene, which expresses an active xylose (glucose) isomerase. Appl Environ Microbiol 62:4648–4651
Waltz E (2008) Cellulosic ethanol booms despite unproven business models. Nat Biotechnol 26:8–9
Wang Y, Seo B, Wang B et al (2020) Fundamentals, materials, and machine learning of polymer electrolyte membrane fuel cell technology. Energy AI 1:100014. https://doi.org/10.1016/J.EGYAI.2020.100014
Zakaria Z, Kamarudin SK (2016) Direct conversion technologies of methane to methanol: an overview. Renew Sust Energ Rev 65:250–261
Zamora F (2009) Biochemistry of alcoholic fermentation. In: Wine chemistry and biochemistry. Springer, New York, pp 3–26. https://doi.org/10.1007/978-0-387-74118-5_1
Zhang M, Yu Y (2013) Dehydration of ethanol to ethylene. Ind Eng Chem Res 52:9505–9514

Chapter 8
Feedstock for Second-Generation Bioethanol Production

Letícia Raquel Paliga, Andressa Janaina Warken, Caroline Dalastra, Maria Luíza Rodrigues Soares, Simone Kubeneck, Taís Rosângela Correia Souza, Sérgio Luiz Alves Jr, and Helen Treichel

Abstract The current broad need to bring new energy sources, especially in the transportation sector due to economic and population growth, causes different sources to be sought to produce fuels. In this sense, second-generation bioethanol from different biomasses has been gaining prominence since it allows the use of nonfood feedstocks, such as lignocellulosic biomass from agricultural and forestry residues, as well as secondary wastes. In addition to these lignocellulosic residues, this chapter will also address pectin- and starch-rich raw materials generated daily on a large scale worldwide. Second-generation bioethanol (2G) has gained space in several countries, known for not competing with cultivars intended for human and animal food, increasing production to replace fossil fuels, and waste recovery. However, there are still some difficulties to be overcome regarding low productivity compared to others. This underperformance may be linked to different factors, such as the quality of waste, selection of the fermenting microorganisms, and presence of inhibitor components during the last production stage. Thus, mastering the knowledge on residual biomasses is imperative to a highly efficient first stage of 2G ethanol production, providing reduction through process optimization.

L. R. Paliga · A. J. Warken · C. Dalastra · S. Kubeneck · T. R. Correia Souza · H. Treichel (✉)
Laboratory of Microbiology and Bioprocesses, Federal University of Fronteira Sul, Erechim, RS, Brazil

M. L. Rodrigues Soares
Postgraduate program in Biotechnology and Biosciences, Federal University of Santa Catarina, Florianópolis, Brazil

S. L. Alves Jr
Postgraduate program in Biotechnology and Biosciences, Federal University of Santa Catarina, Florianópolis, Brazil

Laboratory of Biochemistry and Genetics, Federal University of Fronteira Sul, Chapecó, Brazil

C. R. Soccol et al. (eds.), *Liquid Biofuels: Bioethanol*, Biofuel and Biorefinery Technologies 12, https://doi.org/10.1007/978-3-031-01241-9_8

8.1 Feedstock for Second-Generation Bioethanol Production

The high increasing rates in global warming make biofuels gain more visibility at the same time as they have been proven conditions to replace petroleum-based fuels (Sharma et al. 2020) with lower carbon emissions and cleaner combustion (when added from 3%–20% in gasoline), meeting the goal of cleaner energy in the transportation sector (Demichelis et al. 2020). Unlike first-generation ethanol (1G), which uses feedstocks that can serve as a source of human or animal food, second-generation (2G) bioethanol tends to value byproducts that do not have a direct destination for food and are treated as waste in industrial agriculture, forestry, and industrial systems, being discarded into the environment when not used for another purpose (Bera et al. 2021).

It can be waste from plants, scraps, and foods. These wastes, also called byproducts, are used as biomass. Plants are used for the production of bioethanol through agro-industrial production waste, which generates a large load of this type of waste that is not used in internal processes and ends up being discarded without added value, which justifies the search for their use in secondary processes (Shahid et al. 2021).

The factors that justify bioethanol production are diverse, one of them being its influence on carbon sequestration. In Brazil, for example, approximately 500 kg of soil carbon per hectare is sequestered annually, for a total of 12 million annually (Hogarth 2017). In contrast, agriculture can take this as induction for land-use change, just as the deforestation resulting from the search for new arable areas (Keles et al. 2018).

The US National Intelligence Council analysis demonstrated that food, water, and energy are three axes entirely connected. The union of these axes was because this junction needs protection, and the thought of security must be done collectively. By 2030, demand is expected to increase by 35%, 40%, and 50%, and one of the highlights for this increase is the climate changes occurring year by year (Alzaabi and Mezher 2021; Keyhanpour et al. 2021).

Therefore, one of the ways of control is the circular economy that acts in a way to make a cycle of use with specific components, in which the item in question after reuse and repair will be sent to recycling, reducing waste and decreasing the emission of carbonic gas (Akhimien et al. 2021). With this, the availability of resources is defined together, emphasizing the possible guarantee of sustainability (Keyhanpour et al. 2021). In this regard, the unrestrained consumption of the axes results in the need for high production. The increase in consumption of the axes is due to significant population growth (approximately 80 million people per year (Keyhanpour et al. 2021)). Consequently, large amounts of waste occur, on a global scale, reaching 1.3 billion tons wasted; moreover, high waste causes up to 10% greenhouse gas emissions over four years (Weber et al. 2020).

New energy alternatives are incessantly sought so that all natural resources may have their uses mitigated somehow. Moreover, even though the oil industry

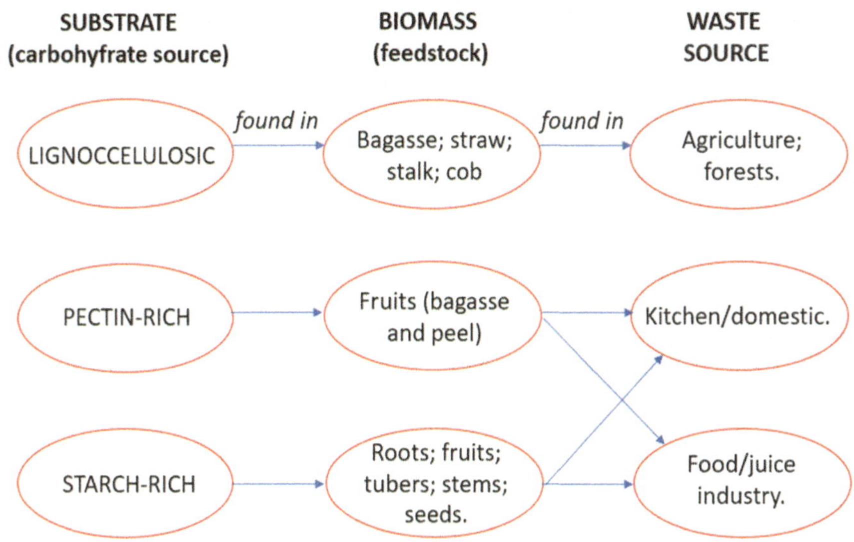

Fig. 8.1 Division of potentially promising biomasses for second-generation bioethanol production. (Source: the authors)

currently has production efficiency, its natural resources can be depleted. Every day, the prices of products increase significantly, which justifies production via valorization of byproducts, mainly of agro-industrial origin (Szklo and Schaeffer 2006).

In this way, bioethanol (2G) production appears to be an alternative due to its lower water, energy, and soil consumption. Figure 8.1 highlights the different types of biomass promising in this production, addressed in this chapter.

8.2 Carbohydrate-Rich Wastes

Rapid population growth and development have increased the need for food and, consequently, the production of edibles and their wastes. New technologies that aim to solve food waste must be developed; for example, Brazil generates approximately 150 million tons of sugarcane bagasse annually (Ebikade et al. 2020; Silva et al. 2014).

The current vision is to look at these residues as treatable wastes. Food waste can be considered a raw material to obtain valuable products. Within this approach, the valorization of carbohydrate-rich residues not only helps to solve food waste but can also contribute to a more sustainable economy, such as the production of ethanol from these substrates (Gálvez-Martos et al. 2021).

Carbohydrate-rich wastes harbor essential polysaccharides: cellulose, hemicellulose, and, in some cases, pectin. Cellulose is a primary composite of plant tissue and consists of up to 50% of the wood composition. Pectin is a soluble fiber in fruits and vegetables, while hemicellulose is a matrix highly branched by pentoses, hexoses, and uronic acids. These polysaccharides, when hydrolyzed, release mainly glucose (in the case of cellulose), xylose (in the case of hemicellulose or xylan), and galacturonic acid (in the case of pectin), where they are used to the formation of new applications (Venkatanagaraju et al. 2020).

The world cereal trade in 2020/21 increased by 3.4% compared to the previous year, driven primarily by a faster-than-expected pace in US corn sales and a continued substantial purchase from China. The global carbohydrate production in 2020 was 2742 million tons. These carbohydrate sources generate large amounts of processing residues that can be transformed into other products with an industrial and sustainable bias. (Cheng et al. 2020).

Foods with a higher concentration of carbohydrates are also components that generate more carbohydrate-rich residues worldwide. Rice processing residues, for example, are mainly rice husks, bran, and broken grains. Most agricultural rice residues are composed of lignocellulosic biomass that has 32% cellulose and 21% lignin. The husk of this grain, when eliminated in nature, can cause environmental imbalances, as its natural absorption is prolonged (approximately five years). Thus, rice processing residues are unnecessarily destroyed since they can be used as feedstock for second-generation biorefineries (Sivaramakrishnan et al. 2021).

According to a survey by Hafid et al. (2017), kitchen waste contains 60% of total carbohydrates compared to other studies. Additionally, food waste containing carbohydrates is estimated at approximately 70% of the complete solid waste generated. In addition to carbohydrates, food residues are rich in proteins, lipids, vitamins, and minerals, making them suitable as raw materials with high added value for the production of biofuels (Xu et al. 2020).

Second-generation ethanol can be obtained from the beneficiation of carbohydrate-rich residues, so waste is minimized. Once left out, this substrate now plays a vital role in the circular economy. The primary raw materials for the production of 2G ethanol are processing wastes from the rice, soy, corn, and sugarcane industries (Lee and Lavoie 2013).

Thus, in this new paradigm, competitive and high-quality secondary materials are essential for the competitiveness of a sustainable economic system. In Europe, for example, where the annual generation of residues is projected to increase up to 70% by 2050, interest in the high production of residues rich in carbohydrates has been increasing significantly (Gálvez-Martos et al. 2021). For this reason, carbohydrate-rich residues have been used not only in the production of paper but also as energy and second-generation ethanol sources, thus reducing environmental pollution (Rosales-Calderon and Arantes 2019).

8.2.1 Lignocellulosic Biomasses

The use of lignocellulosic biomass has been present throughout history, in which straw was used to heat houses. These have been exchanged for coal since 1910 and then boosted the use of ethanol. It is considered a promising feedstock for second-generation bioethanol production since it does not need to increase production areas because it has no direct connection with human consumption (Liu et al. 2019). New fuel-powered vehicles have led to the scarcity of fossil fuels, which has prompted recent research to look for biofuel production allying to preserve the environment and waste valuation (Wang et al. 2021).

Lignocellulosic biomass is the largest natural source of carbohydrates found in natural resources worldwide. Each constituent of lignocellulosic biomass is rich in polysaccharides hydrolyzable into fermentable sugars, composed mainly of cellulose, hemicellulose, and lignin. This type of biomass differs in that it is based on agricultural residues, such as leaves, cuttings, straw, manure, forest biomass, forest residues (sawdust, wood chips), and industrial and urban solid residues (food waste, kraft paper), among others (Santos et al. 2020; Su et al. 2020).

Ethanol since 2013 has already produced 88.69 billion liters of sugarcane and corn in Brazil and the United States, respectively. With this biomass, bioethanol production is possible and used in means of transport as fuel. Lopes et al. (2016) cite that using ethanol can reduce greenhouse gas emissions compared to petroleum fuels. In contrast, 40% of corn production is used for bioethanol (in the US), so when there is a drought in the crop, the agricultural price of corn can vary up to $6.22 to $6.89 per bushel (Stolarski et al. 2015).

Then, new resources are being used so that grain and corn straw, in addition to sugarcane bagasse and straw, are used for cellulosic bioethanol. Studies report that up to 422 billion liters of second-generation bioethanol can be produced from lignocellulosic biomass, and crop waste can yield up to 491 billion liters (Stolarski et al. 2015). In this sense, the main components of lignocellulosic biomass that contribute to second-generation ethanol production via biotechnology using residues from agribusiness will be described.

8.2.1.1 Cellulose, Hemicellulose, and Lignin

Cellulose is inserted in a large part of the plant cell wall and can be considered the most abundant natural polymer in the world. The main characteristics are the ability to stop cell swelling and plasma membrane rupturing when there is an excess of water (Santos et al. 2020).

Cellulose represents up to 50% of the lignocellulosic biomass, a long-chain polymer of glucose units (Wang et al. 2021), connected by β-1,4-glycosidic bonds (Sharma et al. 2020). Moreover, cellulose chains can be united by hydrogen bonds and van der Waals forces; thus, in terms of production, cellulose chains stand out because the hydrolysis of cellulolytic enzymes leads to the generation of glucose, a highly and easily fermented monosaccharide (Paul et al. 2020).

There are three hydroxyl groups (reactive) in each glucose unit, which gives cellulose strong hydrophilic and biodegradable characteristics. Cellulose can be classified into amorphous (best degraded by chemical reagents and enzymes) and crystalline (Wang et al. 2021). Glucose molecules are linked to each other, forming fibrils. The union of these monomers forms dilute-acid soluble cellulose compounds with strong hydrogen interactions but low water solubility. The amorphous regions of cellulose are intercalated by crystalline regions, conferring, due to the order, recalcitrance to the structure when submitted to biotechnological processes via microorganisms (Balat 2011; Silveira et al. 2015). In addition, cellulose fibrils are firmly coated with hemicellulose (Sharma et al. 2020).

Hemicellulose is a heteropolysaccharide located in the primary cell wall, comprising 23% to 32% of lignocellulosic biomass, being a polymer without a determined shape with branched and short chains, with modified heterogeneity of pentoses and hexoses with approximately 500 to 3000 sugar monomers (xylose, arabinose, mannose, galactose), in addition to hexuronic acids and deoxyhexoses. They associate with each other through hydrogen and covalent bonds. Due to the presence of pentoses, hemicellulose becomes viscous when in contact with water. Unlike cellulose, hemicelluloses can be more easily hydrolyzed into sugars via microbes due to their branched and amorphous structure. Cellulose and hemicellulose are bound by lignin between them (Wang et al. 2021; Sharma et al. 2020).

Lignin, which represents 15% to 25% of lignocellulosic biomass, differs from other lignocellulosic biomasses by not having large molecular chains but aromatic rings with manometric compositions of various chemical structures. Its bonds are given by carbon-carbon and ether. This type of connection makes the walls of the plants stable and resistant to infecting agents and insects (Sharma et al. 2020).

The structure is three-dimensional, noncrystalline, and irregular and is composed of syringyl, guaiacyl, and p-hydroxyphenyl units. Each type of lignin varies between these three units. For example, harder woods have larger siringyl and guaiacyl units, differing from softer woods, mostly guaiacyl units, while woody biomass has all three units (Wang et al. 2021). Despite its influence on the hydrolysis efficiency of lignocellulosic biomasses, due to its hydrophobic characteristics, lignin is not directly used for the production of second-generation ethanol (Silveira et al. 2015; Rabelo et al. 2013; Banu et al. 2021).

8.2.1.2 Sources of Lignocellulosic Feedstocks

Different feedstock sources stand out in the production of second-generation bioethanol. When using biomass from firewood, there is a yield of 300 to 350 liters of ethanol per ton, even with the presence of 25% to 30% lignin in the biomass. However, this will only occur if all parts that contain cellulose and hemicellulose are converted into ethanol (Sharma et al. 2020).

The biomass from sawdust wood can accumulate up to a million cubic meters after processing, concentrated in cities with the Ayous (*Triplochiton scleroxylon*)-type tree. For example, most of it is burned due to the large amount to be handled. The main principle for achieving good production is to remove all the lignin from the process, releasing cellulose. However, the high concentration of lignin and the difficulty of breaking it down make the energy consumption higher, a consequence of the longer hydrolysis time required (Assabjeu et al. 2020).

Wheat straw also stands out in second-generation bioethanol production; it can reach 1.3 kg of residue for every 1 kg of grain weight, which shows the large amount of residue that this crop can produce (Ziaei-Rad et al. 2021). The mixture of E85 fuel (85% anhydrous ethanol and 15% pure gasoline) reduces greenhouse gases by 73%. Nevertheless, a life cycle assessment shows that 72% of the energy is consumed in straw preparation and transport, making straw preparation and transport environmentally inefficient. The same has been reported for E85 bioethanol from corn straw, but with a slightly higher CO_2 reduction (86% to 113%) than conventional bioethanol (Sharma et al. 2020).

The removal of all straw from the soil after harvest causes a reduction in soil fertility and acidification, so there may be removal that does not exceed 28% of the total residue (Graham et al. 2007; Sharma et al. 2020). For this issue of waste removal, many researchers indicate that the use of marginal lands, which are unsuitable for planting food crops, can be used without the necessary fertility and with contamination (Brusca et al. 2018).

It is important to note that raw materials play an essential role in the economy. Coffee, for example, has excellent visibility in the economic and industrial sectors; thus, the use of coffee residues, which are generated in large quantities, potentially increases its monetary value. The coffee pulp needs to be processed; for good results, the highest amount of sugar is available due to the low concentration of soluble solids available in the samples. By increasing the number of fermentable sugars and inoculum for fermentation processing, the bioethanol yield can be up to 77.29% (Harsono et al. 2015; Shenoy et al. 2011).

Argania spinosa (L.) *Skeel* forest residue biomass is promising for bioethanol production from Morocco and generates different byproducts; in this case, pulp (43% of total) is used. However, because of the high content of phenolic compounds that inhibit organisms in fermentation, the pulp should contain high carbohydrates and, when possible, fewer phenolic compounds. As this is new research in this area, the proportion of the net amount of bioethanol in the final processing is not yet known (Zouhair et al. 2020).

First-generation bioethanol production is based on food sources such as sugarcane, corn, and beet that are easily used for ethanol production. Nevertheless, excess waste (straw and cob) occurs from these sources, resulting in a new environmental problem (Silva et al. 2020). These residues are rich in cellulose, hemicellulose, and lignin, which can result in second-generation bioethanol. Cobs, for example, are composed of 32.2% cellulose, 29% hemicellulose, and 18.8% lignin. According to the author, when hydrolyzed with acids, these cobs provide a high fermentable sugar concentration.

The best production technologies achieve results of 19% to 22%; in less efficient technologies, this result falls to 11%. Sugarcane, like corn, when it is used, generates large amounts of residues that need to be used in some way to avoid being treated as an environmental problem. In Brazil, sugarcane bagasse and straw are commonly used as fuel sources to produce the steam needed to generate the energy used in the first-generation bioethanol production process (Zhao et al. 2019; Ayodele et al.

2020). Corn straw, which can also be used, has a high lignin content and a solid structure to be converted into bioethanol.

When a way to integrate first- and second-generation production is being sought, with the union of sugar flows, the concentration of ethanol increases in distillation, which reduces energy costs and allows the use of the entire agricultural crop. Thus, this integration consequently increases the productivity per hectare of bioethanol and the industrial production index of the second-generation fuel. Thus, the energy integration between the flows increases the overall energy efficiency, reducing steam and electricity consumption in the processes. Due to the low mass and inhibition of substrates, the recurrent use of lignocellulosic biomass can be reversed by high doses of enzymes, resulting in increased production costs (Ayodele et al. 2020).

8.2.2 Pectin-Rich Biomass

Recently, with the emergence and strengthening of the concept of biorefineries and circular economies, discussions on waste management through chain recovery to minimize impacts on health and the environment have expanded (Morone et al. 2019). Large amounts of pectin residues are discarded daily worldwide, and among them, fruit residues stand out. According to FAO (2021a, b), approximately 1300 billion tons of food residues are generated annually, with fruit residues representing approximately 50% of this total. Food waste is present from the beginning of the production chain, persisting during all stages of food production until it reaches the consumer. In addition to the rapid maturation of fruits, which is one of the leading causes of their disposal by consumers, the processing of these foods can generate an even more significant loss due to the amount of bagasse (Scapini et al. 2019). Biomasses such as sugarcane bagasse, rice straw, and fruit residues are being evaluated as potential raw materials for conversion into biomaterials, bioproducts, and bioenergy (Bonatto et al. 2021; Dagnino et al. 2017).

Lignocellulosic biomass is complex and mainly composed of cellulose, hemicellulose, and lignin. Lignin is associated with cellulose in the cell wall, and its primary function is to provide rigidity, impermeability, and resistance to microbial attack. Biomass rich in pectin has low levels of lignin, which makes the conversion of biomass into fermentable sugars more flexible than lignocellulosic biomass (Edwards and Doran-Peterson 2012; Scapini et al. 2019; Zanivan et al. 2021).

Pectin is a complex carbohydrate composed mainly of covalently linked galacturonic acid (70%). It is also composed of glucose, fructose, arabinose, and galactose. The three most prominent types of pectin present in the cell wall are homogalacturonan (HG), rhamnogalacturonan I (RG-I), and rhamnogalacturonan II (RG-II), accounting for approximately 65%, 20–35%, and 10% of the pectin in the cell wall, respectively (Mohnen 2008; Scapini et al. 2019; Cárdenas-Pérez et al. 2018). When hydrolyzed, pectin's main products are galacturonic acid and fermentable sugars that can produce biofuels, such as second-generation bioethanol (Vaez et al. 2021; Talebnia et al. 2008).

Approximately 115 million tons of citrus fruits are produced annually. After processing by food industries, approximately 50% of the mass that constitutes these fruits becomes residues (Choi et al. 2015; Cohn and Cohn 1997), which can produce biofuels due to their composition (Choi et al. 2013). The skins of these fruits, for example, have approximately 25% pectin in their composition (Vaez et al. 2021). Citrus fruits are made up of large amounts of sugars such as glucose, fructose, and sucrose, acids such as citric and malic, and carbohydrates such as cellulose, hemicellulose, and pectin (Sharma et al. 2017; Rivas et al. 2008).

The use of fruit residues, especially citrus fruits, in second-generation ethanol production reduces dependence on fossil fuels. It provides an alternative for disposing of these residues in the soil. However, it is essential to emphasize that the use of citrus fruits in biotechnological processes requires additional processing before the fermentation process due to the presence of microbial inhibitors, such as D-limonene, which cause stress and inhibition in the glycolytic pathway and can lead to disruption of the cell membrane at high concentrations (Choi et al. 2015). Several studies have been carried out to reuse residues such as fruit processing in the production of biofuels. Therefore, this topic will address the use of raw materials rich in pectin in second-generation ethanol production.

8.2.2.1 Apple

Apple is one of the most produced and consumed fruits worldwide. In 2018 alone, global apple production reached more than 85 million tons (Demiray et al. 2021). During processing for the production of apple juice, approximately 40% of its composition turns into bagasse, which can contain approximately 23% pectin (Gabriel et al. 2013; Dranca et al. 2020; Demiray et al. 2021). To apply this residue in the production of biofuels, it is essential to carry out a pretreatment to release the sugars present in cellulose and hemicellulose that make up the apple pomace (Molinuevo-Salces et al. 2020).

The use of this residue in the production of biofuels is already being explored. Authors such as Demiray et al. (2021) evaluated the potential of apple pomace pretreated with sulfuric acid in the production of second-generation ethanol supplemented with soluble soy protein. The results indicate that apple pomace is a promising residue for biofuel production since it yields 48.7 g L^{-1} ethanol. When the medium was supplemented, there was an increase of up to 8.28%, obtaining yields of approximately 53 g L^{-1} of bioethanol (Demiray et al. 2021).

Molinuevo-Salces et al. (2020) evaluated the valuation of two apple pomaces from second-generation ethanol production. The authors performed a physicochemical pretreatment and enzymatic hydrolysis of the residue to increase the availability of fermentable sugars. As a result, they obtained a concentration of 153 g L^{-1} of total sugars and a production of approximately 51 g L^{-1} ethanol using yeasts of the genera *Kluyveromyces, Lachancea,* and *Saccharomyces cerevisiae*, with the sugar consumption of these yeasts being approximately 84% (Molinuevo-Salces et al. 2020).

8.2.2.2 Orange

More than 48 million tons of oranges were produced worldwide in 2020, with Brazil responsible for more than 11 million tons (USDA 2021), making up most citrus fruit production. A large part of this production is destined for fruit processing, which generates large amounts of waste, such as peels. The composition of orange peel consists of soluble sugars (16.9%) and polysaccharides such as pectin (42.5%), cellulose (9.21%), and hemicellulose (10.5%). The use of these residues has already been studied for the production of biofuels. These compounds, when hydrolyzed, release glucose, galactose, xylose, arabinose, and inhibitor compounds such as limonene (Rivas et al. 2008).

Santi et al. (2014) used orange peel to produce second-generation ethanol with the yeast *S. cerevisiae* F15. The residue obtained was composed of 169.6 g kg^{-1} pectin, 67 g kg^{-1} glucose, 68 g kg^{-1} fructose, 260.9 g kg^{-1} cellulose, and 118.8 g kg^{-1} hemicellulose. To obtain fermentable sugars, the residues underwent a pretreatment based on steam explosion followed by acid catalysis. The solids obtained in the pretreatment were subjected to enzymatic hydrolysis, and with the liquid fraction obtained, alcoholic fermentation was carried out. The fermentation medium was yeast extract, ammonium sulfate, $MgSO_4.7H_2O$, KH_2PO_4, and the inoculum with *S. cerevisiae* F15 cells. As a result, the authors obtained a maximum ethanol yield coefficient equal to 0.495 g g^{-1} in the third fermentation cycle lasting 4 hours.

Another study that used orange processing residues was by Widmer et al. (2010). The authors pretreated the residue with a steam explosion followed by enzymatic hydrolysis. The hydrolysate obtained was used as a fermentation medium together with calcium carbonate and Saccharomyces cerevisiae yeast cells. Widmer et al. (2010) obtained ethanol yields between 76 and 94%. However, the contents of total sugars in the raw and hydrolyzed residue were equal to 24.43 and 57.66%, respectively, contributing to higher ethanol production yields.

8.2.2.3 Lemon

Global production in 2020 is estimated at 8.4 million tons of lemon (USDA 2021). Like orange residues, lemons are rich in sugars and carbohydrates. They belong to the same class of citrus fruits (Boluda-Aguilar and López-Gómez 2013), making them a residue with the potential for producing biofuels. Consequently, the processing of this fruit by juice production industries generates large amounts of waste.

In ethanol production, a recent study by Kundu et al. (2021) was based on residues, such as lemon peel and pulp. The residue used was approximately 3% pectin, 28% cellulose, 18.56% hemicellulose, and 34% free sugars. The residues underwent enzymatic saccharification by hexose and pentose produced by *S. cerevisiae* and *Pichia kudriavzevii*. Subsequently, alcoholic fermentation was carried out using yeast strains, obtaining a maximum ethanol yield equal to 12.16% in 24 h of fermentation (Kundu et al. 2021).

The study by Boluda-Aguilar and López-Gómez (2013) used lemon residues to produce bioethanol. Previously, this residue was pretreated with a steam explosion to remove possible inhibitors after performing enzymatic hydrolysis and with the hydrolysate and the yeast *S. cerevisiae* CECT 1329 for alcoholic fermentation. As a result, Boluda-Aguilar and López-Gómez (2013) obtained ethanol yields of up to 63 L of ethanol per 1000 kg of lemon residue.

8.2.2.4 Tangerines/Mandarins

Global production in 2020 is estimated to be flat at 33.3 million tons (USDA 2021). The rinds of this fruit represent approximately 60% of the dry weight of the fruit. Tangerine residues have great potential in the production of biofuels, as they have low concentrations of lignin (0.56%) and a considerable number of fermentable sugars (31.58%), in addition to approximately 10% cellulose, 4% hemicellulose, and 23% pectin (Oberoi et al. 2011).

Oberoi et al. (2011) evaluated dry, ground, and hydrothermally pretreated tangerine residues in ethanol production via simultaneous saccharification and fermentation. Through the statistical optimization of the process, the authors obtained an ethanol concentration of 42 g L^{-1} and ethanol productivity of 3,50 g L^{-1} h^{-1}. The authors observed that the pretreatment of the residues increased the sugar concentration in the extract by almost 30%, indicating partial hydrolysis of hemicellulose and lignin.

Boluda-Aguilar et al. (2010) used tangerine peel residues pretreated by steam explosion. The pretreatment allowed a higher sugar yield, where ethanol production reached 50–60 L/1000 kg of residue. In addition to ethanol production, the authors evaluated the influence of D-limonene concentration on biofuel production. The authors observed an optimal concentration of up to 0.0065% (v.w^{-1}) D-limonene, for which the highest ethanol production was obtained. On the other hand, when the tests showed a concentration of D-limonene above 0.16% (v.w^{-1}), total inhibition of the fermentation process was observed, where there was no ethanol production.

8.2.2.5 Mango

In 2016, approximately 46 million tons of mango were produced (Perea-Moreno et al. 2018). Mango is mainly cultivated in a subtropical climate, a highly perishable fruit, maturing in approximately 6–7 days. Very ripe fruits, characterized by excessive softening, desiccation, and microbial contamination, are not marketable, and for this reason, large amounts of waste are generated (Buenrostro-Figueroa et al. 2018). The fruit residue comprises the skin, fibrous material, and stone. The mango kernel corresponds to 30–45% of the fruit weight, depending on the variety, composed of lignin, cellulose, hemicellulose, and starch (Perea-Moreno et al. 2018). The mango peel represents approximately 15–20% of the fruit's weight (Maisuthisakul and Gordon 2009; Buenrostro-Figueroa et al. 2018; Scapini et al. 2019). Mango contains a high concentration of sugar (16–18% $w.w^{-1}$), acids with organoleptic properties, and antioxidants such as carotene (Reddy and Reddy 2011), making it a residue with the potential for producing biofuels.

8.2.3 Starch-Rich Wastes

Starch is the second most abundant polysaccharide in nature and is produced by several plants that use it for energy storage. It is formed by two polysaccharides, amylose (18–33%) and amylopectin (72–82%). It is found in roots, fruits, tubers, stems, and seeds. Among the main foods rich in starch are potatoes, cassava, corn, rice, and wheat (Buléon et al. 1998, Le Corre et al. 2010).

Roots and tubers such as cassava, sweet potato, and yam or stalks reached a global production of 237 Mt. (dry matter) from 2018 to 2020 (OECD/FAO 2021). The processing of these foods generates many starch-rich residues, such as husks, pulps, stems, and leaves, that can obtain high value-added products, such as second-generation ethanol. (Zhang et al. 2016; Sivamani et al. 2018).

Corn starch is already successfully used in the production of first-generation ethanol. The largest ethanol producer globally, the USA, has most of its production generated from starch, reaching approximately 59.8 billion liters in 2019 (RFA 2020). However, this type of production creates competition for land use between ethanol production and food production (Zhang et al. 2016; Thatoi et al. 2014).

Studies such as those by Martinez et al. (2018) show alternatives to this problem when using starch residues in ethanol production. The authors investigated the use of cassava bagasse as a substrate. The residues used in the study were obtained from starch manufacturers in Brazil and were analysed for their chemical composition. The residue samples had high starch contents, on average 64%. For alcoholic fermentation, the authors used a strain of Saccharomyces cerevisiae and reported an average ethanol yield of 30%.

Apiwatanapiwat et al. (2011) reported 10 g. L^{-1} ethanol using pretreated cassava pulp as the only carbon source. The authors developed a CBP yeast S. cerevisiae codisplaying α-amylase, glucoamylase, endoglucanase, cellobiohydrolase, and β-glucosidase on the cell surface. According to the authors, the recombinant strain could directly produce ethanol from the cassava pulp without hydrolytic enzymes.

Potatoes are one of the most important food crops for starch-rich tubers worldwide (FAO 2017). It is estimated that in 2019, world potato production was approximately 370 million tons (FAO 2021a, b). However, little is reported about its use in the production of biofuels.

Abanoz et al. (2012) used wastewater from potato processing to produce ethanol. Izmirlioglu and Demirci (2012) made 30.99 g. L^{-1} ethanol after optimizing the hydrolysis process of mashed potato residues. Potato peels also have a high potential, as shown in Arapoglou et al.'s (2010) study, which reported 7.6 g. L^{-1} of ethanol by Saccharomyces cerevisiae var. bayanus after enzymatic hydrolysis with three enzymes. Yamada et al. (2009) obtained 20 g. L^{-1} ethanol from sugars released from fresh potato peel. In the same study, the authors evaluated the addition of mashed potato residue to the medium, which increased ethanol production, reaching 50 g.L^{-1}.

The potential of tuber husks such as sweet potato, elephant foot yam, tannia, yam, and sugar beet for the production of bioethanol was investigated by Mithra et al. (2018). Previous studies show that these residues have starch (27 and 32%), cellulose (13 and 19%), hemicellulose (13 and 20%), and lignin (4 and 8%) in their composition (Mithra and Padmaja 2016, 2017). A comparison was made between production by fed-batch separate hydrolysis and fermentation (F-SHF) and simultaneous saccharification and fermentation (F-SSF) using Saccharomyces cerevisiae. The authors reported that waste pretreated with dilute sulfuric acid under F-SHF had a higher ethanol yield (34 and 43 g. L^{-1}) than steam pretreatment (27 and 36 g.L^{-1}). In comparison, in F-SSF, steam-pretreated sweet potato peel, elephant foot yam, and tannia were higher, with yields from 281 to 302 ml.kg^{-1} ethanol.

Dioscorea composita is a yam species rich in starch that is used as raw material for cofermentation with sugarcane bagasse (Ye et al. 2018). The analysis of the chemical composition of the residues used in the study shows that Dioscorea contained 47.9% starch, 9.4% cellulose, 13.2% xylan, and 10.9% lignin. The authors evaluated various mixture proportions of sugarcane bagasse pretreated with alkali and Dioscorea composita residues via simultaneous saccharification and fermentation (SSF). The results obtained show that the mixture ratio of 1:1 reached the highest concentration (31.77 g. L^{-1}) and yield (84.40%) of ethanol.

Rice byproducts have also been exploited for the production of bioethanol. Rice processing generates various residues, such as rice husk, discolored, unripe, and broken rice bran. These residues have a starch content ranging from 7–85% (Favaro et al. 2017).

The potential of rice bran was investigated by Tiwari et al. (2015). Bran contains many sugars, such as residual starch (10–20%), that can be converted to ethanol (Beaugrand et al. 2004; Sharma et al. 2007; Tiwari et al. 2015). The authors used the Bacillus cereus MCR-3 bacterial strain for fermentation and obtained an ethanol production of approximately 11% under optimized pH (5) and temperature (37 °C) conditions.

In their study, Favaro et al. (2017) showed the feasibility of using rice processing byproducts (rice bran, broken, unripe and discolored rice) as raw materials for ethanol production. The residues were first fermented individually and then mixed to assess ethanol production. The authors reported that the ethanol yields of each raw material exceeded 88% of the theoretical value, while the byproduct mixture reached 92% of the theoretical value (51 g.L^{-1}).

These examples of alternative starchy substrates demonstrate the variety of raw materials used to produce bioethanol. However, using these byproducts presents challenges. For example, the logistics of their collection and transport can be more complicated and more expensive. Furthermore, each type of waste has different structures and compositions, which may require auxiliary enzymes for hydrolysis. (Cripwell et al. 2020). Many studies are still needed to improve the conditions used in generating bioethanol from waste to make its production viable on a large scale at a commercial level.

8.3 It Is no Piece of Cake

Among the most significant challenges present today, replacing fossil fuels with environmentally friendly fuels is in great movement and discussion (Su et al. 2020). As seen above, second-generation bioethanol is already a reality in biofuels, with large world production. As production grows, production logistics and the advancement of technology in the sector are challenged to achieve totally clean production. Thus, this topic will discuss the difficulties that make residues unviable for production since studies show that the expansion of ethanol and biodiesel in 20 years will be 2 to 4 times (Sydney et al. 2019; Sharma et al. 2020).

The enzymes used to hydrolyze cellulose into glucose are very expensive ($0.10–$0.20 per gallon) compared to corn ethanol ($0.03 per gallon); furthermore, the energy yield of cellulosic bioethanol is disadvantaged by the fact that there is many bacterial consumption during the whole process, thus making the cost between cellulosic bioethanol ($18/million BTU) and corn ethanol ($12/million BTU) have approximately a difference of $6/million BTU. It is also known that not all biomasses are available for hydrolysis due to the current difficulty in doing this quickly and in an energy-efficient and economical way (Tan et al. 2008).

Environmentally, second-generation bioethanol stands out because it does not cause as much damage to the expansion of arable land as first-generation bioethanol (Su et al. 2020). However, by occupying the residues left over from crops, especially in monocultures, some types of perennial grasses, which even though they overgrow, when removed in excess from the soil, make them unprotected from natural causes, such as erosion and flooding (Tan et al. 2008).

Nevertheless, on the question of waste removal from the land, Sharma et al. (2020) point out that excessive removal of wheat straw from the soil results in soil with no protective layer for erosion, besides that this straw will bring fertilization and correction needed for the next crop desired for that soil. Nevertheless, studies show that some portions can be removed without the soil being affected, and studies vary this removal between 25% and 60% depending on the product, location, and fertility of the area. Suppose there is no fertility in the soil due to waste removal to improve production. In that case, farmers need to add new chemical additives for remediation, with the likelihood of increased pollution and ill health (Tan et al. 2008).

The life cycle assessment (LCA) method evaluates the inputs, outputs, and environmental impacts throughout its complete cycle to better assess the entire bioethanol production and process cycle. Life cycle assessment pinpoints precisely how each biomass will impact the environment. As a function of agricultural use,

LCA investigates land use and carbon sequestration, as when waste is removed in excess, another carbon sequestration is lower and the land use to generate the byproducts is also higher; chemical additives are added by the lack of natural fertilization that the removed biomass would bring to the soil; effects of waste removal, as seen, the reduction decreases the carbon sequestration, consequently increasing the probability of generating greenhouse gases and transport, which will be used to bring the biomass to the biorefinery, increasing pollution and the use of fossil fuels (Wiloso et al. 2012).

8.4 Is the 2G Ethanol a Reality?

Ethanol is an organic substance obtained from the alcoholic fermentation of sugars (especially yeasts), ethylene hydration, or reduction to acetaldehyde. It is found in beverages, perfumery, and fuel industries. Its physical and chemical properties depend mainly on the hydroxyl group -OH, which implies the molecule's polarity and promotes intermolecular interactions via hydrogen bonds (Jiang et al. 2021).

The new concept of ethanol production is the utilization of lignocellulosic biomasses as a raw material. These feedstocks come from residues of natural products, such as corn, rice, cocoa, straw, sweet potato, and sugarcane bagasse, giving rise to so-called second-generation ethanol (Anderson and Wallington 2020).

In the world, biofuels are related to more than 1.7 million jobs; 845,000 are in Brazil. Commodities for ethanol production, such as sugarcane, are growing and taking place amid food agriculture. There is concern about competition between food production and renewable sources, thus generating food insecurity on the world stage. Today's food system (production, transportation, processing, packaging, storage, retail, consumption, loss, and waste) feeds the vast majority of the world's population and supports the livelihoods of more than 1 billion people. Since 1961, the per capita food supply has increased by more than 30%, accompanied by increased use of sugar cane through fuel production. An estimated 821 million people are currently undernourished, 151 million children under the age of five are stunted, 613 million women and girls aged 15 to 49 are iron deficient, and 2 billion adults are overweight or obese. The food system is under pressure and competition in the production of biofuels. These factors impact the four pillars of food security (availability, access, utilization, and stability) (Gold et al. 2020).

One of the essential facts for alerting these divergences is the prices of agricultural inputs, which increase over the years, generating fear and malnutrition in various parts of the planet. Additionally, with the price of oil at US$66.00 a barrel, the value of corn increases, for example, in addition to other crops. According to the UN (2018), the world population will be over 8 billion people in 2024 and over 9.5 billion people in 2050, requiring a better food supply. Because of this, the guarantee of food in quantity, quality, and diversity is a right of all people and a duty of the state (Anwar Saeed et al. 2018).

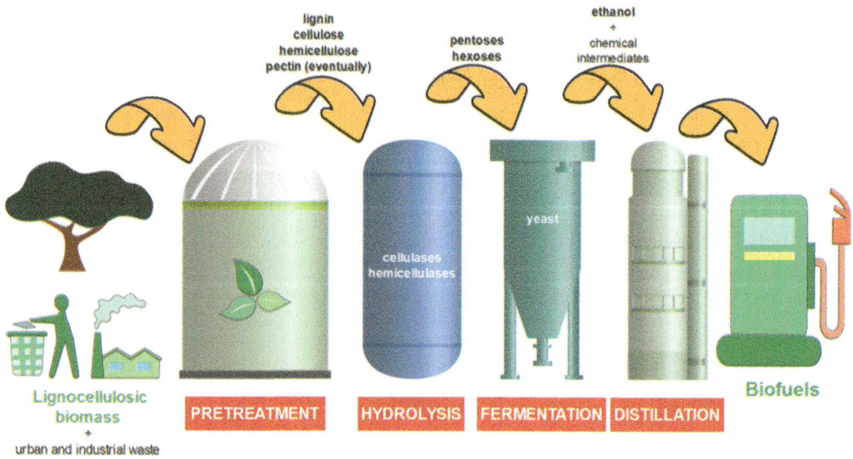

Fig. 8.2 Production of second-generation ethanol. Different residues can be employed as raw materials in second-generation processes. Usually, it requires three steps that may or may not be conducted together before ethanol distillation: pretreatment (mainly chemical or physicochemical), hydrolysis (preferably enzymatic), and fermentation (where yeasts are the main microorganism employed). (Source: the authors)

The need for agricultural spaces for food production competes with the means of producing biofuels. Currently, Brazil ranks among the ten countries that most waste food globally, with approximately 35% of production being wasted every year. The solution to this issue comes with the use of food industry residues as a source for the generation of ethanol and other fuels from renewable sources, thus having a stable balance between food, waste, and ethanol production. Our current way of life and the constant increase in population make energy one of the leading global needs. Fossil resources are being depleted and will be economically unviable soon, so there is an urgent need to develop renewable energy to replace petroleum-derived fossil fuels (Su et al. 2020).

Biomass is an answer to replacing oil. Among the biomass components, lignocellulose is the most abundant, producing 170 billion metric tons. The different sources of lignocellulose are agricultural residues (leaves, greenhouses, and straw), agro-soils (solid cattle manure), and forest residues (Ray et al. 2016). Lignocellulosic compounds and agricultural residues are not associated with food and are available in abundance worldwide. Therefore, the consumption of lignocellulosic biomass on a large scale for ethanol production does not converge with the use of arable land as in first-generation bioethanol (from agricultural cultivars for food) (Robak and Balcerek 2018).

Second-generation ethanol is produced using a process that involves the following main steps: (i) pretreatment, (ii) hydrolysis to sugars, (iii) fermentation, and (iv) product upgrade (Madu and Agboola 2018). Therefore, the 2G ethanol production process has methodologies to develop appropriate bioremediation and is fully viable for industrial use. Figure 8.2 below demonstrates the primary system for the production of 2G ethanol:

Coupled with the sustainable bias and reuse of extracts, second-generation ethanol presents itself as a technology capable of facing contemporary international energy challenges: high oil prices in international markets, dependence on this source by certain countries, global warming, and growing demand for first-generation ethanol in both the domestic and international markets (Parashar et al. 2016).

8.5 Conclusion and Perspectives for Feedstocks for Second-Generation Bioethanol Production

Through the above, research on new sources capable of generating biofuels to replace fossil fuels has been gaining space on the world stage due to the need to replace the latter due to its scarcity and the large load of waste generated daily by industries to generate energy and fuel, and waste from production for food purposes that are discarded into the environment, therefore, second-generation bioethanol, can supply the energy demand and reduce the environmental impacts, besides adding value to the byproducts.

On the other hand, this valorization brings to light the difficulties and challenges faced at the economic and environmental level of future industrial-scale production, the removal of waste from the land, and the high cost of production, for example. In this sense, there is a range of diverse biomasses from different sectors with promising potential for 2G bioethanol production, with various proposals, pretreatment and processing methods, and an integration of first- and second-generation bioethanol. Thus, every production will have pros and cons, as mentioned in the chapter, and each biomass has a different LCA to be analysed. Therefore, we see that second-generation bioethanol is already a reality today and has trends of large future **production.**

Acknowledgements The authors thank CNPq, FAPERGS, and CAPES.

Compliance with Ethical Standards This chapter was written according to ethical standards.

References

Abanoz K, Stark BC, Akbas MY (2012) Enhancement of ethanol production from potato-processing wastewater by engineering Escherichia coli using Vitreoscilla haemoglobin. Lett Appl Microbiol 55:436–443

Alzaabi MSMA, Mezher T (2021) Analysing existing UAE national water, energy and food nexus related strategies. Renew Sust Energ Rev 144:111031

Akhimien NG, Latif E, Hou SS (2021) Application of circular economy principles in buildings: a systematic review. J Build Eng 38:102041

Apiwatanapiwat W, Murata Y, Kosugi A et al (2011) Direct ethanol production from cassava pulp using a surface-engineered yeast strain codisplaying two amylases, two cellulases, and β-glucosidase. Appl Microbiol Biotechnol 90:377–384

Anderson JE, Wallington TJ (2020) Novel method to estimate the octane ratings of ethanol-gasoline mixtures using base fuel properties. Energy Fuel 34:4632–4642

Saeed AM, Hongzhi M, Yue S et al (2018) Concise review on ethanol production from food waste: development and sustainability. Environ Sci Poll Resear 25:28851–28863

Arapoglou D, Varzakas T, Vlyssides A et al (2010) Ethanol production from potato peel waste (PPW). Waste Manag 30:1898–1902

Assabjeu AC, Noubissié E, Desobgo SCZ et al (2020) Optimization of the enzymatic hydrolysis of cellulose of triplochiton scleroxylon sawdust in view of the production of bioethanol. Sci African 8:e00438

Ayodele BV, Alsaffar MA, Mustapa SI (2020) An overview of integration opportunities for sustainable bioethanol production from first- and second-generation sugar-based feedstocks. J Clean Prod 245:118857

Balat M (2011) Production of bioethanol from lignocellulosic materials via the biochemical pathway : a review. Energy Conv Manag 52:858–875

Banu JR, Preethi KS, Tyagi VK et al (2021) Lignocellulosic biomass based biorefinery: a successful platform towards circular bioeconomy. Fuel 302:121086

Beaugrand J, Cronier D, Bebeire P et al (2004) Arabinoxylan and Hydroxycinnamate content of wheat bran in relation to endoxylanase susceptibility. J Cereal Sci 40:223–230

Bera T, Inglett KS, Inglett PW et al (2021) Comparing first- and second-generation bioethanol byproducts from sugarcane: impact on soil carbon and nitrogen dynamics. Geoderma 384: 114818

Boluda-Aguilar M, García-Vidal L, González-Castañeda FP et al (2010) Mandarin peel wastes pretreatment with steam explosion for bioethanol production. Bioresour Technol 101:3506–3513

Boluda-Aguilar M, López-Gómez A (2013) Production of bioethanol by fermentation of lemon (Citrus Limon L.) peel wastes pretreated with steam explosion. Ind Crop Produc 41:188–197

Bonatto C, Scapini T, Zanivan J et al (2021) Utilization of seawater and wastewater from shrimp production in the fermentation of papaya residues to ethanol. Bioresour Technol 321:124501

Brusca S, Cosentino SL, Famoso F et al (2018) Second generation bioethanol production from Arundo donax biomass: an optimization method. Energy Proced 148:728–735

Buenrostro-Figueroa J, Tafolla-Arellano JC, Flores-Gallegos AC et al (2018) Native yeasts for alternative utilization of overripe mango pulp for ethanol production. Rev Arg Microbiol 50: 173–177

Buléon A, Colonna P, Planchot V et al (1998) Starch granules: structure and biosynthesis. Int J Biol Macromol 23:85–112

Cárdenas-Pérez S, Chanona-Pérez JJ, Güemes-Vera N et al (2018) Structural, mechanical and enzymatic study of pectin and cellulose during mango ripening. Carbohydr Polym 196:313–321

Cheng X, Zheng J, Lin A et al (2020) A review: roles of carbohydrates in human diseases through regulation of imbalanced intestinal microbiota. J Funct Food 74:104197

Choi S, Kim J-H, Wi SG et al (2013) Bioethanol production from mandarin (Citrus unshiu) peel waste using popping pretreatment. Appl Energy 102:204–210

Choi S, Lee YG, Khanal SK et al (2015) A low-energy, cost-effective approach to fruit and citrus peel waste processing for bioethanol production. Appl Energy 140:65–74

Citrus: World Markets and Trade USDA (2021). https://www.fas.usda.gov/data/citrus-world-markets-and-trade

Cohn R, Cohn AL (1997) Subproductos del procesado de las frutas. In: Arthey D, Ashurst PR (eds) Procesado de frutas. Acribia, Zaragoza, Spain

Cripwell RA, Favaro L, Viljoen-Bloom M et al (2020) Consolidated bioprocessing of raw starch to ethanol by Saccharomyces cerevisiae: achievements and challenges. Biotechnol Adv 42:107579

Dagnino EP, Felissia FE, Chamorro E et al (2017) Optimization of the soda-ethanol delignification stage for a rice husk biorefinery. Ind Crop Product 97:156–165

Demichelis F, Laghezza M, Chiappero M et al (2020) Technical, economic and environmental assessement of bioethanol biorefinery from waste biomass. J Clean Product 277:124111

Demiray E, Kut A, Karatay SE et al (2021) Usage of soluble soy protein on enzymatically hydrolysis of apple pomace for cost-efficient bioethanol production. Fuel 289:119785

Dranca F, Vargas M, Oroian M (2020) Physicochemical properties of pectin from Malus domestica 'Fălticeni' apple pomace as affected by nonconventional extraction techniques. Food Hydrocol 100:105383

Ebikade E, Athaley A, Fisher B et al (2020) The future is garbage: repurposing of food waste to an integrated biorefinery. ACS Sust Chem Eng 8:8124–8136

Edwards MC, Doran-Peterson J (2012) Pectin-rich biomass as feedstock for fuel ethanol production. Appl Microbiol Biotechnol 95:565–575

Favaro L, Cagnin L, Basaglia M et al (2017) Production of bioethanol from multiple waste streams of rice milling. Bioresour Technol 244:151–159

Food and Agriculture Organization (FAO) (2017) Crops and livestock products. http://www.fao.org/faostat/en/#data/QCL

Food and Agriculture Organization (FAO) (2021a) Crops and livestock products. http://www.fao.org/faostat/en/#data/QCL

Food and Agriculture Organization (FAO) (2021b) Food losses and food waste in Latin America and the Caribbean. http://www.fao.org/americas/noticias/ver/pt/c/239394/

Gabriel LS, Prestes RA, Pinheiro LA et al (2013) Multivariate analysis of the spectroscopic profile of the sugar fraction of apple pomace. Braz Arch Biol Technol 56:439–446

Gálvez-Martos JL, Greses S, Magdalena JA et al (2021) Life cycle assessment of volatile fatty acids production from protein- and carbohydrate-rich organic wastes. Bioresour Technol 321:124528

Gold M, Cassar CM, Zurbrügg C et al (2020) Biowaste treatment with black soldier fly larvae: increasing performance through the formulation of biowastes based on protein and carbohydrates. Waste Manag 102:319–329

Graham RL, Nelson R, Sheehan J et al (2007) Current and potential U.S. Corn Stover Supplies Agron J 99:1–11

Hafid HS, Rahman NA, Mokhtar MN et al (2017) Over production of fermentable sugar for bioethanol production from carbohydrate-rich Malaysian food waste via sequential acid-enzymatic hydrolysis pretreatment. Waste Manag 67:95–105

Harsono SS, Salahuddin FM, Purwono GS et al (2015) Second generation bioethanol from Arabica coffee waste processing at smallholder plantation in Ijen plateau region of East Java. Procedia Chem 14:408–413

Hogarth JR (2017) Evolutionary models of sustainable economic change in Brazil: no-till agriculture, reduced deforestation and ethanol biofuels. Environ Innov Soc Trans 24:130–141

Izmirlioglu G, Demirci A (2012) Ethanol production from waste potato mash by using *saccharomyces cerevisiae*. Appl Sci 2:738–753

Jiang J, Ding X, Isaacson KP et al (2021) Ethanol-based disinfectant sprays drive rapid changes in the chemical composition of indoor air in residential buildings. J Hazar Mat Lett 2:100042

Keles D, Choumert-Nkolo J, Motel PC et al (2018) Does the expansion of biofuels encroach on the forest? J Forest Econom 33:75–82

Keyhanpour MJ, Jahromi SHM, Ebrahimi H (2021) System dynamics model of sustainable water resources management using the nexus water-food-energy approach. Ain Shams Eng J 12: 1267–1281

Kundu D, Banerjee S, Karmakar S et al (2021) Valorization of citrus lemon wastes through biorefinery approach: an industrial symbiosis. Bioresour Technol Report 15:100717

Le Corre D, Bras J, Duresne A (2010) Starch nanoparticles: a review. Biomacromol 11:1139–1153

Lee RA, Lavoie JM (2013) From first- to third-generation biofuels: challenges of producing a commodity from a biomass of increasing complexity. Animal Front 3:6–11

Liu C-G, Li K, Wen Y et al (2019) Bioethanol: new opportunities for an ancient product. Adv Bioen 4:1–34

Lopes ML, Paulillo SCL, Godoy A et al (2016) Ethanol production in Brazil: a bridge between science and industry. Braz J Microbiol 47:64–76

Madu JO, Agboola BO (2018) Bioethanol production from rice husk using different pretreatments and fermentation conditions. 3. Biotech 8:1–6

Maisuthisakul P, Gordon MH (2009) Antioxidant and tyrosinase inhibitory activity of mango seed kernel by product. Food Chem 117:332–341

Martinez AG, Feiden A, Bariccatti R et al (2018) Ethanol production from waste of cassava processing. Appl Sci 8:2158

Mithra MG, Padmaja G (2016) Compositional profile and ultrastructure of steam and dilute sulfuric acid pretreated root and vegetable processing residues. Curr Biotechnol 7:288–301

Mithra MG, Jeeva ML, Sajeev MS et al (2018) Comparison of ethanol yield from pretreated lignocellulos-starch biomass under fed-batch SHF or SSF modes. Heliyon 4:e00885

Mohnen D (2008) Pectin structure and biosynthesis. Curr Opinion Plant Biol 11:266–277

Molinuevo-Salces B, Riano B, Hijosa-Valsero M et al (2020) Valorization of apple pomaces for biofuel production: a biorefinery approach. Biomass Bioenergy 142:105785

Morone P, Koutinas A, Gathergood N et al (2019) Food waste: challenges and opportunities for enhancing the emerging bioeconomy. J Clean Produc 221:10–16

Oberoi HS, Vadlani PV, Nanjundaswamy A et al (2011) Enhanced ethanol production from Kinnow mandarin (Citrus reticulata) waste via a statistically optimized simultaneous saccharification and fermentation process. Bioresour Technol 102:1593–1601

Organization for Economic Cooperation and Development/Food and Agriculture Organization (OECD/FAO) (2021) Agricultural Outlook 2021–2030. https://www.oecd.org/publications/oecd-fao-agricultural-outlook-19991142.htm

Paul M, Panda G, Mohapatra PK et al (2020) Study of structural and molecular interaction for the catalytic activity of cellulases: an insight in cellulose hydrolysis for higher bioethanol yield. J Mol Struc 1204:127547

Parashar A, Jin Y, Mason B et al (2016) Incorporation of whey permeate, a dairy effluent, in ethanol fermentation to provide a zero waste solution for the dairy industry. J Dairy Sci 99:1859–1867

Perea-Moreno A-J, Perea-Moreno M-Á, Dorado MP et al (2018) Mango stone properties as biofuel and its potential for reducing CO2 emissions. J Clean Produc 190:53–62

Rabelo SC, Filho RM, Costa AC (2013) Lime pretreatment and fermentation of Enzymatically Hydrolyzed Sugarcane Bagasse. Appl Biochem Biotechnol 169:1696–1712

Ray S, Raychaudhuri U, Chakraborty R (2016) An overview of encapsulation of active compounds used in food products by drying technology. Food Biosci 13:76–83

Reddy LVA, Reddy OVS (2011) Effect of fermentation conditions on yeast growth and volatile composition of wine produced from mango (Mangifera indica L.) fruit juice. Food Bioproduc Proces 89:487–491

Renewable Fuels Association (RFA) (2020) Focus forward: 2020 pocket guide to ethanol. https://ethanolrfa.org/wp-content/uploads/2020/02/2020-Outlook-Pocket-Guide-for-Web.pdf

Rivas B, Torrado A, Torre P et al (2008) Submerged citric acifd fermentation on orange peel autohydrolysate. J Agric Food Chem 56:2380–2387

Robak K, Balcerek M (2018) Review of second generation bioethanol production from residual biomass. Food Technol Biotechnol 56:174–187

Rosales-Calderon O, Arantes V (2019) A review on commercial-scale high-value products that can be produced alongside cellulosic ethanol. Biotechnol Biofuels 12:240

Santi G, Crognale S, D'annibale A et al (2014) Orange peel pretreatment in a novel lab-scale direct steam-injection apparatus for ethanol production. Biomass Bioenergy 61:146–156

Santos F, Eichler P, Queiroz JH et al (2020) Production of second-generation ethanol from sugarcane. Sugarcane Bioref Technol Perspec:195–228

Scapini T, Favaretto DPC, Camargo AF, et al. (2019) Bioethanol from fruit residues, in: Treichel H, Júnior S, FongaroG., üller, C (Eds.), ethanol as a green alternative fuel: insight and perspectives. pp. 139–176
Shahid MK, Batool A, Kashif A et al (2021) Biofuels and biorefineries: development, application and future perspectives emphasizing the environmental and economic aspects. J Environ Manag 297:113268
Sharma B, Larroche C, Dussap C-G (2020) Comprehensive assessment of 2G bioethanol production. Bioresour Technol 313:123630
Sharma HR, Chauhan GS, Agrawal K (2007) Physico-chemical characteristics of rice bran processed by dry heating and extrusion cooking. Int J Food Prop 7:603–614
Sharma K, Mahato N, Cho MH et al (2017) Converting citrus wastes into value-added products: economic and environmently friendly approaches. Nutrition 34:29–46
Shenoy D, Pai A, Vikas RK et al (2011) A study on bioethanol production from cashew apple pulp and coffee pulp waste. Biomass Bioenergy 35:4107–4111
Silva LF, Taciro MK, Raicher G et al (2014) Perspectives on the production of polyhydroxyalkanoates in biorefineries associated with the production of sugar and ethanol. Int J Biol Macromol 71:2–7
Silva MD, Santos VAQ, Ernandes FMP et al (2020) Acid hydrolysis of corn cob for the production of second generation ethanol by *saccharomyces cerevisiae* ATCC 26602. Int J Develop Resear 10:38871–38878
Silveira MHL, Morais RHC, Lopes AMC et al (2015) Current pretreatment Technologies for the Development of cellulosic ethanol and biorefineries. ChemSusChem 8:3366–3390
Sivamani S, Chandrasekaran AP, Balajii M et al (2018) Evaluation of the potential of cassava-based residues for biofuels production. Rev Environ Sci Biotechnol 17:553–570
Sivaramakrishnan R, Ramprakash B, Ramadoss G et al (2021) High potential of Rhizopus treated rice bran waste for the nutrient-free anaerobic fermentative biohydrogen production. Bioresour Technol 319:124193
Stolarski MJ, Krzyŝaniak M, Łuczyński M et al (2015) Lignocellulosic biomass from short rotation woody crops as a feedstock for second-generation bioethanol production. Ind Crop Produc 75: 66–75
Su T, Zhao D, Khodadadi M et al (2020) Lignocellulosic biomass for bioethanol: recent advances, technology trends, and barriers to industrial development. Curr Opinion Green Sust Chem 24: 56–60
Sydney EB, Letti LAJ, Karp SG et al (2019) Current analysis and future perspective of reduction in worldwide greenhouse gases emissions by using first and second generation bioethanol in the transportation sector. Bioresour Technol Report 7:100234
Szklo A, Schaeffer R (2006) Alternative energy sources or integrated alternative energy systems? Oil as a modern lance of Peleus for the energy transition. Energy 31:2513–2522
Talebnia F, Pourbafrani M, Lundin M et al (2008) Optimization of citrus wastes saccharification by dilute acid hydrolysis. BioResour 3:108–122
Tan KT, Lee KT, Mohamed AR (2008) Role of energy policy in renewable energy accomplishment: the case of second-generation bioethanol. Energy Policy 36:3360–3365
Thatoi H, Dash PK, Mohapatra S, Swain MR (2014) Bioethanol production from tuber crops using fermentation technology: a review. Int J Sust Energy 35:443–468
Tiwari S, Jadhav SK, Tiwari KL (2015) Bioethanol production from rice bran with optimization of parameters by Bacillus cereus strain McR-3. Int J Environ Sci Technol 12:3819–3826
Vaez S, Karimi K, Mirmohamadsadeghi S et al (2021) An optimal biorefinery development for pectin and biofuels production from orange wastes without enzyme consumption. Proces Saf Environ Protect 152:513–526
Venkatanagaraju E, Bharathi N, Rachiraju S et al (2020) Extraction and purification of pectin from agro-industrial wastes. Pectins–Extraction, Purification, Characterization and Applications, pp 1–15

Wang F, Ouyang D, Zhou Z et al (2021) Lignocellulosic biomass as sustainable feedstock and materials for power generation and energy storage. J Energy Chem 57:247–280

Weber CT, Trierweiler LF, Trierweiler JO (2020) Food waste biorefinery advocating circular economy: bioethanol and distilled beverage from sweet potato. J Clean Produc 268:121788

Widmer W, Zhou W, Grohmann K (2010) Pretreatment effects on orange processing waste for making ethanol by simultaneous saccharification and fermentation. Bioresour Technol 101: 5242–5249

Wiloso EI, Heijungs R, Snoo GR (2012) LCA of second generation bioethanol: a review and some issues to be resolved for good lca practice. Renew Sust Energy Rev 16:5295–5308

Xu Q, Liao Y, Cho E, Ko JH (2020) Effects of biochar addition on the anaerobic digestion of carbohydrate-rich, protein-rich, and lipid-rich substrates. J Air Waste Manag Assoc 70:455–467

Yamada S, Shinomiya N, Ohba K, Sekikawa M et al (2009) Enzymatic hydrolysis and ethanol fermentation of by-products from potato processing plants. Food Sci Technol Res 15:653–658

Ye G, Zeng D, Zhang S et al (2018) Ethanol production from mixtures of sugarcane bagasse and Dioscorea composita extracted residue with high solid loading. Bioresour Technol 257:23–29

Zanivan J, Bonatto C, Scapini T et al. (2021) Evaluation of bioethanol production from a mixed fruit waste by Wickerhamomyces sp. UFFS-CE-3.1.2. Bioenerg Resear in press

Zhang M, Xie L, Yin Z et al (2016) Biorefinery approach for cassava-based industrial wastes: current status and opportunities. Bioresour Technol 215:50–62

Zhao Y, Damgaard A, However X, al. (2019) Bioethanol from corn Stover –global warming footprint of alternative biotechnologies. Appl Energy 247:237–253

Ziaei-Rad Z, Fooladi J, Pazouki M et al (2021) Lignocellulosic biomass pretreatment using low-cost ionic liquid for bioethanol production: an economically viable method for wheat straw fractionation. Biomass Bioenergy 151:106140

Zouhair FZ, Benali A, Kabbour MR et al (2020) Typical characterization of argane pulp of various Moroccan areas: a new biomass for the second generation bioethanol production. J Saudi Soc Agric Sci 19:192–198

Chapter 9
Diversity and Use of Genetically Modified Microorganisms for Second-Generation Ethanol Production

Pooja and Sudesh Kumar Yadav

Abstract Realizing the need for alternate sources of energy and fuels as their present source fossil fuels are continuously depleting, exploration of renewable agricultural biomass could be very useful. Recovery of primary produce from agriculture generates a large amount of secondary biomass. Such biomass is composed of lignocellulose and can be converted to bioenergy, biofuels and platform chemicals. Efforts have been initiated for the production of such value-added products from agricultural biomass through chemical pretreatments. However, microorganisms have been identified for their potential role in the transformation of biomass into biofuels. In view of this, the present chapter describes the various microorganisms for their potential in biofuel/ethanol production from agricultural waste biomass. Furthermore, the potential of such microorganisms can be improved through manipulation of their machinery by genetic modifications. Based on these ideas, industries have already been set up and opened the great scope for biofuel production.

9.1 Introduction

The increasing energy demand and negative impact of fossil fuels on the environment have drifted towards the replacement of nonrenewable source-based energy. The worldwide consumption of fossil fuels has been gradually increasing with time, and more than 84% fossil-based carbon emissions have been reported since the 1980s. Recent reports cited an increase of approximately 7% in GHG emissions at the beginning of the current decade (Manochio et al. 2017). Therefore, awareness of alarming global warming issues, as well as depletion of fossil reserves, boosts the attempt to alter the current scenario by exploring sustainable and renewable energy sources more. Biofuels are perceived as viable contenders to replace fossil fuels due to their flexibility, abundant amount of feedstock available and reduction in

Pooja · S. K. Yadav (✉)
Center of Innovative and Applied Bioprocessing (CIAB), Mohali, Punjab, India
e-mail: sudesh@ciab.res.in

C. R. Soccol et al. (eds.), *Liquid Biofuels: Bioethanol*, Biofuel and Biorefinery Technologies 12, https://doi.org/10.1007/978-3-031-01241-9_9

greenhouse gas (GHG) emissions (Momayez et al. 2017). The global scenario is experiencing a continuous increase in biofuel production to combat climate change mitigation and energy security issues. To address the problem of increasing demand for transportation fuel without environmental pollution, blending ethanol with fossil fuels is recommended. The basic idea behind such implementation was promising properties of ethanol, such as anti-knock potential and cleaner combustion (Kannuchamy et al. 2016). Countries across the globe are adopting this strategy and plan to increase the blending of ethanol in a phased manner. Blending the generated ethanol from first-generation substrates has created energy securities concerns addressed in the manner of food versus fuel competition for industrial ethanol production (Karp and Richter 2011). Therefore, to eliminate food security concerns, the utilization of abundant lignocellulosic biomass could serve the potable purpose of industrial-scale ethanol production.

Photosynthesis and the resulting plant biomass are the only substantial sources of organic compounds in the terrestrial biosphere (Cavicchioli et al. 2019). Lignocellulosic biomass serves as a reliable feedstock for renewable energy since it is admittedly not in competition with food. In some countries, biomass, such as poplar, sunflower, and jatropha, is cultivated primarily for biofuel production (Adegboye et al. 2021). They have significant advantages over first-generation biomass feedstock because they are not used as food sources. Moreover, biofuel produced from lignocellulosic feedstock has been proven to be environmentally friendly, helps to reduce dependence on fossil fuel, serves as an alternative for declining petroleum reservoirs, and provides an economic improvement, especially to rural communities (Lin and Lu 2021). Apart from this, the primary product of photosynthesis, cellulosic biomass has disadvantage associated with it such as the biomass has evolved to be recalcitrant as the chemical linkages in the plant cell wall make the biomass impediment to deconstruction and reduce accessibility of enzymes (Sharma and Saini 2020; Sethupathy et al. 2021) Therefore, depolymerisation of the compact biomass can be performed by applying different physical, chemical, biological and combined pretreatment strategies. The production of ethanol from lignocellulosic biomass includes three major steps: pretreatment, hydrolysis and fermentation (Chen et al. 2016). Studies have demonstrated the use of various concentrations of acid and alkali for pretreating biomass to extract the maximum amount of sugars. However, scientists across the globe are searching for the best alternatives to reduce the cost of the process at the industrial scale. These developments are carried out at the pretreatment stage, exploring microbes that secrete economically beneficial enzymes as well as microorganisms that could support consolidated bioprocessing to reduce the overall cost of the system and make the processes feasible and reliable (Jiang et al. 2018).

Microorganisms inhabit different spaces in nature and are capable of producing a broad array of bioactive compounds that are used as fuels, drugs, and other important chemicals (Singh et al. 2005). The wild microorganisms present in the environment are quite difficult to isolate and characterize for a desired product, making the process quite laborious and time consuming. Moreover, these native organisms are generally not able to ferment all kinds of sugars available in the biomass for fermentation purposes, and during the pretreatment step, certain inhibitors are produced that hinder the growth and activity of these native microorganisms. Therefore, designing metabolically engineered microbes that can efficiently produce the desired product is necessary. Various strategies are being employed for designing metabolically engineered organisms for biofuel production, which can boon the economy.

There is a dire need to develop economically feasible industrial-scale methods, as pretreatment for the depolymerization of lignocellulosic biomass and the cost of enzymes are the most capital-consuming practices for the lignocellulose-based biofuel industry (Kumari and Singh 2018). Therefore, intensive research is being carried out globally to increase biofuel production from renewable energy sources while reducing the production cost parameters for sustainable industries. With advancements in technology, such as the development of microbial strains and their genetic engineering or optimization of fermentation parameters, augmentation in biofuel production has been observed over time. Metabolic engineering has become convenient to manipulate microbial metabolic pathways and produce numerous essential chemicals with an increase in the number of whole-genome sequenced organisms (Pirie et al. 2013; Simeonidis and Price 2015). Another approach to increase the number and types of bioactive compounds synthesized by microorganisms is the manipulation and evolution of different pathway enzymes (Lee et al. 2008).

Conventionally, microbial strains for industrial applications have been improvised by induced mutations and random mutagenesis. However, these improvement processes are largely uncontrolled, slow, and random. In some cases, the desired phenotype is not obtained even by mutagenesis and selection if the relevant biochemistry is missing in a microorganism (Park and Lee 2008). Therefore, the development of microorganisms for various purposes increasingly relies on rational engineering, including genetic and metabolic engineering (Liao et al. 2016). With the advent of engineering tools, classical approaches have been upgraded for improved titers and better product yields. Engineering has made it possible to modify the surface of cells by displaying novel proteins that may improve the biocatalytic capability of cells (Liu et al. 2015). For example, enzymatic conversion of starch into sugars for alcohol fermentation may improve if microorganisms display specific enzymes on their surface. Additionally, certain feedstocks can be converted to biofuels if cells display a suite of the relevant surface enzymes and have the relevant intracellular metabolic pathways.

Metabolic engineering complemented with genetic engineering is among the core technologies for the enhanced production of microbial biofuels (Fig. 9.1). It can provide microorganisms with improved titers and yields of products and capacity to utilize a broad range of carbon substrates. Metabolic engineering of microbes can also enhance the ability to better withstand high titers of potentially toxic products and adverse environmental conditions (Zhang et al. 2011). Continuing developments in synthetic biology, genetic engineering and manipulation of enzyme expression allow biofuel-producing pathways to be inserted in a microbial host. Metabolic engineering is a precise modification method that hampers the accumulation of unfavorable mutations.

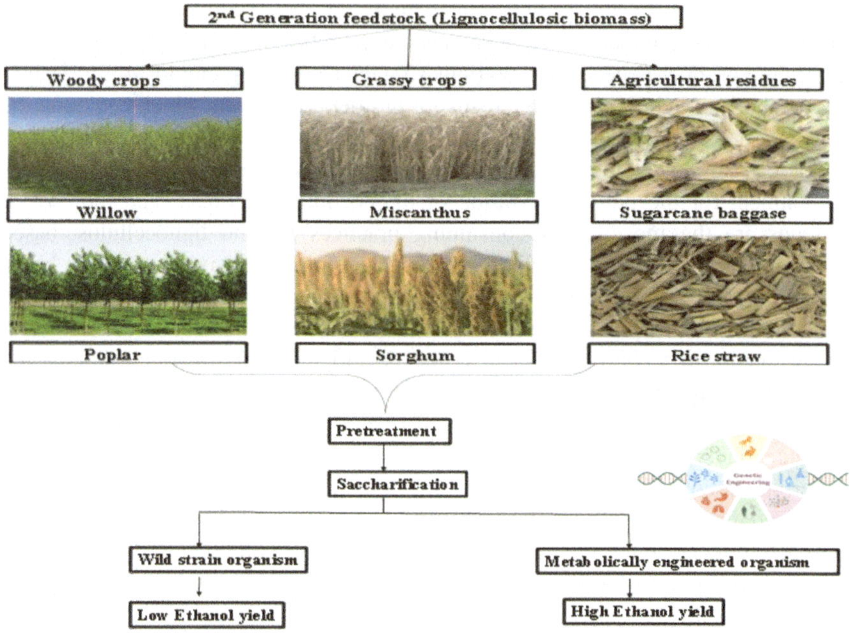

Fig. 9.1 Metabolically engineered organisms yield higher ethanol from different lignocellulosic feed stocks

9.2 Metabolic Engineering for Biofuels Production

Microorganisms are well known for their potential to produce biofuels. Conventional methods of ethanol production were not appropriate, as in some cases, the amount of desired product formed was very low. Some microbes isolated from the natural environment are not efficient at achieving the expected outcomes desired by humans, and it is well known that the identification and characterization of microorganisms with high fermentative performance is laborious and time-consuming. Therefore, to overcome these problems, the advent of metabolic engineering has contributed to the improved genetic machinery of wild microbes to more specific machinery for efficient product formation with sound profits (Lee et al. 2012). Transformation of a microbe metabolically requires some basic fundamentals to be clarified, such as knowledge of the biosynthetic pathway, expression and suppression of certain genes and mechanism of host metabolism to obtain the desired product. Microbial communities such as bacteria, yeast, and fungi are the pioneering hosts for metabolic engineering.

9.2.1 Various Approaches for Metabolic Engineering

Different approaches are being explored for engineering organisms, as shown in Fig. 9.2.

9.2.1.1 Rational and Intuitive Approach

This approach is basically used for diversified substrate utilization by organisms, for rerouting metabolic pathways and to eradicate byproduct formation (Lee et al. 2012). This approach includes *in silico* gene insertion technology, which allows microbes to utilize a broad range of substrates, eventually reducing the cost of the bioprocess with maximum conversion of biomass to biofuel. For example, *S. cerevisiae, a* widely accepted yeast with a high fermentation rate and ethanol tolerance, has been the emphasis of researchers across the globe to improve its pentose fermenting ability for higher ethanol yield. This yeast has been metabolically engineered by inserting two different genes from *Pichia stipitis* encoding xylose reductase and xylitol dehydrogenase to utilize xylose (Chu and Lee 2007). The utilization of diverse sugars by *S. cerevisiae* for ethanol production followed by genetic engineering can be booned to the bioethanol industry. Another approach of transporter engineering helps to improve microbial resistance against toxic compounds as well as target products. Elimination of byproducts also plays an important role in the efficient production of desired compounds. For example, the fermentation efficiency of *E. coli* is improved by minimizing byproduct formation, which could be achieved by disruption of the tricarboxylic acid cycle, which further eliminates the pathway for NADH oxidation (Causey et al. 2003).

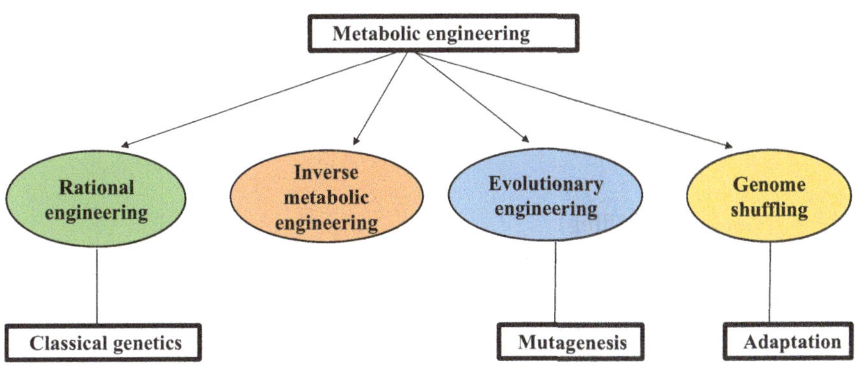

Fig. 9.2 Different approaches of metabolic engineering applied to microorganisms

9.2.1.2 Inverse Metabolic Engineering (IME)

Inverse metabolic engineering is a combinational approach for elucidation of a metabolic engineering strategy. It involves three key steps. The primary step is to identify, construct or calculate a desired phenotype. The secondary step is to determine the genetic or particular environmental factors conferring that phenotype. The third step is providing that phenotype on another strain or organism by directed genetic or environmental manipulation. For example, four endogenous *S. cerevisiae* genes producing improved alcohol tolerance have been identified through a sequence of IME techniques (Hong et al. 2010).

9.2.1.3 Evolutionary Engineering

Evolutionary genetic engineering is carried out using recombinant strains for enhanced ethanol productivity and increased tolerance towards stress. It includes random mutagenesis followed by cell selection with desired traits. Martín et al. (2007) reported enhanced stress tolerance levels in *S. cerevisiae* using an evolutionary engineering approach. In another study, xylose toxicity was removed in *S. cerevisiae* using evolutionary engineering. When the strain was grown on xylose at a concentration of 10 g/L, xylose toxicity was evident, and further subculturing of the strain consecutively for a third time led to enhanced ethanol productivity. (Kim et al. 2013). Therefore, this strategy can be very effective in strain improvement.

9.2.1.3.1 Mutagenesis

Mutagenesis is a process by which the genetic information of an organism is changed by the production of a mutation. It is either random, i.e., not locus specific or a site directed. Random mutations could be introduced through adaptation and protoplast fusion. Even UV can induce random mutations. In the case of *S. cerevisiae*, acetic acid tolerance was enhanced using UV mutagenesis (Zheng et al. 2011). For example, random mutations using the chemical agent ethyl methane sulfonate (EMS) have been introduced in *S. cerevisiae* to improve its ethanol production ability (Mobini-Dehkordi et al. 2008).

9.2.1.4 Genome Shuffling

Genome shuffling is usually carried out for strains that do not possess any genetic information (Heluane et al. 1993). The basics behind gene shuffling involve the creation of random mutations through genome transfer. Moreover, this approach promotes the transfer of larger segments of DNA, in contrast to the classical genetic engineering approach, which solely relies on vectors for gene transfer.

Adaptation of a strain involves exposing the selected strain to various stress conditions for a longer duration. It involves repeated passaging of the cultures to increasing concentrations of inhibitors in lignocellulosic hydrolysates. Generally, adaptive techniques are used to produce inhibitor-tolerant strains (Zhu et al. 2009; Liu et al. 2004).

Rerouting metabolic pathways is a better alternative to native pathways of organisms that cannot provide optimum concentrations of desired products. Other approaches are also very useful, such as classical metabolic engineering, which includes overexpression of genes, gene knockout approaches, construction of synthetic metabolic pathways, and tolerance engineering, also called inverse metabolic engineering, as discussed in the text (Sartaj et al. 2020).

9.3 Bacteria

Abundant bacterial populations have the capability to convert naturally available substrates to biofuels, such as biohydrogen, bioalcohols, and biogas. Scientific communities are continually looking for novel bacterial species in nature with enhanced production potential, particularly in thermophilic environments, as thermophiles are thought to be more robust for cellulose degradation and hydrogen production (Mehta et al. 2016). Certain microorganisms that are extensively used for bioethanol production are discussed below.

9.3.1 Mesophilic Bacteria

9.3.1.1 *Zymomonas mobiliz*

The diversity of microorganisms has been explored for bioethanol production, one of the most commonly produced biofuels. Ethanol has been used as an additive to gasoline due to its clean burning characteristics and significant reduction in greenhouse gas emissions. Bacterial species are well known for their potential to produce value-added products. *Zymomonas mobiliz* is an anaerobic organism that has been best explored for its bioethanol production capacity. *Zymomonas* has many industrially desirable characteristics, such as high ethanol tolerance, high specific activity and a broad range of pH values (3.5–7.5) (Yang et al. 2016). Although yeasts are preferred over bacterial species, the efficiency of conversion of glucose to ethanol by *Z. mobiliz* is much more proficient, with a drawback of its inability to utilize most natural sugars. *Z. mobiliz* uses the Entner-Doudoroff (ED) pathway for ethanol production, and the energy consumption in the form of ATP is much less in the case of the ED pathway, with 50% more efficiency than the EMP pathway used by *S. cerevisiae* during the ethanologenic process. Cells of *Z. mobiliz* are smaller in size; therefore, a large area is available for glucose uptake, which makes it much more convenient for *Z. mobiliz* to consume glucose at a much higher rate than *S. cerevisiae* (Conway 1992). These factors are collectively responsible for higher ethanol productivity by *Z. mobiliz* than by *S. cerevisiae*. In the case of *Z. mobiliz*, ethanol production is mainly controlled by PET operon-encoding enzymes such as pyruvate decarboxylase and alcohol dehydrogenase (Ingram and Conway 1988). One of the major drawbacks in ethanol production by *Z. mobiliz* is that it is unable to utilize pentose sugars, which are cheap and abundantly available in lignocellulosic biomass (Xia et al. 2019). Therefore, analysing the need for pentose sugar utilization metabolic engineering has emerged as an advanced tool for improving the sugar utilization capability in *Z. mobiliz*. Ethanol production using this engineering technology is being commercialized by DuPont, the largest cellulosic ethanol plant in the United States. Moreover, a recombinant *Z. mobiliz* strain TMY-HFPX with multiple improved characteristics was developed and used for very high gravity (VHG) fermentation, which is the mainstream technology in the ethanol industry. VHG fermentation requires the strains to be resistant to multiple stresses, such as high glucose concentration, high ethanol concentration, high temperature and harsh acidic conditions. Therefore, the engineered strain harbored multiple gene modules, such as the metB/yfdZ operon for lysine and methionine biosynthesis, the thioesterase gene tesA to enhance free fatty acid biosynthesis to increase ethanol tolerance, mainly the xylA/xlyB/tktA/talB operon for xylose utilization, and a small heat shock protein operon for heat stress tolerance. This engineered strain has been shown to achieve an ethanol concentration of up to 136 g/L in a solution containing 295 g/L glucose (90% of theoretical yield) without requiring exogenous amino acids and vitamins (Wang et al. 2016). To date, this engineered strain is significantly superior to that produced by all the reported ethanol-producing strains.

9.3.1.2 *Escherichia coli*

Currently, the majority of bioethanol is produced using mesophilic microorganisms. Researchers have tried engineering bacterial species such as *Bacillus* and *E. coli* to develop ethanol-producing bacteria. *Bacillus subtilis* has been most widely used to construct ethanol-producing strains because the molecular biology of these microorganisms is well understood. They are capable of utilizing diverse substrates and are amenable to metabolic engineering for exclusive production of ethanol (Kim et al. 2010; Soo et al. 2017). *Escherichia coli* was the first successful bacterium engineered for ethanol production, as it is able to utilize both pentose and hexose sugars. This initial success of engineering *E. coli* strains was further built upon by the introduction of relevant foreign genes and disrupting pathways for competing products. Thus, *E. coli* W (wild) was transformed into *E. coli* KO11. Later, a novel strain was found to be capable of producing ethanol (Ajit et al. 2017). *E. coli* KO11 contained genes from Z. mobiliz encoding pyruvate decarboxylase and alcohol dehydrogenase (PET operon), as they are responsible for anaerobic production of ethanol. The engineered *E. coli* KO11 was successfully adapted to produce ethanol at nearly 95% of the theoretical yield in a complex medium and showed increased ethanol tolerance relative to the original host. However, the inability of this strain to grow in ethanol at a concentration of 3.5% and its complex nutritional requirements might eventually add up to the cost of ethanol production (Ajit et al. 2017).

In another approach to overcome this tolerance issue in *E. coli*, strain KO11 was subjected to conventional techniques of adaptive evolution and selection through long-term exposure to media supplemented with increasing ethanol content to increase its ethanol tolerance to 10%. This engineered strain, LY01, and its parent strain KO11 were not as economical as they required complex nutritional supplements and, therefore, enhanced the expenses of ethanol production. Further considering the complex media requirements, KO11 was engineered to strain SZ110 to allow ethanol production in less nutritionally demanding media. Strain SZ110 was engineered by eliminating and inserting certain genes. Engineering through transposon-mediated mutagenesis and metabolic evolution provided strain LY168. Functional selections were subsequently performed by serial transfers in mineral salt media without antibiotics. Eventually, the combination of genetic engineering and long-term adaptation resulted in microbial biocatalysts that produced up to 45 g ethanol/L in 48 h in a simple mineral salt medium.

9.3.1.3 *Bacillus subtilis*

Bacillus subtilis is a gram-positive bacterium and is generally recognized as safe (GRAS). *Bacillus subtilis* is amenable to genetic manipulation as well (Liu et al. 2017). This bacterium has an optimal temperature of 37 °C and can grow at temperatures up to 50 °C. The ability of *Bacillus subtilis* to ferment multiple carbohydrates from mono-, di-, oligo-, and polysaccharides makes it a favorable organism for ethanol production. In this context, the ability of *B. subtilis* to utilize starch, xylan, galactan, pullulan, arabinan, rhamnogalacturonan, and pectin is quite fascinating in regard to converting plant biomass wastes into value-added biotechnological products. The capability of *Bacillus subtilis* to produce and efficiently secrete various hydrolytic enzymes makes it feasible to survive on different carbohydrates. Therefore, for complete degradation of plant biomass, it is advantageous to extend the substrate range of *B. subtilis* by the expression of exogenous genes encoding novel enzymes (Chen et al. 2016). To date, the only successful attempt to develop an ethanologenic strain of *B. subtilis* has been reported by Romero et al. 2007. They engineered an exogenous ethanol pathway using heterologous expression of *Z. mobiliz* genes encoding pyruvate decarboxylase (pdc) and alcohol dehydrogenase (adhB) to create a novel strain. The native genes encoding lactate dehydrogenase (ldh) and acetolactate synthase (alsS) were knocked out to obstruct lactate and 2,3-butanediol production, respectively, as major fermentation products of *B. subtilis*. Further insertion and deletion of genes followed by the redirection of fermentative metabolism, the resulting strain was able to produce 8.9 g/L ethanol from 20 g/L glucose during 9 days of fermentation under nonaerated conditions (Soo et al. 2017).

Certain drawbacks associated with mesophilic bacteria such as *E. coli* and *B. subtilis* are their poor ability to hydrolyze carbohydrate polymers. They show poor tolerance towards extreme pH values and an inability to withstand high salt concentrations (Jin et al. 2014). Therefore, bioprocesses carried out using these microorganisms become contaminated easily with undesirable species, making them unsuitable for use in large-scale production operations.

9.3.2 *Thermophilic Bacteria*

Thermophilic bacteria are preferred over mesophilic bacteria, as these microorganisms have a high ability to hydrolyze lignocellulosic materials and simultaneously convert the produced sugars into ethanol. The potential of thermophilic bacteria for enhancing the rate of biomass hydrolysis and fermentation of sugars into ethanol has been well explored. Moreover, an additive advantage of thermophiles over mesophilic bacteria is that they also reduce the chances of contamination of the fermentation process by unwanted microbes (Scully and Orlygsson 2015). Genetically engineered thermophilic bacteria have been extensively investigated for the production of ethanol, as summarized in Table 9.1.

9.3.2.1 *Clostridium* Spp.

Generally, bacteria belonging to the genus *Clostridium* are Gram-positive. They are obligate anaerobes and are spore-forming microorganisms with the capability to utilize both simple and complex carbohydrates. Members in this genus can ferment a wide range of sugars (hexoses, pentoses, and disaccharides) to produce organic acids (Wu et al. 2012). The bacteria in this genus are able to act on recalcitrant structures and degrade lignin-containing materials, such as lignocellulosic waste. These bacteria contain a large extracellular multiprotein complex called a cellulosome that contains multiple cellulases and hemicellulases, helping in the degradation of lignocellulosic components of biomass (Mitchell 1997). Starches and simple sugars derived from sugar cane and corn are the most commonly used feedstocks for the industrial production of biofuels. Different *Clostridium* species have been explored for industrial purposes. *C. acetobutylicum* has already been used successfully for the industrial production of acetone and butanol from starch and molasses. *C. acetobutylicum* ATCC 824 and *C. beijerinckii* BA101 were proven to increase n-butanol tolerance and n-butanol production from carbohydrates. *Clostridium* sp. can efficiently utilize lignocellulosic biomass for biofuel production after pretreatment. One of the key role players among *Clostridium* sp. is *C. thermocellum,* which can carry out bulk conversion of cellulosic wastes into ethanol (Brown et al. 2011). The major obstacle of *Clostridium* biofuel fermentation to an economic process is the production of the side products acetate and lactate, which could limit the conversion yield (Demain et al. 2005). Other limitations

Table 9.1 Genetically engineered thermophilic bacteria

Organisms	Genetic modification	Substrate	Fermentation mode	Pretreatment	Ethanol yields (mM/g)	References
Clostridium thermocellum	*ΔpyrF*	Avicel (2.5 g/L)	Batch	A	5	Lynd et al. (1989)
Clostridium thermocellum	*Δldh, Δpta*	Whatman paper (8.0 g/L)	Batch	None	7.20–8.00	Rani et al. (1997)
Clostridium thermocellum	*Δldh, Δpta*	Paddy straw (8.0 g/L)	Batch	None	6.10–8.00	Rani et al. (1997)
Clostridium strain DBT-IOC-C19	–	Avicel (10.0 g/L)	Batch	None	3.26	Singh et al. (2017)
Clostridium strain AK1	–	Hemp (5.0 g/L)	Batch	A/Alk	3.5	Örlygsson (2012)
Thermoanaerobacter pentosaceus	–	Rapeseed straw (5 0 g/L)	Con	Alk	1.4	Tomas et al. (2013)
Thermoanaerobacter mathranii	*Δldh*	Wheat straw (6.7 g/L)	Batch	WO/E	2.61	Ahring et al. (1999)
Thermoanaerobacter ethanolicus	*Δldh*	Beet molasses (30.0 g/L)	Batch	None	4.81	Avci and Dönmez (2006)
Thermoanaerobacter mathranii BG1L1	*Δldh*	Wheat straw (30.0–120.0 g/L)	Batch	WO/E	8.50–9.20	Georgieva et al. (2007)
T. Ethanolicus	*Δtdk, ΔadhE*	Wood HL (8.0 g/L)	Batch	E	3.30–4.50	Carreira et al. (1983)
Thermoanaerobacter strain AK5	–	Grass (4.5 g/L)	Batch	A/E	4.31	Brynjarsdottir et al. (2012)
Thermoanaerobacter strain J1	–	Hemp (4.5 g/L)	Batch	A/E	4.3	Jessen and Orlygsson (2012)
T. Saccharoylticum	*Δldh*	Xylan (10.0 g/L)	Batch	WO	6.3	Ahring et al. (1996)
Thermoanaerobacterium strain AK17	–	Grass (2.5 g/L)	Batch	A/Alk/E	5.5	Almarsdotti et al. (2012)
T. Saccharolyticum TD1	*Δldh*	Xylose	Batch	None	0.98 Mol/Mol	Biswas et al. (2014)
T. Saccharolyticum ALK2	*Δldh, Δack, Δpta*	Cellobiose	Continuous	None	ND	Shaw et al. (2008)

(continued)

Table 9.1 (continued)

Organisms	Genetic modification	Substrate	Fermentation mode	Pretreatment	Ethanol yields (mM/g)	References
Geobacillus thermoglucosidasius TM242	*Δldh-ΔpflB-*	Glucose	Batch	None	1.73 Mol/Mol	Cripps et al. (2009)

[a]A-acid; Alk-alkaline; E-enzymatic; WO-wet oxidation

Δldh-lactate dehydrogenase, Δack-acetate kinase, Δpta-phosphate acetyl transferase, ΔpflB-pyruvate formate lyase Δtdk-thymidine kinase, ΔadhE-acetaldehyde alcohol dehydrogenase, ΔpyrF-orotidine-5-phosphate decarboxylase include low yields, low productivity, low final product concentration, and inhibition by end products. A combination of recombinant DNA technology and metabolic engineering research could be very useful to overcome these bottlenecks.

9.4 Yeasts

9.4.1 Saccharomyces Cerevisiae

S. cerevisiae is a well-known yeast and has been the mainstay of bioethanol production from sugars. During ethanol production, yeasts are preferred over bacterial species due to their high level of alcohol tolerance in comparison to wild-type bacterial strains such as *E. coli*. It is known that *S. cerevisiae* is one of the renowned potential yeasts for ethanol production across the globe, with an ethanol yield on sugars approaching the theoretical limit of 0.51 g ethanol/g of glucose (Jansen et al. 2017). Ethanol is the largest-volume product in industrial biotechnology (Stewart et al. 1983). *S. cerevisiae* is used for the production of bioethanol mainly by fermentation of sugar-based feed stocks such as sugarcane, sweet sorghum, and sugar beet and hydrolysis of starch. From the past two decades, a large international effort has aimed to access abundantly available agricultural and forest residues for fuel production. Studies have shown that nearly 93% of the glucose provided during fermentation to *Saccharomyces* is efficiently converted to ethanol, and the remaining part is mostly converted to cell biomass (Majidian et al. 2018). *Saccharomyces* uses the glycolysis pathway to produce ethanol; generally, one molecule of glucose is used to produce two pyruvate molecules. Under anaerobic conditions, pyruvate is reduced to 0.511 ethanol by a reaction catalyzed by alcohol dehydrogenase, one which also produces 0.489 carbon dioxide. The generation of two ATPs during glycolysis is helpful in driving yeast cell biosynthesis, which also involves a variety of energy-dependent reactions. Therefore, ethanol production is tightly coupled with yeast cell growth. Apart from CO_2, glycerol, organic acids, and higher alcohols may be produced during fermentation, which may inevitably result in the production of intermediate compounds. The formation of such compounds may ultimately lead to a lower ethanol yield. The main problem with native *S. cerevisiae* is its inability to utilize sugar polymers for ethanol production. *Saccharomyces* cannot utilize starch or cellulose to produce ethanol (Lin et al. 1998). The inability of yeast to convert sugar polymers to ethanol makes prior hydrolysis crucial for efficient ethanol production. Pretreatment (hydrolysis) accounts for more than 30% of the cost of the process, making it industrially unfavorable. The industry favors an extremely cheap process for competitive fuel production. Therefore, to reduce the process cost instead of pretreatment, metabolic engineering plays a role by engineering *S. cerevisiae* to enhance its hydrolytic capability to utilize polymeric substrates directly. Certain studies are being carried out to engineer yeast for efficient ethanol production. An engineered strain capable of co-fermenting mixtures of xylose and cellobiose with an ethanol productivity of nearly 0.65 g/L/h has been reported

(Young et al. 2010). The reason behind engineering of *S. cerevisiae* was to make it capable of fermenting sugar polymers.

A significant transformation could be the introduction of very high gravity (VHG) fermentation technology in industrial ethanol production. In this VHG fermentation, mashes with larger than 27 g dissolved solids/100 g mash can be batch-fermented with all substrates present at zero time and without the use of conditioned or genetically modified *S. cerevisiae*. This technology has led to the production of 23.8% v/v ethanol in the laboratory from wheat mash containing 38% w/v dissolved solids (Bayrock and Michael Ingledew 2001). Various stresses are encountered by yeast cells during VHG fermentations, such as environmental stresses such as nutrient deficiency, high temperature, and contamination, while the others are from yeast cell metabolism, such as ethanol accumulation and its corresponding inhibition of yeast cell growth and ethanol production (Bai et al. 2008).

Considering the current status of ethanol production, a major tilt towards second-generation biofuel production is seen. This could be due to the availability of inexpensive substrates in the form of lignocellulosic biomass, an abundant heterogeneous renewable resource from plants. However, such substrates often contain high levels of inhibitors, either present in the substrate itself or generated during the pretreatment process (Palmqvist and Hahn-Hägerdal 2000). This has necessitated the development of engineered microorganisms for the transformation of biomass with higher concentrations of inhibitors. Moreover, lignocellulosic biomass contains readily fermentable hexoses such as glucose and mannose. It also contains substantial amounts of pentoses D-xylose and L-arabinose. Since wild-type *S. cerevisiae* strains are unable to utilize pentoses derived from hemicellulose and pectin polymers in plant biomass, several strains of *Saccharomyces* and *Clostridium* have been metabolically engineered for higher ethanol yields to further improve tolerance to stresses and inhibitors, as well as pentose fermentation capability (Table 9.2).

9.4.2 Thermotolerant Yeasts

Mesophilic microorganisms are widely used for fermentation purposes. Thermoethanologenic yeasts are receiving considerable interest due to current challenges of increasing temperature, which could possibly overcome numerous difficulties. The use of thermophilic/thermotolerant yeast for bioethanol production has several process advantages, including a broad substrate utilization range, higher saccharification and fermentation rates, minimized contamination risk, lower costs of pumping and stirring, no cooling problems, and less energy requirements for mixing and product recovery (Dung et al. 2012; Kumar et al. 2013; Arora et al. 2015a; Scully and Orlygsson 2015). The use of thermotolerant yeasts offers important advantages in the production of bioethanol (Arora et al. 2015a, b; Caspeta and Nielsen 2015), and the potential of these yeasts can be further enhanced by metabolic engineering. *Kluyveromyces marxianus*, a thermotolerant yeast, ferments glucose to ethanol and can grow at a temperature of nearly 50 °C. In addition, *K. marxianus* has

Table 9.2 Genetically engineered microorganisms for second-generation ethanol production

Organisms	Product	Pathway or key enzymes expressed	Substrate	References
Clostridium cellulovorans	Ethanol, n-butanol	Fatty acyl-ACP reductase-dependent	Cellulose	Yang et al. (2015)
C. Autoethanogenum	Ethanol	Ferredoxin oxidoreductase	Synthetic medium	Liew et al. (2017)
Clostridium thermocellum and *Thermoanaerobacterium saccharolyticum*	Ethanol	Embden-Meyerhof	Cellulose	Argyros et al. (2011)
S. cerevisiae ADAP8	Ethanol	Evolved Orpinomyces XI strain overexpressing XKS1 and SUT1	Xylose	Madhavan et al. (2009a, b)
S. cerevisiae INVSc1/pRS406XKS/pILSUT1/pWOXYLA	Ethanol	Orpinomyces XI strain overexpressing XKS1 and SUT1	Xylose	Madhavan et al. (2009a, b)
S. cerevisiae RWB202-AFX	Ethanol	Piromyces XI evolved isolate	Xylose	Van Maris et al. (2007)
S. cerevisiae TMB3057	Ethanol	XR-XDH strain overexpressing XKS1 and PPP Δgre3	Xylose	Karhumaa et al. (2007)
S. cerevisiae TMB3066	Ethanol	Piromyces XI strain overexpressing XKS1 and PPP Δgre3	Xylose	Hahn-Hägerdal et al. (2007)
S. cerevisiae strain XUSAE57	Ethanol	Xylose-isomerase pathway	Xylose and glucose	Ko et al. (2020)

the ability to use a variety of substrates, such as cellobiose, xylose, xylitol, arabinose, glycerol, lactose, and inulin. It has been claimed that *K. marxianus* provides a significantly higher ethanol yield than native *S. cerevisiae* using inulin hydrolysates. (Hu et al. 2012). Similarly, another thermotolerant yeast, *P. kudriavzevii*, has been reported to ferment cassava hydrolysate to ethanol and achieved an ethanol concentration of 78.6 g/L in approximately 24 h at 40 °C (Yuangsaard et al. 2013). An ethanol productivity of 3.28 g/L/h with a theoretical yield of 85.4% was also reported for this process.

9.5 Industrial Perspective of Ethanol Production

In the era of modern technology, industrial biotech companies are making use of potential microorganisms and specialized proteins to develop neo biobased products from renewable and sustainable agricultural sources. Industrial biotechnology is the key technology for ethanol production from renewable feedstocks. Considering the current scenario, a joint study by USDA/DOE in 2005 concluded that the US could produce 60 billion gallons of ethanol by 2030 using grains and cellulosic feedstock without hampering food, feed and fiber production (Hettenhaus 2006).

The second generation of bioethanol production relies on different feedstocks, primarily paddy straw, wheat straw, and bagasse, among others. Although 2G ethanol production technology from agricultural residues is fairly well known around the globe, India seeks to diversify its energy basket by upgrading from the pilot scale to setting up demonstration plants producing ethanol that are commercially viable. In India, twelve production plants are being set up for processing approximately 400–500 tons of agricultural residues and are converted into ethanol along with byproducts in the form of biogas, biomanure and bioelectricity (Zhou et al. 2021).

Across the globe, pioneers in ethanol production are the United States of America, Canada, Europe and Brazil. Their operational production units have the capacity to handle biomass at 250–1500 tons per day. Continuous efforts have been made to develop efficient and cost-effective technologies that reduce bioethanol production costs in recent decades. After years of research and development, various cellulosic ethanol pilot and demonstration plants have started operations around the world. The first industrial cellulosic ethanol plant in the world was set by Beta Renewable in 2012. Then, by 2015, Beta Renewable was reported to operate on a daily basis in Italy with 40 MMgy plants, shipping cellulosic ethanol to Europe (Prasad et al. 2020). Proceeded by DuPont, it started producing cellulosic ethanol in Nevada, USA, at its 30-MMgy plant. Abengoa celebrated the inaugural 25 MMgy cellulosic ethanol plant in Hugoton, Kansas, USA in 2015. In contrast, Raizen started operations at its 40 MMgy cellulosic ethanol plant in 2014. In 2014, GranBio started up a cellulosic ethanol plant with a capacity of 20 MMgy in Brazil. In 2014, POET-DSM Advanced Biofuels, a 50/50 joint venture between Royal DSM (Netherlands) and POET, LLC (USA), opened its Project Liberty facility in Emmetsburg, Iowa, USA. The cellulosic ethanol facility was set to produce 20 MMgy of ethanol and then ramp up to 25 MMgy. Praj Industries is the leading biobase technology company with 1G ethanol-producing plants established across the globe. Its first 2G ethanol demonstration plant was established in India in 2017 with a production capacity of one million liters of ethanol per year. By 2047, Praj Industries targeted deriving 30% energy from biomass. Praj, Asia's first integrated second-generation biorefinery produces 3000 liters of ethanol per day using 10 tons of raw material. Ethanol production by Praj includes cofermentation using genetically modified organisms that can ferment both C5 and C6 sugars to produce ethanol from potential substrates such as corn cob, corn stover, wheat straw, bamboo, wood chips and cotton stalk (Silveira et al. 2018). Another plant set up by DBT-ICT, operational under the name of India Glycols Limited located in Kashipur, India, is based on zero waste technology that can produce ethanol in just 24 hours from any kind of biomass.

9.6 Conclusion

The advent of metabolic engineering is proving useful for the production of biofuels such as ethanol. The diversity of microorganisms, including bacteria, yeasts, cyanobacteria, and microalgae, can be metabolically engineered for desired products. In comparison to slow methods, such as traditional random mutagenesis or selection for strain improvement, metabolic engineering is rational, quick, and enormously powerful. More specifically, metabolic engineering has made it convenient that entire new pathways can be introduced into suitable microorganisms to enable the production of biofuels and precursors of interest. Lignocellulosic biomass, which is abundantly produced and is left underutilized by some or other means, can be efficiently converted to ethanol without using harsh treatment methods. Either an entire pathway introduction or, alternatively, parts of specific biochemical pathways may be engineered to optimize the production of a biofuel. Recent advantages in metabolic engineering have improved ethanol yields mostly by knocking out pathways leading to the formation of undesirable products. Commercialization of biofuels will inevitably require microbial cell factories that are substantially superior to wild types and some of the engineered strains now available. Genetic and metabolic engineering in combination with synthetic biology and systems biology are the key to generating highly capable cell factories for the production of biofuels.

References

Adegboye MF, Ojuederie OB, Talia PM et al (2021) Bioprospecting of microbial strains for biofuel production: metabolic engineering, applications, and challenges. Biotechnol Biofuels 14:1–21

Ahring BK, Jensen K, Nielsen P et al (1996) Pretreatment of wheat straw and conversion of xylose and xylan to ethanol by thermophilic anaerobic bacteria. Bioresour Technol 58:107–113

Ahring BK, Licht D, Schmidt AS et al (1999) Production of ethanol from wet oxidized wheat straw by *Thermoanaerobacter mathranii*. Bioresour Technol 68:3–9

Ajit A, Sulaiman AZ, Chisti Y (2017) Production of bioethanol by Zymomonas mobiliz in high-gravity extractive fermentations. Food Bioprod Process 102:123–135

Almarsdotti AR, Sigurbjornsdottir MA, Orlygsson J (2012) Effect of various factors on ethanol yields from lignocellulosic biomass by *Thermoanaerobacterium* AK17. Biotechnol Bioeng 109:686–694

Argyros DA, Tripathi SA, Barrett TF et al (2011) High ethanol titers from cellulose by using metabolically engineered thermophilic, anaerobic microbes. Appl Environ Microbiol 77:8288–8294

Arora R, Behera S, Kumar S (2015a) Bioprospecting thermophilic/thermotolerant microbes for production of lignocellulosic ethanol: a future perspective. Renew Sust Energ Rev 51:699–717

Arora R, Behera S, Sharma NK et al (2015b) A new search for thermotolerant yeasts, its characterization and optimization using response surface methodology for ethanol production. Front Microbiol 6:889

Avci A, Dönmez S (2006) Effect of zinc on ethanol production by two *Thermoanaerobacter* strains. Process Biochem 41:984–989

Bai FW, Anderson WA, Moo-Young M (2008) Ethanol fermentation technologies from sugar and starch feedstocks. Biotechnol Adv 26:89–105

Bayrock DP, Michael Ingledew W (2001) Application of multistage continuous fermentation for production of fuel alcohol by very-high-gravity fermentation technology. J Ind Microbiol 27: 87–93

Biswas R, Prabhu S, Lynd LR et al (2014) Increase in ethanol yield via elimination of lactate production in an ethanol-tolerant mutant of *clostridium thermocellum*. PLoS One 9:e86389

Brown SD, Guss AM, Karpinets TV et al (2011) Mutant alcohol dehydrogenase leads to improved PNAS ethanol tolerance in *clostridium thermocellum*. Proc Natl Acad Sci 108:13752–13757

Brynjarsdottir H, Wawiernia B, Orlygsson J (2012) Ethanol production from sugars and complex biomass by *Thermoanaerobacte*r AK5: the effect of electron-scavenging systems on end-product formation. Energy Fuel 26:4568–4574

Carreira LH, Wiegel JÜRGEN, Ljungdahl LG (1983) Production of ethanol from biopolymers by anaerobic, thermophilic, and extreme thermophilic bacteria. I. Regulation of carbohydrate utilization in mutants of *Thermoanaerobacter ethanolicus*. Biotechnol Bioeng Symp 13(830567)

Caspeta L, Nielsen J (2015) Thermotolerant yeast strains adapted by laboratory evolution show trade-off at ancestral temperatures and preadaptation to other stresses. MBio 6:e00431–e00415

Causey TB, Zhou S, Shanmugam KT et al (2003) Engineering the metabolism of *Escherichia coli* W3110 for the conversion of sugar to redox-neutral and oxidized products: homoacetate production. Proc Natl Acad Sci 100:825–832

Cavicchioli R, Ripple WJ, Timmis KN et al (2019) Scientists' warning to humanity: microorganisms and climate change. Nat Rev Microbiol 17:569–586

Chen J, Zhao L, Fu G et al (2016) A novel strategy for protein production using nonclassical secretion pathway in *Bacillus subtilis*. Microb Cell Factories 15:1–16

Chu BC, Lee H (2007) Genetic improvement of *Saccharomyces cerevisiae* for xylose fermentation. Biotechnol Adv 25:425–441

Conway T (1992) The Entner-Doudoroff pathway: history, physiology and molecular biology. FEMS Microbiol Rev 9:1–27

Cripps RE, Eley K, Leak DJ et al (2009) Metabolic engineering of *Geobacillus thermoglucosidasius* for high yield ethanol production. Metabol Eng 11:398–408

Demain AL, Newcomb M, Wu JD (2005) Cellulase, clostridia, and ethanol. Microbiol Mol Biol Rev 69:124–154

Dung NTP, Thanonkeo P, Phong HX (2012) Screening useful isolated yeasts for ethanol fermentation at high temperature. Int J Appl 2:65–71

Georgieva TI, Mikkelsen MJ, Ahring BK (2007) High ethanol tolerance of the thermophilic anaerobicethanol producer Thermoanaerobacter BG1L1. Cent Eur J Biol 2:364–377

Hahn-Hägerdal B, Karhumaa K, Jeppsson M et al (2007) Metabolic engineering for pentose utilization in *Saccharomyces cerevisiae*. Biofuels 108:147–177

Heluane H, Spencer JFT, Spencer D et al (1993) Characterization of hybrids obtained by protoplast fusion, between *Pachysolen tannophilus* and *Saccharomyces cerevisiae*. Appl Microbiol Biotechnol 40:98–100

Hettenhaus J (2006) Achieving sustainable production of agricultural biomass for biorefinery feedstock. Ind Biotechnol 2:257–275

Hong ME, Lee KS, Yu BJ et al (2010) Identification of gene targets eliciting improved alcohol tolerance in *Saccharomyces cerevisiae* through inverse metabolic engineering. J Biotechnol 149:52–59

Hu N, Yuan B, Sun J et al (2012) Thermotolerant *Kluyveromyces marxianus* and *Saccharomyces cerevisiae* strains representing potentials for bioethanol production from Jerusalem artichoke by consolidated bioprocessing. Appl Microbiol Biotechnol 95:1359–1368

Ingram LO, Conway T (1988) Expression of different levels of ethanologenic enzymes from *Zymomonas mobiliz* in recombinant strains of *Escherichia coli*. Appl Environ Microbiol 54: 397–404

Jansen ML, Bracher JM, Papapetridis I et al (2017) *Saccharomyces cerevisiae* strains for second-generation ethanol production: from academic exploration to industrial implementation. FEMS Yeast Res 17(5):fox044

Jessen JE, Orlygsson J (2012) Production of ethanol from sugars and lignocellulosic biomass by *Thermoanaerobacter* J1 isolated from a hot spring in Iceland. J Biomed Biotechnol 2012: 186982

Jiang Y, Zhang T, Lu J et al (2018) Microbial coculturing systems: butanol production from organic wastes through consolidated bioprocessing. Appl Microbiol Biotechnol 102:5419–5425

Jin H, Chen L, Wang J et al (2014) Engineering biofuel tolerance in nonnative producing microorganisms. Biotechnol Adv 32:541–548

Kannuchamy S, Mukund N, Saleena LM (2016) Genetic engineering of *clostridium thermocellum* DSM1313 for enhanced ethanol production. BMC Biotechnol 16:1–6

Karhumaa K, Sanchez RG, Hahn-Hägerdal B et al (2007) Comparison of the xylose reductase-xylitol dehydrogenase and the xylose isomerase pathways for xylose fermentation by recombinant *Saccharomyces cerevisiae*. Microb Cell Factories 6:1–10

Karp A, Richter GM (2011) Meeting the challenge of food and energy security. J Exp Bot 62:3263–3271

Kim JH, Block DE, Shoemaker SP et al (2010) Atypical ethanol production by carbon catabolite derepressed *lactobacilli*. Bioresour Technol 101:8790–8797

Kim SR, Skerker JM, Kang W et al (2013) Rational and evolutionary engineering approaches uncover a small set of genetic changes efficient for rapid xylose fermentation in *Saccharomyces cerevisiae*. PLoS One 8:e57048

Ko JK, Enkh-Amgalan T, Gong G et al (2020) Improved bioconversion of lignocellulosic biomass by *Saccharomyces cerevisiae* engineered for tolerance to acetic acid. GCB Bioen 12:90–100

Kumar S, Dheeran P, Singh SP et al (2013) Cooling system economy in ethanol production using thermotolerant yeast *Kluyveromyces sp.* IIPE453. Am J Microbiol Res 1:39–44

Kumari D, Singh R (2018) Pretreatment of lignocellulosic wastes for biofuel production: a critical review. Renew Sust Energ Rev 90:877–891

Lee JW, Na D, Park JM et al (2012) Systems metabolic engineering of microorganisms for natural and nonnatural chemicals. Nat Chem Biol 8:536–546

Lee SK, Chou H, Ham TS et al (2008) Metabolic engineering of microorganisms for biofuels production: from bugs to synthetic biology to fuels. Curr Opin Biotechnol 19:556–563

Liao JC, Mi L, Pontrelli S et al (2016) Fuelling the future: microbial engineering for the production of sustainable biofuels. Nat Rev Microbiol 14:288–304

Liew F, Henstra AM, Köpke M et al (2017) Metabolic engineering of *clostridium autoethanogenum* for selective alcohol production. Metabol Eng 40:104–114

Lin CY, Lu C (2021) Development perspectives of promising lignocellulose feedstocks for production of advanced generation biofuels: a review. Renew Sust Energ Rev 136:110445

Lin LL, Ma YJ, Chien HR et al (1998) Construction of an amylolytic yeast by multiple integration of the *aspergillus awamori* glucoamylase gene into a *Saccharomyces cerevisiae* chromosome. Enzym Microb Technol 23:360–365

Liu D, Evans T, Zhang F (2015) Applications and advances of metabolite biosensors for metabolic engineering. Metabol Eng 31:35–43

Liu Y, Li J, Du G et al (2017) Metabolic engineering of *Bacillus subtilis* fueled by systems biology: recent advances and future directions. Biotechnol Adv 35:20–30

Liu ZL, Slininger PJ, Dien BS et al (2004) Adaptive response of yeasts to furfural and 5-hydroxymethylfurfural and new chemical evidence for HMF conversion to 2,5-bis-hydroxymethylfuran. J Ind Microbiol Biotechnol 31:345–352

Lynd LR, Grethlein HE, Wolkin RH (1989) Fermentation of cellulosic substrates in batch and continuous culture by *clostridium thermocellum*. Appl Environ Microbiol 55:3131–3139

Madhavan A, Tamaludi S, Srivastava A et al (2009a) Alcoholic fermentation of xylose and mixed sugars using recombinant *Saccharomyces cerevisiae* engineered for xylose utilization. Appl Microbiol Biotechnol 82:1037–1047

Madhavan A, Tamaludi S, Ushida K et al (2009b) Xylose isomerase from polycentric fungus Orpinomyces: gene sequencing, cloning, and expression in *Saccharomyces cerevisiae* for bioconversion of xylose to ethanol. Appl Microbiol Biotechnol 82:1067–1078

Majidian P, Tabatabaei M, Zeinolabedini M et al (2018) Metabolic engineering of microorganisms for biofuel production. Renew Sust Energ Rev 82:3863–3885

Manochio C, Andrade BR, Rodriguez RP et al (2017) Ethanol from biomass: a comparative overview. Renew Sust Energ Rev 80:743–755

Martín C, Marcet M, Almazán O et al (2007) Adaptation of a recombinant xylose-utilizing *Saccharomyces cerevisiae* strain to a sugarcane bagasse hydrolysate with high content of fermentation inhibitors. Bioresour Technol 98:1767–1773

Mehta R, Singhal P, Singh H et al (2016) Insight into thermophiles and their wide-spectrum applications. 3 Biotech 6:81

Mitchell WJ (1997) Physiology of carbohydrate to solvent conversion by clostridia. Adv Microb Physiol 39:31–130

Mobini-Dehkordi M, Nahvi I, Zarkesh-Esfahani H et al (2008) Isolation of a novel mutant strain of Saccharomyces cerevisiae by an ethyl methane sulfonate-induced mutagenesis approach as a high producer of bioethanol. J Biosci Bioeng 105:403–408

Momayez F, Karimi K, Karimi S et al (2017) Efficient hydrolysis and ethanol production from rice straw by pretreatment with organic acids and effluent of biogas plant. RSC Adv 7:50537–50545

Örlygsson J (2012) Ethanol production from biomass by a moderate thermophile *clostridium* AK1. Iceland Agri Sci 25(1):25–35

Palmqvist E, Hahn-Hägerdal B (2000) Fermentation of lignocellulosic hydrolysates II: inhibitors and mechanisms of inhibition. Bioresour Technol 74:25–33

Park JH, Lee SY (2008) Towards systems metabolic engineering of microorganisms for amino acid production. Curr Opin Biotechnol 19:454–460

Pirie CM, De Mey M, Prather KLJ et al (2013) Integrating the protein and metabolic engineering toolkits for next-generation chemical biosynthesis. ACS Chem Biol 8:662–672

Prasad S, Singh A, Korres NE et al (2020) Sustainable utilization of crop residues for energy generation: a life cycle assessment (LCA) perspective. Bioresour Technol 303:122964

Rani KS, Swamy MV, Seenayya G (1997) Increased ethanol production by metabolic modulationof cellulose fermentation in *clostridium thermocellum*. Biotechnol Lett 19:819–823

Romero S, Merino E, Bolívar F et al (2007) Metabolic engineering of *Bacillus subtilis* for ethanol production: lactate dehydrogenase plays a key role in fermentative metabolism. Appl Environ Microbiol 73:5190–5198

Sartaj KM, Pruthi V, Prasad R (2020) Biofuels production using metabolic engineering. In: Engineering of microbial biosynthetic pathways. Springer, Singapore, pp 231–244

Scully SM, Orlygsson J (2015) Recent advances in second generation ethanol production by thermophilic bacteria. Energies 8:1–30

Sethupathy S, Morales GM, Li Y et al (2021) Harnessing microbial wealth for lignocellulose biomass valorization through secretomics: a review. Biotechnol Biofuels 14:1–31

Sharma D, Saini A (2020) Pretreatment technologies for biomass deconstruction. In: Lignocellulosic ethanol production from a biorefinery perspective. Springer, Singapore, pp 65–109

Shaw AJ, Podkaminer KK, Desai SG et al (2008) Metabolic engineering of a thermophilic bacterium to produce ethanol at high yield. Proc Natl Acad Sci 105:13769–13774

Silveira MHL, Chandel AK, Vanelli BA et al (2018) Production of hemicellulosic sugars from sugarcane bagasse via steam explosion employing industrially feasible conditions: pilot scale study. Biores Technol Rep 3:138–146

Simeonidis E, Price ND (2015) Genome-scale modeling for metabolic engineering. J Ind Microbiol Biotechnol 42:327–338

Singh N, Mathur AS, Tuli DK et al (2017) Cellulosic ethanol production via consolidated bioprocessing by a novel thermophilic anaerobic bacterium isolated from a Himalayan hot spring. Biotechnol Biofuels 10:1–18

Singh S, Kate BN, Banerjee UC (2005) Bioactive compounds from cyanobacteria and microalgae: an overview. Crit Rev Biotechnol 25:73–95

Soo CS, Yap WS, Hon WM et al (2017) Improvement of hydrogen yield of ethanol-producing *Escherichia coli* recombinants in acidic conditions. Electron J Biotechnol 26:27–32

Stewart GG, Panchal CJ, Russell I et al (1983) Biology of ethanol-producing microorganisms. Crit Rev Biotechnol 1:161–188

Tomas AF, Karakashev D, Angelidaki I (2013) *Thermoanaerobacter pentosaceus sp.* nov., an anaerobic, extremely thermophilic, high ethanol-yielding bacterium isolated from household waste. Int J Syst Evol Microbiol 63:2396–2404

Van Maris AJ, Winkler AA, Kuyper M et al (2007) Development of efficient xylose fermentation in *Saccharomyces cerevisiae*: xylose isomerase as a key component. Adv Biochem Eng Biotechnol 108:179–204

Wang H, Cao S, Wang WT et al (2016) Very high gravity ethanol and fatty acid production of *Zymomonas mobiliz* without amino acid and vitamin. J Ind Microbiol Biotechnol 43:861–871

Wu H, Chen XP, Liu GP et al (2012) Acetone–butanol–ethanol (ABE) fermentation using *clostridium acetobutylicum* XY16 and *in situ* recovery by PDMS/ceramic composite membrane. Bioprocess Biosyst Eng 35:1057–1065

Xia J, Yang Y, Liu CG et al (2019) Engineering *Zymomonas mobiliz* for robust cellulosic ethanol production. Trend Biotechnol 37:960–972

Yang S, Fei Q, Zhang Y et al (2016) *Zymomonas mobiliz* as a model system for production of biofuels and biochemicals. Microb Biotechnol 9:699–717

Yang X, Xu M, Yang ST (2015) Metabolic and process engineering of *clostridium cellulovorans* for biofuel production from cellulose. Metab Eng 32:39–48

Young E, Lee SM, Alper H (2010) Optimizing pentose utilization in yeast: the need for novel tools and approaches. Biotechnol Biofuels 3:1–12

Yuangsaard N, Yongmanitchai W, Yamada M et al (2013) Selection and characterization of a newly isolated thermotolerant *Pichia kudriavzevii* strain for ethanol production at high temperature from cassava starch hydrolysate. Antonie Van Leeuwenhoek 103:577–588

Zhang F, Rodriguez S, Keasling JD (2011) Metabolic engineering of microbial pathways for advanced biofuels production. Curr Opin Biotechnol 22:775–783

Zheng DQ, Wu XC, Wang PM et al (2011) Drug resistance marker-aided genome shuffling to improve acetic acid tolerance in *Saccharomyces cerevisiae*. J Ind Microbiol Biotechnol 38:415–422

Zhou Y, Searle S, Anup S (2021) Techno-economic analysis of cellulosic ethanol in India using agricultural residues. International Council on Clean Transportation, Washington, DC

Zhu JJ, Yong Q, Xu Y et al (2009) Adaptation fermentation of *Pichia stipitis* and combination detoxification on steam exploded lignocellulosic prehydrolyzate. Nat Sci 1:47

Chapter 10
Pretreatment Technologies for Second-Generation Bioethanol Production

Kristell Atziry Bahena-Molina, Sushant Sunder, Ambarish Ganesan, Rahul Saini, Carlos Saul Osorio-González, and Satinder Kaur Brar

Abstract In the recent scenario, the attention of the industry and the researchers has been directed towards the production of biofuels mainly due to the current depletion of fossil fuels and the environmental problems that have been worsened for the past decade. In this context, the production of second-generation bioethanol has been widely considered. The increased interest in utilization of lignocellulosic biomass is due its renewability, easy availability, and it does not compete for the food security. However, to take advantage of these promising benefits, it is necessary to face different challenges related to feedstock exploitation such as its structural complexity and conversion efficiency into fermentable sugars. Various pretreatment techniques have been investigated over the years and are classified as physical, chemical, biological, and physicochemical methods. In this sense, this chapter aims to bring an overview of the current knowledge regarding the technologies used to treat lignocellulosic biomass as a feedstock for second-generation bioethanol production.

Kristell Atziry Bahena-Molina, Sushant Sunder and Ambarish Ganesan contributed equally with all other contributors.

K. A. Bahena-Molina
Department of Bioengineering, School of Engineering and Sciences, Mexico City, Mexico

S. Sunder
Department of Biotechnology, Delhi Technological University, Shahbad Delhi, India

A. Ganesan
Department of Biotechnology, Kamaraj College of Engineering and Technology, Virudhunagar, Tamil Nadu, India

R. Saini · C. S. Osorio-González
Department of Civil Engineering, Lassonde School of Engineering, Toronto, ON, Canada

S. Kaur Brar (✉)
Department of Civil Engineering, Lassonde School of Engineering, York University, North York, ON, Canada
e-mail: satinder.brar@lassonde.yorku.ca

C. R. Soccol et al. (eds.), *Liquid Biofuels: Bioethanol*, Biofuel and Biorefinery Technologies 12, https://doi.org/10.1007/978-3-031-01241-9_10

10.1 Introduction

Over the past decade, the depletion of fossil resources has promoted an enormous interest in the development and production of biofuels (such as biodiesel, bioethanol, biobutanol or drop-in biofuel) as pollution-free and renewable substitutes (Magda et al. 2021). Currently, bioethanol has been regarded as the most available biofuel in the world market used to partial or total replacement of transportation fuels as gasoline-ethanol mixtures such as E15 and E85 mixtures (Cheng and Timilsina 2011). Bioethanol is a clear, colorless and less toxic alcohol that can be produced from a wide range of lignocellulosic biomass materials such as agriculture and forestry residues as well as residues from other industries (Roberts and Patterson 2014). The worldwide bioethanol production during 2020 was ~26.059 million gallons and the countries with the highest production are United States (53%), Brazil (30%), France (5%), China (3%), India (2%) and Canada (2%) (US Department of Energy 2021).

The bioethanol production methods strongly depend on the nature of the substrate used. However, the common step for all kinds of feedstock is fermentation, one of the oldest biotechnological processes that consist in the use of ethanologenic microorganisms from different species such as *Saccharomyces* sp., *Candida* sp., *Zymomonas*, or *Pichia* sp., to convert monomeric sugars into ethanol. Different feedstocks including sugarcane, sugar beet, fruits and sugar refinery waste have been used. Nevertheless, they require an extraction process followed by purification with lime and filtration steps to get fermentable sugars. In addition, starchy sources including cereals, roots, legumes, and immature fruits are processed either by the dry-grind or wet mill methods. Nonetheless, the most used method is a dry-grind process that consists in the preparation of slurry which is then liquefied with α-amylase to breakdown the starch molecules and saccharified with β-amylases to convert dextrin into glucose (Häggström et al. 2014; Zabed et al. 2017).

Another alternative is lignocellulosic biomass, one of the most attractive feedstocks to produce biofuel such as bioethanol. Among the main advantages of this feedstock, it is renewable, easily available, and non-competitive with food security (Saini et al. 2020). Lignocellulosic biomass can be obtained from different residues considered as wastes such as forest residues, sugarcane bagasse and crop residues (Osorio-González et al. 2019). In general, lignocellulosic-based bioethanol production is composed of four main steps: (i) pretreatment, (ii) saccharification, (iii) fermentation and; (iv) product recovery (Hu et al. 2020). Figure 10.1 shows the main step for bioethanol production using different feedstocks.

Moreover, lignocellulosic biomass is the complex structure between lignin, cellulose and hemicellulose and requires the use of efficient pretreatment methods to improve the release of fermentable sugars. In this sense, various pretreatment methods have been investigated over the years and are classified as physical (microwave radiation, milling, extrusion, and ultrasonication), chemical (acid, alkaline, oxidation and ionic liquids), biological (microbes and enzymes) and physicochemical (steam explosion, ammonia fiber explosion and carbon dioxide explosion)

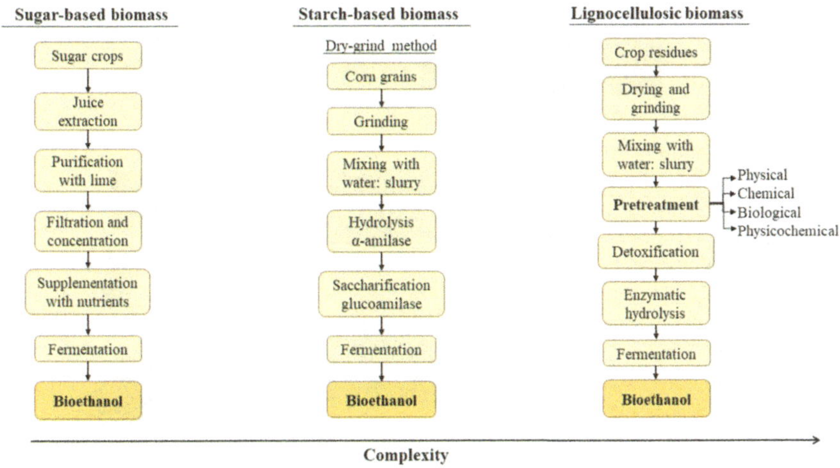

Fig. 10.1 Schematic diagram of bioethanol production using different feedstocks. Modified from (Zabed et al. 2017; Hu et al. 2020)

(Su et al. 2020). Each pretreatment has its approach and action mode as well as a specific interaction with the feedstock. Based on these interactions, the decision can be made to choose or not the appropriate pretreatment strategy. The aim of this chapter is to give an overview of the current technologies used to pretreatment on lignocellulosic biomass for second-generation bioethanol production.

10.2 Feedstock for 2G Bioethanol Production

In the process of bioethanol production, the type of feedstock is one of the major determinants that play a vital role, due to its direct effect on productivity, affordability, and economical feasibility. Generally, feedstocks used on bioethanol production can be classified into four main groups: (i) sugar-containing such as molasses, sugarcane, or cheese whey; (ii) starch-containing such as corn, wheat, and root crops; (iii) lignocellulosic biomass produced from agricultural and forestry residues; and (iv) algal biomass (Baskar et al. 2019). Sugar and starch feedstocks belong to the first-generation biofuels production; lignocellulosic biomass to the second generation and algal biomass is regarded to the third generation. Figure 10.2 illustrates different generations of biofuels. However, production from first-generation biofuels raises a concern about food security, mainly because they are competing directly in the use of basic foods, and arable land. Furthermore, this type of feedstock increases the total production cost due to the competition for their availability (Subramaniam et al. 2019).

To above the competition and compromise food security, other types of feedstocks have been investigated. In this sense, second-generation (2G) feedstocks are

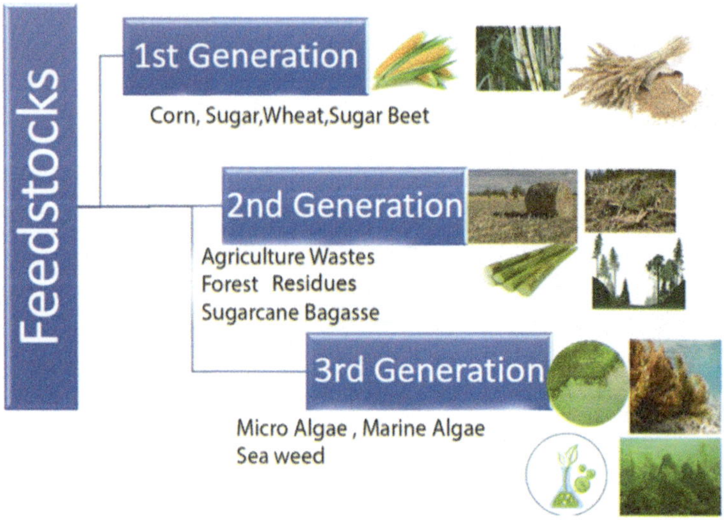

Fig. 10.2 Classification of bioethanol feedstocks according to the generation of biofuels they belong

abundant, cost-effective, and most of them are residues and waste materials with low or null use. 2G feedstocks include agricultural residues, forestry residues, and municipal solid wastes (Lee and Lavoie 2013). The main composition of this type of feedstock is cellulose, hemicellulose, and lignin. Cellulose is the major structural component with 40–50% and provides the mechanical strength, is a linear polymer linked by β-1,4-glycosidic bonds, mainly composed of glucose units (crystalline and amorphous). Hemicellulose is 20 to 30% of the matrix, is the second most abundant heterogeneous polymer, which mainly consists of glucuronoxylan, glucomannan, and other branched polysaccharides composed of xylose, arabinose, mannose, glucose, and galactose units. Finally, lignin is the smallest fraction with 15 to 20% of the total matrix, is the most complex fraction and provides the protective layer to cellulose and hemicellulose fractions. Lignin is an insoluble amorphous compound, conformed by lignin monomers such as coumaric, sinapyl, and coniferyl alcohols. This fraction is difficult to break or degrade and its hardness and composition vary between source, species, age, among others (Osorio-González et al. 2020).

10.3 Pretreatment Technologies Involved in 2G Bioethanol Production

Pretreatment is a salient tool for lignocellulose biomass conversion and is important to make cellulose and hemicellulose available to the enzymes in order to obtain fermentable sugars (Mosier 2005). Different pretreatment processes have been

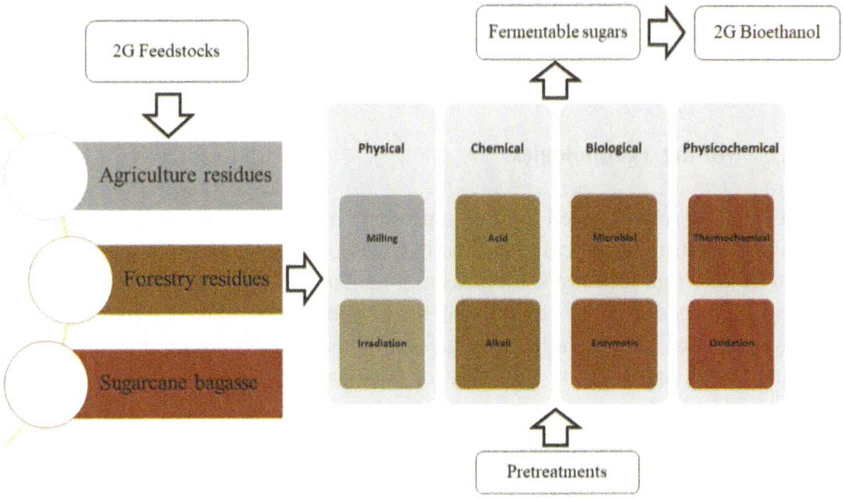

Fig. 10.3 Classification of the main technologies used on lignocellulosic biomass for second-generation bioethanol production

developed to fractionate, solubilize, and separate cellulose, hemicellulose, and lignin components using physical, chemical, physicochemical, and biological treatment methods (Fig. 10.3), which will be discussed in the following section. Furthermore, the pretreatment method contributes to the affordability and total cost of the production process due to its efficiency has a high influence on the product yield and quality of the produced bioethanol (Satari and Jaiswal 2021).

10.3.1 Physical-Based Pretreatment Technologies

Physical-based pretreatment generally decreases the crystallinity of cellulosic structures present in the biomass through mechanical disruption techniques such as milling, grinding, and irradiation to reduce the particle size of lignocellulosic biomass (Tu and Hallett 2019). Factors such as processing temperature, pressure, residence time and type of feedstock determine the efficiency of physical pretreatment. The process is usually carried at 180 to 240 °C temperature complemented with mechanical shearing. The treatment creates particles of different sizes and increases the surface area of feedstock materials for the better hydrolysis of sugars. The resulting fine particles are then sectioned off by sieving to obtain the desired size particle. This type of pretreatment is quite energy-intensive as compared to other pretreatments, which involve more than 20% of the overall operational cost (Kumari and Singh 2018). The major disadvantage of this process is high energy consumption that leads to an increase in production cost. In general, the use of mechanical methods such as milling is not sufficient to release fermentable sugars, but it helps to decrease the particle size. Therefore, the process can be integrated with chemical or physicochemical methods to release sugar monomers.

10.3.1.1 Milling Technologies

It is the primary method used for any large-scale lignocellulosic biomass waste due to the inherent structure of cellulose and extensive degree of crystallinity. Milling is found to render lignocellulosic biomass more amenable to cellulases because it can decrease the crystallinity of cellulose up to 0.2 nm (Kumari and Singh 2018). This process consumes lots of energy which contributes to its limitation. Factors such as type of milling, its duration and type of biomass determine the final degree of polymerization and surface area available for enzymatic hydrolysis. In order to improve the digestibility of biomass, different kinds of milling such as hammer, colloid, and two-roll are performed. Vibratory milling is found to be more effective because the cellulose crystallinity gets improved in digestibility of spruce, straw, bagasse, aspen chips and other lignocellulosic biomass (Kumar and Sharma 2017).

To overcome the challenges of energy consumption, wet milling is more preferred for pretreatment. It is more effective when combined with other pretreatment methods such as grinding, heating, or chemical treatments (Nakagawa et al. 2007) as it helps cellulose and hemicellulose solubilization (Kumari and Singh 2018). Table 10.1 shows the main advantage and disadvantages of some of the most common physical methods used as a pretreatment. Among the main advantages that milling offers is that it does not generate any type of inhibitors or toxic compounds that can be a problem in future production steps. This is one of the reasons why it is a preferred preliminary pretreatment for the range of lignocellulosic biomass. Studies have shown that the better hydrolysis and good yield of reducing sugar for rice straw was obtained when the milling pretreatment is aided with steam

Table 10.1 Advantages and disadvantages of several physical-based pretreatment methods

Pretreatment	Advantages	Disadvantages	References
Milling	The generation of ultrafine particles enhances the surface area for enzymatic treatment. No generation of toxic compounds.	Pretreatment is unable to remove lignin and hemicellulose from the biomass. High energy-demanding process.	(Kumari and Singh 2018)
Sonication	Provides greater accessibility and reactivity of cellulose. Highly efficient process and lower product extraction time. Eco-friendly pretreatment.	Negatively impacts enzymatic hydrolysis. Longer duration of time. High power consumption.	(Saif Ur Rehman et al. 2013)
Irradiation	It enhances enzymatic digestibility. It is an industry-proven process and used at the commercial level.	Excessive irradiation dose reduces the yield of glucose. In thin layer materials, penetration is limited so results in poor yield.	(Chaturvedi and Verma 2013)

explosion or alkaline solution (Abomohra 2019). Lin et al. (2010) reported a 110% increase in enzymatic hydrolysis of lignocellulosic biomass when an alkaline solution of 6% NaOH is used as a wet milling treatment for 3 weeks at room temperature.

10.3.1.2 Sonication Technologies

Sonication is a relatively new physical pretreatment technique used for lignocellulosic biomass. The ultrasonic waves produced in this treatment alter the physical morphology of biomass and its chemical composition. Briefly, the treatment disrupts the cell wall structure leading to an increase in the surface area, which ultimately increases the accessibility of polymeric sugars to hydrolytic enzymes (Bhutto et al. 2017). In general, the pretreatment is carried out at a frequency range of 10KHz to 20 MHz at an energy of 7.2×10^4 J/g to improve the cavity formation in the biomass (Balan et al. 2009; Park et al. 2010). Maximum cavity formation was proven to be at 50 °C which favors the optimal activity of cellulolytic enzymes (Yachmenev et al. 2009). The cavitation is highly dependent on ultrasound frequency, which is proportional to the amount of energy released. The released energy generates localized hot spots of 5000 °C and 5000 atm of pressure for a few microseconds, which improves the lignin disruption and cellulose solubilization in the biomass. The frequency of ultrasonic waves and the amount of released energy are used to determine the degree of delignification in the biomass. For instance, it has been shown that frequency in the range of 10–100 kHz is enough for the decrease in hemicellulose and lignin by 41.45% and 30.16% respectively for soybean straw in the presence of a solution of 10% of ammonia (NH_3) for 24 hours at ambient temperature (Stamatelatou et al. 2014). In addition, it has been observed that the use of lower pretreatment time, enhanced delignification degree and increased accessibility to cellulose are the key advantages to be used at an industrial scale (Kim et al. 2016). Nevertheless, it has been reported that the sonication time, type of solvent used, biomass characteristics, reactor configuration, and rate kinetics have a high influence on the efficacy of this pretreatment. On the other hand, the use of prolonged sonication times that can go beyond the threshold limit could negatively affect the delignification as well as cellulose accessibility. This is because the use of higher sonication power increases cellulose and hemicellulose oxidation, hence decreases the sugar yield during the saccharification process (Saif Ur Rehman et al. 2013). Likewise, Chen et al. (2011) reported an increase in viscosity in a wood cellulose suspension when treated with 1200 W in the presence of 20–25 kHz and 120 ml of a solution containing purified cellulose. The power and duration of sonication need to be optimized depending upon the lignocellulosic biomass used in order to meet the desired pretreatment objectives. To achieve better results, this type of pretreatment needs to be synchronized with lower frequencies, appropriate power dissipation levels and appropriate periods to enhance the mass transfer rate. Furthermore, combined enzyme-sonication hydrolysis of lignocellulosic biomass can accelerate the conversion of lignocellulosic biomass as a feedstock to produce biofuels.

10.3.1.3 Irradiation Technologies

Among the different physical techniques, irradiation such as gamma rays, electron beam, ultrasound, and microwaves are valued as one of the most efficient pretreatment methods. This method accounts for lower energy demand, ease to handle, maximum energy efficiency and selectivity are few properties that have made this process more promising and approachable as discussed in Table 10.2 (Kassim et al. 2016). Highly energetic radiation causes several changes such as loosening of cellulose crystallinity, hydrolysis of hemicellulose and depolymerization of lignin in the lignocellulosic biomass (Chen et al. 2012; Chaturvedi and Verma 2013). In general, exposure time, biomass composition and frequency of radiation are the few parameters that characterize the efficiency of irradiation pretreatment (Saini et al. 2015). Irradiation has been a very constructive approach over the last 30 years and established itself from bench scale to pilot scale (Hassan et al. 2018; Zabed et al. 2019). In order to increase the efficiency of this technique, it has been reported to use the different chemical substances during or after the irradiation process to improve the saccharification process (Lee et al. 2017; Kainthola et al. 2019). For example, this technique is coupled with other substances such as ionic fluids and nanoparticles and has demonstrated excellent results, but unfortunately,

Table 10.2 Assessment of irradiation pretreatments on different biomass

Pretreatment method	Biomass	Treatment conditions	Effect	References
Gamma radiation	Sugarcane bagasse	>100 KGy[a] of gamma radiation	A comparison was made with 45.5% (w/w) in raw biomass. 30% more reduction in crystallinity index as compared to raw material (at 500KGy). 11.2% decrease in cellulose (at 2000KGy).	(Kapoor et al. 2017)
Electron beam	Kenaf core	5000KGy of electron beam dose with 3% sulphuric acid	Comparison was made with 25.4% (w/w) in raw biomass. High dose of the electron beam at (5000KGy) followed by acid hydrolysis resulted in a 45.1% reduction in lignin.	(Lee et al. 2017)
Gamma radiation	Cedarwood	800KGy of gamma radiation followed by milling (4 minute)	The particle size reduction (<180 μm) was obtained by 55.9% with irradiated biomass as compared to 35.5% (w/w) in untreated biomass.	(Liu et al. 2017)
Gamma radiation	Eucalyptus	800KGy of gamma radiation followed by milling (4 minute)	The particle size reduction (<180 μm) obtained by 55.9% with irradiated biomass as compared to 35.5% (w/w) in untreated biomass.	(Liu et al. 2017)

[a]KGy: KiloGray

this leads to an elevation in the total cost of the pretreatment, which limits its scaling at an industrial level (Sankaran et al. 2020).

The impact of irradiation methods differs from type to type of biomass. For the lignocellulosic feedstock pretreatment microwave irradiation is favoured. Microwave has several advantages such as ease to operate, low energy requirement, minimum generation of inhibitors, and high heating capacity in a relatively short period. The efficiency of this pretreatment can be further enhanced with the help of mild-alkali treatment, where sodium hydroxide is a good catalyzer to increase xylose and glucose content after the pretreatment process (Keshwani and Cheng 2010). Another very promising form of irradiation technique is electron beam and gamma radiation (Subhedar et al. 2018; Tsubaki et al. 2018; Sewsynker-Sukai and Gueguim Kana 2018). These two methods are associated with high temperature and no inhibitory compounds generation during those pretreatments (Kumar and Sharma 2017; Sankaran et al. 2020). Likewise, irradiation using an electron beam is a very efficient method to maximize the conversion of hemicellulose and lignin. This technique is implemented through the breakdown of glycosidic bonds present in cellobiose and other oligomers. The cellulose and the crystallinity present in the lignocellulosic biomass were reduced from 85,000 Da to 5000 Da at 1000KGy. Studies found that enzymatic hydrolysis of glucose yield increased up to 44.3% at 270 KGy for 48 hours.

10.3.1.4 Microwave Technology

Microwave is another technique used to pretreat lignocellulosic biomasses. This pretreatment method is a more suitable alternative than conventional heating methods because microwaves penetrate at the molecular level through dipole rotation to heat the biomass and break the complex linkage (Diaz et al. 2015; Kostas et al. 2017). The advantage of microwave is the heat approach, i.e., heating inside-out (Zhu et al. 2015). The microwave-assisted pretreatment approach creates hot spots within the polymeric structures and facilitates an 'explosion' like effect in the biomass. One of the main advantages that this method is the microwave radiation efficacy because the heating is highly dependent on the raw material dielectric properties as well as the used solvent as a catalyzer. Alcohol, acetone, and water are the most popular because of their high ability to absorb radiation and get heated through the waves. This allows the biomass to be selectively heated and at the same time protects it from superheat (i.e., the heat generated by the solvent). Additionally, the generation of hotspots due to instantaneous heating by microwaves is a more effective method to cause explosions within the biomass through the generated pores. The shorter duration of time and energy-efficient properties of this technique makes it stand out among other pretreatments (Zhu et al. 2015).

The addition of acid or base to treat biomass before the microwave exposure helps the biomass to alter the surface area and provide the larger pore size within the particles (Hu et al. 2019). The use of sodium hydroxide is the most suitable alkaline treatment. Ease in operational use, low energy consumption, high heat for a short duration, degradation of complex structural organization of cellulose biomass and low production of inhibitors are few merits of this pretreatment. However, the generation of inhibitors is increased by the usage of acid and alkaline pretreatment (Kumari and Singh 2018). Several studies have found to carry this pretreatment at 130 to 200 °C with 30 minutes of exposure and energy input of 200 to 800 W due it has been observed an explosion-like effect on the biomass with a better pores generation. As a result of the pore generation, an improvement in lignin reduction, and surface area accessibility has been reported (Zhu et al. 2015; Kostas et al. 2017; Kainthola et al. 2019). In summary, lignocellulosic biomass pretreated with microwave technology is more efficient related to energy use as compared to other conventional methods.

10.3.2 Chemical-Based Pretreatment Technologies

Among all the methods used on lignocellulosic biomass, chemical pretreatment has been the most effective and feasible for lignocellulosic-based bioethanol production. The use of pretreatments using acids, alkalis, ozonolysis, ionic liquids and organic solvents belong to this group (Mohapatra et al. 2017). Table 10.3 demonstrates the different chemical-based pretreatment for lignocellulosic biomass.

10.3.2.1 Ionic Liquids-Based Pretreatment

Ionic Liquids (ILs) have recently emerged as an attractive technology with a wide range of uses such as electrolytes, plasticizers, solvents, and catalysis in the chemical industry. ILs are salts usually composed of large organic cations and small inorganic anions that have special properties such as low melting points (<100 °C), negligible vapour pressure, high thermal and chemical stability, high polarity, low toxicity, non-volatile, wide liquid range, and good solvation properties (Olivier-Bourbigou et al. 2010). Because of their unique properties, they have been intensively studied as "green" solvents for lignocellulosic biomass pretreatment breaking the hydrogen bonds and stimulate the dissolution of the lignocellulosic polysaccharides (da Costa Lopes et al. 2015). During this process, cations and anions play a significant role in the cleavage of cellulose, hemicellulose, and lignin bonds, giving the enzymes more accessibility to the substrate during the saccharification. The above mechanism is mainly influenced by the composition and structure of ILs, the nature of lignocellulosic biomass and the process conditions (Sathitsuksanoh et al. 2013).

Several types of ILs cations capable of dissolving cellulose fibers through hydrogen bond cleavage include imidazolium, pyridinium, aliphatic ammonium, alkylated phosphonium, and sulfonium ions, while the anions are hydroxyl groups that can include chlorides, carboxylates (acetate, formate, propionate, lactate), dialkyl phosphates, and amino acid (Hou et al. 2017). Among these, imidazolium is the most frequently used and has been mainly studied as a solvent in the

Table 10.3 Illustrations of chemical-based pretreatment methods for lignocellulosic biomass

Pretreatment	Description	Action mechanism	Reaction conditions	Advantages	Disadvantages
Acid-based	Acids such as H_2SO_4, HCl, H_3PO_4 and HNO_3 at high or low concentrations.	Cleavage of glycosidic bonds for hydrolysis of polysaccharides into fermentable sugars.	Diluted: <5% w/w acid at high temperatures (>140 °C). Concentrated: >30% w/w acid at moderate temperatures (<100 °C).	Availability of reagents, industrial scale, low cost, high sugar conversion.	Additional recovery steps, corrosive equipment, high toxic, high maintenance cost, inhibitors.
Alkali-based	Bases like sodium potassium, ammonium, and calcium hydroxides.	Degradation of esters and glycosides bonds. Alteration of lignin, de-crystallization of cellulose and solvation of hemicellulose.	Low temperatures and pressures but long residence time. Strongly dependent on reagent used.	Safer handling, low formation of inhibitors, minimal sugar degradation, easy recovery, and reuse of reagents.	Expensive reagents, less sugar degradation.
Ionic liquids-based	Salts of organic cations (e.g., imidazolium), and inorganic anions (e.g., carboxylates).	Disruption of the hydrogen bonds in cellulose, hemicellulose, and lignin.	Strongly dependent on reagent used.	Green solvent, recoverable, reusable, easy to synthesize, chemical stability, low toxicity.	High cost, toxicity for hydrolytic enzymes, viscosity, and formation of inhibitors.
Deep eutectic solvents	Fluids (e.g., choline chloride with carboxylic acids) capable to form a eutectic mixture.	Lignin and cellulose dissolution and depolymerization of lignocellulosic biomass.	Strongly dependent on reagent used.	Green solvent, cost-effective, low toxicity, biodegradable, no purification, and no waste disposal.	Lack of knowledge, instability, high viscosity and reactivity that can form inhibitors.
Organosolv-based	Mixture of organic solvents (e.g., ethanol, formic acid), and water with catalyst (e.g., hydrochloric, sulfuric acid).	Disruption of the bonds leaving the lignin and a part of the hemicellulose dissolved, and reactive cellulose suitable for enzymatic hydrolysis.	Temperatures ranging from 100 to 250 °C and high pressures.	High-purity lignin and cellulose, high efficiency, low input of solvents, and easy to recover.	High costly by conditions and recovery process, corrosive equipment.

(continued)

Table 10.3 (continued)

Pretreatment	Description	Action mechanism	Reaction conditions	Advantages	Disadvantages
Ozonolysis	Ozone is used to oxidize, solubilize and degrade the lignin and hemicellulose.	Dissolution of lignin and partial alteration of hemicellulose by cleaving the hydrogen bonds. Cellulose fraction is not affected.	Room temperature and normal pressure.	No toxic inhibitory compounds, selective lignin degradation, low environmental pollution and no chemical additives.	Commercially unfeasible, high energy requirements, high ozone doses. Dangerous process: Toxic, exothermic, and flammable.
Oxidazing agents	Oxidation agents such as H_2O_2, $C_2H_4O_3$, SO_3 and ClO_2.	Solubilization of hemicellulose, amorphous cellulose, and lignin.	Ambient pressure and temperature under mild alkaline or neutral conditions.	Low-cost bioreactors and not residue production in the biomass.	Inhibitors, nonselective, highly-cost oxidants. Not for industrial scale.

pretreatment of lignocellulosic biomass (Valladares-Diestra et al. 2021). For instance, (Karatzos et al. 2012) used 1-ethyl-3-methylimidazolium acetate to pretreated sugarcane bagasse at 130 °C. It resulted in 60% of delignification, 83% of saccharification yield (% cellulose dissolved) and a full recovery of the ILs. As it was mentioned, ILs are recoverable, reusable, and easy to synthesize. However, complete removal of them is an unfeasible and expensive process because it requires the consumption of a large amount of water or anti-solvent, mixing process, and complex recycling systems. Furthermore, they are toxic for the hydrolytic enzymes used during the saccharification process, become more viscous in the processing and generate inhibitors. In addition, the high cost of ionic liquids also remains in these areas that need further exploration (Rastogi and Shrivastava 2017).

10.3.2.2 Deep Eutectic Solvents

To tackle the issues of high price and toxicity of ILs, a new generation of solvents has emerged in recent decades named Deep Eutectic Solvents (DESs). A deep eutectic solvent is a fluid generally composed of two or three components, often linked by hydrogen bonding, capable of forming a eutectic mixture with a lower melting point. This melting point depression can be attributed to the hydrogen bonding network between the components and the resulted charge delocalization (Morais et al. 2020). The application of DESs has been recently described as biomass pretreatment acting in polysaccharides dissolution and the depolymerization of lignocellulosic biomass (Kumar et al. 2016). Choline chloride (ChCl) is the most used organic salt due to its biodegradability, low cost, and its use in combination with urea, glycerol, carboxylic acids, or polyols (Tang et al. 2017). In comparison with ILs, DESs are cost-effective and easy to produce, pose low toxicity and high biodegradability and biocompatibility with enzymes. Additionally, preparation of DESs is by simply mixing the components, thus purification and waste disposal are not required. These advantages make DESs a versatile alternative to be used as a pretreatment method at a large-scale, specifically reinforcing the concept of biorefinery. However, an increase in research and development is still needed to solve challenges such as proper selection of DESs accordingly with the application purpose, instability during electrochemical processes, viscosity during its process and its reactivity that can lead to forming undesirable inhibitors (Zhang et al. 2012).

10.3.2.3 Oxidation-Based

Unlike the methods discussed above, oxidation pretreatment such as ozonolysis has rarely received attention among researchers mainly because of its high cost involved, nonetheless, it has been used as a pretreatment in lignocellulosic biomass.

10.3.2.3.1 Ozonolysis

In the search of green alternatives to decrease the use and generation of toxic chemicals, ozonolysis has gained interest as delignifying agent with different types of lignocellulosic biomass such as sugarcane bagasse, wheat straw, cotton stalks and wood (Barrera-Martínez et al. 2016; de Guilherme et al. 2017; Osuna-Laveaga et al. 2020), (García-Cubero et al. 2009; Jibouri et al. 2015), (Mamleeva et al. 2009; Kaur et al. 2012). This oxidative pretreatment consists of the direct exposition of the lignocellulosic biomass to ozone at ambient temperature and pressure conditions. Other parameters such as moisture content, particle size, and ozone concentration could vary according to the biomass properties (Taherzadeh and Karimi 2008).

When biomass is treated by ozonolysis, ozone acts as a strong oxidizing agent that removes the lignin by oxidizing aromatic groups and attracting oxygen electrons from the lignin structure. By this mechanism, it can remove a high lignin fraction with minimal degradation of the hemicellulose and cellulose contents (Osorio-González et al. 2020). Travaini et al. (2013) showed enhanced enzymatic hydrolysis obtaining 42% and 52% of glucose and xylose yields, respectively, by treating sugarcane bagasse with 3.44% v/v of ozone at 40% moisture. Furthermore, Travaini et al. (2016a) obtained maximum glucose and xylose yields of 77% and 57%, respectively, optimizing parameters to 50% moisture, 2% mol/mol (ozone/oxygen) and 60 L/h (ozone/oxygen flow). These optimal conditions provided an ethanol yield of 76% (glucose to ethanol conversion) in fermentation experiments with *Saccharomyces cerevisiae*. Overall, ozonolysis is an attractive pretreatment method since it does not produce toxic compounds such as furans, phenolic compounds, and organic acids. Other advantages of ozonolysis are its selective lignin degradation, it avoids problems with chemical storage because the on-site ozone generation, reduces the environmental pollution and does not require chemical additives. However, the main drawbacks with this process are that it is not currently used commercially due to high energy requirements of ozone production, high ozone doses and the high costly maintenance of the ozone generator. Moreover, ozone it is also highly reactive, flammable and corrosive hence making it dangerous processes (Travaini et al. 2016b).

10.3.3 Biological-Based Pretreatment Technologies

Generally, pretreatment which is processed with any kind of biological entity (such as microorganisms) or the biological by-products (such as enzymes) comes under one roof called biological pretreatment (Fig. 10.4). In contrast with other pretreatment methods, biological pretreatment methods are known as the green pretreatment technique and do not produce toxic by-products (Narayanaswamy et al. 2013). Even though biological pretreatment has been treatment as an eco-friendly and green technique, there are several drawbacks associated with this strategy. For instance, the biological pretreatment process is time-consuming

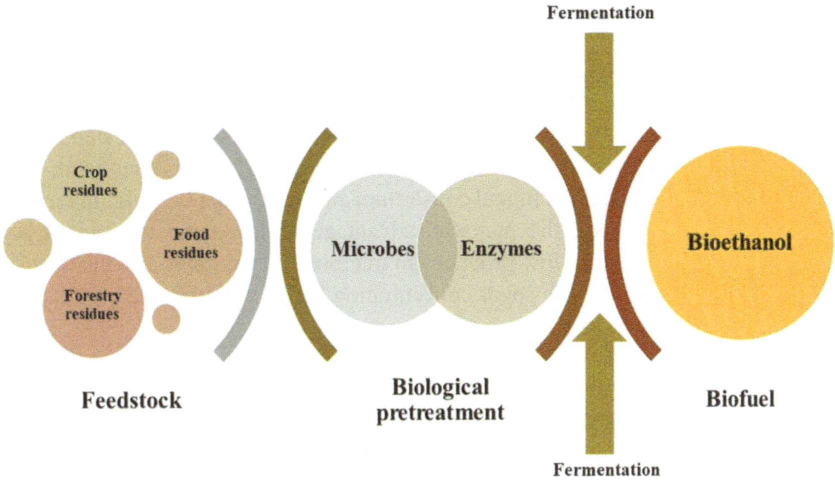

Fig. 10.4 Overview of biological pretreatment

process, reduction in total sugar and extra purification step before implementing the pretreatment biomass for bioethanol production (Vasco-Correa et al. 2016). To overcome these challenges, the integrated strategies have been developed and a lot of research is going on throughout the world (Zabed et al. 2019). Additionally, recombinant technologies are being used to modify microorganisms according to desired treatments, the combination with other methods (such as, physical and chemical ones) with biological pretreatment has also been developed for effective pretreatment (Ummalyma et al. 2019). In the following sections, different types of biological-based pretreatments have been described briefly.

10.3.3.1 Microorganism-Based

The structure of lignocellulosic is complex and heterogeneous, while microorganisms can degrade the complex linkage between lignin, cellulose and hemicellulose resulting in the release of polymeric subunits (Vasco-Correa et al. 2016). Microbial-based depolymerization can be achieved either by fungi or bacteria such as *Punctualaria* sp., *Ceriporiopsis subvermispora, P. ostreatus*, *P. pulmonarius*, *P. chrysosporium,* and *Irpex lacteus* (Sindhu et al. 2016). In general, microbial degradation is one of the natural degradations which occurs naturally for balancing the natural carbon cycle. Microbial depolymerization has a high potential for application in many other industries that utilize lignocellulosic biomass. Microorganisms use different enzymes to depolymerize the biomass such as lignin peroxidase, aryl alcohol oxidase and glycosidase (Sindhu et al. 2016). From the studies in fungal and bacterial pretreatment of lignocellulosic biomass, Vasco-Correa et al. (2016) performed the pretreatment of switchgrass using *Pycnoporus* sp. SYBC-L3 and reported the 30% Lignin reduction, 90% enzymatic digestibility (compared with

75% for untreated material) in 36 days, while micromycete *Myrothecium roridum* was shown to extensively delignify paddy straw and the herbaceous weed *Parthenium* sp. (more than 50% lignin degradation) while leaving most of the cellulose intact and significantly enhancing the enzymatic digestibility in only 7 days of incubation. Likewise, Du et al. (2011) pretreatment the corn stalks using *Irpex lacteus* and reported the 82% hydrolysis of biomass. On the other hand, Song et al., (2013) implemented the fungal consortium and achieved 43.8% lignin removal in corn stover. Moreover, the major disadvantage of using the microbial-based pretreatment method is the slow process and microorganisms could use the released sugar which will decrease the sugar concentration hence will ultimately impact the bioethanol production.

10.3.3.2 Enzyme-Based

Enzymes-based pretreatment involves the direct use of hydrolytic enzymes (such as cellulolytic, hemicellulolytic and ligninolytic enzymes) to dissolve the connecting bonds between lignocellulosic complexes. This method requires less time and does not lead to the reduction in sugar concentration as a major drawback of using microbial-based pretreatment (Narayanaswamy et al. 2013). On the contrary, the extraction and purification process of enzymes is difficult and expensive which would directly affect the overall production process. Recently, many research works have been going on regarding genetically engineered microbes that could specifically produce the desired lytic enzymes for pretreatment purposes (Saini et al. 2020). The factors influencing biological pretreatment include temperature, pH, incubation time, microbial adaptation, moisture, substrate size, and aeration (Sindhu et al. 2016).

10.3.3.2.1 Cellulolytic Enzymes

Cellulase includes endoglucanase, exoglucanases, and β-glucosidase enzymes belong to the glycosyl hydrolase (GH) family 128. GH families are consisting of different cellulase enzymes and the synergistic actions of these hydrolytic enzymes catalyze the cellulose into monomeric sugar units. The endo-glucanases hydrolyze the glycosidic bonds of cellulose to release cellobiose and some glucose. The β-glucosidases finally cleave cellobiose to glucose (Sharma et al. 2019).

10.3.3.2.2 Hemicellulolytic Enzymes

This enzyme includes glycoside hydrolase groups and carbohydrate esterase groups hydrolyze the glycosidic bonds and the ester bonds of acetate or ferulic acid groups. There is a wide range of interdependent hemicellulases involved synergistically during hydrolysis of hemicellulose to form several monomeric sugars (Sindhu et al. 2016). The glycoside hydrolase groups were found in about 29 GH families and carbohydrate esterase (CE) groups were found in about 9 CE families. The interdependent hemicellulases transform synergistically during hydrolysis of hemicellulose to form several monomeric sugars and also liberate cellulase. The enzymes like endo- and exoxylanases hydrolyze the cross-link of hemicelluloses that cleave the xylene to generate oligosaccharides. Where other enzymes like β-xylosidases, α-arabinofuranosidase, and esterases hydrolyze xylooligosaccharides to xylose; arabinose into furanose and pyranose forms; acetyl group into arabinose (Sharma et al. 2019).

10.3.3.2.3 Ligninolytic Enzymes

The ligninolytic enzymes are generally used to degrade the highly complex and recalcitrant lignin. Most of the white-rot fungi possess an enzymatic system to degrade the lignin. They produce laccase and various peroxidases. The white-rot fungi are the major producers of ligninolytic enzymes. Bacteria have a low potential for lignin degradation. However, actinomycetes, α-proteobacteria, and γ-proteobacteria are known to have a ligninolytic system (Sharma et al. 2019). The bacterial ligninolytic enzymes such as laccase, lignin peroxidase, dye decolorizing peroxidases, β-etherases, superoxide dismutases have already been discovered in other strains of bacteria. Among these above enzymes, some of the most significant ligninolytic enzymes are laccase and peroxidases.

10.3.4 Physicochemical-Based Pretreatment Technologies

A physicochemical pretreatment is an integrative approach widely used for the breakdown of hemicellulose and lignin polymers present in the biomass. It is an application of high pressure and temperature in presence of a chemical catalyst. Most of the methods used in this category are performed in the range of 50 °C to 250 °C with pressure up to 40 psi. However, the microbial inhibitory compounds such as furans, organic acids, and phenolic compounds generated by the use of chemical catalyst are released. One of the most popular methods used in this category is steam explosion, wet oxidation, ammonia fiber expansion (AFEX), liquid hot water. A lower amount of energy, high sugar yield during enzyme hydrolysis, lower degradation of sugars and chemical utilization are few properties that favors scaling up for industrial usages.

10.3.4.1 Acid-Based Pretreatment

The acid pretreatment of lignocelluloses is a highly effective and well-known process that consists of the use of acids such as sulphuric, hydrochloric, phosphoric and nitric acid to disrupt the lignocellulosic matrix and obtain pretreated biomass for saccharification (Fennouche et al. 2019). The major mechanism is through the cleavage of glycosidic bonds, which leads to the hydrolysis of polysaccharides into fermentable sugars, and the simultaneous disruption of the composite material linkage. This pretreatment mainly act into the hemicellulose fraction, and to a lesser extent in lignin and cellulose. The most important parameters to consider during the acid pretreatment process are the liquid: solid ratio (wet or dry), acid concentration, temperature, particle size and residence time. From the acid concentration factor, it is derived two different variations of these pretreatments: diluted and concentrated acid treatment.

10.3.4.1.1 Diluted Acid Pretreatment

The dilute acid pretreatment process is very well known and considerably feasible due to its simplicity, low reagents costs and effectiveness. It is usually performed with dilute acid solutions (less than 5% w/w) and at high temperatures (120 to 215 °C) from several minutes to an hour (Galbe and Zacchi 2012). However, due to the low acid concentration, there is only removal of the hemicellulose and acid-soluble lignin fractions (Chiranjeevi et al. 2018). Additionally, the high temperatures tend to produce sugar degradation compounds that have been described as potential microbial inhibitory compounds. Dilute acid pretreatment has been recently studied on lignocellulosic materials, for instance, Shekiro III et al. (2014) used sulphuric acid (H_2SO_4) at 0.14 to 0.30% (w/w) (160 °C, 15 min) for the pretreatment of deacetylated corn stover to produce high monomeric xylose yields (73.5%) and improved cellulose digestibility of the pretreated solids (Shekiro III et al. 2014). Corn stover hydrolysate was also pretreated with 1% H_2SO_4 at 165 °C for 10 min exhibiting an 80% monomeric xylose yield (Sievers et al. 2017). On other hand, Ávila-Lara et al. (2015), pretreated agave bagasse using several variables, such as catalyst loading, retention time, and solids loading to obtain an optimal process. The optimum conditions were 2.1% H_2SO_4 during 33.8 min and 8.5% solids loading (Ávila-Lara et al. 2015).

10.3.4.1.2 Concentrated Acid Pretreatment

The concentrated acid pretreatment consists of the use of acids at high concentrations to disrupt cellulose and hemicellulose fractions. It is usually carried out using over 30% w/w of acid concentration at room or moderate temperatures below 100 °C (Jung and Kim 2015). The primary advantage of the concentrated process is the high sugar recovery efficiency. Jang and Choi (2018), obtained 77% (w/w) sugar yield from the cellulose and hemicellulose fractions in lignocellulosic biomass by using 75% (w/w) H_2SO_4 (Jang and Choi 2018). Although the concentrated acid pretreatment means lower operating cost because of lower temperatures and residence time requirements, the equipment corrosion problems and acid recovery are important drawbacks that imply a greater investment. This adds to the cost makes the total production process more expensive (Shahbazi and Zhang 2010). Hence, dilute acid pretreatment appears to be a more favorable method for large-scale bioethanol production and as a pretreatment on a wide range of lignocellulosic biomass. Among other pretreatment methods, acid treatment is by far the best method because of its availability of highly commercial compounds, flexibility to be applied at the industrial level, relatively low cost, high disruption of hemicellulose and high glucose recovery yield (Solarte-Toro et al. 2019). Nevertheless, some disadvantages of this process are the additional steps of washing to the acid recovery, the high cost of corrosion-resistant equipment, the acid pollution, toxicity to the environment, the high maintenance cost and the formation of inhibitory compounds (Tian et al. 2018). In this sense, due to the partial breakdown of lignin and disruption of cellulose and hemicellulose, some compounds such as furfural, phenolic compounds and hydroxymethylfurfural (HMF) are released and considered inhibitory compounds for microbial growth (Mathew et al. 2016).

10.3.4.2 Alkali-Based Pretreatment

Alkaline pretreatment is another form of chemical pretreatment whereby a base like sodium, potassium, ammonium, and calcium hydroxides are used to improve the enzymatic digestibility of lignocellulosic biomass. The process involves the alteration of the lignin structure, partial solubilization of cellulose and partial solvation of hemicellulose through degradation of the side chains of esters and glycosides. The whole mechanism alters the polymerization degree bringing some changes in the physical properties of the biomass such as surface area, porosity, and crystallinity (Kim 2013). Alkali pretreatment usually utilizes lower temperature and pressure, but residence time is measured in terms of hours or days according to the fixed temperature. The optimization made by Chen et al. (2012) found an optimal temperature range for glucose and xylose release between 103 to 106 °C and 93 to 97 °C, respectively. Although an increase in temperature, high alkali loading, and longer residence time accelerates the delignification. Furthermore, high severity pretreatment conditions may also lead to undesired sugar loss by dissolution and degradation of hemicellulose and reduce glucan/xylan conversion.

Additionally, the reaction conditions also depend on the reagent used, for instance, sodium hydroxide (NaOH) has been applied to wood biomass at higher temperature and concentration, in particular at 150 °C and 5 N of concentration, resulting in the efficient removal of hemicellulose and lignin (~ 60%), while the glucose content was maintained (Oka et al. 2013). Kim (2018) use a thermo-chemical treatment as pretreatment on palm fruit fibers with 0.5 to 3 M of a NaOH solution at 121 °C, 15 psi of pressure for 60 min and observed was a 56.9% of delignification yield. On other hand, (Yang et al. 2012) reported the use of sodium carbonate (Na_2CO_3) and observed the sugar recovery of 71.7% from pretreated rice straw at 140 °C and total titratable alkali charge of 8%. Li and Kim (2011) treat corn stover using 50% (w/w) of ammonia at 30 °C for 12 weeks and observed 55% of delignification and 86.5% glucans digestibility. Finally, high cellulose-to-glucose conversion of 89.5% was achieved using lime ($Ca(OH)_2$) pretreatment carried out at 90 °C for 5 h and a lime loading of 0.4 g of $Ca(OH)_2$/g of corn stover improving the ethanol conversion to 72.4% (28.7 g/L of bioethanol produced) (Fírvida et al. 2021). Unlike acid-based catalyzed pretreatments, alkali-based pretreatment reduces the formation of inhibitory compounds, decreases the need for corrosion-resistant equipment, has minimal sugar degradation and, in some methods, allows the recovery and reuse of alkaline reagents (Kim 2018). The recovery of alkaline reagents is important because they usually are expensive and their disposal can cause serious environmental issues (Badiei et al. 2014). To face these challenges, the industry and researchers have shown great interest in the development of green pretreatment technologies to decrease or eliminate the use and generation of caustic chemicals. Examples of these technologies includes some ionic liquids and organosolv pretreatments, as well as ozonolysis and some irradiation methods, because they are less or non-toxic and have been shown a low production of hazardous wastes (Capolupo and Faraco 2016).

10.3.4.3 Organosolv-Based Pretreatment

Organosolv-based pretreatment has recently attracted great interest in the conversion of lignocellulosic biomass into fermentable sugars and high-purity lignin fractions. This method consists of using organic aqueous solvents such as alcohols, esters, ketones, glycols, organic acids, phenols, and ethers, and a small amount of water (usually 1:1 v/v solvent:water) in presence of a catalyst (hydrochloric, sulfuric, oxalic, or salicylic acids) at temperatures ranging from 100 to 250 °C. When lignocellulose is mixed with this organic liquid and heated, the network of lignin and a part of the hemicellulose is dissolved, leaving reactive cellulose in the solid phase suitable for enzymatic hydrolysis. In this process, organosolv mainly works by disrupting the bonds between lignin and hemicelluloses to cleave off ether, ester and C-C bonds (Zhang et al. 2016). The process produces three main fractions: high purity lignin, a hemicellulose syrup with C5 and C6 sugars, and a pure cellulose fraction. Xylose and high-purity lignin can be extracted from the solvent for their utilization in other industrial applications (Taherzadeh and Karimi 2008). Catalyst such as mineral acids (e.g., hydrochloric acid and sulfuric acids) and organic acids (oxalic and salicylic acids) are added to decrease the operating temperature and to enhance the delignification (Zhao et al. 2009).

The most popular organic solvent are ethanol, methanol, acetone, formic acid, acetic acid, and glycerol have also been studied for lignocellulosic biomass (Dapía et al. 2002; Huijgen et al. 2010; Snelders et al. 2014; Teramura et al. 2016). Ethanol is the most popular solvent in lignocellulosic organosolv-based pretreatment due to the high conversion of cellulose, high availability, and low boiling point (Galbe and Wallberg 2019). For instance, (Salapa et al. 2017) obtained a maximum cellulose conversion of 89% and a bioethanol yield of 67% using a pretreated wheat straw with ethanol at 180 °C in the presence of sulfuric acid as catalyst. When methanol and butanol were used at the same conditions, a decrease in bioethanol yield was observed with 56% and 49%, respectively, and less than 69% conversion of cellulose. In almost all the cases, the use of organosolv as a pretreatment method has shown a higher efficiency compared with other methods such as ionic liquids and deep eutectic solvents due to the enhancement of the saccharification process by increasing the enzymatic accessibility to cellulose and hemicellulose. Even when organosolv-based pretreatment is a promising treatment for the biorefinery sector, but more research and cost optimization to solve the major drawbacks of this method that include the high operation costs due to the high temperature and pressure conditions, the expensive corrosion-resistant equipment and the solvent recovery process is needed. Even when the solvents can be recovered using distillation and recycled, they need to be drained from the reactor, evaporated, and condensed, making this process's costs relatively high (Bensah and Mensah 2013).

10.3.4.4 Steam Explosion

Steam explosion is one of the most widely used physicochemical pretreatment techniques. This method deals with high temperature and pressure that contribute to the disruption of internal hydrogen bonds and alters the ordered structure of cellulose (Chen and Liu 2015; Alvira et al. 2016). High temperature in the treatment removes the hemicellulose from solid fraction and promotes cellulose digestibility. Whereas the high pressure applied to the biomass gets released from the closed pores from the fibers in the form of steam. The high temperature and high pressure applied to the biomass, results in the breakdown of cellulose into monosaccharides and releases new hydroxyl groups. The new hydroxyl group emerged from the sugars, maximizes the adsorption capacity of cellulose and promotes the hemicellulose transformation and lignin hydrolysis (Fernández-Bolaños et al. 2001; Hongzhang and Liying 2007). This method is integrated with chemicals and mechanical forces to hydrolyze the acetyl group present in the hemicellulose. This is because the alkali-treated pretreatment extract results in a higher conversion rate of sugar and ethanol. However, a large number of phenolic compounds are generated during the breakdown of lignin. The production of toxic compounds such as phenolic compounds, furan derivatives and weak organic acids can hamper the yield of enzymatic hydrolysis. This reduces the conversion rate of glucose during the fermentation and de-accelerates the ethanol production (McIntosh et al. 2012; Park et al. 2012; Buratti et al. 2015). Therefore, some detoxification methods are necessary to overcome the challenges of toxic compounds. Several challenges associated with this method are mentioned in Table 10.4. Approaches like genetic modification, evolutionary engineering, and adaptive strategies have very promising results. Apart from the challenges mentioned above, the pretreatment still results in excellent enzymatic hydrolysis yield. The autohydrolysis of acetyl group at the high temperature produces acetic acid and proportionally increases the scale-up process at the industry level. Since a good amount of acetic acid is formed during the autohydrolysis, therefore, an additional supply of acid catalyst is not required for the pretreatment.

Table 10.4 Summary of physicochemical pretreatment

Pretreatment	Advantages	Disadvantages	References
Steam explosion	Minimum energy consumption and short duration time. No environmental and recycling cost. By using the steam explosion technique with wheat straw, 3,000,000 L ethanol is produced. The pretreatment is highly recommended for agricultural residues and hardwoods.	Incomplete or partial decomposition of a lignin-carbohydrate matrix. This results in incomplete enzymatic hydrolysis. Great destruction to hemicellulose (especially xylan). Generation of inhibitors at higher temperatures.	(Chen and Liu 2015; Alvira et al. 2016)
Liquid hot water	Low energy and temperature requirements and minimized degradation products are formed. No usage of chemicals. Higher hemicellulose sugars are obtained and low fermenter inhibitors. Reduced cost of product recovery because of eco-friendly and no usage of chemicals.	Number of solubilized products is higher, but the concentration of products formed is lower. Down-steam processing requires a lot of energy because of the huge volume of water.	(Chandra et al. 2012)
Wet oxidation	Easily availability of oxygen or air as an oxidizing agent. Mineralizes organic components into CO_2, water and inorganic acids.	High temperature and high-pressure demand for this pretreatment make it costly.	(Banerjee et al. 2011)
AFEX	Dry-to-dry process as no wash stream is used. Also, no generation of toxic chemicals. High ethanol production without biomass washing, detoxification and nutrient supplementation. High cellulose retention after the pretreatment.	Higher capital investment as compared to other pretreatments. Since ammonia is hazardous and corrosive in nature. Therefore, pretreatment requires a highly controlled environment.	(Rabemanolontsoa and Saka 2016; Hu et al. 2019)

This makes the overall system more economical. Other factors such as material particle size, humidity, steam temperature and residence time accounts are the few advantages that account for the efficiency of this treatment. Minimum investment, less use of hazardous reagents, lower environmental impact, lower requirement for reaction condition and complete sugar recovery are some of the benefits obtained from this pretreatment. Brazil is well known as one of the most important countries in second generation bioethanol production. An example is Raizen® with ~2.5

billions of bioethanol production during 2019–2020. Raizen® bioethanol production is sugarcane bagasse-based as well as sugar and regular ethanol residues. The company use steam explosion technology followed for enzymatic saccharification to pretreat the bagasse in order to obtain fermentable sugars.

10.3.4.5 Liquid Hot Water

Liquid hot water (LHW) is also known as "Hot Compressed Water". This pretreatment is employed at high temperature (160-220 °C) and high pressure to keep water in the liquid state. This method does not require a catalyst for the pretreatment process. The technique does not require rapid expansion or decompression and the pressure used in this method is meant to avoid evaporation. The application of LHW pretreatment has been efficient to a variety of lignocellulosic biomass due to an increase in cellulose digestibility because of enhanced hemicellulose removal (up to 80%). It has been commonly applied for rice straw, corn stover, wheat straw, corncobs and rye straws (Chandra et al. 2012). In order to prevent sugar degradation and generation of inhibitors, it is preferred to maintain the pH between 4 to 7 at 190 °C for 15 minutes. This controlled pH, temperature and exposure time is well suited for the anaerobic digestion process by the cells. Based on water flow and biomass flow direction in a reactor, LHW can be performed in three ways. The first technique is co-current pretreatment. This involves the heating of biomass slurry at high temperatures (Abomohra 2019). This condition is held for a controlled residence time and finally applied to a cool environment. The second technique is countercurrent pretreatment. This is carried out using the pumping of hot water against biomass in a controlled environment (Abomohra 2019). The third technique is a flow-through pretreatment that is carried out by hot water flow through lignocellulosic biomass. Like other pretreatments, this LHW has also a few advantages and disadvantages that are briefly discussed in Table 10.4. This pretreatment accounts for high energy consumption and the extensive requirement of water during the downstream processing accounts for its limitation. Whereas no usage of chemicals, catalysts and minimum production of inhibitors favors the scale-up processes. Another advantage of this pretreatment is that it does not corrode the bioreactor therefore, this process can be more economically from the industrial perspective. For instance, Granbio® in Brazil, use this technology to pretreat sugarcane baggasse for second generation bioethanol production. Granbio®, produce ~60 million liters of bioethanol per year, being thus the first company in South America with mass production of second-generation bioethanol.

10.3.4.6 Wet Oxidation

Wet oxidation (WO) is another physicochemical pretreatment that uses an oxidizing agent in an aqueous environment to break down lignocellulose (Ravindran and Jaiswal 2016). The WO pretreatment consists in treating lignocellulosic biomass with an aqueous solution (acidic, neutral, and alkaline conditions) via oxygen reaction in the presence of high temperature and pressure conditions (120 to 315 °C and 0.5 to 5 MPa) for a period over 10 minutes (Zhang et al. 2020).

Oxygen, temperature, pressure, and reaction time are the main factors that determine the efficacy of wet oxidation pretreatment. This is because an increase in temperature leads to an increase in hydrogen ion concentration which ultimately decreases the pH. Therefore, in this process at 170 °C the water acts as an acid catalyst and catalyzes the hydrolytic reaction. The mechanism involves the solubilization of hemicellulose and part of the lignin fraction by de-esterifying the acetate groups. This results in the breaking down of hemicellulose into pentose sugar molecules, oxidation of lignin while the cellulose remains ineffective (Zhang et al. 2020). The addition of chemical agents such as alkaline peroxide and sodium carbonate are proven to reduce the reaction temperature, improve the hemicellulose degradation and minimize the formation of inhibitory compounds (Banerjee et al. 2011). Schmidt and Thomsen (1998) found that the solubilization of hemicellulose is directly influenced by the function of temperature, where the concentration of hemicellulose increased up to three-fold with the temperature increased from 150 to 180 °C. In the oxidative mechanism of this process, carbon dioxide and water are released as by-products. The decomposition of polymeric chains from the biomass yields to the production of aldehydes, low-weight carboxylic acids and alcohols during hydrolysis. WO pretreatment has been considered a method with an ideal effect compared with other treatments because of its potential efficiency for fractionating the lignocellulose biomass matrix. Furthermore, this method reduces the formation of enzyme-inhibiting compounds compared with steam explosion or hot water treatment and as it is an exothermic process, the total energy demand is reduced (Refaat 2012). On the other hand, high-pressure oxygen demand, high cost of hydrogen peroxide and combustion nature of pure oxygen during the pretreatment are the demerits of this technique. These disadvantages make the operation more expensive and limits its application at industrial levels (de Jong and Gosselink 2014; Bajpai 2016).

10.3.4.7 Ammonia Fiber Expansion (AFEX)

Ammonia Fiber Expansion (AFEX) is usually operated at lower temperature and concentrated ammonia as a catalyst. The concentration varies from 0.3 to 2 kg of ammonia per Kg of dry biomass. The catalyst (ammonia) is added in a high-pressure reactor containing biomass for a range of 5–45 minutes of pretreatment. This maintains the temperature around 90° C in the bioreactor and after this time the pressure in the reactor is released rapidly to break the lignocellulose matrix. Only a small amount of material was solubilized while hemicellulose and lignin remained intact. Ammonia being volatile was easily eluted out and reused in the process later. The principle of this process is highly coherent with steam explosion pretreatment. In the case of steam explosion, water is used to create pressure whereas, in AFEX, ammonia is used to create high pressure (Rabemanolontsoa and Saka 2016).

For example, in the case of AFEX, the lignocellulosic biomass is heated with liquid ammonia at a 1:1 ratio in a closed reactor at 60-90° C and maintains the pressure above 3 MPa for 60 min (Kumar and Sharma 2017). Once the temperature is reached the desired value it is held for 5 minutes and then the reactor is opened which releases the pressure. This sudden pressure drop reduces the temperature of the system and results in the evaporation of ammonia. When the lignocellulosic biomass is treated with high temperature and pressure complemented with ammonia, it causes swelling and change in the crystallinity of the lignocellulosic matrix. This results in increased reactivity of carbohydrates present after the pretreatment. The method modifies the lignin structure which accounts for increased digestibility and water holding capacity. The advantage of using AFEX (Table 10.4) is that it avoids the generation of toxic inhibitors. This property makes it highly desirable to combat the challenges during downstream processing. Besides this, the cost of AFEX pretreatment is very low as compared to other methods. This is because it does not include additional steps of washing, detoxification, recovery and reuse of water. Almost 90% of the hemicellulose and cellulose is changed to fermentable sugars by this AFEX pretreatment under optimized conditions of temperature, moisture, pressure, ammonia leading and pretreatment duration (Uppugundla et al. 2014).

10.4 Future Perspectives and Conclusions

The chapter aimed to provide a concise summary as well as the advantages and disadvantages of technologies currently used as a pretreatment in lignocellulosic biomass to produce bioethanol. It was further discussed the role of key factors that have a significant impact on the pretreatment. Nevertheless, several challenges have to be addressed to improve the pretreatment process such as by decreasing the use of expensive chemicals, high energy consumption, denaturation of hemicellulose as well as inhibitor generation.

To tackle these issues, additional advancements are being developed to achieve the most effective, efficient, environmentally sustainable, and economically suitable pretreatment processes. The subcritical and supercritical fluids are the new way to look for the effective pretreatment method. The subcritical solvent is the less polar solvent which requires a small amount of water for the pretreatment whereas supercritical solvent is a solvent that can exist above the critical temperature and pressure in a transient phase. The application of sub- and supercritical fluids such as CO_2 has raised in the last years because of its unique properties to penetrate the surface of biomass, increase the permeability of cellulose and reach maximum reducing sugar yields. The above is because the high temperature and pressure increase the permeability and disrupt all the outer crosslinking fibers with lignin and hemicellulose. As a result, this kind of treatment provides the maximum possible surface area to the enzymes for saccharification. For example, Liang et al. (2017) used subcritical CO_2-water hydrolysis to treat sugarcane bagasse producing 45.8% of reducing sugar at the optimal conditions of 200 °C, 40 min, and 1 MPa.

Additionally, a low temperature steep of delignification (LTSD) during the application of supercritical fluids has been proposed to remove and recover more than 90% of lignin from renewable biomass such as hardwood chips. This pretreatment operates at low temperatures, utilizes low concentrations of nontoxic chemicals, and does not produce toxic inhibitors (Bhatia et al. 2020). There are also some variations of current technologies to make them more efficient or economically feasible such as modified organosolv that consists of a combination of the organic compound, tetrahydrofuran water, and dilute sulfuric acid for the fractionation of lignin.

Moreover, more attention and effort are needed on knowledge of biological pretreatment, their mechanisms as well as the implications of advanced tools used to improve this category of pretreatment. Recently, recombinant technologies are very good tools that lead to developments in enhancing or suppressing specific genes that play salient roles in increased lipid production, enhanced inhibitor tolerances, and improved carbon consumption rates (Saini et al. 2020) this could elevate the productivity of the biofuels. Despite these advancements in pretreatment, there is still a huge scope for improved biomass pretreatment that can offer maximum sugar yield with minimum inhibitory compounds, energy, and chemical consumption, as well as to increase its marketable profitability and sustainability. In economic terms, the use of waste residues, less energy demand, and the addition of value-added products to the biofuel production help to reduce the production cost of bioethanol. Hence, the constant development of more economically feasible techniques, isolating or engineering better productive microbes, and eco-friendly pretreatment method based on positive or neutral life cycle assessment could make the bioethanol and biorefinery process much better and efficient than the current trends.

Acknowledgments Financial Support from Mitacs Globalink, Natural Sciences and Engineering Research Council of Canada (No. 284111, Discovery; No. 476649–14, Strategic Research Grant), and James and Joanne Love Chair in Environmental Engineering

References

Abomohra AE-F (2019) Biomass for bioenergy: recent trends and future challenges. BoD–Books on Demand

Alvira P, Negro MJ, Ballesteros I et al (2016) Steam explosion for wheat straw pretreatment for sugars production. Bioethanol 2. https://doi.org/10.1515/bioeth-2016-0003

Ávila-Lara AI, Camberos-Flores JN, Mendoza-Pérez JA et al (2015) Optimization of alkaline and dilute acid pretreatment of agave bagasse by response surface methodology. Front Bioeng Biotechnol 0. https://doi.org/10.3389/fbioe.2015.00146

Badiei M, Asim N, Jahim JM, Sopian K (2014) Comparison of chemical pretreatment methods for cellulosic biomass. APCBEE Procedia 9:170–174. https://doi.org/10.1016/j.apcbee.2014.01.030

Bajpai P (2016) Pretreatment of lignocellulosic biomass. In: Bajpai P (ed) Pretreatment of lignocellulosic biomass for biofuel production. Springer, Singapore, pp 17–70

Balan V, Bals B, Chundawat SPS et al (2009) Lignocellulosic biomass pretreatment using AFEX. In: Mielenz JR (ed) Biofuels: methods and protocols. Humana Press, Totowa, NJ, pp 61–77

Banerjee S, Sen R, Mudliar S et al (2011) Alkaline peroxide assisted wet air oxidation pretreatment approach to enhance enzymatic convertibility of rice husk. Biotechnol Prog 27:691–697. https://doi.org/10.1002/btpr.589

Barrera-Martínez I, Guzmán N, Peña E et al (2016) Ozonolysis of alkaline lignin and sugarcane bagasse: structural changes and their effect on saccharification. Biomass Bioenergy 94:167–172. https://doi.org/10.1016/j.biombioe.2016.08.010

Baskar G, Kalavathy G, Aiswarya R, Abarnaebenezer Selvakumari I (2019) 7–Advances in bio-oil extraction from nonedible oil seeds and algal biomass. In: Azad K (ed) Advances in eco-fuels for a sustainable environment. Woodhead Publishing, pp 187–210

Bensah EC, Mensah M (2013) Chemical pretreatment methods for the production of cellulosic ethanol: technologies and innovations. Int J Chem Eng 2013:e719607. https://doi.org/10.1155/2013/719607

Bhatia SK, Jagtap SS, Bedekar AA et al (2020) Recent developments in pretreatment technologies on lignocellulosic biomass: effect of key parameters, technological improvements, and challenges. Bioresour Technol 300:122724. https://doi.org/10.1016/j.biortech.2019.122724

Bhutto AW, Qureshi K, Harijan K et al (2017) Insight into progress in pre-treatment of lignocellulosic biomass. Energy 122:724–745. https://doi.org/10.1016/j.energy.2017.01.005

Buratti C, Barbanera M, Lascaro E (2015) Ethanol production from vineyard pruning residues with steam explosion pretreatment. Environ Prog Sustain Energy 34:802–809. https://doi.org/10.1002/ep.12043

Capolupo L, Faraco V (2016) Green methods of lignocellulose pretreatment for biorefinery development. Appl Microbiol Biotechnol 100:9451–9467. https://doi.org/10.1007/s00253-016-7884-y

Chandra R, Takeuchi H, Hasegawa T (2012) Methane production from lignocellulosic agricultural crop wastes: a review in context to second generation of biofuel production. Renew Sust Energ Rev 16:1462–1476. https://doi.org/10.1016/j.rser.2011.11.035

Chaturvedi V, Verma P (2013) An overview of key pretreatment processes employed for bioconversion of lignocellulosic biomass into biofuels and value added products. 3 Biotech 3:415–431. https://doi.org/10.1007/s13205-013-0167-8

Chen W-H, Tu Y-J, Sheen H-K (2011) Disruption of sugarcane bagasse lignocellulosic structure by means of dilute sulfuric acid pretreatment with microwave-assisted heating. Appl Energy 88:2726–2734. https://doi.org/10.1016/j.apenergy.2011.02.027

Chen H-Z, Liu Z-H (2015) Steam explosion and its combinatorial pretreatment refining technology of plant biomass to bio-based products. Biotechnol J 10:866–885. https://doi.org/10.1002/biot.201400705

Chen W-H, Ye S-C, Sheen H-K (2012) Hydrolysis characteristics of sugarcane bagasse pretreated by dilute acid solution in a microwave irradiation environment. Appl Energy 93:237–244. https://doi.org/10.1016/j.apenergy.2011.12.014

Cheng JJ, Timilsina GR (2011) Status and barriers of advanced biofuel technologies: a review. Renew Energy 36:3541–3549. https://doi.org/10.1016/j.renene.2011.04.031

Chiranjeevi T, Mattam AJ, Vishwakarma KK et al (2018) Assisted single-step acid pretreatment process for enhanced delignification of Rice straw for bioethanol production. ACS Sustain Chem Eng 6:8762–8774. https://doi.org/10.1021/acssuschemeng.8b01113

Dapía S, Santos V, Parajó JC (2002) Study of formic acid as an agent for biomass fractionation. Biomass Bioenergy 22:213–221. https://doi.org/10.1016/S0961-9534(01)00073-3

de Guilherme A, Dantas PVF, Soares JCJ et al (2017) Pretreatments and enzymatic hydrolysis of sugarcane bagasse aiming at the enhancement of the yield of glucose and xylose. Braz J Chem Eng 34:937–947. https://doi.org/10.1590/0104-6632.20170344s20160225

de Jong E, Gosselink RJA (2014) Chapter 17–Lignocellulose-based chemical products. In: Gupta VK, Tuohy MG, Kubicek CP et al (eds) Bioenergy research: advances and applications. Elsevier, Amsterdam, pp 277–313

Diaz AB, Moretti MM de S, Bezerra-Bussoli C, et al (2015) Evaluation of microwave-assisted pretreatment of lignocellulosic biomass immersed in alkaline glycerol for fermentable sugars production. Bioresour Technol 185:316–323. https://doi.org/10.1016/j.biortech.2015.02.112

Du W, Yu H, Song L et al (2011) The promoting effect of byproducts from Irpex lacteus on subsequent enzymatic hydrolysis of bio-pretreated cornstalks. Biotechnol Biofuels 4:37. https://doi.org/10.1186/1754-6834-4-37

EIA U.S. (2021) Energy Information Administration: independent statistics & analysis. http://www.eia.gov.html. Accessed 5 Oct 2021

Fennouche I, Khellaf N, Djelal H, Amrane A (2019) An effective acid pretreatment of agricultural biomass residues for the production of second-generation bioethanol. SN Appl Sci 1:1460. https://doi.org/10.1007/s42452-019-1517-x

Fernández-Bolaños J, Felizón B, Heredia A et al (2001) Steam-explosion of olive stones: hemicellulose solubilization and enhancement of enzymatic hydrolysis of cellulose. Bioresour Technol 79:53–61. https://doi.org/10.1016/S0960-8524(01)00015-3

Fírvida I, del Río PG, Gullón P et al (2021) Alternative lime pretreatment of corn Stover for second-generation bioethanol production. Agronomy 11:155. https://doi.org/10.3390/agronomy11010155

Galbe M, Zacchi G (2012) Pretreatment: the key to efficient utilization of lignocellulosic materials. Biomass Bioenergy 46:70–78. https://doi.org/10.1016/j.biombioe.2012.03.026

Galbe M, Wallberg O (2019) Pretreatment for biorefineries: a review of common methods for efficient utilisation of lignocellulosic materials. Biotechnol Biofuels 12:294. https://doi.org/10.1186/s13068-019-1634-1

García-Cubero MT, González-Benito G, Indacoechea I et al (2009) Effect of ozonolysis pretreatment on enzymatic digestibility of wheat and rye straw. Bioresour Technol 100:1608–1613. https://doi.org/10.1016/j.biortech.2008.09.012

Häggström C, Rova U, Brandberg T, Hodge DB (2014) Chapter 8–Integration of ethanol fermentation with second generation biofuels technologies. In: Qureshi N, Hodge DB, Vertès AA (eds) Biorefineries. Elsevier, Amsterdam, pp 161–187

Hassan SS, Williams GA, Jaiswal AK (2018) Emerging technologies for the pretreatment of lignocellulosic biomass. Bioresour Technol 262:310–318. https://doi.org/10.1016/j.biortech.2018.04.099

Hongzhang C, Liying L (2007) Unpolluted fractionation of wheat straw by steam explosion and ethanol extraction. Bioresour Technol 98:666–676. https://doi.org/10.1016/j.biortech.2006.02.029

Hou Q, Ju M, Li W et al (2017) Pretreatment of lignocellulosic biomass with ionic liquids and ionic liquid-based solvent systems. Molecules 22:490. https://doi.org/10.3390/molecules22030490

Hu J, Jiang B, Wang J et al (2019) Physicochemical characteristics and pyrolysis performance of corn stalk torrefied in aqueous ammonia by microwave heating. Bioresour Technol 274:83–88. https://doi.org/10.1016/j.biortech.2018.11.076

Hu Y, Bassi A, Xu C (Charles) (2020) 21–Energy from biomass. In: Letcher TM (ed) Future energy (third edition). Elsevier, pp. 447–471

Huijgen WJJ, Reith JH, den Uil H (2010) Pretreatment and fractionation of wheat straw by an acetone-based Organosolv process. Ind Eng Chem Res 49:10132–10140. https://doi.org/10.1021/ie101247w

Jang M-O, Choi G (2018) Techno-economic analysis of butanol production from lignocellulosic biomass by concentrated acid pretreatment and hydrolysis plus continuous fermentation. Biochem Eng J 134:30–43. https://doi.org/10.1016/j.bej.2018.03.002

Jibouri AKHA, Turcotte G, Wu J, Cheng C-H (2015) Ozone pretreatment of humid wheat straw for biofuel production. Energy Science & Engineering 3:541–548. https://doi.org/10.1002/ese3.93

Jung YH, Kim KH (2015) Chapter 3–Acidic pretreatment. In: Pandey A, Negi S, Binod P, Larroche C (eds) Pretreatment of biomass. Elsevier, Amsterdam, pp 27–50

Kainthola J, Shariq M, Kalamdhad AS, Goud VV (2019) Enhanced methane potential of rice straw with microwave assisted pretreatment and its kinetic analysis. J Environ Manag 232:188–196. https://doi.org/10.1016/j.jenvman.2018.11.052

Kapoor K, Garg N, Diwan RK et al (2017) Study the effect of gamma radiation pretreatment of sugarcane bagasse on its physcio-chemical morphological and structural properties. Radiat Phys Chem 141:190–195. https://doi.org/10.1016/j.radphyschem.2017.07.010

Karatzos SK, Edye LA, Doherty WOS (2012) Sugarcane bagasse pretreatment using three imidazolium-based ionic liquids; mass balances and enzyme kinetics. Biotechnol Biofuels 5: 62. https://doi.org/10.1186/1754-6834-5-62

Kassim MA, Khalil HPSA, Serri NA et al (2016) Irradiation pretreatment of tropical biomass and biofiber for biofuel production. In: Monteiro WA (ed) Radiation effects in materials. InTech

Kaur U, Oberoi HS, Bhargav VK et al (2012) Ethanol production from alkali- and ozone-treated cotton stalks using thermotolerant Pichia kudriavzevii HOP-1. Ind Crop Prod 37:219–226. https://doi.org/10.1016/j.indcrop.2011.12.007

Keshwani DR, Cheng JJ (2010) Microwave-based alkali pretreatment of switchgrass and coastal bermudagrass for bioethanol production. Biotechnol Prog 26:644–652. https://doi.org/10.1002/btpr.371

Kim JS, Lee YY, Kim TH (2016) A review on alkaline pretreatment technology for bioconversion of lignocellulosic biomass. Bioresour Technol 199:42–48. https://doi.org/10.1016/j.biortech.2015.08.085

Kim S (2018) Evaluation of alkali-pretreated soybean straw for lignocellulosic bioethanol production. International Journal of Polymer Science 2018:e5241748. https://doi.org/10.1155/2018/5241748

Kim TH (2013) Pretreatment of lignocellulosic biomass. In: Bioprocessing technologies in biorefinery for sustainable production of fuels, chemicals, and polymers. Wiley, pp 91–110

Kostas ET, Beneroso D, Robinson JP (2017) The application of microwave heating in bioenergy: a review on the microwave pre-treatment and upgrading technologies for biomass. Renew Sust Energ Rev 77:12–27. https://doi.org/10.1016/j.rser.2017.03.135

Kumar AK, Parikh BS, Pravakar M (2016) Natural deep eutectic solvent mediated pretreatment of rice straw: bioanalytical characterization of lignin extract and enzymatic hydrolysis of pretreated biomass residue. Environ Sci Pollut Res 23:9265–9275. https://doi.org/10.1007/s11356-015-4780-4

Kumar AK, Sharma S (2017) Recent updates on different methods of pretreatment of lignocellulosic feedstocks: a review. Bioresources and Bioprocessing 4:7. https://doi.org/10.1186/s40643-017-0137-9

Kumari D, Singh R (2018) Pretreatment of lignocellulosic wastes for biofuel production: a critical review. Renew Sust Energ Rev 90:877–891. https://doi.org/10.1016/j.rser.2018.03.111

Lee RA, Lavoie J-M (2013) From first- to third-generation biofuels: challenges of producing a commodity from a biomass of increasing complexity. Anim Front 3:6–11. https://doi.org/10.2527/af.2013-0010

Lee B-M, Jeun J-P, Kang P-H (2017) Enhanced enzymatic hydrolysis of kenaf core using irradiation and dilute acid. Radiat Phys Chem 130:216–220. https://doi.org/10.1016/j.radphyschem.2016.08.026

Li X, Kim TH (2011) Low-liquid pretreatment of corn Stover with aqueous ammonia. Bioresour Technol 102:4779–4786. https://doi.org/10.1016/j.biortech.2011.01.008

Liang J, Chen X, Wang L et al (2017) Subcritical carbon dioxide-water hydrolysis of sugarcane bagasse pith for reducing sugars production. Bioresour Technol 228:147–155. https://doi.org/10.1016/j.biortech.2016.12.080

Lin Z, Huang H, Zhang H et al (2010) Ball milling pretreatment of corn Stover for enhancing the efficiency of enzymatic hydrolysis. Appl Biochem Biotechnol 162:1872–1880. https://doi.org/10.1007/s12010-010-8965-5

Liu Y, Guo L, Wang L et al (2017) Irradiation pretreatment facilitates the achievement of high total sugars concentration from lignocellulose biomass. Bioresour Technol 232:270–277. https://doi.org/10.1016/j.biortech.2017.01.061

da Costa Lopes AM, Roseiro LB, Bogel-Lukasik R (2015) Chapter 5: Relevance of ionic liquids and biomass feedstocks for biomolecule extraction. In: Ionic liquids in the biorefinery concept, pp 121–167

Magda R, Szlovák S, Tóth J (2021) Chapter 7–The role of using bioalcohol fuels in sustainable development. In: Bochtis D, Achillas C, Banias G, Lampridi M (eds) Bio-economy and Agri-production. Academic, pp 133–146

Mamleeva NA, Autlov SA, Bazarnova NG, Lunin VV (2009) Delignification of softwood by ozonation. Pure Appl Chem 81:2081–2091. https://doi.org/10.1351/PAC-CON-08-10-11

Mathew AK, Parameshwaran B, Sukumaran RK, Pandey A (2016) An evaluation of dilute acid and ammonia fiber explosion pretreatment for cellulosic ethanol production. Bioresour Technol 199: 13–20. https://doi.org/10.1016/j.biortech.2015.08.121

McIntosh S, Vancov T, Palmer J, Spain M (2012) Ethanol production from eucalyptus plantation thinnings. Bioresour Technol 110:264–272. https://doi.org/10.1016/j.biortech.2012.01.114

Mohapatra S, Mishra C, Behera SS, Thatoi H (2017) Application of pretreatment, fermentation and molecular techniques for enhancing bioethanol production from grass biomass–a review. Renew Sust Energ Rev 78:1007–1032. https://doi.org/10.1016/j.rser.2017.05.026

Morais ES, da Costa Lopes AM, Freire MG et al (2020) Use of ionic liquids and deep eutectic solvents in polysaccharides dissolution and extraction processes towards sustainable biomass valorization. Molecules 25:3652. https://doi.org/10.3390/molecules25163652

Mosier N (2005) Features of promising technologies for pretreatment of lignocellulosic biomass. Bioresour Technol 96:673–686. https://doi.org/10.1016/j.biortech.2004.06.025

Nakagawa H, Harada T, Ichinose T et al (2007) Biomethanol production and CO_2 emission reduction from forage grasses, trees, and crop residues. Japan Agricultural Research Quarterly: JARQ 41:173–180. https://doi.org/10.6090/jarq.41.173

Narayanaswamy N, Dheeran P, Verma S, Kumar S (2013) Biological pretreatment of lignocellulosic biomass for enzymatic Saccharification. In: Fang Z (ed) Pretreatment techniques for biofuels and biorefineries. Springer, Berlin, Heidelberg, pp 3–34

Oka D, Kobayashi K, Isobe N et al (2013) Enzymatic hydrolysis of wood with alkaline treatment. J Wood Sci 59:484–488. https://doi.org/10.1007/s10086-013-1359-x

Olivier-Bourbigou H, Magna L, Morvan D (2010) Ionic liquids and catalysis: recent progress from knowledge to applications. Appl Catal A Gen 373:1–56. https://doi.org/10.1016/j.apcata.2009.10.008

Osorio-González CS, Hegde K, Brar SK et al (2019) Challenges in lipid production from lignocellulosic biomass using Rhodosporidium sp.; a look at the role of lignocellulosic inhibitors. Biofuels Bioprod Biorefin 13:740–759. https://doi.org/10.1002/bbb.1954

Osorio-González CS, Hegde K, Brar SK et al (2020) Pulsed-ozonolysis assisted oxidative treatment of forestry biomass for lignin fractionation. Bioresour Technol 313:123638. https://doi.org/10.1016/j.biortech.2020.123638

Osuna-Laveaga DR, García-Depraect O, Vallejo-Rodríguez R et al (2020) Integrated ozonation-enzymatic hydrolysis pretreatment of sugarcane bagasse: enhancement of sugars released to expended ozone ratio. PRO 8:1274. https://doi.org/10.3390/pr8101274

Park J, Shiroma R, Al-Haq MI et al (2010) A novel lime pretreatment for subsequent bioethanol production from rice straw–calcium capturing by carbonation (CaCCO) process. Bioresour Technol 101:6805–6811. https://doi.org/10.1016/j.biortech.2010.03.098

Park J-Y, Kang M, Kim JS et al (2012) Enhancement of enzymatic digestibility of Eucalyptus grandis pretreated by NaOH catalyzed steam explosion. Bioresour Technol 123:707–712. https://doi.org/10.1016/j.biortech.2012.07.091

Rabemanolontsoa H, Saka S (2016) Various pretreatments of lignocellulosics. Bioresour Technol 199:83–91. https://doi.org/10.1016/j.biortech.2015.08.029

Rastogi M, Shrivastava S (2017) Recent advances in second generation bioethanol production: an insight to pretreatment, saccharification and fermentation processes. Renew Sust Energ Rev 80: 330–340. https://doi.org/10.1016/j.rser.2017.05.225

Ravindran R, Jaiswal AK (2016) A comprehensive review on pre-treatment strategy for lignocellulosic food industry waste: challenges and opportunities. Bioresour Technol 199:92–102. https://doi.org/10.1016/j.biortech.2015.07.106

Refaat AA (2012) 5.13–Biofuels from waste materials. In: Sayigh A (ed) Comprehensive renewable energy. Elsevier, Oxford, pp 217–261

Roberts LG, Patterson TJ (2014) Biofuels. Encyclopedia of Toxicology. Elsevier, In, pp 469–475

Saif Ur Rehman M, Kim I, Chisti Y, Han J-I (2013) Use of ultrasound in the production of bioethanol from lignocellulosic biomass. Energ Educ Sci Technol 30:1391–1410

Saini A, Aggarwal NK, Sharma A, Yadav A (2015) Prospects for irradiation in cellulosic ethanol production. Biotechnol Res Int 2015:1–13. https://doi.org/10.1155/2015/157139

Saini R, Hegde K, Brar SK, Vezina P (2020) Advanced biofuel production and road to commercialization: an insight into bioconversion potential of Rhodosporidium sp. Biomass Bioenergy 132:105439. https://doi.org/10.1016/j.biombioe.2019.105439

Salapa I, Katsimpouras C, Topakas E, Sidiras D (2017) Organosolv pretreatment of wheat straw for efficient ethanol production using various solvents. Biomass Bioenergy 100:10–16. https://doi.org/10.1016/j.biombioe.2017.03.011

Sankaran R, Parra Cruz RA, Pakalapati H et al (2020) Recent advances in the pretreatment of microalgal and lignocellulosic biomass: a comprehensive review. Bioresour Technol 298: 122476. https://doi.org/10.1016/j.biortech.2019.122476

Satari B, Jaiswal AK (2021) Green fractionation of 2G and 3G feedstocks for ethanol production: advances, incentives and barriers. Curr Opin Food Sci 37:1–9. https://doi.org/10.1016/j.cofs.2020.07.004

Sathitsuksanoh N, George A, Zhang Y-HP (2013) New lignocellulose pretreatments using cellulose solvents: a review: new lignocellulose pretreatments using cellulose solvents. J Chem Technol Biotechnol 88:169–180. https://doi.org/10.1002/jctb.3959

Schmidt AS, Thomsen AB (1998) Optimization of wet oxidation pretreatment of wheat straw. Bioresour Technol 64:139–151. https://doi.org/10.1016/S0960-8524(97)00164-8

Sewsynker-Sukai Y, Gueguim Kana EB (2018) Microwave-assisted alkalic salt pretreatment of corn cob wastes: process optimization for improved sugar recovery. Ind Crop Prod 125:284–292. https://doi.org/10.1016/j.indcrop.2018.08.086

Shahbazi A, Zhang B (2010) 5–Dilute and concentrated acid hydrolysis of lignocellulosic biomass. In: Waldron K (ed) Bioalcohol production. Woodhead Publishing, pp 143–158

Sharma HK, Xu C, Qin W (2019) Biological pretreatment of lignocellulosic biomass for biofuels and bioproducts: an overview. Waste Biomass Valor 10:235–251. https://doi.org/10.1007/s12649-017-0059-y

Shekiro J III, Kuhn EM, Nagle NJ et al (2014) Characterization of pilot-scale dilute acid pretreatment performance using deacetylated corn Stover. Biotechnol Biofuels 7:23. https://doi.org/10.1186/1754-6834-7-23

Sievers DA, Kuhn EM, Tucker MP, McMillan JD (2017) Effects of dilute-acid pretreatment conditions on filtration performance of corn Stover hydrolyzate. Bioresour Technol 243. https://doi.org/10.1016/j.biortech.2017.06.144

Sindhu R, Binod P, Pandey A (2016) Biological pretreatment of lignocellulosic biomass–an overview. Bioresour Technol 199:76–82. https://doi.org/10.1016/j.biortech.2015.08.030

Snelders J, Dornez E, Benjelloun-Mlayah B et al (2014) Biorefining of wheat straw using an acetic and formic acid based organosolv fractionation process. Bioresour Technol 156:275–282. https://doi.org/10.1016/j.biortech.2014.01.069

Solarte-Toro JC, Romero-García JM, Martínez-Patiño JC et al (2019) Acid pretreatment of lignocellulosic biomass for energy vectors production: a review focused on operational conditions and techno-economic assessment for bioethanol production. Renew Sust Energ Rev 107:587–601. https://doi.org/10.1016/j.rser.2019.02.024

Song L, Yu H, Ma F, Zhang X (2013) Biological pretreatment under non-sterile conditions for enzymatic hydrolysis of corn stover. Bioresources 8:3802–3816

Stamatelatou K, Antonopoulou G, Michailides P (2014) 15–Biomethane and biohydrogen production via anaerobic digestion/fermentation. In: Waldron K (ed) Advances in biorefineries. Woodhead Publishing, pp 476–524

Su T, Zhao D, Khodadadi M, Len C (2020) Lignocellulosic biomass for bioethanol: recent advances, technology trends, and barriers to industrial development. Current Opinion in Green and Sustainable Chemistry 24:56–60. https://doi.org/10.1016/j.cogsc.2020.04.005

Subhedar PB, Ray P, Gogate PR (2018) Intensification of delignification and subsequent hydrolysis for the fermentable sugar production from lignocellulosic biomass using ultrasonic irradiation. Ultrason Sonochem 40:140–150. https://doi.org/10.1016/j.ultsonch.2017.01.030

Subramaniam Y, Masron TA, Azman NHN (2019) The impact of biofuels on food security. Int Econ 160:72–83. https://doi.org/10.1016/j.inteco.2019.10.003

Taherzadeh MJ, Karimi K (2008) Pretreatment of lignocellulosic wastes to improve ethanol and biogas production: a review. Int J Mol Sci 9:1621–1651. https://doi.org/10.3390/ijms9091621

Tang X, Zuo M, Li Z et al (2017) Green processing of lignocellulosic biomass and its derivatives in deep eutectic solvents. ChemSusChem 10:2696–2706. https://doi.org/10.1002/cssc.201700457

Teramura H, Sasaki K, Oshima T et al (2016) Organosolv pretreatment of sorghum bagasse using a low concentration of hydrophobic solvents such as 1-butanol or 1-pentanol. Biotechnol Biofuels 9:27. https://doi.org/10.1186/s13068-016-0427-z

Tian S-Q, Zhao R-Y, Chen Z-C (2018) Review of the pretreatment and bioconversion of lignocellulosic biomass from wheat straw materials. Renew Sust Energ Rev 91:483–489. https://doi.org/10.1016/j.rser.2018.03.113

Travaini R, Barrado E, Bolado-Rodríguez S (2016a) Effect of ozonolysis pretreatment parameters on the sugar release, ozone consumption and ethanol production from sugarcane bagasse. Bioresour Technol 214:150–158. https://doi.org/10.1016/j.biortech.2016.04.102

Travaini R, Martín-Juárez J, Lorenzo-Hernando A, Bolado-Rodríguez S (2016b) Ozonolysis: an advantageous pretreatment for lignocellulosic biomass revisited. Bioresour Technol 199:2–12. https://doi.org/10.1016/j.biortech.2015.08.143

Travaini R, Otero MDM, Coca M et al (2013) Sugarcane bagasse ozonolysis pretreatment: effect on enzymatic digestibility and inhibitory compound formation. Bioresour Technol 133:332–339. https://doi.org/10.1016/j.biortech.2013.01.133

Tsubaki S, Azuma J, Fujii S et al (2018) Chapter 13–Microwave-driven biorefinery for utilization of food and agricultural waste biomass. In: Bhaskar T, Pandey A, Mohan SV et al (eds) Waste biorefinery. Elsevier, pp 393–408

Tu W-C, Hallett JP (2019) Recent advances in the pretreatment of lignocellulosic biomass. Current Opinion in Green and Sustainable Chemistry 20:11–17. https://doi.org/10.1016/j.cogsc.2019.07.004

Ummalyma SB, Supriya RD, Sindhu R et al (2019) Biological pretreatment of lignocellulosic biomass–current trends and future perspectives. In: Second and third generation of feedstocks. Elsevier, pp 197–212

Uppugundla N, da Costa SL, Chundawat SP et al (2014) A comparative study of ethanol production using dilute acid, ionic liquid and AFEX™ pretreated corn Stover. Biotechnol Biofuels 7:72. https://doi.org/10.1186/1754-6834-7-72

Valladares-Diestra KK, de Souza P, Vandenberghe L, Zevallos Torres LA et al (2021) Imidazole green solvent pre-treatment as a strategy for second-generation bioethanol production from sugarcane bagasse. Chem Eng J 420:127708. https://doi.org/10.1016/j.cej.2020.127708

Vasco-Correa J, Ge X, Li Y (2016) Biological pretreatment of lignocellulosic biomass. In: Biomass fractionation technologies for a lignocellulosic feedstock based biorefinery. Elsevier, In, pp 561–585

Yachmenev V, Condon B, Klasson T, Lambert A (2009) Acceleration of the enzymatic hydrolysis of corn Stover and sugar cane bagasse celluloses by low intensity uniform ultrasound. J Biobaased Mater Bioenergy 3:25–31. https://doi.org/10.1166/jbmb.2009.1002

Yang L, Cao J, Jin Y et al (2012) Effects of sodium carbonate pretreatment on the chemical compositions and enzymatic saccharification of rice straw. Bioresour Technol 124:283–291. https://doi.org/10.1016/j.biortech.2012.08.041

Zabed H, Sahu JN, Suely A et al (2017) Bioethanol production from renewable sources: current perspectives and technological progress. Renew Sust Energ Rev 71:475–501. https://doi.org/10.1016/j.rser.2016.12.076

Zabed HM, Akter S, Yun J et al (2019) Recent advances in biological pretreatment of microalgae and lignocellulosic biomass for biofuel production. Renew Sust Energ Rev 105:105–128. https://doi.org/10.1016/j.rser.2019.01.048

Zhang J, Zhou H, Liu D, Zhao X (2020) Chapter 2–Pretreatment of lignocellulosic biomass for efficient enzymatic saccharification of cellulose. In: Yousuf A, Pirozzi D, Sannino F (eds) Lignocellulosic Biomass to Liquid Biofuels. Academic, pp 17–65

Zhang K, Pei Z, Wang D (2016) Organic solvent pretreatment of lignocellulosic biomass for biofuels and biochemicals: a review. Bioresour Technol 199:21–33. https://doi.org/10.1016/j.biortech.2015.08.102

Zhang Q, Vigier KDO, Royer S, Jérôme F (2012) Deep eutectic solvents: syntheses, properties and applications. Chem Soc Rev 41:7108–7146. https://doi.org/10.1039/C2CS35178A

Zhao X, Cheng K, Liu D (2009) Organosolv pretreatment of lignocellulosic biomass for enzymatic hydrolysis. Appl Microbiol Biotechnol 82:815–827. https://doi.org/10.1007/s00253-009-1883-1

Zhu Z, Macquarrie DJ, Simister R et al (2015) Microwave assisted chemical pretreatment of Miscanthus under different temperature regimes. Sustain Chem Process 3:15. https://doi.org/10.1186/s40508-015-0041-6

Chapter 11
Capabilities of the Ascomycete Fungus *Penicillium Verruculosum* and its Enzymes for Conversion of Cellulosic Feedstock

Aleksandra M. Rozhkova, Alexander V. Gusakov, Anna S. Dotsenko, Olga A. Sinitsyna, and Arkady P. Sinitsyn

Abstract Cellulolytic enzymes are found in various microorganisms, however, ascomycete fungi are the most effective in cellulose hydrolysis. In world practice, the enzymes of the fungus *Trichoderma*, which is the ancestor of all cellulolytic enzymes producers, are still used as a source of industrial cellulases. At the moment there are other promising cellulase-producing fungi. These include fungal strains belonging to the genera *Aspergillus, Myceliophtora (Chrysosporium), Chaetomium* etc. The most promising strain among other microorganisms is the ascomycete cellulolytic *Penicillium verruculosum*, which contains cellulases with higher molecular activity and better operational characteristics. Using fungus *P. verruculosum* as a host, an expression system has been created that allows one to obtain improved enzyme complexes containing a «finely tuned» set of enzymes capable of effectively destroying plant polysaccharides under specified technological conditions.

11.1 Introduction

On the current moment, there is a trend towards the priority development of alternative energy based on renewable sources and a reduction of fossil hydrocarbons in the global energy balance. In addition, there are serious environmental

A. M. Rozhkova (✉) · A. S. Dotsenko
Fundamentals of Biotechnology Federal Research Center, Russian Academy of Sciences, Moscow, Russia
e-mail: a.rojkova@fbras.ru

A. V. Gusakov · O. A. Sinitsyna
M.V. Lomonosov Moscow State University, Moscow, Russia

A. P. Sinitsyn
Fundamentals of Biotechnology Federal Research Center, Russian Academy of Sciences, Moscow, Russia

M.V. Lomonosov Moscow State University, Moscow, Russia

C. R. Soccol et al. (eds.), *Liquid Biofuels: Bioethanol*, Biofuel and Biorefinery Technologies 12, https://doi.org/10.1007/978-3-031-01241-9_11

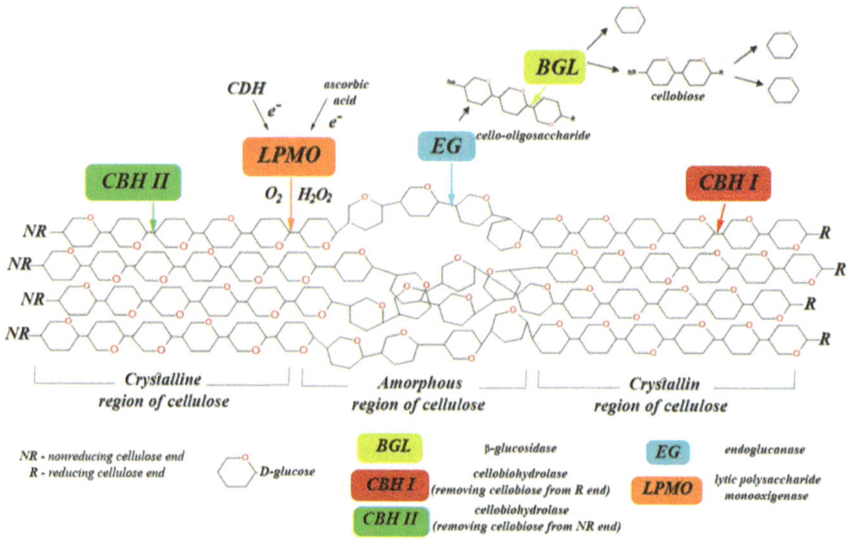

Fig. 11.1 Schematic representation of cellulose degradation

problems associated with an increase of greenhouse gas emissions into the atmosphere.

Biotechnological processing of feedstock which is a renewable and most widespread plant biomass in the world, can significantly reduce the stress on the Earth's ecosystem.

Glucose and other fermentable sugars obtained by enzymatic hydrolysis of feedstocks can be converted by microorganisms into biofuels, ethanol and butanol, as well as other useful products such as organic and amino acids, feed products, adhesives, biosynthesized materials, biodegradable plastics and other demanded products. Thus, in the future, renewable plant biomass will be considered as a source of raw materials, including bioethanol production.

The efficiency of the sugar release depends on a number of factors and, first of all, on the composition of the cellulolytic enzymes complex and the synergistic interaction of individual enzymes within it (Bajaj and Mahajan 2019; Singh et al. 2017).

Cellulolytic enzymes belong to the β-1,4-glucanases, i.e. carbohydrases (glycoside hydrolases), cleaving β-1,4-bonds in O-glycosyl compounds. According to the type of action on substrates, cellulases are divided into endo-β-1,4-glucanases (EC 3.2.1.4), exo-cellobiohydrolase (EC 3.2.1.91 and 3.2.1.176) and β-glucosidase (EC 3.2.1.21) (Lynd et al. 2002; Payne et al. 2015; Liberato et al. 2016) (Fig. 11.1).

Endo-β-1,4-glucanases (EGs) are enzymes that hydrolyze internal β-1,4-glucoside bonds distant from the ends of the polymer chain of cellulose (as well as lichenan, β-glucan, carboxymethylcellulose - CMC) with the formation of polymer fragments substrate and cello-oligosaccharides, which is accompanied by a significant decrease in the degree of substrate polymerization. EGs cleave amorphous regions of cellulose, but EGs are known to hydrolyze and crystalline cellulose (Payne et al. 2015; Vlasenko et al. 1998; Wu and Wu 2020). EG is often divided into enzymes of a less ordered and more ordered type of action (Vlasenko et al. 1998).

Cellobiohydrolases (CBHs) cleave cellobiose residues from the ends of polymer molecules of native or partially hydrolyzed cellulose. CBHs of the first type (CBH I, EC 3.2.1.176) attack the cellulose molecule from the reducing end, while CBHs of the second type (CBH II, EC 3.2.1.91) act from the non-reducing end. Unlike endoglucanases, CBH can hydrolyze both amorphous and microcrystalline cellulose (MCC), however only CBHs are responsible for hydrolysis of MCC (Sidar et al. 2020).

β-Glucosidases (BGL) cleave terminal non-reducing-D-glucose residues from cellobiose and oligosaccharides, and the rate of hydrolysis decreases with an increase in the number of glucoside residues in the oligosaccharide (Korotkova et al. 2009; Teugjas and Valjamae 2013; Singh et al. 2016). In contrast to EGs and CBHs, the specificity of BGL actions is broader and often they are able to cleave not only β-1,4-, but also β-1,2-, β-1,3-, β-1,6-glucosidic bonds (Singh et al. 2016).

Currently, a wide range of enzyme preparations based on the culture fluids secreted by mycelial fungi *Trichoderma* are presented on the market. These preparations are traditionally used for the cellulosic feedstocks hydrolysis, but they have a number of drawbacks, which consists in: 1) the relatively low efficiency of hydrolysis due to the lack of β-glucosidase (cellobiase) in composition of *Trichoderma* enzyme complex and 2) the relatively low molecular activity of CBHs, EGs and BGL. This fact leads to the excessive accumulation of the cellobiose in the reaction, which inhibits CBH I - key enzyme involved in saccharification process (Chekushina et al. 2013).

Therefore, modern fundamental and applied research is very actively developing, aimed at finding and obtaining new super-producing strains of cellulases, hemicellulases and boosting enzymes that have a high hydrolytic capacity and have a "controlled" enzyme composition, so that the complex of produced enzymes was best adapted to saccharification of certain types of feedstock.

Considering the information above, this chapter aims to describe the capabilities of the fungus *Penicillium verruculosum* as another source of cellulases for the hydrolysis of feedstocks to fermentable sugars. Analysis of the properties of cellulolytic enzymes secreted by this ascomycete will make it possible to assess the potential for using *P. verruculosum* in biotechnology as an alternative to *Trichoderma*.

11.2 Enzyme Complex of *Penicillium Verruculosum* for Cellulosic Feedstock Hydrolysis as an Alternative to Commercial Enzymes of *Trichoderma Reesei*

Enzymatic hydrolysis of cellulosic feedstock leads to formation of monosaccharides without significant energy costs and anthropogenic impact on the environment (Fulop and Ecker 2020). Since plant biomass is a complex substrate that includes polysaccharides of various compositions, its deep destruction requires a mixture of enzymes of different specificity, which carry out the cooperative hydrolysis of all its components (Guo et al. 2018).

Currently, mutant or recombinant fungal strains of the genus *Trichoderma* (*T. reesei, T. viride, T. longibrachiatum*) play a leading role among industrial microbial producers of cellulases and hemicellulases (Merino and Cherry 2007; Margeot et al. 2009; Baker et al. 1998; Kubicek et al. 2009), which are commercialized by Novozymes (Denmark), Genencor & Danisco (USA), Iogen (Canada), PrimAlko (Finland), AB Enzymes (Germany), EnMex (Mexico), Meiji Seika Kaisha Ltd. and Shin Nihon Chemical Co. (Japan) and others. It should be noted that fungal strains belonging to the other genera (*Penicillium*, *Acremonium*, *Chrysosporium*, *Myceliophthora*, *Chaetomium* and *Humicola*) can become a worthy alternative to traditional *Trichoderma* strains (Skomarovskii et al. 2005; Martins et al. 2008; Ikeda et al. 2007; Maeda et al. 2013).

Some species of the genus *Penicillium* are very promising producers of highly active cellulase complexes. As a rule, cellulases from *Penicillium* are superior to *T. reesei* enzymes in the rate of hydrolysis and glucose yield from various cellulose-containing substrates at the same dosage in terms of protein concentration or cellulase activity, which has been repeatedly noted by various researchers since the mid-1990s. These data are discussed in detail in reviews (Gusakov 2011; Gusakov and Sinitsyn 2012); in addition, they are supplemented by later publications (Liao et al. 2015; Zhao et al. 2016) (Table 11.1). Almost all authors note a relatively high level of the BGL activity in enzyme preparations based on *Penicilli*. This property is one of the main advantages of *Penicillium* over *Trichoderma*. As a result, the increase in the glucose yields with the use of enzymes from *Penicillium* sometimes reaches a five-fold effect (Table 11.1).

Another reason for the high efficiency of the *Penicillium* cellulase multienzyme cocktails is the extremely high specific activity of their key enzymes-cellobiohydrolases, CBH I and CBH II, in comparison with the corresponding enzymes of *T. reesei* (the difference can reach 2–2.5 times). In particular, this was demonstrated for cellobiohydrolases from *P. funiculosum, P. pulvirrolum, P. verruculosum, P. canescens* (Gusakov and Sinitsyn 2012; Morozova et al. 2010; Volkov et al. 2014). One of the possible reasons for a high specific activity of *P. verruculosum* CBH I and CBH II is the optimal distribution of N- bound glycans on the surface of catalytic domains of these enzymes (Dotsenko et al. 2016b; Gusakov et al. 2017).

Table 11.1 Comparison of the hydrolytic capacity of cellulase complexes from fungi of the genus *Penicillium* and *T. reesei* (Gusakov 2011, Gusakov and Sinitsyn 2012)

Cellulase producer	Substrate and feedstock pretreatment	Magnitude of the effect[a]
Penicillium sp.	Fir wood pretreated with steam explosion or organosolv	1.6–3.6
	Poplar wood pretreated by organosolv	1.4–2.0
	Red maple wood pretreated by organosolv	1.5–2.1
P. brasilianum	Spruce wood pretreated with a steam explosion	2.1
P. echinulatum	Bleached eucalyptus Kraft pulp	1.1
P. Funiculosum	Shredded corn cobs	1.1–1.5
P. Occitanis	Esparto grass	1.8
P. Pinophilum	Spruce wood pretreated with a steam explosion	3.1
P. verruculosum	Microcrystalline cellulose	1.3–1.7
	Eucalyptus pulp	1.3–3.9
	Softwood pretreated with organosolv	1.3–5.1

[a] The ratio of the product concentrations (usually glucose) obtained in hydrolysis of cellulose-containing substrates using enzyme preparations based on the fungi *Penicillium* and *T. reesei*

It is known that unproductive adsorbtion of cellulases to lignin leads to their partial inhibition, which negatively affects the efficiency of enzymatic hydrolysis (Ko et al. 2015; Rahikainen et al. 2013a; Rahikainen et al. 2013b). During the hydrolysis of microcrystalline cellulose, cellulases from *P. verruculosum* were significantly less inhibited by two types of lignin artificially added to the reaction system than cellulases from five different *T. reesei* preparations (Berlin et al. 2006a). Similar results demonstrating less negative effect of lignin on *P. verruculosum* cellulases in comparison with *T. reesei* enzymes were reported by Steffien et al. (2014).

The enzyme complex secreted by *P. verruculosum* B151 contains more than 20 enzymes differing in biochemical and catalytic properties. The main cellulases secreted by the control high-yielding strain B1–221-151 are CBH I (Cel7A), CBH II (Cel6A), EG I (Cel7B), EG IIa (EG II), EG IIb (Cel5B), EG III (Cel12A) and BGL (Cel3A), while the content of CBH I is about 35%, CBH II - 21-33%, the content of each of the mentioned EG varies from 2 to 5%, and the proportion of BGL is 3–4% (Table 11.2) (Dotsenko et al. 2015; Morozova et al. 2010; Sinitsyn et al. 2016). The enzyme complex also includes other carbohydrases (18–30% of the total protein) - these are xyloglucanases, xylanases, α-galactosidase and glucoamylase. Molecular masses of the enzymes vary from 19 to 120 kDa, isoelectric points - from 2.0 to 5.8. For most of these cellulases, the pH optima of activity are in the range of pH 4.0–5.0. As a result of limited proteolysis of the linker, *P. verruculosum* cellulases possessing a CBM (CBH I, CBH II, EG I) are usually represented by high-molecular-weight (full-length) and low-molecular-weight forms, the last representing a catalytic domain of an enzyme without a CBM (Morozova et al. 2010).

Comparative data characterizing the composition of *P. verruculosum* B151 and Cellic CTec-2 enzyme preparations (the last is one of the powerful commercial

Table 11.2 Component composition (%) of enzyme preparation from *P. verruculosum* B151 and commercial preparation Cellic CTec-2 (Chekushina et al. 2013)

Enzymes	*P. verruculosum* B151	Cellic CTec-2
Cellobiohydrolases (CBHs)		
P. verruculosum		
CBH I with CBM (66 kDa)	20	–
CBH I without CBM (55 kDa)	15	–
CBH II with CBM (60 kDa)	17	–
CBH II without CBM (50 kDa)	17	–
T. reesei		
CBH I total (65 and 57 kDa)	–	20–30
CBH II total (60 and 50 kDa)	–	20–30
Endoglucanases (EGs)		
P. verruculosum		
EGs total (70, 52, 57, 36, 33, 25 kDa)	12–15	–
T. reesei		
EGs total (57, 50, 48, 40 kDa)	–	6–8
F. oxysporum		
EG C (46 kDa)	–	2
H. insolens		
EG V (23 kDa)	–	2–5
β-Glucosidases (BGs)		
P. verruculosum		
BG (116 kDa)	4	–
T. reesei		
BG (74 kDa)	–	<2
A. fumigatis and A. oryzae		
BGs (116, 94 kDa)	–	12–15

cellulase cocktails based on *T. reesei*) are presented in Table 11.2. For the Cellic CTec-2, the total content of CBHs is 40–60% while in the *P.verruculosum* B151 preparation the total CBH content is 69%. In the Cellic CTec-2, the content of a high-molecular-weight CBH (with CBM) exceeded the content of a low-molecular-weight form (without CBM) 10 times for the CBH I and 2 times for the CBH II. The *P. verruculosum* B151 preparation was characterized by rather similar content of high-molecular-weight and low-molecular-weight CBH I and CBH II forms. The total content of EGs in *P.verruculosum* B151 and Cellic CTec-2 turned out to be approximately the same (14–16%). However, the *P. verruculosum* B151 contained only its own (homologous) endoglucanases, while the Cellic CTec-2 included both homologous and heterologous enzymes (EGs from *F. oxysporum* and *H. insolens*). The *P. verruculosum* B151 was characterized by a much higher content of the homologous BGL (4%) compared to that in the Cellic CTec-2 preparation (1%). However, the total content of BGLs in the Cellic CTec-2 was 18% due to the presence of the heterologous BGLs from *A. fumigatus* and *A. oryzae*.

Table 11.3 Total and specific activities of *P. verruculosum* B151 and Cellic CTec-2

	FP	MCC	CMC	*p*NPG	Cellobiose	Birch xylan	*p*NPXyl
Enzymatic preparation (EP)	50°C, pH 5.0	40°C, pH 5.0	50°C, pH 5.0	40°C, pH 5.0	40°C, pH 5.0	50°C, pH 5.0	40°C, pH 5.0
Total activity, U/g (B151) or U/ml (Cellic CTec-2)							
P.verruculosum B151	760	578	15,116	1404	603	25,028	3
Cellic CTec-2	135	75	3760	1326	960	6796	<1
Specific activity, U/mg of protein							
P.verruculosum B151	0.9	0.7	18.3	1.7	0.7	30.3	<1
Cellic CTec-2	0.6	0.3	15.6	5.5	4.0	11.3	<1

Thus, the compared *P. verruculosum* B151 and Cellic CTec-2 preparations had approximately the same component composition, with the exception of the increased content of the homologous BGL in the *P. verruculosum* B151, which was replenished by additional heterologous expression of BGLs in the enzyme preparation based on *Trichoderma*.

The most important criterion for comparing *P. verruculosum* B151 and Cellic CTec-2 enzyme preparations are their molecular activities toward a number of specific substrates - soluble and insoluble polysaccharides (FP, MCC, CMC, birch xylan), synthetic substrates (*p*NPG, *p*NPXyl) and oligosaccharides (cellobiose), determined from the initial rates of hydrolysis. The set of these substrates characterizes the activity of both the enzyme complex and its individual components. As follows from Table 11.3, the activities of *P. verruculosum* B151 and Cellic CTec-2 preparations against FP, MCC, CMC and *p*NPXyl were comparable. The activity of *P. verruculosum* B151 against birch xylan exceeded the activity of Cellic CTec-2 by about 3 times (Chekushina et al. 2013).

Thus, B151 was inferior to Cellic CTec-2 only in terms of the BGL activity, which was due to the presence of heterologous BGLs in the composition of Cellic CTec-2 (Table 11.2).

It should be noted that both CBH I and CBH II of *P. verruculosum* significantly surpass the respective CBHs from *T. reesei* in terms of their hydrolytic capacity when acting on crystalline cellulose (Gusakov 2014), while even the low-molecular-weight form of *P. verruculosum* CBH II is comparable in effectiveness to the full-length CBH I of *T. reesei* (Gusakov and Sinitsyn 2012; Sinitsyn and Sinitsyna 2021). This fact, together with a higher BGL activity, determines the huge potential of enzyme preparations based on *P. verruculosum* as compared to classical commercial preparations of *T. reesei* (Steffien et al. 2014; Gusakov 2011; Sinitsyn and Sinitsyna 2021).

However, despite of the high hydrolytic potential, commercial cellulases from *Penicillium sp.* have not yet become widespread. Perhaps this is due to the fact that the existing *Penicillium* strains are still inferior in the level of extracellular protein secretion to the best industrial strains of *T. reesei*. Another reason may be the insufficiently aggressive marketing policy of the producers of these enzymes.

11.3 The Potential of the *P. Verruculosum* Expression System in Obtaining Optimal Enzyme Preparations

In world practice, recombinant DNA technologies are used to improve the composition of the secreted fungal multienzyme systems (Saunders et al. 1989; Kruszewska 1999; Su et al. 2012; Singh et al. 2017). An expression system has been developed for the *P. verruculosum* B151 strain, and its broad capabilities have been demonstrated in the latest review (Sinitsyn et al. 2020b). Over the past decade, more than 100 strains have been obtained that produce demanded enzymes for the food, pulp and paper, textile and other industries (Sinitsyn et al. 2020a).

For genetic engineering manipulations on the basis of the highly productive strain P. *verruculosum* B151, an auxotrophic mutant *P. verruculosum* B1–537 (ΔniaD) was obtained. The resulting host had a defective nitrate reductase gene, which allowed to use it as a basis for obtaining recombinant producers of homologous and heterologous proteins by carrying out the selection of transformants on a medium with sodium nitrate. At the same time, the content of key cellulases in the multienzyme cocktails produced by strains B1–537 (ΔniaD) and B151 is practically the same (Sinitsyn et al. 2020b). Cotransformation with plasmids carrying the gene of the target protein and the gene of native nitrate reductase leads to the return of the recombinants to prototrophy, which is a marker for the selection of transformants.

It is important to note that the expression of the target gene can be controlled by various "promoter-terminator" systems. In most works, the expression of various genes in the *P. verruculosum* B1–537 (niaD-) strain was regulated by a *cbhI* gene promoter (Rubtsova et al. 2015; Sinitsyn et al. 2016, 2017; Denisenko et al. 2017, 2019, 2021; Volkov et al. 2019). So, the *P. verruculosum cbhI* gene encoding CBH I, the main component of the cellulase complex of the *P. verruculosum* fungus, was used in the same manner as the *cbhI* promoter in *Trichoderma* recombinant strains (Mach and Zeilinger 2003; Liu et al. 2008). The CBH I expression system is inducible, which makes its use convenient for obtaining preparations with a high content of the target protein (Gupta et al. 2014).

From the point of view of improving the efficiency of saccharification of cellulosic feedstock, the most successful recombinant strain was *P. verruculosum* F10, a super-producer of the highly active *A. niger* BGL. Enzyme preparations based on this strain contain up to 80% of the heterologous BGL of the total secreted protein (Dotsenko et al. 2015; Chekushina et al. 2013).

In the case of *P. verruculosum* F10 strain, the most BGL-enriched transformant contained up to 18 copies of the heterologous *bgl1* gene under the *cbh1* promoter control, while the expression of own CBH I *P.verruculosum* was largely suppressed. However, careful selection of transformants with controlled copies of the target

Table 11.4 Using the *cbhI* promoter for the production of recombinant strains and enzyme preparations of *P. verruculosum* with homologous or heterologous expressed cellulases (Morozova et al. 2010; Dotsenko et al. 2015; Sinitsyn et al. 2016)

Cloned genes	Expressed cellulases	Content of enzymes, % [a]
egl1 T. reesei	EG I	0 (3)
cbhII T. reesei	CBH II	0 (2)
bgl1 A. niger	BGL	0 (21)
cbh1 P. verruculosum	CBH I	20 (66)
cbh1 P. verruculosum egII P. verruculosum	CBH I + EG II	20 + 3 (56 + 19)
cbh1 P. verruculosum egII P. verruculosum bgl1 A. niger	CBH I + EG II + BGL I	20 + 3 + 0 (37 + 34 + 12)
cbh1 P. verruculosum egII P. verruculosum bgl1 A. niger	CBH I + EG II + BGL I	20 + 3 + 0 (28 + 30 + 8)

[a] The content of enzymes as % of the total protein in the control preparation *P. verruculosum* B151 is given. The content of the expressed enzymes in the preparation obtained using the corresponding recombinant *P. verruculosum* strain is indicated in brackets

genes allowed to use the strong *cbh1* promoter to obtain recombinant *P. verruculosum* strains capable of secreting balanced complex of heterologous and homologous cellulases (Dotsenko et al. 2015). Examples are given in Table 11.4. They also include enzyme preparations in which the content of their own cellulases was increased by homologous expression (for example, CBH I or EG II).

An original approach for adjusting the composition of the *P. verruculosum* cellulolytic complex was to obtain the so-called "fusion construct" consisting of sequentially connected genes encoding CBH I, EG II of *P. verruculosum*, and BGL of *A. niger*, expressed under the *cbh1* promoter (Sinitsyn et al. 2016). The use of this design made it possible to obtain an enzyme preparation enriched with three components most important for cellulose hydrolysis (Table 11.4). Using this preparation together with the original *P. verruculosum* B151 cocktail led to an increase in the yield of sugars during the hydrolysis of pretreated pine and aspen wood up to 70%.

The alternative (non-inducible) expression systems can be used to obtain strains with a moderate level of expression of the target recombinant enzymes. This type includes, for example, a constitutive expression system based on the promoter of the histone *hist4.1* gene. The use of such expression systems makes it possible to increase the production of the target protein without significant changes in the composition of the main secreted enzyme complex. Using the expression system based on the *hist4.1* gene promoter, an enzyme preparation B1_PrHist with a heterologous BGL of *A. niger* was obtained, the content of which was about 13% of the total secreted protein. In this case, together with the heterologous BGL, the expression of *P. verruculosum* own CBH I was increased. This preparation provided a higher glucose yield (by 10–21%) in hydrolysis of MCC and pretreated aspen wood relative to the control preparation *P. verruculosum* B151 (Dotsenko et al. 2015; Sinitsyn et al. 2016).

Thus, both genetically engineered approaches considered above, in which the target protein is expressed either under the control of a strong inducible (*cbh1*) promoter, or under the control of a weaker constitutive (*hist4.1*) promoter of the corresponding gene, can significantly increase the hydrolytic capacity of secreted cellulase complexes. The choice of this or that approach largely depends on the specific application of the final enzyme preparation in a particular biotechnological process, on the type of cellulosic feedstock and the method of its pretreatment.

11.4 The Role of Hydrolysis Boosting Enzymes in Feedstock Degradation

Although the most important enzymes in lignocellulosic biomass degradation are cellulolytic ones, that is, exo-cellobiohydrolases and endo-β-1,4-glucanases, there are other types of enzymes whose presence is necessary for the most complete conversion of the feedstock. These enzymes are usually called accessory or auxiliary, and they typically act as boosters for cellulase activities. Hemicellulases (xylanases, xyloglucanases, mannanases, arabinases, etc.) that catalyze the hydrolysis of xylan, xyloglucan, mannan and other hemicelluloses provide better accessibility of cellulose to cellulases and increase the total yield of soluble sugars derived from the polysaccharides (Berlin et al. 2005; Rashmi and Siddalingamurthy 2018).

Among other accessory enzymes, the most powerful enhancers of glucose yield from cellulose are β-glucosidases (BGLs) and lytic polysaccharide monooxygenases (LPMOs) (Fig. 11.1). This section of the chapter will provide examples of using these enzymes for boosting the enzymatic hydrolysis of cellulosic materials.

As noted above the BGLs (EC 3.2.1.21) convert cellobiose and higher oligosaccharides, formed from cellulose under the action of CBHs and EGs, into glucose, and the mechanism of the boosting effect may be explained by elimination of the inhibitory effect of cellobiose on CBHs due to the BGL-catalyzed hydrolysis of this disaccharide (Lynd et al. 2002; Berlin et al. 2005). The role of BGLs as components of microbial cellulolytic systems has been known for a long time, while it should be noted that some fungi cannot produce enough levels of the BGL activity for efficient conversion of cellulose into glucose. This is particularly true for fungi of the genus *Trichoderma*, including older *T. reesei* industrial strains (Gusakov 2011). Although the basic high-cellulase strains of *P. verruculosum* secrete higher levels of the BGL activity than *T. reesei*, it still may be a limiting factor for efficient cellulose hydrolysis (Gusakov and Sinitsyn 2012).

Berlin et al. demonstrated that the addition of Novozym 188 commercial enzyme preparation, enriched with *A. niger* BGL, to the MSUBC1 sample produced by a *P. verruculosum* laboratory strain boosted the glucose yield from Douglas fir pretreated by organosolv or steam explosion by 70 and 20%, respectively (Berlin

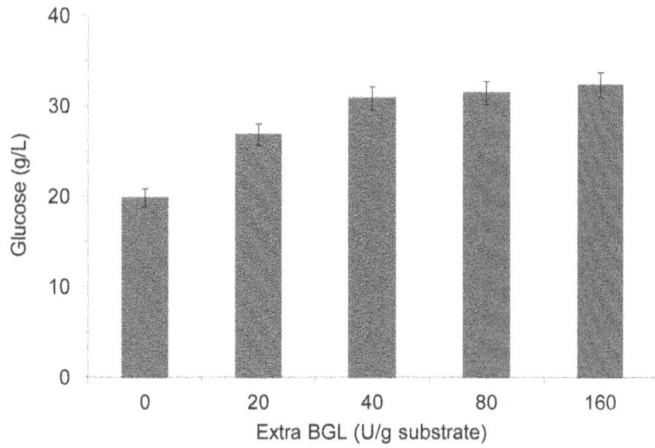

Fig. 11.2 Effect of the addition of purified *A. niger* BGL on the glucose concentration produced after 72-h hydrolysis of milled aspen wood by *P. verruculosum* B151 cellulase preparation, 50 °C, pH 5.0. Substrate concentration 100 g/L, cellulase loading 5 mg/g substrate. Source: (Dotsenko et al. 2015)

et al. 2005). In hydrolysis of organosolv-pretreated yellow poplar and red maple with the same enzyme preparations the boosting effects reached 38 and 57% (Berlin et al. 2006b).

Experiments on the addition of different dosages of purified *A. niger* BGL to the *P. verruculosum* cellulase preparation at 5 mg/g substrate loading in hydrolysis of milled aspen wood showed that after supplementation of 20 and 40 units of BGL per g of substrate the glucose yield increased by 35 and 56% relative to the control (cellulase preparation in the absence of extra added BGL, see Fig. 11.2), while further increase in the BGL loading up to 160 U/g provided rather small additional effect on the product formation (Dotsenko et al. 2015). Thus, 20–40 U/g of the extra BGL activity seemed to be optimal.

Because of the high potential of the *A. niger* BGL in boosting the hydrolysis of cellulosic materials, the corresponding *bgl1* gene was heterologously cloned and expressed in the host *P. verruculosum* B1–537 (niaD-) strain under the control of a strong inducible *cbh1* or a weaker constitutive *hist4* promoter (Chekushina et al. 2013; Dotsenko et al. 2015). As it mentioned above *P. verruculosum* F10 strain obtained based on the *cbh1* promoter provided extremely high expression of the heterologous BGL (up to 80% of the total secreted protein); strain obtained based on the *hist4* gene promoter (B1_PrHist) provided 13% content of the heterologous BGL (Dotsenko et al. 2015).

Then different approaches for boosting the hydrolytic performance of *P. verruculosum* cellulases due to the presence of the *A. niger* BGL in the multienzyme cocktail were tested. The best boosting effect on the yield of sugars in 72-h hydrolysis of milled aspen wood (64% increase relative to the control) provided the enzyme preparation obtained by co-fermentation of a high-cellulase *P. verruculosum* B151 strain with the F10 strain (Fig. 11.3a); in the case of MCC the effect was even higher (two-fold, Fig. 11.3b). When the preparations produced by single recombinant *P. verruculosum* strains (B1_PrCBH1 or B1_PrHist) expressing the *A. niger* BGL were used at the same protein loading, the increase in the sugar yield was also observed, although the boosting effects were less pronounced (Fig. 11.3). It should be noted that for all enzyme samples containing the heterologous BGL glucose made

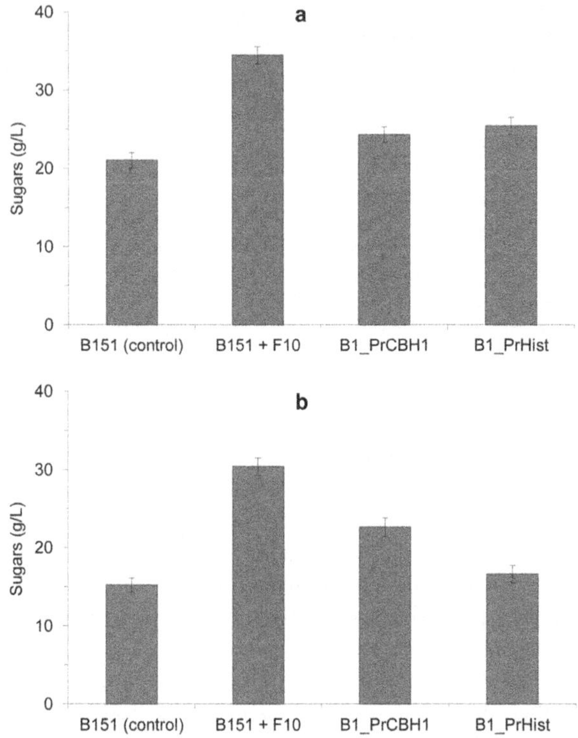

Fig. 11.3 The concentration of sugars (glucose + cellobiose) formed after 72-h hydrolysis of milled aspen wood (**a**) and microcrystalline cellulose (**b**) by various *P. verruculosum* cellulase preparations at 50 °C, pH 5.0. Substrate concentration 100 g/L, enzyme loading 5 mg/g substrate. Source: (Dotsenko et al. 2015)

up not less that 99% of the total sugars, while in the case of control (B151 alone) the final hydrolysates contained cellobiose (6–7%) together with glucose.

LPMOs, as enhancers for cellulases, were discovered rather recently (Vaaje-Kolstad et al. 2010; Quinlan et al. 2011), although they had previously been mistakenly known as EGs belonging to family 61 of glycoside hydrolases (Harris et al. 2010). LPMOs break down the cellulose polymer chain by an oxidative mechanism, the oxidation occurring either at C1 (EC 1.14.99.54) or at C4 atom of the glycoside ring (EC 1.14.99.56) with formation of aldonic acid (its lactone) or 4-ketoaldose, respectively (Phillips et al. 2011; Horn et al. 2012). Some LPMOs are of the mixed type (C1/C4), oxidizing both at C1 and C4 position (Quinlan et al. 2011; Hemsworth et al. 2013). To carry out the catalysis by an LPMO, an electron donor must be present in the reaction system. Different compounds, such as ascorbic or gallic acid, may act as electron donors for LPMOs; in nature, the enzyme cellobiose dehydrogenase (CDH), secreted by some fungi, or building blocks of lignin may play the role of the electron donor (Bulakhov et al. 2017; Horn et al. 2012; Quinlan et al. 2011; Phillips et al. 2011). Most of LPMOs acting on cellulose as a major substrate belong to AA9 family of Auxiliary Activities (Levasseur et al. 2013).

Bulakhov et al. studied the effects of addition of three purified LPMOs with gallic acid or CDH as electron donors on the yield of reducing sugars (RS) in hydrolysis of microcrystalline cellulose by a crude *P. verruculosum* cellulase complex (Bulakhov et al. 2016). Two LPMOs (from *Thielavia terrestris* and *T. reesei*) represented recombinant enzymes heterologously expressed in *P. verruculosum*, while the third, native LPMO, was isolated from a culture liquid of *Myceliophthora thermophila* fungus. After replacing 10% of cellulase protein with equivalent amount of LPMO the final concentration of RS increased by 17–31%, despite the lower content of hydrolytic enzymes in the reaction system. The most significant boosting effect was observed for *T. reesei* LPMO in the presence of CDH electron donor. The data obtained also indicated the possible presence of some amount of intrinsic LPMO in the original *P. verruculosum* enzyme sample. So, in order to increase the content of its own LPMO in the *P. verruculosum* multienzyme preparations, the *lpmo1* gene was sequenced and homologously overexpressed in *P. verruculosum* B1–537 strain under the control of the *cbh1* gene promoter (Semenova et al. 2019). Seven recombinant LPMO-producing strains were tested, for which the target enzyme content in the culture filtrate varied from 9 to 57% of the total secreted protein. Two enzyme preparations, in which the LPMO portion was 9–30% while the cellulase content was substantially preserved, demonstrated the enhanced yields of RS in 48-h saccharification of microcrystalline cellulose and milled aspen wood: by 19–31 and 11–26%, respectively, relative to the reference cellulase cocktail. Although the enzyme samples with LPMO content of 50% or higher demonstrated rather poor performance in saccharification of cellulosic materials, they still could be used as blends to the traditional cellulase preparations not having their own LPMO.

In another work (Bulakhov et al. 2017), genes encoding *A. niger* BGL and *T. reesei* LPMO (TrLPMO) were cloned and expressed by *P. verruculosum* B1–537 strain under the control of the inducible glucoamylase (*gla1*) gene promoter. The heterologous BGL content in the respective culture liquids (hBGL1, hBGL2 and hBGL3) varied from 4 to 10% of the total protein, while the content of TrLPMO in the hLPMO sample was ~3%. The glucose yields in 48-h hydrolysis of microcrystalline cellulose and milled aspen wood by the hBGL1, hBGL2 and hBGL3 preparations increased by up to 99 and 80%, respectively, relative to the reference enzyme preparations (PvC1 and PvC2 obtained on similar growth media as samples under study) without the heterologous BGL (Fig. 11.4a). The TrLPMO in the hLPMO preparation boosted the conversion of the aspen wood by 10–43%; however, in hydrolysis of microcrystalline cellulose the hLPMO sample was less effective than the reference preparations. The highest product yield in hydrolysis of the lignocellulosic substrate was obtained when the hBGL2 and hLPMO preparations were used at the ratio 1:1 (Fig. 11.4b).

Since cellulase systems of fungi, and *P. verruculosum* in particular, consist of various enzymes differing in the mode of action and substrate specificity, LPMOs, oxidizing the polysaccharide chain either at C1 or C4 or both atoms of a glycoside ring, may have different effects on the performance of individual cellulases. So, the peculiarities of the kinetic interaction of three LPMOs, belonging either to the C1-type (from *T. terrestris*) or to the mixed C1/C4-type (from *P. verruculosum* and *T. reesei*), with purified CBH I, CBH II and EG II, the major components of

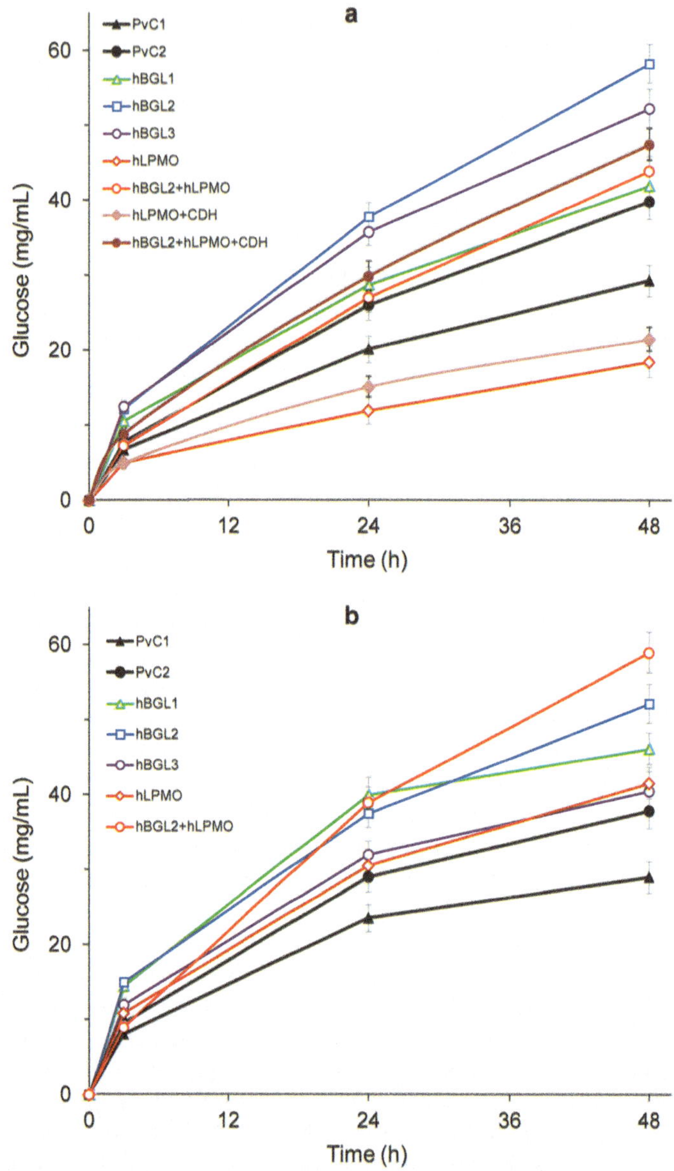

Fig. 11.4 Progress kinetics of hydrolysis of microcrystalline cellulose (**a**) and milled aspen wood (**b**) by different *P. verruculosum* preparations. Conditions: substrate concentration 100 g/L; protein loading 5 mg/g substrate; CDH loading (when applied) 0.1 mg/g substrate; 50 °C; pH 5.0. Source: (Bulakhov et al. 2017)

P. verruculosum cellulase complex, during the destruction of cellulosic substrates were studied (Semenova et al. 2021). All the LPMOs under study boosted the yield of RS in hydrolysis of microcrystalline or amorphous cellulose by CBH II or EG

II. The synergistic effects were expressed to the greatest extent in the initial period of the reaction, the coefficients of synergism reached 1.67–1.89 for the pairs LPMO/CBH II (the last enzyme with 10% of BGL addition) or 1.78–2.27 for the pairs LPMO/EG II. The combinations of *P. verruculosum* and *T. reesei* LPMOs with CBH I (supplemented with 10% of BGL) also displayed synergism (up to 36% increase in the RS yield), while the *T. terrestris* LPMO, on the contrary, induced an inhibitory effect (up to 44%) on CBH I. The observed kinetic antagonism between the C1-oxidizing LPMO and CBH I could be explained by the formation of the inhibitory to CBH I cellulose chain fragments (containing aldonic acid or its lactone instead of the reducing ends) after the LPMO-catalyzed breakdown of the polysaccharide molecules (Semenova et al. 2021).

These results show that LPMOs as boosting agents for cellulases should be selected with caution. Taking into account the fact that CBH I, acting on the reducing ends of cellulose molecules, typically represents the major enzyme of most fungal cellulolytic systems, C4 or C1/C4-oxidizing LPMOs seem to be more preferential than the C1-oxidizing enzymes.

Thus, two key boosting enzymes – BGL and LPMO assist to major cellulases of *P. verruculosum* (CBHs and EGs). The level of naturally secreted BGL for *P. verruculosum* is not high enough and need to be increased to optimal level (to 20–40 U of BGL activity per 1 g of cellulosic feedstock). Recombinant strains of *P. verruculosum* obtained by using *cbh1* or *hist4.1* promoters – producers of BGL (secreted 80 and 13% of BGL from total protein respectively) were generated. That allowed obtaining cellulase preparations enriched by BGL which lead to more than 60% improvement of glucose yield from cellulosic feedstock. The homologous *lpmo1* gene of *P. verruculosum* was also expressed using the *cbhl* promoter. Replacement of cellulase protein with equivalent amount of LPMO protein in the reaction mixture lead up to 30% improvement of RS yield. Genes encoding BGL and LPMO were expressed by *P. verruculosum* B1–537 (niaD-) strain under the control of *gla1* promoter (the obtained recombinant strains provides 4–10% of BGL and 3% of LPMO from total secreted protein); the simultaneous use of enzyme preparations produced by these recombinant strains allowed improve the glucose yield for different cellulosic feedstock up to 80–99%.

11.5 Enzyme Improvement Opportunities for the Effective Enzymatic Hydrolysis

Catalytic activity and stability of enzymes, mainly affect the yields of sugars during the hydrolysis of cellulosic feedstock. Improvement in the catalytic activity of the enzymes means an increase in the quantity of sugars that are released per unit of time at the same dosage of the enzymes. Improvement in the stability of the enzymes in certain conditions of the hydrolysis means prolonged action of the enzymes.

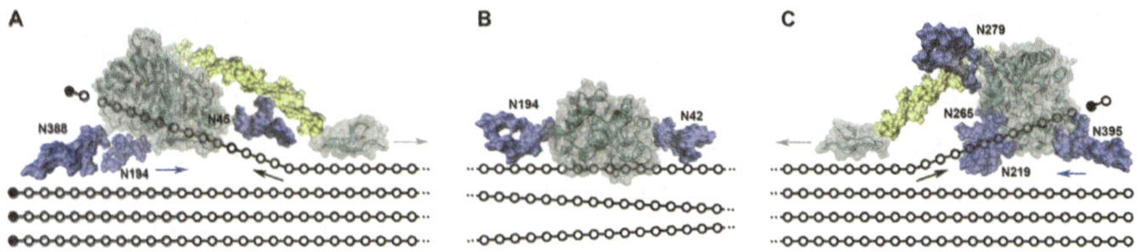

Fig. 11.5 Mechanistic model of action of cellulolytic enzymes CBH I, CBH II, and EG II produced by *P. verruculosum* on cellulose. Cellobiohydrolases CBH I (**a**) and CBH II (**c**) have catalytic and cellulose binding domains and demonstrate exo-type of action toward the crystalline cellulose. Endoglucanase EG II (**b**) contains only a catalytic domain and demonstrates endo-type of action toward the amorphous cellulose. The open circles represent anhydroglucose residues in cellulose chains; the solid circles represent reducing ends. Numbers correspond to sites of N-glycosylation. Arrows indicate the direction of movement of the respective elements during the processive hydrolysis of a cellulose chain

Therefore, the development of cellulolytic enzymes with enhanced activity and stability is important for the wide implementation of the enzymes in feedstocks conversion.

The catalytic activity of the cellulolytic enzymes can be increased through the engineering of the mechanism of the enzyme action. Cellulose in the composition of feedstocks exists in the form of microfibrils containing segments of both crystalline and amorphous cellulose. Cellulolytic enzymes need first effectively bind with the cellulose chain in the complex structure of the microfibrils and then productively hydrolyze the glycosidic linkages. N-linked glycosylation occurring in the molecules of the cellulolytic enzymes produced by *P. verruculosum* was successfully engineered to increase the rate of the enzymatic hydrolysis of feedstocks. The reason is that the N-linked glycosylation affects the binding affinity and also the mechanism of the binding.

To explain the effect of the N-linked glycosylation on the catalytic activity of the cellulolytic enzymes, it is necessary to elucidate the mechanism of action of these enzymes. Main cellulolytic enzymes produced by *P. verruculosum* are CBH I, CBH II, and EG II. Endoglucanase EG II belonging to glycoside hydrolase (GH) family 5 demonstrates endo-type of action and hydrolyzes glycosidic linkages in the amorphous cellulose, while cellobiohydrolases CBH I and CBH II belonging to GH families 7 and 6 have exo-type of action toward the crystalline cellulose. The specificity of action of CBH I, CBH II, and EG II is demonstrated in Fig. 11.5.

A distinctive feature of CBH I and CBH II is a processive type of action. At the initial stage of the hydrolysis, the cellulose chain is threaded into the tunnel of the enzyme active center. After a catalytic act of splitting off a cellobiose residue is performed the cellulose chain is further moved through the tunnel for the next catalytic act. The glycans linked to sites N219 and N265 in CBH II and those linked to N45 in CBH I are located near the entrance of the tunnel and should inevitably interact with the threaded chain thus slowing down its movement through the tunnel. Indeed, deletion of the site N45 from the CBH I structure through N45A substitution increased the yield of reducing sugars in the hydrolysis of MCC and aspen wood in

1.3 and 1.6-times respectively (Dotsenko et al. 2016b). Deletion of the sites N219 and N265 in the CBH II structure through N219A and N265A substitutions increased the yield of reducing sugars in the hydrolysis of Avicel in 1.3 and 1.2 times respectively. In the case of hydrolysis of aspen wood, deletion of the sites N219 and N265 increased the yield of reducing sugars in 1.4 and 1.2 times. Also, the processivity of CBH II was increased in 1.3–1.4 times because of N219A and N265A substitutions (Gusakov et al. 2017). The increase in the degree of processivity for the engineered enzyme variants strongly supports the statement on the interaction of the N-linked glycans located near the tunnel entrance with the cellulose chain.

Two more sites N194 and N388 are located at the "bottom" of the enzyme molecule of CBH I similar to site N395 in CBH II. Deletion of these sites through N to A substitution decreased the yield of reducing sugars in the hydrolysis. Deletion of the sites N194 and N388 in the CBH I structure decreased the sugars yield in 1.1–1.4 times in hydrolysis of Avicel and aspen wood (Dotsenko et al. 2016b). Deletion of the site N395 in CBH II decreased the yield in 1.4 and 1.5 times in the case of Avicel and aspen wood. Deletion of the site N279 led to a dramatic decrease in the yield of reducing sugars, although the site is located on the side surface of the enzyme globule closer to the linker part. The deletion decreased the yield in 6.2 and 2.4 times in the case of Avicel and aspen wood hydrolysis (Gusakov et al. 2017). Deletion of the sites N194, N388, and N395 decreased the sorption of the enzymes on Avicel that confirms the active interaction of the N-linked glycans with cellulose chains.

A decrease in the sugars yield caused by the elimination of glycans from the "bottom" of the CBH I and CBH II molecules corresponds to a considerable influence of these glycans on the processive machinery of the enzymes. At the beginning of the hydrolysis, the glycans provide the enzyme correct orientation through non-specific dynamic interactions with a cellulose microfibril. Then in the processive acts of splitting off cellobiose residues and threading the cellulose chain into the active site tunnel, they work as a wedge, thus helping to detach a single chain from a cellulose microfibril.

EG II *P. verruculosum*, despite CBH I and CBH II, contains only a catalytic domain but not a cellulose binding domain. Two sites of N-glycosylation are located by both sides of the active site cleft. Deletion of the sites N42 and N194 in the EG II structure increased the yield of sugars in aspen wood hydrolysis in 1.1 and 1.2 times respectively. EG II acts toward amorphous cellulose and does not have the property of processivity. EG II binds with the cellulose chain, hydrolyzes the glycosidic linkage, and desorbs from the chain. The substitutions N42A and N194A led to a 1.4–1.7-fold increase in the k_{cat}, at the same time no notable change in the Km was obtained. A possible explanation is that the N-linked glycans affect the dissociation of the two chain parts from the enzyme active site after the bond hydrolysis. Deletion of the glycans accelerates the dissociation and increases the enzymatic turnover and the observed k_{cat} (Dotsenko et al. 2016a).

Similar effects of N-linked glycans on catalytical properties of cellulolytic enzymes were demonstrated for CBH I and CBH II produced by *P. funiculosum*,

T. reesei, and *Myceliophthora thermophila* (*Chrysosporium lucknowense*). Deletion of the N-glycosylation site through N to A substitution in *T. reesei* CBH I resulted in up to a 1.6-fold increase in conversion degree of bacterial crystalline cellulose (Adney et al. 2009). Deletion of the N-glycosylation sites in *P. funiculosum* CBH I increased the conversion degree of bacterial crystalline cellulose in up to 1.4 times. While adding one more site of the glycosylation at the "bottom" of the enzyme molecule boosted the hydrolysis in 1.7 times (Adney et al. 2009). In the case of CBH II, *P. verruculosum*, *T. reesei*, and *M. thermophila* CBH II possess a different number of glycosylation sites in the enzyme molecule and practically identical catalytical sites. *M. thermophila* CBH II with a single N-linked glycan at the "bottom" and *P. verruculosum* Cel6A with one N-linked glycan at the "bottom" and three glycans on the "side" produce up to 1.6 times higher yield of sugars in Avicel hydrolysis compared to *T. reesei* CBH II with N-glycosylation on the "side" and "top" (Gusakov et al. 2017; Gusakov et al. 2007).

As well as the improvement in the catalytic activity the improvement in the stability allows boosting the enzymatic hydrolysis. The improvement in the stability of the cellulolytic enzymes can prolong the enzyme action under high temperatures and the presence of destabilizing compounds. Performing the hydrolysis at elevated temperatures allows obtaining higher sugar yields in a shorter time. Different compounds that destabilize the enzymes are accumulated in the feedstocks after the pretreatment, for example, ionic liquids and deep eutectic solvents. Improved variants of *P. verruculosum* CBHI with amino acid substitutions demonstrated increased thermostability at 60–65 °C. Single substitution G415P in the C-terminus of α-helix provided a 3.4-fold increase in half-life time at 60 °C compared to wild-type enzyme (Dotsenko et al. 2020a). The variant with multiple substitutions A65R-G415R-S181F demonstrated 2.5-fold improved thermostability at 65 °C due to the formation of additional salt bridges and π–π interaction in the enzyme structure (Pramanik et al. 2021). Molecular dynamics simulations revealed that the performed substitutions lead to stabilization of surface-exposed flexible α-helixes and loop in the structure. Moreover, the variant with multiple substitutions exhibited 1.9-fold, 1.4-fold, and 1.6-fold higher specific activities in ionic liquid [Bmim]Cl (50 g/L), deep eutectic solvent [Ch]Cl (50 g/L), and two-fold concentrated seawater compared to wild type enzyme. Using the improved variant provided up to a 1.7-fold increase in sugar yield during hydrolysis of aspen wood in the presence of [Bmim]Cl or in seawater (Pramanik et al. 2021).

Amino acid substitutions in α-helix allowed improvement in the thermostability of *P. verruculosum* EG II. Single substitution S308P in α-helix provided a 2.4–four-fold increase in half-life time at 70–80 °C (Dotsenko et al. 2019). Multiple substitutions F16L/Y293F/Q289G being in α-helix and β-strand provided a 5.5-fold improvement in half-life time at 75 °C. The computational analysis demonstrated the key role of the C-terminal region of the enzyme structure for the improvement in the enzyme thermostability (Contreras et al. 2020). Additional disulfide bridges S127C-A165C binding two α-helixes and Y171C-L201C binding two loops in the enzyme structure provided a 1.5–two-fold increase in the half-life times at 70–80 °C (Bashirova et al. 2019).

The surface of the enzyme molecule of *P. verruculosum* EG II was engineered to improve the enzyme operation in the presence of [Bmim]Cl. Amino acid substitutions were performed in order to fulfill the pockets located on the surface of the enzyme molecule and in distance from the active center. The substitutions E70S and V150L provided improvement both in thermostability and stability in the presence of [Bmim]Cl. The enzyme thermostability was increased in 1.2–1.6 times at temperature 70–80 °C. The enzyme stability in the presence of [Bmim]Cl (50 g/L) was improved in 1.7–1.9 times. The substitutions provided up to a 1.2-fold improvement in sugars yield during hydrolysis of aspen wood (Dotsenko et al. 2020b).

Therefore, independent amino acid substitutions provided considerable improvement in the activity and stability of the cellulolytic enzymes produced by *P. verruculosum.* The improvement supported up to a 1.7-fold increase in the yield of sugars obtained in the enzymatic hydrolysis. However, combining all beneficial substitutions within one enzyme structure will further improve the properties and enhance the hydrolysis. Moreover, the construction of the enzymatic mixtures for the hydrolysis from several improved enzymes will further enhance the hydrolysis.

Thus, protein engineering gives the possibility to improve the properties of *P. verruculosum* enzymes such as catalytic activity and stability. Deletion of N-linked glycans (deletion of N45 site through N45A substitution) improved the catalytic activity of CBH I toward different cellulosic feedstock in 1.3–1.6-times, for CBH II – in 1.3–1.4-times (deletion of the sites N219 and N2655 by N219A and N265A substitutions). In the case of EG II deletion of N-linked glycans increased k_{cat} in 1.4–1.7-times (deletion of N42 and N194 sites by N42A and N194A substitutions). CBH I with amino acid substitutions demonstrated increased thermostability – single substitution G415P in the C-terminus of α-helix provided a 3.4-fold increase in half-life time at 60 °C compared to wild-type enzyme; single substitution G415P in the C-terminus of α-helix provided 3.4-fold increase in half-life time at 60 °C.

11.6 Conclusions

P. verruculosum fungus is a promising producer of highly active cellulases – cellobiohydrolases, endoglucanases. Cellulases from *P. verruculosum* outperform *T. reesei* enzymes in the hydrolytic capacity toward cellulosic feedstock at the same protein or the same cellulase activity loading. Higher specific activity of key enzymes – CBH I and CBH II of *P. verruculosum* in comparison with the corresponding enzymes of *T. reesei*, higher BGL content and less inhibition by lignin of cellulases from *P. verruculosum* explain advantages in terms biomass hydrolysis.

Bioethanol is produced in general by starch fermentation from corn or sugar beets. An alternative is the use of cellulosic feedstocks by digesting cellulose to glucose. In addition to *Trichoderma*, fungi of the genus *Penicillium* such as *P. janthinellum* (Singhania et al. 2014), *P. decumbens* (Gao et al. 2011),

P. oxalicum (Ye et al. 2017) are also successfully used for the synergistic saccharification of cellulosic feedstocks. For example, in the production of bioethanol in the process of simultaneous saccharification and fermentation (SSF), preparations from *P. janthinellum* show greater efficiency in comparison with known commercial preparations. Often, an increase in the yield of bioethanol during SSF is associated with the tolerance of the biochemical properties of the *Penicillium* enzymes, which are more adaptable to the process of cellulose hydrolysis.

Acknowledgments This research was funded by the Russian Foundation for Basic Research (RFBR No. 18-54-80027) and Ministry of Science and High Education.

References

Adney WS, Jeoh T, Beckham GT et al (2009) Probing the role of N-linked glycans in the stability and activity of fungal cellobiohydrolases by mutational analysis. Cellulose 16:699–709

Bajaj P, Mahajan R (2019) Cellulase and xylanase synergism in industrial biotechnology. Appl Microbiol Biot 103:8711–8724. https://doi.org/10.1007/s00253-019-10146-0

Baker JO, Ehrman CI, Adney WS et al (1998) Hydrolysis of cellulose using ternary mixtures of purified cellulases. Appl Biochem Biotechnol 70-72:395–403

Bashirova A, Pramanik S, Volkov P et al (2019) Disulfide bond engineering of an endoglucanase from Penicillium verruculosum to improve its Thermostability. Int J Mol Sci 20:1602. https://doi.org/10.3390/ijms20071602

Berlin A, Balakshin M, Gilkes N et al (2006a) Inhibition of cellulase, xylanase and beta-glucosidase activities by softwood lignin preparations. J Biotechnol 125:198–209. https://doi.org/10.1016/j.jbiotec.2006.02.021

Berlin A, Gilkes N, Kilburn D et al (2005) Evaluation of novel fungal cellulase preparations for ability to hydrolyze softwood substrates - evidence for the role of accessory enzymes. Enzyme Microb Technol 37:175–184

Berlin A, Gilkes N, Kilburn D et al (2006b) Evaluation of cellulase preparations for hydrolysis of hardwood substrates. Appl Biochem Biotech 130:528–545

Bulakhov AG, Gusakov AV, Chekushina AV et al (2016) Isolation of homogeneous polysaccharide monooxygenases from fungal sources and investigation of their synergism with cellulases when acting on cellulose. Biochemistry-Moscow 81:530–537

Bulakhov AG, Volkov PV, Rozhkova AM et al (2017) Using an inducible promoter of a gene encoding Penicillium verruculosum Glucoamylase for production of enzyme preparations with enhanced Cellulase performance. PLoS One 12:e0170404. https://doi.org/10.1371/journal.pone.0170404

Chekushina AV, Dotsenko GS, Sinitsyn AP (2013) Comparing the efficiency of plant material bioconversion processes using biocatalysts based on Trichoderma and Penicillium verruculosum enzyme preparations. Catal Ind 5:98–104. https://doi.org/10.1134/S2070050413010042

Contreras F, Pramanik S, Rozhkova AM et al (2020) Engineering Robust Cellulases for Tailored Lignocellulosic Degradation Cocktails. Int J Mol Sci:21. https://doi.org/10.3390/ijms21051589

Denisenko YA, Gusakov AV, Rozhkova AM et al (2017) Site-directed mutagenesis of GH10 xylanase a from Penicillium canescens for determining factors affecting the enzyme thermostability. Int J Biol Macromol 104:665–671. https://doi.org/10.1016/j.ijbiomac.2017.06.079

Denisenko YA, Gusakov AV, Rozhkova AM et al (2019) Protein engineering of GH10 family xylanases for gaining a resistance to cereal proteinaceous inhibitors. Biocatal Agric Biote 17: 690–695. https://doi.org/10.1016/j.bcab.2019.01.042

Denisenko YA, Korotkova OG, Zorov IN et al (2021) Heterologous Expression of Thermogutta terrifontis Endo-Xanthanase in Penicillium verruculosum, Isolation and Primary Characterization of the Enzyme. Biochemistry-Moscow 86:489–495. https://doi.org/10.1134/S000629792104009x

Dotsenko AS, Dotsenko GS, Rozhkova AM et al (2020a) Rational design and structure insights for thermostability improvement of Penicillium verruculosum Cel7A cellobiohydrolase. Biochimie 176:103–109. https://doi.org/10.1016/j.biochi.2020.06.007

Dotsenko AS, Gusakov AV, Rozhkova AM et al (2016a) Effect of N-linked glycosylation on the activity and other properties of recombinant endoglucanase IIa (Cel5A) from Penicillium verruculosum. Protein Eng Des Sel 29:495–501. https://doi.org/10.1093/protein/gzw030

Dotsenko AS, Gusakov AV, Volkov PV et al (2016b) N-linked glycosylation of recombinant cellobiohydrolase I (Cel7A) from Penicillium verruculosum and its effect on the enzyme activity. Biotechnol Bioeng 113:283–291. https://doi.org/10.1002/bit.25812

Dotsenko AS, Pramanik S, Gusakov AV et al (2019) Critical effect of proline on thermostability of endoglucanase II from Penicillium verruculosum. Biochem Eng J 152:107397. https://doi.org/10.1016/j.bej.2019.107395

Dotsenko AS, Rozhkova AM, Zorov IN et al (2020b) Protein surface engineering of endoglucanase Penicillium verruculosum for improvement in thermostability and stability in the presence of 1-butyl-3-methylimidazolium chloride ionic liquid. Bioresour Technol 296:122370. https://doi.org/10.1016/j.biortech.2019.122370

Dotsenko GS, Gusakov AV, Rozhkova AM et al (2015) Heterologous beta-glucosidase in a fungal cellulase system: comparison of different methods for development of multienzyme cocktails. Process Biochem 50:1258–1263. https://doi.org/10.1016/j.procbio.2015.05.008

Fulop L, Ecker J (2020) An overview of biomass conversion: exploring new opportunities. Peerj 8: e9586. https://doi.org/10.7717/peerj.9586

Gao L, Wang F, Gao F et al (2011) Purification and characterization of a novel cellobiohydrolase (PdCel6A) from Penicillium decumbens JU-A10 for bioethanol production. Bioresour Technol 102:8339–8342. https://doi.org/10.1016/j.biortech.2011.06.033

Guo HL, Chang YJ, Lee DJ (2018) Enzymatic saccharification of lignocellulosic biorefinery: research focuses. Bioresour Technol 252:198–215. https://doi.org/10.1016/j.biortech.2017.12.062

Gupta VK, Schmoll M, Herreraestrella A et al (2014) Biotechnology and biology of Trichoderma. Biotechnology and Biology of Trichoderma:1–549

Gusakov AV (2011) Alternatives to Trichoderma reesei in biofuel production. Trends Biotechnol 29:419–425. https://doi.org/10.1016/j.tibtech.2011.04.004

Gusakov AV (2014) Comment on "revealing Nature's Cellulase diversity: the digestion mechanism of Caldicellulosiruptor bescii CelA". Science 344:578-a. https://doi.org/10.1126/science.1251248

Gusakov AV, Dotsenko AS, Rozhkova AM et al (2017) N-linked glycans are an important component of the processive machinery of cellobiohydrolases. Biochimie 132:102–108. https://doi.org/10.1016/j.biochi.2016.11.004

Gusakov AV, Salanovich TN, Antonov AI et al (2007) Design of highly efficient cellulase mixtures for enzymatic hydrolysis of cellulose. Biotechnol Bioeng 97:1028–1038. https://doi.org/10.1002/bit.21329

Gusakov AV, Sinitsyn AP (2012) Cellulases from Penicillium species for producing fuels from biomass. Adv Biochem Eng Biot 3:463–477. https://doi.org/10.4155/bfs.12.41

Harris PV, Welner D, Mcfarland KC et al (2010) Stimulation of lignocellulosic biomass hydrolysis by proteins of glycoside hydrolase family 61: structure and function of a large, enigmatic family. Biochemistry 49:3305–3316

Hemsworth GR, Davies GJ, Walton PH (2013) Recent insights into copper-containing lytic polysaccharide mono-oxygenases. Curr Opin Struc Biol 23:660–668

Horn SJ, Vaaje-Kolstad G, Westereng B et al (2012) Novel enzymes for the degradation of cellulose. Biotechnol Biofuels 5:45. https://doi.org/10.1186/1754-6834-5-45

Ikeda Y, Hayashi H, Okuda N et al (2007) Efficient cellulase production by the filamentous fungus Acremonium cellulolyticus. Biotechnol Prog 23:333–338. https://doi.org/10.1021/bp060201s

Ko JK, Ximenes E, Kim Y et al (2015) Adsorption of enzyme onto lignins of liquid hot water pretreated hardwoods. Biotechnol Bioeng 112:447–456. https://doi.org/10.1002/bit.25359

Korotkova OG, Semenova MV, Morozova VV et al (2009) Isolation and properties of fungal beta-glucosidases. Biochemistry (Mosc) 74:569–577. https://doi.org/10.1134/s0006297909050137

Kruszewska JS (1999) Heterologous expression of genes in filamentous fungi. Acta Biochim Pol 46:181–195. https://doi.org/10.18388/abp.1999_4196

Kubicek CP, Mikus M, Schuster A et al (2009) Metabolic engineering strategies for the improvement of cellulase production by Hypocrea jecorina. Biotechnol Biofuels 2:19. https://doi.org/10.1186/1754-6834-2-19

Levasseur A, Drula E, Lombard V et al (2013) Expansion of the enzymatic repertoire of the CAZy database to integrate auxiliary redox enzymes. Biotechnol Biofuels 6:41. https://doi.org/10.1186/1754-6834-6-41

Liao HP, Fan XT, Mei XL et al (2015) Production and characterization of cellulolytic enzyme from Penicillium oxalicum GZ-2 and its application in lignocellulose saccharification. Biomass Bioenergy 74:122–134

Liberato MV, Silveira RL, Prates ET et al (2016) Molecular characterization of a family 5 glycoside hydrolase suggests an induced-fit enzymatic mechanism (vol 6, 23473, 2016). Sci Rep-Uk 6

Liu T, Wang TH, Li X et al (2008) Improved heterologous gene expression in Trichoderma reesei by cellobiohydrolase I gene (cbh1) promoter optimization. Acta Bioch Bioph Sin 40:158–165. https://doi.org/10.1111/j.1745-7270.2008.00388.x

Lynd LR, Weimer PJ, Van Zyl WH et al (2002) Microbial cellulose utilization: fundamentals and biotechnology. Microbiol Mol Biol R 66:506-+

Mach RL, Zeilinger S (2003) Regulation of gene expression in industrial fungi: Trichoderma. Appl Microbiol Biot 60:515–522. https://doi.org/10.1007/s00253-002-1162-x

Maeda RN, Barcelos CA, Santa Anna LM et al (2013) Cellulase production by Penicillium funiculosum and its application in the hydrolysis of sugar cane bagasse for second generation ethanol production by fed batch operation. J Biotechnol 163:38–44. https://doi.org/10.1016/j.jbiotec.2012.10.014

Margeot A, Hahn-Hagerdal B, Edlund M et al (2009) New improvements for lignocellulosic ethanol. Curr Opin Biotechnol 20:372–380. https://doi.org/10.1016/j.copbio.2009.05.009

Martins LF, Kolling D, Camassola M et al (2008) Comparison of Penicillium echinulatum and Trichoderma reesei cellulases in relation to their activity against various cellulosic substrates. Bioresour Technol 99:1417–1424. https://doi.org/10.1016/j.biortech.2007.01.060

Merino ST, Cherry J (2007) Progress and challenges in enzyme development for biomass utilization. Adv Biochem Eng Biotechnol 108:95–120. https://doi.org/10.1007/10_2007_066

Morozova VV, Gusakov AV, Andrianov RM et al (2010) Cellulases of Penicillium verruculosum. Biotechnol J 5:871–880. https://doi.org/10.1002/biot.201000050

Payne CM, Knott BC, Mayes HB et al (2015) Fungal Cellulases. Chem Rev 115:1308–1448

Phillips CM, Beeson WT, Cate JH et al (2011) Cellobiose dehydrogenase and a copper-dependent polysaccharide monooxygenase potentiate cellulose degradation by Neurospora crassa. ACS Chem Biol 6:1399–1406

Pramanik S, Semenova MV, Rozhkova AM et al (2021) An engineered cellobiohydrolase I for sustainable degradation of lignocellulosic biomass. Biotechnol Bioeng 118:4014–4027. https://doi.org/10.1002/bit.27877

Quinlan RJ, Sweeney MD, Lo Leggio L et al (2011) Insights into the oxidative degradation of cellulose by a copper metalloenzyme that exploits biomass components. Proc Natl Acad Sci U S A 108:15079–15084. https://doi.org/10.1073/pnas.1105776108

Rahikainen JL, Martin-Sampedro R, Heikkinen H et al (2013a) Inhibitory effect of lignin during cellulose bioconversion: the effect of lignin chemistry on non-productive enzyme adsorption. Bioresour Technol 133:270–278. https://doi.org/10.1016/j.biortech.2013.01.075

Rahikainen JL, Moilanen U, Nurmi-Rantala S et al (2013b) Effect of temperature on lignin-derived inhibition studied with three structurally different cellobiohydrolases. Bioresour Technol 146: 118–125. https://doi.org/10.1016/j.biortech.2013.07.069

Rashmi R, Siddalingamurthy KR (2018) Microbial xyloglucanases: a comprehensive review. Biocatal Biotransformation 36:280–295. https://doi.org/10.1080/10242422.2017.1417394

Rubtsova EA, Bushina EV, Rozhkova AM et al (2015) Novel enzyme preparations with high pectinase and Hemicellulase activity based on Penicillium canescens strains. Appl Biochem Micro 51:591–599. https://doi.org/10.1134/S0003683815050142

Saunders G, Picknett TM, Tuite MF et al (1989) Heterologous gene-expression in filamentous fungi. Trends Biotechnol 7:283–287. https://doi.org/10.1016/0167-7799(89)90048-6

Semenova MV, Gusakov AV, Telitsin VD et al (2021) Enzymatic destruction of cellulose: characteristics of the kinetic interaction of lytic polysaccharide monooxygenases and individual Cellulases. Appl Biochem Micro 57:618–625. https://doi.org/10.1134/S0003683821050136

Semenova MV, Gusakov AV, Volkov PV et al (2019) Enhancement of the enzymatic cellulose saccharification by Penicillium verruculosum multienzyme cocktails containing homologously overexpressed lytic polysaccharide monooxygenase. Mol Biol Rep 46:2363–2370. https://doi.org/10.1007/s11033-019-04693-y

Sidar A, Albuquerque ED, Voshol GP et al (2020) Carbohydrate binding modules: diversity of domain architecture in amylases and Cellulases from filamentous microorganisms. Front Bioeng Biotech 8

Singh A, Patel AK, Adsul M et al (2017) Genetic modification: a tool for enhancing cellulase secretion. Biofuel Res J 4:600–610. https://doi.org/10.18331/Brj2017.4.2.5

Singh G, Verma AK, Kumar V (2016) Catalytic properties, functional attributes and industrial applications of beta-glucosidases. 3 Biotech 6:3. https://doi.org/10.1007/s13205-015-0328-z

Singhania RR, Saini JK, Saini R et al (2014) Bioethanol production from wheat straw via enzymatic route employing Penicillium janthinellum cellulases. Bioresour Technol 169:490–495. https://doi.org/10.1016/j.biortech.2014.07.011

Sinitsyn AP, Korotkova OG, Rubtsova EA et al (2020a) Construction of recombinant producers of enzyme preparations for feed production with an expression system based on Penicillium verruculosum fungus. Appl Biochem Micro 56:875–880

Sinitsyn AP, Korotkova OG, Sinitsyna OA et al (2016) Optimizing the composition of Cellulase enzyme complex from Penicillium verruculosum: enhancing hydrolytic capabilities via genetic engineering. Catal Ind 8:101–106. https://doi.org/10.1134/S2070050416010128

Sinitsyn AP, Rubtsova EA, Shashkov IA et al (2017) Preparation and properties of new biocatalysts for the degradation of nonstarch plant polysaccharides. Catal Ind 9:349–356. https://doi.org/10.1134/S2070050417040092

Sinitsyn AP, Sinitsyna OA (2021) Bioconversion of renewable plant biomass. Second-generation biofuels: raw materials, biomass pretreatment, enzymes, processes, and cost. Analysis. Biochemistry-Moscow 86:S166–S195. https://doi.org/10.1134/S0006297921140121

Sinitsyn AP, Sinitsyna OA, Zorov IN et al (2020b) Exploring the capabilities of the Penicillium verruculosum expression system for the development of producers of enzymes for the effective degradation of renewable plant biomass: a review. Appl Biochem Micro 56:638–646

Skomarovskii AA, Gusakov AV, Okunev ON et al (2005) Studies of hydrolytic activity of enzyme preparations of Penicillium and Trychoderma fungi. Prikl Biokhim Mikrobiol 41:210–212

Steffien D, Aubel I, Bertau M (2014) Enzymatic hydrolysis of pre-treated lignocellulose with Penicillium verruculosum cellulases. J Mol Catal B-Enzym 103:29–35

Su XY, Schmitz G, Zhang ML et al (2012) Heterologous gene expression in filamentous fungi. Adv Appl Microbiol 81(81):1–61. https://doi.org/10.1016/B978-0-12-394382-8.00001-0

Teugjas H, Valjamae P (2013) Selecting beta-glucosidases to support cellulases in cellulose saccharification. Biotechnol Biofuels 6

Vaaje-Kolstad G, Westereng B, Horn SJ et al (2010) An oxidative enzyme boosting the enzymatic conversion of recalcitrant polysaccharides. Science 330:219–222. https://doi.org/10.1126/science.1192231

Vlasenko EY, Ryan AI, Shoemaker CF et al (1998) The use of capillary viscometry, reducing end-group analysis, and size exclusion chromatography combined with multi-angle laser light scattering to characterize endo-1,4-beta-D-glucanases on carboxymethylcellulose: a comparative evaluation of the three methods. Enzyme Microb Technol 23:350–359

Volkov PV, Gusakov AV, Rubtsova EA et al (2019) Properties of a recombinant GH49 family dextranase heterologously expressed in two recipient strains of Penicillium species. Biochimie 157:123–130. https://doi.org/10.1016/j.biochi.2018.11.010

Volkov PV, Rozhkova AM, Gusakov AV et al (2014) Homologous cloning, purification and characterization of highly active cellobiohydrolase I (Cel7A) from Penicillium canescens. Protein Expr Purif 103:1–7. https://doi.org/10.1016/j.pep.2014.08.011

Wu SS, Wu SF (2020) Processivity and the mechanisms of Processive endoglucanases. Appl Biochem Biotech 190:448–463

Ye Y, Li X, Cao Y et al (2017) A beta-xylosidase hyper-production Penicillium oxalicum mutant enhanced ethanol production from alkali-pretreated corn Stover. Bioresour Technol 245:734–742. https://doi.org/10.1016/j.biortech.2017.08.155

Zhao XH, Wang W, Tong B et al (2016) A newly isolated Penicillium oxalicum 16 Cellulase with high efficient synergism and high tolerance of monosaccharide. Appl Biochem Biotech 178: 173–183. https://doi.org/10.1007/s12010-015-1866-x

Chapter 12
Third-Generation Bioethanol Production Technologies

N. Dlangamandla and K. Permaul

Abstract The depletion of fossil fuels has increased demand for alternative energy resources globally. Currently, bioethanol and biodiesel produced via first and second generation technologies are the most attractive biofuels, which have shown sustainability as renewable energy sources. The challenge with first and second generation biofuels is that feedstocks are associated with food security and there is lower yield of the process. Recently, third generation bioethanol from microalgae and macroalgae has been shown to be an emerging technology for the biofuel industry globally. The advantage is that this system does not require large amounts of land and pure water. Moreover, bioethanol that has been produced from algae has been shown to have higher yield compared to the second generation production process. Therefore, the main aim of this review is to take a detailed look at the third generation bioethanol technologies and the prospective future of the process. The pretreatment processes that are associated with processing microalgae and macroalgae to generate fermentable sugars for bioethanol production are also discussed.

12.1 Introduction

The demand fossil fuels has been growing in the past decade due to population growth and improved socioeconomic conditions (Bhore 2014). Therefore, alternative energy sources are required and being researched to overcome this demand for energy as well as to satisfy environmental regulations in various countries. Currently, solar and wind power and lignocellulosic biomass are the major renewable

Conceptualization, draft preparation, N.D.; review and editing, K.P; final draft review and editing, all authors. N.D. and K.P. contributed equally to this work.

N. Dlangamandla · K. Permaul (✉)
Department of Biotechnology and Food Science, Faculty of Applied Sciences, Durban University of Technology, Durban, South Africa
e-mail: nkosikhod@dut.ac.za; kugen@dut.ac.za

C. R. Soccol et al. (eds.), *Liquid Biofuels: Bioethanol*, Biofuel and Biorefinery Technologies 12, https://doi.org/10.1007/978-3-031-01241-9_12

energy sources. A fundamental property of renewable energy sources is their recyclability over time (Tan et al. 2020). Therefore, a variety of attractive renewable feedstocks for the biofuel industry is necessary. The most widely produced and researched biofuel is bioethanol as it can be used either directly in automobiles or combined with gasoline. It can take advantage of existing infrastructure to reduce greenhouse gas emissions (Edeh 2020). Due to the process's predicted economic competivity, global bioethanol production between 2016–2022 has been forecasted to climb from 110 to 140 billion litres, with an annual growth rate of 8% (Sharma et al. 2020). Lignocellulosic biomasses are the most attractive available feedstock for renewable energy and least expensive for second generation biofuels. Algal biomass, on the other hand, is classified as a third generation biofuel feedstock and comprises of microalgae and macroalgae. Microalgae are regarded as a promising feedstock for biodiesel production due to their rich lipid concentrations and fast biomass growth (Sun et al. 2016; Choi et al. 2019; Khan et al. 2018). Recent studies have described a biorefinery using microalgae as the feedstock for the third-generation bioethanol production (Pereira et al. 2019). Algae comprise of polysaccharides, which are suitable and beneficial for ethanol production. Moreover, cellulose can be derived from polysaccharides, which is the major source for biofuel industries. Finally, microalgae are well known for carbon mitigation and bioenergy production (Choi et al. 2019).

The primary goal of this chapter is to examine third generation bioethanol technologies, as well as the process's potential future. The chapter also describes the pre-treatment steps involved in processing microalgae and macroalgae to produce fermentable sugars for bioethanol production.

12.2 Feedstock/Substrates for Bioethanol Production

The majority of currently produced (first generation) bioethanol is derived from edible crops (sugarcane, sugar beet, corn, wheat, potato, barley, etc.) which contain high sucrose and starch content which can be further hydrolysed to fermentable sugars. Furthermore, bioethanol can also be produced from lignocellulosic biomass (polysaccharides), which requires pre-treatment and then enzymatic hydrolysis to release fermentable sugars for bioethanol production. Even though the process is still in development, lignocellulosic biomass is the most appealing feedstock for second generation bioethanol production, due to its wide availability and low cost. For the development of a sustainable third generation bioethanol production process, an abundant and cheaper feedstock is required since high process costs are associated with the raw materials used. The advantage of the third generation bioethanol is that feedstock that can be obtained more easily. In recent years, researchers have evaluated the feedstocks for bioethanol production much more closely (Laopaiboon et al. 2007; Khalil et al. 2015). Therefore, 3G production has a promising future due to availability and abundance of feedstocks. Table 12.1 provides a summary of some common feedstocks and their defining characteristics.

Table 12.1 Bioethanol production from various feedstocks (Ahmed et al. 2021; Teixeira et al. 2016; Dalla Marta et al. 2014; Macedo et al. 2020; Panahi et al. 2019)

Systems	Biomass	Carbohydrate content	Bioethanol production technologies	Bioethanol yield
First generation	Edible crops [sugar (from sugarcane, sugar beet, etc), and starch crops (from corn, wheat, potato, barley and etc)]	High starch/ sucrose content	Sucrose and starch extraction, fermentation, purification	1824–2572 L/ha
Second generation	Non-edible crops (*Jatropha*, cassava and *Miscanthus*), lignocellulosic biomass (straw, bagasse, wood, grass, and agricultural waste)	High cellulose content with hemicellulose and lignin	Pretreatment, saccharification, fermentation, purification	7,861,074 L/ ha
Third generation	Microalgae and macroalgae	High cellulose, hemicellulose and low lignin content	Pretreatment (mild conditions), saccharification, fermentation, purification	143–153 ton/ha

12.2.1 Sucrose-Containing Feedstocks

Sucrose is obtained mainly from energy-rich edible crop resources such as sugarcane and sweet sorghum. In both these edible crops, the juice is extracted and can be used directly in fermentations. Sugarcane (juice or molasses) has been extensively used for bioethanol production in Brazil, the largest sugarcane growing country. In sweet sorghum, the stem and grains are harvested. A juice with a high content of edible sucrose is extracted while the grains contain a high amount of edible starch that can be further processed to yield fermentable sugars. Bagasse forms a component of the lignocellulosic biomass from the plant (Laopaiboon et al. 2007; Khalil et al. 2015). The advantage of sweet sorghum as a feedstock is the ability to grow the plant at least twice a year, with a harvesting cycle of 4 months, and a 2- and four-fold lower fertilizer and water requirement (Tinôco et al. 2021; Bedzo et al. 2021).

12.2.2 Algae as the Feedstock

Algal feedstock (macroalgae or microalgae) has been shown to be the most promising feedstock for third generation bioethanol production. The investigation of algae as feedstock started early in the 1950s and further progressed into the early 1970s due to the oil crisis. Therefore, over the years, algae have been extensive researched as feedstock for biofuels production (Moser 2010; Carriquiry et al. 2011; Kumar et al. 2020). Studies have shown that algae can produce oil, which can be further processed into biodiesel by different types of microorganisms. Studies have also reported the conversion of macroalgae and microalgae to bioethanol and butanol. Algae are organisms that usually grow in aquatic ecosystems, utilizing light and carbon dioxide (CO_2) to form biomass. Recent studies have shown several ways of converting microalgal biomass into energy sources which include: biological conversion; chemical reactions; physical conversion; and thermal conversion (Tan et al. 2020; Hong and Wu 2020). The development of third generation biorefineries can improve and add value to the bio-renewables industry. Currently, only first generation biorefineries are commercially successful, while second generation and third generation bioethanol production processes have not yet been commercialized due to challenges that are related to production cost, scalability and technical issues (Bhatia et al. 2020). To overcome these challenges, recent studies have shown the potential of algae as the feedstock for third generation bioethanol to reduce the commercialization challenges (Tan et al. 2020; Chong et al. 2020; Chung et al. 2021). Algal biomass has the advantage of having a faster growth rate than plants and a higher photosynthetic efficiency too (Abdul Latif et al. 2019), while they can grow in various liquid media with significantly less land use., Therefore, algae can be grown with low-cost nutrients and municipal and agricultural wastewater (Chung et al. 2021; Abdul Latif et al. 2019; Liang et al. 2017).

12.2.3 The Technology of Algal Cultivation

Algal farming technology promises considerable economic benefit for biofuels (biodiesel) and other added-value biochemicals (Kumar et al. 2015). Pilot-scale cultivation of microalgae was developed in the early 1950s using *Chlorella* sp. in Japan for high quality production (Oswald and Golueke 1960). Since then, new technology has been changing, from the open pond to photobioreactors (PBRs) (Gupta et al. 2015). Open pond has been utilized for many years since they are less expensive and easier to build than PBR systems, but they are associated with contamination. The first commercial production of microalgae was initiated in the late 1970s in Japan, Europe and Israel. The main aim of the open ponds was to grow healthy food. The process was further used for the production of fine chemicals and health supplements. The biofuels industry has been subsequently developing this process to use algae biomass as a feedstock. Currently, the two most widely used cultivation systems are open ponds and closed PBRs (comprising mainly stirred tanks, vertical columns, horizontal tubular and flat panels). During the process a variety of organic and inorganic carbon sources are employed, including high-productivity energy resources like lipids, omega-3 and other microalgal oils (Kumar et al. 2021).

Furthermore, microalgae cultivation requires a variety of extra supplements as well as favourable environmental conditions (light intensity, temperature and pH) for growth (García-López et al. 2020). Different supplements such as bicarbonate, nitrogen and carbon dioxide are utilized during microalgal cultivation (Menegazzo and Fonseca 2019). Based on external colour differences, macroalgae (seaweeds) are categorized into three broad types: red (*Chlorophyta*), brown (*Phaeophyta*) and green (*Rhodophyta*) (Santos et al. 2018). Seaweeds consist of a diverse range of species (25,000 species) with differing morphologies and bioactive properties. Macroalgae can be further divided into 10 taxonomic divisions: *Cyanophyta, Prochlorophyta, Phaeophyta, Chlorophyta, Charophyta, Euglenophyta, Chrysophyta, Pyrrhophyta, Cryptophyta* and *Rhodophyta (*Bold 1978*)*. For growth, macroalgae minimally require sunlight, nutrients and water. Breeding, and genetic engineering have been used as key technologies to develop commercial strains for the enhanced production of macroalgal species. Genetic improvement has shown some success but breeding experiments have not progressed significantly since their early development.

12.3 Bioethanol Conversion Technologies

Bioethanol has been the most frequently utilized biofuel globally for the first and second generation biofuels. Bioethanol is one of the most researched biofuels and is currently produced on a commercial scale using corn (USA) and sugarcane (Brazil) (Tan et al. 2020; Gohain et al. 2021). The challenge of the current bioethanol production is related to the cost of the production process, both upstream and downstream, as well as the impact on food security. For years, yeast (e.g., *Saccharomyces cerevisiae*) has been utilized as the main bioethanol producer, offering unique benefits such as low cultivation costs, high ethanol productivity and tolerance, and ease of handling. The main feedstock for bioethanol includes edible crops and lignocellulosic biomass (Mussatto et al. 2010). Second generation bioethanol produced from lignocellulosic biomass has been shown as a promising technology for future production of bioethanol. However, the issue with the second-generation process is the by-products that are generated during pre-treatment. These technical issues are due to the presence of lignin in the feedstock and the delignification process which are cost-intensive for scaling up the process to commercialization. Third generation bioethanol produced from macroalgae has advantages since it does not require fresh water and large tracts of land for cultivation, while having a fast growth rate with high carbohydrate and lower lignin content. Furthermore, seaweed biomass has proved to be a suitable feedstock for third generation bioethanol and the benefits of this type of biomass provides a new avenue for biofuel research to reduce the dependence on fossil fuels (Hafting et al. 2015; Marquez et al. 2014; Salvi et al. 2021; Pablo et al. 2020; Satari and Jaiswal 2021; Qarri and Israel 2020). Third generation bioethanol production research has been dominated by the use of macroalgal biomass as the feedstock. Compositional analysis of these feedstocks are shown in Table 12.2. The overall process consists of three main stages: the collection of raw materials; pre-treatment; and finally fermentation (Dave et al. 2019). However, macroalgal organisms with sustainable or economically sound growth conditions for third generation bioethanol production are required (Balina

Table 12.2 Macromolecular composition of macroalgae species (% dry weight)

Macroalgae	Carbohydrate	Protein	Lipids	Reference
Brown algae	16.8–61.6	5.4–24	0.3–4.9	(Hong et al. 2014; Lee et al. 2014; Matanjun et al. 2009)
Red algae	26–66	9–22.9	0.2–1.3	(Kostas et al. 2016, 2020; Cho et al. 2014)
Green algae	42.8–59	17–31.6	1.8–4.8	(Choi et al. 2012; Kim et al. 2014)

et al. 2017). Recent studies have identified strains and optimised conditions to improve the growth of feedstock for bioethanol production. Promising techniques for developing macroalgal cultivation such as *ex situ* and *in situ* techniques have been developed. Artificial or onshore cultivation is the most utilized method for growth of macroalgae. Strains such *Ulva lactuca* and *Cystoseira amentacea* were cultivated by *ex-situ* methods, while *Kappaphycus alvarezii* and *Gracilaria* sp. were cultivated *in-situ* off-shore. Moreover, the on-shore cultivation technique has been the most utilized method and is most likely to be used for the future high production of macroalgal biomass to meet the global demand (Kim et al. 2017). The advantages of macroalgal production systems include consistent quantities of biomass per ton of dry weight (Dave et al. 2019). The methods have several drawbacks, including high production costs and contamination by other marine epiphytes. The transformation of the energy-rich crops or lignocellulosic biomass into bioethanol relies on their pre-treatment before the fermentative microorganisms can play their role (Cardona and Sánchez 2007). Several technologies have been developed to overcome the engineering challenges arising in the process.

12.3.1 Macroalgae Pre-Treatment Process

The macroalgae pretreatment process consists of five different processes, i.e., physical, chemical, thermal, mechanical, and biological processes. The main aim of chemical pre-treatment is the depolymerization of cellulose, while solubilizing hemicellulose, and delignifying the structure of algae (Ahmed et al. 2021). Acidic and alkaline solutions are used under mild pre-treatment conditions followed by enzymatic hydrolysis to further convert cellulose/glucans into fermentable sugars. Alkaline pre-treatment often uses potassium and sodium salts (Kostas et al. 2016). However, the limitation of the alkaline pre-treatment process is the excessive water consumption during the desalting process. The diluted sulfuric acid pre-treatment process is the most widely used method for pre-treatment of macroalgae. However, the challenge with acid pre-treatment are the inhibitors that are generated during the process. Recent studies have shown that the inhibitors can be reduced by neutralizing the hydrolysate to a pH of 7 with salts. Mechanical pre-treatment is a widely used method to breakdown the solid particles into smaller pieces. This process consists of steps such as washing, drying, and milling. The washing process is used to remove salts, sand,, and unrequired materials (Yuhendra et al. 2021; Tedesco et al. 2013). Thermal pre-treatment has been traditionally used for enhancing polysaccharide extraction and release of fermentable sugars from macroalgae. However, the process has limitations since it does not degrade the structure of the macroalgal cortex. The biological pretreatment process uses microorganisms such as fungi and bacteria, and enzymes to solubilize lignin and hemicellulose. The advantage of the biological process is that toxic compounds are not generated during the process and there is also less energy consumption (Tedesco et al. 2013). The process is conducted under either aerobic or anaerobic conditions. White rot fungi have been the most utilised microbes for the biological pre-treatment process. Studies have shown that white rot fungi have the ability to produce enzymes such as lignin peroxidase, manganese peroxidase and laccase, which can biodegrade lignin (Cui et al. 2021; Omoni et al. 2021; Tapia-Tussell et al. 2018; Thompson et al. 2019). Enzymatic hydrolysis is one of the final processes before fermentation, and is conducted by using enzymes (cellulase, agarase, and amylase) to hydrolyse macroalgal polysaccharides into fermentable sugars (Yun et al. 2016; Ekborg et al. 2006). Cellulase and amylase have been widely used for saccharification, while, the β-agarase system has been employed since the early 2000s (Al Abdallah et al. 2016). The disadvantage of enzymatic hydrolysis is the long duration of the process, high costs of enzymes production, and complex purification and recovery of the enzymes.

12.3.2 Microalgae Processes for Bioethanol Production

Microalgae have been investigated as a carbon source for biofuel industry through biomass transformation technologies such as biochemical conversion, chemical reactions, direct combustion, and thermochemical conversion. The most well-developed biochemical conversion processes are anaerobic digestion for biogas production and yeast fermentation for bioethanol production. These processes have been commercially successful. Microalgal biomass also requires a hydrolysis process before fermentation to convert carbohydrates into simple fermentable sugars (monomers) for bioethanol production. The hydrolysis process can be carried out independently of fermentation or in tandem with saccharification and fermentation (Smachetti et al. 2018). Ho et al. (Ho et al. 2013) investigated the feasibility of microalgal biomass as a feedstock for bioethanol production. The results obtained in their study showed a high yield of bioethanol at 79.9% and 92.3% theoretical yield using separate hydrolysis and fermentation (SHF) and simultaneous saccharification and fermentation (SSF), respectively, while using enzymatic hydrolysis. A final concentration of 11.7 g/L at 87.6% theoretical yield was obtained using the

Table 12.3 Production of bioethanol using microalgae and macroalgae as the feedstock

Microalgae/Macroalgae	Residual sugars (g/L)	Ethanol (g/L)	Ethanol yield (g/g)	Reference
Chlorella vulgaris FSP-E	50.00	11.66	0.233	(Ho et al. 2013)
Chlorella sp. ABC-001	50.00	2.45	0.430	(Seo et al. 2020)
Spirulina platensis	70.45	41.20	Nd	(Bader et al. 2020)
Chlamydomonas sp. JSC4 pigment extracted	Nd	73.00	Nd	(Huang et al. 2020)
Chlamydomonas reinhardtii CC125	22.40	10.75	0.510	(Bader et al. 2020)
Scenedesmus raciborskii WZKMT	72.50	79.38	Nd	(Alam et al. 2019)
Chaetomorpha linum	50.00	8.00	0.440	(Schultz-Jensen et al. 2013)
Ulva linza	50.00	48.24	0.329	(Greetham et al. 2020)
Sargassum spp.	45.66	18.14	0.760	(Aparicio et al. 2021)
Sargassum spp.	44.70	20.50	0.960	(del Río et al. 2021)

carbohydrate-rich microalga *Chlorella vulgaris* FSP-E as feedstock. A recent study by Seon et al. (Ho et al. 2013) reported an effective post-treatment process for bioethanol production using *Chlorella* sp. ABC-001 as the feed stock. The hydrolysis process using H_2SO_4 and $Ca(OH)_2$ solution showed the highest ethanol yield of 0.43 g/g. More studies on the third-generation bioethanol production using algae are presented in Table 12.3.

12.3.3 Third Generation Bioethanol Production from Macroalgae

Macroalgae are one of the most promising feedstock for the production of bioethanol, due to its growth conditions, since it require less land, freshwater, nutrients, and fumigation (Greetham et al. 2020). Recent studies have reported that the bioethanol yield per unit land area is higher for third generation production, when macroalgae is used as the feedstock (Cardona and Sánchez 2007; Ramachandra and Hebbale 2020). Macroalgae are classified into three categories, i.e., red, brown, and green, according to their pigments. The carbohydrates that are available in macroalgae are processed to bioethanol with selective enzymatic hydrolysis, and fermentation. The third-generation bioethanol production process requires milder pre-treatment, hydrolysis, fermentation, and distillation as the final stage. Aparacio et al. (Aparicio et al. 2021), reported the impact of a high-pressure system on delignifying the macroalgae before the enzymatic process. The system was validated with significant conversion of glucose to bioethanol, where 18.14 g/L of bioethanol was produced with a glucose to bioethanol conversion yield of 76.23%. In their study, they show the feasibility of bioethanol production from macroalgal biomass, which can be further developed into a biorefinery.

12.4 Macroalgal Biorefineries

The biorefinery industry, with macroalgal species has diverse high-value end products which can add value to the manufacturing units of bio-economies. The integrated macroalgal biorefinery concept can support production of high-value products along with biofuels (Pang et al. 2019). The biorefinery has been using crop biomass to produce liquid biofuels. However, the challenges of the first and second generation biorefinery was the competition for energy and food. The advantage of the macroalgae biorefinery is that due to low lignin content, the delignification pre-treatment step is reduced or not necessary, leading to higher bioethanol production (Rajak et al. 2020). The lignin content and the associated drawbacks (recalcitrant structure and low degradation and breakdown products acting as inhibitors for fermentable sugars) has been a critical factor for the slow adoption of the lignocellulosic bioethanol production (Kumar and Sharma 2017). The macroalgal biorefinery can be developed from macroalgal biomass, which adds value with diverse feedstock for different conditions for the industry. For example, *Rhodophyta* account for approximately 61% of the total global seaweed production, which can be used as feedstock for third generation bioethanol (Peñuela et al. 2018). Apart from third generation bioethanol production, macroalgae can be used for the production of agar, carrageenans and other biochemicals of nutritional and pharmaceutical importance like polyunsaturated fatty acids, proteins, antitumour, antioxidant and antiviral compounds. A summary of the steps in an algal biorefinery is shown in Fig. 12.1.

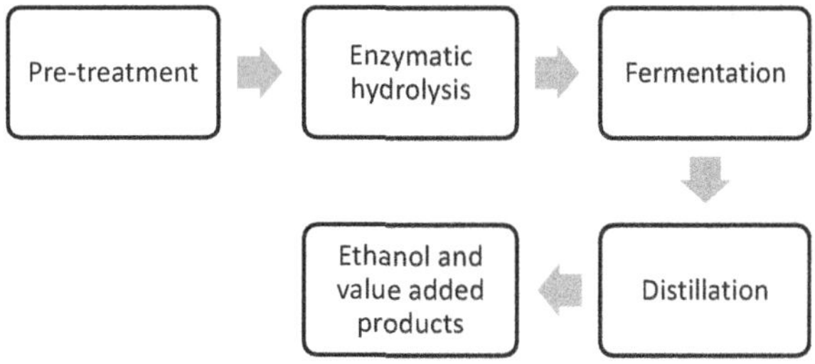

Fig. 12.1 Third-generation bioethanol production from macroalgal biomass

12.5 Future Perspectives of Third Generation of Bioethanol Production

The future third generation bioethanol is well promising, due to the demand on gasoline and the decline on the fossil fuels. Therefore, an alternative source of energy is required with the ability to reduce greenhouse emissions. Recent studies have shown the ability of macroalgae as a feedstock for bioethanol production. However, the technologies are still underdeveloped in large-scale lab or pilot studies. Moreover, certain parameters such as hydrolysis and fermentation still require more extensive research to improve the yield of bioethanol production and enhance fermentable sugars, and to reduce the duration time of the enzymatic process to reduce the costs that are associated with it. Therefore, low-cost macroalgae harvesting methods and simple technologies are required to add value to the future of third generation bioethanol production.

12.6 Conclusions

Third-generation bioethanol production has promise as an alternative source of energy. However, the process has not yet fully developed to a commercial level. Recent research studies have revealed that algal biofuels offer various advantages, including less land requirements for feedstock production and high oil content with high productivity. The macroalgal feedstock has been shown to have low lignin content and a less recalcitrant structure, as discussed in this review. In conclusion, the third-generation bioethanol produced from macroalgae has the ability to yield high ethanol production levels with a suitable pre-treatment processes, enzymatic hydrolysis, and fermenting strains.

Funding This research was funded by the Durban University of Technology and the NRF/BRICS STI Grant No: 2017–418.

Acknowledgments The authors would like to acknowledge funding from the Durban University of Technology (DUT). The financial assistance of the NRF/BRICS STI Grant is hereby acknowledged as well as the consortium partners. We also acknowledge the technical support of the Enzyme Technology Research Group.

Conflicts of Interest The authors declare no conflict of interest.

References

Abdul Latif NIS, Ong MY, Nomanbhay S (2019) Hydrothermal liquefaction of Malaysia's algal biomass for high-quality bio-oil production. Eng Life Sci 19:246–269

Ahmed N, Dhar BR, Pramanik BK, Forehead H, Price WE, Hai FI (2021) A cookbook for bioethanol from macroalgae: review of selecting and combining processes to enhance bioethanol production. Curr Pollution Rep. https://doi.org/10.1007/s40726-021-00202-7

Al Abdallah Q, Nixon BT, Fortwendel JR (2016) The enzymatic conversion of major algal and cyanobacterial carbohydrates to bioethanol. Front Energ Res 4:1–15

Alam MA, Yuan T, Xiong W, Zhang B, Lv Y, Xu J (2019) Process optimization for the production of high-concentration ethanol with *Scenedesmus raciborskii* biomass. Bioresour Technol 294: 122219

Aparicio E, Rodríguez-Jasso RM, Pinales-Márquez CD, Loredo-Treviño A, Robledo-Olivo A, Aguilar CN, Kostas ET, Ruiz HA (2021) High-pressure technology for *Sargassum* spp biomass pretreatment and fractionation in the third generation of bioethanol production. Bioresour Technol 329:1–10

Bader AN, Rizza LS, Consolo VF, Curatti L (2020) Efficient saccharification of microalgal biomass by *Trichoderma harzianum* enzymes for the production of ethanol. Algal Res 48:1–9

Balina K, Lika A, Romagnoli F, Blumberga D (2017) Seaweed cultivation laboratory testing: effects of nutrients on growth rate of *Ulva intestinalis*. Energy Procedia 113:454–459

Bedzo OK, Dreyer CB, van Rensburg E, Görgens JF (2021) Optimisation of pretreatment catalyst, enzyme cocktail and solid loading for improved ethanol production from sweet sorghum bagasse. Bioenergy Res. https://doi.org/10.1007/s12155-021-10298-w

Bhatia L, Bachheti RK, Garlapati VK, Chandel AK (2020) Third-generation biorefineries: a sustainable platform for food, clean energy, and nutraceuticals production. Biomass Convers Biorefin. https://doi.org/10.1007/s13399-020-00843-6

Bhore N (2014) Energy outlook: a view to 2040. Detroit Automotive Petroleum Forum, Detroit, MI, USA, pp 1–35

Bold HC (1978) Introduction to the algae. Prentice-Hall, Hoboken

Cardona CA, Sánchez ÓJ (2007) Fuel ethanol production: process design trends and integration opportunities. Bioresour Technol 98:2415–2457

Carriquiry MA, Du X, Timilsina GR (2011) Second generation biofuels: economics and policies. Energy Policy 39:4222–4234

Cho H, Ra C-H, Kim S-K (2014) Ethanol production from the seaweed *Gelidium amansii*, using specific sugar acclimated yeasts. J Microbiol Biotechnol 24:264–269

Choi W, Han J, Lee C, Song C, Kim J, Seo Y, Lee S, Jung K, Kang D, Heo S (2012) Bioethanol production from *Ulva pertusa* Kjellman by high-temperature liquefaction. Chem Biochem Eng Q 26:15–21

Choi YY, Patel AK, Hong ME, Chang WS, Sim SJ (2019) Microalgae bioenergy with carbon capture and storage (BECCS): an emerging sustainable bioprocess for reduced CO_2 emission and biofuel production. Bioresour Technol Rep 7:100256–100270

Chong TY, Cheah SA, Ong CT, Wong LY, Goh CR, Tan IS, Foo HCY, Lam MK, Lim S (2020) Techno-economic evaluation of third-generation bioethanol production utilizing the macroalgae waste: a case study in Malaysia. Energy 210:118483–118491

Chung MRWY, Tan IS, Foo HCY, Lam MK, Lim S (2021) Potential of macroalgae-based biorefinery for lactic acid production from exergy aspect. Biomass Convers Biorefin. https://doi.org/10.1007/s13399-021-01375-3

Cui T, Yuan B, Guo H, Tian H, Wang W, Ma Y, Li C, Fei Q (2021) Enhanced lignin biodegradation by consortium of white rot fungi: microbial synergistic effects and product mapping. Biotechnol Biofuels 14:162

Dalla Marta A, Mancini M, Orlando F, Natali F, Capecchi L, Orlandini S (2014) Sweet sorghum for bioethanol production: crop responses to different water stress levels. Biomass Bioenergy 64: 211–219

Dave N, Selvaraj R, Varadavenkatesan T, Vinayagam R (2019) A critical review on production of bioethanol from macroalgal biomass. Algal Res 42:101606 del Río PG, Gullón B, Pérez-Pérez A, Romaní A, Garrote G (2021) Microwave hydrothermal processing of the invasive macroalgae Sargassum muticum within a green biorefinery scheme. Bioresour Technol 340:1–10

Edeh I (2020) Bioethanol production: an overview. In: Inambao F (ed) Bioethanol technologies. IntechOpen, pp 1–22

Ekborg NA, Taylor LE, Longmire AG, Henrissat B, Weiner RM, Hutcheson SW (2006) Genomic and proteomic analyses of the agarolytic system expressed by *Saccharophagus degradans* 2-40. Appl Environ Microbiol 72:3396–3405

García-López D, Olguín E, González-Portela R, Sánchez-Galván G, De Philippis R, Lovitt R, Llewellyn C, Fuentes-Grünewald C, Saldívar RP (2020) A novel two-phase bioprocess for the production of *Arthrospira* (spirulina) *maxima* LJGR1 at pilot plant scale during different seasons and for phycocyanin induction under controlled conditions. Bioresour Technol 298: 1–11

Gohain M, Hasin M, Eldiehy KS, Bardhan P, Laskar K, Phukon H, Mandal M, Kalita D, Deka D (2021) Bio-ethanol production: a route to sustainability of fuels using bio-based heterogeneous catalyst derived from waste. Process Saf Environment Prot 146:190–200

Greetham D, Adams JM, Du C (2020) The utilization of seawater for the hydrolysis of macroalgae and subsequent bioethanol fermentation. Sci Rep 10:1–15

Gupta PL, Lee S-M, Choi H-J (2015) A mini review: photobioreactors for large scale algal cultivation. World J Microbiol Biotechnol 31:1409–1417

Hafting JT, Craigie JS, Stengel DB, Loureiro RR, Buschmann AH, Yarish C, Edwards MD, Critchley AT (2015) Prospects and challenges for industrial production of seaweed bioactives. J Phycol 51:821–837

Ho S-H, Huang S-W, Chen C-Y, Hasunuma T, Kondo A, Chang J-S (2013) Bioethanol production using carbohydrate-rich microalgae biomass as feedstock. Bioresour Technol 135:191–198

Hong IK, Jeon H, Lee SB (2014) Comparison of red, brown and green seaweeds on enzymatic saccharification process. J Ind Eng Chem 20:2687–2691

Hong Y, Wu Y-R (2020) Acidolysis as a biorefinery approach to producing advanced bioenergy from macroalgal biomass: a state-of-the-art review. Bioresour Technol 318:124080

Huang X, Bai S, Liu Z, Hasunuma T, Kondo A, Ho S-H (2020) Fermentation of pigment-extracted microalgal residue using yeast cell-surface display: direct high-density ethanol production with competitive life cycle impacts. Green Chem 22:153–162

Khalil SRA, Abdelhafez AA, Amer EAM (2015) Evaluation of bioethanol production from juice and bagasse of some sweet sorghum varieties. Ann Agric Sci 60:317–324

Khan MI, Shin JH, Kim JD (2018) The promising future of microalgae: current status, challenges, and optimization of a sustainable and renewable industry for biofuels, feed, and other products. Microb Cell Factories 17:36

Kim D-H, Lee S-B, Jeong G-T (2014) Production of reducing sugar from *Enteromorpha intestinalis* by hydrothermal and enzymatic hydrolysis. Bioresour Technol 161:348–353

Kim JK, Yarish C, Hwang EK, Park M, Kim Y (2017) Seaweed aquaculture: cultivation technologies, challenges and its ecosystem services. Algae 32:1–13

Kostas ET, White DA, Cook DJ (2020) Bioethanol production from UK seaweeds: investigating variable pre-treatment and enzyme hydrolysis parameters. Bioenergy Res 13:271–285

Kostas ET, White DA, Du C, Cook DJ (2016) Selection of yeast strains for bioethanol production from UK seaweeds. J Appl Phycol 28:1427–1441

Kumar BR, Mathimani T, Sudhakar MP, Rajendran K, Nizami A-S, Brindhadevi K, Pugazhendhi A (2021) A state of the art review on the cultivation of algae for energy and other valuable products: application, challenges, and opportunities. Renew Sust Energ Rev 138:110620–110649

Kumar K, Mishra SK, Shrivastav A, Park MS, Yang J-W (2015) Recent trends in the mass cultivation of algae in raceway ponds. Renew Sust Energ Rev 51:875–885

Kumar AK, Sharma S (2017) Recent updates on different methods of pretreatment of lignocellulosic feedstocks: a review. Bioresour and Bioprocess 4:1–19

Kumar M, Sun Y, Rathour R, Pandey A, Thakur IS, Tsang DC (2020) Algae as potential feedstock for the production of biofuels and value-added products: opportunities and challenges. Sci Total Environ 716:1–17

Laopaiboon L, Thanonkeo P, Jaisil P, Laopaiboon P (2007) Ethanol production from sweet sorghum juice in batch and fed-batch fermentations by *Saccharomyces cerevisiae*. World J Microbiol Biotechnol 23:1497–1501

Lee SY, Chang JH, Lee SB (2014) Chemical composition, saccharification yield, and the potential of the green seaweed *Ulva pertusa*. Biotechnol Bioprocess Eng 19:1022–1033

Liang S, Wei L, Passero ML, Feris K, McDonald AG (2017) Hydrothermal liquefaction of laboratory cultivated and commercial algal biomass into crude bio-oil. Environ Prog Sustain Energy 36:781–787

Macedo AA, Medeiros RG, Silvério TAB, Nelson DL, Oliveira DCS, dos Reis AB (2020) Possibilities and prospects regarding ethanol production from saccharin sorghum [*Sorghum bicolor* (L.) Moench]. SN Appl Sci 2:1–12

Marquez GPB, Santiañez WJE, Trono GC Jr, Montaño MNE, Araki H, Takeuchi H, Hasegawa T (2014) Seaweed biomass of the Philippines: sustainable feedstock for biogas production. Renew Sustain Energ Rev 38:1056–1068

Matanjun P, Mohamed S, Mustapha NM, Muhammad K (2009) Nutrient content of tropical edible seaweeds, *Eucheuma cottonii, Caulerpa lentillifera* and *Sargassum polycystum*. J Appl Phycol 21:75–80

Menegazzo ML, Fonseca GG (2019) Biomass recovery and lipid extraction processes for microalgae biofuels production: a review. Renew Sustain Energ Rev 107:87–107

Moser BR (2010) Camelina (*Camelina sativa* L.) oil as a biofuels feedstock: Golden opportunity or false hope? Lipid Technol 22:270–273

Mussatto SI, Dragone G, Guimarães PM, Silva JPA, Carneiro LM, Roberto IC, Vicente A, Domingues L, Teixeira JA (2010) Technological trends, global market, and challenges of bio-ethanol production. Biotechnol Adv 28:817–830

Omoni VT, Lag-Brotons AJ, Ibeto CN, Semple KT (2021) Effects of biological pre-treatment of lignocellulosic waste with white-rot fungi on the stimulation of 14C-phenanthrene catabolism in soils. Int Biodeterior Biodegradation 165:105324

Oswald WJ, Golueke CG (1960) Biological transformation of solar energy. Adv Appl Microbiol 2: 223–262

Pablo G, Gomes-Dias JS, Rocha CM, Romaní A, Garrote G, Domingues L (2020) Recent trends on seaweed fractionation for liquid biofuels production. Bioresour Technol 299:1–15

Panahi HKS, Dehhaghi M, Aghbashlo M, Karimi K, Tabatabaei M (2019) Shifting fuel feedstock from oil wells to sea: Iran outlook and potential for biofuel production from brown macroalgae (ochrophyta; phaeophyceae). Renew Sust Energ Rev 112:626–642

Pang N, Gu X, Chen S, Kirchhoff H, Lei H, Roje S (2019) Exploiting mixotrophy for improving productivities of biomass and co-products of microalgae. Renew Sustain Energ Rev 112:450–460

Peñuela A, Robledo D, Bourgougnon N, Bedoux G, Hernández-Núñez E, Freile-Pelegrín Y (2018) Environmentally friendly valorization of *Solieria filiformis* (Gigartinales, Rhodophyta) from IMTA using a biorefinery concept. Mar Drugs 16:1–19

Pereira H, Silva J, Santos T, Gangadhar KN, Raposo A, Nunes C, Coimbra MA, Gouveia L, Barreira L, Varela J (2019) Nutritional potential and toxicological evaluation of *Tetraselmis* sp. CTP4 microalgal biomass produced in industrial photobioreactors. Molecules 24:1–18

Qarri A, Israel A (2020) Seasonal biomass production, fermentable saccharification and potential ethanol yields in the marine macroalga *Ulva* sp.(Chlorophyta). Renew Energ 145:2101–2107

Rajak RC, Jacob S, Kim BS (2020) A holistic zero waste biorefinery approach for macroalgal biomass utilization: a review. Sci Total Environ 716:1–17

Ramachandra TV, Hebbale D (2020) Bioethanol from macroalgae: prospects and challenges. Renew Sustain Energ Rev 117:1–18

Salvi KP, da Silva OW, Horta PA, Rörig LR, de Oliveira BE (2021) A new model of algal turf scrubber for bioremediation and biomass production using seaweed aquaculture principles. J Appl Phycol. https://doi.org/10.1007/s10811-021-02430-2

Santos SCR, Ungureanu G, Volf I, Boaventura RAR, Botelho CMS (2018) Macroalgae biomass as sorbent for metal ions. In: Popa V, Volf I (eds) Biomass as renewable raw material to obtain bioproducts of high-tech value. Elsevier, Amsterdam, pp 69–112

Satari B, Jaiswal AK (2021) Green fractionation of 2G and 3G feedstocks for ethanol production: advances, incentives and barriers. Curr Opin Food Sci 37:1–9

Schultz-Jensen N, Thygesen A, Leipold F, Thomsen ST, Roslander C, Lilholt H, Bjerre AB (2013) Pretreatment of the macroalgae *Chaetomorpha linum* for the production of bioethanol–comparison of five pretreatment technologies. Bioresour Technol 140:36–42

Seo G, Kim HS, Cho JM, Kim M, Park W-K, Chang YK (2020) Effect of post-treatment process of microalgal hydrolysate on bioethanol production. Sci Rep 10:1–12

Sharma B, Larroche C, Dussap C-G (2020) Comprehensive assessment of 2G bioethanol production. Bioresour Technol 313:123630

Smachetti MES, Rizza LS, Coronel CD, Do Nascimento M, Curatti L (2018) Microalgal biomass as an alternative source of sugars for the production of bioethanol. In: Kulia A, Sharma V (eds) Principles and applications of fermentation technology. John Wiley & Sons, Hoboken, pp 351–386

Sun Z, Liu J, Zhou Z-G (2016) Algae for biofuels: an emerging feedstock. In: Luque R, Lin CSK, Wilson K, Clark J (eds) Handbook of biofuels production, 2nd edn. Woodhead Publishing, Cambridge, pp 673–698

Tan IS, Lam MK, Foo HCY, Lim S, Lee KT (2020) Advances of macroalgae biomass for the third generation of bioethanol production. Chin J Chem Eng 28:502–517

Tapia-Tussell R, Avila-Arias J, Domínguez Maldonado J, Valero D, Olguin-Maciel E, Pérez-Brito D, Alzate-Gaviria L (2018) Biological pretreatment of mexican caribbean macroalgae consortiums using Bm-2 strain (*Trametes hirsuta*) and its enzymatic broth to improve biomethane potential. Energies 11:1–11

Tedesco S, Benyounis K, Olabi A (2013) Mechanical pretreatment effects on macroalgae-derived biogas production in co-digestion with sludge in Ireland. Energy 61:27–33

Teixeira ACR, Sodré JR, Guarieiro LLN, Vieira ED, De Medeiros FF, Alves CT (2016) A review on second and third generation bioethanol production. SAE Technical Paper 2016-36-0515. https://doi.org/10.4271/2016-36-0515

Thompson TM, Young BR, Baroutian S (2019) Advances in the pretreatment of brown macroalgae for biogas production. Fuel Process Technol 195:106151

Tinôco D, Genier HLA, da Silveira WB (2021) Technology valuation of cellulosic ethanol production by *Kluyveromyces marxianus* CCT 7735 from sweet sorghum bagasse at elevated temperatures. Renew Energy 173:188–196

Yuhendra A, Farghali M, Mohamed IM, Iwasaki M, Tangtaweewipat S, Ihara I, Sakai R, Umetsu K (2021) Potential of biogas production from the anaerobic digestion of *Sargassum fulvellum* macroalgae: influences of mechanical, chemical, and biological pretreatments. Biochem Eng J 175:108140

Yun EJ, Kim HT, Cho KM, Yu S, Kim S, Choi I-G, Kim KH (2016) Pretreatment and saccharification of red macroalgae to produce fermentable sugars. Bioresour Technol 199:311–318

Chapter 13
Feedstocks and Pre-Treatment Techniques for Third-Generation Bioethanol Production

Gurpreet Kaur and Satinder Kaur Brar

Abstract Biofuels are introduced as an effective replacement for conventional fuels to reduce the environmental impacts and to meet the growing demand for energy, raised nowadays because of the growing population and industrialization. Due to running shortage and various economical and environmental drawbacks of conventional carbon-based fuels, biofuels are developed. Among biofuels, bioethanol is a demanding biofuel that could be an effective fuel and could be used in conventional engines of automobiles in combination with gasoline without any alterations. Depending upon the use of feedstock used to produce bioethanol, 1G, 2G, 3G, and most recent 4G bioethanol came into light. In this chapter, the focus is on 3G bioethanol produced from macroalgae and microalgae. 3G bioethanol has been developed to overcome the limitations associated with 1G and 2G bioethanol. 3G bioethanol is a recent emerging technique that is still in its development phase, while the previous research shows that it is a promising way to cope with several environmental issues related to foregoing generations.

13.1 Introduction

In this modern era, fuel consumption has been increased abruptly due to the development of industrialization and growing population, which leads to a high energy demand world widely. The primary source for fuel production is carbon-based fossil fuels, which are associated with a huge environmental issue. And the fast depletion rate of such conventional sources leads to find an alternative to producing fuels from other sources (Dudley 2018; Sommer 2015). There has been worldwide interest in technologies based on renewable energy sources, as a means of decreasing reliance on fossil fuels, minimizing climate change effects, and reducing greenhouse gas emissions. The concentration of global greenhouse gases reached a

---

G. Kaur · S. Kaur Brar (✉)
Department of Civil Engineering, Lassonde School of Engineering, York University, North York, ON, Canada
e-mail: satinder.brar@lassonde.yorku.ca

C. R. Soccol et al. (eds.), *Liquid Biofuels: Bioethanol*, Biofuel and Biorefinery Technologies 12, https://doi.org/10.1007/978-3-031-01241-9_13

concentration of approximately 400 ppm in 2016. This increase represents a growth of 146% when compared to pre-industrial phase, and is a result of anthropogenic activities and use of fossil fuel such as transportation fuels and energy demands (Tang et al. 2020). As the global population is increasing continuously, the demand for energy sources will also increase; therefore a non-polluting, green, and renewable approach is required to meet the growing demand for energy (Singh et al. 2017).

One of the innovative and sustainable technologies helping towards energy production with low adverse effects is biofuels. Industrial biotechnology manufactures biofuels (biodiesel, bioethanol, biobutanol) from renewable feedstocks by employing and integrating biotechnology-based tools with conventional industrial processes.

Here, the researchers come with an alternative to use corps (first-generation bioethanol, 1G) and industrial, forestry, agricultural waste (second-generation bioethanol, 2G) as feedstock to produce fuel to overcome the limitations associated with conventional sources. Biofuel production has been in process from the last four decades in the USA and European countries after the catastrophic oil crisis in 1973. Firstly, the biodiesel was extracted from oil crops for instance corn, hemp, soybean, etc.; but these crops had some drawbacks such as require a large space for cultivation and compete with food crops. Therefore, these oil crops failed to replace fossil fuels (Schenk et al. 2008).

The first- and second-generation bioethanol proved to be an interesting and beneficial replacement method, however, some impediments related to these 1G and 2G bioethanol came to light. For instance, in the case of 1G bioethanol, a large quantity of crops rich in starch as a sugar source for ethanol production like sugarcane, wheat, sorghum, barley is required, huge area to cultivate such crops, and competition between fuel and food supply chain. Such limitations of 1G bioethanol trigger the usage of lignocellulosic biomass from various industrial and agricultural waste like fruit peels, forestry waste, molasses, sugarcane bagasse, etc. having a high content of cellulose and hemicellulose as a sugar source. Although, 2G bioethanol provides numeral benefits like low-cost feedstock, reduces the demand to manage waste; however, the high content of lignin and hemicellulose in such lignocellulosic biomass requires appropriate pre-treatment techniques to breakdown the complex bond to release sugar from them. The pre-treatment steps depend on the type of lignocellulosic biomass used and are the costlier part among the whole bioethanol process (Dave et al. 2019; Harun et al. 2010; Singh et al. 2017). Hence to overcome the drawbacks corresponding to 1G and 2G bioethanol, algae become the recent interest as feedstock for bioethanol production which is known as 3G bioethanol production which is a more viable and sustainable option. Algae is a large group of aquatic and oxygenic photoautotrophs, found in diverse varieties from unicellular to multicellular forms, and have the potential for biofuel production. They are particularly divided into two broad groups as macroalgae (phytoplankton) and microalgae (seaweeds) (Barsanti and Gualtieri 2005; Hoek et al. 1995).

In 3G bioethanol, the carbohydrate content of the algae is used as a source to produce fermentable sugar molecules such as glucose or fructose, which are later converted to ethanol through fermentation process. This process requires less space

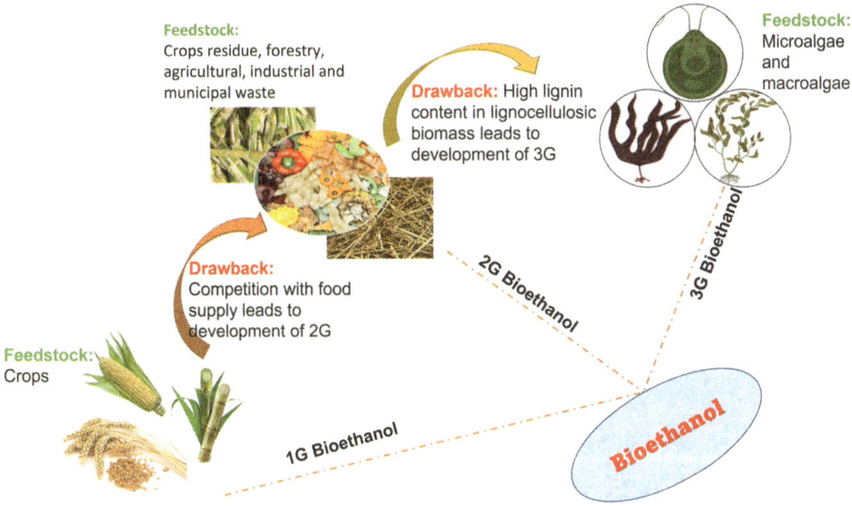

Fig. 13.1 Comparison between three different generations of bioethanol production

and also helps in reducing the carbon dioxide concentration from the atmosphere (Wang et al. 2019) (Fig. 13.1).

13.2 Feedstock for 3G Bioethanol: Algae

With respective to 1G and 2G bioethanol, which are utilizing starch-rich crops and lignocellulosic biomass respectively, 3G bioethanol explores the usage of carbohydrates-rich algae as their feedstock to obtain biofuels. Algae offer several dominant factors over the feedstocks used in 1G and 2G biofuels, making it more economically, socially, and environmentally convenient.

Algae possess the capability to produce lipids, carbohydrates, and proteins in large abundance within a short period of growth, which are valuable initiatives for several products which can serve mankind, for instance, biodiesel, bioethanol, and other co-products. In anaerobic and dark conditions, some algae are known to possess self-biorefinery system and generate ethanol in natural conditions using organic carbon sources (Khoo et al. 2019; Laurens et al. 2017; Zachleder and Brányiková 2014). There are many reports on the role of algae in the production of biofuels. Among biofuels, biodiesel is considered to be the commonest product, because algae are an oleaginous species, rich in lipid content (Mahendran et al. 2020; Nematian et al. 2020; Pugliese et al. 2020). Besides biodiesel, algae are exploited as

Table 13.1 Summary of some previous studies on pre-treatment and bioethanol yield from Algae

Type	Biomass	Carbohydrate content	Pretreatment	Ethanol yield	Reference
Microalgae	*Scenedesmus dimorphus*	49% w/w	Enzymatic hydrolysis	84%	(Chng et al. 2017)
			Acid-hydrolysis	80% theoretical yield	
	Hindakia tetrachotoma ME03	52.2%	Acid (H_2SO_4, HCI, H_3PO_4, and NaOH) hydrolysis	94%	(Onay 2019)
	Chlorella sp.	–	Hydrothermal pretreatment and enzymatic hydrolysis	52.88 ± 1.69 mg/g	(Ngamsirisomsakul et al. 2019)
Macroalgae (Brown seaweeds)	*Nizimuddinia zanardini*	25%w/w	Dilute sulfuric acid and hot water	34.6 g/kg	(Yazdani et al. 2015)
	Saccharina latissima	55%w/w laminarin and mannitol	No pretreatment	0.45% (v/v)	(Adams et al. 2008)
			Thermal (70 °C) and acidic (pH-2) treatment	0.13% (v/v)	
(Green seaweeds)	*Chaetomorphalinum*	–	Ball milling	18 g Ethanol/100 g dry biomass	(Schultz-Jensen et al. 2013)
	Ulva lactuca	75%–93%	Sulfuric acid treatment	40%	(van der Wal et al. 2013)
(Red seaweeds)	*Gelidium amansii*	35 g/L galactose, 8 g/L glucose	1.5% sulfuric acid, 140 °C, 15% Solid load, 60 min	39.2% (27.6 g/L)	(Park et al. 2012)
	Eucheuma cottonii	99.8% glucose after saccharification	Solid acid catalyst pretreatment	42.4%	(Tan and Lee 2015)
	Gracilaria verrucose waste biomass after agar extraction	56.65%	Enzymatic hydrolysis	27.2 g/L	(Shukla et al. 2016)

feedstock for bioethanol production (Table 13.1) using its carbohydrate content present in the form of starch, cellulose, or other polysaccharides.

To obtain bioethanol from algae, the polysaccharide carbohydrates are first hydrolyzed to simple fermentable sugars which can be then used by microorganisms during the fermentation step (Alfonsín et al. 2019; Bibi et al. 2017; Mushlihah et al. 2020; Ungureanu et al. 2020). With the advancement in recent biotechnological approaches such as genetic engineering and metabolic engineering, many technical improvements are being made to optimize the biofuel production from algae using these tools. Genetic engineering applications are exploited to manipulate the genetic makeup and central carbon metabolic pathways using algal model systems to improve biofuel yield (da Maia et al. 2020; Lakatos et al. 2019; Radakovits et al. 2010; Surendhiran and Sirajunnisa 2019). For successful genetic manipulation, expressed sequence tag (EST) databases and genome sequence databases of several algal species have been developed. To date many successful genetic transformations in microalgae, green algae, brown algae, and red algae has been reported for optimized biofuel production (Jiang et al. 2003; Kumar et al. 2004; Radakovits et al. 2010).

For instance, the starch content increased four to six folds in mutant strain in comparison to wild type with manipulation of GWD (*sex1* phenotype) in A. thaliana (Yu et al. 2001). Likewise, metabolic pathways are modified to produce easily accessible substrate, for instance, smaller and soluble sugar molecules (maltose) are likely to ferment more easily than the higher polysaccharides (starch). Maltose export protein (MEX1) has been introduced to the cytosol of green microalgae to directly produce maltose in cytosol for higher bioethanol yield (Deschamps et al. 2008).

13.2.1 Types of Algae

The algae are either unicellular or multicellular organisms having the ability to perform photosynthesis. They contain chlorophyll pigment to convert light into food using simple substances like carbon, nitrogen, and phosphorus. The major requirements for algae growth are aquatic media, light, and carbon sources. Based on the mode of synthesis of food, they are either phototrophic or heterotrophic organisms. The algae, which utilize atmospheric carbon dioxide to produce nutrients as food by harnessing sunlight is referred to as phototrophic algae while which use organic-based carbon sources for synthesizing the building blocks mainly protein, fats for their food are referred to as heterotrophic algae. Another type of algae also exists which can use either inorganic carbon dioxide or small organic carbon sources for their growth are termed as mixotrophs (Barsanti and Gualtieri 2005; Harun et al. 2010).

However, based on the size and cellular arrangement, algae are broadly classified into two major groups:

13.2.1.1 Microalgae

The microscopic unicellular algae which could be prokaryotic or eukaryotic depending on the presence of cell wall in cells are known as Microalgae. There are several species of microalgae that possess the ability to provide the opportunity to obtain bioethanol from them. Various studies by several different research groups focus on the potential of microalgae to produce biofuel and high valuable bio-product at a commercial scale. Microalgae are oil-amassing species that help to reduce the load on fossil fuels for energy supply (Tang et al. 2020).

Characteristics of Microalgae Microalgae are microscopic in size with high oil content ranging from 25%–77% of dry weight. Approximate, 10,000 ton dry algal biomass is produced globally in a year and 20,000–80,000 L of biofuel could be produced per acre per year (Tang et al. 2020). They grow as small colonies and have the capacity to cope with stressful conditions. Various factors control the rate of bio-oil production in third-generation biofuels; for instance strain species, available light for photosynthesis of microalgae, carbon dioxide, organic carbon source, surrounding environment, nutrients (nitrogen and phosphorus), and cultivation conditions (Chen et al. 2018). The microalgal strains mostly used for bioethanol production are *Chlorella vulgaris, Dunaliella tertiolecta, Spirulina* sp., *Botryococcus braunii, Chlamydomonas reinhardtii, Neochloris oleoabundans,* and *Nannochloris* sp. (Harun et al. 2010).

The polysaccharides produced in microalgal species are converted to monosaccharides such as glucose, xylose, arabinose, and galactose, etc. with the potential to produce 0.234 g/g bioethanol per dry biomass. Biobutanol, biogas, biodiesel, acetone, cosmetics, omega-3 oil, and livestock feed are additional co-products that could also be produced from algae. The production of these co-products helps to reduce the overall cost of bioethanol production (Hossain et al. 2019; Lakatos et al. 2019; Velazquez-Lucio et al. 2018).

13.2.1.2 Macroalgae (Seaweeds)

The species of algae that are multicellular and seen by the naked eye, are known as macroalgae or commonly as seaweed. Such species of algae are present since the evolution in marine habitats (Jiang et al. 2016). The major species of macroalgae which are used for biofuel production are as follows:

13.2.1.2.1 Red Algae (Rhodophyceae, Red Seaweed

In literature, approximately 10,000 species of red algae are recorded, however, a certain number is not known. Red algae comprise eukaryotic cells with chloroplast with different pigments. Most of the species have chlorophyll a, and carotenoids likely phycoerythrin and phycocyanin in high proportions. Red algae feature unique characteristics among the other phylum and division such as chloroplast with unstacked thylakoids, with no external endoplasmic reticulum and absence of flagella, while other species of algae have flagella, aggregated thylakoids, and phycoerythrin and phycocyanin are not found except Cryptophyte (Barsanti and Gualtieri 2005; Ramachandra and Hebbale 2020). These species mostly reserve their food in the form of Floridian starch in the cytoplasm, while polysaccharides such as Carrageenan, Agar, Cellulose are also present. *Kappaphycus alvarezii, Eucheuma denticulatum, Eucheuma* sp., *Gracilaria verrucose* are some examples that are often cultivated for bioethanol production with an annual production of 8,978,535 tons per year (Lee and Lee 2016). In the process of bioethanol production, the stored starch is hydrolyzed to simple fermentable sugars and later used for ethanol production.

13.2.1.2.2 Brown Algae (Phaeophyceae, Brown Seaweed)

Brown algae, commonly known as Kelp, is the fastest-growing algae among all macroalgae, which possess approximately 60% of carbohydrates content in dry biomass (Kraan 2013). The photosynthetic pigment frequently found in these seaweeds is Fucoxanthin and stores reserved food as Laminarin and mannitol, while xanthophylls are present as Fucoxanthin, Violaxanthin, Diadinoxanthin, Heteroxanthin, and Vacheriaxanthin. Unlike red seaweeds, these species have two flagella only in reproductive cells in unequal, lateral whiplash and tinsel shape (Ramachandra and Hebbale 2020).

Some frequently used strains for bioethanol production are *Laminaria japonica, Undaria pinnatifida, Sargassum fusiforme*. In these macroalgae, carbohydrate content is present in the form of Laminarin, Mannitol, Alginate, Fucoidan, Cellulose. According to FAO, the estimated production of brown algae per year is 678,493 tons (Lee and Lee 2016).

13.2.1.2.3 Green Algae (Chlorophyceae, Green Seaweed)

Towards bioethanol production, green algae play the least role among the other seaweeds. These species contain Chlorophyll *a, b*, carotene and Xanthophyll as photosynthetic pigments and have 2 or 4, equal anterior, and whiplash flagella (Ramachandra and Hebbale 2020). The overall commercial production of green algae is around 21,545 tons per year, which is comparatively low concerning red and brown algae. The common strains studied for bioethanol production are *Codium fragile, Caulerpa* sp., *Monostroma nitidum,* and *Enteromorpha clathrate.* The carbohydrate content is nearby 50% of dry biomass weight and is found in the form of *Starc*h and Cellulose (Lee and Lee 2016).

13.2.2 Advantage of Algae as Feedstock

Numerous properties make algae a sustainable and effective feedstock for biofuels production. A few of the major advantages of using algae biomass are as follow:

1. The algal species are rich in carbohydrate content as compared to terrestrial crops (Aizawa et al. 2007),
2. There is no or very less amount of lignin is present in comparison to terrestrial lignocellulosic biomass used in 2G bioethanol (Jiang et al. 2016),
3. Algae is the cheapest feedstock, and it does not compete with human food supply,
4. Algae could be cultivated on wastewater, which in turn assist in energy production along with wastewater management sustainably (Onay 2018; Ungureanu et al. 2020)
5. Algae use atmospheric carbon dioxide as an inorganic carbon source and convert it to sugar molecules, helping to reduce greenhouse gases and this phenomenon is known as CO_2 sequestration (Ramachandra and Hebbale 2020; Rosenberg et al. 2011),
6. The higher growth rate in a short period and lesser water requirement as compared to crops,
7. Could be cultivated in artificial bioreactors or on the seashore, nearshore and offshore in open ponds (Ghadiryanfar et al. 2016), and
8. Short cultivation time helps to produce more biomass annually.

13.3 Commercial Production

To produce biofuels, especially in the form of biodiesel or bioethanol, the algae are produced on large scale to meet the global fuel demand sustainably. Various simple to complex steps are followed to retrieve bio-products like ethanol from their feedstocks. These steps can vary depending upon the type of feedstock used for bioproduction, however, the major procedure is similar. In the case of algae, either in the form of microalgae (phytoplankton) or macroalgae (seaweeds), the process used for bioethanol generation is summarised in Fig. 13.2.

13.3.1 Cultivation

To generate 3G bioethanol, a large quantity of algal feedstock is required. It can either be attained from naturally occurring seaweeds and phytoplankton or could be achieved by artificially culturing them in bioreactors and open raceway ponds. The common type of bioreactors used are flat-bed and tubular bioreactors in horizontal and vertical geometry (Bibi et al. 2017; Dave et al. 2019).

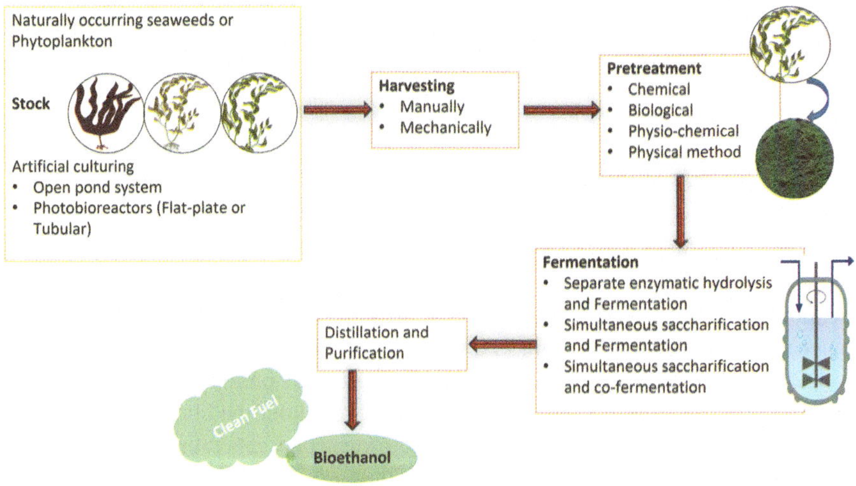

Fig. 13.2 Overall presentation of the bioethanol production process from algae as a 3G feedstock

Another approach is also used where, seeds are produced in the artificial laboratory conditions, and then they are cultivated near the seashore, on-shore, and off-shore in open ponds. Both the approaches have some pros and cons associated with them. For instance, the cultivation on seashore reduces the cultivation cost and provides the natural growth condition while there are chances of being affected by climate, weather, and sea waves (Daroch et al. 2013; Ghadiryanfar et al. 2016). However, in the case of an artificial culturing system, there is no damage from climatic circumstances, but there is a requirement for a proper supply of growth nutrients and light for the higher growth yield in bioreactors (Jiang et al. 2016). A medium rich in nutrients like carbon, nitrogen, and phosphorus is required to sustain the growth of microalgae. Few species absorb nitrogen directly from the atmosphere; however, in most cases, urea or ammonia is used as a nitrogen source (Tang et al. 2020). To make the final product cost-effective, natural sunlight is used instead of an artificial source for the photosynthesis of microalgae at the large-scale production of algae (Koyande et al. 2019).

13.3.2 Harvesting

After the cultivation period, the grown biomass of algae is harvested mostly in two ways: manually or mechanically. In manual harvesting, the labor is used to collect the algal biomass by using sickle, blades, or fork and is mainly used in artificial culturing. In the mechanical method, various technical motorboats with cutting blades and storage areas are used especially in seashore cultivation and for harvesting marine seaweeds (Ghadiryanfar et al. 2016; Roesijadi et al. 2010).

13.3.3 Pre-Treatment Techniques for 3G Algal Feedstock

The harvested algal biomass is transported to the production sites employing transportation, where the harvested biomass is pre-processed to enable them for saccharification. The initial steps involved in the processing are very simple like removal of unnecessary parts (stones, snails, plastic), dewatering, drying, etc. (Ghadiryanfar et al. 2016). After this pre-processing, the next part is the pre-treatment of feedstock, which is the most challenging and costly step throughout the complete bioethanol production process. Although, in comparison to second-generation feedstock (lignocellulosic biomass), pre-treatment of algal biomass is easier and cheaper because algal contains almost no or very less lignin content, while lignocellulosic biomass accommodates a high proportion of lignin which is difficult to break down (Bibi et al. 2017; Dave et al. 2019). The main purpose of pre-treatment for algal biomass is to rupture the cell wall of microalgae and macroalgae to release and recover the carbohydrate content for saccharification in enzymatic hydrolysis, to obtain the sugars directly, to prevent the damage to recovered carbohydrates, and to inhibit the toxic effect of inhibitors on bioethanol production.

To accomplish the pre-treatment process, numerous methods are studied and evaluated for their efficiency. The pre-treatment methods are broadly classified into four categories, named physical pre-treatment, physicochemical pre-treatment, chemical pre-treatment, and biological pretreatment methods as shown in Fig. 13.3 (Tang et al. 2020; Velazquez-Lucio et al. 2018).

13.3.3.1 Physical Method

The physical treatment of feedstock aims mainly towards size reduction to enhance the surface area. The increased surface area allows the escalated transport of acid, base, or catalyst inside the cells to release polysaccharides (Lee et al. 2014). Physical treatment can be performed in numerous ways as follows:

Mechanical comminution

These methods employ mechanical forces to decompose the algal cell wall and matrix such as milling, chipping, and grinding. These steps reduce the degree of polymerization, which in turn accelerates the efficiency of the further process (Bibi et al. 2017; Harun et al. 2014).

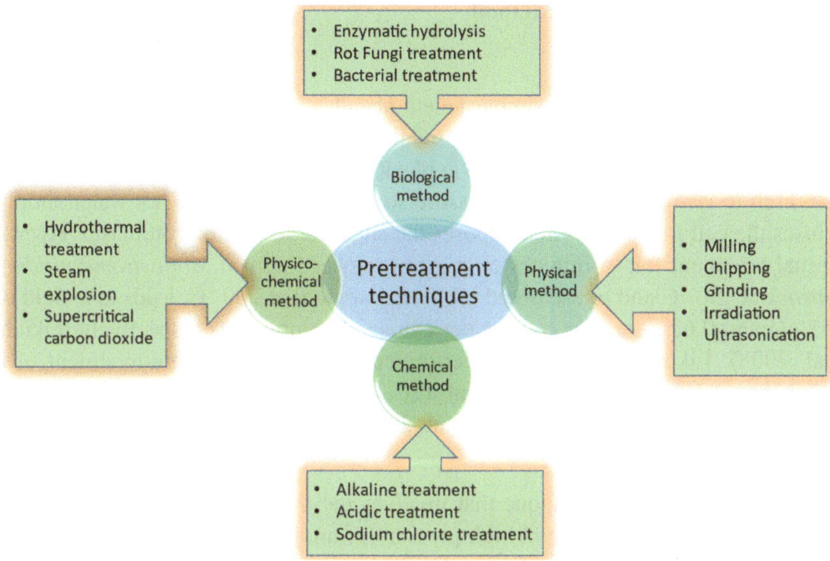

Fig. 13.3 Different pre-treatment techniques used to rupture the rigid cell wall of algae prior to fermentation

Ultrasonication

In ultrasonication, the high-frequency waves with continuous compression and rarefaction induce the microbubbles, which produce a shear force on collapsing (Khanal et al. 2007). The generated shear force disintegrates the cell membrane and cell wall, releasing the polysaccharides. For instance, in a study by Choi et al. 2011, the biomass of *Scenedesmus obliquus* was pretreated with ultrasonication results in an excellent release of sugars to the surface (Choi et al. 2011).

Irradiation

The use of gamma radiations is also introduced to pre-treat the biomass to disrupt the cell matrix and crystallinity of polysaccharides. The highly reactive radicals are produced in the chain scission process with radiation, which helps to decompose the original algal structure. In a study, *Undaria* biomass was treated with 0, 10, 50, 100, 200, and 500 kGy irradiation followed by acidic hydrolysis, which showed that irradiation causes cell wall breakage (Yoon et al. 2012).

Microwave heating

It is a simple, short, and effective pre-treatment method, requires a short reaction time. The microwave heating generates polar bond vibration in the liquid phase, which results in an explosion within the particles. The resultant explosion reduces the polymerization of large polysaccharides in both algal and lignocellulosic biomass (Maceiras et al. 2021; Ocreto et al. 2021).

13.3.3.2 Physicochemical Method

Hydrothermal Treatment

Hydrothermal treatment is the simplest method and is widely used for lignocellulosic biomass. In this method biomass is exposed to high temperature in liquid medium for 15 minutes, however unlike the steam explosion method, there is no sudden shifting from high to normal atmospheric pressure so, no expansion. The hot water simply diffuses the cell membranes and exposes the sugar molecule to protoplasm. Hydrothermal treatment for 15 minutes on a green seaweed named *Monostroma nitidum Wittrock* at 150 °C and red seaweed (*Solieria pacifica*) at 200 °C, leads to a yield of 0.51 g/g and 0.62 g/g sugar content per extracted components respectively (Okuda et al. 2008). Likewise, Choi et al. performed this method for pre-treatment of a macroalgae *Ulva pertusa* Kjellman at 150 °C and 15 MPa and achieve 9.08% glucose and 21% xylose yield (Choi et al. 2012).

Steam Explosion

It is a physicochemical technique that involves the treatment of biomass at a high temperature of 160–260 °C and high pressure of around 20–50 bar for a period of few seconds to few minutes. Afterward, the biomass is suddenly exposed to atmospheric pressure resulting in explosive decompression, lead to depolymerization of complex carbohydrates. This method has been used in several studies as a pre-treatment method and shows the good result (Kumar et al. 2009; Manzanares et al. 2012; Martín-Sampedro et al. 2012; Tomás-Pejó et al. 2011).

Supercritical Carbon Dioxide Treatment

In this method, the temperature and pressure of a liquid are increased above its critical value, where it exhibits properties of both the liquid and gas phase and is known as the supercritical phase. In such fluid extraction methods, various solvents can be used, however, carbon dioxide is widely preferred, because carbon dioxide has a low critical temperature of 32 °C and pressure of 72 bar, which is easy to achieve. Its non-reactive, less toxic, and less flammable nature helps to extract the sugars and lipids from biomass. It is a cost-effective treatment as it does not require much higher temperature like the steam explosion and also gives higher yield (Halim et al. 2013; Tabil et al. 2011).

13.3.3.3 Chemical Method

Acidic Treatment

Acids are the effective and commonly used chemical for the delignification of both algal biomass and lignocellulosic biomass. The acids catalyze the decomposition of the cell wall, hydrolyze the cellulose, and help to expose the polysaccharides to enzymatic hydrolysis to recover more fermentable sugar molecules. Acid can be used in concentrated and diluted form depending upon the crystallinity of biomass, however; dilute acidic solutions are preferred because of less harmful effects (Spiden et al. 2015). The widely used acids for pre-treatment are sulfuric acid, hydrochloric acid, phosphoric acid, and nitric acid. It is the most adapted pretreatment method according to the literature, several studies used different combination of acids and different temperatures, ranging from 40 °C to 150 °C for different treatment period varying from few minutes to hours. Instead of its effectiveness and wide-scale applications, there are some disadvantages associated with this pre-treatment, for instance; harmful chemicals and generation of some toxic chemicals like furfural, which act as an inhibitor for ethanol (Balat et al. 2008). Shokrkar et al., use the combination of different dilute acids with Magnesium sulfate for the pre-treatment of mixed culture of microalgae and observed that the combination of dilute sulfuric acid with $MgSO_4$, give the highest sugar yield (Shokrkar et al. 2017).

Alkaline Treatment

In comparison to acidic pre-treatment, the use of an alkaline solution such as NaOH, is quite rare for the pre-treatment of algal biomass. The alkaline solution breaks the intermolecular bonds within polymeric carbohydrates, releasing fermentable sugars for fermentation. This process is carried out at a low temperature and low pressure as compared to other pretreatment methods. Algal biomass was first exposed to alkaline pre-treatment by Harun et al. (2011) using 0.75% NaOH for 30 minutes and at 120 °C, resulted in 350 mg glucose yield per g of *Chlorococcum infusionum* biomass, with 0.26 g ethanol/g algae bioethanol yield (Harun et al. 2011). In contrast, unsatisfied results were recorded for 0.05–0.2 N $Ca(OH)_2$ (121 °C, 15 min) on *Ulva lactuca* (Kim et al. 2011).

13.3.3.4 Biological Method

In recent years, microbe-mediated pretreatment of biomass is carried out to avoid post-process environment hazards. The white-rot fungi such as *Phanerochaete chrysosporium, Ceriporiopsis subvermispora*, and *Pleurotus florida*; and soft rot fungi like *Trichoderma reesei* are used for this purpose (Sara et al. 2016; Zabed et al. 2016). These fungi release enzyme system includes laccase, manganese peroxidase, and lignin peroxidase for the disintegration of complex structures. The wet biomass of *S. occidentalis* subjected to NaOH pretreatment is easily degraded using enzyme extract from *Trichoderma harzianum*. This enzyme extract has contained a mixture of amidases, proteases, glucanases, mannosidases, and sarcosine oxidases (Heshof et al. 2020). Although this pretreatment method is eco-friendly, it requires longer residence time (several hours to few days). Even, fungi also consume some part of sugars for their growth, which imposes a negative impact on the process. Mushlihah et al. (2020) suggested the fungal pretreatment as a cheap and effective pretreatment method for marine algae and observe 2.3 folds more sugar yields in comparison to untreated biomass (Mushlihah et al. 2020). However, to make this pretreatment method more convenient, natural depolymerization processes are being investigated as well as various genetic engineering technologies are also explored to modify the microbes to overexpress lignin/cellulose-degrading enzymes (Ning et al. 2021).

13.3.4 Hydrolysis

The pre-treatment is followed by hydrolysis of polysaccharides to fermentable monosaccharide sugar molecules like glucose, fructose, sucrose which can be performed either by a conventional method like acidic hydrolysis using acids like sulfuric acid, hydrochloric acid, etc. (Harun and Danquah 2011; Shokrkar et al. 2017; Spiden et al. 2015; Wang et al. 2014) or by using enzymatic hydrolysis like cellulase (Ismail et al. 2020; Kwon et al. 2016; Okuda et al. 2008; Shokrkar et al. 2018; Trivedi et al. 2013; Ungureanu et al. 2020). The latter one is gaining much attraction due to greater yield and use of biological components which are non-hazardous, unlike acids. A company, Novozymes has cellulase preparations known as Cellic CTec2, Cellic CTec3, and Cellic HTec3 which potentially achieve higher levels of saccharification (Brar et al. 2020).

13.3.5 Fermentation

After depolymerization of complex polysaccharides into fermentable sugar molecules, the algal feedstock is used as fermentation media for the growth of various microbes for instance *Saccharomyces cerevisiae* (El-Sayed et al. 2016; Lee et al. 2015), *Saccharomyces bayanus* (El-Mekkawi et al. 2019), *Escherichia coli* (Kim et al. 2011; Lee et al. 2011), *Pichia stipites* (Sukwong et al. 2018; Yeon et al. 2010), *Pichia angophorae* (Khammee et al. 2021), and *Clostridium* sp. (Abomohra and Elshobary 2019; Hong et al. 2020) to convert hydrolyzed sugar into ethanol using their metabolic activities. This step can be performed in three different ways as separate hydrolysis and fermentation (SHF), simultaneous saccharification and fermentation (SSF), and consolidated bioprocessing (CBP). When hydrolysis of polysaccharides is done before fermentation in a different chamber, and then followed by fermentation separately is know as SHF; in SSF, both the hydrolysis and fermentation are performed in a single chamber; while in CBP, in addition to hydrolysis and fermentation, enzyme production is also done simultaneously. The former two methods are most common and most studied for ethanol production from algal biomass (Harun et al. 2010; Ho et al. 2013; Taherzadeh and Karimi 2007). Various parameters like temperature, pH, selection of microbe, feedstock content, nutrients, fermentation time, inhibitors content affects the bioethanol yield. For the high efficiency and high bioethanol yield content, these parameters are needed to be considered while the process of fermentation (El-Mekkawi et al. 2019; Harun et al. 2014).

After completion of fermentation, the end-products are recovered, and ethanol is separated by distillation process using water-ethanol mixture and purified for market application.

13.4 Future Prospective and Conclusion

Nowadays, the need to explore new renewable resources for fuel production is required as it is expected that the fossil fuels like coal, natural gas, and crude oil will exhaust in the coming 50 years (Dudley 2018; Tang et al. 2020). In such a scenario, there is a great need to explore all possible sustainable options to produce fuel at a large scale. Among the biofuels, which are deemed to be an effective replacement option, third-generation biofuels like bioethanol or biodiesel from algal biomass overcome the limitations associated with first- and second-generation biofuels. However, the developments are still in continuity to improve the yield from such green biomass. Cultivation of pure culture at a large scale is a challenging process as it needs aseptic conditions, however, mixed cultures of algae are explored nowadays for their efficiency of ethanol production (Shokrkar et al. 2017). Likewise, genetic engineering and biochemical engineering approaches should be investigated to enhance the synthesis in near future. Several companies such as Algaetech International, AlgaeBiotech, Sapphire Energy, Algae Tec, EcoFuel Laboratories, IBV Biotech, etc. want to replace fossil fuels with biofuels with an aim to meet energy demand (Katiyar and Arora 2020).

To conclude, algal biomass is considered to be an eco-friendly, sustainable and effective feedstock owing to its high growth rate, more productivity, easy and cheaper cultivation, and harvesting processes, ability to grow in wastewater. Concerning pre-treatment methods, acidic pretreatment and biological treatment are seemed to be promising methods among all others for algal biomass with high efficiency. Algal biomass possesses numerous advantages as a biofuel feedstock over the crops and lignocellulosic biomass, making it a preferable choice. However, few drawbacks which limit the use of algal biomass can prevail with the help of recent technologies like genetic and biochemical engineering.

Acknowledgments The financial support provided by Natural Sciences and Engineering Research Council of Canada (Discovery Grant 355254, CRD Grant, and Strategic Grant 447075) is sincerely acknowledged. The support of James and Joanne Love Chair in Environmental Engineering at York University is appreciated as well.

Conflict of Interest The Authors Declare no Conflict of Interest

References

Abomohra AE-F, Elshobary M (2019) Biodiesel, bioethanol, and biobutanol production from microalgae. Microalgae Biotechnology for Development of Biofuel and Wastewater Treatment, 293–321

Adams JM, Gallagher JA, Donnison IS (2008) Fermentation study on Saccharina latissima for bioethanol production considering variable pre-treatments. J Appl Phycol 21(5):569. https://doi.org/10.1007/s10811-008-9384-7

Aizawa M, Asaoka K, Atsumi M, Sakou T (2007) Seaweed bioethanol production in Japan-The Ocean sunrise project. Oceans 2007:1–5

Alfonsín V, Maceiras R, Gutiérrez C (2019) Bioethanol production from industrial algae waste. Waste Manag 87:791–797

Balat M, Balat H, Öz C (2008) Progress in bioethanol processing. Prog Energy Combust Sci 34(5): 551–573

Barsanti L, Gualtieri P (2005) Algae: anatomy, biochemistry, and biotechnology. CRC press

Bibi R, Ahmad Z, Imran M, Hussain S, Ditta A, Mahmood S, Khalid A (2017) Algal bioethanol production technology: a trend towards sustainable development. Renew Sust Energ Rev 71: 976–985

Brar KK, Chadha BS, Brar SK, Singh P (2020) Biotechnological strategies for enhanced production of biofuels from lignocellulosic biomass. In: *Valorization of biomass to value-added commodities*. Springer, pp 521–551

Chen Z, Wang L, Qiu S, Ge S (2018) Determination of microalgal lipid content and fatty acid for biofuel production. BioMed Research International 2018

Chng LM, Lee KT, Chan DJC (2017) Synergistic effect of pretreatment and fermentation process on carbohydrate-rich Scenedesmus dimorphus for bioethanol production. Energy Convers Manag 141:410–419

Choi J-A, Hwang J-H, Dempsey BA, Abou-Shanab RA, Min B, Song H, Lee DS, Kim JR, Cho Y, Hong S (2011) Enhancement of fermentative bioenergy (ethanol/hydrogen) production using ultrasonication of Scenedesmus obliquus YSW15 cultivated in swine wastewater effluent. Energy Environ Sci 4(9):3513–3520

Choi WY, Han JG, Lee CG, Song CH, Kim JS, Seo YC, Lee SE, Jung KH, Kang DH, Heo SJ (2012) Bioethanol production from Ulva pertusa Kjellman by high-temperature liquefaction. Chem Biochem Eng Q 26(1):15–21

da Maia JL, Cardoso JS, da Silveira Mastrantonio DJ, Bierhals CK, Moreira JB, Costa JAV, de Morais MG (2020) Microalgae starch: a promising raw material for the bioethanol production. Int J Biol Macromol

Daroch M, Geng S, Wang G (2013) Recent advances in liquid biofuel production from algal feedstocks. Appl Energy 102:1371–1381

Dave N, Selvaraj R, Varadavenkatesan T, Vinayagam R (2019) A critical review on production of bioethanol from macroalgal biomass. Algal Res 42:101606

Deschamps P, Haferkamp I, d'Hulst C, Neuhaus HE, Ball SG (2008) The relocation of starch metabolism to chloroplasts: when, why and how. Trends Plant Sci 13(11):574–582. https://doi.org/10.1016/j.tplants.2008.08.009

Dudley B (2018) BP statistical review of world energy. BP Statistical Review, London, UK

El-Mekkawi SA, Abdo SM, Samhan FA, Ali GH (2019) Optimization of some fermentation conditions for bioethanol production from microalgae using response surface method. Bulletin of the National Research Centre 43(1):1–8

El-Sayed WMM, Ibrahim HAH, Abdul-Raouf UM, El-Nagar MM (2016) Evaluation of bioethanol production from Ulva lactuca by Saccharomyces cerevisiae. J Biotechnol Biomater 6(226):2

Ghadiryanfar M, Rosentrater KA, Keyhani A, Omid M (2016) A review of macroalgae production, with potential applications in biofuels and bioenergy. Renew Sust Energ Rev 54:473–481

Halim R, Harun R, Webley PA, Danquah MK (2013) Bioprocess engineering aspects of biodiesel and bioethanol production from microalgae. In: Advanced Biofuels and Bioproducts. Springer, pp 601–628

Harun R, Danquah MK (2011) Influence of acid pre-treatment on microalgal biomass for bioethanol production. Process Biochem 46(1):304–309

Harun R, Danquah MK, Forde GM (2010) Microalgal biomass as a fermentation feedstock for bioethanol production. J Chem Technol Biotechnol 85(2):199–203

Harun R, Jason WSY, Cherrington T, Danquah MK (2011) Exploring alkaline pre-treatment of microalgal biomass for bioethanol production. Appl Energy 88(10):3464–3467

Harun R, Yip JW, Thiruvenkadam S, Ghani WA, Cherrington T, Danquah MK (2014) Algal biomass conversion to bioethanol–a step-by-step assessment. Biotechnol J 9(1):73–86

Heshof R, Visscher B, van de Zilver E, van de Vondervoort R, van Keulen F, Delahaije RJ, Wind RD (2020) Production of tailor-made enzymes to facilitate lipid extraction from the oleaginous yeast Schwanniomyces occidentalis. AMB Express 10(1):1–11

Ho S-H, Huang S-W, Chen C-Y, Hasunuma T, Kondo A, Chang J-S (2013) Bioethanol production using carbohydrate-rich microalgae biomass as feedstock. Bioresour Technol 135:191–198

Hoek C, Mann D, Jahns HM, Jahns M (1995) Algae: an introduction to phycology. Cambridge University Press

Hong Y, Chen C, Wu Y-R (2020) Biobutanol production from sulfuric acid-pretreated red algal biomass by a newly isolated clostridium sp. Strain WK. Biotechnology and Applied Biochemistry 67(5):738–743

Hossain N, Zaini J, Mahlia TMI (2019) Life cycle assessment, energy balance and sensitivity analysis of bioethanol production from microalgae in a tropical country. Renew Sust Energ Rev 115:109371

Ismail MM, Ismail GA, El-Sheekh MM (2020) Potential assessment of some micro-and macroalgal species for bioethanol and biodiesel production. Energy Sources, Part A: Recovery, Utilization, and Environmental Effects, 1–17

Jiang P, Qin S, Tseng CK (2003) Expression of the lacZ reporter gene in sporophytes of the seaweed Laminaria japonica (Phaeophyceae) by gametophyte-targeted transformation. Plant Cell Rep 21(12):1211–1216. https://doi.org/10.1007/s00299-003-0645-2

Jiang R, Ingle KN, Golberg A (2016) Macroalgae (seaweed) for liquid transportation biofuel production: what is next? Algal Res 14:48–57

Katiyar R, Arora A (2020) Health promoting functional lipids from microalgae pool: a review. Algal Res 46:101800

Khammee P, Ramaraj R, Whangchai N, Bhuyar P, Unpaprom Y (2021) The immobilization of yeast for fermentation of macroalgae Rhizoclonium sp. for efficient conversion into bioethanol. Biomass Conversion and Biorefinery 11:827–835

Khanal SK, Grewell D, Sung S, Van Leeuwen J (2007) Ultrasound applications in wastewater sludge pretreatment: a review. Crit Rev Environ Sci Technol 37(4):277–313

Khoo CG, Dasan YK, Lam MK, Lee KT (2019) Algae biorefinery: review on a broad spectrum of downstream processes and products. Bioresour Technol 292:121964

Kim N-J, Li H, Jung K, Chang HN, Lee PC (2011) Ethanol production from marine algal hydrolysates using Escherichia coli KO11. Bioresour Technol 102(16):7466–7469. https://doi.org/10.1016/j.biortech.2011.04.071

Koyande AK, Show P-L, Guo R, Tang B, Ogino C, Chang J-S (2019) Bio-processing of algal bio-refinery: a review on current advances and future perspectives. Bioengineered 10(1): 574–592

Kraan S (2013) Mass-cultivation of carbohydrate rich macroalgae, a possible solution for sustainable biofuel production. Mitig Adapt Strateg Glob Chang 18(1):27–46

Kumar P, Barrett DM, Delwiche MJ, Stroeve P (2009) Methods for pretreatment of lignocellulosic biomass for efficient hydrolysis and biofuel production. Ind Eng Chem Res 48(8):3713–3729

Kumar SV, Misquitta RW, Reddy VS, Rao BJ, Rajam MV (2004) Genetic transformation of the green alga—Chlamydomonas reinhardtii by agrobacterium tumefaciens. Plant Sci 166(3): 731–738. https://doi.org/10.1016/j.plantsci.2003.11.012

Kwon O-M, Kim D-H, Kim S-K, Jeong G-T (2016) Production of sugars from macro-algae Gracilaria verrucosa using combined process of citric acid-catalyzed pretreatment and enzymatic hydrolysis. Algal Res 13:293–297

Lakatos GE, Ranglová K, Manoel JC, Grivalský T, Kopecký J, Masojídek J (2019) Bioethanol production from microalgae polysaccharides. Folia Microbiol 64(5):627–644

Laurens LM, Markham J, Templeton DW, Christensen ED, Van Wychen S, Vadelius EW, Chen-Glasser M, Dong T, Davis R, Pienkos PT (2017) Development of algae biorefinery concepts for biofuels and bioproducts; a perspective on process-compatible products and their impact on cost-reduction. Energy Environ Sci 10(8):1716–1738

Lee HV, Hamid SBA, Zain SK (2014) Conversion of lignocellulosic biomass to nanocellulose: structure and chemical process. Sci World J 2014

Lee H-J, Kim S-J, Yoon J-J, Kim KH, Seo J-H, Park Y-C (2015) Evolutionary engineering of Saccharomyces cerevisiae for efficient conversion of red algal biosugars to bioethanol. Bioresour Technol 191:445–451

Lee OK, Lee EY (2016) Sustainable production of bioethanol from renewable brown algae biomass. Biomass Bioenergy 92:70–75. https://doi.org/10.1016/j.biombioe.2016.03.038

Lee S, Oh Y, Kim D, Kwon D, Lee C, Lee J (2011) Converting carbohydrates extracted from marine algae into ethanol using various ethanolic Escherichia coli strains. Appl Biochem Biotechnol 164(6):878–888

Maceiras R, Alfonsín V, Seguí L, González JF (2021) Microwave assisted alkaline pretreatment of algae waste in the production of cellulosic bioethanol. Energies 14(18):5891

Mahendran J, Saravanan K, Ragulnath D (2020) Performance and emission characteristics of algae derived biodiesel processes. Materials Today: Proceedings 21:268–271

Manzanares P, Ballesteros I, Negro MJ, Oliva JM, Gonzalez A, Ballesteros M (2012) Biological conversion of forage sorghum biomass to ethanol by steam explosion pretreatment and simultaneous hydrolysis and fermentation at high solid content. Biomass Conversion and Biorefinery 2(2):123–132

Martín-Sampedro R, Eugenio ME, García JC, Lopez F, Villar JC, Diaz MJ (2012) Steam explosion and enzymatic pre-treatments as an approach to improve the enzymatic hydrolysis of Eucalyptus globulus. Biomass Bioenergy 42:97–106

Mushlihah S, Husain DR, Langford A, Tassakka ACMA (2020) Fungal pretreatment as a sustainable and low cost option for bioethanol production from marine algae. J Clean Prod 265:121763

Nematian T, Salehi Z, Shakeri A (2020) Conversion of bio-oil extracted from Chlorella vulgaris micro algae to biodiesel via modified superparamagnetic nano-biocatalyst. Renew Energy 146: 1796–1804

Ngamsirisomsakul M, Reungsang A, Liao Q, Kongkeitkajorn MB (2019) Enhanced bio-ethanol production from chlorella sp. biomass by hydrothermal pretreatment and enzymatic hydrolysis. Renew Energy 141:482–492. https://doi.org/10.1016/j.renene.2019.04.008

Ning P, Yang G, Hu L, Sun J, Shi L, Zhou Y, Wang Z, Yang J (2021) Recent advances in the valorization of plant biomass. Biotechnol Biofuels 14(1):1–22

Ocreto JB, Chen W-H, Ubando AT, Park Y-K, Sharma AK, Ashokkumar V, Ok YS, Kwon EE, Rollon AP, De Luna MDG (2021) A critical review on second-and third-generation bioethanol production using microwaved-assisted heating (MAH) pretreatment. Renew Sust Energ Rev 152:111679

Okuda K, Oka K, Onda A, Kajiyoshi K, Hiraoka M, Yanagisawa K (2008) Hydrothermal fractional pretreatment of sea algae and its enhanced enzymatic hydrolysis. Journal of Chemical Technology & Biotechnology: International Research in Process, Environmental & Clean Technology 83(6):836–841

Onay M (2018) Bioethanol production from Nannochloropsis gaditana in municipal wastewater. Energy Procedia 153:253–257

Onay M (2019) Bioethanol production via different saccharification strategies from H. tetrachotoma ME03 grown at various concentrations of municipal wastewater in a flat-photobioreactor. Fuel 239:1315–1323. https://doi.org/10.1016/j.fuel.2018.11.126

Park J-H, Hong J-Y, Jang HC, Oh SG, Kim S-H, Yoon J-J, Kim YJ (2012) Use of Gelidium amansii as a promising resource for bioethanol: a practical approach for continuous dilute-acid hydrolysis and fermentation. Bioresour Technol 108:83–88

Pugliese A, Biondi L, Bartocci P, Fantozzi F (2020) Selenastrum capricornutum a new strain of algae for biodiesel production. Fermentation 6(2):46

Radakovits R, Jinkerson RE, Darzins A, Posewitz MC (2010) Genetic engineering of algae for enhanced biofuel production. Eukaryot Cell 9(4):486–501. https://doi.org/10.1128/EC.00364-09

Ramachandra TV, Hebbale D (2020) Bioethanol from macroalgae: prospects and challenges. Renew Sust Energ Rev 117:109479

Roesijadi G, Jones SB, Snowden-Swan LJ, Zhu Y (2010). Macroalgae as a biomass feedstock: a preliminary analysis. Pacific Northwest National Lab. (PNNL), Richland, WA (United States)

Rosenberg JN, Mathias A, Korth K, Betenbaugh MJ, Oyler GA (2011) Microalgal biomass production and carbon dioxide sequestration from an integrated ethanol biorefinery in Iowa: a technical appraisal and economic feasibility evaluation. Biomass Bioenergy 35(9):3865–3876

Sara M, Rouissi T, Brar SK, Blais JF (2016) Life cycle analysis of potential substrates of sustainable biorefinery. In: Platform chemical biorefinery. Elsevier, pp 55–76

Schenk PM, Thomas-Hall SR, Stephens E, Marx UC, Mussgnug JH, Posten C, Kruse O, Hankamer B (2008) Second generation biofuels: high-efficiency microalgae for biodiesel production. Bioenergy Res 1(1):20–43

Schultz-Jensen N, Thygesen A, Leipold F, Thomsen ST, Roslander C, Lilholt H, Bjerre AB (2013) Pretreatment of the macroalgae Chaetomorpha linum for the production of bioethanol–comparison of five pretreatment technologies. Bioresour Technol 140:36–42

Shokrkar H, Ebrahimi S, Zamani M (2017) Bioethanol production from acidic and enzymatic hydrolysates of mixed microalgae culture. Fuel 200:380–386. https://doi.org/10.1016/j.fuel.2017.03.090

Shokrkar H, Ebrahimi S, Zamani M (2018) Enzymatic hydrolysis of microalgal cellulose for bioethanol production, modeling and sensitivity analysis. Fuel 228:30–38

Shukla R, Kumar M, Chakraborty S, Gupta R, Kumar S, Sahoo D, Kuhad RC (2016) Process development for the production of bioethanol from waste algal biomass of Gracilaria verrucosa. Bioresour Technol 220:584–589

Singh K, Kaloni D, Gaur S, Kushwaha S, Mathur G (2017) Current research and perspectives on microalgae-derived biodiesel. Biofuels

Sommer W (2015) Modelling and monitoring of aquifer thermal energy storage: impacts of heterogeneity, thermal interference and bioremediation. Modelling and Monitoring of Aquifer Thermal Energy Storage: Impacts of Heterogeneity, Thermal Interference and Bioremediation. https://www.cabdirect.org/cabdirect/abstract/20153237389

Spiden EM, Scales PJ, Yap BHJ, Kentish SE, Hill DRA, Martin GJO (2015) The effects of acidic and thermal pretreatment on the mechanical rupture of two industrially relevant microalgae: chlorella sp. and Navicula sp. Algal Res 7:5–10. https://doi.org/10.1016/j.algal.2014.11.006

Sukwong P, Ra CH, Sunwoo IY, Tantratian S, Jeong G-T, Kim S-K (2018) Improved fermentation performance to produce bioethanol from Gelidium amansii using Pichia stipitis adapted to galactose. Bioprocess Biosyst Eng 41(7):953–960

Surendhiran D, Sirajunnisa AR (2019) Chapter 18—Role of genetic engineering in bioethanol production from algae. In: Ray RC, Ramachandran S (eds) Bioethanol production from food crops. Academic, pp 361–381. https://doi.org/10.1016/B978-0-12-813766-6.00018-7

Tabil L, Adapa P, Kashaninejad M (2011) Biomass feedstock pre-processing-part 1: pre-treatment. Biofuel's Engineering Process Technology 18:411–439

Taherzadeh MJ, Karimi K (2007) Enzymatic-based hydrolysis processes for ethanol from lignocellulosic materials: a review. Bioresources 2(4):707–738

Tan IS, Lee KT (2015) Solid acid catalysts pretreatment and enzymatic hydrolysis of macroalgae cellulosic residue for the production of bioethanol. Carbohydr Polym 124:311–321

Tang DYY, Yew GY, Koyande AK, Chew KW, Vo D-VN, Show PL (2020) Green technology for the industrial production of biofuels and bioproducts from microalgae: a review. Environ Chem Lett:1–19

Tomás-Pejó E, Alvira P, Ballesteros M, Negro MJ (2011) Pretreatment technologies for lignocellulose-to-bioethanol conversion. In: Biofuels. Elsevier, pp 149–176

Trivedi N, Gupta V, Reddy CRK, Jha B (2013) Enzymatic hydrolysis and production of bioethanol from common macrophytic green alga Ulva fasciata Delile. Bioresour Technol 150:106–112

Ungureanu N, Vladut V, Biris S-S (2020) Capitalization of wastewater-grown algae in bioethanol production. Engineering for Rural Development: Jelgava, Latvia, 1859–1864

van der Wal H, Sperber BL, Houweling-Tan B, Bakker RR, Brandenburg W, López-Contreras AM (2013) Production of acetone, butanol, and ethanol from biomass of the green seaweed Ulva lactuca. Bioresour Technol 128:431–437

Velazquez-Lucio, J., Rodríguez-Jasso, R. M., Colla, L. M., Sáenz-Galindo, A., Cervantes-Cisneros, D. E., Aguilar, C. N., Fernandes, B. D., & Ruiz, H. A. (2018). Microalgal biomass pretreatment for bioethanol production: a review

Wang H, Ji C, Bi S, Zhou P, Chen L, Liu T (2014) Joint production of biodiesel and bioethanol from filamentous oleaginous microalgae Tribonema sp. Bioresour Technol 172:169–173

Wang J-X, Cao J-P, Zhao X-Y, Liu S-N, Ren X-Y, Zhao M, Cui X, Chen Q, Wei X-Y (2019) Enhancement of light aromatics from catalytic fast pyrolysis of cellulose over bifunctional hierarchical HZSM-5 modified by hydrogen fluoride and nickel/hydrogen fluoride. Bioresour Technol 278:116–123

Yazdani P, Zamani A, Karimi K, Taherzadeh MJ (2015) Characterization of Nizimuddinia zanardini macroalgae biomass composition and its potential for biofuel production. Bioresour Technol 176:196–202

Yeon J-H, Seo H-B, Oh S-H, Choi W-S, Kang D-H, Lee H-Y, Jung K-H (2010) Bioethanol production from hydrolysate of seaweed Sargassum sagamianum. KSBB Journal 25(3): 283–288

Yoon M, Choi J, Lee J-W, Park D-H (2012) Improvement of saccharification process for bioethanol production from Undaria sp. by gamma irradiation. Radiat Phys Chem 81(8):999–1002

Yu T-S, Kofler H, Häusler RE, Hille D, Flügge U-I, Zeeman SC, Smith AM, Kossmann J, Lloyd J, Ritte G, Steup M, Lue W-L, Chen J, Weber A (2001) The Arabidopsis sex1 mutant is defective in the R1 protein, a general regulator of starch degradation in plants, and not in the chloroplast hexose transporter. Plant Cell 13(8):1907–1918. https://doi.org/10.1105/TPC.010091

Zabed H, Sahu JN, Boyce AN, Faruq G (2016) Fuel ethanol production from lignocellulosic biomass: an overview on feedstocks and technological approaches. Renew Sust Energ Rev 66:751–774

Zachleder V, Brányiková I (2014) Starch overproduction by means of algae. In: Algal biorefineries. Springer, pp 217–240

Chapter 14
Microalgae and Macroalgae for Third-Generation Bioethanol Production

Ibham Veza, Anh Tuan Hoang, Muhammad Mujtaba Abbas, Noreffendy Tamaldin, Muhammad Idris, Djati Wibowo Djamari, Ahmed Sule, Eka Maulana, Nicky Rahmana Putra, and A. C. Opia

Abstract First-generation bioethanol, made from edible feedstocks, are currently not regarded as a sustainable source due to the food versus fuel dilemma. Second-generation bioethanol, despite being made from non-edible sources, are not cost-effective owing to their high production cost. To avoid the drawbacks of its predecessor, third-generation bioethanol from microalgae and macroalgae have

---

I. Veza (✉) · N. Tamaldin (✉)
Faculty of Mechanical Engineering, Universiti Teknikal Malaysia Melaka, Durian Tunggal, Melaka, Malaysia
e-mail: noreffendy@utem.edu.my

A. T. Hoang
Institute of Engineering, HUTECH University, Ho Chi Minh City, Vietnam

M. M. Abbas
Department of Mechanical Engineering, University of Engineering and Technology (New Campus), Lahore, Pakistan

M. Idris
PT PLN (Persero), Engineering and Technology Division, Jakarta, Indonesia

D. W. Djamari
Mechanical Engineering Study Program, Sampoerna University, South Jakarta, Indonesia

A. Sule
Faculty of Engineering, Automotive Development Centre, School of Mechanical Engineering, Universiti Teknologi Malaysia, Johor Bahru, Malaysia

Mechanical Technology Education Section, Technical Education Department, Kogi State College of Education, Ankpa, Nigeria

E. Maulana
Department of Mechanical Engineering, Universitas Pancasila, Jakarta, Indonesia

N. R. Putra
Centre of Lipid Engineering and Applied Research (CLEAR), Ibnu Sina Institute for Scientific and Industrial Research, Johor Bahru, Malaysia

A. C. Opia
Department of Marine Engineering, Niger Delta University, Amassama, Nigeria

C. R. Soccol et al. (eds.), *Liquid Biofuels: Bioethanol*, Biofuel and Biorefinery Technologies 12, https://doi.org/10.1007/978-3-031-01241-9_14

been considered a promising replacement to depleting petroleum fuels. In this chapter, the progress of research on micro and macroalgae for third-generation bioethanol production is discussed. This chapter thoroughly explains the use of microalgae and macroalgae for bioethanol production, starting from strains selection, cultivation, harvesting and drying, to hydrolysis, fermentation and distillation. To become a competitive source for bioethanol, the production of microalgae should be cheap and highly efficient. Therefore, each of the above processes should be improved and optimised. At the end of this chapter, the future direction on bioethanol production from microalgae and macroalgae is highlighted.

14.1 Introduction

The major transportation fuels presently in use are predominantly from non-renewable fossil resources (Katijan et al. 2019; Veza et al. 2019b, 2020a). Biofuels have drawn considerable attention as a promising resource of future energy due to their renewability and comparable fuel properties to conventional petrol fuel (Veza et al. 2020b, 2021d). Today, the most technologically viable and commercialised biofuels available in the global market are bioethanol (Prasad et al. 2019, 2020; Malode et al. 2021) and biodiesel (Mohammed et al. 2021; Rusli et al. 2021; Veza et al. 2021a). Both biofuels are currently being supplied in increasing quantities, but their mass-production is not environmentally sustainable. Shahid et al. (2021) provided the projected data on the worldwide production-consumption of both bioethanol and biodiesel (Fig. 14.1). Note that biodiesel is more suitable to replace diesel engines, while alcohol fuels such as bioethanol and biobutanol are preferable for gasoline engines (Veza et al. 2019a, 2020c; 2021b; Roslan et al. 2020).

Bioethanol can be made from three different raw materials: (i) first-generation bioethanol (starch-based crops); (ii) second-generation bioethanol (lignocellulosic substrates such as agricultural waste); and (iii) third-generation bioethanol (microalgae). First-generation (1G) bioethanol uses agricultural land to grow crops. In the second-generation (2G) bioethanol, a lignocellulosic substrate is employed, which often needs a pre-treatment owing to the high lignin content (Rajak and Banerjee 2020; Mankar et al. 2021; Devi et al. 2021). The pre-treatment process is time-consuming and contributes to the increase in the production cost (Onumaegbu et al. 2018; Sharma et al. 2020; Pandey et al. 2020; Veza et al. 2021c). In the third-generation (3G) bioethanol, photosynthetic microorganisms like microalgae are used.

Algae have been extensively examined as potential feedstocks for biofuel production. They have relatively simple cellular structures, higher growth rates and lower lignin content (Zabed et al. 2019; Sankaran et al. 2020). Other intrinsic benefits include faster and uninterrupted production as opposed to regular crops, high lipid content, being carbon neutral due to their atmospheric CO_2 sequestering ability, and they can be cultivated in brackish waters (Sudhakar and Viswanaathan 2019; Özçimen et al. 2020a).

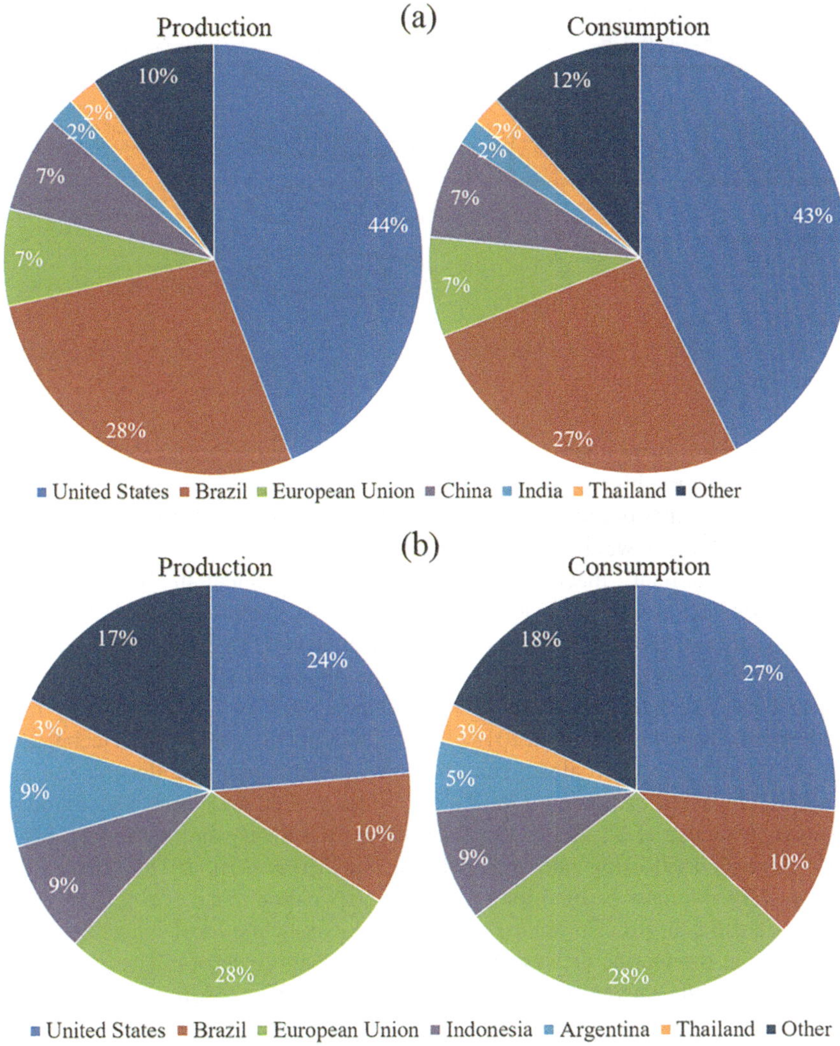

Fig. 14.1 Forecast of bioethanol (**a**) and biodiesel (**b**) global production-consumption in 2025 (Shahid et al. 2021)

Algae have been industrially produced for centuries. A great number of commercial microalgae production processes have been established to produce pharmaceutical supplies, food supplements and bioenergy (Bhalamurugan et al. 2018; Bhattacharya and Goswami 2020). Bioenergy from microalgae is produced as liquid bioethanol (Rizza et al. 2017; de Farias Silva et al. 2019), solid biochar (Lee et al. 2020; Bolognesi et al. 2021), or syngas (Hong et al. 2017; Adnan et al. 2020). However, the commercial feasibility of microalgal biofuels continues to be uncertain. Major enhancements are required to produce more economical microalgal

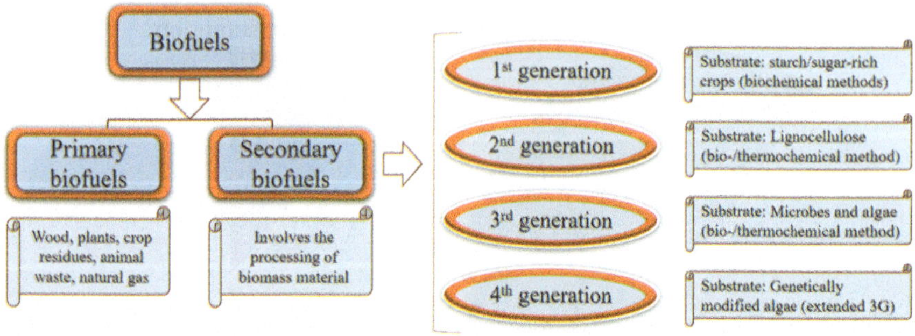

Fig. 14.2 Classification of biofuels according to feedstock and conversion method (Shahid et al. 2021)

biofuels. These include improvement in strain selection, cultivation techniques, and harvesting. Despite this, microalgae and cyanobacteria remain as promising feedstocks for bioethanol production. This is because their carbohydrate content can be as many as 50% (dry weight) (Möllers et al. 2014).

Equally important, macroalgae have comparable potential. Seaweeds, for instance, are rich in carbohydrate (20%–50%) without lignin content, thus making their polysaccharides easy to be hydrolysed and fermented (Yanagisawa et al. 2011). Also, after the utilisation of starch for bioethanol, the algae cake residue is protein-rich and has the potential to produce co-products such as fish and poultry feed supplements (Kawai and Murata 2016). To harvest algal biomass, several techniques can be applied. For microalgae, centrifugation, precipitation and flocculation are normally employed (Abomohra et al. 2017) along with sonication and filtration flotation (Milledge et al. 2014). For macroalgae, machinery and harvesting boats are commonly used using rake or trawler methods (Bak et al. 2018). After being harvested, the biomass is dried for commercial application (Saad et al. 2019).

Microalgae can grow in almost all water habitats, while most macroalgae can only grow in the marine ecosystem (Hanifzadeh et al. 2018a, b). The benefits of macroalgae are that they possess comparable features to plants (Alfonsín et al. 2019). The harvesting process is therefore more straightforward (Sudhakar et al. 2018). There are various methods to produce bioethanol from biochemical and thermochemical processes. Bioethanol production from macroalgae is a promising biofuel route owing to their high carbohydrate content. However, hydrolysis followed by fermentation using bacteria or yeast is required (Sudhakar et al. 2018).

It is critical to understand the position of algae in the classification of biofuels. Figure 14.2 shows the grouping of biofuels according to the substrate and conversion technique. Fourth-generation biofuel is the expansion of 3G biofuel which encompasses the use of sophisticated biological methods (Kumar et al. 2020b; Zhou et al. 2020). Bioethanol from genetically modified algae, for example, belongs to this category. The frequently used methods for algal genetic modification includes light penetration improvement, photosynthetic efficiency enhancement and photoinhibition reduction (Tandon and Jin 2017). Although metabolic engineering could drastically improve the carbohydrates content of algae (Abdullah et al. 2019), the present book chapter only focuses on the progress of 3G bioethanol production from micro and macroalgae.

14.2 Selection of Algal Species

The selection of algal strains for maximum biofuel production requires numerous considerations. For biodiesel production, lipids and lipid composition are critically important, whereas starch and carbohydrate composition are the two most significant factors for bioethanol production. To decrease the costs of the overall process, it is essential to determine if the microorganism could produce by-products that can offer added value. Certain algal species are therefore selected due to the high quantity of their desired components. Note that the species selection should consider advantages at different process stages (both upstream and downstream) to make it competitive.

In general, the selected algal species must have a high level of photosynthetic efficiency, a high CO_2 consumption rate, capability to be cultivated in various types of water, and a rapid growth rate, in addition to being flexible and robust to tolerate diverse environmental circumstances. Strain isolation to produce particular compounds can be performed from its natural habitat. More recently, genetic manipulation is carried out to improve photosynthesis ability and carbon dioxide utilisation, in addition to enhancing the accumulation of the desired component at greater concentrations, as well as the capability to grow in extreme conditions. Nevertheless, genetic manipulation falls under the category of 4G bioethanol and still has numerous challenges for large-scale commercial applications.

A number of published works have reported production of bioethanol from different algal species to produce high carbohydrates. Due to microalgae's low lignin and hemicellulose content compared to lignocellulosic biomass, they have been regarded as preferable for bioethanol production. To produce high polysaccharides levels e.g., starch and cellulose, microalgae require light, nutrients, and CO_2. By means of hydrolysis, fermentation and distillation, the polysaccharides can then be converted into bioethanol.

The selection of microalgal species relies on the preferred final product. For instance, *Dunaliella salina* was selected to produce β-carotene, while *Haematococcus pluvialis* was chosen to obtain astaxanthin (Rammuni et al. 2019). *Chlorella protothecoides* is more suitable to produce biodiesel owing to high lipid content (Al-lwayzy and Yusaf 2017; Yan et al. 2019). In contrast, higher carbohydrate content is preferred for bioethanol production. *Chlorella*, *Dunaliella* and *Scenedesmus*, for example, contain over 50% carbohydrate content (Özçimen et al. 2020b). Rizza et al. (2017) carried out screening of seventeen carbohydrate-rich microalgal strains in South America and found a novel strain known as SP2–3 that could accumulate as much as 70% (w/w) carbohydrates under environments of nitrogen deficiency with constant temperature and light. *Desmodesmus* sp. was also selected due to its fermentable sugars productivity (Rizza et al. 2017).

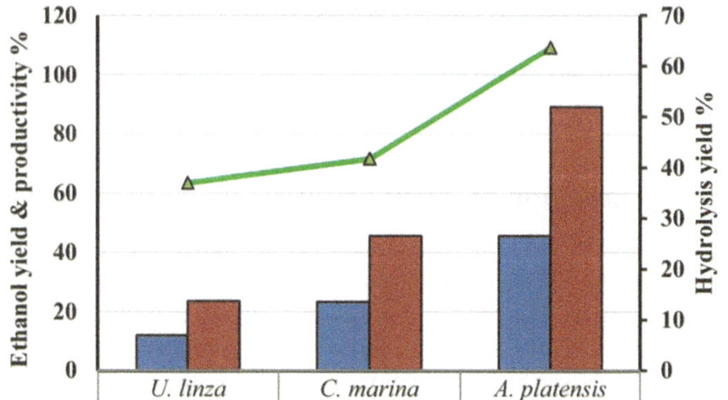

	U. linza	*C. marina*	*A. platensis*
Ethanol yeild %	*12.01*	*23.24*	*45.49*
Ethanol productivity %	*23.5*	*45.49*	*89.02*
Hydrolysis Yeild %	*37.04*	*41.79*	*63.68*

Fig. 14.3 Ethanol yield, productivity and hydrolysis yield of carbohydrate-rich algal species biomass (Ismail et al. 2020)

Macroalgae have low lipid and high carbohydrate content (Nhat et al. 2018). Like microalgae, macroalgae such as seaweed do not require freshwater (Xu et al. 2014). Apart from being used as a food, various strains of seaweed have been utilised for bioethanol production. Macroalgae can be categorised into three different groups according to their photosynthetic pigmentation differences: (i) red (*Rhodophyta*); (ii) brown (*Phaeophyta*); and (iii) green (*Chlorophyta*) (Chen et al. 2015). Cultivation of macroalgae is dependent on a number of aspects, such as macroalgal species, the season of collection, and the carbohydrate conversion through fermentation.

Ismail et al. (2020) screened and compared different micro- and macroalgae strains. Five macroalgae, four microalgae and seven blue-green algae were isolated from various environments. To differentiate the carbohydrate-rich strains, the sugar yield was evaluated. Sulphuric acid (3%) yielded the highest level of reducing sugars. For each strain, the reducing sugars yield relies on the hydrolysis technique instead of on the total sugars quantity. After 48 hours being fermented, *Ulva linza, Chlorella marina* and *Arthrospira platensis* showed promising results with 12.01%, 23.24% and 45.49% of ethanol yield respectively (Fig. 14.3) and were recommended for large-scale cultivation for bioethanol production.

Normally, bioethanol produced from macroalgae are predominantly from the phycocolloid seaweed. The use of green seaweed is inadequately investigated (El Harchi et al. 2018). Green seaweeds, especially *Ulva* species, are regarded as adaptable seaweeds (Teichberg et al. 2010). They feature in daily diets owing to high their of nutritious value (Bobin-Dubigeon et al. 1997). In fact, *Ulva* spp. have been traditionally consumed as food in Japan known as "Aonori" (Nisizawa et al. 1987). *Ulva* species have also attracted numerous interests for farming (Fakihi Kachkach et al. 2014).

Ulva species represent a substantially complex polysaccharides source (Hernández-Garibay et al. 2011). Since the majority of sugars are not easily available, the pre-treatment process is therefore required. Acid hydrolysis is extensively utilised for fermentable sugars release (Mosier et al. 2005). To prevent the drawbacks of acid pre-treatment, yeast adaptation has been suggested for bioethanol fermentation improvement. Adaptation is more viable and environmentally friendly with fewer negative effects and energy needs. Also, in-situ detoxification will eliminate the requirement for a separate detoxification phase.

El Harchi et al. (2018) examined the potential of the green seaweed *Ulva rigida* using pre-treatment that was optimised through thermal acid hydrolysis in order to achieve a high level yield of sugars. Using the non-conventional yeast *Pachysolen tannophilus* for hydrolysate fermentation, it was found that the ethanol from dry seaweed was produced at 0.12 g/g and 0.09 g/g for the adapted and non-adapted yeast. Maximum ethanol yield reached 0.37 g/g or equal to almost 75% efficiency. Overall, acid hydrolysis was found to be effective with mild conditions. Biological adaptation could increase bioethanol production from *Ulva* hydrolysate.

Macroalgal biomass containing mannitol are also promising sources to produce bioethanol. Chades et al. (2018) investigated the mannitol extract-fermentation of brown macroalgae by thermophilic *Clostridia*. Screening of the type strains was performed on 20 mM mannitol. The results showed that 11 of the examined 41 strains could use mannitol to produce ethanol.

To conclude, the selection of microalgae strains relies on the preferred final product. Microorganism for bioethanol production relies on a number of factors. These include pH range, temperature, osmotic tolerance, alcohol tolerance, inhibitors resistance, genetic stability and growth rate. Temperature is considered the most important aspect that influences ethanol production.

14.3 Microalgae for 3G Bioethanol Production

The production of bioethanol comprises five major steps: (i) cultivation; (ii) harvesting and drying; (iii) hydrolysis; (iv) fermentation; and (v) distillation. Note that in some studies, enzymatic hydrolysis is considered as pre-treatment, while enzymatic saccharification is categorised as a separate and extra stage after the pre-treatments. The entire process of microalgae bioethanol production is simplified in Fig. 14.4.

14.3.1 Cultivation

Cultivation is the main method to produce microalgal biomass. It is a crucial process as high carbohydrate is compulsory for cost-effective bioethanol production. The cultivation process can be conducted in two systems: open and closed.

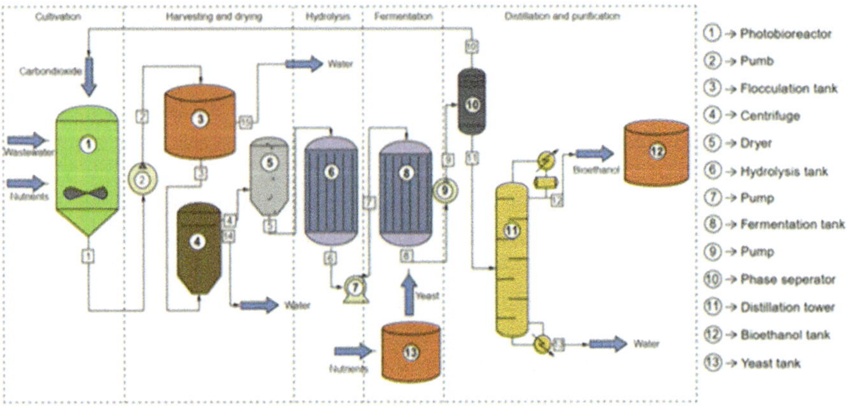

Fig. 14.4 Overview of bioethanol production from microalgae (Özçimen et al. 2020b)

Currently, the industrial cultivation of microalgae is predominantly performed in the open system owing to their affordable structure and low investment costs. Yet, it is difficult to control its operational conditions, resulting in water evaporation, external contaminations, inefficient microalgal cells exposure to sunlight and carbon dioxide as a result of poor mixing, thus producing low yields of biomass. Despite having free and abundant sunlight in the outdoor system, cyclical differences in sunlight can considerably restrict productivity. In contrast, the utilisation of closed systems tends to have better control of the culture, as well as its environment. A closed system also has a larger surface to volume ratios, relatively lower water evaporation, improved isolation as well as greater productivity.

Microalgae can be grown in numerous types of water like salt, fresh, waste, or brackish water and with or without a source of organic carbon. Based on their carbon metabolism, they can be categorised into three different groups: (i) phototrophic (Ho et al. 2018); (ii) heterotrophic; (Morales-Sánchez et al. 2017); and (iii) mixotrophic (Patel et al. 2020). Microalgae from the phototrophic group consume light as the energy source and carbon dioxide from the atmosphere or flue gases as an inorganic carbon source. In contrast, heterotrophic microalgae are independent of light. The mixotrophic microalgal type can grow either photo- or heterotrophically. Thus far, only phototrophic microalgae are technologically and economically feasible for large scale microalgae cultivation, preferably using seawater. This is because high salinity could prevent the culture media contamination, enabling seawater to be utilised directly, as opposed to using freshwater.

While recent technologies may enhance the economical aspect of microalgae biofuels, significant economic drawbacks can be overcome for the time being by integrating wastewater treatment with bioethanol production for supplementary environmental and economic advantages. Compared to traditional wastewater techniques, microalgal wastewater treatment is a relatively inexpensive, unsophisticated process having half the energy consumption of typical mechanical methods. Water and nutrients are the major costs in microalgal biofuel production, but both can be supplied by wastewater.

Wastewater has great potential for algal cultivation. Wastewater contains abundant carbon, nitrogen and phosphorous, which can lead to eutrophication, thus increasing algae production (Ramachandra et al. 2013). These elements are needed to produce carbohydrates for bioethanol production (He et al. 2013). Currently, the wastewater cultivation of microalgae is considered the most-effective method for biofuel production (Cheah et al. 2016). Also, the utilisation of microalgae for wastewater treatment and recycling has drawn interest owing to their capability of CO_2 fixation.

For macroalgae, cultivation plays an important environmental role as opposed to harvesting wild populations. The majority of land areas have been already exploited for terrestrial plant agriculture. Yet, the oceans, covering over 70% of the earth, has great potential. Macroalgae are a rarely exploited resource in Europe and North America, despite their promise for numerous uses like food, cosmetics, animal feed, bioactive components, bioethanol and biodiesel (Marinho et al. 2015; van den Burg et al. 2016; Kraan 2016; Bruhn et al. 2016; Fernand et al. 2017; Charrier et al. 2017). Directed by a market demand (Marinho et al. 2015; van den Burg et al. 2016; Buschmann et al. 2017) and environmental concerns of macroalgal wild harvest (Jensen 1993; Troell et al. 1999; Titlyanov and Titlyanova 2010), the cultivation of macroalgae has started attracting considerable attention.

A feasibility study carried by Zuniga-Jara et al. (2016) found that the cultivation of offshore commercial kelp was unprofitable since the investment and production costs could not be covered by the sale of biomass. To reduce operation costs, van den Burg et al. (2016) emphasised the significance of a cultivation that could facilitate multiple partial harvests. Also, biofouling appears to be an important factor in Europe (Handå et al. 2013; Marinho et al. 2015), and it seems to be combined with sheltered locations, thus stopping the utilisation of multiple partial harvesting (McNeary and Erickson 2013; Bruhn et al. 2016). For that reason, offshore cultivation can be the solution for a cost-effective macroalgal industry (Fei 2004; Sulaiman et al. 2013; Bruhn et al. 2016).

Cultivation in offshore is described as the site's activities that are dependent on ocean waves, which lack shelter from topographical features, thus mitigating the ocean force and waves caused by the wind with considerable wave heights as high as two meters or beyond (Bak et al. 2018). Producing offshore macroalgae is promising due to its sustainability and market potential but is challenging (Zhang et al. 2012; Kraan 2016). Some cultivation trials for macroalgae have been carried out in the Atlantic Ocean, utilising *S. latissima* (Buck and Buchholz 2005; Peteiro and Freire 2009; Mols-Mortensen et al. 2017). Yet, none of such cultivation trials has led to large-scale cost-effective cultivation (van den Burg et al. 2016). Bak et al. (2018) raised the possibility of large-scale cultivation of offshore kelp, but further development is still required to lower the production cost.

The major objective of cultivation is to obtain high biomass production with preferred qualities. Controlling the growth of microalgae is a complex process owing to the influence of photosynthesis, carbon dioxide concentration and various

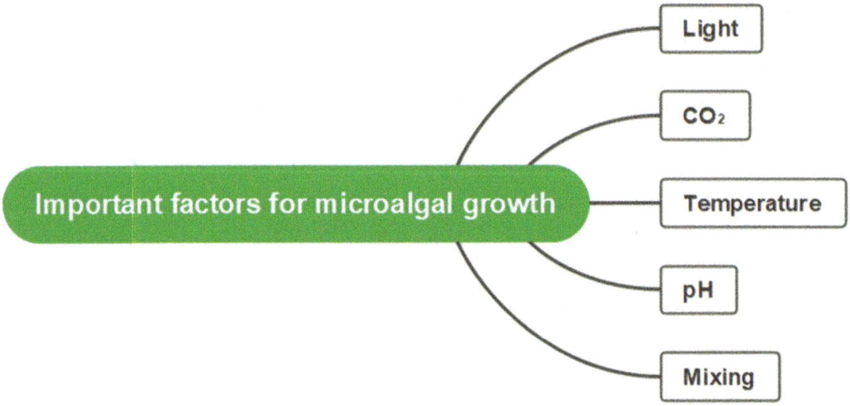

Fig. 14.5 Parameters to be considered for microalgal growth

ecological conditions. Therefore, information on growth requirements such as temperature, light and photoautotrophic organism nutrients have critical roles in obtaining the highest possible biomass output in the cultivation step. A number of important parameters should be considered for microalgal growth as shown in Fig. 14.5. These include temperature, light, pH, nutrient availability, and culture mixing.

14.3.1.1 Light

For growth, microalgae need light to create lipids, carbohydrates, and proteins from CO_2 and H_2O. Thanks to its unsophisticated morphology, metabolism and development, microalgae can achieve greater photosynthetic efficiency compared to terrestrial plants. Natural light sources are free, but the light intensity has a substantial role in microalgal photosynthesis because exposure to extended periods of high light intensity can lead to photoinhibition, which could result in free radical formation, thus triggering photo-oxidative damage. For that reason, the efficiency of light use should be determined by saturation light intensity. Sunlight between 400 and 700 nm is considered photosynthetic active radiation (Zhen and Bugbee 2020), which makes up 40–50% of total sunlight.

To improve the efficient distribution of light in the photobioreactor, a number of techniques can be employed. These include: enhanced gas exchange (carbon dioxide supply and oxygen removal) (Sforza et al. 2018); nutrients uniform distribution (Yan et al. 2020); and optimising the reactor size (Deprá et al. 2019). However, the high costs of a photobioreactor are usually a constraint, thus limiting its application only to laboratory applications. Another significant factor is the influence of the diurnal cycle for outdoor cultivation (Jin et al. 2020), which have outstanding results on the solar energy capture overall efficiency. Nevertheless, it may be restricted by accessible sunlight owing to diurnal cycles, as well as seasonal disparities, thus restricting the feasibility of commercial production to high-level solar radiation regions.

Photo-inhibition and low-level light stress of photosynthesis affects biomass production. Consequently, almost half of the biomass from daytime production could be lost throughout the following night. To overcome this problem, artificial light using fluorescent lamps are normally employed for large-scale purposes, which enables uninterrupted production (Ramanna et al. 2017; Brzychczyk et al. 2020; Erbland et al. 2020). Still, such method needs a higher level of energy input. Light limitation owing to high volume to surface ratios may lead to biomass productivity reduction. For that reason, better light supply using reduced layer thickness or thin layer photobioreactors with optimised culture mixing could be employed. Recently, light-emitting diodes (LEDs) were recommended as a substitute for fluorescent lamps to illuminate photobioreactors to increase microalgal productivity (Wishkerman and Wishkerman 2017; Schulze et al. 2017; Ajayan et al. 2019; Jung et al. 2019).

14.3.1.2 Carbon Dioxide (CO_2)

Carbon is an essential nutrient for microalgae to grow; almost half of the microalgae biomass consists of carbon. Microalgae can be utilised to sequester CO_2 from numerous sources, such as the atmosphere, organic compounds, inorganic carbonates and flue gas from power plants and the industrial sector. Several microalgae strains can be cultivated in dark environments and utilise organic carbons as energy and carbon sources, but the use of organic carbons is expensive for the growth of microalgae. Therefore, a more affordable CO_2 source for photosynthesis is required. Atmospheric CO_2, which is only around 0.03% (Castellote et al. 2009), is not enough to supply the rates of microalgal growth for full-scale bioethanol production, thus making it economically non-viable. Fossil fuel combustion is believed to be the greatest source of CO_2 throughout the globe. The use of CO_2 from flue gas is a promising method to yield greater biomass productivities owing to its higher CO_2 concentration (Collotta et al. 2018; Kothari et al. 2019). It is crucial to remember that a large supply of CO_2 requires sufficient light for efficient microalgae growth.

14.3.1.3 Temperature

Another restricting factor in microalgae cultivation is temperature as it affects oxygen evolution and production efficiency (Ras et al. 2013). The optimal growth temperature for most microalgae strains is within the range of 20-25 °C (Ras et al. 2013). A small number of microalgae can withstand temperatures below 16 °C, but this will slow the growth rates and decrease biomass production (Malcata et al. 2018). Likewise, temperatures above 35 °C are avoided (Andersen and Andersen 2006). However, several cyanobacteria like *Arthrospira* sp. can be cultivated optimally at temperatures between 30 and 38 °C (Vonshak and Tomaselli 2002; Ogbonda et al. 2007).

14.3.1.4 pH

The majority of microalgal strains are sensitive to changes in pH. Controlling pH is therefore important to obtain high growth rates. As microalgae can capture CO_2, the pH tends to increase. Most microalgae generally grow faster at pH values from 8.0 to 9.0 (Bartley et al. 2014). However, some strains such as *Spirulina* can withstand and thrive at lower or higher pH values (Schenk et al. 2008).

14.3.1.5 Mixing

Mixing is a critical aspect that: (i) facilitates gas exchange in microalgal cultures; (ii) uniformly distributes CO_2; (iii) avoids cell sedimentation; (iv) optimises light and nutrients exposure; and (v) assists in excess oxygen removal. A high amount of dissolved oxygen is toxic to the microalgal cells, and non-stop light exposure can generate oxygen radicals that can result in cell damage and growth inhibition (Malcata et al. 2018).

14.3.2 Harvesting and Drying

Harvesting is defined as the concentration of diluted algae to a thick paste, which may correspond to 20–30% of the entire production costs (Mata et al. 2010). It is an essential stage in bioethanol production from microalgae. Efficient and affordable harvesting methods are considerably important. Microalgae with high dry weight need effective harvesting techniques. One or a combination of harvesting approaches can be used. Also, physical, biological and chemical processes can be added to obtain the preferred separation efficiency.

Harvesting microalgae involves two major steps. The first stage is microalgal separation. Flocculation and flotation are normally employed for this process. The second stage is thickening of the algal slurry, where centrifugation and filtration techniques are typically utilised. The harvesting approach differs in accordance with microalgal properties like density and size.

Flocculation is a common method employed in bulk harvesting for great quantities of algal suspensions. To avoid suspension self-agglomeration, the microalgae are negatively charged. With the addition of flocculants to the suspension, the negatively-charged cells then become neutralised and are precipitated. The flocculants should be affordable, renewable, manageable, and accessible in small dosages. They can be inorganic or organic. Inorganic flocculants such as metal salts lead to a high inorganic material concentration of the biomass, resulting in problems in the subsequent stages. Therefore, natural polymeric materials like chitosan are often employed to prevent such problems. However, the flocculation cost increases with the use of natural polymeric materials.

Microalgal biomass is more easily separated from large amounts of water by flocculation and sedimentation than by filtration. The use of flocculation and sedimentation is cost effective owing to reduced power consumption. Cells of microalgae carry a negative charge that inhibits cells aggregation in suspension. These agents are widely used in harvesting processes as they increase the effective particle size and facilitate aggregation, which is a step before other harvesting techniques.

Centrifugation is another frequently used technique for harvesting microalgae since no extra chemicals are required. Nevertheless, the major drawbacks of such method are high energy costs and maintenance owing to freely moving parts. The centrifugation is ideal for microalgae with a thick cell wall and a diameter above 5 μm, but this technique is not appropriate for large-scale purposes owing to the increased energy consumption. Centrifugation is also not the preferred choice for bulk harvesting, but it is suitable for thickening.

Large cells or filamentous microalgae are able to solve harvesting problem, as they can be easily harvested using filtration. The cells or filaments of approximately 200 μm in size are more suitable for cultivation and harvesting than smaller microalgae (0.5–30 μm) (Pereira et al. 2016). Two-stage harvesting with dewatering and thickening can be relatively expensive, especially if an extra drying step is required (Barros et al. 2015). Settling is a suitable harvesting method for strains with large cells/filaments like *Tetraselmis* sp. CTP4 to minimise the cost of biomass dewatering (Pereira et al. 2016). Since there is no universal harvesting method, research and development are being conducted to obtain the appropriate processes for particular microalgae.

After being harvested, the algae paste will be sent for the drying process before the hydrolysis step can be started. Harvested microalgae biomass comprise around 70%–90% water. Therefore, the drying process is required to make the process more efficient. The most frequently employed methods for water removal from microalgae biomass include solar drying and fluidised bed drying. Solar drying is considered the most effective drying method for large-scale purposes.

14.3.3 Hydrolysis

Carbohydrates available in algae are predominantly starch and cellulose with no lignin, making them easier to hydrolyse to monosaccharides than lignocellulosic feedstocks (Ho et al. 2013). However, the carbohydrates found in algae are not directly fermentable for bioethanol production, therefore pre-treatment processes are required (Hahn-Hägerdal et al. 2007; Gírio et al. 2010). The pre-treatment process have room for improvement to convert biomass into bioethanol (Talebnia et al. 2010).

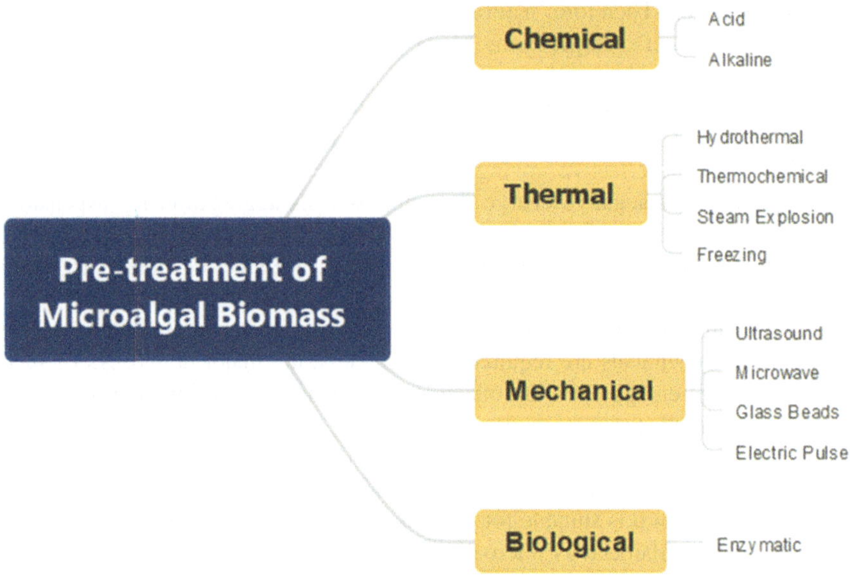

Fig. 14.6 Microalgal biomass pre-treatments for cell disruption

Pre-treatment is a critical process that allows biomass accessibility to enzymes for monosaccharides release. Thus far, pre-treatment is considered one of the most expensive processes for biomass conversion into bioethanol and biohydrogen. Typically, the pre-treatment cost contributes up to 30% of the total cost in the biomass conversion process (Shokrkar et al. 2017). Several physicochemical techniques have been examined for the microalgal biomass pre-treatment. Nevertheless, most of the methods are energy intensive in addition to the issue of by-products formation.

Microalgal biomass requires a pre-treatment process to disrupt algal cell walls to release carbohydrates. After the treatment process, it is also necessary to hydrolyse polysaccharides to release monosaccharides for further utilisation in the fermentation process (Hernández et al. 2015). As a consequence of their simpler cell structure and the absence of lignin, pre-treatment and saccharification can be applied to microalgal biomass simultaneously.

Various pre-treatments can be utilised for cell wall disruption and polysaccharides liberation before hydrolysis. The pre-treatment efficiency is greatly influenced by biomass type and composition. Therefore, the optimum method must be selected for the highest sugar yields with minimum degradation and operational costs. A number of pre-treatment techniques for microalgae are available including chemical, thermal, mechanical and biological pre-treatment (Velazquez-Lucio et al. 2018) as shown in Fig. 14.6.

Pre-treatment methods for microalgae can be grouped into chemical, thermal, mechanical and biological methods. Chemical pre-treatment is applied to the microalgal biomass with acids (HCl, H_2SO_4, HNO_3) and bases (KOH, NaOH,

Na_2CO_3). It is amongst the most frequently used method to disrupt the cell wall or membrane for releasing simple sugars. Cell disruption using acid pre-treatment is accompanied by sugar polymers hydrolysis in the cell walls of microalgae. Chemical pre-treatments are fast, inexpensive and easy. Yet, they need extreme operating conditions like high pressure, temperature and acid concentration, which leads to carbohydrate structure degradation and toxic compound formation. These by-product structures can decrease ethanol yield by yeast metabolism inhibition in the fermentation stage. Therefore, the optimisation of chemical pre-treatment is critical.

Another well-known method is biological pre-treatment such as enzymatic degradation. It provides several advantages over chemical methods. Enzymatic pre-treatment does not generate toxic by-products with higher hydrolysis conversion ratios. Also, extreme operating conditions in physicochemical pre-treatment like high pressure and temperature are not compulsory, thus reducing operating cost. Enzymatic pre-treatment does not require specific enzymes to break down the structures. In addition to chemical and biological pre-treatment, mechanical techniques have also been extensively used for different microalgal species. Despite being easy to scale-up and having a high-level of disruption efficiency, heat used during physical pre-treatment can harm fragile bioactive compounds.

Each of the above pre-treatment techniques has its own benefits and drawbacks. Chemical pre-treatment is effective for polysaccharides degradation, but the resulting sugar yields are relatively low. Furthermore, the acid or base concentration requires optimisation to prevent the formation of inhibitors (Prajapati et al. 2015). Mechanical pre-treatment is more straightforward to operate but also suffers from low sugar yields. Mechanical pre-treatment techniques were employed to disrupt the cellular of algae biomass (Postma et al. 2015). Freezing and thawing techniques produce ice crystals that can break the algal cell walls to release water soluble intracellular compounds (Yang et al. 2015). Ultrasound with low frequency along with various solvents was utilised for extraction of carbohydrates, proteins, lipids and pigments (Ferreira et al. 2016). Ultrasound treatment enhanced enzymatic hydrolysis of the carbohydrates to produce bioethanol by modifying the crystalline regions (Zheng et al. 2013). Microwave pre-treatment could perform 80% of cell lysis (Ali and Watson 2016). The enzymatic hydrolysis has no or low inhibitors formation, but it requires a longer reaction time in addition to the high enzymes cost. Hence, it is important to find an efficient pre-treatment technique to obtain maximum sugar extraction.

Extracting maximum sugars from deoiled algal biomass could be done using different pre-treatment approaches followed by sugar-rich hydrolysate utilization (Kumar et al. 2020a). It was found that the hybrid pre-treatment technique led to increased sugar solubilisation compared to the individual physico-chemical and enzymatic techniques. The use of sugars from hybrid pre-treatment with *Saccharomyces cerevisiae* at pH 5.5 led to maximum bioethanol production. Hydrothermal or liquid hot water pre-treatment is also interesting approach. In this technique, water temperatures should be higher than 100 °C under high pressure with steady pressure being released following the pre-treatment process (Chen and Oswald 1998). Such high-temperature water could lead to cell wall structure disruption (Yuan et al. 2016). Ngamsirisomsakul et al. (2019) found that the acid-hydrothermal was an effective technique for *Chlorella* sp. to produce bioethanol. The production of bioethanol could be additionally improved by hydrolysing the crude slurry from pre-treatment using amylase enzymes. The pre-treated biomass hydrolysis by glucoamylase was able to improve bioethanol production by two-fold.

Thus far, bioethanol production using enzymatic hydrolysis of microalgae using thermostable enzymes is difficult to find (Choi et al. 2010; Shokrkar et al. 2017). The enzymatic hydrolysis utilising thermostable enzymes has benefits in decreasing the requirement for initial acid pre-treatment. Although several studies have examined the pure culture of macroalgae (John et al. 2011; Wargacki et al. 2012; Kumar et al. 2013; Golberg et al. 2014; Yazdani et al. 2015) and microalgae (Harun et al. 2010; Harun and Danquah 2011; Ho et al. 2013), the use of algae mixed cultures needs further investigation. Microalgae cultivation in pure culture will lead to higher operating costs owing to the requirement for sterile conditions. Hence, the application of mixed microalgal culture could overcome such a problem (Hassanpour et al. 2015; Shokrkar et al. 2017). The use of mixed culture is a desirable solution for microalgae in order to dominate contamination risk and improve economic feasibility. Mooij et al. (2013) proposed the term "survival of the fattest", an approach for species enrichment with high storage productivity in the culture of mixed microalgae. Afterwards, a gravimetric enrichment technique could be used to screen lipid- and carbohydrate-accumulating species (Hassanpour et al. 2015).

The extraction of the sugar in a mixed microalgae culture through various pre-treatment could be performed by means of thermostable enzymes (Fig. 14.7) (Shokrkar et al. 2017). The bioethanol yields from different pre-treatment methods were compared using *Saccharomyces cerevisiae* yeast. It was found that the combination of dilute sulfuric acid and $MgSO_4$ showed an increased sugar yield compared to dilute acid. Of all the tested pre-treatments, the enzymatic method using thermostable enzymes exhibited the highest recovery. Furthermore, the wet microalgae enzymatic pre-treatment was compared with dried microalgae under similar conditions and concentrations of dried biomass (50 g/l). It was reported that comparable sugar yields were obtained, showing the potential to decrease the requirement for the drying process. The enzymatic and acidic pre-treatments exhibited yields of 0.46 and 0.38 g/g glucose, equivalent to 92% and 76% of the theoretical values. This finding indicated that bioethanol yield from enzymatic hydrolysis was relatively higher compared to acid hydrolysis.

Hydrolysis of carbohydrates plays a critical role in bioethanol production. To yield higher reducing sugars, it is essential to develop a cost-effective hydrolysis process. Hydrolysis of algal carbohydrates can be performed utilising acids or enzymes. Enzymatic hydrolysis offers a number of benefits as opposed to acid hydrolysis, such as low utility consumption; fewer corrosion issues; increased glucose yield with no sugar degradation; and inhibitory generation (Shokrkar et al. 2018). However, enzymatic hydrolysis may increase the cost of bioethanol production (Hamelinck et al. 2005; Kumar et al. 2015). Shokrkar et al. (2018) investigated the enzymatic hydrolysis of microalgal cellulose to produce bioethanol using

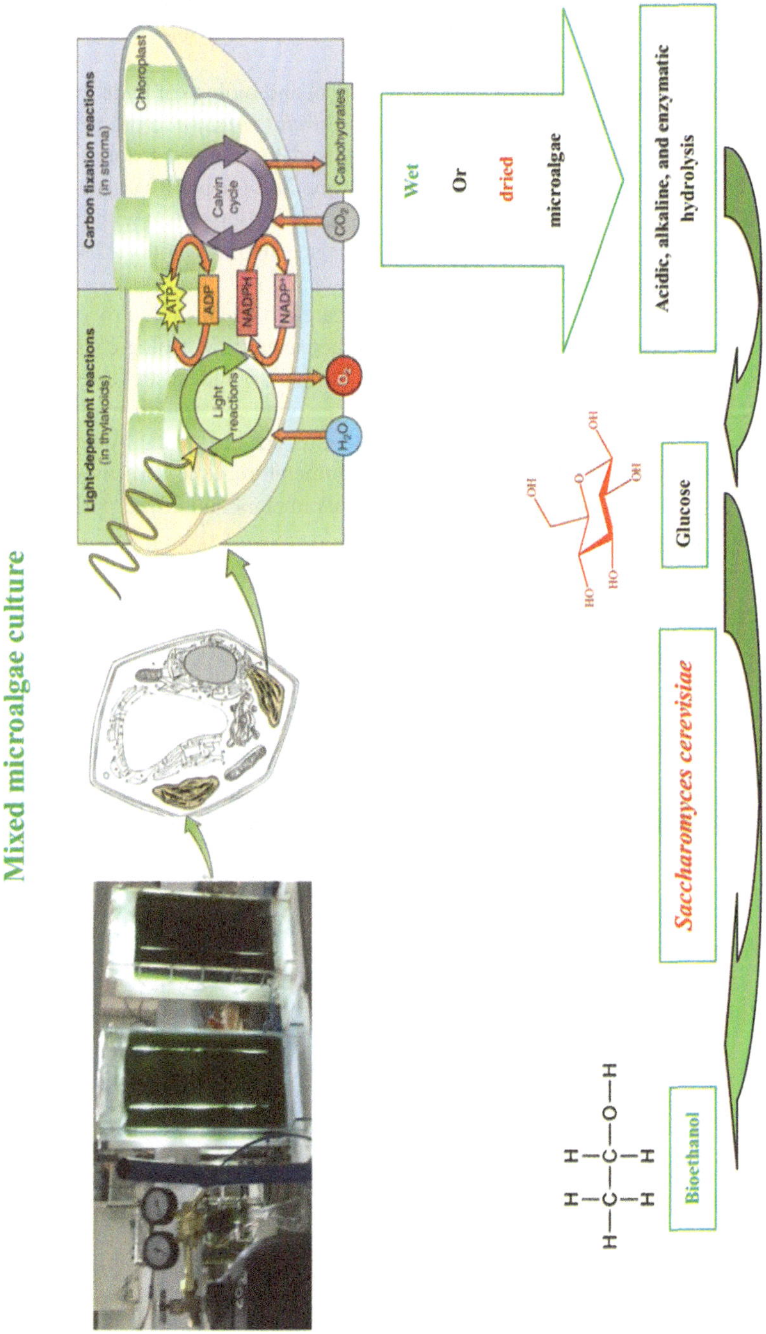

Fig. 14.7 Sugars extraction in a mixed microalgae culture through various pre-treatment methods (Shokrkar et al. 2017)

modelling and sensitivity analysis. They developed a kinetic model by taking into account several factors such as product inhibition, pH and temperature. It was found that the highest glucose yield of almost 60% was obtained at 50 g/l biomass concentration.

Hamouda et al. (2018) investigated the chemical and biological hydrolysis of the macroalgae *Ulva fasciata* and the microalgae *Chlorella vulgaris* to pre-treatment of cell wall and sugar production. *U. fasciata* showed the highest amount of sugar. Chemical hydrolysis was also found as the best method for pre-treatment of algae. Furthermore, *S. cerevisiae* SH02 was reported as the microorganism with the highest ethanol yield at 40%, while that of *Pseudomonas* sp. SH03 was only 22%, attained by means of 5% algal sugar fermentation.

Enzymatic hydrolysis has more benefits compared to chemical hydrolysis. These include mild operation (lower energy needs), higher selectivity, biological specificity (greater conversion yields-reduced by-product formation) and easier scale-up. Nevertheless, it also has some disadvantages, such as enzyme cost and difficult recovery, making it economically impractical. The effectiveness of enzymatic hydrolysis was mainly subject to operational parameters (pH, temperature, time, enzymes type and concentration). The optimisation for different parameters should be performed to achieve the highest yields with minimum costs.

14.3.4 Fermentation

The next major stage in bioethanol production is fermentation. In this stage, bacteria or yeast converts 6-carbon sugars into ethanol. Fermentation mainly comprises the monomeric sugars conversion from the previous stages into alcohols. The highest possible ethanol amount that can be produced from 1 g of glucose is 0:511 g, as can be seen in the stoichiometry reaction presented below.

$$C_6H_{12}O_6 \rightarrow 2CH_3CH_2OH + 2CO_2$$

$$\text{glucose} \rightarrow \text{ethanol} + \text{carbon dioxide}$$

$$1\ \text{g} \rightarrow 0.511\ \text{g} + 0.489\ \text{g}$$

For bioethanol production using yeast, *Saccharomyces cerevisiae* is among the most frequently employed microorganisms (El-Mekkawi et al. 2019; Cripwell et al. 2020). It can tolerate high amounts of ethanol and inhibitory ingredients from the pre-treatment process. Other advantages of *S. cerevisiae* are that it has high osmotic resistance, growth at low pH and high production efficiency. Another commonly used microorganism is *Zymomonas mobilis* (Onay 2019). It has fast growth and is highly efficient but cannot tolerate the presence of many phenolic compounds.

For bioethanol production from microalgae, the fermentation stage comprises two common techniques: separate hydrolysis and fermentation (SHF) and simultaneous saccharification and fermentation (SSF). To start the fermentation process, *Saccharomyces cerevisiae* is added to the fermenter. The released CO_2 is then transferred to the microalgae cultivation system. Sugars that cannot be transformed into bioethanol are kept in a reservoir for another process. The released process water is also stored for future use.

A few studies have been performed investigating the fermentation of macroalgae, but suitable techniques for efficient saccharification with subsequent fermentation has not yet been obtained (Sudhakar et al. 2017). Physical and/or chemical treatments are not considered effective as they might consume relatively higher energy and generate unwanted products that can hamper the yeast growth (Tsuji et al. 2013). Hence, an exclusive and effective saccharification technique for each macroalgal biomass should be developed to obtain effective hydrolysis. Also, unlike sugar hydrolysates fermentation of terrestrial biomass, the fermentation of macroalgal hydrolysate needs salinity tolerant yeast (Khambhaty et al. 2013).

Sudhakar et al. (2017) developed a method from the spent seaweeds biomass via saccharification. The use of marine yeast aimed to achieve zero waste discharge for bioethanol production (Fig. 14.8). It was found that the total carbohydrate was reported at the highest in the fresh seaweeds, while the production of reducing sugar was reported at the highest in the industrial spent samples. The maximum bioethanol from spent biomass was produced by *Meyerozyma guilliermondii* AY17 KJ754141. The spent residues produced after pigment extraction were potential biomass sources for bioethanol production.

14.3.5 Distillation

The last step in the production of algal bioethanol is the recovery or distillation process to separate ethanol from the mixture. This is because ethanol from the fermentation process still consists of water and ethanol, thus a separation process is required. Firstly, the fermentation broth should be filtered for solid removal before the filtrate can be distilled (Hasin et al. 2021). The water–bioethanol mixture is separated according to the difference in boiling point. Water has a boiling point of 100 °C, which is higher compared to that of ethanol (78 °C), thus the ethanol vaporises earlier than water.

After fermentation, bioethanol needs to be separated from the broth. The algal fermentation broths may also contain high quantities of salts like sodium chloride. Air Gap Membrane Distillation (AGMD) can be used to extract bioethanol from such complex compounds (Loulergue et al. 2019). The SEM analysis as presented in Fig. 14.9a showed that the pristine membrane fibrous structure could not be seen anymore. Rather, a dense cake layer was seen on the membrane surface (Fig. 14.9b

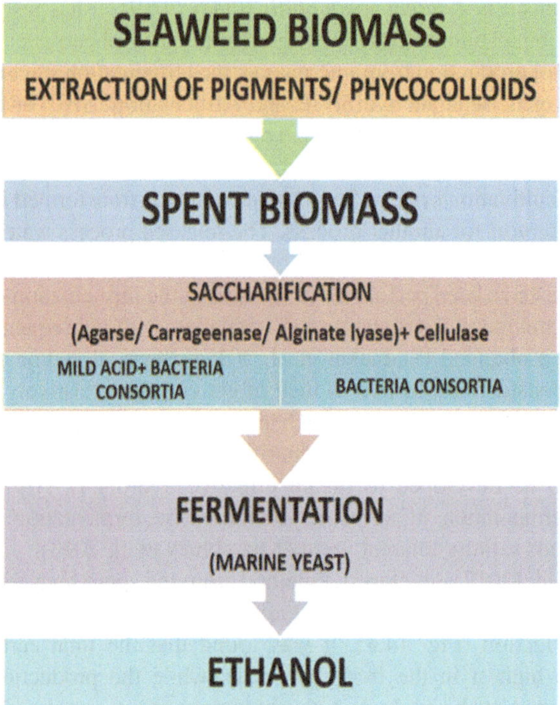

Fig. 14.8 Biosaccharification and bioethanol production from spent macroalgae biomass (Sudhakar et al. 2017)

and c) with its thickness was around a few tens of micrometres (Fig. 14.9d). AGMD could achieve a bioethanol-enriched permeate from macroalgae, proving that such method is suitable for bioethanol extraction.

14.4 Future Outlook

Overall, bioethanol production from algae comprises five major steps: cultivation; harvesting and drying; hydrolysis; fermentation; and distillation. Algae have great potential for bioethanol production, but further investigations should be carried out, particularly in the cultivation and pre-treatment steps. Furthermore, numerous limitations related to harvesting, drying and extraction have also hindered the commercial production of algal bioethanol. Hence, innovative technologies need to be fostered to produce large-scale and sustainable bioethanol from microalgae.

Regarding the cultivation step, accumulating more carbohydrates in algae can be performed by optimising growth conditions as well as exposing algae to light and temperature. The most difficult stage in the bioethanol production is the

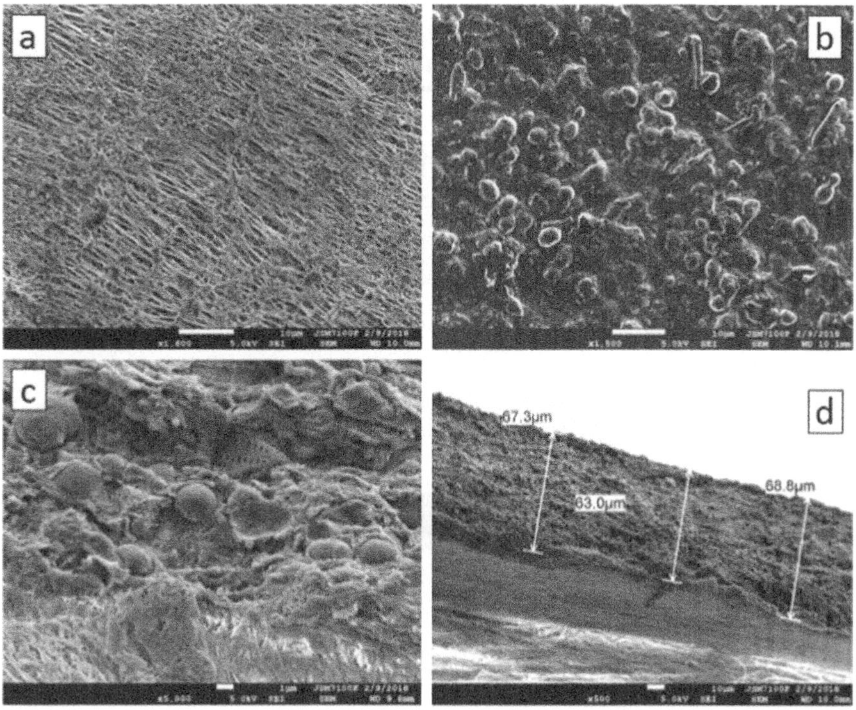

Fig. 14.9 SEM of the membrane surface (**a**) pristine (×1600), (**b**) fouled (×1500), (**c**) fouled (×5000) and (**d**) cross-section of fouled membrane (×500) (Loulergue et al. 2019)

pre-treatment, where polysaccharides are broken down into smaller monomers, making them fermentable. Thus far, current pre-treatment methods are not efficient enough and need further improvement to reduce the cost and increase their performance. Genetic modifications can also be done to improve the production rates of carbohydrates or starch accumulation, but this is the focus of 4G bioethanol production.

It is also important to screen new algae strains for higher ethanol yield. The ideal algae species should be able to develop and grow at high-level biomass concentration and severe circumstances to produce a high-value final product. The precise number of algal species is not known. It is estimated between several hundred thousand and several million unique species—with other additional types being recognised from time to time. Only several thousand microalgal species can grow in cultures, and merely a few of them have been successfully cultivated for commercial purposes.

The utilisation of micro and macroalgae for 3G bioethanol production can be a promising solution for the increasing world energy demand by solving the inherent problems associated with previous biofuels generations. To maximise resource recovery with economical and environmental benefits, microalgal biomass biorefining can be performed in a circular loop. This could be done by integrating

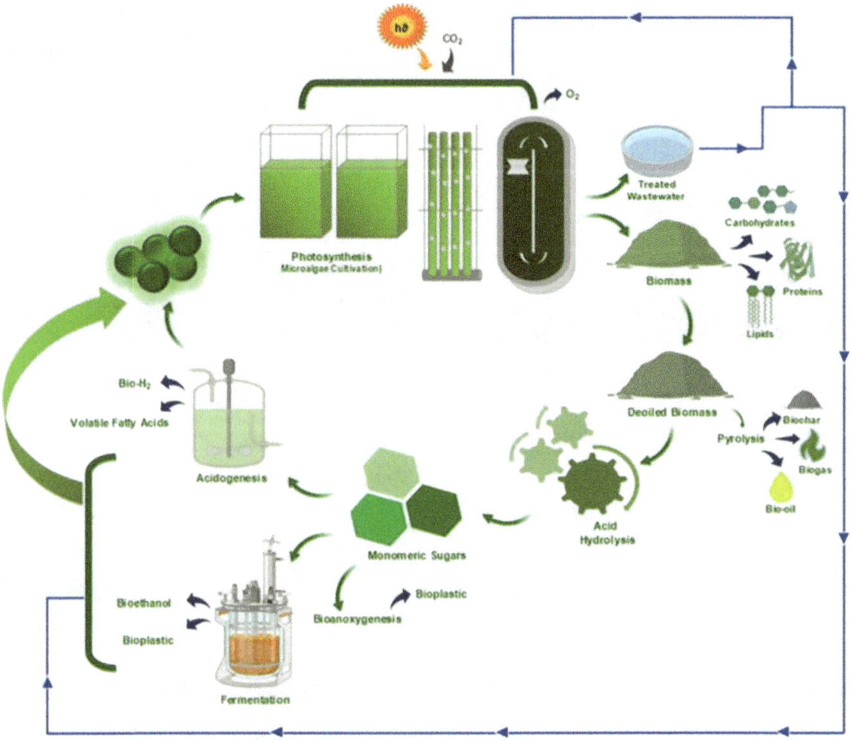

Fig. 14.10 A model of an algal biorefinery with a self-sustainable closed-loop strategy (Mohan et al. 2020)

biorefinery to generate multi-based products as proposed (Fig. 14.10) by Mohan et al. (2020).

Integrating microalgae-based bioethanol production to the sugarcane biorefinery has attracted recent attention. The microalgae growth integration to first and second generation bioethanol sugarcane biorefinery was investigated by means of computer simulation (Albarelli et al. 2018). The CO_2 amount that was employed as a solvent to the supercritical fluid extraction was the major component affecting the feasibility of the economic process. However, with microalgae pre-treatment by co-solvent extraction or cell disruption, the amount of CO_2 could be reduced, thus increasing the process yields. The utilisation of a co-solvent was found to increase lipids and carotenoids extraction by 1.4 and 2.4 times, respectively. Also, a lower investment was achieved compared to microalgal extraction with no cell disruption.

As mentioned above, the integration between two different generations of biofuels is rarely investigated in the literature. One interesting study was carried out by Li et al. (2021) who proposed an innovative industrial symbiosis design by integrating the 2G and 3G biofuels as demonstrated in Fig. 14.11. The design involved three major stakeholders: cellulosic ethanol production plant; microalgae

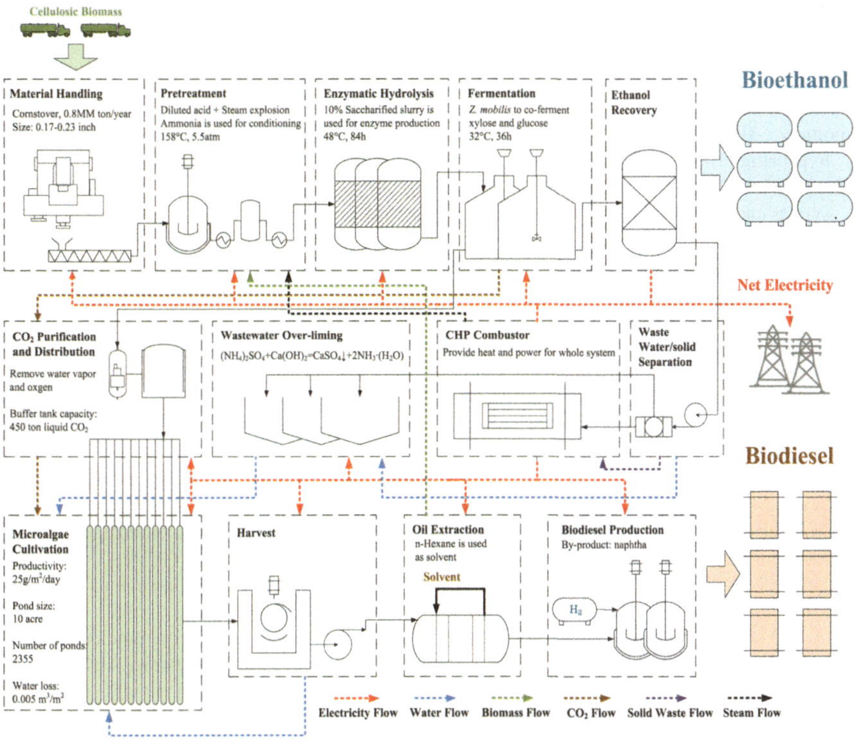

Fig. 14.11 The production system of industrial symbiosis bioenergy proposed by Li et al. (2021)

biodiesel production plant; and utility system. To produce bioethanol and provide carbon dioxide for algal cultivation, the cellulosic ethanol production plant; used cellulosic as well as recycled lipid biomass from microalgal biodiesel production. The wastewater from cellulosic ethanol production was sent to an over-liming treatment to remove sulphate and utilised as backup water and a nitrogen supply for microalgae. A large amount of wastewater was produced; thus, this recycling approach could significantly improve the use efficiency of water resources. Also, the waste solid from cellulosic ethanol production and microalgae biodiesel production could be combusted for power and heat generation to run the entire system.

Micro and macroalgae for bioethanol production should not only involve novel methods to increase the carbohydrate content but also innovative techniques to increase biomass productivity. Furthermore, detailed techno-economic analysis is required to validate the advantages and disadvantages of energy involved in bioethanol production. It is also interesting to examine by-products resulted from algal production. Liquid by-product, for instance, can be valorised for biogas or for animal feed. Despite the challenges, micro and macroalgae have the potential to become the source of future bioethanol production. More detailed investigations are, however, required.

References

Abdullah B, SaFaS M, Shokravi Z et al (2019) Fourth generation biofuel: a review on risks and mitigation strategies. Renew Sust Energ Rev 107:37–50

Abomohra AE-F, El-Sheekh M, Hanelt D (2017) Screening of marine microalgae isolated from the hypersaline Bardawil lagoon for biodiesel feedstock. Renew Energy 101:1266–1272

Adnan MA, Xiong Q, Muraza O et al (2020) Gasification of wet microalgae to produce H_2-rich syngas and electricity: a thermodynamic study considering exergy analysis. Renew Energy 147: 2195–2205

Ajayan K, Harilal C, Gani P (2019) Performance of reflector coated LED bio-box on the augmentation of growth and lipid production in aerophytic trebouxiophyceaen algae *Coccomyxa* sp. Algal Res 38:101401

Al-Lwayzy SH, Yusaf T (2017) Diesel engine performance and exhaust gas emissions using microalgae *Chlorella protothecoides* biodiesel. Renew Energy 101:690–701

Albarelli JQ, Santos DT, Ensinas AV et al (2018) Product diversification in the sugarcane biorefinery through algae growth and supercritical CO_2 extraction: thermal and economic analysis. Renew Energy 129:776–785

Alfonsín V, Maceiras R, Gutiérrez C (2019) Bioethanol production from industrial algae waste. Waste Manag 87:791–797

Ali M, Watson IA (2016) Microwave thermolysis and lipid recovery from dried microalgae powder for biodiesel production. Energ Technol 4:319–330

Andersen T, Andersen FØ (2006) Effects of CO_2 concentration on growth of filamentous algae and *Littorella uniflora* in a Danish softwater lake. Aquat Bot 84:267–271

Bak UG, Mols-Mortensen A, Gregersen O (2018) Production method and cost of commercial-scale offshore cultivation of kelp in the Faroe Islands using multiple partial harvesting. Algal Res 33: 36–47

Barros AI, Gonçalves AL, Simões M et al (2015) Harvesting techniques applied to microalgae: a review. Renew Sust Energ Rev 41:1489–1500

Bartley ML, Boeing WJ, Dungan BN et al (2014) pH effects on growth and lipid accumulation of the biofuel microalgae *Nannochloropsis Salina* and invading organisms. J Appl Phycol 26: 1431–1437

Bhalamurugan GL, Valerie O, Mark L (2018) Valuable bioproducts obtained from microalgal biomass and their commercial applications: a review. Environmental Engineering Research 23: 229–241

Bhattacharya M, Goswami S (2020) Microalgae–a green multi-product biorefinery for future industrial prospects. Biocatal Agric Biotechnol 25:101580

Bobin-Dubigeon C, Lahaye M, Barry JL (1997) Human colonic bacterial degradability of dietary fibres from sea-lettuce (*Ulva* sp). J Sci Food Agric 73:149–159

Bolognesi S, Bernardi G, Callegari A et al (2021) Biochar production from sewage sludge and microalgae mixtures: properties, sustainability and possible role in circular economy. Biomass Conversion and Biorefinery 11:289–299

Bruhn A, Tørring DB, Thomsen M et al (2016) Impact of environmental conditions on biomass yield, quality, and bio-mitigation capacity of *Saccharina latissima*. Aquac Environ Interact 8: 619–636

Brzychczyk B, Hebda T, Pedryc N (2020) The influence of artificial lighting systems on the cultivation of algae: the example of *Chlorella vulgaris*. Energies 13:5994

Buck BH, Buchholz CM (2005) Response of offshore cultivated *Laminaria saccharina* to hydrodynamic forcing in the North Sea. Aquaculture 250:674–691

Buschmann AH, Camus C, Infante J et al (2017) Seaweed production: overview of the global state of exploitation, farming and emerging research activity. Eur J Phycol 52:391–406

Castellote M, Fernandez L, Andrade C et al (2009) Chemical changes and phase analysis of OPC pastes carbonated at different CO_2 concentrations. Mater Struct 42:515–525

Chades T, Scully SM, Ingvadottir EM et al (2018) Fermentation of mannitol extracts from brown macro algae by *thermophilic clostridia*. Front Microbiol 9:1931

Charrier B, Abreu MH, Araujo R et al (2017) Furthering knowledge of seaweed growth and development to facilitate sustainable aquaculture. New Phytol 216:967–975

Cheah WY, Ling TC, Show PL et al (2016) Cultivation in wastewaters for energy: a microalgae platform. Appl Energy 179:609–625

Chen H, Zhou D, Luo G et al (2015) Macroalgae for biofuels production: progress and perspectives. Renew Sust Energ Rev 47:427–437

Chen PH, Oswald WJ (1998) Thermochemical treatment for algal fermentation. Environ Int 24: 889–897

Choi SP, Nguyen MT, Sim SJ (2010) Enzymatic pretreatment of *Chlamydomonas reinhardtii* biomass for ethanol production. Bioresour Technol 101:5330–5336

Collotta M, Champagne P, Mabee W et al (2018) Wastewater and waste CO_2 for sustainable biofuels from microalgae. Algal Res 29:12–21

Cripwell RA, Favaro L, Viljoen-Bloom M et al (2020) Consolidated bioprocessing of raw starch to ethanol by *Saccharomyces cerevisiae*: achievements and challenges. Biotechnol Adv 42: 107579

De Farias Silva CE, Barbera E, Bertucco A (2019) Biorefinery as a promising approach to promote ethanol industry from microalgae and cyanobacteria. Bioethanol production from food crops. Elsevier, pp 343–359

Deprá MC, Mérida LG, De Menezes CR et al (2019) A new hybrid photobioreactor design for microalgae culture. Chem Eng Res Des 144:1–10

Devi A, Singh A, Bajar S et al (2021) Ethanol from lignocellulosic biomass: an in-depth analysis of pre-treatment methods, fermentation approaches and detoxification processes. J Environ Chem Eng 105798

El-Mekkawi SA, Abdo SM, Samhan FA et al (2019) Optimization of some fermentation conditions for bioethanol production from microalgae using response surface method. Bulletin of the National Research Centre 43:1–8

El Harchi M, Fakihi Kachkach FZ, El Mtili N (2018) Optimization of thermal acid hydrolysis for bioethanol production from *Ulva rigida* with yeast *Pachysolen tannophilus*. S Afr J Bot 115: 161–169

Erbland P, Caron S, Peterson M et al (2020) Design and performance of a low-cost, automated, large-scale photobioreactor for microalgae production. Aquac Eng 90:102103

Fakihi Kachkach F, El Harchi M, El Mtili N (2014) In vitro effect of *Ulva rigida* extract on the growth of *Lepidium sativum* and *Allium cepa*. Maroccan Journal of Biology 11:26–31

Fei X (2004) Solving the coastal eutrophication problem by large scale seaweed cultivation. Asian Pacific phycology in the 21st century: prospects and challenges. Springer, pp 145–151

Fernand F, Israel A, Skjermo J et al (2017) Offshore macroalgae biomass for bioenergy production: environmental aspects, technological achievements and challenges. Renew Sust Energ Rev 75: 35–45

Ferreira AF, Dias APS, Silva CM et al (2016) Effect of low frequency ultrasound on microalgae solvent extraction: analysis of products, energy consumption and emissions. Algal Res 14:9–16

Gírio FM, Fonseca C, Carvalheiro F et al (2010) Hemicelluloses for fuel ethanol: a review. Bioresour Technol 101:4775–4800

Golberg A, Vitkin E, Linshiz G et al (2014) Proposed design of distributed macroalgal biorefineries: thermodynamics, bioconversion technology, and sustainability implications for developing economies. Biofuels Bioprod Biorefin 8:67–82

Hahn-Hägerdal B, Karhumaa K, Fonseca C et al (2007) Towards industrial pentose-fermenting yeast strains. Appl Microbiol Biotechnol 74:937–953

Hamelinck CN, Van Hooijdonk G, Faaij AP (2005) Ethanol from lignocellulosic biomass: techno-economic performance in short-, middle-and long-term. Biomass Bioenergy 28:384–410

Hamouda RA, Sherif SA, Ghareeb MM (2018) Bioethanol production by various hydrolysis and fermentation processes with micro and macro green algae. Waste and Biomass Valorization 9: 1495–1501

Handå A, Forbord S, Wang X et al (2013) Seasonal-and depth-dependent growth of cultivated kelp (*Saccharina latissima*) in close proximity to salmon (*Salmo salar*) aquaculture in Norway. Aquaculture 414:191–201

Hanifzadeh M, Garcia EC, Viamajala S (2018a) Production of lipid and carbohydrate from microalgae without compromising biomass productivities: role of ca and mg. Renew Energy 127:989–997

Hanifzadeh M, Sarrafzadeh M-H, Nabati Z et al (2018b) Technical, economic and energy assessment of an alternative strategy for mass production of biomass and lipid from microalgae. J Environ Chem Eng 6:866–873

Harun R, Danquah MK (2011) Influence of acid pre-treatment on microalgal biomass for bioethanol production. Process Biochem 46:304–309

Harun R, Danquah MK, Forde GM (2010) Microalgal biomass as a fermentation feedstock for bioethanol production. J Chem Technol Biotechnol 85:199–203

Hasin M, Gohain M, Deka D (2021) Bio-ethanol production from carbohydrate-rich microalgal biomass: Scenedesmus obliquus. Singapore, Springer Singapore

Hassanpour M, Abbasabadi M, Ebrahimi S et al (2015) Gravimetric enrichment of high lipid and starch accumulating microalgae. Bioresour Technol 196:17–21

He P, Mao B, Shen C et al (2013) Cultivation of *Chlorella vulgaris* on wastewater containing high levels of ammonia for biodiesel production. Bioresour Technol 129:177–181

Hernández-Garibay E, Zertuche-González JA, Pacheco-Ruíz I (2011) Isolation and chemical characterization of algal polysaccharides from the green seaweed *Ulva clathrata (Roth) C. Agardh*. J Appl Phycol 23:537–542

Hernández D, Riaño B, Coca M et al (2015) Saccharification of carbohydrates in microalgal biomass by physical, chemical and enzymatic pre-treatments as a previous step for bioethanol production. Chem Eng J 262:939–945

Ho S-H, Huang S-W, Chen C-Y et al (2013) Bioethanol production using carbohydrate-rich microalgae biomass as feedstock. Bioresour Technol 135:191–198

Ho S-H, Nagarajan D, Ren N-Q et al (2018) Waste biorefineries—integrating anaerobic digestion and microalgae cultivation for bioenergy production. Curr Opin Biotechnol 50:101–110

Hong Y, Chen W, Luo X et al (2017) Microwave-enhanced pyrolysis of macroalgae and microalgae for syngas production. Bioresour Technol 237:47–56

Ismail MM, Ismail GA, El-Sheekh MM (2020) Potential assessment of some micro-and macroalgal species for bioethanol and biodiesel production. Energy Sources, Part A: Recovery, Utilization, and Environmental Effects: 1–17

Jensen A Present and future needs for algae and algal products. Fourteenth international seaweed symposium, 1993. Springer

Jin D, Kotar J, Silvester E et al (2020) Diurnal variations in the motility of populations of biflagellate microalgae. Biophys J 119:2055–2062

John RP, Anisha G, Nampoothiri KM et al (2011) Micro and macroalgal biomass: a renewable source for bioethanol. Bioresour Technol 102:186–193

Jung J-H, Sirisuk P, Ra CH et al (2019) Effects of green LED light and three stresses on biomass and lipid accumulation with two-phase culture of microalgae. Process Biochem 77:93–99

Katijan A, Latif MFA, Zahmani QF et al (2019) An experimental study for emission of four stroke carbureted and fuel injection motorcycle engine. J Adv Res Fluid Mechanic Thermal Sci 62: 256–264

Kawai S, Murata K (2016) Biofuel production based on carbohydrates from both brown and red macroalgae: recent developments in key biotechnologies. Int J Mol Sci 17

Khambhaty Y, Upadhyay D, Kriplani Y et al (2013) Bioethanol from macroalgal biomass: utilization of marine yeast for production of the same. Bioenergy Res 6:188–195

Kothari R, Ahmad S, Pathak VV et al (2019) Algal-based biofuel generation through flue gas and wastewater utilization: a sustainable prospective approach. Biomass Conversion and Biorefinery:1–24

Kraan S (2016) Seaweed and alcohol: biofuel or booze? Seaweed in health and disease prevention. Elsevier:169–184

Kumar AN, Chatterjee S, Hemalatha M et al (2020a) Deoiled algal biomass derived renewable sugars for bioethanol and biopolymer production in biorefinery framework. Bioresour Technol 296:122315

Kumar R, Ghosh AK, Pal P (2020b) Synergy of biofuel production with waste remediation along with value-added co-products recovery through microalgae cultivation: a review of membrane-integrated green approach. Sci Total Environ 698:134169

Kumar S, Dheeran P, Singh SP et al (2015) Kinetic studies of two-stage sulphuric acid hydrolysis of sugarcane bagasse. Renew Energy 83:850–858

Kumar S, Gupta R, Kumar G et al (2013) Bioethanol production from *Gracilaria verrucosa*, a red alga, in a biorefinery approach. Bioresour Technol 135:150–156

Lee XJ, Ong HC, Gan YY et al (2020) State of art review on conventional and advanced pyrolysis of macroalgae and microalgae for biochar, bio-oil and bio-syngas production. Energy Convers Manag 210:112707

Li L, Ge Y, Xiao M (2021) Towards biofuel generation III+: a sustainable industrial symbiosis design of co-producing algal and cellulosic biofuels. J Clean Prod 306:127144

Loulergue P, Balannec B, Fouchard-Le Graët L et al (2019) Air-gap membrane distillation for the separation of bioethanol from algal-based fermentation broth. Sep Purif Technol 213:255–263

Malcata FX, Pinto IS, Guedes AC (2018) Marine macro-and microalgae: an overview. CRC Press

Malode SJ, Prabhu KK, Mascarenhas RJ et al (2021) Recent advances and viability in biofuel production. Energy Conversion and Management: X 10:100070

Mankar AR, Pandey A, Modak A et al (2021) Pre-treatment of lignocellulosic biomass: a review on recent advances. Bioresour Technol 125235

Marinho GS, Holdt SL, Angelidaki I (2015) Seasonal variations in the amino acid profile and protein nutritional value of *Saccharina latissima* cultivated in a commercial IMTA system. J Appl Phycol 27:1991–2000

Mata TM, Martins AA, Caetano NS (2010) Microalgae for biodiesel production and other applications: a review. Renewable and Eustainable Energy Reviews 14:217–232

Mcneary WW, Erickson LE (2013) Sustainable management of algae in eutrophic ecosystems. J Environ Prot 4:9

Milledge JJ, Smith B, Dyer PW et al (2014) Macroalgae-derived biofuel: a review of methods of energy exxtraction from seaweed biomass. Energies 7

Mohammed AT, Jaafar MNM, Othman N et al (2021) Soil fertility enrichment potential of Jatropha curcas for sustainable agricultural production: a case study of Birnin Kebbi, Nigeria. Annals of the Romanian Society for Cell Biology:21061–21073

Mohan SV, Hemalatha M, Chakraborty D et al (2020) Algal biorefinery models with self-sustainable closed loop approach: trends and prospective for blue-bioeconomy. Bioresour Technol 295:122128

Möllers KB, Cannella D, Jørgensen H et al (2014) Cyanobacterial biomass as carbohydrate and nutrient feedstock for bioethanol production by yeast fermentation. Biotechnol Biofuels 7:64

Mols-Mortensen A, Jacobsen C, Holdt SL (2017) Variation in growth, yield and protein concentration in *Saccharina latissima* (*Laminariales, Phaeophyceae*) cultivated with different wave and current exposures in the Faroe Islands. J Appl Phycol 29:2277–2286

Mooij PR, Stouten GR, Tamis J et al (2013) Survival of the fattest. Energy Environ Sci 6:3404–3406

Morales-Sánchez D, Martinez-Rodriguez OA, Martinez A (2017) Heterotrophic cultivation of microalgae: production of metabolites of commercial interest. J Chem Technol Biotechnol 92: 925–936

Mosier N, Wyman C, Dale B et al (2005) Features of promising technologies for pretreatment of lignocellulosic biomass. Bioresour Technol 96:673–686

Ngamsirisomsakul M, Reungsang A, Liao Q et al (2019) Enhanced bio-ethanol production from *chlorella* sp. biomass by hydrothermal pretreatment and enzymatic hydrolysis. Renew Energy 141:482–492

Nhat PVH, Ngo H, Guo W et al (2018) Can algae-based technologies be an affordable green process for biofuel production and wastewater remediation? Bioresour Technol 256:491–501

Nisizawa K, Noda H, Kikuchi R et al (1987) The main seaweed foods in Japan. Hydrobiologia 151: 5–29

Ogbonda KH, Aminigo RE, Abu GO (2007) Influence of temperature and pH on biomass production and protein biosynthesis in a putative *spirulina* sp. Bioresour Technol 98:2207–2211

Onay M (2019) Bioethanol production via different saccharification strategies from H. tetrachotoma ME03 grown at various concentrations of municipal wastewater in a flat-photobioreactor. Fuel 239:1315–1323

Onumaegbu C, Mooney J, Alaswad A et al (2018) Pre-treatment methods for production of biofuel from microalgae biomass. Renew Sust Energ Rev 93:16–26

Özçimen D, Koçer AT, İnan B et al (2020a) Bioethanol production from microalgae. Handbook of microalgae-based processes and products. Elsevier, pp 373–389

Özçimen D, Koçer AT, İnan B et al (2020b) Chapter 14 - Bioethanol production from microalgae. In: JACOB-LOPES E, MARONEZE MM, QUEIROZ MI, ZEPKA LQ (eds) Handbook of Microalgae-Based Processes and Products. Academic, pp 373–389

Pandey A, Srivastava S, Kumar S (2020) Development and cost-benefit analysis of a novel process for biofuel production from microalgae using pre-treated high-strength fresh cheese whey wastewater. Environ Sci Pollut Res 27:23963–23980

Patel AK, Choi YY, Sim SJ (2020) Emerging prospects of mixotrophic microalgae: way forward to sustainable bioprocess for environmental remediation and cost-effective biofuels. Bioresour Technol 300:122741

Pereira H, Gangadhar KN, Schulze PS et al (2016) Isolation of a euryhaline microalgal strain, *Tetraselmis* sp. CTP4, as a robust feedstock for biodiesel production. Sci Rep 6:1–11

Peteiro C, Freire Ó (2009) Effect of outplanting time on commercial cultivation of kelp *Laminaria saccharina* at the southern limit in the Atlantic coast, NW Spain. Chin J Oceanol Limnol 27:54

Postma P, Miron T, Olivieri G et al (2015) Mild disintegration of the green microalgae *Chlorella vulgaris* using bead milling. Bioresour Technol 184:297–304

Prajapati SK, Bhattacharya A, Malik A et al (2015) Pretreatment of algal biomass using fungal crude enzymes. Algal Res 8:8–14

Prasad S, Kumar S, Sheetal K et al (2020) Global climate change and biofuels policy: Indian perspectives. Global Climate Change and Environmental Policy. Springer, pp 207–226

Prasad S, Venkatramanan V, Kumar S et al (2019) Biofuels: a clean technology for environment management. Sustainable green technologies for environmental management. Springer, pp 219–240

Rajak RC, Banerjee R (2020) An innovative approach of mixed enzymatic venture for 2G ethanol production from lignocellulosic feedstock. Energy Convers Manag 207:112504

Ramachandra T, Madhab MD, Shilpi S et al (2013) Algal biofuel from urban wastewater in India: scope and challenges. Renew Sust Energ Rev 21:767–777

Ramanna L, Rawat I, Bux F (2017) Light enhancement strategies improve microalgal biomass productivity. Renew Sust Energ Rev 80:765–773

Rammuni MN, Ariyadasa TU, Nimarshana PHV et al (2019) Comparative assessment on the extraction of carotenoids from microalgal sources: Astaxanthin from *H. pluvialis* and *β-carotene* from *D. salina*. Food Chem 277:128–134

Ras M, Steyer J-P, Bernard O (2013) Temperature effect on microalgae: a crucial factor for outdoor production. Rev Environ Sci Biotechnol 12:153–164

Rizza LS, Smachetti MES, Do Nascimento M et al (2017) Bioprospecting for native microalgae as an alternative source of sugars for the production of bioethanol. Algal Res 22:140–147

Roslan MF, Veza I, Said MFM (2020) Predictive simulation of single cylinder n-butanol HCCI engine. IOP Conference Series: Materials Science and Engineering, 2020. IOP Publishing

Rusli M, Said MFM, Sulaiman A et al. (2021) Performance and emission measurement of a single cylinder diesel engine fueled with palm oil biodiesel fuel blends. IOP Conference Series: Materials Science and Engineering, 2021. IOP Publishing

Saad MG, Dosoky NS, Zoromba MS et al. (2019) Algal biofuels: current status and key challenges. Energies 12

Sankaran R, RaP C, Pakalapati H et al (2020) Recent advances in the pretreatment of microalgal and lignocellulosic biomass: a comprehensive review. Bioresour Technol 298:122476

Schenk PM, Thomas-Hall SR, Stephens E et al (2008) Second generation biofuels: high-efficiency microalgae for biodiesel production. Bioenergy Res 1:20–43

Schulze PS, Guerra R, Pereira H et al (2017) Flashing LEDs for microalgal production. Trends Biotechnol 35:1088–1101

Sforza E, Pastore M, Spagni A et al (2018) Microalgae-bacteria gas exchange in wastewater: how mixotrophy may reduce the oxygen supply for bacteria. Environ Sci Pollut Res 25:28004–28014

Shahid MK, Batool A, Kashif A et al (2021) Biofuels and biorefineries: development, application and future perspectives emphasizing the environmental and economic aspects. J Environ Manag 297:113268

Sharma B, Larroche C, Dussap C-G (2020) Comprehensive assessment of 2G bioethanol production. Bioresour Technol 313:123630

Shokrkar H, Ebrahimi S, Zamani M (2017) Bioethanol production from acidic and enzymatic hydrolysates of mixed microalgae culture. Fuel 200:380–386

Shokrkar H, Ebrahimi S, Zamani M (2018) Enzymatic hydrolysis of microalgal cellulose for bioethanol production, modeling and sensitivity analysis. Fuel 228:30–38

Sudhakar K, Mamat R, Samykano M et al (2018) An overview of marine macroalgae as bioresource. Renew Sust Energ Rev 91:165–179

Sudhakar M, Viswanaathan S (2019) Algae as a sustainable and renewable bioresource for biofuel production. New and Future Developments in Microbial Biotechnology and Bioengineering. Elsevier, p 77–84

Sudhakar MP, Jegatheesan A, Poonam C et al (2017) Biosaccharification and ethanol production from spent seaweed biomass using marine bacteria and yeast. Renew Energy 105:133–139

Sulaiman O, Magee A, Bahrain Z et al (2013) Mooring analysis for very large offshore aquaculture ocean plantation floating structure. Ocean & Coastal Management 80:80–88

Talebnia F, Karakashev D, Angelidaki I (2010) Production of bioethanol from wheat straw: an overview on pretreatment, hydrolysis and fermentation. Bioresour Technol 101:4744–4753

Tandon P, Jin Q (2017) Microalgae culture enhancement through key microbial approaches. Renew Sust Energ Rev 80:1089–1099

Teichberg M, Fox SE, Olsen YS et al (2010) Eutrophication and macroalgal blooms in temperate and tropical coastal waters: nutrient enrichment experiments with *Ulva* spp. Glob Chang Biol 16:2624–2637

Titlyanov E, Titlyanova T (2010) Seaweed cultivation: methods and problems. Russ J Mar Biol 36: 227–242

Troell M, Rönnbäck P, Halling C et al. Ecological engineering in aquaculture: use of seaweeds for removing nutrients from intensive mariculture. Sixteenth International Seaweed Symposium, 1999. Springer

Tsuji A, Tominaga K, Nishiyama N et al (2013) Comprehensive enzymatic analysis of the cellulolytic system in digestive fluid of the sea hare Aplysia kurodai. Efficient glucose release from sea lettuce by synergistic action of 45 kDa endoglucanase and 210 kDa ß-glucosidase. PLoS One 8:e65418

Van Den Burg SW, Van Duijn AP, Bartelings H et al (2016) The economic feasibility of seaweed production in the North Sea. Aquaculture Economics & Management 20:235–252

Velazquez-Lucio J, Rodríguez-Jasso RM, Colla LM et al. (2018) Microalgal biomass pretreatment for bioethanol production: a review

Veza I, Muhammad V, Oktavian R et al (2021a) Effect of COVID-19 on biodiesel industry: a case study in Indonesia and Malaysia. Int J Automotive Mech Eng 18:8637–8646

Veza I, Roslan MF, Said MFM et al. Potential of range extender electric vehicles (REEVS). IOP Conference Series: Materials Science and Engineering, 2020a. IOP Publishing

Veza I, Roslan MF, Said MFM et al. (2020b) Cetane index prediction of ABE-diesel blends using empirical and artificial neural network models. Energy Sources, Part A: Recovery, Utilization, and Environmental Effects: 1–18

Veza I, Roslan MF, Said MFM et al (2021b) Physico-chemical properties of acetone-butanol-ethanol (ABE)-diesel blends: blending strategies and mathematical correlations. Fuel 286: 119467

Veza I, Said MFM, Latiff ZA (2019a) Progress of acetone-butanol-ethanol (ABE) as biofuel in gasoline and diesel engine: a review. Fuel Process Technol 196:106179

Veza I, Said MFM, Latiff ZA (2020c) Improved performance, combustion and emissions of SI engine fuelled with butanol: a review. Int J Automotive Mech Eng 17:7648–7666

Veza I, Said MFM, Latiff ZA (2021c) Recent advances in butanol production by acetone-butanol-ethanol (ABE) fermentation. Biomass Bioenergy 144:105919

Veza I, Said MFM, Latiff ZA et al (2021d) Application of Elman and Cascade neural network (ENN and CNN) in comparison with adaptive neuro fuzzy inference system (ANFIS) to predict key fuel properties of ABE-diesel blends. Int J Green Energy:1–13

Veza I, Said MFM, Latiff ZA et al. Simulation of predictive kinetic combustion of single cylinder HCCI engine. AIP Conference Proceedings, 2019b Pahang, Malaysia. AIP Publishing

Vonshak A, Tomaselli L (2002) Arthrospira (spirulina): systematics and Ecophysiology. In: WHITTON BA, POTTS M (eds) The ecology of cyanobacteria: their diversity in time and space. Springer Netherlands, Dordrecht, pp 505–522

Wargacki AJ, Leonard E, Win MN et al (2012) An engineered microbial platform for direct biofuel production from brown macroalgae. Science 335:308–313

Wishkerman A, Wishkerman E (2017) Application note: a novel low-cost open-source LED system for microalgae cultivation. Comput Electron Agric 132:56–62

Xu X, Kim JY, Oh YR et al (2014) Production of biodiesel from carbon sources of macroalgae, Laminaria japonica. Bioresour Technol 169:455–461

Yan J, Kuang Y, Gui X et al (2019) Engineering a malic enzyme to enhance lipid accumulation in *Chlorella protothecoides* and direct production of biodiesel from the microalgal biomass. Biomass Bioenergy 122:298–304

Yan Z-J, Liu J, Qian L et al (2020) Development and validation of a photobioreactor for uniform distribution of light intensity along the optical path based on numerical simulation. Environ Sci Pollut Res 27:42230–42241

Yanagisawa M, Nakamura K, Ariga O et al (2011) Production of high concentrations of bioethanol from seaweeds that contain easily hydrolyzable polysaccharides. Process Biochem 46:2111–2116

Yang C, Liu W, He Z et al (2015) Freezing/thawing pretreatment coupled with biological process of thermophilic *Geobacillus* sp. G1: acceleration on waste activated sludge hydrolysis and acidification. Bioresour Technol 175:509–516

Yazdani P, Zamani A, Karimi K et al (2015) Characterization of *Nizimuddinia zanardini* macroalgae biomass composition and its potential for biofuel production. Bioresour Technol 176:196–202

Yuan T, Li X, Xiao S et al (2016) Microalgae pretreatment with liquid hot water to enhance enzymatic hydrolysis efficiency. Bioresour Technol 220:530–536

Zabed HM, Akter S, Yun J et al (2019) Recent advances in biological pretreatment of microalgae and lignocellulosic biomass for biofuel production. Renew Sust Energ Rev 105:105–128

Zhang J, Fang J, Wang W et al (2012) Growth and loss of mariculture kelp *Saccharina japonica* in Sungo Bay, China. J Appl Phycol 24:1209–1216

Zhen S, Bugbee B (2020) Far-red photons have equivalent efficiency to traditional photosynthetic photons: implications for redefining photosynthetically active radiation. Plant Cell Environ 43: 1259–1272

Zheng J, Li Q, Hu A et al (2013) Dual-frequency ultrasound effect on structure and properties of sweet potato starch. Starch-Stärke 65:621–627

Zhou Y, Chen Y, Li M et al (2020) Production of high-quality biofuel via ethanol liquefaction of pretreated natural microalgae. Renew Energy 147:293–301

Zuniga-Jara S, Marín-Riffo MC, Bulboa-Contador C (2016) Bioeconomic analysis of giant kelp *Macrocystis pyrifera* cultivation (*Laminariales; Phaeophyceae*) in northern Chile. J Appl Phycol 28:405–416

Chapter 15
Evaluating Decarbonisation Pathways in Road Transportation via Life Cycle Assessment

Jorge E. Velandia Vargas, Rafael S. Capaz, Simone P. Souza, Otávio Cavalett, and Joaquim E. A. Seabra

Abstract This study aimed to quantify the carbon footprint of different private cars and urban buses technologies. For private cars, the highest carbon intensities were associated to internal combustion engines propelled by gasoline. Moreover, when fueled by bioethanol both technologies displayed a competitive carbon footprint, associated to the low footprint of the vehicle. Fuel cell-based technologies were less competitive when compared to other technologies as their environmental performance was heavily influenced by the vehicle burden. Thus, if technical and financial hassles were resolved, the solid-oxide fuel-cell vehicle would represent a promising technology in order to tackle greenhouse gas emissions while providing an option for synergy between electrification and biofuels. Regarding urban busses, biodiesel was found to be an option for carbon mitigation; however, the environmental and health hazards of other emissions remain a concern, which makes the case for the solid-oxide fuel-cell bus and the electric bus. Moreover, the large uncertainties related to land use change of soybean biodiesel could seriously hamper the potential carbon mitigation of biodiesel. All-in-all, the strategies for climate change mitigation in the transport sector should be designed taking into account the local conditions and all the available options at a given time.

J. E. Velandia Vargas · S. P. Souza (✉) · J. E. A. Seabra
University of Campinas, Campinas, Brazil
e-mail: spsouza@unicamp.br; jseabra@fem.unicamp.br

R. S. Capaz
Federal University of Itajubá, Itajubá, Brazil

O. Cavalett
Norwegian University of Science and Technology, Trondheim, Norway

C. R. Soccol et al. (eds.), *Liquid Biofuels: Bioethanol*, Biofuel and Biorefinery Technologies 12, https://doi.org/10.1007/978-3-031-01241-9_15

15.1 Introduction

The decarbonisation of the transport sector is frequently addressed in the context of tackling the climate change and the dependency of fossil fuels. In 2018, the global transport operations were responsible for around 8.2 Gt CO_{2e}, *i.e.* 17% of the total greenhouse gases (GHG) emissions, excluding land use change (LUC) and forestry aspects (Climate Watch 2018). On the same year, the energy demand for load and passenger transportation corresponded to around 25% (*i.e.* 2.9 billion toe) of the global energy use (IEA 2020).

The road transportation accounts for more than 70% of the GHG emissions and energy demand of the transport sector, which justifies the interests in decarbonising this segment. Biofuels are often advocated as one of the key global initiatives to replace fossil fuels and mitigate GHG emissions. Even so, the current consumption of biofuels, mostly composed by biodiesel and ethanol, represented less than 5% of the total energy consumption in the global road transportation, as observed in Fig. 15.1 (IEA 2020).

The previous Intergovernmental Panel Climat Change (IPCC) Special Report on Climate Change and Land (SRCCL) has indicated that global scenarios limiting warming would involve land-based mitigation options that include bioenergy deployment and expansion of forests. As all the sectors are connected, the less we reduce emissions overall, the more we will have to rely on bioenergy and other mitigation strategies to fight climate change (IPCC 2019). However, the report also highlights that depending on the scale necessary for bioenergy deployment, there is an increasing risk of negative consequences for food security, biodiversity

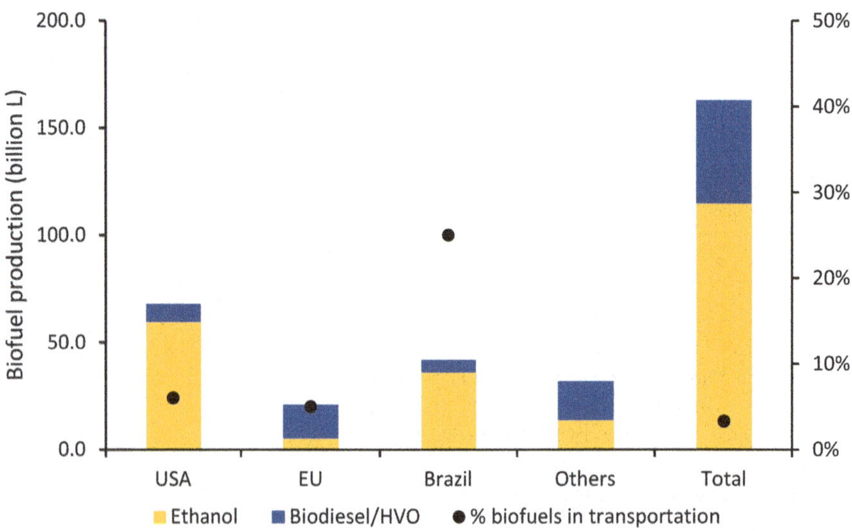

Fig. 15.1 Biofuels production and contribution to the energy demand by transportation. Based on IEA (2020)

conservation and land degradation. Therefore, bioenergy needs to be carefully managed to avoid these risks. In fact, if properly managed, bioenergy can also bring several positive co-benefits for biodiversity and ecosystems quality (Robertson et al. 2017; Field et al. 2020). This is a question of locally appropriate policies and governance systems to maximize the positive outcomes expected by bioenergy deployment.

The European Commission has proposed ambitious targets for the decarbonisation of the transportation sector. The target is 50–55% reduction of the GHG emissions from cars and vans by 2030 (European Commission, 2021), in additional to promoting the carbon market, stimulating cleaner fuel use, and re-investing in clean technologies. Crop-based biofuels are the current leaders in the decarbonisation of the EU transportation sector, but biofuels shall assure a low risk for indirect land use change (iLUC) to enable the drive to carbon-neutrality. Alternative feedstocks also include waste, energy-crops, recycled carbon fuel, renewable fuel of non-biological origin, and green hydrogen. However, the phase-out of the internal combustion engine in favor of the battery electric vehicles has been strongly defended by the Commission and might limit the use of advanced fuels. Advocates of liquid fuels claim that the objectives of the EU will hardly be achieved without advanced liquid fuels due to the lower costs of these alternatives (ePURE, 2021).

In the same way, according to the scenarios for decarbonisation proposed by IEA (2021a), the electrification of end-users, such as transportation, would provide the largest contribution to net-zero carbon emission target by 2070. From a global perspective, a massive shifting of the fleet to electric vehicles—including light-vehicles, buses and trucks—could lead to an electricity contribution of around 50% of the energy demand by road transportation in 2070, while biofuels would provide less than 20%. On the other hand, is the global perspective for decarbonizing the transportation suitable for all national contexts?

Historically, Brazil is one of the global leaders in renewable energy deployment, especially regarding biofuels use (see Fig. 15.1). About half of the total energy supplied in the country comes from renewable energy sources—such as biomass, hydropower, and wind—mostly led by sugarcane products, which were responsible for more than 16% of the national energy supply (52.8 Mtoe) in 2019 (IEA 2020).

The contribution of bioenergy is especially relevant in the Brazilian transportation sector, where biofuels have constituted roughly 25% (or 21.3 Mtoe in 2019) of the energy consumed in this sector. The accumulated learning and recognized expertise have decisively contributed to the prominent role of liquid biofuels in the road transportation sector over the last decades (Nogueira and Silva Capaz 2013; Nogueira et al. 2016). While 33.8 million m^3 of ethanol were consumed in Brazil last year, directly or blended with gasoline, around 4.7 million m^3 of biodiesel were used in mandatory blends with fossil diesel (10% v/v) (EPE 2020a). Both biofuel supply-chains are supported by a well-consolidated agroindustry of roughly 380 sugarcane mills and 110 biodiesel plants (MapBiomas Brasil 2020).

Considering the tough challenge in techno-economic terms on promoting the effective transition to low-carbon and sustainable transportation systems, some

Low-Carbon Policies (LCP) have played an important role in boosting the production and use of biofuels. For instance, the current Brazilian program *Renovabio* (Brasil 2018) seeks to reduce the carbon intensity of the national fuel matrix by up to 11% until 2029 by trading decarbonisation credits (CBIO). Likewise, the United States set forth a target of 36 billion gallons for biofuels by 2022 through the *Renewable Fuel Standard* (RFS) (EPA 2010), setting specific targets for different fuel categories. And, in Europe, the *Renewable Energy Directive* (RED) has introduced targets for renewable energy consumed in the transportation that are periodically updated (EU Science Hub 2020).

Under all these regulatory schemes, the potential GHG reduction for biofuels in comparison to their fossil counterparts is a crucial indicator for the decision-making process. This issue has been estimated using the Life Cycle Assessment (LCA) approaches, where the GHG emissions along the whole biofuel life cycle—*i.e.* from the feedstock procurement to the fuel production and/or use—are accounted for (Gerbrandt et al. 2016; Pereira et al. 2019; de Souza et al. 2021).

The LCA, which is standardized by International Organization for Standardization (ISO) (ISO 2006a, b), is carried out through four steps comprising the definition of the system boundaries, the detailed description of each process inputs and emissions inside the defined boundaries (*i.e.*, the life cycle inventory, LCI), the characterization of the environmental flows according to the impact category evaluated, and the interpretation of the results. Notwithstanding, different approaches and assumptions are normally used in the LCA.

Then, considering the potential of sugarcane ethanol in Brazil to decarbonize the transport sector, a comprehensive and harmonized LCA was carried out in this chapter comprising light and heavy-duty vehicles, more specifically busses, supplied by different energy sources, from fossil fuels, to conventional biofuels (ethanol and biodiesel) and future alternatives, such as green diesel, hydrogen, and electrification.

15.2 Methodology

15.2.1 Goal and Scope

The objective of this study is to quantify the life-cycle GHG emissions of biofuel alternatives used for transportation in the Brazilian context. A *cradle-to-grave* LCA analysis is performed comprising vehicle manufacturing to end-of-life, including fuel production and use-phase. In addition to light-duty vehicles (LDVs), we included urban busses to our scope, aiming to create a broader analysis. Within this document, we refer to urban busses as heavy-duty vehicles (HDVs).

Furthermore, the LCA was performed under an attributional approach since the analysis focused on the environmentally-relevant physical flows described by average data to and from the product system (JCR 2010). In general, energy allocation was assumed for multiple-products systems. The functional unit was one kilometer (km) travelled. The characterization factors for climate change impacts are taken from the fifth IPCC report (2014). The carbon emissions related to the biofuel use were disregarded as they are considered biogenic.

15.2.2 Product-System Description

As the LCA methodology is employed to quantify the use of resources and emissions related to a product or service, it is necessary to define the scope of the data acquisition. According to ISO 14044 (2006b) the product system is the collection of processes containing elementary and product flows, which perform one or more defined functions, and which models the life cycle of a product. However, as the data may include information from multiple supply chains including hundreds of flows it is fundamental to define what processes are the most relevant for the study. This screening is necessary and could save resources as the LCI construction is often a time-intensive task and investing time or computational resources in processes with a low environmental burden could be not worth the effort, depending on the scope of the study.

Once the significance of a process has been recognized, often due to its potential contribution to the overall results or the scope of the study, the LCA practitioner could classify it as a *foreground* process. In contrast, less relevant processes could use generic data found in LCA databases; such processes are identified as *background* processes. Foreground processes construction is under direct control of the LCA practitioner whereas background processes are not usually modified by the LCA practitioner as they often describe the surrounding supply chains for the process inputs. For this case, our system boundaries were outlined to include the carbon emissions associated with the use of diverse vehicle technologies and fuels along their entire life cycle, also known as a cradle-to-grave approach.

The vehicle energy chain in different stages, such as vehicle manufacturing, Well-to-Tank (WTT), Tank-to-Wheel (TTW), and End-of-Life (EoL). Vehicle manufacturing refers to vehicle production, including the raw materials acquisition and processing but also the assembly of the components. WTT includes the fuels/energy carriers/electricity production and the logistics and infrastructures associated to them (transportation, required infrastructures, etc.). TTW stage refers to the use phase, which includes the energy required for the vehicle to travel any given distance; in addition, it includes the emissions linked to the use of the vehicle. EoL refers to the disposal of the vehicle once the life has ended. This classification is valid for both, light-duty vehicles (LDVs) and heavy-duty vehicles (HDVs). For this study, HDVs refer exclusively to busses. The product system for passenger vehicles and busses is depicted in Figs. 15.2 and 15.3. The EoL stage is not specified in either of the figures, but it was considered as well.

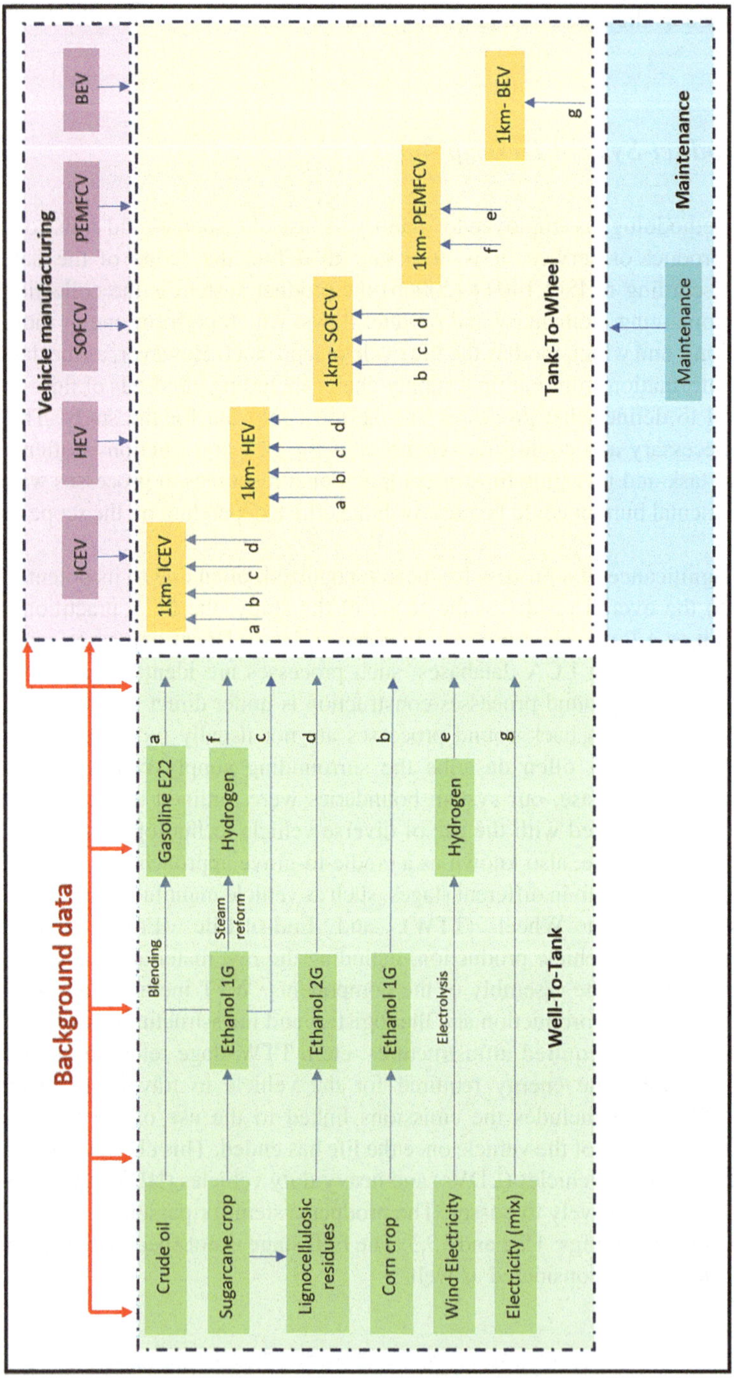

Fig. 15.2 Product system for light-duty vehicles

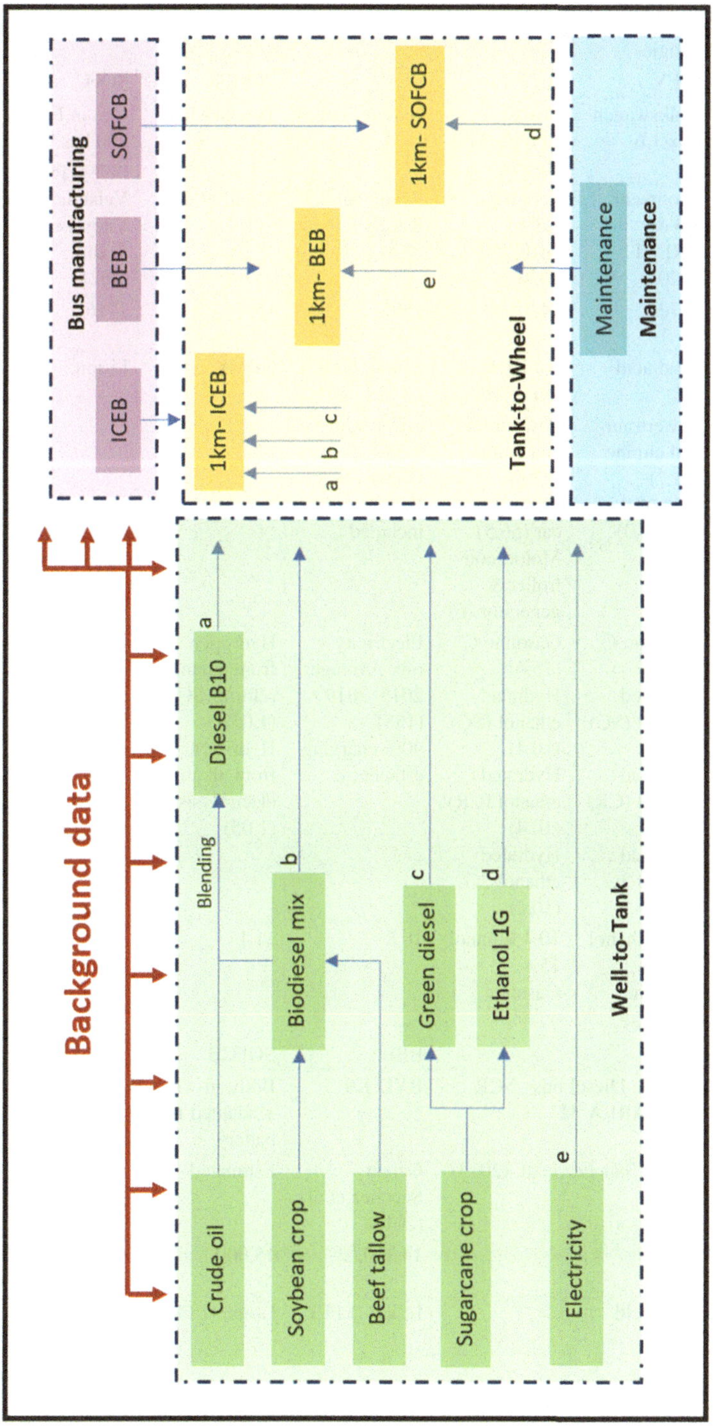

Fig. 15.3 Product system for heavy-duty vehicles (buses)

Table 15.1 Vehicle modelling parameters

Lightweight technologies					
	ICEV	HEV	BEV	PEMFC	SOFC
Model	Volkswagen Fox 1.6	Toyota Corolla Altis Hybrid 1.8	Nissan Leaf S	Toyota Mirai	Nissan EV 200 based prototype[a]
Data source	Ecoinvent 3.3 & GREET 2020	Ecoinvent 3.3 & GREET 2020	Ecoinvent 3.3 & GREET 2020	Miotti et al. (2017)	Velandia Vargas and Seabra (2021)
Curb weight (kg)	1072	1440	1595	1537	1716
Battery chemistry	Lead acid	NiMH & Li ion (NMC)	Li-ion (NMC)	NiMH	Li-ion[b]
Inventory adaptation (weight %)	Powertrain and engine (21.7) Rest of the car (78.3)	Powertrain and engine (27.5) Rest of the car (66.5) Motor, controller & generator (6)	Powertrain (23.6) Glider (78.3)[a] Battery not included	–	–
Fuel /energy consumption (km L^{-1}) (Wh km^{-1}) (kg 100 km^{-1})	Gasoline C (12.3) Hydrated ethanol (SC) (8) Hydrated ethanol (CR) (8) Hydrated ethanol 2G (8)	Gasoline C (15.4) Hydrated ethanol (SC) (10.4) Hydrated ethanol (CR) (10.4) Hydrated ethanol 2G (10.4)	Electricity mix Average 2015–2019 (188) 90% charging efficiency	Hydrogen from ethanol reform (SC) (1.05) Hydrogen from alkaline electrolysis (1.05)	Hydrated ethanol (SC) (20) Hydrated ethanol (CR) (20) Hydrated ethanol 2G (20)
Weight to power ratio (kg HP^{-1})	10.2 Ethanol 10.8 Gasoline	10.4 Ethanol 15.4 Gasoline	10.8	11.1	–

Heavyweight technologies			
	ICEB	BEB	SOFCB
Model	Generic Diesel bus+ SCR using ARLA 32	BYD K9	Body-in-white BYD K9 + adapted fuel cell + Li-ion battery
Reference	García Sánchez et al. (2013)	García Sánchez et al. (2013)	Estimated
Curb weight (kg)	12,180	14,300	15,000
Battery chemistry	Lead acid	Li-ion (LFP)	Li-ion (LFP)

(continued)

Table 15.1 (continued)

Lightweight technologies					
	ICEV	HEV	BEV	PEMFC	SOFC
Fuel/energy consumption (km L^{-1}) ICEB & SOFCB (Wh km^{-1}) BEB	Diesel B10 (1.58) Biodiesel mix (1.67) Green diesel (1.58)		1660 90% charging efficiency	Hydrated ethanol (SC) (0.83)	

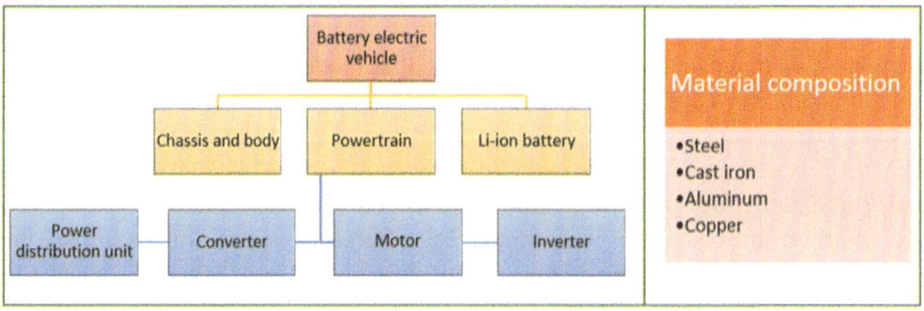

Fig. 15.4 Depiction of unit process (left) and system process (right) schemes

15.2.2.1 Vehicle Manufacturing and Tank-to-Wheel Stage

For the vehicle powertrain technologies analyzed in this chapter, we strived to model the raw materials extraction and the manufacturing stage. Light-duty technologies include the internal combustion engine vehicles (ICEVs) and the hybrid electric vehicles (HEVs); for this case, both powertrains use *flex-fuel* engines, which allow them to operate on both gasoline and hydrated ethanol. Furthermore, in order to create a more comprehensive study, we included fuel cell vehicle technologies. The first assessed fuel cell vehicle was the polymer-electrolyte-membrane fuel cell vehicle (PEMFCV), a technology commercially available in several markets. In fact, the Japanese Ministry of Economy, Trade and Industry put forth a goal to have 800,000 vehicles on the road by 2030 (AsiaTimes 2021). PEMFCV vehicles are propelled on hydrogen; for this study, we considered hydrogen obtained by bioethanol reforming. Moreover, a different kind of fuel cell vehicle technology, the solid-oxide fuel cell vehicle (SOFCV), was also considered for analysis. Although such technology is still under development and only prototypes have been manufactured, it bears potential to be a genuine breakthrough for biofuels use since the vehicles are able to operate on a mixture of bioethanol and water. Finally, as a way to offer a broader comparison, we incorporated a battery electric vehicle (BEV).

For heavy-duty technologies, we included internal combustion engine busses (ICEBs) propelled by blended diesel B10, biodiesel, and renewable diesel. Furthermore, we modelled a theoretical solid-oxide fuel cell bus (SOFCB) according to the best of our findings in scientific literature and lastly, for the sake of comparison, we included a battery electric bus (BEB). In order to cover the Tank-to-Wheel stage, the use-phase of the vehicles was also included; this stage is anticipated to display a large contribution to the overall carbon footprint for combustion engine technologies. Thus, the vehicle energy consumption—in the form of fuels, hydrogen or electricity—is incorporated here along with the emissions linked to the use phase of each vehicle. Braking and tires wearing emissions were not considered for this analysis. Table 15.1 exhibits basic modelling parameters for LDVs and HDVs.

15.2.2.1.1 LCI Construction and Adaptation

There are two main approaches to build an LCI, as modelled in Ecoinvent databases (Swiss Centre for Life Cycle Inventories 2014). In the first one, known as unit process, each product is described as the collection of sub-systems, for instance, the BEV LCI is constituted by individual car systems namely, Li-ion battery (LIB), powertrain and glider—which includes chassis and body-in-white. In its turn, each one of those subsystems is constituted by other subsystems, for example, the powertrain comprises electric motor, power distribution unit, inverter, cables and converter. Subsequently, each one of those subsystems is composed by other subsystems, and so on, until it reaches a bottom level in which the subcomponents are described by raw materials. Figure 15.4 shows a schematic depiction of the product systems. In contrast to the unit process approach, system process LCIs display a list of material flows and emissions. Thus, the BEV LCI appears as a list of steel, aluminum, copper, polymers, magnesium, and every other material in the composition of the car.

The modelling of our LDVs was based on the unit process datasets included in Ecoinvent 3.3. Ecoinvent designed the vehicle LCIs aiming to represent 1 kg of vehicle; this means that the components of the LCI for 1 kg of the vehicle are representative of the entire car. In other words, the LCI for 1 kg of vehicle represents the components—or mass shares—of the entire vehicle but scaled down to 1 kg. For instance, for each 1 kg of a gasoline-driven ICEV, the engine represents 0.26 kg while the rest of the vehicle is 0.73 kg. This approach allows practitioners to easily model vehicles with different curb weight. Considering that the original vehicle data in Ecoinvent is based on average passenger cars spanning from 2000 to 2010, which were expected to be representative only up to 2015, a further adaptation was made considering the mass fractions for ICEVs, HEVs, and BEVs found in GREET 2019 (Argonne National Laboratory 2020), based on more recent data. See Table 15.1. LCIs for PEMFCV and SOFCV were taken from scientific literature. For the BEV we incorporated a NMC (Lithium-Nickel-Manganese-Cobalt-Oxide) battery as described by Ellingsen et al. (2013).

A word should be given to our car selection. As LCA enables the practitioners to compare the environmental performance of diverse technologies and processes, it is crucial to ensure the product system does not benefit any product—or technology—over the others. Since our functional unit was already defined as travelling 1 km, a

Table 15.2 Parameters for LCI adaptation

Raw material	Data year	Production pathway (per kg)	Notes
Steel	2018	Basic oxygen furnace (primary): 77.26%; Electric furnace (recycled): 22.4% (Instituto Aço Brasil 2019)	Glider and powertrain production is assumed to be entirely supplied by Brazilian steel. Brazilian LCI for 43% iron ore concentrate (Ferreira and Leite 2015) replaced the LCI found in Ecoinvent 3.3. An additional beneficiation process for iron ore, to 65%, was included, from Ecoinvent 3.3 Hard coal imports taken as in (DNPM 2017)
Aluminum	2018	Søderberg route (primary): 28.7%; Prebaked anode route (primary): 25.1%; Recycled aluminum: 53.9% (Associação Brasileira do Alumínio 2019)	Aluminum oxide production and bauxite extraction processes were adapted to Brazil. All aluminum oxide was assumed to be obtained from primary production.
Copper	2017	18.1% Imported from Chile (primary): 49.7% (Associação Brasileira do Cobre 2018)	Latin-American LCI included in Ecoinvent 3.3 used to model imported copper.

fair comparison would include vehicles displaying the same curb weight; in fact, vehicle manufacturing is expected to present a significant contribution for the LDV results (Ma et al. 2012; Faria et al. 2013; Hawkins et al. 2013; Messagie et al. 2014). However, curb weight varies notoriously for different vehicle powertrain technologies as seen in Table 15.1. For instance, for a similar weight-to-power ratio, BEVs tend to be heavier than their ICEVs counterparts, mainly due to the battery weight; PEMFCVs are heavier also due to the hydrogen tank and the balance of plant. Thus, car selection was based on the presence of the car in the market—Brazilian or global—while keeping an eye on the weight-to-power ratio to guarantee we are not using vehicles with too different settings.

For the BEV we based our modelling in the 40 kW Nissan Leaf, the second best-selling BEV in the world. Battery weight was estimated as 228.6 kg by assuming a 175 Wh kg^{-1} specific energy (Löbberding et al. 2020). Our ICEV was based on a Volkswagen Fox, which, despite being considered as a compact, presents a similar weight-to-power ratio to Nissan Leaf. The HEV was based on Toyota Corolla Altis hybrid 1.8, a *flex-fuel* hybrid that runs on either gasoline or ethanol. For the HEV modelling, besides of adapting the ICEV LCIs available in Ecoinvent 3.3 to the GREET 2020 weight shares, we added an electric motor, power distribution unit and converter. Additionally, we included a Ni-MH battery, common in HEV models. For the PEMFCV we used the vehicle modelling as in Miotti et al. (2017), which portrays an 80-kW fuel cell system composed of an array of single fuel cells known as the fuel cell stack, the hydrogen tank and the balance of plant (BoP), the set of components in charge of the hydrogen flow within the vehicle.

Additionally, a vehicle powered by a solid oxide fuel cell (SOFC) was included. The potential of SOFC technology for automotive applications has been only recently considered after being long discarded due to technical issues such as high working temperatures and long start-up times. SOFC offers a technically feasible option for on-board reform of biofuels, despite of the technological challenges ahead. In 2016, Nissan unveiled the world's first prototype SOFC car designed to run on bioethanol. Unfortunately, as Nissan disclosed little information about the prototype—we know it contains a 5 kW SOFC, a 24 kWh Li-ion battery and exhibits a 600 km range (Elsevier 2016)—we had to appeal to scientific literature in order to create an adequate approach for vehicle modelling, as discussed in Velandia Vargas and Seabra (2021).

In contrast to the LDVs, the HDVs LCIs were modelled based on the material composition of the vehicles; hence, such scheme is more alike to a system process LCI than to a unit process LCI. The modelling for the ICEB and BEB was based in the material composition displayed by García Sánchez et al. (2013). Each one of the materials in the composition, *e.g.*, steel, aluminum, glass, cast iron, etc. were modelled by Ecoinvent LCIs. Nonetheless, as Ecoinvent v.3.3 only contains a few processes describing Brazilian products, and aiming to accurately reproduce local conditions, a group of priority datasets was adapted. We focused on steel, aluminum and copper production as these materials are expected to significantly contribute to the environmental burden of the vehicles (Velandia Vargas et al. 2019).

The LCI adaptation was carried out as follows: firstly, the LCIs for steel, aluminum and copper had their electricity inputs switched to the Brazilian mix, replacing the global average mix. Secondly, the shares of primary and recycled materials were adapted to match Brazilian supply chains. For instance, our primary aluminum LCIs match the production routes shares found in the country, see Table 15.2. Finally, upstream raw materials, *e.g.*, pig iron, iron pellets and iron ore concentrates were adapted as well.

Thus, the ICEB list of materials had its inputs adapted to Brazilian LCIs entirely while the BEB had the entire bus body and chassis adapted to Brazilian conditions. The BEB battery was assumed to have a $LiFePO_4$ cathode which was modelled after Majeau-Bettez et al. (2011). Within the battery, all inputs of steel, aluminum and copper were changed for Brazilian LCIs, except for the battery cell components, which were assumed completely imported. We should highlight that adapted LCIs were applied to both HDVs and LDVs.

Regarding the modelling of the solid-oxide fuel cell bus (SOFCB), we had to make several assumptions and rely on the best guesses found on scientific literature and therefore, the modelling is bound to uncertainties. Firstly, we considered the chassis and body to be the same as in the fuel cell bus depicted by García Sánchez et al. (2013). Similarly, we assumed that a 120 kW fuel cell would be enough to propel a bus of around 15 t. Assuming our SOFC to have the same power density as the fuel cell proposed by Strazza et al. (2010) which was also employed for the SOFC vehicle—20 kW and 112 kg—then the bus SOFC weight would be 672 kg.

Moreover, as the theoretical SOFCB would run on ethanol, the fuel tank provides the energy stock, allowing the battery to specialize on delivering power. Akin to

Table 15.3 Parameters used for well-to-tank stage

Biofuel/fuel/ energy carrier	Fuel material (Feedstock)	Data sources	Coproducts	Observations
Fuels for Light-duty vehicles				
Hydrated ethanol	Sugarcane	Agricultural stage & refinery: Cavalett et al. (2013)	Per $TonCane^{-1}$: Hydrated ethanol 69.3 kg (85.66 L) Electricity 30.0 kWh	Ashes, vinasse and filtercake were not considered as coproducts as the product system is outlined to include all waste products used for irrigation. Electricity output adjusted to 30 kWh $TonCane^{-1}$. Energy allocation. Transportation distance: 340 km.
Hydrated ethanol	Corn	Agricultural stage: Donke et al. (2016) Refinery: Moreira et al. (2020) adapted.	Per $TonCorn^{-1}$: Hydrated ethanol 446 L Corn oil 13 kg DDGs (high fiber) 82 kg DDGs (high protein) 113 kg DDGs (wet) 169 kg	Energy allocation.
Anhydrous ethanol	Sugarcane	Cavalett et al. (2013) (adapted)	Per $TonCane^{-1}$: Anhydrous ethanol 64.7 kg (81.8 L) Electricity 30.0 kWh	Same electricity output as in hydrated ethanol production. Zeolite 0,026 kg Ton $cane^{-1}$. Energy allocation. Transportation distance: 340 km.
Hydrated ethanol (2G)	Sugarcane	Capaz et al. (2020)	Hydrated ethanol 289.1 kg $TonCane^{-1}$ Electricity 232 kWh $TonCane^{-1}$	Enzymatic hydrolysis route. Energy allocation. Electricity output adjusted to 30 kWh $TonCane^{-1}$
Low sulfur gasoline "Gasoline A"	Crude oil	Ecoinvent	Not specified	Ecoinvent dataset: '*petrol production, low sulfur*. Transportation distance: 50 km.

(continued)

Table 15.3 (continued)

Biofuel/fuel/ energy carrier	Fuel material (Feedstock)	Data sources	Coproducts	Observations
Blended gasoline-E22. "Gasoline C"	Anhydrous ethanol+ Low-sulfur gasoline			Gasoline C was modelled as the blending of 22% anhydrous ethanol and 78% fossil gasoline. This is in consonance with automotive validation tests.
Hydrogen (alkaline electrolysis)	Water + wind electricity	Bekel and Pauliuk (2019)	Hydrogen 1 kg Oxygen 8 kg	51 kWh kg H_2^{-1}. Exergy allocation.
Hydrogen (ethanol reform)	Sugarcane ethanol	Velandia Vargas and Seabra (2021)	No coproducts.	Feedstock use: 6.72 kg kg-^{1}H_2 hydrated ethanol. Electricity use: 0.49 kWh kg^{-1} H_2.
Fuels for Heavy-duty vehicles				
Low sulfur diesel	Crude oil	Ecoinvent		
Biodiesel	Soybean oil	Agricultural stage: Ibict (2019) Transesterification: Garcilasso (2014), Cerri et al. (2017)	Per 1.07 kg^{-1} soybean oil. Biodiesel 1 kg Glycerin 0.1255 kg Fatty acid 0.0404 kg	Energy allocation.
Biodiesel	Tallow	Garcilasso (2014), Dufour and Iribarren (2012)	Per 1.07 kg^{-1} beef tallow. Biodiesel 1 kg Glycerin 0.1255 kg Fatty acid 0.0509 kg	Energy allocation.
Blended diesel (B10)	Soybean biodiesel+ beef tallow biodiesel + Low-sulfur biodiesel			B10 Diesel is a blending of Low-sulfur diesel (90%) and biodiesel mix (10%). In its turn, the Biodiesel mix includes 84% of soybean biodiesel and 16% of beef tallow biodiesel.

(continued)

Table 15.3 (continued)

Biofuel/fuel/ energy carrier	Fuel material (Feedstock)	Data sources	Coproducts	Observations
Green Diesel	Sugarcane ethanol	Capaz et al. (2021)	Renewable jet fuel 0.0227 kg Green diesel 0.0017 kg Naphtha 0.011 kg	Hydrogen required for
Electricity				
Electricity		National energy balance (EPE 2020b)	N/A	Mix 2014–2019. Transmission and distribution losses included as in Ecoinvent.

Table 15.4 2015–2019 average electricity mix

Source	Hydro	Natural gas	Wind	Biomass[a]	Nuclear	Coal	Oil derivatives	Solar[b]
%	65.1	10.2	6.5	8.3	2.5	3.4	2.6	1.3

[a]Biomass generation share is composed of 57% sugarcane bagasse, 32% woodchips burning and 11% biogas burning
[b]Sources specified as "others" were included into solar

Miotti et al. (2017), who proposed a 28 kW battery for a 1530 kg vehicle, we assumed that a 280 kW LIB would be enough to drive the bus whether a 1.2 kW kg^{-1} power density was considered. Applying a 1.5 security factor, the battery weight would be 350 kg. Given the uncertainty regarding the chassis and body weight, we took a conservative stance and considered that the 120 kW fuel cell, the 280 kW battery and the chassis and body would weight 15,000 kg overall. For the fuel consumption, and due to the lack of data, we linearly scaled the SOFCV consumption to match the new SOFC size of 120 kW, resulting in 1.2 L km^{-1}. We had no evidence to support that either the SOFC bus or the car would be substantially heavier than their counterparts; nonetheless, we adopted this strategy in face of the lack of data. The implications of this are analyzed in the sensitivity analysis.

15.2.2.2 Well-to-Tank Stage

This section presents the overall assumptions used in the construction of the Well-to-Tank inventories for the electricity and each one of the fuels considered in the analysis. Table 15.3 displays the parameters used for fuels production and electricity generation. For the fossil fuels, the system boundaries comprised the extraction of commodities and their transformation processes, *e.g.*, crude oil extraction and refining. In the case of the biofuels, the boundaries included the agricultural and industrial stage of production. For energy carriers, such as gaseous hydrogen, the use of primary materials and the energy required for transformation was included; the same principle was used for electricity generation.

15.2.2.2.1 Electricity Generation

The electricity generation mix in Brazil varies over time in response to yearly variations, such as dry and wet seasons and demand, *e.g.*, peak and off-peak hours. Given the fluctuations in demand caused by the COVID-19 pandemic and the changes in the generation mix triggered by the current drought in Brazil (Reuters 2021), the authors adopted a 5 years average (2015–2019) according to the Brazilian energy research company (EPE 2020b). Each annual generation mix is representative of the average dispatch in the National Interconnected System (SIN). Thus, isolated systems (non-connected municipalities), self-production units not connected to the SIN (such as industries with their own generation plants) and distributed generation systems (e.g., photovoltaic panels installed by consumers themselves) were not quantified by our modelling. The electricity mix is shown in Table 15.4.

Furthermore, we appealed to Ecoinvent datasets for modelling each generation source. The electricity mix model comprises of high-voltage generation LCIs that undergo a series of voltage transformations, *i.e.*, from high to medium and from medium to low. In a general way, the LCI structure consisted of the individual contribution of diverse generation sources (hydro, wind, gas, coal, etc.) required for the generation of 1 kWh and its subsequent injection to the grid, for transmission and distribution. The losses linked to distribution and generation are kept as in the Ecoinvent datasets.

15.2.2.2.2 Sugarcane Ethanol

The foreground data for sugarcane ethanol production were obtained from Cavalett et al. (2013), which depict average conditions, found in mills in the Center-South region. The system boundaries include the inputs and emissions related to agricultural and industrial processes: 100% mechanized harvest, no burning and no straw collection. The emission factors from the burning of bagasse in boilers were taken from the software GREET 2019 (Argonne National Laboratory 2020).

The inventory was adjusted by assuming that ethanol production at an autonomous distillery would produce a surplus electricity generation of 30 $kWh.t^{-1}$ sugarcane, which the authors agree to better represent the national average. Energy allocation was applied. For background data, Ecoinvent 3.3 was used, prioritizing inventories for Brazil, whenever possible. For ethanol transportation, trucks with a capacity greater than 32 t were used, with distances of 290 km between the ethanol plant and the distributor and 50 km between the distributor and the gas station.

For 2G ethanol, which uses a mix of residues—bagasse and straw—from 1G ethanol production as feedstock we considered a plant physically separated from the 1G process. The technology is based on enzymatic hydrolysis.

15.2.2.2.3 Corn Ethanol

Corn ethanol is evaluated in this study due to the grain's importance as a feedstock for ethanol production in the United States and the recent irruption in the Brazilian biofuel's landscape. The geographic scope consists of the corn growing and harvesting in the Brazilian state of Mato Grosso; ethanol production takes places in the same region. The foreground data—corn agricultural stage and ethanol production—were obtained from Donke et al. (2016) and Moreira et al. (2020) respectively. The product system consists of a dry mill processing with co-production of distiller's dried grains (DDGS) and corn oil. The background data were taken from Ecoinvent 3.3.

15.2.2.2.4 Fossil Gasoline

Although fossil fuels are still dominant in Brazil, relatively few studies have evaluated their environmental performances in a national context. Publications began to appear in the second half of the 2000s, but often based on international data. In fact, the available data refers mainly to the oil exploration and extraction stage (but also incorporates refining). Thus, in this study we used the Ecoinvent 3.3 *'petroleum production, low sulfur | Alloc Rec, U'* LCI, which incorporates the extraction and transport to a refinery in Europe. A similar strategy was adopted for the background data by the national biofuels policy RenovaBio (Brasil 2018).

As the addition of anhydrous ethanol to gasoline is mandatory in Brazil, we adopted the blending shares employed for vehicle test and validation purposes, *i.e.*, 22% anhydrous ethanol and 78% fossil gasoline. The anhydrous ethanol was adapted from the hydrated ethanol source, only with minor adjustments due to zeolite use and the difference of density. Transportation of anhydrous ethanol is 340 km (as described above: 290 km + 50 km).

15.2.2.2.5 Hydrogen Production

This study explored two pathways for hydrogen production in Brazil. One for sugarcane ethanol steam reform and a second route, included for comparison purposes, which uses alkaline electrolysis powered by wind-based electricity.

The ethanol steam reform is a process in which the ethanol is exposed to high temperature water steam that dissociates the alcohol molecules into CO_2 and H_2. The process includes a water-gas shift and a pressure swing adsorption stage. Electricity and feedstock inputs are considered as in Velandia Vargas and Seabra (2021), which is based on the H2A data (National Renewable Energy Laboratory 2018), see Table 15.3. H2A data portrays the hydrogen production at a forecourt, instead of a central plant, a logistically advantageous scheme; however, unable to take advantage from large economies of scale. The plant is able to supply hydrogen at 99.99% purity at a rate of 1500 kg day^{-1}. The data for the sugarcane crop and the ethanol production stages were obtained from Cavalett et al. (2013).

15.2.2.2.6 Biodiesel

The biodiesel produced in Brazil comes from several oilseed sources and animal fats. Soybean and beef tallow are the most representative sources. We thus assumed a mix of 86% soy and 16% beef tallow, which corresponds to the normalized contributions for 2018 (ANP 2018).

The data for soybean production refers to production conditions in the state of Mato Grosso, as included in the National Bank of Life Cycle Inventary SICV Brazil database (Ibict 2019). For soybean processing and biodiesel transesterification, values from Garcilasso (2014) and Cerri et al. (2017) were adopted. Tallow biodiesel is based on Garcilasso (2014) and Dufour and Iribarren (2012). Tallow was treated as a waste product from meat production; thus, no environmental burden was bestowed at the point of generation. For the transportation of tallow biodiesel from the plant to the distributor, we assumed the distance of 363 km between Lins, São Paulo (one of the main tallow biodiesel production centers in Brazil) and the refinery in Paulínia, São Paulo, where the blending would occur.

Soybean biodiesel processing was assumed to take place in Mato Grosso state. Transportation from Mato Grosso to Paulínia refinery occurs by trucks over a distance of 1200 km (Cerri et al. 2017). Transportation from the refinery to the gas station, for both biodiesel products, is carried out by truck over a distance of 50 km.

15.2.2.2.7 Low-Sulfur Diesel

For low-sulfur diesel data we appealed to Ecoinvent 3.3 '*diesel, low-sulfur, RoW | Alloc Rec, U'* process. This process is based in "Rest of the world" geography, which models conditions for countries outside Europe and the United States. Such process was used due to the absence of peer-reviewed data depicting the Brazilian conditions for diesel production.

Diesel-biodiesel blending is mandatory in Brazil. For this analysis, we assumed a 10% blend of biodiesel in diesel. Transportation from the factory to the gas station occurs by truck for a distance of 50 km.

15.2.2.2.8 Green Diesel

The production of alternative diesel from sugarcane ethanol is explored given the recognition of this pathway by the recent Brazilian regulation for green diesel production (EPE 2020c; ANP 2021). In this pathway—which is called *Alcohol-to-Jet* (ATJ) and has been well discussed for producing alternative jet fuel (Staples et al. 2014; Klein et al. 2018; Santos et al. 2018)—alcohol molecules are dehydrated, oligomerized, and finally hydrogenated to suitable hydrocarbon chains (Staples et al. 2014; Atsonios et al. 2015; Klein et al. 2018), such as kerosene, diesel, and naphtha. The ATJ technology was mostly based on Klein et al. (2018), who designed the overall process for maximizing the kerosene production with a small slate of drop-in diesel. In this case, the light streams production (*e.g.*, propane) is assumed to be used for process self-supply; furthermore, the green diesel production process is fed by hydrated ethanol and hydrogen at 11.0 kg $H_2/t_{ethanol}$. For the latter, we assumed hydrogen is produced by wind energy-fed alkaline electrolysis.

15.2.2.3 Maintenance and End-of-Life

Input data for the maintenance of LDVs were retrieved from Ecoinvent 3.3 as well. The dataset *Maintenance, passenger car, electric, without battery| Alloc Rec, U* was used for the BEV and the SOFCV; such LCI includes information for the maintenance of non-ICEV vehicles. Furthermore, for the LIB in the BEV and the FC in the SOFCV, we assumed that the manufacturing techniques enable such devices to cover the entire lifetime—150,000 km—of the cars. These assumptions for the BEV are supported by empirical evidence (Rodrigues et al. 2015; D'Angelo 2017), while for the SOFCV we accepted it for the sake of simplicity and due to the lack of data. For the PEMFCV we included the maintenance information provided by Miotti et al. (2017). For ICEVs and HEVs the maintenance information was obtained from Ecoinvent dataset *Maintenance, passenger car| Alloc Rec, U.*

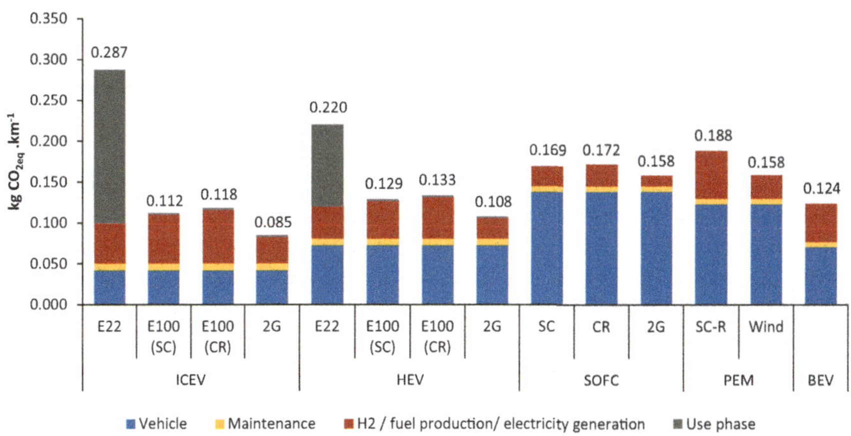

Fig. 15.5 The highest carbon intensities of the evaluated pathways are associated to the fossil fuel-related options, namely ICEV and HEV propelled by E22 gasoline. For those cases, the use phase exhibits the dominant environmental burden, highlighting a central issue of combustion-based technologies: the tailpipe emissions. Internal combustion engine technologies using biofuels also present a contribution to global warming impacts, mostly due to methane formation; however, such contribution is not significant and it is overshadowed by the other lifecycle stages. As perceived, the most competitive results were displayed by bioethanol driven ICEVs, closely followed by HEVs

Regarding the EoL stage, we incorporated the mean conditions included in Ecoinvent 3.3 LCIs, which include the BEV, the ICEV and the HEV. For the SOFCV, we adopted the chassis and body data from a BEV, while for the SOFC itself the modelling of EoL stage was impracticable, due to large uncertainties. In fact, there is absence of information about EoL for the majority of materials in the fuel cell stack (HyTechCycling 2019); however, EoL stage is expected to have a marginal contribution to the overall results. For the PEMFCV we adopted the data from Miotti et al. (2017).

For the HDVs maintenance stage, a proxy LCI was created due to the lack of data. This proxy LCI is based on the assumption that the maintenance stage along the entire lifecycle of the bus would require approximately 17% of the materials and energy required for the bus manufacturing, as stated by García Sánchez et al. (2013). The FC in the SOFCB was considered not to require any maintenance analogous to the SOFC vehicle. In spite of the maintenance LCI being constructed based on approximations, we do not expect it could significantly alter the results since the maintenance is expected to present a low contribution to the overall environmental burden per km (Cooney et al. 2013; Velandia Vargas et al. 2019). For the EoL of the HDVs chassis and body, our modelling was based on the Brazilian conditions represented by Velandia Vargas et al. (2019). For the LIB we maintained the EoL conditions included in the Ecoinvent datasets.

15.3 Lyfe Cycle Impact Assessment for LDVs

The life cycle impact assessment results for the LDVs using one travelled kilometer as functional unit are shown in Fig. 15.5.

For LDVs, the carbon footprint associated to the vehicle manufacturing vary. The environmental advantages, per km, of ICEVs and HEVs options are backed by the lower environmental burden associated to the vehicle. This advantage is notorious when compared to new powertrain technologies, such as SOFCVs and PEMFCVs.

This should not be surprising as decades of investments to reduce curb weight and fuel consumption put the ICEVs in an advantageous position in terms of research and development. HEVs can also benefit from the expertise inherited from their ICEV components; however, the partial electrification of their powertrains exposes HEVs to new challenges. Taking into account the lower vehicle mass and the biogenic nature of the CO_2 emissions linked to biofuels, the ethanol-fueled ICEV and HEV are expected to be competitive considering its low global warming impacts. However, the use of biofuels in combustion-based technologies may only tackle the greenhouse gas emissions and would not solve other environmental problems related to other emissions, such as CO, SO_x, NO_x and particulates.

Fuel cell-based technologies were found to be the less competitive, per km, among the studied cases. Nonetheless, the results must be evaluated keeping in mind the contribution of the vehicle manufacturing to the entire life cycle. As seen in Table 15.1, the weight of vehicles varies significantly. For instance, the ICEV weight

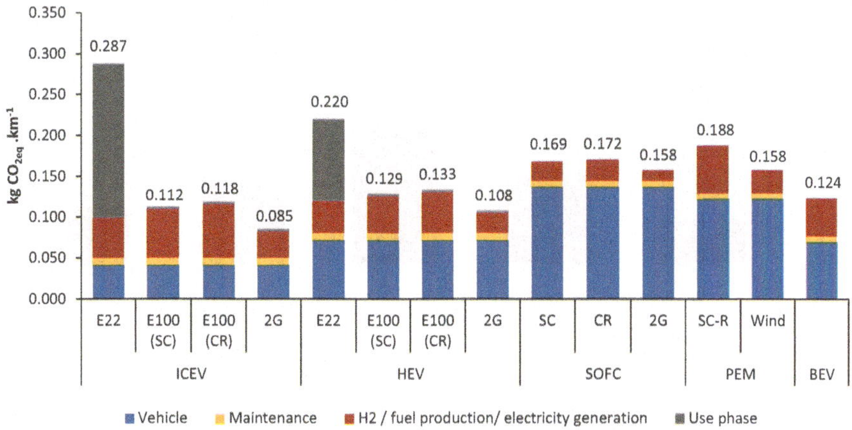

Fig. 15.6 Results of the characterization for light-duty vehicles. *2G* second-generation ethanol, *BEV* battery electric vehicle, *CR* corn ethanol, *HEV* hybrid electric vehicles, *ICEV* internal combustion engine vehicles, *PEM* polymer-electrolyte-membrane fuel cell vehicle, *SC* sugarcane ethanol, *SC-R* sugarcane ethanol reform, *SOFC* solid-oxide fuel cell vehicle

is only 62% of the modelled SOFCV, which puts the mature ICEV way ahead in terms of environmental competitiveness, as more materials imply, inextricably, larger environmental impacts. Analogously, the PEMFCV environmental performance is also affected by the vehicle; however, hydrogen production via sugarcane ethanol reform also displays a significant footprint even considering the biogenic nature of its CO_2 emissions. Moreover, BEVs are poised to reduce their weight in the following decades, which would render them more competitive in terms of carbon footprint. Brazil and other biofuel producing countries should engage in discussions on how to electrify the LDV and HDV fleets without resigning their technological, economic and environmental leadership (Vasconcelos 2017; CNPEM 2018; Teixeira 2018; Velandia Vargas et al. 2020).

As the validity of the results of an LCA study are reliant on the selection of an adequate functional unit to be compared, our results are well founded, considering that we compare the transportation service of the vehicles travelling 1 km. Nonetheless, the comparison of different powertrain technologies could be unfair considering that the vehicle contribution to the overall footprint could be substantial. Specially, the comparison tends to be biased by the large curb weight differences between the different vehicle technologies, for instance, the SOFCV is largely penalized by their weight. Although, it must be said that given the large uncertainties inherent to the modelling of a vehicle that does not even exist commercially yet, our premises were conservative and necessary to avoid creating expectations that could not be met by this technology, still in prototype tests.

However, if the SOFCV technical and economic challenges were to be overcome, we would see no reasons for the technology to be significantly heavier than the other evaluated technologies. In such scenario, the SOFCV vehicle would indeed represent a promising technology in tackling greenhouse gas emissions and additionally,

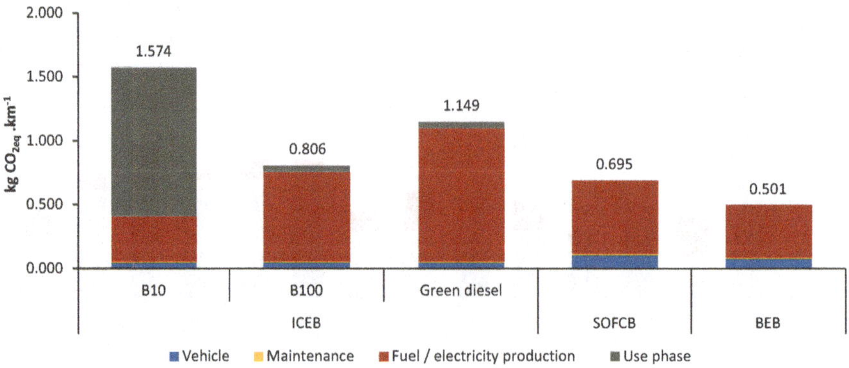

Fig. 15.7 Results of the characterization for heavy-duty vehicles. *B10* biodiesel blended with diesel (10/90, % volume), *B100* pure biodiesel, *BEB* battery electric bus, *ICEB* internal combustion engine busses, *SOFC* solid-oxide fuel cell bus

it would provide the automotive market with a magnificent option to create a synergy between transport electrification and biofuels (Velandia Vargas and Seabra 2021).

Figure 15.6 depicts the results for HDVs. At first sight, there is a notorious divergence between HDVs and LDVs characterization results per lifecycle stage. The contributions of the vehicle—or bus in this case—in the total per km are considerably less significant than in the case of LDVs. This is due to several factors, such as the large fuel consumption per km and the larger life expectancy that was presumed to be 880,000 km. The use phase in the ICEB is responsible for the largest contribution to total results. Remarkably, the blended diesel production—B10—by itself presents much less global warming potential burden compared to other fuels and even electricity; in other words, it is the fossil nature of the carbon released into the atmosphere what causes the most concerning impacts and not the energy or emissions related to fossil fuel extraction.

In fact, despite of the tailpipe emissions of biodiesel-fueled and green diesel-fueled busses being characterized as biogenic carbon, the fuel production stage remains the largest source of GHG emissions. Actually, although the Alcohol-to-Jet route for green diesel in this study was assumed to be supplied with hydrogen totally produced from renewable energy—*i.e.,* 100% wind electricity via alkaline electrolysis—, the carbon footprint is only 30% lower than the current scenario (B10). Nevertheless, we ought to state that green diesel was assumed here as a coproduct of jet-fuel (Capaz et al. 2020). Specific process design for maximizing green diesel production from hydrotreatment of ethanol could provide a different environmental performance (Fig. 15.7).

Furthermore, although the use of biofuels on busses could mitigate GHG emissions when compared to B10 diesel, the environmental (EPE 2021) and health hazards (Dallmann 2019) of other emissions remain a concern, which founds the case for transport electrification, which eliminates tailpipe emissions. BEBs and the theoretical SOFCBs display the best mitigation results despite of being much

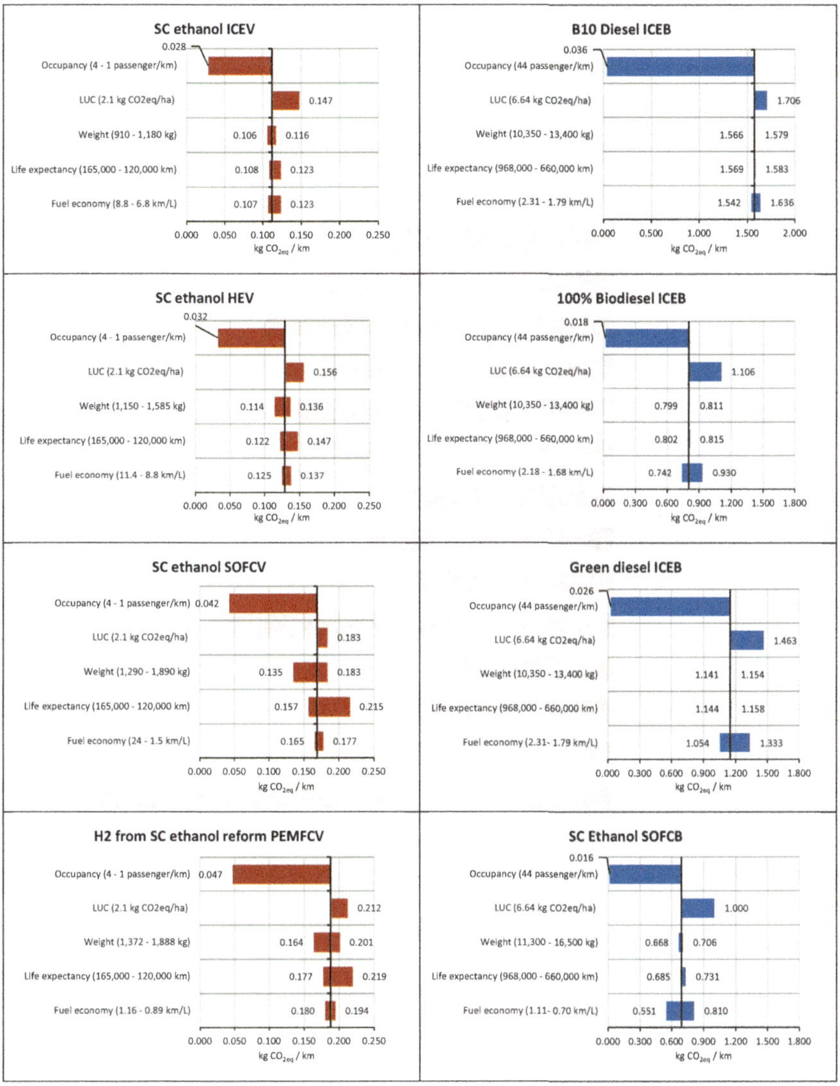

Fig. 15.8 Results of the sensitivity analysis

heavier. In fact, as the HDVs face intense use throughout their lifecycles and are intended to transport a large amount of people, the contribution of the vehicle is diluted, rendering the fuel burden more significant when evaluated per km.

In a global context, many scenarios evaluated, using integrated assessment models, suggest an increasing importance of electrification in the transport sector (van Vuuren et al. 2017; Rochedo et al. 2018; Daioglou et al. 2020). At the same time, the use of biofuels combined or not with technologies for carbon capture and storage are also projected to be a key climate change mitigation strategy, together

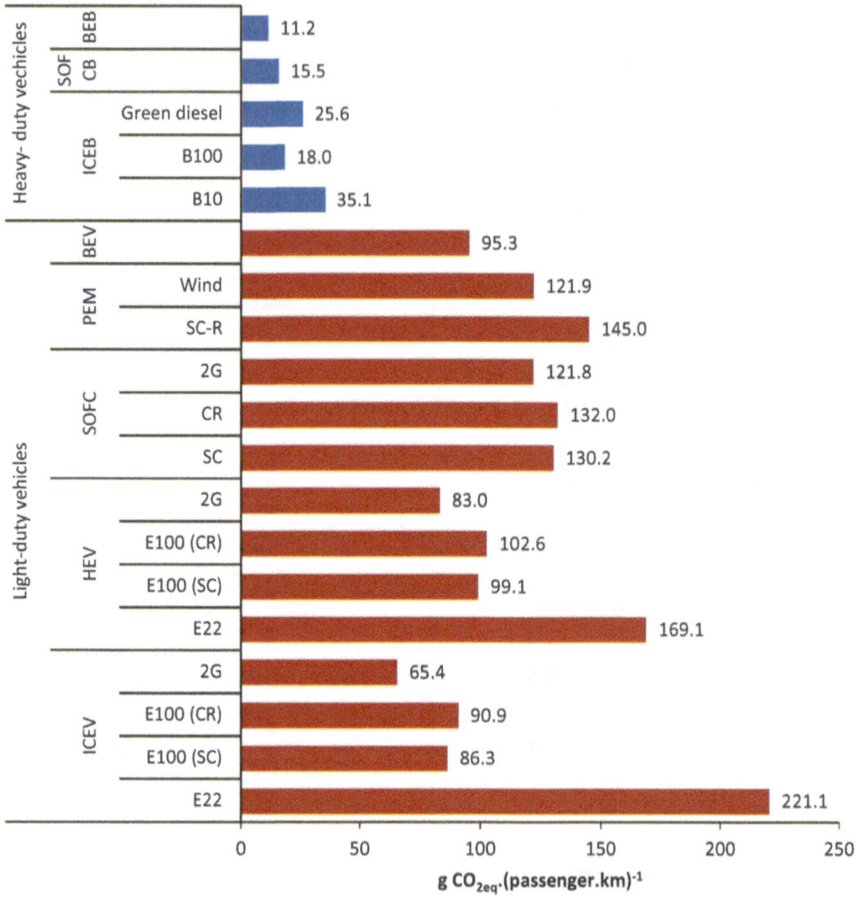

Fig. 15.9 GHG emissions considering average occupancy rates for light and heavy-duty vehicles

with other options including hydrogen, ammonia, and e-fuels (synthetic hydrocarbons produced from renewable electricity) and reductions in the demand side (IEA 2021a, b). While it is increasingly argued that the priority use of biofuels is shifting towards the aviation and marine sectors (where drop-in fuels or other high energy density liquid fuels may be required as electrification may be more complex), the use of biofuels is still expected to provide significant climate change mitigation in the road transport sector (IEA Bioenergy 2020; Bauer et al. 2020).

15.3.1 Sensitivity Analysis

The results of the sensitivity analysis are displayed in Fig. 15.8. In the face of the large number of cases evaluated in this study, we decided to perform the sensitivity analysis only for four LDV and four HDV options. All of the evaluated LDV cases are those of vehicles propelled by sugarcane ethanol.

Occupancy rate is, by an overwhelming difference, the parameter able to create the most significant mitigation of impacts. For HDVs the mitigation potential is even larger, emphasizing the environmental advantages of public transportation over private cars. However, it could be considered that occupancy rate should not be included in the same category of other evaluated parameters, such as life expectancy and fuel consumption, as it is not related to vehicle performance itself. However, we maintained the occupancy rate in the sensitivity analysis in order to create a broader discussion for the reader considering that it could effectively affect the lifecycle results per passenger transported.

The average occupancy rate of private vehicles and busses in Sorocaba, a city in São Paulo state, is 1.3 and 44 passengers per trip, respectively (D'Agosto et al. 2018). Assuming such values are representative for our analysis, we were able to graphically represent the potential reduction of GHG emissions when evaluating the emissions per passenger as displayed in Fig. 15.9. There is a clear appeal for public transportation and carsharing as it allows the reduction of private vehicle trips and their related emissions, not to mention the reduction of vehicles in the streets which enhances traffic speed.

Regarding HDVs, LUC is also a significant parameter, inextricably linked to the use of soybean for biodiesel production. In fact, carbon emissions linked to LUC are a key issue in crop-based fuels, having the potential to reduce or even nullify the benefits of replacing fossil fuels by biofuels (Wong 2008; Bailis and Baka 2010; Stratton et al. 2010; Moreira et al. 2014). In light of the ample debate on the topic, this study evaluated the sensitivity of the results considering only the direct LUC, *i.e.* addressing changes only within the assessed boundaries of a study (ISO 2018). Indirect or rebound effects, which are typically estimated using economic models, are not addressed here. From the average case reported by Novaes et al. (2017), the emission factors related to sugarcane and soybean expansion in Brazil (1999–2018) would be 2.10 and 6.64 tCO_{2e} ha $year^{-1}$, respectively. According to Novaes et al. (2017), which considered only CO_2 emissions, the sugarcane crop expanded over pasture areas, while soybean did it on natural vegetation.

Life expectancy and curb weight present a larger potential for GHG mitigation in LDVs while fuel consumption presents the lowest potential. In contrast, HDVs display the opposite behavior. This is mostly due to the significant fuel consumption associated to busses, as seen in Fig. 15.6. In other words, the environmental burden, per km, of private cars is notoriously more influenced by the vehicle whereas for busses, considering their significant fuel consumption and larger lifespan, the fuel production is the largest contributor.

The large uncertainties related to LUC of soybean biodiesel could seriously hamper the carbon emissions mitigation of Brazilian biodiesel. In short, whether soybean biodiesel comes from deforested areas, there is no point in making the substitution as the overall carbon emissions could be higher than for B10 diesel propelled ICEBs (Dallmann 2019). In contrast, SOFCBs would not face LUC concerns of the same magnitude as the considered sugarcane emission factors are lesser and sugarcane traceability in Brazil is less complex than that of soybean. Furthermore, a higher occupancy rate, could be a large source of GHG emissions mitigation as long as public transportation and carsharing is encouraged.

15.4 Conclusions

There is not one-size-fits-all measure for decarbonizing the transport sector. Even as electrification arrives faster than anticipated in some regions, the use of liquid biofuels for decarbonizing the road transport sector is also recognized as an important strategy, in the short to medium term, in many developing countries (IEA Bioenergy 2020; Milovanoff et al. 2020). Therefore, the strategies for climate change mitigation in the transport sector should be designed taking into account the local conditions and all the available options at a given time; they should also be periodically revisited in face of the novel technology developments and sustainability implications. In this context, this work has demonstrated that in the case of Brazil, both first- and second-generation ethanol from sugarcane are options that offer a high climate change mitigation.

References

ANP (2018) Biocombustíveis, Biodiesel: Boletim Mensal Do Biodiesel, Jan-Dez 2018. Agência Nacional de Petróleo, Gás Natural e Biocombustíveis [Brazilian National Agency of Petroleum, Natural Gas and Biofuels], Brasília. http://www.anp.gov.br/publicacoes/boletins-anp/2395-boletim-mensal-da-producao-de-petroleo-e-gas-natural

ANP (2021) Resolução n° 842, de 14 de maio de 2021. In: Estabelece a especificação do diesel verde, bem como as obrigações quanto ao Control. da Qual. a serem atendidas pelos agentes econômicos que o Comer. em Territ. Nac

Argonne National Laboratory (2020) GREET 2020. https://greet.es.anl.gov/net. Accessed 20 Jan 2021

AsiaTimes (2021) Japan may be on road to nowhere with hydrogen vehicles. https://asiatimes.com/2021/09/japan-may-be-on-road-to-nowhere-with-hydrogen-vehicles/. Accessed 11 Oct 2021

Associação Brasileira do Alumínio (2019) Anuário estatístico do alumínio. http://abal.org.br/estatisticas/nacionais/perfil-da-industria/. Accessed 12 Nov 2020

Associação Brasileira do Cobre (2018) Anuário brasileiro do cobre. http://www.virapagina.com.br/abcobre2018/. Accessed 10 Nov 2020

Atsonios K, Kougioumtzis M, Panopoulos KD, Kakaras E (2015) Alternative thermochemical routes for aviation biofuels via alcohols synthesis: process modeling, techno-economic assessment and comparison. Appl Energy 138:346–366. https://doi.org/10.1016/j.apenergy.2014.10.056

Bailis RE, Baka JE (2010) Greenhouse gas emissions and land use change from Jatropha curcas-based jet fuel in Brazil. Environ Sci Technol 44:8684–8691. https://doi.org/10.1021/es1019178

Bauer N, Rose SK, Fujimori S et al (2020) Global energy sector emission reductions and bioenergy use: overview of the bioenergy demand phase of the EMF-33 model comparison. Clim Chang 163:1553–1568. https://doi.org/10.1007/s10584-018-2226-y

Bekel K, Pauliuk S (2019) Prospective cost and environmental impact assessment of battery and fuel cell electric vehicles in Germany. Int J Life Cycle Assess 24:2220–2237. https://doi.org/10.1007/s11367-019-01640-8

Brasil (2018) Lei 13576. http://www.planalto.gov.br/ccivil_03/_ato2015-2018/2017/lei/L13576.htm. Accessed 28 Oct 2021

Capaz RS, de Medeiros EM, Falco DG et al (2020) Environmental trade-offs of renewable jet fuels in Brazil: beyond the carbon footprint. Sci Total Environ 714:136696. https://doi.org/10.1016/j.scitotenv.2020.136696

Capaz RS, Posada JA, Osseweijer P, Seabra JEA (2021) The carbon footprint of alternative jet fuels produced in Brazil: exploring different approaches. Resour Conserv Recycl 166:105260. https://doi.org/10.1016/j.resconrec.2020.105260

Cavalett O, Chagas MF, Seabra JEA, Bonomi A (2013) Comparative LCA of ethanol versus gasoline in Brazil using different LCIA methods. Int J Life Cycle Assess 18:647–658. https://doi.org/10.1007/s11367-012-0465-0

Cerri CEP, You X, Cherubin MR et al (2017) Assessing the greenhouse gas emissions of Brazilian soybean biodiesel production. PLoS One 12:1–14. https://doi.org/10.1371/journal.pone.0176948

Climate Watch (2018) Climate Watch data. In: Emissions Policies https://www.climatewatchdata.org/. Accessed 28 Oct 2021

CNPEM (2018) Eletrificação de veículos e o futuro do etanol combustível no Brasil. In: Separatas CNPEM. http://cnpem.br/acesso-a-informacao/separatas-cnpem/. Accessed 6 Aug 2019

Cooney G, Hawkins TR, Marriott J (2013) Life cycle assessment of diesel and electric public transportation buses. J Ind Ecol 17:689–699. https://doi.org/10.1111/jiec.12024

D'Agosto DAM, Gonçalves D, de Almeida I, et al (2018) Relatório de Resultados obtidos após aplicação de metodologias nas cidades piloto de Sorocaba e Uberlândia. https://antigo.mdr.gov.br/images/stories/ArquivosSEMOB/ArquivosPDF/eficiencia/publicacoes/estudoeemu3aplicacaodemetodologiasnascidadespiloto.pdf. Accessed 16 Sep 2021

D'Angelo M (2017) Finnish Tesla Model S taxi driver crosses 400,000 km, 93% of battery life remains. In: Teslarati. https://www.teslarati.com/tesla-model-s-400k-km-250k-mi-7-percent-battery-degradation/. Accessed 1 Aug 2018

Daioglou V, Rose SK, Bauer N et al (2020) Bioenergy technologies in long-run climate change mitigation: results from the EMF-33 study. Clim Chang 163:1603–1620. https://doi.org/10.1007/S10584-020-02799-Y

Dallmann T (2019) Climate and air pollutant emissions benefits of bus technology options in São Paulo. http://www.promobe.com.br/library/climate-and-air-pollutant-emissions-benefits-of-bus-technology-options-in-sao-paulo/. Accessed 6 Jan 2020

de Souza NRD, Klein BC, Chagas MF et al (2021) Towards comparable carbon credits: harmonization of LCA models of cellulosic biofuels. Sustain 13:10371. https://doi.org/10.3390/SU131810371

DNPM (2017) Sumário Mineral 2016. https://www.google.com/url?sa=t&rct=j&q=&esrc=s&source=web&cd=&ved=2ahUKEwiKxcLT2cruAhWPHrkGHcUeB6AQFjACegQIAxAC&url=https%3A%2F%2Fwww.gov.br%2Fanm%2Fpt-br%2Fcentrais-de-conteudo%2Fpublicacoes%2Fserie-estatisticas-e-economia-mineral%2Fsumario-mineral%2Fsu. Accessed 1 Jan 2021

Donke A, Nogueira A, Matai P, Kulay L (2016) Environmental and energy performance of ethanol production from the integration of sugarcane, corn, and grain sorghum in a multipurpose plant. Resources 6:1. https://doi.org/10.3390/resources6010001

Dufour J, Iribarren D (2012) Life cycle assessment of biodiesel production from free fatty acid-rich wastes. Renew Energy 38:155–162. https://doi.org/10.1016/j.renene.2011.07.016

Ellingsen LAW, Majeau-Bettez G, Singh B et al (2013) Life cycle assessment of a lithium-ion battery vehicle pack. J Ind Ecol 18:113–124. https://doi.org/10.1111/jiec.12072

Elsevier (2016) Nissan unveils world's first FCEV with SOFC running on bioethanol. Fuel Cells Bull 2016. https://doi.org/10.1016/S1464-2859(16)30231-0

EPA (2010) Renewable fuel standard (RFS2): final rule additional resources. https://www.epa.gov/renewable-fuel-standard-program/renewable-fuel-standard-rfs2-final-rule-additional-resources. Accessed 28 Oct 2021

EPE (2020a) National energy balance. Brasilia

EPE (2020b) Balanço energético nacional. Relatório Síntese BEN 2020. https://www.epe.gov.br/sites-pt/publicacoes-dados-abertos/publicacoes/PublicacoesArquivos/publicacao-479/topico-521/RelatórioSíntese BEN 2020-ab 2019_Final.pdf. Accessed 16 Nov 2020

EPE (2020c) Combustíveis renováveis para uso em motores do ciclo Diesel. https://www.epe.gov.br/sites-pt/publicacoes-dados-abertos/publicacoes/PublicacoesArquivos/publicacao-467/NT_Combustiveis_renovaveis_em_motores_ciclo_Diesel.pdf. Accessed 8 Aug 2021

EPE (2021) Impacto na saúde humana pelo uso de biocombustíveis na Região Metropolitana de São Paulo. In: Publicações. https://www.epe.gov.br/pt/publicacoes-dados-abertos/publicacoes/impacto-na-saude-humana-pelo-uso-de-biocombustiveis-na-regiao-metropolitana-de-sao-paulo. Accessed 28 Sep 2021

EU Science Hub (2020) Renewable Energy – Recast to 2030 (RED II). https://ec.europa.eu/jrc/en/jec/renewable-energy-recast-2030-red-ii. Accessed 2 Dec 2020

Faria R, Marques P, Moura P et al (2013) Impact of the electricity mix and use profile in the life-cycle assessment of electric vehicles. Renew Sust Energ Rev 24:271–287. https://doi.org/10.1016/j.rser.2013.03.063

Ferreira H, Leite MGP (2015) A life cycle assessment study of iron ore mining. J Clean Prod 108:1081–1091. https://doi.org/10.1016/j.jclepro.2015.05.140

Field JL, Richard TL, Smithwick EAH et al (2020) Robust paths to net greenhouse gas mitigation and negative emissions via advanced biofuels. Proc Natl Acad Sci U S A 117:21968–21977. https://doi.org/10.1073/pnas.1920877117

García Sánchez JA, López Martínez JM, Lumbreras Martín J et al (2013) Impact of Spanish electricity mix, over the period 2008–2030, on the life cycle energy consumption and GHG emissions of electric, hybrid diesel-electric, fuel cell hybrid and diesel bus of the madrid transportation system. Energy Convers Manag 74:332–343. https://doi.org/10.1016/j.enconman.2013.05.023

Garcilasso VP (2014) Análise entre processos e matérias-primas para a prodção de biodiesel. In: University of São Paulo. PhD Thesis. https://teses.usp.br/teses/disponiveis/106/106131/tde-24022015-102338/publico/garcilasso.pdf. Accessed 6 Sep 2021

Gerbrandt K, Chu PL, Simmonds A et al (2016) Life cycle assessment of lignocellulosic ethanol: a review of key factors and methods affecting calculated GHG emissions and energy use. Curr Opin Biotechnol 38:63–70. https://doi.org/10.1016/J.COPBIO.2015.12.021

Hawkins TR, Singh B, Majeau-Bettez G, Strømman AH (2013) Comparative environmental life cycle assessment of conventional and electric vehicles. J Ind Ecol 17:53–64. https://doi.org/10.1111/j.1530-9290.2012.00532.x

HyTechCycling (2019) Case studies with new strategies in dismantling and recycling stage D4.3. In: LCA FCH Technologies considering new Strategies and Technologies in the phase recycling and dismantling. http://hytechcycling.eu/wp-content/uploads/d4-3-case-studies-with-new-strategies-in-dismantling-and-recycling-stage.pdf. Accessed 29 May 2020

Ibict (2019) SICV - Banco Nacional de Inventários Do Ciclo de Vida. http://sicv.acv.ibict.br. Accessed 1 Sep 2021

IEA (2020) Data and statistics. https://www.iea.org/data-and-statistics. Accessed 28 Oct 2021

IEA (2021a) Energy Technology Perspectives 2020. https://www.iea.org/reports/energy-technology-perspectives-2020. Accessed 28 Oct 2021

IEA (2021b) Net Zero by 2050. https://www.iea.org/reports/net-zero-by-2050. Accessed 28 Oct 2021

IEA Bioenergy (2020) The role of renewable transport fuels in decarbonizing road transport | Bioenergy. https://www.ieabioenergy.com/blog/publications/new-publication-the-role-of-renewable-transport-fuels-in-decarbonizing-road-transport/. Accessed 28 Oct 2021

Instituto Aço Brasil (2019) Anuário estatístico 2019. https://acobrasil.org.br/site/wp-content/uploads/2019/10/AcoBrasil_Anuario_2019.pdf. Accessed 12 Nov 2020

IPCC (2014) Climate change 2014: Synthesis report. Contribution of working groups i, ii and iii to the fifth assessment report of the intergovernmental panel on climate change. Geneva

IPCC (2019) Special report on climate change and land. https://www.ipcc.ch/srccl/cite-report/. Accessed 28 Oct 2021

ISO (2006a) ISO 14040:2006 Environmental management – Life cycle assessment – Principles and framework

ISO (2006b) ISO 14044:2006 Environmental management – Life cycle assessment – Requirements and guidelines

ISO (2018) ISO 14067:2018 – Greenhouse gases – Carbon footprint of products - Requirements and guidelines for quantification

JCR (2010) International reference life cycle data system (ILCD) handbook: general guide for life cycle assessment - detailed guidance (1st edn). Luxembourg

Klein BC, Chagas MF, Junqueira TL et al (2018) Techno-economic and environmental assessment of renewable jet fuel production in integrated Brazilian sugarcane biorefineries. Appl Energy 209:290–305. https://doi.org/10.1016/j.apenergy.2017.10.079

Löbberding H, Wessel S, Christian O et al (2020) From cell to battery system in BEVs: analysis of system packing efficiency and cell types. World Electr Veh J 1–15. https://doi.org/10.3390/wevj11040077

Ma H, Balthasar F, Tait N et al (2012) A new comparison between the life cycle greenhouse gas emissions of battery electric vehicles and internal combustion vehicles. Energy Policy 44:160–173. https://doi.org/10.1016/j.enpol.2012.01.034

Majeau-Bettez G, Hawkins TR, StrØmman AH (2011) Life cycle environmental assessment of lithium-ion and nickel metal hydride batteries for plug-in hybrid and battery electric vehicles. Environ Sci Technol 45:4548–4554. https://doi.org/10.1021/es103607c

MapBiomas Brasil (2020) Dados de Infraestrutura. https://mapbiomas.org/dados-de-infraestrutura?cama_set_language=pt-BR. Accessed 28 Oct 2021

Messagie M, Boureima F-S, Coosemans T et al (2014) A range-based vehicle life cycle assessment incorporating variability in the environmental assessment of different vehicle technologies and fuels. Energies 7:1467–1482. https://doi.org/10.3390/en7031467

Milovanoff A, Posen ID, MacLean HL (2020) Electrification of light-duty vehicle fleet alone will not meet mitigation targets. Nat Clim Chang 1012(10):1102–1107. https://doi.org/10.1038/s41558-020-00921-7

Miotti M, Hofer J, Bauer C (2017) Integrated environmental and economic assessment of current and future fuel cell vehicles. Int J Life Cycle Assess 22:94–110. https://doi.org/10.1007/s11367-015-0986-4

Moreira M, Gurgel AC, Seabra JEA (2014) Life cycle greenhouse gas emissions of sugar cane renewable jet fuel. Environ Sci Technol 48:14756–14763. https://doi.org/10.1021/es503217g

Moreira MMR, Seabra JEA, Lynd LR et al (2020) Socio-environmental and land-use impacts of double-cropped maize ethanol in Brazil. Nat Sustain. https://doi.org/10.1038/s41893-019-0456-2

National Renewable Energy Laboratory (2018) H2A: hydrogen analysis production case studies. https://www.nrel.gov/hydrogen/h2a-production-case-studies.html. Accessed 9 Jan 2021

Nogueira LAH, Silva Capaz R (2013) Biofuels in Brazil: evolution, achievements and perspectives on food security. Glob Food Sec 2:117–125

Nogueira LAH, Capaz RS, Souza SP, Seabra JEA (2016) Biodiesel program in Brazil: learning curve over ten years (2005–2015). Biofuels Bioprod Biorefin 10:728–737. https://doi.org/10.1002/BBB.1718

Novaes RML, Pazianotto RAA, Brandão M et al (2017) Estimating 20-year land-use change and derived CO2 emissions associated with crops, pasture and forestry in Brazil and each of its 27 states. Glob Chang Biol 23:3716–3728. https://doi.org/10.1111/gcb.13708

Pereira LG, Cavalett O, Bonomi A et al (2019) Comparison of biofuel life-cycle GHG emissions assessment tools: the case studies of ethanol produced from sugarcane, corn, and wheat. Renew Sust Energ Rev 110:1–12. https://doi.org/10.1016/J.RSER.2019.04.043

Reuters (2021) Drought-struck Brazil expects Sept rainfall well below average. In: Reutes.com. https://www.reuters.com/world/americas/drought-struck-brazil-expects-sept-rainfall-well-below-average-2021-08-27/. Accessed 21 Sep 2021

Robertson GP, Hamilton SK, Barham BL et al (2017) Cellulosic biofuel contributions to a sustainable energy future: choices and outcomes. Science (80-) 356:eaal2324. https://doi.org/10.1126/SCIENCE.AAL2324

Rochedo PRR, Soares-Filho B, Schaeffer R et al (2018) The threat of political bargaining to climate mitigation in Brazil. Nat Clim Chang 88(8):695–698. https://doi.org/10.1038/s41558-018-0213-y

Rodrigues A, Cooper T, Watkins M (2015) Driving in the wrong lane: towards a longer life-span of cars. In: Plate conference. Nottingham

Santos CI, Silva CC, Mussatto SI et al (2018) Integrated 1st and 2nd generation sugarcane bio-refinery for jet fuel production in Brazil: Techno-economic and greenhouse gas emissions assessment. Renew Energy 129:733–747. https://doi.org/10.1016/j.renene.2017.05.011

Staples MD, Malina R, Olcay H et al (2014) Lifecycle greenhouse gas footprint and minimum selling price of renewable diesel and jet fuel from fermentation and advanced fermentation production technologies. Energy Environ Sci 7:1545. https://doi.org/10.1039/c3ee43655a

Stratton RW, Wong HM, Hileman JI (2010) Life cycle greenhouse gas emissions from alternative jet fuels. Partnership for Air Transportation Noise and Emission Reduction, Cambridge

Strazza C, Del Borghi A, Costamagna P et al (2010) Comparative LCA of methanol-fuelled SOFCs as auxiliary power systems on-board ships. Appl Energy 87:1670–1678. https://doi.org/10.1016/j.apenergy.2009.10.012

Swiss Centre for Life Cycle Inventories (2014) Ecoinvent database. Version 3.3

Teixeira L (2018) O que falta para os carros elétricos invadirem o Brasil. In: Forbes. https://forbes.uol.com.br/negocios/2018/10/o-que-falta-para-os-carros-eletricos-invadirem-o-brasil/. Accessed 25 Oct 2019

van Vuuren DP, Stehfest E, Gernaat DEHJ et al (2017) Energy, land-use and greenhouse gas emissions trajectories under a green growth paradigm. Glob Environ Chang 42:237–250. https://doi.org/10.1016/J.GLOENVCHA.2016.05.008

Vasconcelos Y (2017) The challenges faced in Brazil. In: Pesquisa Fapesp. http://revistapesquisa.fapesp.br/en/2018/01/12/the-challenges-faced-in-brazil/. Accessed 1 May 2019

Velandia Vargas JE, Seabra JEA (2021) Fuel-cell technologies for private vehicles in Brazil: environmental mirage or prospective romance? A comparative life cycle assessment of PEMFC and SOFC light-duty vehicles. Sci Total Environ 798:149265. https://doi.org/10.1016/j.scitotenv.2021.149265

Velandia Vargas JE, Falco DG, da Silva Walter AC et al (2019) Life cycle assessment of electric vehicles and buses in Brazil: effects of local manufacturing, mass reduction, and energy consumption evolution. Int J Life Cycle Assess. https://doi.org/10.1007/s11367-019-01615-9

Velandia Vargas JE, Seabra JEA, Cavaliero CKN et al (2020) The new neighbor across the street: an outlook for battery electric vehicles adoption in Brazil. World Electr Veh J 11:60. https://doi.org/10.3390/wevj11030060

Wong HM (2008) Life-cycle assessment of greenhouse gas emissions from alternative jet fuels. Massachusetts Institute of Technology

Chapter 16
Carbon Credits and the Bioethanol Industry: Governmental Programs and Incentives

Renato Godinho, Miguel Ivan Lacerda de Oliveira, Luciano Rodrigues, Marcelo Moreira, and Joaquim E. A. Seabra

Abstract Strong policy mechanisms are one of the most important factors to enable biofuels production in a country. While technology-push policies are important to allow faster development in new technologies and to create supporting mechanisms, market-pull instruments appear to be central for biofuels development worldwide, given the presence of positive externalities that are not autonomously recognized by the market system. This chapter brings an overview of the most prevalent market-pull instruments in place (mandates, exemptions, and carbon related mechanisms) and few examples of policies implemented in major markets, outlining their main mechanisms. It presents a more detailed discussion about the new biofuels policy in Brazil (RenovaBio) as an example of carbon-related mechanism that could be explored in other developing countries with substantial bioenergy potential, recognizing the contributions that biofuels can give to climate change mitigation. As land use change is arguably the most important (and contentious) aspect to be considered in any biofuels policy, the chapter briefly addresses the concept behind this subject

R. Godinho
Ministry of Foreign Affairs (MRE), Brasília (DF), Brazil
e-mail: renato.godinho@itamaraty.gov.br

M. I. Lacerda de Oliveira
Instituto Nacional de Meteorologia (INMET), Brasília (DF), Brazil
e-mail: Miguel.oliveira@inmet.gov.br

L. Rodrigues
Agribusiness Studies Center at FGV and Brazilian Sugarcane Industry Association (UNICA), São Paulo (SP), Brazil
e-mail: luciano.rodrigues@fgv.br

M. Moreira
Agroicone, São Paulo (SP), Brazil
e-mail: marcelo@agroicone.com.br

J. E. A. Seabra (✉)
University of Campinas (UNICAMP), Campinas (SP), Brazil
e-mail: jeseabra@unicamp.br

C. R. Soccol et al. (eds.), *Liquid Biofuels: Bioethanol*, Biofuel and Biorefinery Technologies 12, https://doi.org/10.1007/978-3-031-01241-9_16

as well as its potential implications for the environmental benefits perceived for biofuels.

16.1 Introduction

Transport is the least diversified energy end-use sector. It consumes about two thirds of global oil final energy demand, with more than 90% of the final energy demand consisting of oil products (IEA 2017). In terms of CO_2, transport is responsible for roughly one-fourth of global energy-related emissions, mainly due to the contributions from road transport, mostly for passenger travel (IEA 2019a).

Biofuels are the main renewable energy source that can be directly used in the transport sector. Although the recognition of nonmarket benefits is often the driving force behind efforts to increase their use, especially with respect to climate change, biofuels can also bring economic benefits to rural communities and enhance energy security (Souza et al. 2015). Nevertheless, biofuels still struggle to reach cost-effectiveness against fossil fuels, and in 2018 they contributed to only around 90 Mtoe (or almost two million barrels of oil equivalent per day) (IEA 2019b).

A number of factors concur to make biofuels less economically competitive than the fossil-based alternatives, including: (1) much less established industry than the petrochemical industry, with smaller scale and much fewer sunk capital in the form of upstream, midstream and downstream assets (production/extraction, refineries, distribution centers, etc.); (2) wider variety and complexity of feedstocks, depending on natural factors not under control of producers, including rain patterns and commodity price fluctuations; (3) fossil-fuel subsidies, which are still very significant around the world, despite movements towards a phase-out; (4) still-maturing state of some key conversion pathway technologies used for some feedstock and final products, which in some cases are just now leaving demonstration stages and reaching market-ready scales.

This condition poses difficulties for the expansion of biofuels based on price competitiveness. As with other renewable energies, in the biofuels market it is observed the presence of positive externalities associated especially with the reduction of greenhouse gases emissions (GHG) promoted by these products. These externalities are not priced autonomously by the supply and demand system, creating a market failure that requires regulation and public policies. This explains why biofuels is currently a policy-driven industry all around the world, growing around a number of state-led mechanisms in various forms.

Strong policy mechanisms are arguably the single most important factor to enable biofuels production in a country. The Biofuture Platform[1] classifies biofuels policies under three areas (Biofuture Platform 2018), thus defined:

[1] The Biofuture Platform is a government-led, multistakeholder coalition of 21 countries, launched in November 2016 during UNFCCC COP 22. It seeks to promote increased policy alignment and convergence around the most efficient and impactful programs and instruments to promote

- Technology-push—where policies help reduce the cost of research and development to drive new ideas and reduce the cost of technology, taking early-stage technologies through the valley of death that exists between early development and demonstration.
- Market-pull—where the policy helps to create or increase market demand for the technology.
- Enabling support—where the policy addresses the barriers existent in the institutional environment to enable further innovation and deployment.

Technology-push policies are important to induce faster development in new technologies and to create supporting mechanisms, which are in many cases necessary to overcome specific natural or institutional constraints. On the other hand, the "market pull", or also called "demand-pull" category of instruments is the one most central to biofuels development worldwide. By forcing or stimulating demand, rewarding positive externalities, or establishing targets or carbon credit trading mechanisms, market pull instruments are capable to bridge the competitivity gap with fossil fuels and allow for scaling up the industry.

This chapter brings an overview of the most prevalent market-pull instruments (mandates, exemptions, and carbon related mechanisms) and few examples of policies implemented in major markets, outlining their main mechanisms. Section 16.4 presents a more detailed discussion about the new biofuels policy in Brazil (RenovaBio) as an example of carbon-related mechanism that could be explored in other developing countries with substantial bioenergy potential, recognizing the contributions that biofuels can give to climate change mitigation. As land use change is arguably the most important (and contentious) aspect to be considered in any biofuels policy, the concept behind this subject is addressed in Sect. 16.5 along with its potential implications for the environmental benefits perceived for biofuels.

16.2 Market-Pull Instruments

Market-pull instruments are popular across the main international players on bioenergy, and they particularly focus on biofuels rather than on bioproducts. Among the instruments, there is a clear preference for volumetric based mandates, quotas, tax incentives and investment support mechanisms. These are all broadly effective options to support technologies that are relatively mature, as they create a demand for biofuels, which is typically met with commercial conversion

---

sustainable bioenergy deployment. The most recent effort from the coalition is the Policy Blueprint, which "aims to accelerate the growth of the sustainable bioeconomy by providing countries with the methodologies, tools and practical guidance to evaluate and improve the impacts and effectiveness of their bioenergy and bioeconomy policies" (Biofuture Platform 2016).

technologies. The following paragraphs give a brief description of the most prevalent market-pull instruments.

Biofuels mandates: Biofuels mandates are the most traditional, by far the most widely disseminated, and arguably still the most relevant instrument supporting biofuels development worldwide, reaching 65 countries in 2021 (Biofuels Digest 2021). Those mandates can assume different forms, including the most simple and direct blending mandate of a minimum blend level nationwide (i.e., Brazil's separate ethanol and biodiesel mandates), target-based mandates, including the EU's RED II targets for renewable energy use in transport, and nationwide volumetric-based mandates, such as the USA's Renewable Fuel Standard's.

Tax exemptions: Biofuels benefit from a wide number of federal and provincial-level tax exemptions and benefits, frequently justified on an environmental, climate, or energy security basis. In particular contexts, those instruments do help to partially rebalance the competitiveness disadvantage against fossil fuels. Those exemptions, however, do not come close to match the total subsidies conceded to fossil fuels, which amounted to US$ 494 billion in 2019, according to the International Energy Agency (IEA 2021).

Carbon-related mechanisms: A more recent, but growing trend is a new generation of biofuels-related policies that directly focus on those fuel's key advantage: their inferior carbon footprint. Made from recent-growth biomass, the carbon-content biofuels can be considered as part of the natural carbon cycle, and thus not adding to the fossil-based carbon emissions that are the most important cause of human-driven climate change. Carbon-related biofuel policies seek to assess and quantify specific emissions-reductions values for different biofuels, and attribute them a corresponding economic incentive. Carbon life cycle assessment methodologies vary widely, as do the actual market and policy mechanisms used to provide the incentive. Some of those policies and programs also have an embedded carbon credit market, that can or cannot be fungible with other kinds of carbon markets.

16.3 Examples of Biofuels Policies

16.3.1 Renewable Fuel Standard 2

The United States is currently the world's largest biofuels producer, and the Renewable Fuel Standard 2 (RFS2) policy is the main instrument responsible for that production. RFS2 is in essence a volumetric nationwide blending mandate for biofuels, with the mandate applying to fossil fuel refiners (US EPA 2015). The Environmental Protection Agency determines the mandate level every year, within statutory bounds, based on, among other factors, the availability of sufficient supply of the several biofuel categories stipulated by the program, as a series of concentric sets (biofuels of any kind; "advanced" biofuels; and cellulosic biofuels). As the mandate is a total volumetric mandate, and not a blend level mandate, its achievement depend on total fuel volumes sold, especially because there are either regulatory or practical blending limits on biofuels (biodiesel is largely kept within 7% levels across many states, with some notable exceptions, while ethanol is mostly "stuck" at a 10% practical blending level, as there are few petrol stations carrying E15 throughout the country. Refiners that manage to blend above their target requirements can sell credits to those that underperformed or chose to not blend biofuels at all, resulting in a market for the Renewable Identification Numbers (RINs) which are bought and sold among refiners. The market price of the RIN reflects the implicit price competitiveness boost of biofuels. If this price is high, biofuels can be sold at a larger premium in relation to fossil fuels.

16.3.2 European Union Renewable Energy Directive

The European Union Renewable Energy Directive (EU RED II) is a complex piece of legislation to promote increased uptake of renewable energy across the Union. The main mechanism are mandatory renewable energy targets expressed as a percentual share of total energy used in different sectors. As a Directive, the target is mandatory for Member States, but the specific instruments and mechanisms to meet the target are left to each State. For the transport sector, RED II directs States to "require fuel suppliers to supply a minimum of 14% of the energy consumed in road and rail transport by 2030 as renewable energy" (European Union 2018). RED II in transport is famous for a number of policy complications and individualizations, including the constantly changing sustainability criteria associated with biofuels; the blunt classifications of the types of biofuels in broad categories that face distinct treatment within the target, be it a cap (such as the 7% cap imposed on food/feed based biofuels) or a double counting advantage.

Since it is a State-implemented Directive, RED II itself does not institute any tradeable credit system, and most States have largely sought to implement RED II by means of simple blending mandates, tax incentives, and nationwide targets or "quotas", usually separated by the kind of renewable energy used according to different classifications. Some countries go beyond the RED II's "conventional" and "advanced" biofuels categories to provide further technologies and categories, either capped or incentivized with double or even triple counting mechanisms, raising criticisms for creating "favorites" and for jeopardizing the actual renewable energy target by allowing some forms to count as double or triple for the purposes of the law. An example is given in Table 16.1 from Germany's 2021 "greenhouse gas reduction quota act", used to implement REDII in that country (Appunn 2021).

Table 16.1 German greenhouse gas reduction quota act 2021, based on the EU RED II directive

	2022	2023	2024	2025	2026	2027	2028	2029	2030
GHG quota (CO_2-reduction)	7%	8%	9.25%	10.5%	12%	14.5%	17.5	21%	25%
Food and feed crop (maximum share, energy)	4.4%								
Used cooking oils and animal fats (max. share, energy)	1.9%								
Advanced biofuels (minimum share, energy)	0.2%	0.3%	0.4%	0.7%	1%		1.7%		2.6%
	Quantities above the minimum share are counted with a factor of 2								
Aviation, power-to-liquid kerosene					0.5%		1%		2%
Hydrogen and power-to-x fuels	Quantities will be counted with a factor of 2 (refineries and road transport)								
Electricity	Quantities will be counted with a factor of 3 (electricity from public charging points, private e-cars, e-car fleets)								

From Appunn (2021)

16.3.3 Low Carbon Fuel Standard

California's Low Carbon Fuel Standard (LCFS) is perhaps the most well-known state-level biofuels policy, and with a reputation for being a successful policy in bringing down the State of California's transport emissions while helping to bring efficient, sustainable and innovative pathways and technologies to market. The LCFS's main difference with relation to most other biofuel policies is the centrality of carbon Life Cycle Assessment to its workings. Instead of establishing a more or less arbitrary set of categories, feedstocks and/or technologies to receive favorable or unfavorable treatments, the program has a defined methodology for calculating both default and actual carbon emissions values for any given low carbon fuel pathway, under a well-to-wheel approach. This technology-agnostic setup favors actual performance, letting markets and technologies compete freely for the higher scores, under the same broad methodology. This model, originally adopted in 2009, has later inspired similar policies in other US states and in other countries, such as Brazil's RenovaBio and Canada's Clean Fuel Standard (CFS).

LCFS's main goal is to reduce the carbon intensity of the transportation fuel pool in the state by 20% in 2030. It sets annual carbon intensity (CI) benchmarks for all fuels, which reduce over time. Similar to RFS, it also has an established market for credit transactions, which has exceeded US$ 2 billion since 2018 (California Air Resources Board 2020).

16.3.4 Carbon Offsetting and Reduction Scheme for International Aviation

The Carbon Offsetting and Reduction Scheme for International Aviation (Corsia) is a scheme adopted by the International Civil Aviation Organization (ICAO) to help it meets its Member State determined goal of a carbon neutral growth in the international aviation industry starting in 2020. The goal is to be met by airline operators using a "basket of measures", including operational and aviation efficiency measures, use of aviation biofuels (called "sustainable aviation fuels", SAFs, in ICAO parlance) and—as a transitional measure for when the formers are not enough—the buying of carbon offset credits from ICAO-approved carbon markets, issued from non-aviation sectors (industry, agriculture, others) (ICAO 2021). In this scheme, the implied economic incentive of aviation biofuels, from the airline operator standpoint, is equal to the price of buying offsets equivalent to the listed emissions reductions of a particular biofuel.

In order to quantify the emissions savings provided by different aviation fuel pathways in relation to fossil aviation kerosene, CORSIA sets out a complex full life cycle analysis (LCA) methodology, similar to that used in LCFS, containing both attributional (direct emissions calculations from the value chain) and consequential (inferred or estimated emissions resulting from higher biofuels demand on land use and other factors). CORSIA's two main sides—offsets and direct emissions reductions from biofuel usage—are not always in sync and employ different control mechanisms and criteria.

16.3.5 Other Carbon Markets

Carbon pricing, either via a direct carbon tax or through cap-and-trade market mechanisms, is also thought to be a policy instrument that can help drive competitiveness of biofuels, by creating a tax-based difference between those, which would be exempt from the tax, and fossil-based fuels. While there are very few countries in the world applying a carbon tax and corresponding exemption in this way, Sweden is a notable example. The country has a longstanding carbon tax on all fossil fuels, which in 2021 has surpassed US$ 110/ton (Pomerleau 2020), being therefore the world's highest. Biofuels are exempt from the tax, a benefit that Sweden has had to defend every year to the European Commission, which has been extending year-long waivers for several years (the last one being valid until December 31, 2022). The carbon tax has indeed helped Sweden to achieve a high penetration of biofuels in its transport mix (18% in 2020, according to the Swedish Energy Agency), the highest in Europe, and 54% renewable energy penetration in its whole energy mix (IEA 2019c; Swedish Energy Agency 2021).

As for national or international cap-and-trade carbon markets, there are few documented instances in which they had an impact on biofuels deployment. Offsets in internationally recognized schemes such as the Clean Development Mechanism (CDM), based on the Kyoto Protocol, are usually project-based, require compliance with additionality criteria, among others, and have therefore not been well suited to interface with broad based deployment of biofuels, which usually require a set of national policies which, almost by definition, would exclude the possibility of biofuels production being considered as "additional mitigation" against a contrafactual scenario in absence of an offset project. On the other hand, cap-and-trade schemes could theoretically benefit biofuel deployment in the same way that Sweden's carbon tax did by putting an emissions cap on fossil fuel producers/importers/consumers, of which biofuels would be exempted. However, research found no notice of such an arrangement in place.

16.4 The Brazilian Policy on Biofuels and the Decarbonization Credit Market

16.4.1 Fundamentals, Objectives, and Conceptual Structure

The concept of energy security, usually based on guaranteeing energy supply at affordable prices has gained a new element in recent years associated with the need of reducing GHG emissions. It is precisely in this new component of the world energy field that the Brazilian National Biofuels Policy, also known as RenovaBio, bases itself. The Program was established by Law No. 13,576, of December 26, 2017, after a broad debate involving public and private agents in Brazil. Inspired by successful initiatives in other countries, RenovaBio aims to:

1. Stimulate the reduction of GHG emissions in the transport segment, contributing to the fulfillment of the commitments made by Brazil in the Paris Agreement.
2. Induce economic and environmental efficiency gains in the production of biofuels.
3. Define rules for expanding the supply of clean energy, replacing fossil fuels with biofuels.
4. Recognize positive externalities associated with biofuels—both those already in the market on a commercial scale in Brazil (notably, ethanol and biodiesel) and those under development (e.g., biogas and sustainable aviation fuels).

RenovaBio innovates by creating a mechanism that assigns economic value to the reduction of GHG emissions promoted by biofuels, without setting volumetric targets, changing existing mandates, delimiting a captive market, or setting any type of subsidy.

The Program is based on three main pillars. The first of them refers to the proposition of a 10-year decarbonization target for the transport sector. This instrument aims to define a maximum limit for the carbon intensity of the Brazilian fuel matrix. To meet this limit, it is necessary to expand the participation of biofuels and produce them with lower levels of emissions. The second pillar refers to the valuation mechanism of the carbon that is no longer emitted due to the displacement of fossil energy with renewable energy. This remuneration takes place through the sales of emission reduction certificates known as Decarbonization Credit (CBIO), which is issued upon the sale of the biofuel by the producer. The system requires the purchase of the certificate by distributors to meet the matrix's emission reduction targets each year. The price of CBIO, in turn, is determined by market conditions, with immediate adjustments made in a transparent process of commercialization in an organized market.

This system aims to correct an important market failure that characterizes the renewable energy world: the presence of externalities that result in a suboptimal level of production and consumption of biofuels, and an overinvestment in fossil sources. In essence, in the mechanism proposed by RenovaBio, positive externalities become an economic return to the biofuel producer, and negative externalities, on the other hand, become an additional private cost to fossils. It is up to consumers to make their choice, based on the relative prices, now adjusted, of each fuel.

Finally, the third and last element of RenovaBio establishes a link between the energy-environmental efficiency of production and the revenue that can be earned from the sale of CBIOs by the biofuel producer. By quantifying emissions according to the characteristics of each biofuel producer, the mechanism recognizes the different stages of the production and marketing process, defining a greater number of CBIOs to be issued by the most efficient producers in terms of life cycle GHG emissions. Producers with reduced diesel consumption in the production process, for example, will have a higher energy-environmental efficiency rating and, therefore, will be able to issue a greater number of CBIOs for each volume of biofuel sold. As the certificate represents one tonne of CO_2eq that has not being emitted, it is natural that the most efficient producers present greater income from the sale of CBIOs.

This system valorizes efficiency gains in production, inducing investments in new emission-reducing practices and products. These efficiency gains, in turn, may be transferred to the final consumer through competition among different fuels and even among different biofuels. The following sections bring further details about the three mentioned instruments and present a reflection on the Program's potential and opportunities for improvement.

16.4.2 Decarbonization Target

The decarbonization target specifies the reduction in carbon intensity (CI) required for the Brazilian transport matrix. This carbon intensity, expressed in grams of carbon dioxide equivalent per megajoule (g CO_2eq/MJ), is determined from an average of the carbon footprint of each fuel composing the matrix, measured by the energy contribution of the respective products in the total energy supply in the transport sector. Table 16.2 details this calculation considering the fuel consumption forecast for 2022.

Table 16.2 Example of calculation of the estimated average carbon intensity for the Brazilian fuel matrix in 2022

	(A) Fuel consumption (billion liters)	(B) Fuel energy content (MJ/liter)	(AxB) Fuel energy supply (MJ)	Share of each fuel	Carbon intensity (g CO_2eq/MJ)
Gasoline	30.1	32.2	969	24.8%	87.4
Anhydrous ethanol	11.1	22.4	249	6.4%	19.71
Hydrous ethanol	16.7	21.4	357	9.2%	19.97
Diesel	48.3	35.5	1715	43.9%	86.5
Biodiesel	7.9	35.5	280	7.2%	21.25
Aviation fuel	7.6	34.4	261	6.7%	87.5
Aviation biofuel	0	34.4	0	0.0%	33.29
Natural gas vehicle	2.4	28.9	69	1.8%	78.1
Biogas/ biomethane	0.1	28.9	3	0.1%	7.15
		Average CI—Brazilian fuel matrix			71.56

From MME (2021)

The maximum CI of the fuel matrix expected by RenovaBio decarbonization target is defined for a period of 10 years. It is a parameter established by the National Energy Policy Council (CNPE), based on a technical study conducted by an inter-ministerial committee. This Committee, called the RenovaBio Comittee[2] is coordinated by the Ministry of Mines and Energy (MME). Figure 16.1 shows the current annual targets proposed for the Brazilian fuel matrix.

From Table 16.2 and Fig. 16.1 it derives that the average carbon intensity of the matrix can be reduced by taking two distinct and complementary paths. The first of them is the expansion of the participation of biofuels in the transport sector. As these products have less than the average carbon intensity, their expansion promotes reductions in the matrix's aggregate CI.

The second possibility refers to more efficient production of current biofuels or the inclusion of new biofuels with a lower carbon footprint. As shown below, the Program recognizes the most efficient producers from the point of view of GHG emissions and, therefore, creates incentives for the pursuit of energy-environmental efficiency gains in biofuels production.

[2]The RenovaBio Committee includes representatives of the MME (coordination); Civil House of the Presidency of the Republic; Ministry of Economy; Ministry of Infrastructure; Ministry of Agriculture, Cattle and Supplying; Ministry of Science, Technology, Innovations and Communications; and Ministry of the Environment (Art. 13 of Decree 9888/2019). The regulation allows the participation of other public and private institutions, technicians, and specialists in the biofuels sector.

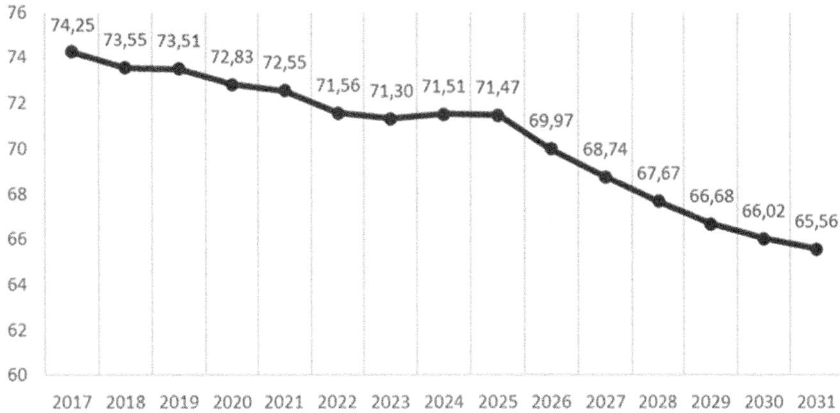

Fig. 16.1 Annual CI targets intended for the Brazilian fuel matrix (values in g CO_2eq/MJ)

To achieve the decarbonization target, the carbon intensity of the matrix is converted into numbers of Decarbonization Credits (CBIOs) to be acquired by fuel distributors (which is an obligated part of the program). CBIO is a certificate that represents the reduction of GHG emissions or the removal of one tonne of CO_2 equivalent from the atmosphere.

As the matrix's carbon intensity is related to the emission of an amount of greenhouse gases annually, it is possible to convert the desired reduction of CI in the matrix into tonnes of carbon dioxide equivalent that must not be emitted. In other words, as CBIO represents one tonne of GHG that is no longer emitted, it is possible to make a relationship between the desired CI for the fuel matrix and the number of CBIOs that would be needed annually to reach this intensity. That is the logic used by Renovabio. Thus, in the CNPE Resolution, the CI defined annually for the fuel matrix is also expressed in CBIOs numbers that should be purchased by distributors (Fig. 16.2).

The allocation of the national target for purchasing CBIOs among distribution companies is carried out by the National Agency for Petroleum, Natural Gas and Biofuels (ANP) and is based on the market share of companies in the sales of fossil fuels of the previous year. Distributors who commercialize a greater amount of fossil fuels are required to purchase a greater number of CBIOs.

In summary, the fundamental dynamics of the model can be summarized as follows:

1. The government, through CNPE, defines a limit for the carbon intensity of the transport matrix annually, considering a period of 10 years.
2. This limit is associated with a specific amount of reduction in tonnes of CO_2eq and, therefore, defines the number of CBIOs that must be acquired and retired (removed from the market) by the distributors.

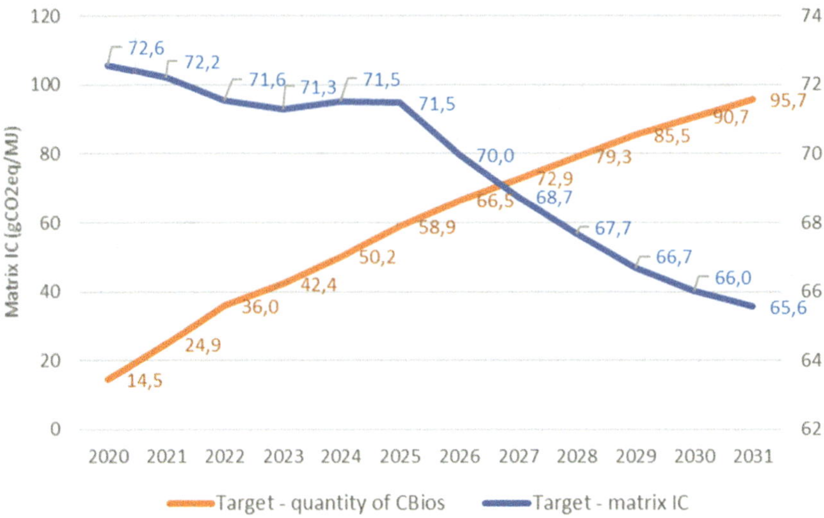

Fig. 16.2 Annual targets of CI intended for the Brazilian fuel matrix and number of CBIOs to be acquired by distributors. (Values for the 2020–2021 period from CNPE 2020 and for the 2022–2031 period from MME 2021)

3. For distributors with a greater share in the fossil fuel market, there is a greater obligation to purchase the CBIOs required for a given year.
4. The CBIOs supply is carried out by certified biofuel producers.
5. The CBIO price represents the economic value of the tonne of CO_2eq captured or avoided.
6. To increase the amount of CBIOs, it is necessary to expand the production of biofuels or to improve the energy-environmental efficiency with which these products are produced.
7. A more intense reduction of CI for the matrix will require a greater amount of CBIOs and, therefore, it will accelerate the incentive for the supply of biofuels and the search for energy-environmental efficiency gains among the producers of these renewables.

16.4.3 Certification of Biofuel Producers

The methodology used to define the carbon intensity of the biofuel manufactured by each producer is the Life Cycle Assessment (LCA) technique. From this perspective, all emissions caused by the biofuel from its production to consumption are accounted for. This process includes the GHG emission from the extraction of natural resources, acquisition or production and treatment of biomass, its conversion into biofuel and its combustion in engines, considering all the transport stages.

The assessment of biofuels' CI is performed through a tool called RenovaCalc, which uses specific information from the biofuel production system and calculates the carbon intensity of this product considering the aforementioned life cycle analysis. To guarantee traceability and reliability of the process, producers wishing to participate in the program must undergo a certification and audit procedure for the data and indicators included in RenovaCalc.

The first stage of certification consists of verifying the eligibility of the rural property where the feedstock is originated. To participate in the Program, each property must essentially meet two criteria:

(a) The non-suppression of native vegetation after the publication date of the Resolution that regulates the certification process—this rule defines that an agricultural property can only be incorporated as a supplier of the feedstock in RenovaBio if it has not registered any type of deforestation.
(b) The Cadastro Ambiental Rural—CAR (Rural Environmental Registry) must be regularized. CAR is a national electronic public registry, which is mandatory for all rural properties with the purpose of integrating environmental information of rural properties and possessions referring to the situation of the Areas of Permanent Preservation (APP), of Legal Reserve areas, forests and remnants of native vegetation, Restricted Use Areas and consolidated areas, composing a database for control, monitoring, environmental and economic planning and combating deforestation.

These criteria were designed to avoid the conversion of native vegetation or the expansion of energy crops into environmentally sensitive areas. It is, therefore, a measure that goes beyond the Brazilian environmental legislation, establishing a zero-deforestation rule on properties included in the Program.

In the second stage of certification, the producer hires a company accredited by the ANP, called "Inspector Firm". This company audits all information provided by the producer. Companies able to audit the values reported by the production units present extensive technical knowledge and internationally recognized competence. The documentation used in the audit is available for public consultation for a period of 30 days. If no inconsistency becomes evident, the certification process is approved by ANP. On the ANP website, it is possible to follow the public consultations in progress and the certification reports issued by the inspection firms. It is, therefore, a process with wide dissemination and publicity of all information and production indicators used.

Once certification is completed, it is possible to identify the carbon intensity of the biofuel manufactured by each producer (the carbon intensity of the biofuel is measured in g CO_2eq/MJ). The difference between the biofuel's life cycle emission and the emissions of its fossil counterpart determines the producer's Environmental Energy Efficiency Score (NEEA). In other words, NEAA identifies the GHG emissions avoided when the biofuel replaces the reference fossil fuel consumption. Thus, NEAA determines the amount of biofuel needed to avoid the emission of one tonne of CO_2eq and, thereby, the Decarbonization Credit (CBIOs).

This process ensures that the plan of emissions reduction of CBIO has a solid governance structure, with a systematic validation by external companies, by the society through public consultation and by ANP. The baseline used for comparing emissions is also consistent when using the fossil fuel as a reference. Furthermore, there are no issues associated with non-permanence or double counting of reductions. Finally, it is important to highlight that this logic innovates by creating incentives for the search for low-emission technologies and products. This happens because the most efficient producers can issue a greater amount of CBIOs per volume of biofuel produced.

16.4.4 Emission and the Decarbonization Credit Market

The importer or producer of biofuel certified by RenovaBio must request authorization to issue the CBIOs within a maximum period of sixty days after the issuance of the invoice attesting the biofuel commercialization. Invoices for the sale of biofuels are submitted to a platform maintained by ANP and validated in the database maintained by the Federal Revenue of Brazil to certify their veracity.

Once the biofuel sales invoice is validated, the producer must hire a financial institution for bookkeeping the credit. This financial institution will be responsible for registering the credit with the recording entity, maintaining the account where the credits are registered, controlling the ownership of credits, writing off retired credits, among others. Each primary issuer may have contract with one registrar.

CBIO trading takes place in an organized market, and regulations prevent the identification of the buyer and seller. Clients send buy and sell orders to intermediaries (it may be the institution that carries out the custody) who launch them in the trading environment. The entire process takes place through the Brazilian Payment System. The custodians of both parties are responsible for transferring the ownership of the credits and carrying out the credit and debit operations of the financial resources arising from the negotiation of the CBIOs. At the end of each day, all trades are registered on the Brazilian stock exchange (registration entity). The position of traded, registered, retired credits and their prices are published daily. In this context, the distributor needs to acquire the credits in the organized market and request their retirement to meet their goals. Retirement is the removal of the credit from circulation.

Speculators and investors wishing to buy CBIOs for future trading can also participate in this market. In addition, companies from other sectors that wish to offset their emissions can purchase CBIOs and request their retirement. This condition makes the CBIO market the first carbon credit market in the country.

In 2019, in its first year of operation, the CBIOs market handled around 15 million decarbonization credits. Decarbonization targets indicate that this number should reach at least 95 million CBIOs in 2030 only with mandatory purchases from distributors, in addition to the expected demand by other companies for use as carbon offset.

16.4.5 Considerations, Improvements, and Opportunities

Despite its recent implementation, RenovaBio has brought a new perspective to the biofuel sector by recognizing the country's differentiated position, which has conditions to sustainably expand. In the last two decades, the Brazilian biofuel industry has experienced cyclical moments, alternating periods of euphoria and crisis. This situation, to a large extent, resulted from changes in public policies and the Brazilian government's positioning on the topics of energy and fuels.

The uncertainty generated by this dynamic required a solution based on greater predictability and recognition of the need for regulation, considering the presence of externalities in this market. In this context, the mechanism proposed by RenovaBio established a guideline for the national fuel matrix, without any type of discrimination for a specific biofuel. In fact, the proposal created greater competition between biofuels, as the most efficient, from an economic and environmental point of view, should prevail.

The imposed system also started to valorize the externalities present in this market, and even created a new perspective associated with the possibility of trading carbon credits by the bioenergy sector. The Program established a first step in the use of biofuels as one of the complementary solutions for the decarbonization of the Brazilian economy. In spite of this condition, there are aspects to be improved in the Program in order to take advantage of the opportunities that arise for years to come.

In the institutional field, it is essential that the defined decarbonization targets are stable and minimally affected by conjunctural situations. Also, it is worthy highlighting the relevance of integrating RenovaBio with other programs linked to the reduction of emissions in the transport sector, such as the rules for the automotive sector in the country.

In the tax sphere, CBIO is still impacted with a higher charge than other financial securities traded in Brazil. In the commercialization stage, it is evident that new opportunities can be created from the development of a derivative market for CBIO, with a future price curve.

Communication actions to disseminate CBIO as an asset to offset emissions in other sectors are important. The integration of this market with other international initiatives or even with an eventual Brazilian emissions trading market is also fundamental for the expansion of bioenergy as a generator of carbon credits. In terms of governance, continuous actions to reduce transaction costs and improve the methodology and certification system stand out.

Finally, the effort to increase economic, energy and environmental efficiency in the production of biofuels must be maintained. Some of these actions are already being carried out and should guarantee the consolidation of RenovaBio as an instrument for decarbonizing the Brazilian transport matrix in the coming years.

16.5 Land Use and Biofuels Policies

16.5.1 The Land Use Change Concept

Given the wide variation in cultivation conditions, as well as accounting divergences, the estimates on biofuels' life cycle emissions can vary over a wide range. Yet, it is possible to say that today there is a relative consensus on how to evaluate the core life cycle GHG emissions of biofuels. However, this is not applicable when it comes to the assessment of the indirect effects associated with biofuels (Edenhofer et al. 2012; Souza et al. 2015).

The Land Use Change (LUC) emissions concept establishes that the carbon footprint calculation of biofuels should also consider the emissions from land use change caused by the conversion of areas previously destined for other purposes, to produce biofuels. Such a concept is aligned with carbon footprint assessment approaches.

Based on carbon cycle modeling, researchers found that the production of biofuels from low-performance inputs, in places with high carbon stocks, can emit more GHGs than the biofuel would be able to mitigate by replacing fossil fuels (Tilman et al. 2006; Righelato and Spracklen 2007; Fargione et al. 2008). Fargione et al. (2008), for example, pointed out that the conversion of forests, peatlands and savannas into agricultural commodity areas for the production of biofuels can lead to a carbon "debt". According to the authors, the conversion of Cerrado to sugarcane would lead to an initial GHG emission 17 times greater than the amount of GHG saved annually by the substitution of ethanol for gasoline. In the case of converting grassland areas for corn in the United States, the result would be a debt 98 times greater than the annual amount saved by the consumption of corn ethanol.

Using economic models integrated to LCA models, Searchinger et al. (2008) challenged the contribution of various biofuels, even when the feedstocks are not produced from areas directly converted from natural vegetation. The rationale (that has been eventually known as indirect land use change, iLUC) considers that the effect of land use change caused by the production of biofuels is not restricted to the area effectively replaced by the feedstock used to produce the biofuel. The abstraction of the iLUC concept leads to the conclusion that if a biofuel feedstock displaces another agricultural activity in one location, market forces (governed by supply and demand) would lead to an area compensation elsewhere. When dealing with agricultural commodities, international market forces mean that the indirect effect can occur in very distant areas, including other continents. In the authors' theoretical experiment, corn ethanol would increase GHG emissions by 50% when replacing gasoline (instead of reducing emissions by 20%) mainly due to iLUC.

With the spread of the iLUC concept, added to the food versus biofuel controversy, the international civil and scientific communities entered into intense debates, generating antagonisms between groups in favor and against biofuels (Kline and Dale 2008; Searchinger 2008; Wang and Haq 2008). Although there is a consensus that iLUC cannot be effectively measured, this concept had enormous international repercussion in the scientific and policy arenas. Therefore, with the debate about the real contributions of biofuels to the reduction of GHG emissions, and with the adoption of policies that conditioned production, an irrevocable link was created between climate policy, energy policy and land use.

16.5.2 iLUC Methods

Given its theoretical abstraction, iLUC cannot be measured for a land-use change that has occurred, nor can its effects be effectively quantified or predicted (Nassar et al. 2011; European Commission 2012; Valin et al. 2015). Even though estimations have been made using historical data (Fritsche and Wiegmann 2011; Moreira et al. 2012), or via simple projections (Lapola et al. 2010), the consequential approach based on the identification of marginal expansion effects of biofuels is the most suitable way for isolating and understanding the effects of iLUC (EPA 2010; CARB 2015; Valin et al. 2015). The following approach is thus recommended:

1. A reference scenario is projected;
2. A change in the reference scenario (exogenous shock) leads to the expansion of consumption and/or production of biofuels;
3. GHG emissions from land use change are estimated by the difference between scenarios (reference and shock);
4. The difference in GHG emissions between the scenarios is attributed to the additional consumption of biofuels (measured in energy terms).

iLUC has been estimated in g CO_2eq/MJ (by applying temporal adjustments), so that this unit can be added to the carbon footprint of the agricultural and industrial steps of the biofuel production and compared with the emissions of the fossil fuel. More recently, the indicator of kg CO_2eq/km has been also adopted to incorporate the efficiency of the vehicles.

As iLUC models did not exist before, modelers aimed at combining existing socioeconomic and biophysical models. Socioeconomic models provide simulations of trends in production, consumption, and land use. They provide matrixes that defines how much the area (usually hectares) of crops, forests pastures have shifted among each other. These models typically deal with large geographic areas such as states, regions, and countries. Biophysical models provide emission factors for each type of lad use change (usually in t CO_2eq/ha) defined by the socioeconomic models.

Both socioeconomic and biophysical models have significant uncertainties and limitations on their own. Broad geographical areas, uncertainties of human behavior, feedbacks, uncertainties regarding carbon stocks and fluxes in different locations are a non-exhaustive list of examples. It is a common understanding among experts that quantitative values resulting from the combination of the two types of models should be considered with caution.

Table 16.3 Land use change in selected policies

	iLUC calculation	Policy risk management
RFS2	Yes	Yes
LCFS	Yes	No
REDII	No	Yes
CORSIA	Yes	Yes
RenovaBio	No	Yes

16.5.3 iLUC in Main Biofuels Policies

The diffusion of the iLUC concept took place at a time when the main consuming regions of fossil fuels began to adopt policies to replace gasoline and diesel with biofuels. Even though science was not ready to rigorously contemplate such assessment, US and European Union legislators determined that domestic policies relating to biofuels should consider iLUC in their formulation. CORSIA, which was latter implemented, also incorporated land use emissions in its design. Table 16.3 summarizes how the main international regulations address the land use change issue.

RFS2 managed iLUC by a combination of restrictions on direct land use change and iLUC estimation. At sectorial level, an iLUC value was estimated within ranges of uncertainty. This iLUC value was added to the "core LCA" so it could be compared to fossil fuel counterparts. Biofuel producers must also demonstrate that the biomass used for biofuels production was not originated from deforestation areas.

The Draft Regulatory Impact Analysis (DRIA), published by the EPA in May 2009 (EPA 2009), featured several problems in capturing the characteristics of land use in Brazil, for example. In particular, the analysis was too aggregated (Brazil was treated as a single region), pasture areas were not modeled following economic assumptions, livestock intensification capacity was not properly considered, and the analysis of satellite images was not accurate enough to differentiate pasture areas and native vegetation in some Brazilian biomes. Improvements were implemented disaggregating Brazil into six regions and incorporating substantially more complete land use change analysis than in the original DRIA (Nassar et al. 2009). The final examination indicated some iLUC for sugarcane ethanol, but significantly lower than original proposal. The technical collaboration between research institutions from USA and abroad (namely FAPRI/CAR and ICONE-Brasil) was of great relevance for the improvement of the modeling results.

LCFS, in turn, does not require producers to demonstrate that biomass did not originate from deforestation areas. Instead, there are estimates for iLUC by feedstock, which are added to (individual) core LCA values to produce the total carbon intensity of each biofuel route. Likewise the RFS2 original assessment, the first version of the regulation was subjected to criticism and suggestions for revision.[3]

[3]Particularly for Brazil, several indices and parameters in the GTAP model, and even productive structures, were inadequate. Hundreds of letters with criticisms and recommendations were sent to US regulators, most of them containing comments regarding iLUC (CARB 2009). With respect to sugarcane ethanol, an extensive letter compiled by the Sugarcane Industry Association (UNICA)

CARB then set up a group of experts responsible for reviewing the modeling, and several improvements were implemented in the regulatory review, completed in 2015. A new version of the GTAP-Bio model, with a competition structure between land uses better suited to deforestation analysis, and a much more detailed model for conversion of land use change soil GHG emissions are the main highlights (Plevin et al. 2011; Taheripour and Tyner 2013; Farina and Philips 2014a, b; CARB 2015). Although iLUC estimations have improved significantly, there are still difficulties in specific points where the general equilibrium modeling bumps into structural limitations.[4]

Unlike the North American case, the European Union (EU) does not adopt a single procedure for estimating iLUC. The European regulation preferred to work with the Precautionary Principle, whereby the "risk" of iLUC is assessed through a variety of methodologies and methodological reviews, coordinated by the EU's Joint Research Center (JRC). The current text of RED-FQD reports values obtained by the MIRAGE model, prepared by the ATLASS consortium (Laborde 2011; European Commission 2012).

There is, therefore, no complete and ideal model for all uses. Each model/approach has its pros and cons. Deciding on the most appropriate model to be used usually leads to choices between precision, consistency, and comprehensiveness of the results, which can have significant implications on the benefits perceived for biofuels within each regulatory context.

16.6 Final Remarks

The production and use of biofuels have grown rapidly during the 2000s decade due to high oil prices and especially a variety of government policies, such as feed-in-tariffs, tax exemptions and biofuel mandates. Policies on the promotion on biofuels have been grounded on multiple arguments, such as reducing dependence on non-renewable resources and increasing energy security, mitigating climate change effects, enhancing economic growth and creating jobs, improving the balance of trade, ensuring food security and the sustainable management of natural resources.

(Jank and Velasco 2009) mentioned the inadequacy of general equilibrium models to obtain precise values; issues regarding the shock size; underestimation of the livestock intensification capacity in Brazil; incorrect elasticities; poorly dimensioned and unequal scenarios for the cases studied; outdated sugarcane productivity; inaccurate biomass stocks for Brazil; and incorrect emission factors for sugarcane.

[4]Modelling limitations relate essentially to the difficulty of incorporating real observations (i.e., land use pattern via satellite images) that are suitable to the regional characteristics in Brazil. Further, models still fail to identify and simulate effective land use change (effective substitution between uses, not resulting land use amounts) and to properly incorporate the livestock dynamics (technological levels, production systems and regional systems), the evolution of second and third harvest technology and its dynamics, the new technologies in the sugarcane industry, and to consider more accurate GHG emission coefficients associated with land use.

However, the justification of these policies depends on their specific goals. In that sense, policies directly connected to a target tend to be more effective.

As discussed in this chapter, different market-pull instruments are in place for the promotion of biofuels. Even though mandates are the most traditional instrument for supporting biofuels worldwide, carbon-related mechanisms are expected to be more effective in capturing the benefits of biofuels in terms climate change mitigation. This chapter addressed the recently launched Brazil's RenovaBio Program as an example of such carbon-related policy, which can be explored by other potential producers without the need for volume targets nor subsidies.

In all cases, regulation will be key to guarantee the achievement of the policy goals, while preventing undesired side effects. As land use change is a topic of special concern in the biofuels industry, the elaboration of robust assessment tools remains as a challenge, as well as the development of harmonized approaches that could make carbon credits more fungible, and thereby allowing the expansion of the market.

References

Appunn K (2021) CO2 reduction and biofuels in Germany's transport sector - implementing the RED II directive. In: Clean energy wire. https://www.cleanenergywire.org/factsheets/co2-reduction-and-biofuels-germanys-transport-sector-implementing-red-ii-directive. Accessed 4 Oct 2021

Biofuels Digest (2021) The digest's biofuels mandates around the world 2021. In: Biofuels digest. https://www.biofuelsdigest.com/bdigest/2021/01/06/the-digests-biofuels-mandates-around-the-world-2021/4/. Accessed 4 Oct 2021

Biofuture Platform (2016) Biofuture platform - kickstarting a global, advanced bioeconomy. http://www.biofutureplatform.org/

Biofuture Platform (2018) Creating the biofuture: a report on the state of the low carbon bioeconomy. Biofuture Platform, Brasília

California Air Resources Board (2020) LCFS basics. https://ww2.arb.ca.gov/resources/documents/lcfs-basics. Accessed 4 Oct 2021

CARB (2009) Board meeting comments. California Air Resources Board, Sacramento, CA

CARB (2015) Staff report: initial statement of reasons for rulemaking: proposed re-adoption of the Low Carbon Fuel Standard regulation. California Air Resources Board

CNPE (2020) Resolução N° 8, de 18 de agosto de 2020. Define as metas compulsórias anuais de redução de emissões de gases causadores do efeito estufa para a comercialização de combustíveis

Edenhofer O, Pichs Madruga R, Sokona Y et al (eds) (2012) Renewable energy sources and climate change mitigation: special report of the Intergovernmental Panel on Climate Change. Cambridge University Press, New York

EPA (2009) Draft regulatory impact analysis: changes to renewable fuel standard program. Environmental Protection Agency

EPA (2010) Renewable fuel standard program (RFS2) regulatory impact analysis. Assessment and Standards Division, Office of Transportation and Air Quality, U.S. Environmental Protection Agency, Washington

European Commission (2012) Commission staff working document - impact assessment. Accompanying the document Proposal for a Directive of the European Parliament and of the Council, amending Directive 98/70/EC relating to the quality of petrol and diesel fuels and amending Directive 2009/28/EC on the promotion of the use of energy from renewable sources

European Union (2018) Directive (EU) 2018/2001 of the European Parliament and of the Council of 11 December 2018 on the promotion of the use of energy from renewable sources (recast)

Fargione J, Hill J, Tilman D et al (2008) Land clearing and the biofuel carbon debt. Science 319: 1235–1238. https://doi.org/10.1126/science.1152747

Farina E, Philips L (2014a) UNICA's preliminary comments on revised indirect land use change values

Farina E, Philips L (2014b) UNICA's comments on the updated indirect land use change analysis of the low carbon fuel standard

Fritsche U, Wiegmann K (2011) Indirect land use change and biofuels. Directorate General for Internal Policies, Department A: Economic and Scientific Policy; Environment, Public Health and Food Safety

ICAO (2021) Carbon offsetting and reduction scheme for international aviation (CORSIA). In: CORSIA. https://www.icao.int/environmental-protection/CORSIA/Pages/default.aspx. Accessed 4 Oct 2021

IEA (2017) World energy outlook 2017. OECD Publishing, Paris

IEA (2019a) CO2 emissions from fuel combustion 2019. International Energy Agency, France

IEA (2019b) World energy outlook 2019. OECD/IEA, Paris

IEA (2019c) Energy policies of IEA countries: sweden 2019 review. International Energy Agency (IEA), Paris

IEA (2021) Fossil fuel subsidies database

Jank M, Velasco J (2009) UNICA's comments on proposed new pathways for Brazil sugarcane ethanol

Kline KL, Dale VH (2008) Biofuels: effects on land and fire. Science 321:199–201. https://doi.org/10.1126/science.321.5886.199

Laborde D (2011) Assessing the land use change consequences of european biofuel policies. International Food Policy Research Institute (IFPRI)

Lapola DM, Schaldach R, Alcamo J et al (2010) Indirect land-use changes can overcome carbon savings from biofuels in Brazil. Proc Natl Acad Sci U S A 107:3388–3393. https://doi.org/10.1073/pnas.0907318107

MME (2021) Nota Técnica No 41/2021/DBIO/SPG, Consulta Pública MME No 12/2021

Moreira M, Nassar A, Antoniazzi L et al (2012) Direct and indirect land use change assessment. In: Cortez LAB, Poppe A (eds) Sustainability of sugarcane bioenergy, Updated edition. Center for Strategic Studies and Management (CGEE), Brasília, DF

Nassar A, Harfuch L, Moreira M, et al (2009) Impacts on land use and GHG emissions from a shock on brazilian sugarcane ethanol exports to the United States using the Brazilian land use model (BLUM). ICONE

Nassar AM, Harfuch L, Bachion LC, Moreira MR (2011) Biofuels and land-use changes: searching for the top model. Interface Focus 1:224–232. https://doi.org/10.1098/rsfs.2010.0043

Plevin RJ, Gibbs HK, Duffy J, Yeh S (2011) Agro-ecological Zone Emission Factor Model. University of California (Berkeley); University of Wisconsin; California Air Resources Board

Pomerleau S (2020) What can we learn from Sweden's carbon tax? Niskanen Center

Righelato R, Spracklen DV (2007) Carbon mitigation by biofuels or by saving and restoring forests? Science 317:902–902. https://doi.org/10.1126/science.1141361

Searchinger TD (2008) Response to M. Wang and Z. Haq's E-Letter

Searchinger T, Heimlich R, Houghton RA et al (2008) Use of U.S. croplands for biofuels increases greenhouse gases through emissions from land-use change. Science 319:1238–1240. https://doi.org/10.1126/science.1151861

Souza GM, Victoria RL, Joly CA, Verdade LM (2015) Bioenergy & sustainability: bridging the gaps. Scientific Committee on Problems of the Environment (SCOPE), Paris Cedex

Swedish Energy Agency (2021) Energy in Sweden 2021: an overview. SEA, Stockholm

Taheripour F, Tyner W (2013) Biofuels and land use change: applying recent evidence to model estimates. Appl Sci 3:14–38. https://doi.org/10.3390/app3010014

Tilman D, Hill J, Lehman C (2006) Carbon-negative biofuels from low-input high-diversity grassland biomass. Science 314:1598–1600. https://doi.org/10.1126/science.1133306

US EPA O (2015) Renewable fuel annual standards. https://www.epa.gov/renewable-fuel-standard-program/renewable-fuel-annual-standards. Accessed 4 Oct 2021

Valin H, Peters D, van den Berg M, et al (2015) The land use change impact of biofuels consumed in the EU. Quantification of area and greenhouse gas impacts. Ecofys, IIASA and E4tech

Wang M, Haq Z (2008) Ethanol's effects on greenhouse gas emissions

Chapter 17
How Would Solid Oxide Fuel Cells and Bioethanol Impact in Electric Mobility Transition?

Fábio Coutinho Antunes, Raissa Venâncio, Gustavo Doubek, and Hudson Zanin

Abstract We live in a transition economy, from fossil fuels based to a more sustainable one, targeting emission neutrality of carbon dioxide (CO_2). The CO_2 gas is the primary greenhouse gas, and it is recognized as the leading cause of global warming and subsequent climate change events. Several alternative technologies are under study, and substantial funding has been applied to reduce emissions and eliminate greenhouse gases. Most advanced technologies so far are sustainable and economically competitive and need a push from governments to fly even higher such as solar and wind renewable energy generation and bioethanol (C_2H_5OH) and hydrogen (H_2) for fuels. The third generation of solid oxide fuel cells (MS-SOFCs: metal-supported solid oxide fuel cells) can use H_2, CO or biofuels for onboarding energy generation at lower temperatures (ca. 600 °C), offering high electrical efficiency and none or low polluting gas emissions. Combined with biofuels such as bioethanol, MS-SOFCs would offer a solution for the transports electrification. This solution uses the infrastructure already existing in many countries and would be a pivotal component in sustaining the transformation required and keeping alive the net of values constructed over the years by biofuels agroindustry.

The original version of the chapter has been revised. A correction to this chapter can be found at https://doi.org/10.1007/978-3-031-01241-9_21

F. Coutinho Antunes (✉) · R. Venâncio · G. Doubek · H. Zanin
Advanced Energy Storage Division, Center for Innovation on New Energies, University of Campinas – Unicamp, São Paulo, Brazil
e-mail: fabioant@unicamp.br; hzanin@unicamp.br

C. R. Soccol et al. (eds.), *Liquid Biofuels: Bioethanol*, Biofuel and Biorefinery Technologies 12, https://doi.org/10.1007/978-3-031-01241-9_17

17.1 How Does Bioethanol Fit in Electric or Hybrid Vehicles?

A conventional fuel cell (FC) directly converts fuel into electrical energy, providing higher energy conversion efficiency than conventional thermodynamic systems (Carnot-Rankine cycles). The operating principle of fuel cells (FCs) is similar to batteries because electrochemical surface reactions generate electricity. This combination may occur between a fuel and an oxidizing gas, for example, H_2 and oxygen (O_2), respectively, to produce electricity, heat and water vapor (steam) (Stambouli and Traversa 2002). H_2 gas is highly costly to store and safely transport, where high pressure and expensive reservoirs are required. Vehicle collision could also be potentialized if H_2 is the fuel. However, in liquid form, all those issues are gone. Hydrogen is stored in the form of several hydrocarbons. Biofuels store hydrogen in liquid form. For instance, C_2H_5OH stores considerable energy per unit of volume (ca. 18.4 MJ.L^{-1}, while H_2 is ca. 0.01 MJ.L^{-1} at Standard Temperature and Pressure-STP) and logistics is much simpler and cheaper than H_2 (Yang et al. 2019a). Inside the FCs or in the bioethanol reformer, hydrogen is obtained from bioethanol and combined with, O^{2-}, OH^- or $(CO_3)^{2-}$ ions to generate electricity, heat, steam, and CO_2. Please, remember that sugarcane plants capture CO_2 as part of photosynthesis processes, and their life cycle has therefore a net-zero emission i.e., carbon neutrality = CO_2 emission and capture sum tends to zero. Among the FCs, solid oxide fuel cells (SOFCs) are the best option to use hydrocarbon as a fuel.

17.2 Why Does SOFC Is the Best Option?

Many FCs are at different development stages (Stonten and Emonts 2012). The FCs are classified according to the type of electrolyte applied: Proton Exchange Membrane FC (PEMFC), Alkaline FC (AFC), Phosphoric Acid FC (PAFC), Molten Carbonate FC (MCFC) and Solid Oxide FC (SOFC). Table 17.1 gathers a brief overview of the FCs features. The SOFC is the only FCs that could use bioethanol directly as fuel and operate in an environment of reformed H_2 fuel i.e., H_2 and CO (syngas), with high efficiency. That is why SOFC is the most relevant technology. PEMFC is the closest second option but does not tolerate traces (10 ppm) of CO and sulfur compounds (H_2S). In SOFC, CO is also fuel, which combined with O^{2-} forms CO_2 and low levels of H_2S are tolerable.

Let us have a closer look at SOFC.

17.3 Explaining SOFC

The main SOFCs components are electrolyte, anode, cathode, interconnectors (ICs) and sealing. Figure 17.1 presents the historical concepts of cells supported by an electrolyte (first generation), anode or cathode (second generation) and metal (third generation).

Table 17.1 Brief overview of the FCs types. Adapted from Fuel Cells Handbook (2004), Carrete et al. (2001, 2000) and Devi et al. (2004)

	PEMFC	AFC	PAFC	MCFC	SOFC
Electrolyte	Hydrated polymeric ion exchange membranes	Mobilised or immobilised potassium hydroxide alkaline in asbestos matrix	Immobilised liquid phosphoric acid in SiC matrix	Immobilised liquid molten carbonate in $LiAlO_2$ matrix	Cubic or perovskites based solid oxide ceramics
Electrodes	Carbon	Transition metals	Carbon	Nickel and nickel oxide	Perovskite or perovskite-cubic/metal cermet
Catalyst	Platinum	Platinum	Platinum	Electrode material	Electrode material
Interconnect	Carbon or metal	Metal	Graphite	Stainless steel or nickel	Nickel-ceramic composites or ferritic stainless steel
$T_{op.}$ (°C)	60–120 °C	<100 °C	160–220 °C	600–800 °C	500–1000 °C
Anode reaction	$H_2 \rightarrow 2H^+ + 2e^-$	$H_2 + 2OH^- \rightarrow 2H_2O + 2e^-$	$H_2 \rightarrow 2H^+ + 2e^-$	$H_2 + (CO_3)^{2-} \rightarrow H_2O + CO_2 + 2e^-$	$H_2 + O^{2-} \rightarrow H_2O + 2e^-$
Cathode reaction	$\frac{1}{2} O_2 + 2H^+ + 2e^- \rightarrow H_2O$	$\frac{1}{2} O_2 + H_2O + 2e^- \rightarrow 2OH^-$	$\frac{1}{2} O_2 + 2H^+ + 2e^- \rightarrow H_2O$	$\frac{1}{2} O_2 + CO_2 + 2e^- \rightarrow (CO_3)^{2-}$	$\frac{1}{2} O_2 + 2e^- \rightarrow O^{2-}$
Charge carrier in the electrolyte	H^+	OH^-	H^+	$(CO_3)^{2-}$	O^{2-}
External reformer for hydrocarbon fuels	Yes	Yes	Yes	No, for some fuels	No, for some fuels and cell designs
External shift conversion of CO to hydrogen	Yes, plus purification to remove trace CO	Yes, plus purification to remove CO and CO_2	Yes	No	No
Prime cell components	Carbon-based	Carbon-based	Graphite-based	Stainless-based	Ceramic

(continued)

Table 17.1 (continued)

	PEMFC	AFC	PAFC	MCFC	SOFC
Product water management	Evaporative	Evaporative	Evaporative	Gaseous product	Gaseous product
Product heat management	Process gas + liquid cooling medium	Process gas + electrolyte circulation	Process gas + liquid cooling medium or steam generation	Internal reforming + process gas	Internal reforming + process gas
Applications	Transportation Space Military Energy storage systems		Combined heat and power for decentralised stationary power systems	Combined heat and power for decentralised stationary systems and for mobility (trains, boats, vehicles, etc.)	
Power Supply	Small plants 5–250 KW modular	Small plants 5–150 KW modular	Small–medium sized plants 50 KW–11 MW	Small power plants 100 KW–2 MW	Small power plants 100–250 KW
Fuel	H_2 from hydrocarbons/methanol	Pure H_2, hydrazine	H_2 from hydrocarbons	Hydrocarbons/CO and H_2	Hydrocarbons/CO, H_2 and Bioethanol
Efficiency (%)	40–50	–	40–50	50–60	45–70

The evolution of the cell was based on reducing the working/operating temperature ($T_{op.}$) to apply low-cost ICs and sealing materials (Binelli et al. 2016, 2017).

Among all SOFC components, the electrolyte is the most important to define the $T_{op.}$. State of the art in SOFC electrolyte is 8mol% yttria (Y_2O_3)-stabilized zirconia (ZrO_2) known as 8YSZ (Subbarao 1981). That is the best option because of its excellent cost-benefit, appropriated ionic conductivity (σ_{ionic}), match in coefficient of thermal expansion (CTE) with the electrodes and ease to manufacture. However, there are materials with suitable σ_{ionic} for electrolyte in SOFCs, as depicted in Fig. 17.2b. Figure 17.2 presents the comparison of the total electrical conductivity (σ) as a function of the inverse of the absolute temperature (bottom x-axis) and temperature (top x-axis) for different solid oxide electrolytes. Special attention has been given to 8YSZ and gadolinium-doped ceria (GDC). Doped ceria-based electrolyte materials, such as GDC and samarium-doped ceria (SDC), exhibit higher ionic conductivity than 8YSZ between $600<T(°C)<750$. Although doped ceria-based electrolytes show phase stability in these intermediate temperatures, Ce^{4+} tend to reduce and form Ce^{3+} at low oxygen partial pressure (PP_{O_2}) in the fueling environment typical of a SOFC device. This leads to several disadvantages, such as high electronic conductivity by small polaron mechanisms, chemical expansion due to larger size of Ce^{3+} when compared to that of Ce^{4+}, and inferior mechanical properties (Mahato et al. 2015; Gautam et al. 2020).

The anode is usually a composite of metallic nickel (Ni) and 8YSZ, and the cathode is a mixed conductor (MIEC: mix ionic and electronic conductor) of the type strontium (Sr) doped-lanthanum manganite ($LaMnO_3$), known as LSM (Subbarao 1981; Mahato et al. 2015; Ormerod 2003; Boldrin and Brandon 2019; Mogensen

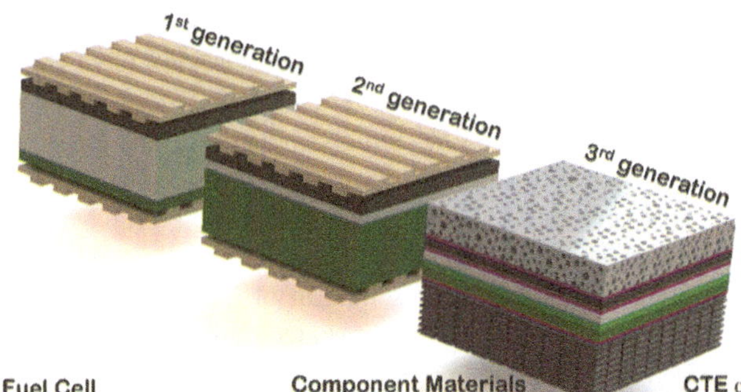

Fuel Cell	Component Materials	CTE α_l (10^{-6} K^{-1})
ceramic interconnector	Sr doped $LaCrO_3$	9
porous support	ferritic stainless steel (FSS)	10 - 12
cathode	LSM, LSC, CMA	12; 22; 18
electrolyte	8YSZ, 10GDC	10; 12.7
active anode	Ni/10GDC, Cu/10GDC	12
base anode	Ni/8YSZ, Cu/8YSZ	12
diffusion barrier layer	10GDC, $(Mn,Co,Cr)_3O_4$ + LSM	12; 7
micro-porous plate	FSS	10 - 12
micro-reformer support	micro-channels FSS + Catalyst	10 - 12

Fig. 17.1 Schematic representation of the three SOFCs generations. Components are unscaled for better visualisation of internal cell components. Special attention will be given to modified metal-supported by micro-reformers. CMA: $CoMnAlO_4$. Source: prepared by the authors

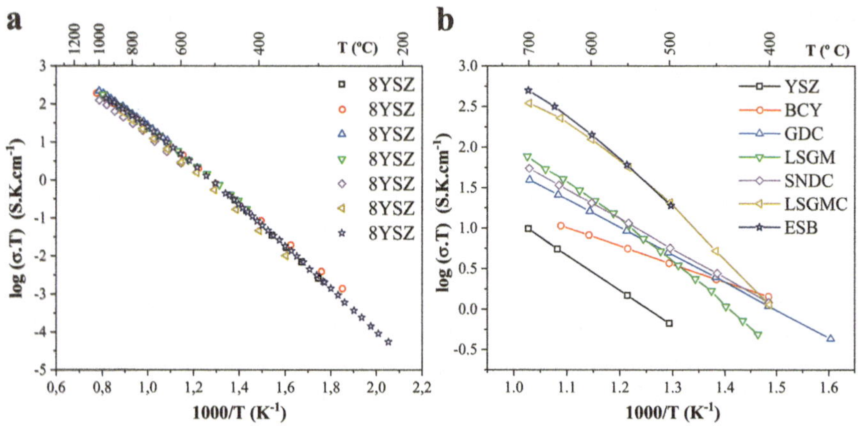

Fig. 17.2 (**a**) Comparing σ of 8YSZ from several works evidencing reproducibility (Violet diamond: Mahato et al. 2015; Blue triangle: Ivers-Tiffée et al. 2001; Black square: Antunes et al. 2018; Red circle: Araki and Arai 2011; Green triangle: Kharton et al. 2004; Yellow triangle: Yamahara 2003; Navy blue star: Ahamer et al. 2017) and (**b**) the σ of several SOFC electrolytes. YSZ: $Y_{0.16}Zr_{0.84}O_{1.92}$; BCY: $BaCe_{0.92}Y_{0.08}O_3$; ESB: $Er_{0.4}Bi_{1.6}O_3$; DWSB: $Dy_{0.08}W_{0.04}Bi_{0.88}O_{1.56}$; GDC: $Gd_{0.2}Ce_{0.8}O_{1.90}$; SNDC: $Sm_{0.075}Nd_{0.075}Ce_{0.85}O_2$; LSGM: $La_{0.8}Sr_{0.2}Ga_{0.8}Mg_{0.2}O_3$; LSGMC: $La_{0.8}Sr_{0.2}Ga_{0.8}Mg_{0.115}Co_{0.085}O_3$. Adapted from Wachsman et al. (2014)

et al. 2002; Mogensen and Chemistry 2012; Ivers-Tiffée et al. 2001; Hammou 2008; Jacobson 2010; Hussain and Yangping 2020; Shabri et al. 2015; Zakaria et al. 2020a; Dwiwedi 2020).

In this system configuration with 8YSZ as an electrolyte, the second generation SOFCs operates between 800<T(°C)<1000, to achieve good performance (0.5–1.9 W cm^{-2}) and high efficiency (40–60%) using hydrocarbons fuels (Mahato et al. 2015; Muccillo et al. 2008; Yaman and Kucukaga 2020). These lower power peak density values are due to the low 8YSZ ionic conductivity and the low electrode electrocatalytic activity (Lyu et al. 2020; Singhal and Kendall 2003; Minh and Takahashi 1995; Huijsmans 2001). Temperature is still too high for a long lifespan, low-cost material application, and biofuel as fuel.

Figure 17.3a shows a schematic representation of SOFC in three layers: (1) a dense 8YSZ electrolyte separated by two porous electrodes, (2) Ni/8YSZ composite anode and (3) LSM-8YSZ composite cathode. The fuel passes through the ceramic IC and the anode support, where may suffer a reform or not and then combine with O^{2-} ions, which were reduced in the cathode and conducted by the electrolyte. The insert figure evidences the $H_2 + O^{2-} \leftrightarrow H_2O + 2e^-$ where fuel (reductant) is combined with oxygen ion (oxidant) forming steam plus electricity.

Figure 17.3 shows additionally a schematic representation of SOFC in (**b**) planar and (**c**) tubular cell shapes. The planar shape is easier to stack the cell to achieve higher voltage, energy and power. Tubular shape cells are interesting from a stack (Sect. 17.10) and construction point of view because they can act under high pressures increasing the cell power at different current densities but face electrode cracking and delamination due to thermal expansions (Singhal 2002).

a

b

c

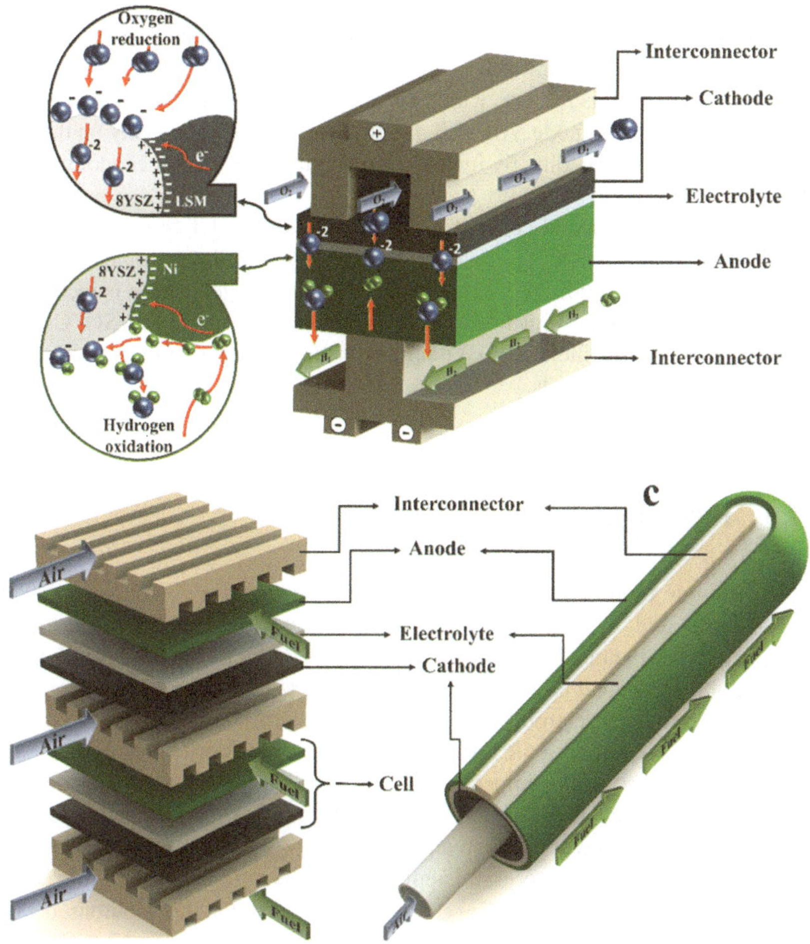

Fig. 17.3 Schematic representation of (**a**) second generation cell formed by the porous anode and cathode separated by the dense 8YSZ membrane; (**b**) stack and (**c**) tubular cells. Components are unscaled, and diffusion barrier layers (DBLs) are hidden for better visualisation of the cell. Source: prepared by the authors

17.4 Low-Temperature SOFC

Low-temperature solid oxide fuel cells (LT-SOFCs) operate below 600 °C and may be low-cost and have a long lifespan compared to conventional SOFCs, which operate between $800<T(°C)<1000$ (Yang et al. 2019a). If LT-SOFCs also directly operate with bioethanol as fuel, FCs give a big step onwards applications on an electric vehicle with onboarding generation and would use the infrastructure already existing in most countries.

The SOFC may have cheaper ICs and sealings by reducing the T_{op}, facilitating starting heating up (quick start-up) and stopping cooling down (light-off), which expands the application of SOFC as a portable electricity generator. However, the main issue of LT-SOFCs that directly operates with bioethanol is the low performance compared to conventional H_2 fueled SOFC. The carbon (C) coking is also a significant issue in this system, which causes deactivation of Ni catalyst by thin C layer formation at the anode. After all, the lifespan and power density on the current state of the art of SOFC directly fueled with bioethanol is not yet economical viable (Yang et al. 2019a).

Here we will detail and contrast all components from LT-SOFCs with conventional SOFC to understand the evolution and what else needs to evolve to apply them on onboarding generation in electric vehicles. The third generation is recognize for lower $T_{op.}$ and ceramic thin films on porous metal supports (MS-SOFCs: metal-supported solid oxide fuel cells). We highlight the electrolyte, functional anode and cathode and DBLs onwards.

17.5 Electrolytes

The electrolyte is an ionic conductor. The ionic conductivity is strongly affected by the thickness and $T_{op.}$ due to the ionic conduction process nature been thermally activated i.e., Arrhenius Law. The charge carriers in thick electrolytes at lower $T_{op.}$ suffers polarization and ohmic resistances, causing a drop in σ_{ionic}. Electrolyte ionic ohmic transport and electrodes activation losses sharply increase at lower temperatures. These were due to ionic conductivity and charge transfer reaction rate exponential dependence with temperature (Yang et al. 2019a). Reducing electrolyte thickness, will diminish the overall pathway for ions transport, thus decreasing cell ohmic drop even at lower T_{op} (Lin and Beale 2006; Zouhri and Lee 2016).

The 8YSZ is the most successful applied electrolyte. The stoichiometric chemical composition is approximately $Zr_{0.84}Y_{0.16}O_{1.92}$, where the Zr^{4+} cations occupy a face-centered cubic (FCC) crystal structure as depicted in Fig. 17.4. This cubic structure is stable at room temperature and up to 2500 °C (Lima and Marple 2017).

Figure 17.4a depicted a draft from the cubic crystal structure of ZrO_2, evidencing in green the Zr^{4+} cations and in red the O^{2-} anions. The Zr^{4+} cations form a FCC crystal lattice with the O^{2-} anions occupying all interstitial tetrahedral positions. There are four Zr^{4+} cations and eight O^{2-} anions per unit cell of the crystal in this crystal structure, thus respecting the ZrO_2 stoichiometric ratio of 1:2 (Skinner and Kilner 2003; Asadikiya et al. 2016; Brodnikovska et al. 2019; Selvaraj et al. 2019; Kingery and Bowen 1976; Barsoum 2002; Götsch et al. 2016; Shimizu et al. 2016; Huang and Goodenough 2009; Li et al. 2020a). As doping with Y_2O_3 creates defects such as oxygen vacancies ($V_O^{\bullet\bullet}$) that are formed when Y^{3+} cations (in purple at Fig. 17.4b) partially replace Zr^{4+} cations. Equation 17.1 shows the chemical reaction for forming this kind of defect according to the Kröger-Vink notation (Kroger and Vink 1958; Moulson and Herbert 2003).

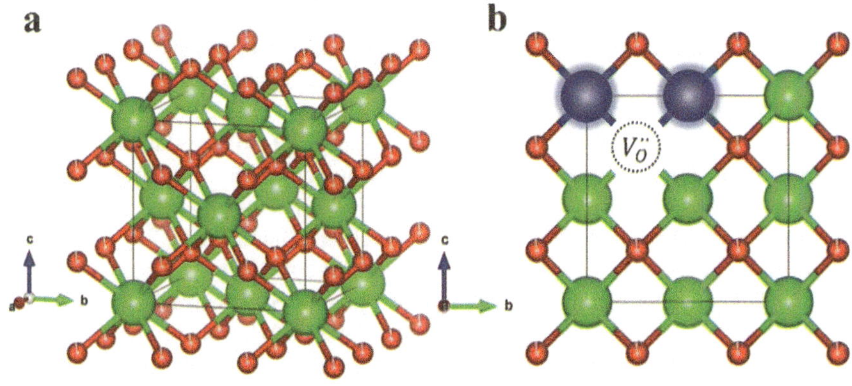

Fig. 17.4 Schematic of the cubic crystal structure of (**a**) ZrO_2 and for (**b**) 8YSZ. In green Zr^{4+}, red O^{2-} and purple Y^{3+} ions. Source: prepared by the authors

$$Y_2O_3 + 2Zr_{Zr} \overset{ZrO_2}{\rightarrow} 2Y'_{Zr} + 3O_O + V_O^{\bullet\bullet} \tag{17.1}$$

$$\left[Y'_{Zr}\right] = 2\left[V_O^{\bullet\bullet}\right] \tag{17.2}$$

According to Eq. (17.1), when doping with Y_2O_3, two Zr^{4+} cations are replaced by two Y^{3+} cations leaving two negative charges in the crystal. These two negative charges are balanced by the formation of an $V_O^{\bullet\bullet}$ that has two positive charges. In this way, electroneutrality is maintained in the crystal.

At high temperatures (ca. 1000 °C), ZrO_2 presents an intrinsic formation of charge carriers such as $V_O^{\bullet\bullet}$, however, it is insufficient to produce high ionic conductivity that guarantees excellent performance in the $T_{op.}$ range between $800 < T(°C) < 1000$. In addition, the pure ZrO_2 ceramics are unstable at room temperature, and they undergo cubic-tetragonal phase transformation at ca. 2320 °C and tetragonal-monoclinic at ca. 1170 °C. This last phase transformation, known as martensitic, is accompanied by a volumetric expansion between 3 and 4%, causing cracks and fractures of the ceramic. To overcome this issue, ZrO_2 may be doped with alliovalent elements, such as Y^{3+} and Ca^{2+}, which stabilize the cubic phase onto room temperature (Mamivand et al. 2013a, b; Basu et al. 2004; Deville et al. 2004a, b; Kelly and Rose 2002; Shibata et al. 2001).

By doping ZrO_2 with trivalent cations such as Y^{3+}, Gd^{3+} or Sm^{3+} the electrical neutrality can be maintained according to Eq. (17.2), in addition to enabling the formation of solid solutions with a controlled concentration of $V_O^{\bullet\bullet}$ defects in the crystalline structure. This equation indicates linearity depending on the concentration of dopants and oxygen vacancies. If all vacancies are free to move, the concentration of mobile vacancies is equal to the stoichiometric fraction of vacancies. If the concentration of dopants is high, interactions between defects (clusters) may arise through electrostatic forces such as Coulomb, through the formation of defects associations of the kind $\left(Y'_{Zr}V_O^{\bullet\bullet}\right)^{\bullet}$ or $\left(Y'_{Zr}V_O^{\bullet\bullet}Y'_{Zr}\right)^{x}$, and these in large quantities can reduce the electrical conductivity (Mogensen et al. 2004). This ordered and relatively open cubic crystal structure allows higher electrical conductivity, as it facilitates the mobility of oxygen vacancies charge carriers (Skinner and Kilner 2003; Cormack 1986).

Figure 17.2a shows the Arrhenius curves of the 8YSZ polycrystalline ceramic σ as a function of the inverse of the absolute temperature extracted from different references (Mahato et al. 2015; Ivers-Tiffée et al. 2001; Antunes et al. 2018; Araki and Arai 2011; Kharton et al. 2004; Yamahara 2003; Ahamer et al. 2017). A dense sample of 8YSZ ceramic electrolyte has an electrical conductivity of 0.05 S cm^{-1} at 800 °C. This σ value guarantees high efficiency and performance of the SOFC at $T_{op.}$=800 °C. Beyond of the 8YSZ electrolyte σ, must also have: (1) a density above 95% of the theoretical (~5.9 g/cm^3) (Panthi et al. 2018) to avoid mixing between fuel and oxidising gas and good mechanical properties; (2) ionic conductivity with a broad electrolytic domain with an ion transfer number greater than 0.99 (Figuereido and Marques 2013; Wincewicz and Cooper 2005) and (3) exhibit good chemical and mechanical stability under operating conditions, which involve wide ranges of PP_{O_2} and thermal cycling (Steele and Heinzel 2001).

All these requirements are necessary for a SOFC electrolyte, however, the σ_{ionic} of this material is only viable at elevated temperatures. The $T_{op.}$ higher than 800 °C restricts the materials applied in its manufacture, increasing the costs of other components (Ormerod 2003; Jacobson 2010; Ralph et al. 2001; Irshad et al. 2016), generating several problems related to materials such as sintering and coarsening of electrodes, interfacial diffusion between electrode and electrolyte materials, thermal instability, thermomechanical stresses due to the different CTEs of each material (Mahato et al. 2015; Ivers-Tiffée et al. 2001; Jacobson 2010; Montross et al. 2002; Tietz 1999). One of the major problems of this high temperature is the Cr volatilization from ICs based on Sr-doped $LaCrO_3$ ceramics and Fe-Cr metal alloys (Sect. 17.9) (Hilpert et al. 1996; Wu and Liu 2010; Sachitanand et al. 2013; Aznam et al. 2019). These conditions reduce the lifespan of the components and require advanced materials, which have high cost, making it difficult to spread this technology on a larger scale (Mahato et al. 2015; Wachsman and Lee 2011; Tu and Stimming 2004).

Several researchers have sought improvements to this device. The main challenge is to reduce the $T_{op.}$ of SOFCs: (1) decreasing the thickness of the 8YSZ electrolyte in order to reduce ohmic losses due to polarization employing cells supported by the electrodes (De Souza et al. 1997; Aguiar et al. 2004; Chen and Wei 2006; Chen et al. 2006; Moon et al. 2008; Ding and Liu 2008; Bailly et al. 2012; Shi et al. 2017) and (2) replacing the 8YSZ electrolyte with more conductive ones, such as doped CeO_2 (Mahato et al. 2015; Kharton et al. 2004, 2001; Mogensen et al. 2000), lanthanum gallates (Mahato et al. 2015; Kharton et al. 2004; Ishihara 2009a; Gomes et al. 2009; Yaremchenko et al. 2004) and doped barium cerates-type proton conductors (Mahato et al. 2015; Kreuer et al. 2004; Ishihara 2009b). However, none of these higher conductivity electrolytes has thermodynamic stability like YSZ electrolytes and competitive commercial prices yet (Ng et al. 2018; Polat 2008).

In this context, many researchers focus to reduce electrolytes into thin films. All these efforts seek creative solutions to improve the performance of SOFCs by $T_{op.}$ ca. 600 °C and apply low-cost materials to using bioethanol as fuel (Antunes et al. 2018; da Silva and de Souza 2017; Mat et al. 2018; Zakaria et al. 2020b; Sanna et al. 2010; Karthikeyan et al. 2006; Lee et al. 2014a; Yamamoto et al. 2015; Duan et al. 2015; Panthi and Tsutsumi 2014; Li et al. 2020b).

17.6 Bioethanol as Fuels

Biofuels may be applied as a direct fuel by internal reforming or indirect fuel by external reforming. Bioethanol also can be applied alone or with steam. We start here with internal reforming.

Here, we consider the most applied case of oxide ion-conducting electrolyte and fuel reaction at the anode, which depends on $T_{op.}$, flow conditions, current density, PP_{O_2}, and catalytic behaviour for this thought exercise. It is well known that bioethanol faces pyrolysis at first, forming H_2 and CO and byproducts (syngas). The H_2, CO and undecomposed fuel molecules can be oxidised at the anode-electrolyte interface in contact with oxygen ions in the triple-phase boundaries (TPBs). The H_2O and CO_2 molecules will be the byproducts of that reaction. The coking formation and removal can occur concomitantly (Yang et al. 2019a).

Bioethanol reform on SOFC anode can occur in the presence of H_2O (ESR: Ethanol Steam Reforming) to produce H_2 according to Eqs. (17.3) and (17.4):

$$C_2H_5OH_{(g)} + H_2O_{(g)} \rightarrow 2CO_{(g)} + 4H_{2(g)} \tag{17.3}$$

$$\Delta H^0_{298\ K} = 255,5\ \text{kJ/mol}$$

$$C_2H_5OH_{(g)} + 3H_2O_{(g)} \rightarrow 2CO_{2(g)} + 6H_{2(g)} \tag{17.4}$$

$$\Delta H^0_{298\ K} = 173,3\ \text{kJ/mol}$$

bioethanol can also react with O_2 by partial oxidation of ethanol (POE) in auto reforming (ATR: AutoThermal Reforming) to produce H_2O, H_2, CO_2 and CO according to Eqs. (17.5), (17.6) and (17.7):

$$C_2H_5OH_{(g)} + 3O_{2(g)} \rightarrow 2CO_{2(g)} + 3H_2O_{(g)} \tag{17.5}$$

$$\Delta H^0_{298\ K} = -1293\ \text{kJ/mol}$$

$$C_2H_5OH_{(g)} + \frac{3}{2}O_{2(g)} \rightarrow 2CO_{2(g)} + 3H_{2(g)} \tag{17.6}$$

$$\Delta H^0_{298\ K} = -552\ \text{kJ/mol}$$

$$C_2H_5OH_{(g)} + \frac{1}{2}O_{2(g)} \rightarrow 2CO_{(g)} + 3H_{2(g)} \tag{17.7}$$
$$\Delta H^0_{298\ K} = -14\ \mathrm{kJ/mol}$$

At low molar ratio of Oxygen/Ethanol (O/E) can produce CO, as presented in Eq. (17.7), but this reaction is thermodynamically less favored at the temperature of ESR and POE ca. 600 °C. Finally, H_2 and CO are converted by O^{2-} anions in the SOFC anode according to Eqs. (17.8) and (17.9):

$$H_{2(g)} + O^{2-}_{(crystal)} \rightarrow H_2O_{(g)} + 2e^- \tag{17.8}$$
$$\Delta H^0_{298\ K} = -241\ \mathrm{kJ/mol}$$

$$CO_{(g)} + O^{2-}_{(crystal)} \rightarrow CO_{2(g)} + 2e^- \tag{17.9}$$
$$\Delta H^0_{298\ K} = -283\ \mathrm{kJ/mol}$$

Coke formation on the SOFC anode can also occur according to Eq. (17.10):

$$2CO_{(g)} \rightarrow C_{(s)} + CO_{2(g)} \tag{17.10}$$
$$\Delta H^0_{298\ K} = -171,5\ \mathrm{kJ/mol}$$

however, reactions between CO and H_2O can generate more H_2 and CO_2 through the displacement reaction of the H_2O (WGS: water-gas shift reaction) reducing the formation of coke on the SOFC anode according to Eq. (17.11):

$$CO_{(g)} + H_2O_{(g)} \rightarrow CO_{2(g)} + H_{2(g)} \tag{17.11}$$
$$\Delta H^0_{298\ K} = -40,4\ \mathrm{kJ/mol}$$

All endothermic reactions are favored by the temperature increase and the exothermic ones are favored by the temperature reduction and that is the reason why syngas production in external reformers are made up of two reformers: (1) one for the ESR and POE at ~600 °C based on the state of the art Ni-Al_2O_3 or Ni-CeO_2 and another (2) to selective oxidation by WGS reaction of CO with H_2O transforming it into CO_2 plus H_2 at ca. 300 °C based on the state of the art Fe_xO_y or PrO_x (Llorca et al. 2013). The use of steam and O_2 are promising in the coke formation suppression by external reforming (Wongsakulphasatch et al. 2013; Ni et al. 2007).

The thermodynamic balance between ESR, POE and ATR are studied by minimizing the Gibbs free energy including the possibility of coke formation for molar ratios between 0<H_2O/Ethanol(S/E)<10 and 0<O_2/Ethanol(O/E)<3 and catalysis temperatures ($T_{cat.}$) between 200<$T_{cat.}$(°C)<1000. The main conclusions are the following: (1) ethanol processed with steam and/or air decomposes throughout the $T_{cat.}$ range, with $T_{cat.}$<400 °C could being produces methane (CH_4) and CO_2, however above $T_{cat.}$>400 °C the concentration of CH_4 decreases while that of H_2 and CO strongly increases; (2) high amounts of H_2 is possible in ESR compared to POE between 550<$T_{cat.}$(°C)<650 and S/E>4 being possible to produce 4 mol H_2/mol C_2H_5OH with molar fraction of CO<0.1 without formation of coke; (3) in POE, high amounts of H_2 and CO are produced and a reasonable amount of H_2 can be reached at O/E<1.5 and $T_{cat.}$>600 °C but to avoid coke formation it is necessary to maintain the O/E >0.8; (4) the ATR, in addition to reducing the energy demand for ESR, reduces the rate of coke formation over the entire S/E range and increasing the O/E ratio from 0 to 0.75 does not greatly affect the H_2 and CO formation at $T_{cat.}$<600 °C (Llorca et al. 2013).

Although, the last process involves many fundamental steps and the C-C bond catalytic breaking is more difficult and also may result into adsorbed C as coke. In counterpoint to internal reforming, the external reforming consists of the passage of hydrocarbons in a system that contains a catalyst based, usually Ni or Ru on heated support, generally, metallic AISI or ceramics based on Al_2O_3, La_2O_3, CeO_2, ZnO and MgO. The $T_{cat.}$ depends on each system and can vary between 250<T(°C)<950. For example, it is possible to maximize the reaction parameters such as hydrocarbon conversion, H_2 production and selectivity by proper catalyst preparation, i.e., wet impregnation technique of the support and optimization of operational parameters. The parameters can be optimized by choosing the appropriate catalyst and its support to improve chemical stability and tolerance for C formation besides optimize reagents flow, the S/E molar ratio and $T_{cat.}$. Catalysis promoters to oxidize bioethanol to H_2 can be the metals Co, Cu, Sn, Nb, Pt, Pd, Mn, Rh, Ru and Au that combined to Ni form bimetallic alloys, improving the production of H_2 and decreasing C deposition (Sect. 17.7) (Sengodan et al. 2018; Zanchet et al. 2015).

Different catalyst-support systems for hydrocarbons external reformers have been proposed such as Ni-Al_2O_3 (Aboudheir et al. 2006; Comas et al. 2004). The ESR works well in Co/Al_2O_3 and Co/SiO_2 (Batista et al. 2004), Rh/Al_2O_3 (Cavallaro et al. 2003), Rh/CeO_2-ZrO_2 (Diagne et al. 2004), Ni/Al_2O_3-La_2O_3 (Fatsikostas and Verykios 2004), Rh, Ru, Pt or Pd onto Al_2O_3, MgO or TiO_2 (Liguras et al. 2003), Ni/$BaZr_{0.1}Ce_{0.7}Y_{0.1}Yb_{0.1}O_{3-\delta}$ (Ma et al. 2020) and Na-Co/ZnO (Llorca et al. 2004). The ATR could occur in bimetallic alloy systems Ni-Rh on CeO_2 (Kugai et al. 2006), Pt-(Ce)/Al_2O_3 and Pt-(La)/Al_2O_3 (Navarro et al. 2005) and Cu-Ni on ZnO and Al_2O_3 (Velu et al. 2005).

Generally, bioethanol is reformed by ESR improving the conversion rate to H_2 and reducing coke formation. From this reforming, C_2H_5OH decomposes into 2CO and $4H_2$ between 400 and 600 °C on the surface of the catalyst particles, usually Ni or a Ni-Rh alloy on an Al_2O_3 or CeO_2 ceramic support that can also help in catalysis (Llorca et al. 2013).

Table 17.2 shows the main catalysts-support reformer systems from literature on ESR and ATR external reforming.

Among all the systems depicted, those presenting Pd, Rh and Ni onto MgO, ZnO, CeO_2 and La_2O_3 exhibit the best performance in terms of bioethanol conversion and H_2 selectivity. The ESR and ATR are promising in the suppression of C formation on the catalyst (Wongsakulphasatch et al. 2013; Ni et al. 2007). However, for the economy of scale of SOFC technology embedded in vehicles, the complexity, size, weight and above all, the cost of the external reformer are also limiting factors and, therefore, internal reform directly in the SOFC anode is preferred (Singhal 2002; Dogdibegovic et al. 2020).

17.7 Functional Anode and Cathode

In the third SOFCS generation, the anode and cathode are divided into two parts: support and functional. The support provides physical robustness to the electrode and the functional give electrochemical activity.

17.7.1 Anode

Many researchers have been optimizing the materials of SOFC components, especially the anode, to use more versatile and cheaper fuels than H_2, for example, bioethanol (Yang et al. 2019a). The Ni/8YSZ composite is the most applied material as an anode in LT-SOFCs due to: (1) the exceptional control of the size, distribution and connection of pores through which the reactants and products are transported; (2) the high surface area of the TPBs; (3) electrocatalytic activity where reduction reactions occur; (4) the unique properties of high electrical conductivity, structural stability and (5) CTEs matching with other materials ($CTE_{52vol\%\ Ni/8YSZ}$ = 12.10^{-6} °C^{-1} between 300<T(°C)<800 which is close to CTE_{8YSZ} = 14.10^{-6} °C^{-1} between 300<T(°C) <900) (Mori et al. 1998).

However, the three major problems of using hydrocarbon fuels in Ni/8YSZ composite anodes in SOFCs are: (1) the low kinetic performance of the redox reaction cycles; (2) instability in prolonged use and, mainly, (3) C deposition on the anode surface. Many researchers have been studying other systems and the addition of other materials to the Ni/YSZ anode to reduce C formation and achieve good SOFC performance and stability (Yang et al. 2019a; Ni et al. 2007; Sengodan et al. 2018; Shi et al. 2020). Table 17.3 presents selected studies of various anodes and cell configuration systems for the different fuels internal reforming.

The studies on anodes are divided according to the use or not of noble metals. The objective of these studies is to improve performance in three aspects: (1) suppress C deposition on the anode; (2) increase the oxidation kinetics of bioethanol and (3) improve mechanical and chemical stability. Among the non-noble metals, Ni is the most used element in anodes. It has high electrical conductivity and excellent activity for H_2 electrocatalytic oxidation. However, Ni is vulnerable to C deposition, coking as it accelerates/catalyzes hydrocarbons thermal decomposition, causing a drop in SOFC performance if hydrocarbon is the fuel. To mitigate this problem, many studies combine Ni and Cu as a catalyst in anode materials. Cu has a poor catalytic activity to breaking down hydrocarbon molecules, and no C deposition

Table 17.2 Different reformer systems for bioethanol ESR, ATR and catalyst preparation methods. Adapted from Sengodan et al. (2018) and Ni et al. (2007)

Reformer system	Preparation method	$T_{cat.}$ (°C)	Conversion rate $\left(L.h^{-1}.g_{cat.}^{-1}\right)$	Molar ratio H_2O/C (ESR) $O_2/H_2O/C$ (ATR)	Conversion (%)	H_2 selectivity (%)	Refs.
ESR							
1 wt% Ru/CeO_2	Impregnation	450	–	3:1	>90	57	Erdohelyi et al. (2006)
15.3 wt% Ni/La_2O_3	Impregnation	250	–	3:1	81	49	Sun et al. (2005)
3.8 wt% Ni/Al_2O_3	Precipitation/ Impregnation	650	–	3:1	100	89	Akande et al. (2005)
$CuNi/CeO_2$	Impregnation	450	57	10:1	60	70	Dancini-Pontes et al. (2015)
Ni/La_2O_3-ZrO_2	Impregnation	350	9.2	30:1	100	55	Dan et al. (2015)
Ni/CeO_2-ZrO_2	Impregnation	650	5.5	8:1	100	>75	Biswas and Kunzru (2007)
Ni/GDC	Hydrothermal combustion	850	80	3:1	100	65	Augusto et al. (2014)
Ni/TiO_2	Precipitation/ Impregnation	500	29.2	3:1	99	–	Rossetti et al. (2014)
$Ni_{0,95}Mo_{0,05}/SiO_2$	Hydrothermal combustion	700	8.6	3:1	80	80	Kim et al. (2012)
$CuNi/SiO_2$	Impregnation	350	17.2	6:1	100	–	Chen and Lin (2014)
Pt-Ni/CeO_2	Hydrothermal combustion/Impregnation	500	36	3:1	69	–	Moraes et al. (2015)
$Pt_{1,5}Co_{98,5}/ZrO_2$	Sol gel/Impregnation	300	22	13:1	100	–	Chiou et al. (2014)
$Ni_3Mg_2AlO_y$	Precipitation	300	168	3:1	100	60	Fang et al. (2015)
Co_2MnO_4	Combustion	700	8.8	0.33:1	100	71	Kwak et al. (2015)
$2KCo/ZrO_2$	Impregnation	420	60	9:1	100	74	Greluk et al. (2015)
$La_{0,6}Sr_{0,4}CoO_3$	Sol gel citrate	700	40	3:1	100	70	Morales and Segarra (2015)

(continued)

Table 17.2 (continued)

Reformer system	Preparation method	$T_{cat.}$ (°C)	Conversion rate $\left(L.h^{-1}.g_{cat.}^{-1}\right)$	Molar ratio H_2O/C (ESR) $O_2/H_2O/C$ (ATR)	Conversion (%)	H_2 selectivity (%)	Refs.
$La_{1-x}Ca_xFe_{0,7}Ni_{0,3}O_3$	Pechini	650	400	3:1	100	65	Chen et al. (2011)
$La_{1-x}K_xFe_{0,7}Ni_{0,3}O_3$	Pechini	450	60	3:1	100	70	Zhao et al. (2016)
$LaCo_xNi_{1-x}O_3$	Pechini	550	60	3:1	100	60	Liu et al. (2015)
$Ni/LaFe_{0,7}Co_{0,3}O_3$	Pechini	550	80	3:1	100	67	Wang et al. (2014a)
$Mn/Co_{0,1}Si_{0,9}$-aluminosilicate	Hydrothermal combustion/impregnation	600	8.6	0.33:1	100	97	Lee et al. (2014b)
$Rh/5\%La_2O_3$-CeO_2-Al_2O_3	Impregnation	500	82.2	3:1	70	70	Osorio-Vargas et al. (2015)
Rh/CeO_2	Deposition/impregnation	650	382	3:1	100	72	Hou et al. (2015)
Ir/CeO_2	Deposition/impregnation	650	6	1.8:1	100	55	Zou et al. (2015)
Ir/CeO_2	Deposition/impregnation	450	22	4:1	100	–	Chiou et al. (2012)
$Ir/Ce_{0,9}Pr_{0,1}O_2$	Impregnation	450	18	3:1	95	60	Wang et al. (2011)
Rh-Co/CeO_2	Impregnation	450	72	3:1	90	60	Ferencz et al. (2014)
Rh/MgO	Impregnation	650	24	8.4:1	100	92	Frusteri et al. (2004)
Pt/CeO_2	Impregnation	350	25.4	3:1	82	–	He et al. (2012)
$Pt/CeZrO_2$	Precipitation/Impregnation	500	80	0	100	>50	de Lima et al. (2009)
ATR							
Pd/ZnO	Impregnation	450	–	0.5:13:1	100	61	Casanovas et al. (2006)
$Ni_{19,4}Cu_{0,6}/Al_2O_3$	Impregnation	727	–	0.68:1.6:1	100	98	Fierro et al. (2005)

Table 17.3 Summary of structures, materials, and performances of SOFCs based on H_2 and bioethanol fuels. Adapted from Yang et al. (2019a)

Anode	Electrolyte	Cathode	Power Density Peak (mW/cm^2)	Fuel (mol: mol) (vol.%) (Pa/Pa)	Main Points	Refs.
Reports mainly on anode and electrolyte supported solid oxide fuel cells (SOFCs) – first and second generations						
Ni/YSZ (50vol.%Ni) (250 μm)	YSZ colloidal thin-film deposition on anode (10 μm)	LSM-YSZ composite (50vol.%LSM)	1935 at 800 °C	H_2/4vol.%H_2O	Inexpensive and scalable button cell with 700 h test exhibited high power density without aging or degradation	De Souza et al. (1997)
Ni/YSZ (anode support: 1 mm) (anode interlayer: ~20 μm)	YSZ (10 μm)	$La_{0.8}Sr_{0.2}MnO_{3-\delta}$ (80 μm) cathode interlayer (20 μm)	800 at 800 °C	1:1 Ethanol/H_2O	The FC performance is governed by electrodes polarizations as well as the cell ohmic resistance	Jiang and Virkar (2001)
NiO-10YSZ with $NiAl_2O_3$ catalyst (800 μm)	10YSZ (10 μm)	$Ba_{1-x}Sr_xCo_yFe_{1-y}O_{3-\delta}$ (BSCF)-$Sm_{0.5}Sr_{0.5}CoO_{3-\delta}$ (SSC)-$Sm_{0.2}Ce_{0.8}O_{1.9}$ (SDC) (25 μm)	300 at 600 °C	Ethanol/pyridine	Ethanol and pyridine fuel blend suppressed coke formation	Wang et al. (2014b)
NiO-10YSZ with Ni/$Ce_{0.8}Zr_{0.2}O_{2-\delta}$ catalyst layer	10YSZ (10 μm)	BSCF-SSC	179 at 550 °C 324 at 600 °C	Ethanol/O_2	Improvement of H_2 selectivity and coking resistance due to catalyst layer	Wang et al. (2012)
NiO + 10YSZ with Ni/$Ce_{0.8}Zr_{0.2}O_{2-\delta}$ catalyst layer	10YSZ (10 μm)	BSCF	162 at 600 °C 74 at 550 °C	Ethanol/H_2O	Good coking resistance and catalytic activity due to the anode interaction	Liao et al. (2011)
Ru-coated (<10 nm) Pt (150 nm)	$Gd_{0.1}Ce_{0.9}O_{2-\delta}$ (5GDC) (350 μm)	Pt	13 at 500 °C	Ethanol	Reduced anode impedance using Ru coating by	Jeong et al. (2015)

(continued)

Table 17.3 (continued)

Anode	Electrolyte	Cathode	Power Density Peak (mW/cm^2)	Fuel (mol: mol) (vol.%) (Pa/Pa)	Main Points	Refs.
					atomic layer deposition (ALD) without coking	
Pt	YSZ	Pt	4 at 550 °C	3.5 kPa/ 7.5 kPa Ethanol/ H_2O	Ag anode shows inferior performance compared to Pt anode	Poulianitis et al. (2006)
			3 at 550 °C	5 kPa/ 7.5 kPa Ethanol/ H_2O		
Ag		Ag	2 at 550 °C			
Pd	Y-$BaZrO_3$ (BYZ) (130 nm)	Pt	15.3 at 400 °C	Ethanol	Increase of bond cleavage energy	Li et al. (2017)
Reports mainly on metal supported solid oxide fuel cells (MS-SOFCs)—third generation						
Ni/5GDC by screen printing on 200 μm of stainless steel stabilised with Ti-Nb-17wt%Cr	10–30 μm of GDC DBL on anode + YSZ blocking layer	$La_{1-x}Sr_xCo_{1-y}Fe_yO_{3-\delta}$ (LSCF)-GDC composite	240 at 550 °C	H_2/3vol. %H_2O	2500 h test delivery 0.25 $A.cm^2$ at 0.7 V on 16 cm^2 active area	Brandon et al. (2004), Leah et al. (2017)
Ni/$Sm_{0,2}Ce_{0,8}O_{2-\delta}$ (40vol.%Ni) on porous metal support	10 mol%Sc_2O_3-1 mol% CeO_2-stabilised ZrO_2 (10Sc1CeSZ) (7 μm)	Pr_6O_{11}	250 at 700 °C	0.65:1 Ethanol/ H_2O	High surface area of Ni nanoparticles in anode and metal support obtaining excellent power density without coking delivery 0.44 $A.cm^{-2}$ at 0.7 V on 5 cm^2 active area	Dogdibegovic et al. (2020, 2021)

Lamination of Fe22Cr stainless steel support (240 μm) + YSZ anode backbone structure by tape casting and subsequent infiltration of Ni/10GDC (50vol.%Ni) as an active layer	ScYSZ (9 μm) 10GDC DBL deposited by PVD	$(La_{0.6}Sr_{0.4})_{0.99}CoO_{3-\delta}$ (LSC) (15 μm)	720 at 700 °C	H_2/ 20vol.% H_2O	Lamination by tape casting, sintering and screen printing of the cathode delivery 1.44 $A.cm^2$ at 0.7 V on 16 cm^2 active area. Performance tests shows stability up to 1000 hs at 650 °C	Nielsen et al. (2018), Blennow et al. (2011a)
NiO-$Ce_{0.55}La_{0.45}O_{2-\delta}$ anode on porous Ni-Mo support (20–40 μm) by plasma spraying (PS)	$La_{0.75}Sr_{0.25}Cr_{0.50}Mn_{0.50}O_{3-\delta}$ (LSCM) DBL by PS on $La_{1-x}Sr_xGa_{1-y}Mg_yO_{3-\delta}$ (LSGM) electrolyte (40–50 μm)	15 wt% C pore formers in SSC by PS	1013 at 700 °C	H_2	Ni-Mo porous metal alloy support with low tortuosity gas flow channels demonstrated that power density and fuel consumption increase with channel densities and delivery 1.4 $A.cm^{-2}$ at 0.8 V on 16 cm^2 active area using 46 vol.% of the fuel	Hwang et al. (2016)
Ni/GDC (active layer: 8 μm) Ni/GDC (intermediate layer: 15 μm) Ni/8YSZ (base layer: 25 μm) on Fe26Cr (support layer: 300 μm)	8YSZ (4 μm) 20GDC DBL (0.3 μm)	LSCF	1250 at 700 °C	N_2 / 20vol.% H_2	Gradually functionalized anodes were optimized by changing the active anode Ni/8YSZ by Ni/GDC delivery 1.79 $A.cm^{-2}$ at 0.7 V on 0.064 cm^2 active area result 38% increase in power density	Bischof et al. (2019)

occurs on its surface when applied as an anode material (Yang et al. 2019a; Atkinson et al. 2004; Pieta et al. 2021). However, power is usually compromised, and a better solution is still to come.

Noble metal-based catalysts such as Pt, Pt-Ru alloy, Ag and Pd can be applied as functional anodes in SOFCs fed with alcohols. Pt is the best-known catalyst for the deprotonation of methanol and bioethanol molecules. A large amount of CO is generated on the surface of the Pt catalyst when fuel alcohols are applied in SOFCs, which does not happen with H_2. Ru added to Pt anode can improve the anode's non-carbon deposition and thermal stability. Ru is more easily oxidized and provides oxygen O species at the Pt vicinity, helping to further oxidize the adsorbed CO (Yang et al. 2019a; Atkinson et al. 2004).

In summary, the catalyst materials applied as anodes for bioethanol oxidation reactions are: (1) non-noble metals such as Ni, Cu or Ni-Cu alloys mixed with the materials applied in the electrolytes or (2) noble metals such as Pt, Pd, Ag or Pt-Ru alloys.

17.7.2 Cathode

The cathode is responsible for the chemical oxygen reduction reaction (ORR), which is also affected by LT-SOFC. Cathode materials must be stable and active in the full-scale of $T_{op.}$ and operational parameter of the cell. It should present thermo-chemical stability for the reaction to occur uniformly throughout the active sites, compatibility with other components avoiding intermediate reactions, thermal shocks due to electrolyte expansion and delamination. High electronic conductivity helps the transport of electrons through the external circuit. Porosity ensures gas diffusion through the electrode and high catalytic activity to breaking down oxygen molecules (Cesário and de Macedo 2017).

Non-noble metals used as catalyst materials for the O_2 reduction reaction in the cathodes are MIECs oxides type $La_{1-x}Sr_xCoO_{3-\delta}$ (LSC), $La_{1-x}Sr_xCo_yFe_{1-y}O_{3-\delta}$ (LSCF) and $Ba_{1-x}Sr_xCo_yFe_{1-y}O_{3-\delta}$ (BSCF). For noble metals, Pt and Ag type are the most relevant example of success, but they are expensive (e.g., Pt: $30–60/g) and have low thermal stability (Yang et al. 2019a; Dogdibegovic et al. 2020; Atkinson et al. 2004; Liu et al. 2021).

The procedure for cathode deposition and stabilization is the main challenge in MS-SOFCs. The metal support, anode, and electrolyte are usually sintered together in a reducing atmosphere to prevent ferritic stainless steel (FSS) oxidation. Next, the cathode is deposited and sintered in the air while the rest of the cell (mainly the FSS support) must remain in a reducing atmosphere. The cathode can be sintered in situ between $900<T(°C)<1000$ for a short soaking time, i.e., before the MS-SOFC works at $T_{op.}$. Thus, cathode activation and their adhesion on electrolytes occur, while the anode region is protected by a reducing atmosphere.

New materials for MS-SOFCs cathodes have been investigated to ensure stability in reducing atmospheres and high electrocatalytic activity. Vibhu et al. (2015) have reported studies on characterization and stability of based-cathodes on $La_2NiO_{4+\delta}$ (LNO) and $Pr_2NiO_{4+\delta}$ (PNO). $La_{2-x}Pr_xNiO_{4+\delta}$ (LPNO) composites were successfully synthesized. The half-cells electrochemical performance with $x<0.5$ exhibited better stability, while $0.5<x<1$ shows better performance. These nickelates are over stoichiometric in oxygen up to ~1000 °C and have suitable electrical conductivities, CTEs, oxygen diffusion and exchange coefficients for application in MS-SOFCs. However, performance tests were not performed in that work.

LSCF-based cathodes with $C_{graphite}$ pore formers have been deposited by PS on samaria-doped ceria DBLs that prevents the formation of the insulating phase between the cathode and the electrolyte while maintaining area-specific resistance (ASR) values as low as 62 mΩ cm^2 (Fan et al. 2016).

BSCF-based cathodes have been sintered in situ at lower temperatures than $T_{op.}$. Kim et al. (2011) have reported a MS-SOFC power density of 0.74 W cm^{-2} when supported by ferritic stainless steel (FSS) 430L with BSCF cathode on gadolinium-doped ceria DBL.

Tucker et al. (2010) reported the use of cathode and anode infiltration techniques on backbone structures. Zhou et al. (2014a) have deposited ~20 μm of YSZ electrolyte on porous FSS 430L support followed by a YSZ backbone structure layer as cathode support. They used $SrFe_{0.75}Mo_{0.25}O_{3-\delta}$ (SFMO) as a catalyst for the porous FSS 430L anode and the YSZ backbone cathode. Immediately after infiltration, all layers were sintered in reducing atmosphere. These MS-SOFCs reached a power density of 0.4 W .cm^{-2} at 700 °C using H_2/H_2O.

17.8 The MS-SOFCs: A Brief Overview

The third generation MS-SOFCs shows exceptional performance and qualify to use bioethanol as fuel in electrified vehicles, allowing onboarding electricity generation powertrain systems. The ~0.5–30 μm thinnest cell components fabricated by scalable processing techniques applied on lower $T_{op.}$ allows cost reduction of overall cell. This achievement was only possible due to low-cost porous FSS as metallic supports under the anode and over the cathode and cells ICs welding in stacks, ensuring tightness and high thermal conductivity, mechanical vibrations resistance and high redox cyclability. These characteristics warrant companies such as Ceres Power (England), Plansee SE (Austria), AVL List GmbH and Bosch (Germany), Nissan (Japan), Weichai Power and BYD (China) starting scale production of MS-SOFCs for automotive industries.

Recently, exciting studies have been reported using different functional anodes layers with high electrocatalytic performance on porous FSS support. Summary, a gadolinium-doped ceria DBL is deposited on FSS (Fe26Cr, ~300 μm) by physical vapor deposition (PVD) like magnetron or radiofrequency sputtering, which prevents diffusion of Fe, Ni and Cr to electrodes avoiding oxides insulating phases formation. Then, a stable terpineol and ethyl cellulose-based ink containing suitable particle size and distribution of Ni/8YSZ anode is applied by screen printing and dried to shape the first layer.

On this first base anode, an intermediate layer with smaller particle size and distribution of Ni/8YSZ anode is also applied by screen printing using a polyester sieve with a tiny mesh aperture, then dried. On top of this layer, a relatively dense thinner active layer of Ni/10GDC anode is deposited using a lower mesh sieve than previous ones, followed by drying. After the drying step, each anode layer is sintered between $1100<T(°C)<1200$ for 3 h of soaking time in reducing atmosphere.

After sintering, the base, intermediate and active anodes will have ca. 25, 25 and 8 μm of thickness, respectively. The set of functional anodes is necessary so that the active layer presents adequate roughness and microstructure to avoid electrolyte delamination. Then, a dense ca. 4 μm layer of 8SYZ or 10GDC based electrolyte is deposited on the active anode by PVD techniques, and another DBL is applied over the electrolyte.

Finally, a high-performance cathode, typically $La_{0.6}Sr_{0.4}CoO_{3-\delta}$ (LSC), is deposited upon the DBL-electrolyte by screen printing, dried and heat treated. This procedure allows to obtain high power and current densities between 1–2 $W.cm^{-2}$ and 1.4–2.8 $A.cm^{-2}$, respectively, at 0.7 V and 650 °C using H_2. These values were obtained mainly due to the lower polarizability in the active anode and low Ni's tortuosity covered by GDC thin particles interconnected from the intermediate anode layer to the electrolyte. Thus, the metallic Ni phase is responsible for the long-range electrons transport while both Ni and GDC phases act in the H_2 catalysis (Bischof et al. 2019; Haydn et al. 2014; Udomsilp et al. 2020).

However, FSS support and Ni (base anode) oxidation were found near the FSS support interface and GDC barrier layer despite the high performance. Even with the protective layer, Fe and Cr have been found on Ni particles creating insulating oxides such as Cr_2O_3, $NiCr_2O_4$ and Fe_2O_3. The Ni diffusion onto support also occurs, causing the austenitic transformation of FSS, increasing the CTE and reducing its corrosion resistance. These kinds of degradations are responsible for reducing performance in the long term. This happens because the deposition of DBL by PVD is not entirely homogeneous on the porous FSS support. Other researchers deposit Co-Mn DBLs to form spinel oxides like $(Co, Mn\ or/and\ Cr)_3O_4$ on the FSS support and applied a LSM or LSC contact layer to prevent Cr diffusion from the spinel to the anode (Larring and Fontaine 2013). These DBLs significantly mitigate this issue but do not prevent diffusion between anode and FSS support.

The FSS support is usually obtained by powder metallurgy using tape casting. This technique makes it possible to produce pores with randomly distributed irregular shapes and small sintering necks between FSS metallic particles. Due to many pores, a high specific surface area impairs corrosion resistance. A layer of $LaCrO_3$ electron conductor is deposited on the inner surface of the FSS support by dip coating (DCoat) to improve corrosion resistance. However, it is necessary to optimize this process to ensure a homogeneous layer over the inner surface of the pores (Molin et al. 2008; Antepara et al. 2005; Deng et al. 2006; Jeong et al. 2020).

Further, besides the FSS support and Ni oxidation, Ni and YSZ or GDC particles coarsening also occur, reducing the specific surface area and TPBs where the fuel electrochemical oxidation reactions arise. This degradation mechanism occurs through the interdiffusion of Ni, which causes the growth of some larger Ni particles causing the smaller ones to vanish, increasing the volume of defects such as pores. To reduce this issue, researchers use nanoparticulate anodes by wet infiltration of Ni and YSZ or GDC nanoparticles in porous YSZ or GDC ceramics backbones structures. However, due to the high specific surface area of these materials, the Ni particles coarsening still occurs, reducing the TPBs (Dogdibegovic et al. 2020, 2019a; Blennow et al. 2011a, 2009; Tucker 2017; Tan et al. 2018).

The cathode and electrolyte usually react at high temperatures forming resistive phases, e.g., $La_2Zr_2O_7$ pyrochlores caused by the reaction between Sr doped $LaMnO_3$ and YSZ. DBLs can mitigate this issue but do not prevent this kind of degradation. Even so, during the co-sintering process in reducing atmosphere to protect the support against oxidation, most cathodes decompose in this atmosphere and also in the high $T_{op.}$. Therefore, other cathodes such as LSC, LSCF and $Sm_{1-x}Sr_xCoO_3$ (SSC) can survive in reducing atmosphere under relatively lower temperatures, but the CTEs of these materials are much higher than the YSZ one, increasing the risk of delamination (Brugnoni et al. 1995; Mitterdorfer and Gauckler 1998; Wang et al. 2008).

To avoid the decomposition of these cathodes and oxidation of FSS support, the entire cell can be sintered in an argon atmosphere at 950 °C. High $T_{op.}$ also causes cathode materials coarsening and this issue can be minimized likewise previously discussed for anodes. The Cr from the FSS support or metallic ICs in H_2O and O_2 presence can evaporate, diffuse, and be deposited on the FSS support-cathode interface, forming Cr_2O_3. $LaNi_{0.6}Fe_{0.4}O_3$ (LNF) cathodes are more resistant to Cr poisoning when compared to LSM and LSCF (Zhen et al. 2007; Komatsu et al. 2008). The CoO_x layers were deposited on cathode by atomic layer deposition (ALD) to suppress their Cr-poisoning from the FSS support or ICs (Dogdibegovic et al. 2019b).

Last but not least, ICs are essential components to connect the cells providing electrical conductivity and mechanical support in the stacks assembly. State of the art is Fe16Cr FSS. The issue with these materials is that the Cr evaporation poisoned the cathode and spoiled their electrical conductivity due to oxidation. To reduce Cr evaporation, spinel and perovskites are applied by PVD or atmospheric plasma spraying (APS) over the ICs to inhibit Cr evaporation and oxidation. Spinels based on Mn-Co are most used because they have an electrical conductivity of 60 $S.cm^{-1}$ at 800 °C and CTE of $9.7 \times 10^{-6}\ K^{-1}$, very close to $10.5 \times 10^{-6}\ K^{-1}$ ICs besides the low cost compared to perovskites ones (Antepara et al. 2005; Shaigan et al. 2010).

To assure the stacks tightness, mica and alumina ceramic glass-based sealants are used in the first and second SOFCs generation. In third SOFCs generation, these sealants are replaced by low-cost laser welding of FSS ICs (Sudireddy et al. 2017). The MS-SOFC materials degradation mechanisms discussed above are already known in the literature, and there are few studies using bioethanol (Dogdibegovic et al. 2020).

The great challenge to be investigated is the FSS support-anode interaction with bioethanol. Coking occurs on Ni catalyst surface, reducing the performance and cell lifespan. The precipitation of the graphitic phase takes place by catalytic graphitization mechanisms on the Ni surface. Ni^{2+} cations transform graphite into channels and precipitate at the end of the surface, creating thin Ni particles. These particles catalyze the formation of carbon fibers and nanotubes reducing the active sites for hydrocarbons conversion into H_2 as well as pulverizing the Ni (Chun et al. 2000, 2002; Chun and Ramanarayanan 2007). To mitigate this issue, researchers have been using Ni-Cu or Cu alloys as anodes. Cu prevents graphite deposition, but has low catalytic activity reducing the cell performance (Costa-Nunes et al. 2005).

In order to use bioethanol in MS-SOFCs, it is crucial to study the new degradation mechanisms and redesign the FSS support and anode. In the last two decades, both structure of FSS support and the method of obtaining it has not changed. Conventional powder metallurgy has been used for manufacturing randomly distributed pores with high curvature/tortuosity FSS supports channels. Three aspects must be considered when optimizing the support: (1) high efficiency of gas diffusion channels; (2) pores with a size gradient and (3) decrease the sensitivity to sintering necks corrosion of FSS support metal particles.

Nielsen et al. (2018) achieved a 40% increase in MS-SOFC power density using a FSS support with straight-line channels, which was attributed to the significant efficiency of gas diffusion due to high porosity and low tortuosity. If small sinter necks can be avoided, FSS support degradation caused by oxidation can be reduced. Although interconnected porosity can be improved by adjusting particle size and morphology, sintering temperature, pore formers and organic additives, evenly distributed pores and high-efficiency gas channels are difficult to achieve by powder metallurgy. Laser drilling has been used by companies such as Ceres Power to improve the directionality of diffusion channels from the FSS support to the anode (Leah et al. 2015, 2011). However, crossways interconnected pores are difficult to obtain by laser drilling, whereas this technique is based on fast-melting metal thin sheets.

Nowadays, metal additive manufacturing has made progress in porous structural components fabrication. The organized 3D interconnected pores with a few hundred microns can be made by powder metallurgy using an electron beam or laser (Yang et al. 2019b,c; Downing et al. 2021; Yadroitsev et al. 2009).

Few studies of modified FSS supports to act as catalysts have been reported, among them, Zhou et al. (2014b) produced MS-SOFCs by laminating the support, anode and electrolyte structured layers that were infiltrated by the catalyst. In that work, FSS (300 μm) was used as support. The electrolyte used was YSZ (30 μm). As the anode, 10wt% of Ni was infiltrated as a catalyst on the metallic support backbone structure. As the cathode, 30wt% of LSCF ($La_{0.6}Sr_{0.4}Fe_{0.9}Sc_{0.1}O_{3-\delta}$) was also infiltrated into the porous YSZ backbone structure. This architecture made it possible to achieve a power density and current density of 0.42 W.cm^{-2} and 0.5 W.cm^{-2}, respectively, at 700 °C using H_2/3vol.%H_2O in an active area of 0.35 cm^2.

Even so, no study has been reported yet on the use of bioethanol in FSS supports modified with micro-reformers produced by additive manufacture. Figure 17.5

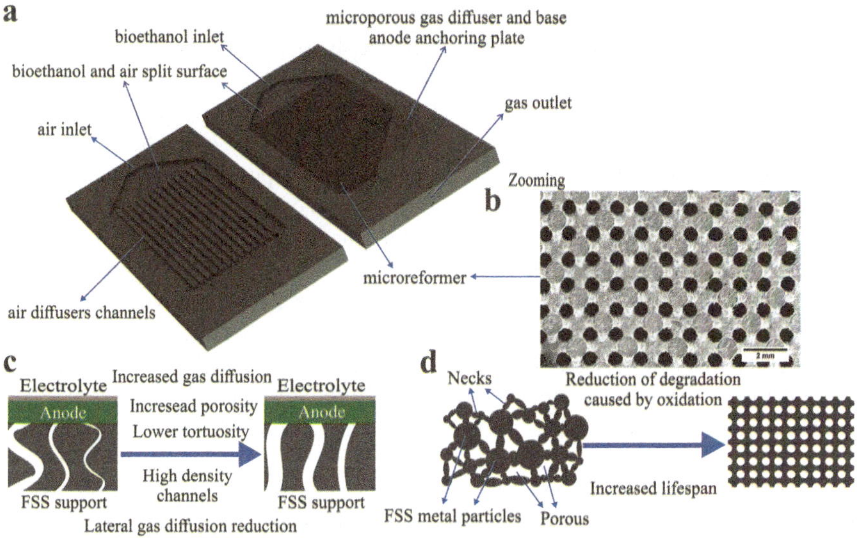

Fig. 17.5 (**a**) project developing for FSS micro-reformer support integrated to the ICs for cell anchoring, containing the organized microchannels produced by metal additive manufacturing; (**b**) micro-reformer volume element zooming (channels with ~200 μm of diameter); (**c**) engineering FSS support for enhancement performance obtained by Nielsen et al. (2018) and (**d**) the impacts that this proposal can bring prototyping micro-channels organized structure as micro-reformers. Adapted from Nielsen et al. (2018) and Zhou et al. (2021)

shows the organized micro-channels in a micro-reformer developed by metal additive manufacturing as a possible engineering technology FSS support project (Binelli et al. 2016, 2017). This new design project can be used for enhancing the performance of MS-SOFC, allowing indirect internal reforming of bioethanol. This figure also shows the impact this technology can have by increasing the MS-SOFCs corrosion resistance and lifespan when a robust, organized micro-channels sinter necks-free of FSS support is used instead of commonly produced by tape casting conventional powder metallurgy.

17.9 DBLs, Interconnectors and Current Collectors

The third SOFCS generation, more commonly namely metal supported-SOFC, enable bioethanol's internal reforming at temperatures $T_{op.}$<800 °C without significant reductions in the performance. That is possible due to thin electrolyte and functional anode and cathode layers. As previously described, that allows the use of stainless steel as a material for ICs and current collectors, enabling the sealing of these components by laser welding. Although FSS supports have low cost and good CTE matching (Zhou et al. 2021), considerable ion diffusions must be controlled. Even far from the austenitic phase transformation at ~1200 °C and the melting point at 1450 °C of 18wt%Cr-0.06wt%C FSS alloy, at $T_{op.}$ Fe, Ni and Cr can migrate and poisoning the cell. Fe and Cr poisoning the anode, Ni poisoning the FSS support and Cr the cathode (Banerjee 2017).

A DBL of ~1–2 μm is essential to block the elements diffusion between the anode and cathode with theirs FSS metal support and ICs. The DBL could be prepared by several techniques such as ALD, APS, pulsed laser deposition (PLD), and PVD (Zhou et al. 2021) with advantages and disadvantages on each technique concerning scalability and quality. The main DBLs requirements are high densities and compatible CTEs. Techniques such as reactive spray deposition (RSD) and electron beam evaporation (EB-PVD) both in air can produce <2 and 0.5–1 μm thicknesses of gadolinium-doped ceria DBLs, respectively (Nédélec et al. 2011; Krishnan 2017).

Materials that can be used as DBLs at interfaces between the anode/FSS support and electrolyte/cathode are CeO_2, 10SDC and 10GDC (Fan et al. 2016; Nédélec et al. 2011; Hwang et al. 2011; Klemensø et al. 2011a; Brandner et al. 2008). Lanthanum chromates and Cr_2O_3/Cr_2MnO_4 have been used only between the anode/FSS or anode/Ni-Al supports (Brandner et al. 2008; Solovyev et al. 2015). At cathode/electrolyte and cathode/ICs interfaces, GDC or SDC and CoO_x are most commonly used, respectively (Fan et al. 2016; Kim et al. 2010). The gadolinium-doped ceria MIEC with CTE ca. 12×10^{-6} K^{-1} is the most employed DBL (Fan et al. 2016; Nédélec et al. 2011; Solovyev et al. 2015).

With proper DBL, the FSS support may be applied as an IC and current collector material. This metal has excellent oxidation resistance and, because of that, have a long lifespan on $T_{op.}$. The oxidation of FSS support could also be improved, i.e., by applying a $LaCrO_3$ thin film by DCoat as discussed above.

FSS-based supports such as 430, 441 and Crofer 22 APU differ by Cr amount in alloy and CTEs ranging from (10–12) $\times$ 10^{-6} K^{-1} which are compatible with YSZ and GDC ceramics. FSS ($2/kg) are cheaper than Ni ($18/kg) and NiCrAlY ($63/kg), according to a review by Tucker (2010) published in 2010. 8YSZ and 10GDC ceramic powders costs are ~$150/kg (Tosoh) and ~$3200/kg (Fuel Cell Materials), respectively.

As mentioned above, metallic support must be porous for gas diffusion, provide mechanical and structural support, be a good electron conductor, CTEs compatible with ceramic materials, good resistance to oxidation in the presence of fuels and steam at $T_{op.}$ and be inexpensive (Tucker 2010). These porous structures could have the morphology of foams, meshes, and holes drilled by laser melting of FSS sheets as produced by Ceres Power or obtained by powder metallurgy routes involving uniaxial and isostatic pressing (pelletizing), spraying or tape casting of ceramic suspensions (Rose et al. 2009).

Antepara et al. (2005) have reported the importance of quantifying some properties of metal supports that should be integrated into MS-SOFCs. These properties are mass gain ($mg.cm^{-2}$) and ASR measured by the 4-point probe in direct current (DC—$m\Omega.cm^2$) after oxidation in air, CTEs, mechanical tensile strength and creep, thermodynamic stability with ceramics in long-term and Cr vaporization in steam atmospheres.

To perform the ASR measurement, samples are air oxidized at $T_{op.}$ leading to Cr oxide scale formation. The FSS Crofer 22 APU (ThyssenKrupp) and ITM (Plansee SE) supports has ASRs of 5 and 2.5 $m\Omega.cm^2$ at 700 °C, respectively. For LSM coated on ITM FSS support by PVD this value decreases to 1 $m\Omega.cm^2$ (Antepara et al. 2005).

Yan et al. (2015) have used sol-gel protective layers on FSS supports to improve oxidation resistance. Precursor solutions containing ethylene glycol, HNO_3, La, Y and Ce cations were infiltrated into porous FSS supports by DCoat stepwise procedure, drying at 400 °C followed by final calcination at 650 °C to form cations oxides on porous FSS support. These oxides and Cr from FSS support react to improve oxidation resistance with appropriate electrical conductivity.

Although Crofer 22 APU and ITM are used as porous supports, they can also be used as dense ICs. When stacking MS-SOFC is preferred, both are made from the same material to ensure welding quality and reduce leakage risks (Krishnan 2017).

The FSS-based ICs issue is the formation of Cr(O)OH vapor in moisture air poisoning the cathode. CTE mismatch between ceramics and ICs is not a big problem since the sealing process and stack design ensure that ICs are not in contact with ceramics. Generally, ICs are separated by glass sealing or welded/brazed-on FSS support which is the same material. The most commonly dense alloys for ICs are Crofer 22 APU, Crofer H, ITM, ZMG 232 and FSS 430 among the ASS series 300. Crofer 22 APU is widely used as IC coated on the cathode to prevent Cr poisoning. ICs are evaluated in the same way as porous FSS supports (Nielsen et al. 2018; Leah et al. 2011, 2013; Yang et al. 2019b; Bance et al. 2004; Ansar et al. 2011; Franco et al. 2011, 2013; Roehrens et al. 2015; Klemensø et al. 2011b; Christiansen et al. 2013; McKenna et al. 2013; Christiansen 2014; Technology 2021).

Sudireddy et al. (2017) have been employed protective layers on anode and cathode ICs. The coating process is a pre-coated PVD process-based developed by Sandvik Materials Technology (2021) that allows scalability and lower cost. In this process, ~600 nm Co layers are applied on the FSS support and cathode ICs. During stack operation, a Co_3O_4 layer forms on coated FSS surface, which reacts with the Mn from FSS IC, creating a $(Co,Mn)_3O_4$ spinel layer. This thin spinel layer prevents Cr evaporation over 3000 h at 850 °C (Froitzheim et al. 2012).

Other ICs and porous supports based on ASS 300 series, Ni-Al, and Ni-Mo alloys have been used, but they exhibit mismatch CTEs and lower corrosion resistance compared to FSS 400 series supports and ICs (Tucker 2010).

17.10 MS-SOFCs Stacks: New Concept, Design and Manufacturing

The Ceres Power produces MS-SOFCs with FSS sheet supports that are laser drilled to create a central inner gas diffusion surrounded by an impermeable one (where no holes are drilled). A thin film of porous cermet anode is deposited over the FSS support drilled region. Afterwards, the electrolyte is layered on the entire anode area overlapping even the undrilled FSS support to seal the cell. This design ensures that a dense electrolyte completely seals the porous anode. In this design, the electrolyte has a complex architecture comprising three layers: (1) a thick GDC film that ensures tightness; (2) a thin electron blocking YSZ film on GDC film and (3) another GDC film that provides a contact layer between YSZ and cathode. The cathode has a double layer: (1) a thin active layer close to the electrolyte where the O_2 reduction reactions take place and (2) another thick layer to collect the current (Leah et al. 2015, 2011, 2013; Bance et al. 2004).

Except for the DBLs and electrolytes, cathode and anode films are deposited by screen printing providing a low cost and high scalability for the manufacturing process. The ceramic films set is sintered in air at ca. 1000 °C without significant FSS support oxidation. It is well known that GDC has low sinterability and densifies above 1400 °C. This low sintering temperature ($T_{sint.}$) is only possible by synthesizing ceramic powders with sintering additives such as CoO and ZnO. Each repetitive unit cell in Ceres Power' stack is comprised of: (1) a two-layer cathode; (2) three-layer electrolyte; (3) anode monolayer; (4) FSS support where the cell is co-sintered; (5) insulating spacer and ICs (Krishnan 2017).

Ceres had reported MS-SOFC single-cells with a power density of ca. 300 mW.cm^{-2} at 600 °C using H_2. These MS-SOFC single-cells had been used to manufacturing stacks. Currently, this company provides 10–15 kW power stacks that can be used as APUs in vehicles due to the advantages of fast light-off, mechanical shocks resistance and vibration absorption (Bance et al. 2004).

The consortium comprises the companies Plansee SE (PSE), Sulzer Metco AG (SM AG) and ElringKlinger AG (EK AG) together they produce MS-SOFC stacks as APUs. On the ~1mm porous FSS support, all catalytic and electrolytic layers have been deposited by PS. DBL on porous FSS support also has been applied by APS and in some cases by PVD. This DBL is a perovskite-based coat developed by the German Aerospace Center (GAC) and Plansee SE that increased the durability by more than 2000 h. At the electrolyte/cathode interface, a gadolinium-doped ceria DBL has been deposited by EB-PVD (Ansar et al. 2011). This consortium manufacture MS-SOFCs by PS modified into low-pressure PS (LPPS) and vacuum PS (VPS). Before the MS-SOFC were assembled, each one received the following functional layers: (1) on the ~1 mm FSS ITM support obtained by powder metallurgy, (2) a 10–30 μm $La_{0.6}Sr_{0.2}Ca_{0.2}CrO_3$ (LSCaCr) plus 2–3 μm of LSM DBLs layers were deposited by APS and PVD, respectively; in addition (3) 40–60 μm of NiO-YSZ anode (50vol.%NiO) by APS; (4) 35–50 μm 8YSZ electrolyte by VPS and LPPS upon which a (5) 20–30 μm of LSM plus LSCF cathode bilayer by APS, PS suspension, colloidal spray or screen printing.

The PS technique has the advantage of being a fast process which can be automated by robots. However, obtaining dense or porous microstructures of functional layers is much more complex than conventional sintering routes. Electrolytes obtained by PS exhibit gas leakage, reducing the open-circuit voltage (OCV) and power density (Ansar et al. 2011).

Two stacks, each one with ten MS-SOFCs, were manufactured by LPPS and VPS. The stacks obtained by LPPS and VPS exhibited power and current density of 160 W at 240 mA.cm^{-2} and 200 W at 300 mA.cm^{-2}, respectively, corresponding to 8 V (0.8 V per cell) using H_2 as fuel. However, after applying the LSCF cathode by screen printing and sintering *in situ* using ITM FSS support, gas tightness tests show leakage rates from seals and electrolytes. This leakage impaired not only the OCV but also the MS-SOFC performance. MS-SOFCs manufacturers face technological challenges in obtaining and operating leak-free stacks (Krishnan 2017).

The AVL List GmbH (AVL) and PSE companies, Forschungszentrum Jülich (FJ) and Karlsruhe Institute of Technology (KIT) developed MS-SOFC stack system as APU to reliably deliver 3 KW. In this system, the 8YSZ electrolyte thickness was decreased to 3–4 μm while 1–2 μm of gadolinium-doped ceria DBL was used. This MS-SOFC shows power and current densities of 1.06 W.cm^{-2} and 1.52 A.cm^{-2}, respectively, in 0.7 V at 820 °C. However, the high OCV of 0.97 V indicated a high gas leakage rate. This is the main challenge for developing new dense electrolyte deposition technologies (Franco et al. 2011; Roehrens et al. 2015). New advances in MS-SOFC stacks achieved by PSE, AVL, FJ and KIT consortium have improved the quality of materials used as the cathode, such as LSC, LSCF and LSC-LSCF composites with C in graphite phase. The $C_{graphite}$ polymorph has been used to improve the adhesion between cathode and electrolyte. Oxidation of $C_{graphite}$ is responsible for in situ cathode activation resulting in improved performances (Franco et al. 2013).

The Energy Conversion and Storage (ECS) and Risø National Laboratory for Sustainable Energy (Riso), both of the Technical University of Denmark (DTU) and Topsoe FCs (TFCs) company supported by the METSOFC and METSAPP consortiums manufactured MS-SOFCs stacks. These consortiums have developed new scalable processes to produce MS-SOFCs stacks that include nanostructured electrodes by infiltration and Sc_2O_3-Y_2O_3 co-doped ZrO_2 (Sc-YSZ) electrolytes-based to work between $600<T(°C)<700$ (Blennow et al. 2011a).

These MS-SOFCs comprises the following steps: (1) co-lamination of the porous Sc-YSZ anode on a 5GDC DBL upon Fe22Cr FSS support by tape casting; (2) co-sintering above 1000 °C in a H_2/Argon atmosphere; (3) Sc-YSZ thin film electrolyte deposition on the anode by PVD; (4) precursor solution of salts and Ni/GDC nanoparticles infiltration on anode and FSS supports followed by calcination at 350 °C–2 h; (5) 5GDC DBL deposition onto electrolyte by PVD; (6) LSCF porous cathode deposition followed by Ni/GDC infiltration and calcination and (7) LSC IC deposition. Cathode and IC were deposited by screen printing and sintered in situ during the sealing of the MS-SOFC stacks (Blennow et al. 2011a).

These consortiums have also been using laser and air-brazing for joining and sealing the MS-SOFCs stacks. These stacks with 12 × 12 cm^2 footprint area of MS-SOFCs have been tested since 2013. Long-term tests up to 3000 h have been performed, obtaining a current density of 0.25 A.cm^{-2} at 650 °C with low fuel utilization (H_2/4vol.%H_2O). Low degradation rates of 36 mΩ cm^2 and 0.9% of cell voltage drop per 1000 h were observed on this $T_{op.}$. Although the degradation rate increases with increasing the fuel, this fact is more due to the corrosive effect of moisture in the anode region. TFCs and DTU had been used Nb-doped $SrTiO_3$ as alternative anode for high oxidation resistance under high moisture levels (Christiansen et al. 2013).

These stacks with 25 MS-SOFCs manufactured by TFCs delivered 430 W and 0.7 V per cell at 700 °C. Long-term tests up to 250 h demonstrated a degradation rate of 160 mΩ cm^2/kh. This degradation is attributed to oxidation due to the high moisture levels in the anode, since the operating H_2 fuel utilization is about 60vol.% and air utilization is around 40vol.%. Likewise, the Ni infiltration process introduces a minimal concentration of 0.3wt%Ni into the anode, almost all nanoparticles undergo Ni→NiO phase transition with volumetric expansion. This volumetric expansion causes leakage through the seals or electrolyte imperfections. Oxidation of Ni and FSS support at high moisture levels is the main degradation source of these MS-SOFCs stacks (Blennow et al. 2011a).

The European RAMSES consortium, comprising by CEA, LITEN, CNRS, Baikowski (France) and Hogänas AB (Sweden), SOFC Power (Italy), SINTEF (Norway) and IKERLAN (Spain) presents each one a speciality, e.g., Hogänas AB supplies the metal powders for the support, SINTEF supplies La-Mn coating solutions to minimize support oxidation, Baiakowsky supplies 8YSZ and ScSZ both with sinterability at 1200 °C. The developed designs provides solutions in planar (better performance) and tubular (better cyclability) geometries (Mougin et al. 2013). The goal of this consortium has been to develop FSS supports with high anti-oxidation properties. Fe and Cr alloys containing different chemical compositions were optimized to obtain CTE compatible with ceramics and good oxidation resistance. The Fe22Cr alloy with <0.15wt% Si was chosen as the material for FSS support (Mougin et al. 2013).

Anode-supported MS-SOFCs (AMSC) stacks are being used in planar and tubular geometries, while another cathode-supported MS-SOFCs (CMSC) stacks have been tested only in planar geometries. The porous FSS support received protective coating layers such as $La(Mn_{0.5}Co_{0.5})_{0.8}$ and $LaMn_{0.8}$ by DCoat to prevent air oxidation. Each layer was calcined at 900 °C for 5 min of soaking time. These materials form perovskite oxides and protect the cathode from Cr poisoning. The authors also reported that the developed Fe22Cr FSS support is stable in air oxidation in the cathode region even without the protective layer, although a protective layer is usually needed in the anode region with H_2 and steam. A long-term test up to 500 h reveals a 3 μm scale oxide on FSS support (Mougin et al. 2013). Green sheets from metal powders were pre-sintered between $1100<T(°C)<1150$ for 30 min of soaking time to be used in planar AMSC and CMSC stacks. These FSS supports achieved porosity between 30 and 40% (Mougin et al. 2013).

Tubular AMSC (Crofer 22 APU) stacks of 50 mm in length and 14 mm in diameter have been used to depositing an yttria-doped ceria (YDC) DBL by DCoat. Afterwards, NiO-YSZ anode and ScSZ electrolyte were also deposited by DCoat and PS, respectively. This single-cell has been sintered in 10vol.%H_2/90vol.%Air ca. 1350 °C. Then, LSF-SDC composite cathode was applied by DCoat and sintered in situ up to 950 °C (Mougin et al. 2013). The goal of this consortium has also been to develop high-performance DBLs. With these technologies, the MS-SOFCs developed power densities greater than 500 mW.cm^{-2} at 0.7 V at 800 °C (Mougin et al. 2013).

Finally, the group Fuel Cell Materials and Manufacturing Laboratory of the University of Toronto has been manufactured MS-SOFCs in a scalable way by APS. In this process, there is non-sintering of the components because the layers densification occurs during the deposition of ceramic suspensions containing the materials. This group, coordinated by Kesler et al. (2013) obtained MS-SOFCs with 700 mW.cm^{-2} at 750 °C using hydrocarbons and H_2 as fuels. This group developed a 21–23wt%Cr of FSS support pelletized with 1 cm in diameter sintered in reducing atmosphere. This FSS support has a larger small pore volume due to the PVB binder and PMMA-based pore formers. The FSS supports were DCoat using La and Y nitrates to form rare earth oxides after calcination (Kesler et al. 2013).

Recently, the EU METSAPP consortium developed more than 200 MS-SOFCs with a 150 cm^2 footprint area by scalable manufacturing processes. Quality control of stacks manufacturing process and assembly have been validated by MS-SOFCs electrochemical properties and microstructure characterizations. Numerical models have been developed and validated to understand the oxidation behavior of MS-SOFCs. Finally, a low-cost DBL on metal ICs allows to reduce Cr evaporation by 90% and Cr_2O_3 scale thickness by three times, increasing the stacks lifespan (Blennow et al. 2011a, b; Sudireddy et al. 2017). These MS-SOFCs were obtained by co-laminating the FSS support, FeCr-YSZ anode backbone structure and electrolyte by tape casting. These layers were sintered in a reducing atmosphere. A DBL was deposited on the electrolyte by PVD before the cathode is applied by screen printing. After cathode deposition, Ni and GDC salts in a precursor solution was impregnated into the backbone anode structure. The cathode was sintered in situ throughout the MS-SOFC stack startup and testing at operating conditions (Sudireddy et al. 2017).

In this consortium, chromite spinel DBLs were developed in order to protect the interface of FeCr-YSZ backbone structure anode functional layer (AFL) and FSS support from Ni, Fe and Cr interdiffusion limiting of Cr_2O_3 scale on FeCr particles (Knibbe et al. 2013). The main challenge was to obtain a homogeneous DBL across the surface of the porous FSS support as the quality is strongly dependent on impurities of the metal surface and the wettability of the precursor solution. The metal surface cleaning process consists of silicon (Si) removal. Si is responsible for the poor adhesion of the spinel protective layer (Sudireddy et al. 2017).

Figure 17.6 shows a stack CAD project concept proof with ten MS-SOFC unity cells comprising dense FSS ICs (bipolar plates) and SOFC thin films on the microporous FSS support cell frame (Binelli et al. 2016, 2017).

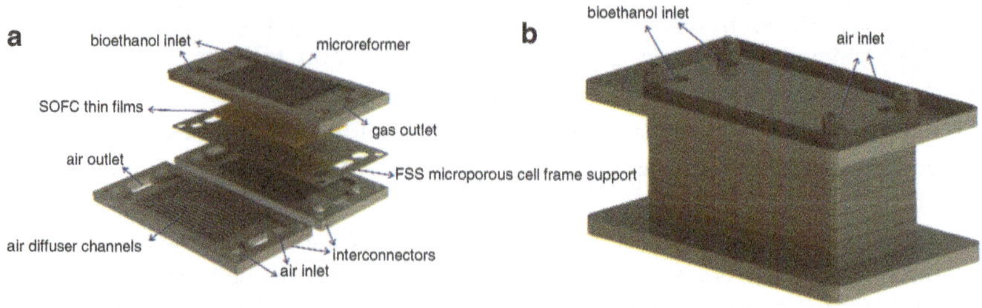

Fig. 17.6 The concept CAD project of (**a**) SOFC unity cell on a FSS microporous cell frame support upon a bioethanol micro-reformer showing the ICs, bioethanol and air inlets besides the air and gas outlets, (**b**) a stack CAD project with ten MS-SOFCs unity cells depicted the bioethanol and air inlet. Source: prepared by the authors

17.11 Final Thoughts and Perspective

Within 20 years, electric vehicles will be commonplace and have superseded petrol vehicles. Currently, Tesla is leading the technology drive, with a market value as significant as those of the nine largest automakers combined (Wayland and Kolodny 2020). Electrification of transport is a reality for all automakers, many of which will probably not produce combustion engine cars after 2030. Indeed, many automakers have already publicly declared their decommissioning of internal combustion engine manufacture. The reasons for this are not only CO_2 emission reduction but also to remove the dependence upon petroleum that comes from politically unstable countries in the Middle East. Here it is important to clarify two main points:

1. CO_2 and particles emitted by the exhaust of gasoline and diesel combustion vehicles have dangerous environmental impacts that must be overcome. An electric car is a good solution only if the country has clean, reliable, and renewable electricity for charging batteries. The infrastructure that generates, transmits, and stores clean energy has massive costs (trillions of dollars) but this will be done anyway at worldwide level. It is essential to emphasize the immense challenge here to meet demand with production because renewable energy production is intermittent i.e., they depend on wind and sunlight. In that sense, large battery containers for energy storage and other forms of energy production such as hydroelectric or nuclear energy are critical to shape peak demand.
2. Countries like Brazil (and likewise India and USA) have an alternative/peculiar solution. Brazil produces large quantities of biofuel as ethanol, which is already close to CO_2 neutral. If all cars in Brazil were powered by ethanol, we would already reach $\sim CO_2$ neutrality in mobility. That is partially true because the amount of CO_2 gases emitted by ethanol-powered car engine are absorbed by sugarcane plants from the air, thereby closing the cycle. Car developers are discussing an electrified-car model for Brazil that uses SOFCs to convert ethanol into electricity which powers electric motors and/or recharges batteries and

powers the motor during the SOFC warming-up process and when extra power is required. This solution uses infrastructure that already exists here in Brazil, such as gas stations, ethanol production, and allow cars to evolve from combustion engine cars to hybrid / e-vehicles, i.e., hybrid car that has only one electric engine. This strategy also reduces noise, increases the range of driving (by developing a more efficient engine), and is aligned with the technology that will be powering cars around the rest of the world. It is worth noting here by reference (Nassar et al. 2008), which demystifies the idea that ethanol production competes with food production in Brazil. This country produces more than enough food and is one of the largest food exporters in the world. The issue of hunger in Brazil is related to economic income distribution and to politics, which prioritize exportation to adjust the commercial balance of trade, over feeding internal markets. For example, meat, soil beans, corn among others, are more expensive in the internal market because of exportation.

The COVID-19 pandemic has taught us several lessons, caused severe debt, and reinforced the production of local goods. In that context/scenario, SOFCs are strategic goods that will play an even more critical role in the near future than today for electric mobility in several countries like Brazil, India, USA and African countries. As such, local development and production will be fundamental for the Brazilian and several countries near-future economy.

From SOFCs perspective, the FSS supports in MS-SOFC warrant scalability. The advanced manufacturing techniques have improved the performance of SOFC cells powered with liquid fuels. However, new degradation mechanisms have emerged, limiting MS-SOFCs performance and lifespan (Zhou et al. 2021). In the last decade, there has been a significant increase in reported works improving the electrochemical kinetics of redox cycles and also reducing coking at the anode when hydrocarbons and alcohols are used as fuels in SOFCs (Bischof et al. 2019; Tucker 2010; Haydn et al. 2014; Udomsilp et al. 2020; Larring and Fontaine 2013; Zhou et al. 2021; Rojek-Wöckner et al. 2016). These achievements were possible through efforts of the international scientific community linking the science and engineering materials used in SOFCs and electrocatalytic properties of electrodes. Studies on diffusion, solid solution formation, nucleation and growth of carbon on metals and ceramic anodes catalysts with bioethanol highlight the possibility of future works to improve MS-SOFCs performance as the potential electric powertrains system for hybrid vehicles (Berry 1973; Bleu et al. 2019; Yu et al. 2021). Another exciting research field could be effective catalysts materials against coking that has been studied in converting alcohols from high molar biomass tar (Gao et al. 2020).

These advances have been improving the cell lifespan enabling the technology transfer from laboratory to scale production. The great challenge consists of developing new materials and architectures that allow $T_{op.}$ reducing between $600<T_{op.}(°C)<800$, maintaining the same performance without losing bioethanol internal reforming. In this context, MS-SOFCs combined with batteries are strong candidates as a hybrid technology applied in vehicles and could be a solution for automotive market. This strategic scenario can drive in bioethanol internal reforming for fuelling the MS-SOFC. Brazil, India, African countries, and the United States are responsible for 99% of the world's bioethanol production (Bajpai 2020).

References

Aboudheir A, Akande A, Idem R, Dalai A (2006) Experimental studies and comprehensive reactor modeling of hydrogen production by the catalytic reforming of crude ethanol in a packed bed tubular reactor over a Ni/Al 2O 3 catalyst. Int J Hydrog Energy 31(6):752–761. https://doi.org/10.1016/j.ijhydene.2005.06.020

Aguiar P, Adjiman CS, Brandon NP (2004) Anode-supported intermediate temperature direct internal reforming solid oxide fuel cell. I: model-based steady-state performance. J Power Sources 138(1–2):120–136. https://doi.org/10.1016/j.jpowsour.2004.06.040

Ahamer C, Opitz AK, Rupp GM, Fleig J (2017) Revisiting the temperature dependent ionic conductivity of yttria stabilized zirconia (YSZ). J Electrochem Soc 164(7):F790–F803. https://doi.org/10.1149/2.0641707jes

Akande AJ, Idem RO, Dalai AK (2005) Synthesis, characterization and performance evaluation of Ni/Al 2O3 catalysts for reforming of crude ethanol for hydrogen production. Appl Catal A Gen 287(2):159–175. https://doi.org/10.1016/j.apcata.2005.03.046

Ansar A et al (2011) Metal supported solid oxide fuel cells and stacks for auxiliary power units - progress, challenges and lessons learned. ECS Trans 35(1):147–155

Antepara I, Villarreal I, Rodríguez-Martínez LM, Lecanda N, Castro U, Laresgoiti A (2005) Evaluation of ferritic steels for use as interconnects and porous metal supports in IT-SOFCs. J Power Sources 151(1–2):103–107. https://doi.org/10.1016/j.jpowsour.2005.02.084

Antunes FC, Goulart CA, Andreeta MRB, De Souza DPF (2018) YSZ/Al2O3 multilayer thick films deposited by spin coating using ceramic suspensions on Al2O3 polycrystalline substrate. Mater Sci Eng B Solid-State Mater Adv Technol 228:60–66. https://doi.org/10.1016/j.mseb.2017.11.007

Araki W, Arai Y (2011) Optimum strain state for oxygen diffusion in yttria-stabilised zirconia. Solid State Ionics 190:75–81

Asadikiya M, Sabarou H, Chen M, Zhong Y (2016) Phase diagram for a nano-yttria-stabilized zirconia system. RSC Adv 6(21):17438–17445. https://doi.org/10.1039/c5ra24330k

Atkinson A et al (2004) Advanced anodes for high-temperature fuel cells. Nat Mater 3:17–27

Augusto BL, Noronha FB, Fonseca FC, Tabuti FN, Colman RC, Mattos LV (2014) Nickel/gadolinium-doped ceria anode for direct ethanol solid oxide fuel cell. Int J Hydrog Energy 39(21):11196–11209. https://doi.org/10.1016/j.ijhydene.2014.05.088

Aznam I, Mah JCW, Muchtar A, Somalu MR, Ghazali MJ (2019) Review on oxidation behavior and chromium volatilization of Fe-Cr-based interconnects at high operation temperatures of solid oxide fuel cells. J Adv Res Fluid Mech Therm Sci 59(1):148–155

Bailly N, Georges S, Djurado E (2012) Elaboration and electrical characterization of electrosprayed YSZ thin films for intermediate temperature-solid oxide fuel cells (IT-SOFC). Solid State Ionics 222–223:1–7. https://doi.org/10.1016/j.ssi.2012.06.020

Bajpai P (2020) Global production of bioethanol. In: Green energy and technology. SpringerLink, pp 177–196

Bance P, Brandon NP, Girvan B, Holbeche P, O'Dea S, Steele BCH (2004) Spinning-out a fuel cell company from a UK University - 2 years of progress at Ceres Power. J Power Sources 131(1–2):86–90. https://doi.org/10.1016/j.jpowsour.2003.11.077

Banerjee MK (2017) Heat treatment of commercial steels for engineering applications, vol 2–3. Elsevier Ltd

Barsoum MW (2002) Fundamentals of ceramics. Fundam Ceram:1–612. https://doi.org/10.1887/0750309024

Basu B, Vleugels J, Van Der Biest O (2004) Transformation behaviour of tetragonal zirconia: role of dopant content and distribution. Mater Sci Eng A 366(2):338–347. https://doi.org/10.1016/j.msea.2003.08.063

Batista MS, Santos RKS, Assaf EM, Assaf JM, Ticianelli EA (2004) High efficiency steam reforming of ethanol by cobalt-based catalysts. J Power Sources 134(1):27–32. https://doi.org/10.1016/j.jpowsour.2004.01.052

Berry BS (1973) Diffusion of carbon in nickel. J Appl Phys 44(8):3792–3793. https://doi.org/10.1063/1.1662846

Binelli ARR, Tasić MB, Filho RM (2016) Catalytic steam reforming of ethanol for hydrogen production: brief status. Chem Ind Chem Eng Q 22(4):327–332. https://doi.org/10.2298/CICEQ160216017B

Binelli ARR, Maciel Filho R, Jardini AL (2017) "Processo para Obtenção de Placas de Microcanais para Microreatores Químicos a Placas Assim Obtidas," BR10201203232

Bischof C et al (2019) Microstructure optimization of nickel/gadolinium-doped ceria anodes as key to significantly increasing power density of metal-supported solid oxide fuel cells. Int J Hydrog Energy 44(59):31475–31487. https://doi.org/10.1016/j.ijhydene.2019.10.010

Biswas P, Kunzru D (2007) Steam reforming of ethanol on Ni-CeO2-ZrO2 catalysts: effect of doping with copper, cobalt and calcium. Catal Lett 118(1–2):36–49. https://doi.org/10.1007/s10562-007-9133-6

Blennow P et al (2009) Development of planar metal supported sofc with novel cermet anode. ECS Trans 25(2):701–710. https://doi.org/10.1149/1.3205585

Blennow P et al (2011a) Manufacturing and characterization of metal-supported solid oxide fuel cells. J Power Sources 196(17):7117–7125. https://doi.org/10.1016/j.jpowsour.2010.08.088

Blennow P, Hjelm J, Klemensø T, Persson ÅH, Ramousse S, Mogensen M (2011b) Planar metal-supported SOFC with novel cermet anode. Fuel Cells 11(5):661–668. https://doi.org/10.1002/fuce.201100029

Bleu Y et al (2019) Dynamics of carbon diffusion and segregation through nickel catalyst, investigated by in situ XPS, during growth of nitrogen doped graphene. Carbon N Y 155: 410–420. https://doi.org/10.1016/j.carbon.2019.08.084

Boldrin P, Brandon NP (2019) Progress and outlook for solid oxide fuel cells for transportation applications. Nat Catal 2(7):571–577. https://doi.org/10.1038/s41929-019-0310-y

Brandner M, Bram M, Froitzheim J, Buchkremer HP, Stöver D (2008) Electrically conductive diffusion barrier layers for metal-supported SOFC. Solid State Ionics 179(27–32):1501–1504. https://doi.org/10.1016/j.ssi.2008.03.002

Brandon NP et al (2004) Development of metal supported solid oxide fuel cells for operation at 500-600°C. J Mater Eng Perform 13(3):253–256. https://doi.org/10.1361/10599490419135

Brodnikovska I et al (2019) Grains, grain boundaries and total ionic conductivity of 10Sc1CeSZ and 8YSZ solid electrolytes affected by crystalline structure and dopant content. Mater Today Proc 6:79–85. https://doi.org/10.1016/j.matpr.2018.10.078

Brugnoni C, Ducati U, Scagliotti M (1995) SOFC cathode/electrolyte interface. Part I: reactivity between La0.85Sr0.15MnO3 and ZrO2-Y2O3. Solid State Ionics 76(3–4):177–182. https://doi.org/10.1016/0167-2738(94)00299-8

Carrete L, Friedrich KA, Stimming U (2001) Fuel cells - fundamentals and applications. Fuel Cells 1(1):5–39. https://doi.org/10.1002/1615-6854(200105)1:1<5::aid-fuce5>3.0.co;2-g

Carrette L, Friedrich KA, Stimming U (2000) Fuel cells: principles, types, fuels, and applications. ChemPhysChem 1(4):162–193. https://doi.org/10.1002/1439-7641(20001215)1:4<162::aid-cphc162>3.0.co;2-z

Casanovas A, Llorca J, Homs N, Fierro JLG, Ramírez de la Piscina P (2006) Ethanol reforming processes over ZnO-supported palladium catalysts: Effect of alloy formation. J Mol Catal A Chem 250(1–2):44–49. https://doi.org/10.1016/j.molcata.2006.01.033

Cavallaro S, Chiodo V, Freni S, Mondello N, Frusteri F (2003) Performance of Rh/Al2O3 catalyst in the steam reforming of ethanol: H2 production for MCFC. Appl Catal A Gen 249(1): 119–128. https://doi.org/10.1016/S0926-860X(03)00189-3

Cesário MR, de Macedo A (2017) Functional materials for solid oxide fuel cells: processing, microstructure and performance. Bentham Science Publishers, Sharjah

Chen LC, Lin SD (2014) Effects of the pretreatment of CuNi/SiO2 on ethanol steam reforming: influence of bimetal morphology. Appl Catal B Environ 148–149:509–519. https://doi.org/10.1016/j.apcatb.2013.11.031

Chen YY, Wei WCJ (2006) Processing and characterization of ultra-thin yttria-stabilized zirconia (YSZ) electrolytic films for SOFC. Solid State Ionics 177(3–4):351–357. https://doi.org/10.1016/j.ssi.2005.10.010

Chen K et al (2006) Development of yttria-stabilized zirconia thin films via slurry spin coating for intermediate-to-low temperature solid oxide fuel cells. J Power Sources 160:436–438

Chen SQ, Li YD, Liu Y, Bai X (2011) Regenerable and durable catalyst for hydrogen production from ethanol steam reforming. Int J Hydrog Energy 36(10):5849–5856. https://doi.org/10.1016/j.ijhydene.2011.01.126

Chiou JYZ et al (2012) Pathways of ethanol steam reforming over ceria-supported catalysts. Int J Hydrog Energy 37(18):13667–13673. https://doi.org/10.1016/j.ijhydene.2012.02.081

Chiou JYZ, Lee CL, Ho KF, Huang HH, Yu SW, Bin Wang C (2014) Catalytic performance of Pt-promoted cobalt-based catalysts for the steam reforming of ethanol. Int J Hydrog Energy 39(11):5653–5662. https://doi.org/10.1016/j.ijhydene.2014.01.156

Christiansen N (2014) METSAPP - metal supported SOFC technology for stationary and mobile applications (GA number 278257), pp 1–19. https://www.fch.europa.eu/sites/default/files/ReviewMETSAPP-FCH JU_2012.pdf

Christiansen N, Primdahl S, Wandel M, Ramousse S, Hagen A (2013) Status of the solid oxide fuel cell development at topsoe fuel cell A/S and DTU energy conversion. ECS Trans 57(1):43–52. https://doi.org/10.1149/05701.0043ecst

Chun CM, Ramanarayanan TA (2007) Mechanism and control of carbon deposition on high temperature alloys. J Electrochem Soc 154(9):C465. https://doi.org/10.1149/1.2750447

Chun CM, Mumford JD, Ramanarayanan TA (2000) Carbon-induced corrosion of nickel anode. J Electrochem Soc 147(10):3680. https://doi.org/10.1149/1.1393958

Chun CM, Mumford JD, Ramanarayanan TA (2002) Mechanisms of metal dusting corrosion of iron. J Electrochem Soc 149(7):B348. https://doi.org/10.1149/1.1483099

Comas J, Marino F, Laborde M, Amadeo N (2004) Bio-ethanol steam reforming on Ni/Al2O3 catalyst. Chem Eng J 98(1–2):61–68. https://doi.org/10.1016/S1385-8947(03)00186-4

Cormack AN (1986) Mass transport in anion deficient fluorite oxides. Kinet Mass Transp Silic Oxide Syst 7:177–186. https://doi.org/10.4028/www.scientific.net/msf.7.177

Costa-Nunes O, Gorte RJ, Vohs JM (2005) Comparison of the performance of Cu-CeO2-YSZ and Ni-YSZ composite SOFC anodes with H2, CO, and syngas. J Power Sources 141(2):241–249. https://doi.org/10.1016/j.jpowsour.2004.09.022

da Silva FS, de Souza TM (2017) Novel materials for solid oxide fuel cell technologies: A literature review. Int J Hydrog Energy 42(41):26020–26036. https://doi.org/10.1016/j.ijhydene.2017.08.105

Dan M, Mihet M, Tasnadi-Asztalos Z, Imre-Lucaci A, Katona G, Lazar MD (2015) Hydrogen production by ethanol steam reforming on nickel catalysts: Effect of support modification by CeO2 and La2O3. Fuel 147:260–268. https://doi.org/10.1016/j.fuel.2015.01.050

Dancini-Pontes I et al (2015) Influence of the CeO2 and Nb2O5 supports and the inert gas in ethanol steam reforming for H2 production. Chem Eng J 273:66–74. https://doi.org/10.1016/j.cej.2015.03.032

de Lima SM et al (2009) Ethanol decomposition and steam reforming of ethanol over CeZrO2 and Pt/CeZrO2 catalyst: reaction mechanism and deactivation. Appl Catal A Gen 352(1–2):95–113. https://doi.org/10.1016/j.apcata.2008.09.040

De Souza S, Visco SJ, De Jonghe LC (1997) Thin-film solid oxide fuel cell with high performance at low-temperature. Solid State Ionics 98(1–2):57–61. https://doi.org/10.1016/s0167-2738(96)00525-5

Deng X, Wei P, Bateni MR, Petric A (2006) Cobalt plating of high temperature stainless steel interconnects. J Power Sources 160(2 Spec. Iss.), pp 1225–1229. doi:https://doi.org/10.1016/j.jpowsour.2006.03.024

Devi PS, Das Sharma A, Maiti HS (2004) Solid oxide fuel cell materials: a review. Trans Indian Ceram Soc 63(2):75–98. https://doi.org/10.1080/0371750X.2004.11012140

Deville S, Guénin G, Chevalier J (2004a) Martensitic transformation in zirconia: Part I. Nanometer scale prediction and measurement of transformation induced relief. Acta Mater 52:5697–5707

Deville S, Guénin G, Chevalier J (2004b) Martensitic transformation in zirconia: Part II. Martensite growth. Acta Mater 52:5709–5721

Diagne C, Idriss H, Pearson K, Gómez-García MA, Kiennemann A (2004) Efficient hydrogen production by ethanol reforming over Rh catalysts. Effect of addition of Zr on CeO2 for the oxidation of CO to CO2. Comptes Rendus Chim 7(6–7, 617):–622. https://doi.org/10.1016/j.crci.2004.03.004

Ding J, Liu J (2008) An anode-supported solid oxide fuel cell with spray-coated yttria-stabilized zirconia (YSZ) electrolyte film. Solid State Ionics 179(21–26):1246–1249. https://doi.org/10.1016/j.ssi.2008.01.094

Dogdibegovic E, Shen F, Wang R, Robinson I, Lau GY, Tucker MC (2019a) Progress in metal-supported solid oxide fuel cells and electrolyzers with symmetric metal supports and infiltrated electrodes. ECS Trans 91(1):877–885

Dogdibegovic E, Wang R, Lau GY, Karimaghaloo A, Lee MH, Tucker MC (2019b) Progress in durability of metal-supported solid oxide fuel cells with infiltrated electrodes. J Power Sources 437:226935. https://doi.org/10.1016/j.jpowsour.2019.226935

Dogdibegovic E, Fukuyama Y, Tucker MC (2020) Ethanol internal reforming in solid oxide fuel cells: A path toward high performance metal-supported cells for vehicular applications. J Power Sources 449:227598

Dogdibegovic E, Fukuyama Y, Tucker MC (2021) Erratum: Ethanol internal reforming in solid oxide fuel cells: A path toward high performance metal-supported cells for vehicular applications [Journal of Power Sources 449 (2020) 227598]. J Power Sources 492:4–6. https://doi.org/10.1016/j.jpowsour.2021.229644

Downing D, Jones A, Brandt M, Leary M (2021) Increased efficiency gyroid structures by tailored material distribution. Mater Des 197:109096. https://doi.org/10.1016/j.matdes.2020.109096

Duan NQ, Yan D, Chi B, Pu J, Jian L (2015) High performance anode-supported tubular solid oxide fuel cells fabricated by a novel slurry-casting method. Sci Rep 5:5–8. https://doi.org/10.1038/srep08174

Dwiwedi S (2020) Solid oxide fuel cell: materials for anode, cathode and electrolyte. Int J Hydrog Energy 45(44):23988–24013. https://doi.org/10.1016/j.ijhydene.2019.11.234

Erdohelyi A, Raskó J, Kecskés T, Tóth M, Dömök M, Baán K (2006) Hydrogen formation in ethanol reforming on supported noble metal catalysts. Catal Today 116(3):367–376. https://doi.org/10.1016/j.cattod.2006.05.073

Fan ESC, Kuhn J, Kesler O (2016) Suspension plasma spraying of La0.6Sr0.4Co0.2Fe0.8O3-δ cathodes: Influence of carbon black pore former on performance and degradation. J Power Sources 316:72–84. https://doi.org/10.1016/j.jpowsour.2016.02.075

Fang W et al (2015) Highly loaded well dispersed stable Ni species in NiXMg2AlOY nanocomposites: application to hydrogen production from bioethanol. Appl Catal B Environ 166–167:485–496. https://doi.org/10.1016/j.apcatb.2014.11.052

Fatsikostas AN, Verykios XE (2004) Reaction network of steam reforming of ethanol over Ni-based catalysts. J Catal 225(2):439–452. https://doi.org/10.1016/j.jcat.2004.04.034

Ferencz Z et al (2014) Effects of support and Rh additive on co-based catalysts in the ethanol steam reforming reaction. ACS Catal 4(4):1205–1218. https://doi.org/10.1021/cs500045z

Fierro V, Akdim O, Provendier H, Mirodatos C (2005) Ethanol oxidative steam reforming over Ni-based catalysts. J Power Sources 145(2):659–666. https://doi.org/10.1016/j.jpowsour.2005.02.041

Figuereido FML, Marques FMB (2013) Electrolytes for solid oxide fuel cells. Wiley Interdiscip Rev Energy Environ 2(1):52–72. https://doi.org/10.1002/wene.23

Franco T et al (2011) Development of metal-supported solid oxide fuel cells. ECS Trans 35(1): 343–349. https://doi.org/10.1149/1.3570009

Franco T et al (2013) The status of metal-supported SOFC development and industrialization at plansee. ECS Trans 57(1):471–480. https://doi.org/10.1149/05701.0471ecst

Froitzheim J, Canovic S, Nikumaa M, Sachitanand R, Johansson LG, Svensson JE (2012) Long term study of Cr evaporation and high temperature corrosion behaviour of Co coated ferritic steel for solid oxide fuel cell interconnects. J Power Sources 220:217–227. https://doi.org/10.1016/j.jpowsour.2012.06.092

Frusteri F et al (2004) H2 production for MC fuel cell by steam reforming of ethanol over MgO supported Pd, Rh, Ni and Co catalysts. Catal Commun 5(10):611–615. https://doi.org/10.1016/j.catcom.2004.07.015

Gao X, Wang Z, Ashok J, Kawi S (2020) A comprehensive review of anti-coking, anti-poisoning and anti-sintering catalysts for biomass tar reforming reaction. Chem Eng Sci X 7:100065. https://doi.org/10.1016/j.cesx.2020.100065

Gautam M, Ahuja A, Sinha A, Sharma J, Patro PK, Venkatasubramanian A (2020) Synthesis and characterization of gadolinium-doped ceria and barium cerate-based composite electrolyte material for IT-SOFC. Bull Mater Sci 43(1). https://doi.org/10.1007/s12034-020-02195-3

Gomes E, Mather GC, Figueiredo FM, Marques FMB (2009) Microstructure and electrical properties of aluminium-substituted La(Sr)Ga(Mg)O3–δ-based solid electrolytes. Monatshefte für Chemie - Chem Mon 140:1041–1052

Götsch T, Wallisch W, Stöger-Pollach M, Klötzer B, Penner S (2016) From zirconia to yttria: Sampling the YSZ phase diagram using sputter-deposited thin films. AIP Adv 6(2):025119-1-025119–20. https://doi.org/10.1063/1.4942818

Greluk M, Rybak P, Słowik G, Rotko M, Machocki A (2015) Comparative study on steam and oxidative steam reforming of ethanol over 2KCo/ZrO2 catalyst. Catal Today 242(Part A):50–59. https://doi.org/10.1016/j.cattod.2014.07.032

Hammou A (2008) Solid oxide fuel cells. Adv Electrochem Sci Eng 2:88–139. https://doi.org/10.1002/9783527616763.ch2

Haydn M et al (2014) Multi-layer thin-film electrolytes for metal supported solid oxide fuel cells. J Power Sources 256:52–60. https://doi.org/10.1016/j.jpowsour.2014.01.043

He Z, Yang M, Wang X, Zhao Z, Duan A (2012) Effect of the transition metal oxide supports on hydrogen production from bio-ethanol reforming. Catal Today 194(1):2–8. https://doi.org/10.1016/j.cattod.2012.05.004

Hilpert K, Das D, Miller M, Peck DH, Weiß R (1996) Chromium vapor species over solid oxide fuel cell interconnect materials and their potential for degradation processes. J Electrochem Soc 143(11):3642–3647. https://doi.org/10.1149/1.1837264

Hou T, Yu B, Zhang S, Xu T, Wang D, Cai W (2015) Hydrogen production from ethanol steam reforming over Rh/CeO2 catalyst. Catal Commun 58:137–140. https://doi.org/10.1016/j.catcom.2014.09.020

Huang K, Goodenough JB (2009) Solid oxide fuel cell technology: principles, performance and operations. Woodhead publishing limited, CRC Press; Boston, New York

Huijsmans JPP (2001) Ceramics in solid oxide fuel cells. Curr Opin Solid State Mater Sci 5(4): 317–323. https://doi.org/10.1016/S1359-0286(00)00034-6

Hussain S, Yangping L (2020) Review of solid oxide fuel cell materials: cathode, anode, and electrolyte. Energy Transit 4:113–126

Hwang CS et al (2011) High performance metal-supported intermediate temperature solid oxide fuel cells fabricated by atmospheric plasma spraying. J Power Sources 196(4):1932–1939. https://doi.org/10.1016/j.jpowsour.2010.10.029

Hwang CS, Tsai CH, Hwang TJ, Chang CL, Yang SF, Lin JK (2016) Novel metal substrates for high power metal-supported solid oxide fuel cells. Fuel Cells 16(2):244–251. https://doi.org/10.1002/fuce.201500216

I. EG&G Technical Services (2004) Fuel cell handbook (7 edn). Morgantown, WV

Irshad M et al (2016) A brief description of high temperature solid oxide fuel cell's operation, materials, design, fabrication technologies and performance. Appl Sci 6(3). https://doi.org/10.3390/app6030075

Ishihara T (2009a) Perovskite oxide for solid oxide fuel cells. Springer, London

Ishihara T (2009b) Perovskite oxide for solid oxide fuel cells fuel cells and hydrogen energy

Ivers-Tiffée E, Weber A, Herbstritt D (2001) Materials and technologies for SOFC-components. J Eur Ceram Soc 21:1805–1811

Jacobson AJ (2010) Materials for solid oxide fuel cells. Chem Mater 22(3):660–674. https://doi.org/10.1021/cm902640j

Jeong HJ, Kim JW, Jang DY, Shim JH (2015) Atomic layer deposition of ruthenium surface-coating on porous platinum catalysts for high-performance direct ethanol solid oxide fuel cells. J Power Sources 291:239–245. https://doi.org/10.1016/j.jpowsour.2015.05.005

Jeong H, Roehrens D, Bram M (2020) Facile route for reactive coating of LaCrO3 on high-chromium steels as protective layer for solid oxide fuel cell applications. Mater Lett 258: 126794. https://doi.org/10.1016/j.matlet.2019.126794

Jiang Y, Virkar AV (2001) A high performance, anode-supported solid oxide fuel cell operating on direct alcohol. J Electrochem Soc 148(7):A706. https://doi.org/10.1149/1.1375166

Karthikeyan A, Chang CL, Ramanathan S (2006) High temperature conductivity studies on nanoscale yttria-doped zirconia thin films and size effects. Appl Phys Lett 89(18):10–13. https://doi.org/10.1063/1.2385211

Kelly PM, Rose LRF (2002) The martensitic transformation in ceramics – its role in transformation toughening. Prog Mater Sci 47:463–557

Kesler O, Cuglietta M, Harris J, Kuhn J, Marr M, Metcalfe C (2013) Progress in metal-supported SOFCs using hydrogen and methane fuels. ECS Trans 57(1):491–501

Kharton VV et al (2001) Ceria-based materials for solid oxide fuel cells. J Mater Sci 36:1105–1117

Kharton V, Marques F, Atkinson A (2004) Transport properties of solid oxide electrolyte ceramics: a brief review. Solid State Ionics 174:135–149

Kim YM, Kim-Lohsoontorn P, Bae J (2010) Effect of unsintered gadolinium-doped ceria buffer layer on performance of metal-supported solid oxide fuel cells using unsintered barium strontium cobalt ferrite cathode. J Power Sources 195(19):6420–6427. https://doi.org/10.1016/j.jpowsour.2010.03.095

Kim YM, Kim-Lohsoontorn P, Baek SW, Bae J (2011) Electrochemical performance of unsintered Ba0.5Sr 0.5Co0.8Fe0.2O3-δ, La 0.6Sr0.4Co0.8Fe0.2O 3-δ, and La0.8Sr0.2MnO 3-δ cathodes for metal-supported solid oxide fuel cells. Int J Hydrog Energy 36(4):3138–3146. https://doi.org/10.1016/j.ijhydene.2010.10.065

Kim D, Kwak BS, Park N-K, Han GB, Kang M (2012) Dynamic hydrogen production from ethanol steam- reforming reaction on NixMoy/SBA-15 catalytic system. Int J Energy Res 39:279–292

Kingery WD, Bowen HK (1976) D, 2nd edn. R. Uhlmann, Introduction to ceramics

Klemensø T et al (2011a) High performance metal-supported solid oxide fuel cells with Gd-doped ceria barrier layers. J Power Sources 196(22):9459–9466. https://doi.org/10.1016/j.jpowsour.2011.07.014

Klemensø T et al (2011b) Development of long-term stable and high-performing metal-supported SOFCs. ECS Trans 35(1):369–378. https://doi.org/10.1149/ma2011-01/12/750

Knibbe R et al (2013) Oxidation in ceria infiltrated metal supported SOFCs-A TEM investigation. J Power Sources 228:75–82. https://doi.org/10.1016/j.jpowsour.2012.11.051

Komatsu T, Chiba R, Arai H, Sato K (2008) Chemical compatibility and electrochemical property of intermediate-temperature SOFC cathodes under Cr poisoning condition. J Power Sources 176(1):132–137. https://doi.org/10.1016/j.jpowsour.2007.10.068

Kreuer K-D, Paddison SJ, Spohr E, Schuster M (2004) Transport in proton conductors for fuel-cell applications: simulations, elementary reactions, and phenomenology. Chem Rev 104:4637–4678

Krishnan VV (2017) Recent developments in metal-supported solid oxide fuel cells. Wiley Interdiscip Rev Energy Environ 6(5). https://doi.org/10.1002/wene.246

Kroger FA, Vink HJ (1958) Relations between the concentrations of imperfections in solids. J Phys Chem Solids 5(3):208–223

Kugai J, Subramani V, Song C, Engelhard MH, Chin YH (2006) Effects of nanocrystalline CeO2 supports on the properties and performance of Ni-Rh bimetallic catalyst for oxidative steam reforming of ethanol. J Catal 238(2):430–440. https://doi.org/10.1016/j.jcat.2006.01.001

Kwak BS, Lee G, Park SM, Kang M (2015) Effect of MnOx in the catalytic stabilization of Co2MnO4 spinel during the ethanol steam reforming reaction. Appl Catal A Gen 503:165–175. https://doi.org/10.1016/j.apcata.2015.06.037

Larring Y, Fontaine M-L (2013) critical issues of metal-supported fuel cell. In: Irvine J, Connor P (eds) Solid oxide fuels cells: facts and figures. Green energy and technology. Springer, London, pp 71–93

Leah R et al (2011) Development of highly robust, volume-manufacturable metal-supported sofcs for operation below 600°C. ECS Trans 35(1):351–367

Leah R et al (2013) Low-cost, REDOX-stable, low-temperature sofc developed by ceres power for multiple applications: latest development update. Pap Knowl Towar a Media Hist Doc 57(1): 461–470

Leah R et al (2015) Ceres power steel cell technology: rapid progress towards a truly commercially viable SOFC. ECS Trans 68(1):95–107

Leah R et al (2017) Development progress on the ceres power steel cell technology platform: further progress towards commercialization. ECS Trans 78(1):87–95

Lee JG, Park JH, Shul YG (2014a) Tailoring gadolinium-doped ceria-based solid oxide fuel cells to achieve 2Wcm-2 at 550 °C. Nat Commun 5. https://doi.org/10.1038/ncomms5045

Lee G, Kim D, Kwak BS, Kang M (2014b) Hydrogen rich production by ethanol steam reforming reaction over Mn/Co10Si90MCM-48 catalysts. Catal Today 232:139–150. https://doi.org/10.1016/j.cattod.2014.03.037

Li Y, Wong LM, Xie H, Wang S, Su PC (2017) Nanoporous palladium anode for direct ethanol solid oxide fuel cells with nanoscale proton-conducting ceramic electrolyte. J Power Sources 340:98–103. https://doi.org/10.1016/j.jpowsour.2016.11.064

Li X, Kuang X, Sun J (2020a) Rare earth elements based oxide ion conductors. Inorg Chem Front 00(1–3):1–22. https://doi.org/10.1039/x0xx00000x

Li S, Lu X, Shi S, Chen L, Wang Z, Zhao Y (2020b) Europium-doped ceria nanowires as anode for solid oxide fuel cells. Front Chem 8(May):1–10. https://doi.org/10.3389/fchem.2020.00348

Liao M, Wang W, Ran R, Shao Z (2011) Development of a Ni-Ce0.8Zr0.2O2 catalyst for solid oxide fuel cells operating on ethanol through internal reforming. J Power Sources 196(15): 6177–6185. https://doi.org/10.1016/j.jpowsour.2011.03.018

Liguras DK, Kondarides DI, Verykios XE (2003) Production of hydrogen for fuel cells by steam reforming of ethanol over supported noble metal catalysts. Appl Catal B Environ 43(4): 345–354. https://doi.org/10.1016/S0926-3373(02)00327-2

Lima RS, Marple BR (2017) Insights on the high-temperature operational limits of ZrO2-Y2O3 TBCs manufactured via air plasma spray. J Mater Eng Perform 26(3):1272–1282. https://doi.org/10.1007/s11665-017-2562-5

Lin Y, Beale SB (2006) Performance predictions in solid oxide fuel cells. Appl Math Model 30(11): 1485–1496. https://doi.org/10.1016/j.apm.2006.03.009

Liu F, Qu Y, Yue Y, Liu G, Liu Y (2015) Nano bimetallic alloy of Ni-Co obtained from LaCoxNi1-xO3 and its catalytic performance for steam reforming of ethanol. RSC Adv 5(22): 16837–16846. https://doi.org/10.1039/c4ra14131h

Liu Y, Shao Z, Mori T, Jiang SP (2021) Development of nickel based cermet anode materials in solid oxide fuel cells – now and future. Mater Reports Energy 1(1):1–77

Llorca J, Homs N, Sales J, Fierro JLG, De La Piscina PR (2004) Effect of sodium addition on the performance of Co-ZnO-based catalysts for hydrogen production from bioethanol. J Catal 222(2):470–480. https://doi.org/10.1016/j.jcat.2003.12.008

Llorca J, Corberán VC, Divins NJ, Fraile RO, Taboada E (2013) Hydrogen from bioethanol

Lyu Y, Xie J, Wang D, Wang J (2020) Review of cell performance in solid oxide fuel cells. J Mater Sci 55(17):7184–7207. https://doi.org/10.1007/s10853-020-04497-7

Ma Z, Wang X, Yang Y, Zhang H, Ou X, Ling Y (2020) Numerical modeling of ethanol-fueled solid oxide fuel cells with a Ni-BaZr0.1Ce0.7 Y0.1Yb0.1O3–δ external reformer. Ionics (Kiel). 26:4587–4598

Mahato N, Banerjee A, Gupta A, Omar S, Balani K (2015) Progress in material selection for solid oxide fuel cell technology: a review. Prog Mater Sci 72:141–337

Mamivand M, Zaeem MA, El Kadiri H (2013a) A review on phase field modeling of martensitic phase transformation. Comput Mater Sci 77:304–311. https://doi.org/10.1016/j.commatsci.2013.04.059

Mamivand M, Zaeem MA, El Kadiri H, Chen LQ (2013b) Phase field modeling of the tetragonal-to-monoclinic phase transformation in zirconia. Acta Mater 61(14):5223–5235. https://doi.org/10.1016/j.actamat.2013.05.015

Mat ZA, Kar YB, Yong TC, Hassan SHA (2018) A short review of material combination in Bilayer Electrolyte of IT-SOFC. Int J Eng Technol 7(4):513–515. https://doi.org/10.14419/ijet.v7i4.35.22901

McKenna BJ et al (2013) Advances in metal supported cells in the METSOFC EU consortium. Fuel Cells 13(4):592–597. https://doi.org/10.1002/fuce.201200185

Minh NQ, Takahashi T (1995) Science and technology of ceramic fuel cells. Oxford, Elsevier Sci Ltd

Mitterdorfer A, Gauckler LJ (1998) La2Zr2O7 formation and oxygen reduction kinetics of the La0.85Sr0.15MnyO3, O2(g)|YSZ system. Solid State Ionics 111(3–4):185–218. https://doi.org/10.1016/s0167-2738(98)00195-7

Mogensen M, Chemistry SS (2012) Introduction to fuel cells: Fundamentals of electrochemical kinetics, thermodynamics and solid state chemistry (II) for the experienced. Dtu, no Ii

Mogensen M, Sammes NM, Tompsett GA (2000) Physical, chemical and electrochemical properties of pure and doped ceria. Solid State Ionics 129:63–94

Mogensen M, Jensen KV, Jørgensen MJ, Primdahl S (2002) Progress in understanding SOFC electrodes. Solid State Ionics 150:123–129

Mogensen M, Lybye D, Bonanos N, Hendriksen PV, Poulsen FW (2004) Factors controlling the oxide ion conductivity of fluorite and perovskite structured oxides. Solid State Ionics 174(1–4):279–286. https://doi.org/10.1016/j.ssi.2004.07.036

Molin S, Kusz B, Gazda M, Jasinski P (2008) Evaluation of porous 430L stainless steel for SOFC operation at intermediate temperatures. J Power Sources 181(1):31–37. https://doi.org/10.1016/j.jpowsour.2007.10.009

Montross CS, Yokokawa H, Dokiya M (2002) Thermal stresses in planar solid oxide fuel cells due to thermal expansion differences. Br Ceram Trans 101(3):85–93. https://doi.org/10.1179/096797802225003956

Moon H, Kim SD, Hyun SH, Kim HS (2008) Development of IT-SOFC unit cells with anode-supported thin electrolytes via tape casting and co-firing. Int J Hydrog Energy 33(6):1758–1768. https://doi.org/10.1016/j.ijhydene.2007.12.062

Moraes TS et al (2015) The study of the performance of PtNi/CeO2-nanocube catalysts for low temperature steam reforming of ethanol. Catal Today 242(Part A):35–49. https://doi.org/10.1016/j.cattod.2014.05.045

Morales M, Segarra M (2015) Steam reforming and oxidative steam reforming of ethanol over La0.6Sr0.4CoO3-δ perovskite as catalyst precursor for hydrogen production. Appl Catal A Gen 502:305–311. https://doi.org/10.1016/j.apcata.2015.05.036

Mori M, Yamamoto T, Itoh H, Inaba H, Tagawa H (1998) Thermal expansion of nickel-zirconia anodes in solid oxide fuel cells during fabrication and operation. J Electrochem Soc 145(4):1374–1381. https://doi.org/10.1149/1.1838468

Mougin J et al (2013) Metal supported solid oxide fuel cells: from materials development to single cell performance and durability tests. ECS Trans 57(1):481–490. https://doi.org/10.1149/05701.0481ecst

Moulson AJ, Herbert JM (2003) Electroceramics: materials, properties, applications, 2nd edn. Wiley, Sussex

Muccillo R, Muccillo ENS, Fonseca FC, de Florio DZ (2008) Characteristics and performance of electrolyte-supported solid oxide fuel cells under ethanol and hydrogen. J Electrochem Soc 155(3):B232. https://doi.org/10.1149/1.2828024

Nassar AM, Rudorff BFT, Antoniazzi LB, De Aguiar DA, Bacchi MRP, Adami M (2008) Prospects of the sugarcane expansion in Brazil: impacts on direct and indirect land use changes. Sugarcane Ethanol Contrib Clim Chang Mitig Environ, no. May 2014, pp 63–95. doi: https://doi.org/10.3920/978-90-8686-652-6

Navarro RM, Álvarez-Galván MC, Sánchez-Sánchez MC, Rosa F, Fierro JLG (2005) Production of hydrogen by oxidative reforming of ethanol over Pt catalysts supported on Al 2O 3 modified with Ce and La. Appl Catal B Environ 55(4):229–241. https://doi.org/10.1016/j.apcatb.2004.09.002

Nédélec R et al (2011) Gas phase deposition of diffusion barriers for metal substrates in solid oxide fuel cells. Surf Coatings Technol 205(16):3999–4004. https://doi.org/10.1016/j.surfcoat.2011.02.021

Ng KH, Rahman HA, Somalu MR (2018) WITHDRAWN: review: enhancement of composite anode materials for low-temperature solid oxide fuels. Int J Hydrog Energy:1–14. https://doi.org/10.1016/j.ijhydene.2018.01.073

Ni M, Leung DYC, Leung MKH (2007) A review on reforming bio-ethanol for hydrogen production. Int J Hydrog Energy 32:3238–3247

Nielsen J, Persson ÅH, Muhl TT, Brodersen K (2018) Towards high power density metal supported solid oxide fuel cell for mobile applications. J Electrochem Soc 165(2):F90–F96. https://doi.org/10.1149/2.0741802jes

Ormerod RM (2003) Solid oxide fuel cells. Chem Soc Rev 32:17–28

Osorio-Vargas P, Campos CH, Navarro RM, Fierro JLG, Reyes P (2015) Rh/Al2O3-La2O3 catalysts promoted with CeO2 for ethanol steam reforming reaction. J Mol Catal A Chem 407, no. M:169–181. https://doi.org/10.1016/j.molcata.2015.06.031

Panthi D, Tsutsumi A (2014) Micro-tubular solid oxide fuel cell based on a porous yttria-stabilized zirconia support. Sci Rep 4:1–6. https://doi.org/10.1038/srep05754

Panthi D, Hedayat N, Du Y (2018) Densification behavior of yttria-stabilized zirconia powders for solid oxide fuel cell electrolytes. J Adv Ceram 7(4):325–335. https://doi.org/10.1007/s40145-018-0282-4

Pieta IS et al (2021) The hallmarks of copper single atom catalysts in direct alcohol fuel cells and electrochemical CO2 fixation. Adv Mater Interfaces 8(8). https://doi.org/10.1002/admi.202001822

Polat C (2008) Market opportunities for hydrogen solid oxide fuel cells (SOFC): A review of the literature and the future market trends. EABR TLC Conference Proceedings, no. June 2008, pp 1–17

Poulianitis C, Maragou V, Yan A, Song S, Tsiakaras P (2006) Investigation of the reaction of ethanol-steam mixtures in a YSZ electrochemical reactor operated in a fuel cell mode. J Fuel Cell Sci Technol 3(4):459–463. https://doi.org/10.1115/1.2349529

Ralph JM, Schoeler AC, Krumpelt M (2001) Materials for lower temperature solid oxide fuel cells. J Mater Sci 36(5):1161–1172. https://doi.org/10.1023/A:1004881825710

Roehrens D et al (2015) Advances beyond traditional SOFC cell designs. Int J Hydrog Energy 40(35):11538–11542. https://doi.org/10.1016/j.ijhydene.2015.01.155

Rojek-Wöckner VA, Opitz AK, Brandner M, Mathé J, Bram M (2016) A novel Ni/ceria-based anode for metal-supported solid oxide fuel cells. J Power Sources 328:65–74. https://doi.org/10.1016/j.jpowsour.2016.07.075

Rose L, Kesler O, Decès-Petit C, Troczynski T, Maric R (2009) Characterization of porous stainless steel 430 for low- and intermediate-temperature solid oxide fuel cell (SOFC) substrates. Int J Green Energy 6(6):638–645. https://doi.org/10.1080/15435070903372510

Rossetti I et al (2014) TiO2-supported catalysts for the steam reforming of ethanol. Appl Catal A Gen 477:42–53. https://doi.org/10.1016/j.apcata.2014.03.004

Sachitanand R, Sattari M, Svensson JE, Froitzheim J (2013) Evaluation of the oxidation and Cr evaporation properties of selected FeCr alloys used as SOFC interconnects. Int J Hydrog Energy 38(35):15328–15334. https://doi.org/10.1016/j.ijhydene.2013.09.044

Sanna S, Esposito V, Tebano A, Licoccia S, Traversa E, Balestrino G (2010) Enhancement of ionic conductivity in sm-doped ceria/yttria-stabilized zirconia heteroepitaxial structures. Small 6(17): 1863–1867. https://doi.org/10.1002/smll.200902348

Selvaraj T, Johar B, Khor SF (2019) Iron/zinc doped 8 mol% yttria stabilized zirconia electrolytes for the green fuel cell technology: A comparative study of thermal analysis, crystalline structure, microstructure, mechanical and electrochemical properties. Mater Chem Phys 222:309–320. https://doi.org/10.1016/j.matchemphys.2018.10.019

Sengodan S et al (2018) Advances in reforming and partial oxidation of hydrocarbons for hydrogen production and fuel cell applications. Renew Sust Energ Rev 82:761–780

Shabri HA, Othman MHD, Mohamed MA, Kurniawan TA, Jamil SM (2015) Recent progress in metal-ceramic anode of solid oxide fuel cell for direct hydrocarbon fuel utilization: a review. Fuel Process Technol 51:1–8. https://doi.org/10.1016/j.fuproc.2020.106626

Shaigan N, Qu W, Ivey DG, Chen W (2010) A review of recent progress in coatings, surface modifications and alloy developments for solid oxide fuel cell ferritic stainless steel interconnects. J Power Sources 195(6):1529–1542. https://doi.org/10.1016/j.jpowsour.2009.09.069

Shi N, Yu S, Chen S, Ge L, Chen H, Guo L (2017) Dense thin YSZ electrolyte films prepared by a vacuum slurry deposition technique for SOFCs. Ceram Int 43(1):182–186. https://doi.org/10.1016/j.ceramint.2016.09.131

Shi N et al (2020) Review of anodic reactions in hydrocarbon fueled solid oxide fuel cells and strategies to improve anode performance and stability. Mater Renew Sustain Energy 9(6):3–18

Shibata N, Katamura J, Kuwabara A, Ikuhara Y, Sakuma T (2001) The instability and resulting phase transition of cubic zirconia. Mater Sci Eng A 312:90–98

Shimizu T et al (2016) The demonstration of significant ferroelectricity in epitaxial Y-doped HfO2 film. Sci Rep 6(April):1–8. https://doi.org/10.1038/srep32931

Singhal SC (2002) Solid oxide fuel cells for stationary, mobile, and military applications. Solid State Ionics 152–153:405–410. https://doi.org/10.1016/S0167-2738(02)00349-1

Singhal SC, Kendall K (2003) High-temperature solid oxide fuel cells: fundamentals, design and applications, 1st edn. Elsevier Science

Skinner SJ, Kilner JA (2003) Oxygen ion conductors. Mater Today 6(3):30–37. https://doi.org/10.1016/S1369-7021(03)00332-8

Solovyev AA et al (2015) Solid oxide fuel cell with Ni-Al support. Int J Hydrog Energy 40(40): 14077–14084. https://doi.org/10.1016/j.ijhydene.2015.07.151

Stambouli AB, Traversa E (2002) Solid oxide fuel cells (SOFCs): a review of an environmentally clean and efficient source of energy. Renew Sust Energ Rev 6:433–455

Steele BC, Heinzel A (2001) Materials for fuel-cell technologies. Nature 414:345–352

Stonten D, Emonts B (2012) Fuel cell science and engineering materials, processes, systems and technology. Wiley-VCH Verlag & Co. KGaA, Weinheim, Boschstr. 12 69469

Subbarao EC (1981) Zirconia - an overview. Adv Ceram 1:1–24

Sudireddy BR et al (2017) Development of robust metal-supported SOFCs and stack components in EU METSAPP Consortium. Fuel Cells 17(4):508–516. https://doi.org/10.1002/fuce.201600191

Sun J, Qiu XP, Wu F, Zhu WT (2005) H2 from steam reforming of ethanol at low temperature over Ni/Y2O3 and Ni/La2O3 catalysts for fuel-cell application. Int J Hydrog Energy 30(4):437–445. https://doi.org/10.1016/j.ijhydene.2004.11.005

Tan J, Lee D, Ahn J, Kim B, Kim J, Moon J (2018) Thermally driven in situ exsolution of Ni nanoparticles from (Ni, Gd)CeO2 for high-performance solid oxide fuel cells. J Mater Chem A 00(31):1–11. https://doi.org/10.1039/c.d.oxide

Technology SM (2021) Pre-coated PVD process technology. https://www.materials.sandvik/en/products/coated-strip-steel/production-process/

Tietz F (1999) Thermal expansion of SOFC materials. Ionics (Kiel) 5(1–2):129–139. https://doi.org/10.1007/BF02375916

Tu H, Stimming U (2004) Advances, aging mechanisms and lifetime in solid-oxide fuel cells. J Power Sources 127(1–2):284–293. https://doi.org/10.1016/j.jpowsour.2003.09.025

Tucker MC (2010) Progress in metal-supported solid oxide fuel cells: a review. J Power Sources 195(15):4570–4582. https://doi.org/10.1016/j.jpowsour.2010.02.035

Tucker MC (2017) Development of high power density metal-supported solid oxide fuel cells. Energ Technol 5(12):2175–2181. https://doi.org/10.1002/ente.201700242

Udomsilp D et al (2020) Metal-supported solid oxide fuel cells with exceptionally high power density for range extender systems. Cell Reports Phys Sci 1(6). https://doi.org/10.1016/j.xcrp.2020.100072

Velu S, Suzuki K, Vijayaraj M, Barman S, Gopinath CS (2005) In situ XPS investigations of Cu 1-xNi xZnAl-mixed metal oxide catalysts used in the oxidative steam reforming of bio-ethanol. Appl Catal B Environ 55(4):287–299. https://doi.org/10.1016/j.apcatb.2004.09.007

Vibhu V, Rougier A, Nicollet C, Flura A, Grenier JC, Bassat JM (2015) La2 - XPrxNiO4 + δ as suitable cathodes for metal supported SOFCs. Solid State Ionics 278:32–37. https://doi.org/10.1016/j.ssi.2015.05.005

Wachsman ED, Lee KT (2011) Lowering the temperature of solid oxide fuel cells. Science (80-) 334(6058):935–939. https://doi.org/10.1126/science.1204090

Wachsman E, Ishihara T, Kilner J (2014) Low-temperature solid-oxide fuel cells. MRS Bull 39(9): 773–779. https://doi.org/10.1557/mrs.2014.192

Wang Z et al (2008) Dynamic evaluation of low-temperature metal-supported solid oxide fuel cell oriented to auxiliary power units. J Power Sources 176(1):90–95. https://doi.org/10.1016/j.jpowsour.2007.10.002

Wang F et al (2011) Hydrogen production from ethanol steam reforming over Ir/CeO2 catalysts: enhanced stability by PrOx promotion. Int J Hydrog Energy 36(21):13566–13574. https://doi.org/10.1016/j.ijhydene.2011.07.091

Wang W, Su C, Zheng T, Liao M, Shao Z (2012) Nickel zirconia cerate cermet for catalytic partial oxidation of ethanol in a solid oxide fuel cell system. Int J Hydrog Energy 37(10):8603–8612. https://doi.org/10.1016/j.ijhydene.2012.02.138

Wang Z, Wang C, Chen S, Liu Y (2014a) Co-Ni bimetal catalyst supported on perovskite-type oxide for steam reforming of ethanol to produce hydrogen. Int J Hydrog Energy 39(11): 5644–5652. https://doi.org/10.1016/j.ijhydene.2014.01.151

Wang W et al (2014b) Coking suppression in solid oxide fuel cells operating on ethanol by applying pyridine as fuel additive. J Power Sources 265:20–29. https://doi.org/10.1016/j.jpowsour.2014.04.111

Wayland M, Kolodny L (2020) Tesla's market cap tops the 9 largest automakers combined – experts disagree about if that can last. https://www.cnbc.com/2020/12/14/tesla-valuation-more-than-nine-largest-carmakers-combined-why.html. Accessed Nov 11 2021

Wincewicz KC, Cooper JS (2005) Taxonomies of SOFC material and manufacturing alternatives. J Power Sources 140(2):280–296. https://doi.org/10.1016/j.jpowsour.2004.08.032

Wongsakulphasatch S, Kiatkittipong W, Assabumrungrat S (2013) Comparative study of fuel gas production for SOFC from steam and supercritical-water reforming of bioethanol. Int J Hydrog Energy 38(14):5555–5562. https://doi.org/10.1016/j.ijhydene.2013.02.125

Wu J, Liu X (2010) Recent development of SOFC metallic interconnect. J Mater Sci Technol 26(4): 293–305. https://doi.org/10.1016/S1005-0302(10)60049-7

Yadroitsev I, Shishkovsky I, Bertrand P, Smurov I (2009) Manufacturing of fine-structured 3D porous filter elements by selective laser melting. Appl Surf Sci 255(10):5523–5527. https://doi.org/10.1016/j.apsusc.2008.07.154

Yamahara K (2003) Influence of powders on ionic conductivity of polycrystalline zirconias. ECS Proc 2003–07(1):187–195. https://doi.org/10.1149/200307.0187pv

Yamamoto K, Qiu N, Ohara S (2015) In situ fabrication of high-performance Ni-GDC-nanocube core-shell anode for low-temperature solid-oxide fuel cells. Sci Rep 5:1–6. https://doi.org/10.1038/srep17433

Yaman C, Kucukaga Y (2020) Performance of NiO/YSZ anode-supported solid oxide fuel cell fueled with landfill gas stream. E3S Web Conf 166:0–5. https://doi.org/10.1051/e3sconf/202016604007

Yan Y, Bateni R, Harris J, Kesler O (2015) Fabrication of reactive element oxide coatings on porous ferritic stainless steel for use in metal-supported solid oxide fuel cells. Surf Coat Technol 272: 415–427

Yang BC, Koo J, Shin JW, Go D, Shim JH, Ah J (2019a) Direct alcohol-fueled low- temperature solid oxide fuel cells: a review. Energ Technol 7(1):5–19

Yang L et al (2019b) Compression–compression fatigue behaviour of gyroid-type triply periodic minimal surface porous structures fabricated by selective laser melting. Acta Mater 181:49–66. https://doi.org/10.1016/j.actamat.2019.09.042

Yang L, Mertens R, Ferrucci M, Yan C, Shi Y, Yang S (2019c) Continuous graded Gyroid cellular structures fabricated by selective laser melting: design, manufacturing and mechanical properties. Mater Des 162:394–404. https://doi.org/10.1016/j.matdes.2018.12.007

Yaremchenko AA et al (2004) Ionic and electronic conductivity of La9.83-xPrxSi4.5Fe1.5O26-d apatites. Solid State Ionics 171:51–59

Yu F et al (2021) Recent progress in direct carbon solid oxide fuel cell: advanced anode catalysts, diversified carbon fuels, and heat management. Int J Hydrog Energy 46(5):4283–4300. https://doi.org/10.1016/j.ijhydene.2020.10.259

Zakaria Z, Mat ZA, Hassan SHA, Kar YB (2020a) A review of solid oxide fuel cell component fabrication methods toward lowering temperature. Int J Energy Res 44(2):594–611. https://doi.org/10.1002/er.4907

Zakaria Z, Abu Hassan SH, Shaari N, Yahaya AZ, Boon Kar Y (2020b) A review on recent status and challenges of yttria stabilized zirconia modification to lowering the temperature of solid oxide fuel cells operation. Int J Energy Res 44(2):631–650. https://doi.org/10.1002/er.4944

Zanchet D, Santos JBO, Damyanova S, Gallo JMR, Bueno JMC (2015) Toward understanding metal-catalyzed ethanol reforming. ACS Catal 5(6):3841–3863. https://doi.org/10.1021/cs5020755

Zhao L, Wei Y, Huang Y, Liu Y (2016) La1−xKxFe0.7Ni0.3O3 catalyst for ethanol steam reforming—The effect of K-doping. Catal Today 2:430–437

Zhen YD, Tok AIY, Jiang SP, Boey FYC (2007) La(Ni,Fe)O3 as a cathode material with high tolerance to chromium poisoning for solid oxide fuel cells. J Power Sources 170(1):61–66. https://doi.org/10.1016/j.jpowsour.2007.03.079

Zhou Y et al (2014a) Novel architectured metal-supported solid oxide fuel cells with Mo-doped SrFeO3-δ electrocatalysts. J Power Sources 267:148–154. https://doi.org/10.1016/j.jpowsour.2014.04.157

Zhou Y et al (2014b) Performance and degradation of metal-supported solid oxide fuel cells with impregnated electrodes. Int J Hydrog Energy 39(5):2279–2285. https://doi.org/10.1016/j.ijhydene.2013.11.086

Zhou Z, Nadimpalli VK, Pedersen DB, Esposito V (2021) Degradation mechanisms of metal-supported solid oxide cells and countermeasures: a review. Materials (Basel) 14(11). https://doi.org/10.3390/ma14113139

Zou J et al (2015) Hydrogen production from ethanol over Ir/CeO2 catalyst: effect of the calcination temperature. Fuel 159:741–750. https://doi.org/10.1016/j.fuel.2015.07.040

Zouhri K, Lee SY (2016) Exergy study on the effect of material parameters and operating conditions on the anode diffusion polarization of the SOFC. Int J Energy Environ Eng 7(2): 211–224. https://doi.org/10.1007/s40095-015-0201-1

Chapter 18
New Feedstocks for Bioethanol Production: Energy Cane and Agave

Fábio Trigo Raya, Luís Guilherme Furlan de Abreu, Marina Pupke Marone, Mozar de Araújo Salvador, José Antônio Bressiani, José Ignacio del Real Laborde, and Gonçalo Amarante Guimarães Pereira

Abstract Bioethanol consumption is projected to increase to 140 billion liters by 2029. However, climate change threatens the global production of the main bioethanol feedstocks, like sugarcane and corn. The search for new sources of biomass with greater resistance to these adverse conditions is essential to minimize impacts on biofuels production. In this context, two crops stand out: energy cane and *Agave*. Originally bred to improve the *Saccharum* genus feasibility to lignocellulosic bioethanol production, energy cane has physiological and biochemical characteristics that enhance robustness and could act as a direct substitute for sugarcane in traditional areas already suffering from irregular climate. On the other hand, *Agave* is a desert crop that has been domesticated as early as corn and has traditionally been used for alcoholic beverages and fibers but has never been applied for bioenergy purposes. With high water use efficiency, agaves can yield as much as sugarcane and can be used to slacken land competition and improve food and energy security. Both these crops present promising productivities and traits, such as high carbohydrates accumulation and drought resistance. However, they still face similar challenges to unravel its potential as new feedstocks for bioethanol in an Era of Climate Change.

F. Trigo Raya · L. G. Furlan de Abreu · M. Pupke Marone · G. Amarante Guimarães Pereira (✉)
Laboratory of Genomics and BioEnergy, Institute of Biology, Department of Genetics, Evolution and Bioagents, UNICAMP, Campinas-SP, Brazil
e-mail: goncalo@unicamp.br

M. de Araújo Salvador
Instituto Nacional de Meteorologia – INMET, Eixo Monumental Sul Via S1–Sudoeste, Brasília, DF, Brazil

J. A. Bressiani
GranBio Investimentos SA, Av. Brigadeiro Faria Lima, São Paulo, SP, Brazil

J. I. del Real Laborde
Agaves, Fibras y Ceras De México, SAPI de CV, Costa Real 586, Fraccionamiento Valle Real, Saltillo, Coahuila, Mexico

C. R. Soccol et al. (eds.), *Liquid Biofuels: Bioethanol*, Biofuel and Biorefinery Technologies 12, https://doi.org/10.1007/978-3-031-01241-9_18

18.1 Introduction

Earth's population is projected to exceed 9 billion by 2050. Consequently, the demand for agricultural products, such as food, fibers, and fuels, will grow, as well as the need for new cropping areas and fertilizers (Long et al. 2015a). Considering biofuels, by 2029 global ethanol production is projected to increase to 140 billion liters (OECD/FAO 2021). It is estimated that bioethanol demand, for Brazil, could increase 37.4 to 70.7 billion liters in 2030, which could potentially drive expansion into pastureland and natural vegetation mosaics (de Andrade Junior et al. 2019). Also, changes in temperature and precipitation are now a reality as is their effect on agricultural production (Trebicki 2020; Läderach et al. 2017; da Silva et al. 2021; Baum et al. 2020). In tropical and temperate regions, an increase of 2 °C above the average temperature levels recorded in the late twentieth century can have negative impacts on production of food crops such as rice, wheat and corn (IPCC 2014). Therefore, more biomass will have to be produced with limited resources and land competition between food and energy crops will increase (Swaminathan and Kesavan 2012).

Today, ethanol production as a biofuel is dominated by two crops: sugarcane and corn (OECD/FAO 2021; da Silva and Castañeda-Ayarza 2021). It is projected that 25% and 14% of global sugarcane and corn production, respectively, will be used to supply the increasing ethanol demand by 2029 (OECD/FAO 2021). However, climate change projections indicate effects in both cultures. Responsible for 40% of global sugarcane production, Brazil is the world's largest producer. In this country, it is estimated that 45% of the sugarcane productivity areas will require rescue irrigation (da Silva et al. 2021). In São Paulo, a Brazilian state that is responsible for more than half of this crop production, the effect of climate change can already be observed. An analysis of precipitation in the state (from 1950 to 1999) showed an increase in intensive rainfall combined with longer periods of drought (Dufek and Ambrizzi 2008). Other regions of the country have also been affected. During the last decades, rainfall is decreasing in the austral summer of Northern Brazil, which affect the rainfall regimes in the Midwest and Southeast regions, diminishing rain precipitation (Reboita et al. 2010; Silva et al. 2019). Also, an increase in the surfaces minimum and maximum air temperature in all regions of the country is predicted, which will lead to higher frequencies of extreme heat, drought, and rainfall deficits in the North and Northeast regions (dos Santos et al. 2020).

Recently, Brazil started using corn as a feedstock for ethanol production (da Silva and Castañeda-Ayarza 2021), a strategy used for many years in the United States (Moore et al. 2017). Corn production in Brazil is mostly concentrated in the Midwest region and it has proven to be a good alternative for reducing operational idleness during the sugarcane off-season (da Silva and Castañeda-Ayarza 2021). Today, corn is the main feedstock for bioethanol and it accounts for 60% of the world's ethanol production (OECD/FAO 2021). Using corn as its main biofuel feedstock, the US maintain its post as the world's largest ethanol producer (OECD/FAO 2021). Yet, by

2030 it is predicted that 99% of the US ethanol will be produced from corn (OECD/FAO 2021). Like sugarcane, corn has already been affected by climate change. In the US optimum planting date is advancing 0.13 days/year (Baum et al. 2020). Although many projections indicate higher corn yields in temperate regions (Baum et al. 2020; Bhattarai et al. 2017), corn is considered highly susceptible to drought (Somerville et al. 2010). Therefore, its productions areas worldwide might need to be shifted and its global productivity is predicted to decrease (Kogo et al. 2021; Leng and Huang 2017).

In the eminence of climate change, global agricultural areas will be reshaped, consequently, food and energy production will be affected across the planet (Daher and Mohtar 2015). The search for new sources of biomass with greater resistance to adverse conditions is essential to minimize impacts on the production of bioethanol. In this context, two crops stand out: energy cane and *Agave*. Originally bred to improve the *Saccharum* genus feasibility to lignocellulosic bioethanol production, energy cane has physiological and biochemical characteristics that enhance robustness and could act as a direct substitute for sugarcane (Carvalho-Netto et al. 2014). On the other hand, *Agave* is a desert crop that has been domesticated as early as corn and has traditionally been used for alcoholic beverages (Nobel 1994) but has never been applied for bioenergy purposes. Both these crops present promising productivities and traits, such as high carbohydrates accumulation and drought resistance. However, they still face similar challenges to unravel its potential as new feedstocks for bioethanol in an Era of Climate Change.

18.2 Energy Cane

18.2.1 Agroindustry History Driving New Cultivars Development

Sugarcane (*Saccharum* spp.) is the main bioenergy feedstock, especially in the tropical and subtropical regions of the world (Long et al. 2015b). It is a plant with high photosynthetic metabolism (C4 plant), being very efficient in converting radiant energy into chemicals. Sugarcane has been exploited for centuries for sugar production. In addition, because of its adaptability for use in the production of sugar and ethanol, straw and bagasse have been increasingly used in the cogeneration of electricity (Cheavegatti-Gianotto et al. 2011). However, it was from the mid-1970s onwards that the oil embargo by Organization of the Petroleum Exporting Countries (OPEC) demonstrated the economic fragility of a non-renewable energy source. In this context, Brazil gains a prominent role in the large-scale production of ethanol as a renewable fuel from fermented sugarcane sucrose through National Alcohol Program (PROALCOOL), opening a new horizon for Brazil's energy matrix, with the objective of stimulating the production of ethanol in the domestic market, anhydrous ethanol became mixed the gasoline as an additive, substitute the tetraethyl lead. Next, hydrated ethanol allowed the first car entirely powered by biofuel to reach the market in 1979. In parallel, the similar strategy is carried out by the United States, using corn as a source of biomass (Goldemberg 2008; Matsuoka et al. 2009; Mumm et al. 2014).

Interestingly, other by-products are generated and can be developed from sugarcane. Yet, the country's production is still insufficient to supply the domestic market demand (Carvalho-Netto et al. 2014). Recently, new advances were made to minimize the loss of productivity and increase the energy gain in the sugar in the mill process (Matsuoka et al. 2009). The bagasse used to be an undesirable residue, but it has become an important substitute to produce sugar and has also been used to generate the electricity necessary in the process. Until the 2000s, sugarcane fields were burned for manual harvesting, which resulted in two problems: (1) increased emission of gases in the atmosphere; (2) loss of juice extraction quality, however, due of this environmental problem resulted in laws and regulations that are intended to limit cane burning, as Federal Decree No. 2661 of July 9, 1998 and São Paulo State Law No. 11241 of September 19, 2002 (Carvalho-Netto et al. 2014).

However, the growing demand and increased interest in bioenergy encouraged scientists to better understand the physiological processes of sugarcane, such as increasing the efficiency of solar energy capture and the increase in the accumulation of sugar and biomass (Waclawovsky et al. 2010). The dry matter of *Saccharum* spp. is composed of sugars, mostly sucrose, and fiber, cellulose, hemicelluloses, and lignin (Kim and Day 2011). Between 80 and 85% of the total biomass of sugarcane is present in the stalks, and the remainder not used in the sugarcane industry consists of leaves and immature top (Landell et al. 2013). In addition, the cellulose digestion technology for ethanol production made it possible to collect the entire plant, a process that is already a reality in Brazil, with two industrial plants by Granbio and Raizen.

In this context, advances in industrial technologies and biotechnology have made possible the fermentation of soluble sugars caused by the deconstruction of biomass, producing the so-called lignocellulosic bioethanol (Dias et al. 2011; dos Santos et al. 2016). In the last decades, genetic improvement programs had as their main objective the search for the increase of sucrose in sugarcane stalks, however, with the emergence of new technologies, the increase in biomass has become the focus of these programs (Creste et al. 2014). Because of this increased demand for biomass to attend the new cellulosic ethanol projects, sugarcane breeders across the world started to select a new type of hybrids in their breeding programs. The selection of hybrids with high amounts of biomass resulted in individuals twice as productive as sugarcane, which are known as energy cane (Carvalho-Netto et al. 2014; Viator and Richard 2012; Matsuoka et al. 2014). In the 1980s, Alexander was the pioneer in the conceptualization of the use of energy cane in Puerto Rico (Alexander 1985). Energy cane is very similar to sugarcane, with a well-developed exploration system, not only in the field but also in the industry (Viator and Richard 2012). Energy cane was bred especially for lignocellulosic bioethanol production, changing the paradigm of sugarcane improvement programs: from cultivars with high sucrose content to cultivars with high fiber content (20–30%) (Matsuoka et al. 2014).

18.2.2 Cultivars and Productivity

Energy cane is an interspecific hybrid resulting from the crossing of two species, *S. spontaneum* (high fiber content) and *S. officinarum* (high sugar content), thus producing a plant with higher fiber content and lower sugar content than sugarcane (Matsuoka et al. 2014). A greater number of alleles of *S. spontaneum* species increases disease resistance and increases ratoon sprouting (Giamalva et al. 1984). In general, sugar and fiber content in sugarcane averages 13% and 12%, respectively. According to the sugar and fiber content present in its stalks, energy cane can be classified into types I and II. For a type I energy cane, the sugar and fiber contents are 13% and 17%, respectively, which would be more energy efficient than current sugarcane cultivars. For a type II energy cane, the sugar and fiber contents are 5% and 30%, respectively. Type II energy cane are more robust cultivars, which adapt well to more restrictive environments and are ideal for ethanol second-generation production (Carvalho-Netto et al. 2014; dos Santos et al. 2016; Matsuoka et al. 2014; Tew and Cobill 2008). Fiber in energy cane consists of 43% cellulose, 24% hemicellulose, and 22% lignin, while for sugarcane these values are 42% cellulose, 25% hemicellulose, and 20% lignin (Table 18.1) (Kim and Day 2011). In addition, fiber-producing plants are gaining more importance, as they can fix carbon in shoot and root structures, thus mitigating the effects of the excess gases in the atmosphere (Byrt et al. 2011; de Siqueira et al. 2013).

Remarkably, type II energy cane has become a promising cultivar with the potential to increase agricultural productivity (Tew and Cobill 2008). The total biomass production of energy cane is twice that of sugarcane, with 180 t ha^{-1} of wet weight, while sugarcane generally produces 92 t ha^{-1} of wet weight (dos Santos et al. 2016). In addition, in the first cycle of plant cane, the production of dry biomass

Table 18.1 Composition of the main energy crops

	Corn	Sugarcane	Energy cane	Agave[a]
Brix %	16 (Machado Filho et al. 2018)	15[b]	16[b]	27–38 (Subedi et al. 2017)
Pol %	n.a.	15[b]	12[b]	n.a.
Cellulose (%)	31–38 (Corbin et al. 2015)	42 (Kim and Day 2011)	43 (Kim and Day 2011)	26–65 (Davis and Long 2015; Yang et al. 2015a)
Hemicellulose (%)	19–25 (Corbin et al. 2015)	25 (Kim and Day 2011)	24 (Kim and Day 2011)	5–22 (Davis and Long 2015; Yang et al. 2015a)
Lignin (%)	17–21 (Corbin et al. 2015)	20 (Kim and Day 2011)	22 (Kim and Day 2011)	13–15 (Davis and Long 2015; Yang et al. 2015a)
Storage carbohydrate	Starch	Sucrose	Sucrose	Fructans
Direct fermentation?	No	Yes	Yes	No

n.a. Data not available
[a] Data available for *Agave tequilana*
[b]Source: Granbio

from energy cane was on average 75% higher than the average for sugarcane, with productions of 43 and 24.6 t ha^{-1} for energy cane and sugarcane, respectively (Boschiero et al. 2019). Concerning the harvesting cycles, energy cane has a life cycle of up to 10 years, therefore, in the long term, it requires fewer expenses in cultivation operations than sugarcane, which has a shorter life cycle (5 years) (Matsuoka et al. 2014; Grassi and Pereira 2019).

Although sucrose (Pol) levels are lower in energy cane (12.67%) than in sugarcane (15.4%), Brix values are higher in energy cane than in sugarcane (Table 18.1), 16.3% and 15.6%, respectively (Carvalho-Netto et al. 2014). In energy cane, the carbon partition is oriented towards the production of fibers rather than the accumulation of soluble sugars, resulting in a high biomass index. However, due to its high productivity, energy cane surpasses sugarcane in the production of sugar per hectare, being a promising raw material to support the development of the bioeconomy (Carvalho-Netto et al. 2014; Alexander 1985).

In 2012, GranBio, a Brazilian company focused on lignocellulosic bioethanol, started an energy cane breeding program focused on maximizing the productivity of the total sugars present in the cane, originating from juice and fibers present in the stalks, tips, and dry leaves, more robust, less demanding on the soil, climate, water and nutrients and more resistant to pests and diseases. The resulting cultivars would have greater energy efficiency in their cultivation and less competition with food production, considering the feasibility of planting these cultivars in marginal agricultural areas. After 8 years of research, GranBio developed 11 cultivars of energy cane, distributed into two groups: (1) Type I varieties, like Vertix 3, recommended for the first-generation industry of sugar, ethanol, and electricity. It can be planted in restrictive environments, for harvesting throughout all the harvest season and can even be used to expand this season permitting its harvest during rainy periods (Fig. 18.1). Because of its well-developed root system, this energy cane cultivar contributes to improving the organic matter content in the soil throughout the cuts. The higher yield of biomass and number of cuts, increase the production of sugars in the juice and fiber per area reducing drastically the production costs. (2) Type II varieties, Vertix 2, recommended to industries of lignocellulosic bioethanol and/or biochemicals, biogas and biomethane, power generation of electricity, miming and steel, pellets and briquettes, corn ethanol as a source of biomass to boilers, among others that need low-cost biomass as feedstock. It can be grown in restrictive environments for year-round harvest without competing for food. The ethanol obtained from the juice is almost enough to pay the production costs, leaving a large amount of bagasse produced with a residual cost significantly lower than other biomass sources available in the market.

18.2.3 Biological and Agronomic Traits

Energy cane presents a series of characteristics that contribute to the high productivity and success of this plant in adverse conditions (Matsuoka et al. 2014). The

Fig. 18.1 Energy cane Type I (Vertix 3) variety from Granbio, plant cane, 8 months, growing in São Paulo, Brazil (Source Granbio)

energy cane field has a high density, erect growth of plants, higher sprouting rate and tillering, resistance to biotic and abiotic stress, rhizome presence, high volume of the root system, high biomass accumulation, and resistance to mechanical damage (Carvalho-Netto et al. 2014; Viator and Richard 2012; Matsuoka et al. 2014; Alexander 1985; Tew and Cobill 2008; de Abreu et al. 2020). The rapid growth of energy cane can be attributed to two specific characteristics of these hybrids, such as: (1) first shoot development and then root formation; (2) high nocturnal growth rate (de Abreu et al. 2020). Another important feature, and contributes to reduces environmental impact, is the presence of fasciculate roots, which prevent soil erosion. In addition, the greater number of ratoons harvested generates a lower cost to renew a sugarcane field, providing for a minimum period of 10 years for the ratoon (Carvalho-Netto et al. 2014).

Recently, several efforts have been made to understand the physiological, morphological, and biochemical differences between energy cane and sugarcane (Boschiero et al. 2019; de Abreu et al. 2020, 2021; Cruz et al. 2021). The high sprouting rate of energy cane demonstrates different strategies in the first hours of axillary bud outgrowth compared to sugarcane. In energy cane, high metabolic activity was demonstrated in axillary bud outgrowth with only 48 h of planting, there was an increase in the levels of reducing sugars, organic compounds related to the phenylpropanoids and lipids pathway, despite a reduction in the levels of amino acids. Taken together, these results lead to increased energy production for cell division and growth processes (de Abreu et al. 2021). After planting sugarcane, root formation occurs after 24 h (Aude 1993), but energy cane has shown to have a different development strategy, with shoot development first and root formation happening 10 days after planting (de Abreu et al. 2020). Also, energy cane setts treated with auxin (bud sprouting inhibitor) showed little response to the treatment during the axillary bud outgrowth, in addition to showing greater resistance to water and cold stress (Cunha et al. 2020).

During the cycle of the plant cane and the first ratoon, a higher tillering rate, green leaves, and total leaf area were demonstrated in energy cane cultivars type I and II, compared to sugarcane (Cruz et al. 2021). Moreover, the average water use efficiency during the period of the plant cane and first ratoon was the same for the energy cane and sugarcane (Cruz et al. 2021). However, the energy cane root system is more developed, which allows for greater uptake of water in the soil (Carvalho-Netto et al. 2014). Interestingly, during the plant cycle, the type II energy cane cultivar had lower angulation of green leaves and a high number of tillers, resulting in lower light interception, consequently lower photosynthetic rate due to self-shading (Cruz et al. 2021). One of the most important plants life survival strategies is the use of some carbohydrates for storage and others for structural growth (Lambers et al. 2008) and this carbohydrate use strategy demonstrates the difference between sugarcane and energy cane productivity. In sugarcane, sucrose is prioritized as a reserve carbohydrate, while in energy cane, the photosynthesized product is mainly diverted to the structural function, that is, cellulose, hemicellulose, and lignin (Tew and Cobill 2008). Lignocellulose is the most abundant product of carbon-based photosynthesis in nature (Cosgrove 2005). Thus, the energy cane strategy is to increase the fiber content at a higher level to withstand adverse growing conditions (Alexander 1985; Tew and Cobill 2008).

Furthermore, it has been shown that energy cane has a greater capacity to access macronutrients from the soil (Boschiero et al. 2019), which implies that energy cane can respond in more restricted environments (low soil fertility). Therefore, the fertilization requirement for this crop is lower than sugarcane (Boschiero et al. 2019). Interestingly, in Brazil, energy cane is cultivated in marginal lands using the sugarcane agronomical practices. This system could cause the depletion of soil fertility, due to energy cane's high capacity of removing macronutrients from the soil. Therefore, it is still necessary to investigate nutrient requirements and management of energy cane to improve this crop sustainability and production.

18.2.4 Challenges and Benefits of Energy Cane as Bioethanol Feedstock

Although energy cane has shown high productivity, resistance to adverse conditions, among other characteristics that make it ideal for planting, some bottlenecks need to

Table 18.2 Estimated productivity and rainfall requirements of bioenergy crops

	Corn	Sugarcane	Energy cane	Agave[a]
Water requirement (mm year^{-1})	500–800 (Somerville et al. 2010)	1200–1800 (Carr and Knox 2011)	n.a.	300–800 (Somerville et al. 2010; Nobel 1988)
Productivity (t (wet) ha-1 year^{-1})	12–18.5 (Novak et al. 2019)	92 (Grassi and Pereira 2019)	180 (Grassi and Pereira 2019)	136–144 (Yan et al. 2020)
Plant density per hectare	59,000 (Novak et al. 2019)	70,000[b]	277,000[b]	4000 (Yan et al. 2020)
Growth Period	175 days	12 months	12 months	3–5 years (Yan et al. 2020)
Bioethanol productivity (L ha^{-1} year^{-1})	2900 (Somerville et al. 2010)	6900 (Somerville et al. 2010)	6000–7000[b]	4854–6673 (Yan et al. 2020)
Lignocellulosic Bioethanol productivity (L ha^{-1} year^{-1})	900 (Somerville et al. 2010)	3000 (Somerville et al. 2010)	3000–6000[b]	490–741 (Yan et al. 2020)

n.a. Data not available

[a] Data available for *Agave tequilana*

[b]Source: Granbio

be surpassed. Some energy cane cultivars are susceptible to smut disease (*S. scitamineum*) and have a high incidence of flowering, which generates remobilization and loss of sugar levels in the stalks (de Souza Barbosa et al. 2020). The biggest challenge is the genetic improvement programs for the development of new cultivars that are more resistant to this pest and with a lower incidence of flowering. Furthermore, there are also challenges in the agro-industrial sector. For agriculture, it is still necessary to determine crop management issues for planting, such as soil preparation, row spacing, fertilizing, irrigation, planting season, and machinery—collection and transport. For industry, it is still necessary to optimize the efficiency of mills for juice extraction (conditions for milling and pretreatment, sugar extraction—water consumption, energy requirements and extraction efficiency and steam consumption during the processing) (de Souza Barbosa et al. 2020; Leal 2007). Due to the high productivity of energy cane per hectare (Table 18.2), this cultivar has the potential to produce more non-cellulosic sugars than conventional sugarcane cultivars and can replace sugarcane areas with energy cane, even to produce ethanol from sucrose fermentation (dos Santos et al. 2016; Matsuoka et al. 2014). Energy cane has not been widely cultivated until recently (Carvalho-Netto et al. 2014), so these challenges will only be surpassed through specific technologies for this crop, both in industrial processing, field management, and adequate mechanization.

In the current climate change scenario, energy cane has proven to be an alternative to sugarcane, although some challenges need to be surpassed. Also, energy cane can be cultivated in marginal areas, avoiding competition with food production, and with fewer fertilizers, herbicides, and pesticides, whose products are among the greatest aggressors to the environment and human health. In summary, several factors contribute to the use of energy cane: it is more robust, has a higher multiplication rate, excellent efficiency in the use of nutrients, high root volume, resistance to pests, and resistance to mechanical damage (Carvalho-Netto et al. 2014; Matsuoka et al. 2014; Grassi and Pereira 2019; de Souza Barbosa et al. 2020).

18.3 Agave

18.3.1 Agaveculture: A Nine-Millennium Relationship Between Humans and Agaves

Agaves are monocot plants native to the semiarid regions of North and Central America. More than 200 species are known, and Mexico, which contains about 70% of the species, is the main biodiversity center of this genus (Eguiarte et al. 2021). These well-adapted plants thrive under hot and dry conditions and were essential for the development of the Mesoamerican native civilizations. With a singular combination of drought resistance and productivity, agaves molded the lifestyle of these indigenous people and allow them to prosper (Ortiz-Cano et al. 2020). This strong interaction was later known as the man-agave symbiosis (Gentry 1982).

Remarkably, archaeological studies indicate that the interaction between humans and agaves has been going on for more than 9 millennia (Radding 2012). In the pre-Columbian era, agaves were used as a feedstock of fibers, sugars, food, beverages, soap, medicines and even as needles (Nobel 1994; Ortiz-Cano et al. 2020; Valenzuela-Zapata and Nabhan 2004). Similar to sugarcane, the indigenous people use *Agave* to prepare non-alcoholic beverages (*aguamiel*), and a fermented alcoholic version, known as pulque (Isabel Enríquez-Salazar et al. 2017). Also, *Agave* was used as a staple crop to ensure food supply in droughts, and its fiber was used for cordage and cloth-making (Ortiz-Cano et al. 2020).

The pre-Columbian people were responsible for the domestication of this crop, generating cultivars dedicated for fiber production, *Agave sisalana* (sisal) (Fig. 18.2a) and *A. fourcroydes* (henequen), and for beverages, like *A. tequilana* (Fig. 18.3a), *A. salmiana* (Fig. 18.2d) and *A. americana* (Ortiz-Cano et al. 2020; Nobel 2010). The selection pressure applied by them was so intense that, until today, the wild ancestors of cultivated species, such as *A. americana*, *A. sisalana*, *A. fourcroydes* and *A. tequilana*, are still up to debate. However, most of the species used for industrial applications are, now, considered to be originated from the *A. angustifolia* and *A. rhodacantha* complexes (Colunga-García Marín and May-Pat 1997; Lledías et al. 2020; Rodríguez-Garay et al. 2009; Álvarez-Ríos et al. 2020).

With the Europeans arrival, agaveculture enters a new cycle of commercial exploration and industrialization. It was during the colonial period that the

Fig. 18.2 (**a**) *Agave sisalana* (Sisal) growing in São Paulo, Brazil. (**b**) Harvested *A. sisalana* leaves waiting for defibering. (**c**) Sisal fibers drying in the sun in Bahia, Brazil. (**d**) Cold tolerant (−16 °C) *A. salmiana* var. *salmiana* growing in Coahuila, Mexico. (**e**) *A. tequilana* piña being harvested for tequila production. (**f**) Shredded Agave's piña before juice extraction

Fig. 18.3 (**a**) Five years-old *Agave tequilana* growing in Queensland, Australia. (**b**) A backhoe lifting the plant for sampling. (**c**) Scale weighing the sampled *Agave tequilana*. Image: courtesy of Don Chambers—AusAgave

distillation processes were introduced in Mexico. This eventually gave rise to the spirits that symbolize Mexican identity, mezcal and tequila (Valenzuela-Zapata and Nabhan 2004). Moreover, the spread of agaves around the world started shortly after Europeans arrived in the Americas. In 1561, the *Agave americana* was the first exotic species introduced in the Padua Botanical Garden (Italy), considered to be the first botanical garden in the world (Medina 1954; UNESCO 2020). However, it is from the beginning of the twentieth century that the agaves started to have great commercial importance, both in Mexico and in other semi-arid regions of the globe (Eguiarte et al. 2021; Monja-Mio et al. 2019).

Outside Mexico, agaves have been most explored in Brazil and East Africa. In Brazil, the first *Agave sisalana* seedlings were introduced to the state of Bahia only in 1903 (Medina 1954). From the post-war period onwards, Brazil became one of the most important exporter of sisal fiber, and, in the 1970s, it became the largest producer in the world, a position it maintains to this day (Silva and Beltrão 1999; FAO 2020). *Agave sisalana* was first introduced to East Africa in 1893, and, a few years later, sisal became one of the most important commodities in the region (Sabea 2008). Accordingly, the region held one of the most successful *Agave* breeding programs. At the former East African Agricultural Research Station in Amani (Tanzania), was developed the first and only commercially exploited *Agave* hybrid cultivar, the hybrid 11,648 ((*A. amaniensis* × *A. angustifolia*) × *A. amaniensis*). Interestingly, among the agaves used for industrial purposes, only those developed for fiber production were heavily disseminated across the globe (Nobel 1994; Monja-Mio et al. 2019). Though, there are a couple of examples of small-scale production of *Agave* spirits outside Mexico (LADB 1997; Smith 2017; Gross 2021).

Due to competition with synthetic products from the late 1960s forward, sisal production has been globally decreasing (Davis and Long 2015). In contrast, in Mexico, since 1970, the production of *Agave* spirits, mainly tequila, has been growing steadily. Today, Mexico produces about 374 million liters of tequila per year (Consejo Regulador del Tequila 2021) that feeds a $1.7 billion niche market in the United States (Distilled Spirits Council of the United States 2011).

Recently, agaveculture is going through a revolution. Lead by biotechnology, agaves are becoming a promising feedstock for biorefineries, especially in areas with limited water supply (Corbin et al. 2015; Pérez-Pimienta et al. 2017; Davis et al. 2011). Agaves are now being used to produce agavins (a probiotic like inulin), syrup, pharmaceuticals, cosmetics and nanocellulose (Morán et al. 2008; Jeong et al. 2020; Jean 2015; Alejandra et al. 2013; Laborde Aguirre et al. 2010, 2013), and field trials are being performed to explore its potential as a biofuel feedstock (Yan et al. 2020; Parascanu et al. 2021; Holtum et al. 2011; Davis et al. 2017).

18.3.2 A Desert Crop Built for Industry Processing

Among the main reasons that *Agave* has been considered as a new bioenergy crop is its unique chemical composition. Unlike other crops, agaves accumulate fructose polymers as storage carbohydrates (Pérez-López and Simpson 2020). Throughout the whole vegetative phase, this crop accumulates these polymers at a fibrous central stem, commonly known as *piña* (Fig. 18.2e). Usually, farmers harvest the *piña* once carbohydrate accumulation reaches 21–30°Brix (Table 18.1), but some producers have achieved 38°Brix (Subedi et al. 2017). For *A. tequilana*, for instance, this occurs by the fifth year of cultivation.

Known as fructans, these fructose polymers are synthesized by adding fructose monomers to sucrose. As the *Agave* plant matures, new fructose monomers are added, which enhances complexity and branching degree (Arrizon et al. 2010). This trait enables fructans to be metabolically stable and act as a substitute for starch for long-term carbohydrates storage. From an industrial perspective, this characteristic releases farmers from harvesting seasons and makes industrial operations smaller and more efficient. In sugarcane, for instance, yield can be strongly affected by harvest time, and its season limits mills' operation through the year (Marin et al. 2021). *Agave* can be used to provide a constant supply of sugary feedstocks through the year or even be used to complement the sugarcane season. Also, farmers could stock these sugars in the field and process them at any moment during a harvest window of more than 3 years managing inventories to only process them when prices are favorable.

Like starch, fructans usually need to be processed before alcoholic fermentation. Traditionally, tequila producers use acid thermal hydrolysis to transform fructans into fermentable sugars (Mancilla-Margalli and López 2006; Michel-Cuello et al. 2008). This process is done in ovens autoclaves and diffusers, and not only hydrolyze fructans, but also softens the *Agave's piña*, which facilitates further processing operations like milling and extraction. However, this traditional approach does not completely convert all the fructans in monomers and these residues contribute to the organoleptic characteristics of tequila (Ávila-Fernández et al. 2009). Nonetheless, for bioethanol production, there are more efficient strategies, like enzymatic hydrolysis and direct fermentation. Commercially available enzymatic preparations for industrial hydrolysis of fructans have been developed by enterprises, like Novozymes and Megazyme, and proven to be efficient in *Agave* juices (Ávila-Fernández et al. 2009). Direct fermentation can be performed by *Kluyveromyces marxianus* and *Torulaspora delbrueckii,* but there is much room for improvement, especially in genetic engineering industrial strains of *Saccharomyces cerevisiae* for fructans consumption (Corbin et al. 2015; Liu et al. 2014).

Like sugarcane, agaves can also be used to produce lignocellulosic bioethanol. However, the *Agave* cell wall presents a singular architecture and composition that allow this crop to be more efficient. The main factor that limits biomass hydrolysis is recalcitrance caused by lignin (Ragauskas et al. 2014). The *Agave* biomass has significantly lower lignin (Table 18.1). Depending on species, lignin mass fraction represents 7–13% of Agave leaves and 10–15% of stems (Davis and Long 2015; Yang et al. 2015a; Raya et al. 2021). Herbaceous crops, such as *Miscanthus*, *Panicum*, and sugarcane, present 9–27% of lignin, while in woody biomasses it represents 21–32% (*e.g.*, *Populus*, *Eucalyptus* and *Pinus*) (Ragauskas et al. 2014; Morgan et al. 2016). Compared to energy cane, which has 17–27% of lignin (Carvalho-Netto et al. 2014), agaves can be 47% more hydrolyzable (Raya et al. 2021). This difference in recalcitrance has a direct impact on yields, energy consumption, and economic viability of lignocellulosic fuels (Davis et al. 2011; Klein-Marcuschamer et al. 2012).

Desert plants carry the stigma of being extremely drought resistant, but small and slow-growing. However, plants of the *Agave* genus can achieve impressive yields

(Garcia-Moya et al. 2011). *Agave* productivity is specie/cultivar dependent. Commercially cultivated agaves are reported to produce 4–22 t of shoot dry weight per hectare in a year. Theoretical analysis indicates that *A. tequilana* could achieve even higher potential yields (38 t ha^{-1} year^{-1} of dry biomass) (Owen et al. 2016). In Australia, a private company is reporting *A. tequilana* yielding 880 t ha^{-1} of net biomass and individual plants weighing 426 kg (Fig. 18.3; Chambers, Don personal communication). Other agave species like *A. salmiana* outyield biomass production of *A. tequilana* with individual plants accumulating more than 1000 kg during their cycle and have a larger geographical potential as they are tolerant to temperatures below −5 °C (AFyC and Frías, Juan T. personal communication). Additionally, only 10% of the plant whole biomass is root (Nobel 1988; Borland et al. 2009). This high shoot to root ratio allows farmers to harvest almost all the biomass produced.

According to the Owen and Griffiths (2016) model (Owen et al. 2016), the estimated yields of *A. tequilana* under Brazilian semiarid conditions could reach 17–19 t ha^{-1} year^{-1} of dry biomass. When considering the conversion rate of biomass to ethanol (Owen and Griffiths 2014), it is estimated that the Brazilian production of bioethanol from *Agave* could reach up to 6000 L ha^{-1} year^{-1}, which is compatible with the productivity obtained in Australia (6673 L ha^{-1} year^{-1}) and from sugarcane in Brazil, 5943 L ha^{-1} year^{-1} (Yan et al. 2020; CONAB 2018). However, sugarcane and *Agave* meet contrasting water requirements. Sugarcane ideally needs 1100–1800 mm year^{-1} of water, while agaves are much less demanding, with optimal development at 460 and 680 mm year^{-1} and tolerating even 300 mm year^{-1} (Table 18.2) (Carr and Knox 2011; Nobel 1988). The climatic conditions in the Brazilian semiarid are milder than those found in Mexico, in this sense, the predicted productivity of *A. tequilana* in Brazil has the potential to surpass the Mexican one (Owen et al. 2016).

18.3.3 Biological and Agronomical Traits

Most *Agave* species are monocarpic plants with life cycles of 5 to over 50 years. Until today, it is not clear the process that regulates its developmental phases. As agaves are unresponsive to photoperiod or temperature, age-determined signals involving carbohydrate regulation could be the factors that triggers the transition from the vegetative to the reproductive phase (Pérez-López and Simpson 2020). In this context, fructan metabolism is being proposed to be a kye regulator. Fructan branching and accumulation remains throughout the whole *Agave* vegetative phase and specifically the presence of agavins, the most complex plant fructans described, may serve as age related molecular signals (Pérez-López and Simpson 2020; Salinas et al. 2016).

Upon entering the reproductive phase, fructans are redirected to generate a huge inflorescence stalk that produces flowers, fruits and bulbils (Pérez-López and Simpson 2020). Bulbils are a form of asexual reproduction by the development of fully formed plants originated from floral meristems. The most common form of reproduction is through rhizomes. Also called suckers, rhizomes are modified underground stems that grow near the soil surface and appear during the first 3 years of plant growth. Once the apical meristem emerges from the soil, it starts forming a new plant that remains attached to the older one. Eventually, the new plant develops a root system and becomes photosynthetically self-sufficient. However, the sucker can have a great impact on plant productivity, and, therefore, are usually removed (Nobel et al. 1992). In *Agave*, sexual reproduction is difficulted by low rates, long life span, and monocarpy. Traditional planting uses the seedlings formed by the offsets, which are a cheap source of planting material but can be very physiologically heterogeneous and often contaminated with pathological agents. Bigger tequila producers use uniformly sized plugs from tissue-cultured plants of high-yielding lines that are compatible with mechanical planting (Holtum et al. 2011).

From photosynthesis to special carbohydrates reserves, agaves have an arsenal of anatomical and biochemical strategies to thrive in desert environments, making them capable of withstanding prolonged drought periods. All described agaves are known to perform the crassulacean acid metabolism (CAM) (Davis and Long 2015; Nobel 1988). CAM is a photosynthetic pathway that allows the plants to close their stomata during the day and open at night, when evapotranspiration rates are lower. To this metabolism work, CO_2 assimilation is performed at nighttime through a biochemical process similar to C4 photosynthesis. CO_2 is temporally converted into an organic acid, like malic acid, and stored in the vacuole. At daytime, CO_2 is released for the photosynthetic carbon fixation (Winter and Smith 1996; Keeley and Rundel 2003). This strategy maximizes water use efficiency resulting in 80–95% less water consumption than other photosynthetic pathways (Yang et al. 2015b). Additionally to the carbon cycle, agaves can re-shift their photochemical metabolism to endure drought (Campos et al. 2014).

Other remarkable strategies are sunken stomata, large water reserves in the mesophyll (Blunden et al. 1973), deposition of waxes and pectin in the epidermis (Raya et al. 2021), which increase the rate of water absorption in the leaves (Boanares et al. 2018), retractable roots that respond to the amount of water in the soil (Nobel 1988) and temporary rain-induced roots that maximize water uptake after a long dry period (Hunt et al. 1987). Also, agaves' growth and development are limited by seasonal water availability. If given a constant supply of water through the year, agaves keep growing and producing new leaves (Novel and De Cortazar 1987). Under constant watering, agaves shift their physiology to constant growth in detriment of sugar storage in the stem.

Plants accumulate compatible osmolytes to withstand drought and high salinity stress through osmoprotection. In *Agave*, an important osmolyte is raffinose. This oligosaccharide is usually found in seeds. For instance, in monocots seeds its concentrations can range from 2.6 to 7.9 mg g^{-1} (Kuo et al. 1988). In contrast, in *Agave* raffinose was detected in vegetative tissues at a much higher concentration, ranging from 10.96 to 23.38 mg g^{-1} (Raya et al. 2021). In addition to acting on drought, raffinose also has the potential to protect against oxidative stress (Nishizawa et al. 2008). Also, raffinose could be used in *Agave* biorefinery system as is a well-adopted and valuable sweetener (Carocho et al. 2017; Silva et al. 2021). Other important carbohydrates involved in stress responses are the fructans. These polymers are not only the main storage carbohydrate in *Agave*, but can still play roles as membrane stabilizers, resistance to oxidative stress and dehydration, and even as molecular signals (Pérez-López and Simpson 2020; Van den Ende 2013; Singh et al. 2020). The evolution of fructan accumulation can be considered an important adaptation of *Agave* to arid conditions (Pérez-López and Simpson 2020).

18.3.4 Challenges and Opportunities for Agave as Bioethanol Feedstock

The potential of *Agave* as a new bioenergy feedstock has been demonstrated throughout this chapter. However, to spread *Agave* as a new global energy crop there are some challenges to overcome. Although *Agave* cultivation is well established in Mexico, most of the tequila production still relies on an agro-extractivism system (Tetreault et al. 2021), which can only work for the production of valuable alcoholic beverages and discards all the foliar parts of the plant. To produce *Agave* bioethanol as a large-scale commodity some adjustments would be required. We must lay our efforts in five aspects: plant breeding, mass propagation of certified disease-free stocks, intensive agriculture with a focus on stress management, mechanization, and renewable energy policies.

Most industrial crops have established plant breeding programs, as they present small genomes and short life/flowering cycles. Because agaves are monocarpic with long life cycles (up to 50 years) and with high genetic complexity, breeding programs are scarce (Monja-Mio et al. 2019). Similar to *Eucalyptus* and other woody crops (Rezende et al. 2014), *Agave* cultivars are selected based on clonal lines that express desirable traits, such as precocity or higher sugar accumulation. As the tequila producers might never forget, the indiscriminate use of a single clonal line can be devastating when a new disease appears. In the late 90s, one-fifth of all agaves planted in Jalisco were decimated by two diseases that can act simultaneously on plants weakened by a severe frost in December of 1997 (Valenzuela-Zapata and Nabhan 2004; Jiménez-Hidalgo et al. 2004). To establish a new large-scale production, it is paramount for farmers to be able to count on a broader variety of species/genotypes. New species, such as *A. mapisaga* and *A. salmiana* (Nobel et al. 1992), must be explored and genetic diversity in existing lines and germplasms must be increased.

To modernize breeding programs and accomplish faster outcomes, the development of molecular markers and genetic engineering tools are fundamental. Molecular markers are specific previously identified DNA sequences that can be used to test whether an organism carries the desirable trait, for instance, high productivity and/or disease resistance. A few molecular markers have been used in *Agave* to evaluate genetic diversity or phylogenetic relationships (Monja-Mio et al. 2019).

However, their use for assisted selection in *Agave* breeding programs is still incipient. Similarly, genetic engineering strategies have been applied in *Agave*, but without any commercial applications (Flores-Benítez et al. 2007; Gao et al. 2014). Genetic engineering can achieve important outcomes in *Agave*, like herbicide resistance and/or pests protection.

Well-established and widespread tissue culture procedures will be a central step for the development of *Agave* as a global feedstock. *In vitro* micropropagation will enable large-scale production in small spaces of high-quality, standardized, and phytosanitary plants. Also, tissue culture is used for germplasm preservation, genetic engineering protocols, and is the most efficient strategy for the multiplication and dissemination of new cultivars (Monja-Mio et al. 2019; Flores-Benítez et al. 2007; Balch et al. 2012). There are many studies about *Agave* tissue culture and this crop is proven to be responsive to those techniques (Rodríguez-Garay 2016). Recently, propagation by temporary immersion systems is being addressed (Vázquez-Martínez et al. 2022; Monja-Mio et al. 2020), these propagation systems will significantly drop production costs and will allow truly large-scale production of micropropagated plants.

Another important aspect to address is agaveculture mechanization. In tequila plantations, planting and harvesting are done by manual labor, which is not only slower but can also be dangerous, because of agave's sharp thorns that can be harmful and even cause blindness. Taking this into account, a planter and a mower to trim the leaves have been developed by a tequila industry (Cortés 2018). Also, the only *Agave* chopper/harvester prototype has been developed and patented recently, and a preliminary analysis demonstrate that this harvester could save U$ 2277 per hectare (Vaca-Navarro et al. 2019). More research on developing proper machinery for agaveculture is needed, considering that different harvesting techniques could be used for different industries' pipelines (Corbin et al. 2015; Yan et al. 2020). *Agave* production worldwide is very labor-intensive, therefore, mechanized operations, especially harvesting, will be essential for industrializing its production.

Although there is much speculation in the literature about the potential of agaves for biorefineries, few effective initiatives have been taken to achieve this potential (Yan et al. 2020; Holtum et al. 2011; Davis et al. 2017). Among the main reasons for this scenario is the low availability of elite cultivars, agricultural technology, and especially solid biofuels policies. Recent studies on the life cycle assessment (LCA) of ethanol from *Agave tequilana* reported that, both in Australia and Mexico, *Agave* bioethanol seems to be more promising from an environmental perspective, but still need government support and renewable energy policies for its production to be economically viable (Yan et al. 2020; Parascanu et al. 2021). In this sense, the Brazilian experience with sugarcane and its new biofuels national policy (Renovabio) (Grassi and Pereira 2019) places the country in a position to become a leader in the production of *Agave* for bioenergy. Today, it is common for mills in northeastern Brazil to cease their operations due to a lack of sugarcane in consequence of drought (FCStone 2018; Vital 2021). By introducing elite *Agave* cultivars and minimal adaptation, these sugarcane mills could operate using *Agave* to extend their production season.

Approximately 40% of the world's land area is considered arid, semi-arid, or dry sub-humid, with precipitation amounts that are inadequate for conventional agriculture. These abundant and cheaper areas could be used to produce bioethanol if we use an appropriate crop such as some *Agave* species (Somerville et al. 2010; Owen and Griffiths 2014). Also, as these marginal ecosystems are unsuitable for food production, agaves can be used to slacken land competition, and, in the water-energy-food-environment nexus, to improve food and energy security (Yan et al. 2020; Parascanu et al. 2021). To establish this crop as a new feedstock in those areas, it is necessary to perform more field trials with different species/cultivars to better define recommended agricultural practices.

18.4 Conclusion

Throughout this chapter, we discussed two different monocot crops with great potential as bioethanol feedstocks. Global ethanol production is threatened with increasing drought events and higher temperatures caused by climate change. Energy cane and *Agave* appear as valid alternatives to complement and increase ethanol production worldwide. As energy cane presents moderate resistance to drought and sugarcane-like agronomical traits it could be used as a fast alternative for traditional sugarcane areas that are facing water shortage, therefore, avoiding irrigation. Also, this crop can be suitable for transition areas, in which the ethanol industry is expanding. For regions with less precipitation or extended drought periods, like areas threatened by desertification or with semiarid climates, *Agave* can be the best choice. This semiarid crop can yield as much as sugarcane with significantly less water. Furthermore, as agaves grow in areas unsuitable for most conventional crops, they can be used to slacken land competition and improve food and energy security. As with all new industrial crops, to establish energy cane and *Agave* as new feedstocks it is necessary to invest in agronomical practices, mechanization, mass propagation of phytosanitary plants, and characterization of their behavior in the field. However, by overcoming these challenges, the bioethanol industry could reach a new level and be able to supply the increasing demand for this biofuel in a hotter and dryer world.

Acknowledgments The authors would like to thank Don Chambers from AusAgave for sharing information and images. This work was supported in part by the *Coordenação de Aperfeiçoamento de Pessoal de Nível Superior*—Brazil (CAPES)—Finance Code 001, and *Fundação de Amparo à Pesquisa do Estado de São Paulo*—Brazil (FAPESP)—process number 2018/10315-2.

Declaration of Competing Interest The authors report no declarations of interest.

References

Alejandra GUJ, Liliana SZ, Othon SSSR (2013) Agave syrup extract having anticancer activity

Alexander AG (1985) The energy cane alternative. Elsevier, Amsterdam

Álvarez-Ríos GD, Pacheco-Torres F, Figueredo-Urbina CJ, Casas A (2020) Management, morphological and genetic diversity of domesticated agaves in Michoacán, México. J Ethnobiol Ethnomed 16:3. https://doi.org/10.1186/s13002-020-0353-9

Arrizon J, Morel S, Gschaedler A, Monsan P (2010) Comparison of the water-soluble carbohydrate composition and fructan structures of Agave tequilana plants of different ages. Food Chem 122: 123–130. https://doi.org/10.1016/j.foodchem.2010.02.028

Aude MIDS (1993) Estádios De Desenvolvimento Da Cana-De-Açúcar E Suas Relações Com A Produtividade. Ciência Rural 23:241–248. https://doi.org/10.1590/S0103-84781993000200022

Ávila-Fernández Á, Rendón-Poujol X, Olvera C et al (2009) Enzymatic hydrolysis of fructans in the tequila production process. J Agric Food Chem 57:5578–5585. https://doi.org/10.1021/jf900691r

Balch EPM, Araiza MJE, Reyes MEP (2012) Conservación in vitro de germoplasma de Agave spp. bajo condiciones de crecimiento retardado. Rev Fitotec Mex 35:279–287

Baum ME, Licht MA, Huber I, Archontoulis SV (2020) Impacts of climate change on the optimum planting date of different maize cultivars in the central US Corn Belt. Eur J Agron 119:126101. https://doi.org/10.1016/j.eja.2020.126101

Bhattarai MD, Secchi S, Schoof J (2017) Projecting corn and soybeans yields under climate change in a Corn Belt watershed. Agric Syst 152:90–99. https://doi.org/10.1016/j.agsy.2016.12.013

Blunden G, Yi Y, Jewers K (1973) The comparative leaf anatomy of Agave, Beschorneria, Doryanthes and Furcraea species (Agavaceae: Agaveae). Bot J Linn Soc 66:157–179. https://doi.org/10.1111/j.1095-8339.1973.tb02167.x

Boanares D, Ferreira BG, Kozovits AR et al (2018) Pectin and cellulose cell wall composition enables different strategies to leaf water uptake in plants from tropical fog mountain. Plant Physiol Biochem 122:57–64. https://doi.org/10.1016/j.plaphy.2017.11.005

Borland AM, Griffiths H, Hartwell J, Smith JAC (2009) Exploiting the potential of plants with crassulacean acid metabolism for bioenergy production on marginal lands. J Exp Bot 60:2879–2896. https://doi.org/10.1093/jxb/erp118

Boschiero BN, de Castro SGQ, da Rocha AEQ et al (2019) Biomass production and nutrient removal of energy cane genotypes in Northeastern Brazil. Crop Sci 59:379–391. https://doi.org/10.2135/cropsci2018.07.0458

Byrt CS, Grof CPL, Furbank RT (2011) C4 plants as biofuel feedstocks: optimising biomass production and feedstock quality from a lignocellulosic perspective free access. J Integr Plant Biol 53:120–135. https://doi.org/10.1111/j.1744-7909.2010.01023.x

Campos H, Trejo C, Peña-Valdivia CB et al (2014) Photosynthetic acclimation to drought stress in Agave salmiana Otto ex Salm-Dyck seedlings is largely dependent on thermal dissipation and enhanced electron flux to photosystem I. Photosynth Res 122:23–39. https://doi.org/10.1007/s11120-014-0008-6

Carocho M, Morales P, Ferreira ICFR (2017) Sweeteners as food additives in the XXI century: a review of what is known, and what is to come. Food Chem Toxicol 107:302–317. https://doi.org/10.1016/j.fct.2017.06.046

Carr MKV, Knox JW (2011) The water relations and irrigation requirements of sugar cane (Saccharum officinarum): a review the water relations and irrigation requirements of sugar cane (Saccharum officinarum): a review. Exp Agric 47:1–25. https://doi.org/10.1017/S0014479710000645

Carvalho-Netto OV, Bressiani JA, Soriano HL et al (2014) The potential of the energy cane as the main biomass crop for the cellulosic industry. Chem Biol Technol Agric 1:20. https://doi.org/10.1186/s40538-014-0020-2

Cheavegatti-Gianotto A, de Abreu HMC, Arruda P et al (2011) Sugarcane (Saccharum X officinarum): a reference study for the regulation of genetically modified cultivars in Brazil. Trop Plant Biol 4:62–89. https://doi.org/10.1007/s12042-011-9068-3

Colunga-García Marín P, May-Pat F (1997) Morphological variation of henequen (Agave fourcroydes, Agavaceae) germplasm and its wild ancestor (A. Angustifolia) under uniform growth conditions: diversity and domestication. Am J Bot 84:1449–1465. https://doi.org/10.2307/2446608

CONAB (2018) Companhia Nacional de Abastecimento. Acompanhamento da safra brasileira. Cana-de-açúcar - Terceiro levantamento, safra 2018/19, dezembro/2018. 73

Consejo Regulador del Tequila (2021) Producción total: Tequila y Tequila 100%. https://www.crt.org.mx/EstadisticasCRTweb/. Accessed 30 Sep 2021

Corbin KR, Byrt CS, Bauer S et al (2015) Prospecting for energy-rich renewable raw materials: agave leaf case study. PLoS One 10:e0135382. https://doi.org/10.1371/journal.pone.0135382

Cortés RR (2018) Automation process of agricultural work. https://www.casasauza.com/tequila-process/automation-process-agricultural-work. Accessed 3 Sep 2021

Cosgrove DJ (2005) Growth of the plant cell wall. Nat Rev Mol Cell Biol 6:850–861. https://doi.org/10.1038/nrm1746

Creste S, Pinto LR, Xavier MA, et al (2014) The importance of the germplasm in developing agro-energetic profile sugarcane cultivars. In: Sugarcane bioethanol – R & D for productivity and sustainability. Editora Edgard Blücher, pp 353–358

Cruz LP, Pacheco VS, Silva LM et al (2021) Morpho-physiological bases of biomass production by energy cane and sugarcane: a comparative study. Ind Crop Prod 171:113884. https://doi.org/10.1016/j.indcrop.2021.113884

Cunha CP, de Abreu LGF, Grassi MCB et al (2020) Metabolic regulation and development of energy cane setts upon auxin stimulus. Plant Cell Physiol 61:606–615. https://doi.org/10.1093/pcp/pcz229

da Silva AL, Castañeda-Ayarza JA (2021) Macro-environment analysis of the corn ethanol fuel development in Brazil. Renew Sust Energ Rev 135:110387. https://doi.org/10.1016/j.rser.2020.110387

da Silva GJ, Berg EC, Calijuri ML et al (2021) Aptitude of areas planned for sugarcane cultivation expansion in the state of São Paulo, Brazil: a study based on climate change effects. Agric Ecosyst Environ 305:107164. https://doi.org/10.1016/j.agee.2020.107164

Daher BT, Mohtar RH (2015) Water–energy–food (WEF) Nexus Tool 2.0: guiding integrative resource planning and decision-making. Water Int 40:748–771. https://doi.org/10.1080/02508060.2015.1074148

Davis SC, Long SP (2015) Sisal/Agave. In: Cruz VMV, Dierig DA (eds) Handbook of plant breeding. Springer, New York, NY, pp 335–349

Davis SC, Dohleman FG, Long SP (2011) The global potential for Agave as a biofuel feedstock. GCB Bioenergy 3:68–78. https://doi.org/10.1111/j.1757-1707.2010.01077.x

Davis SC, Kuzmick ER, Niechayev N, Hunsaker DJ (2017) Productivity and water use efficiency of Agave americana in the first field trial as bioenergy feedstock on arid lands. GCB Bioenergy 9: 314–325. https://doi.org/10.1111/gcbb.12324

de Abreu LGF, Grassi MCB, de Carvalho LM et al (2020) Energy cane vs sugarcane: watching the race in plant development. Ind Crop Prod 156:112868. https://doi.org/10.1016/j.indcrop.2020.112868

de Abreu LGF, Silva NV, Ferrari AJR et al (2021) Metabolite profiles of energy cane and sugarcane reveal different strategies during the axillary bud outgrowth. Plant Physiol Biochem 167:504–516. https://doi.org/10.1016/j.plaphy.2021.08.022

de Andrade Junior MAU, Valin H, Soterroni AC et al (2019) Exploring future scenarios of ethanol demand in Brazil and their land-use implications. Energy Policy 134:110958. https://doi.org/10.1016/j.enpol.2019.110958

de Siqueira FS, Nishiyama MY, Paterson AH, Souza GM (2013) Biofuel and energy crops: high-yield Saccharinae take center stage in the post-genomics era. Genome Biol 14:210. https://doi.org/10.1186/gb-2013-14-6-210

de Souza Barbosa GV, dos Santos JM, Diniz CA et al (2020) Energy cane breeding. In: Sugarcane biorefinery, technology and perspectives. Elsevier, pp 103–116

Dias MOS, Cunha MP, Jesus CDF et al (2011) Second generation ethanol in Brazil: Can it compete with electricity production? Bioresour Technol 102:8964–8971. https://doi.org/10.1016/j.biortech.2011.06.098

Distilled Spirits Council of the United States (2011) U.S. tequila market at a glance. Distill Spirits Counc United States

dos Santos LV, de Barros Grassi MC, Gallardo JCM et al (2016) Second-generation ethanol: the need is becoming a reality. Ind Biotechnol 12:40–57. https://doi.org/10.1089/ind.2015.0017

dos Santos DJ, Pedra GU, da Silva MGB et al (2020) Future rainfall and temperature changes in Brazil under global warming levels of 1.5°C, 2°C and 4°C. Sustentabilidade em Debate 11:57–90. https://doi.org/10.18472/SustDeb.v11n3.2020.33933

Dufek AS, Ambrizzi T (2008) Precipitation variability in São Paulo State, Brazil. Theor Appl Climatol 93:167–178. https://doi.org/10.1007/s00704-007-0348-7

Eguiarte LE, Jiménez Barrón OA, Aguirre-Planter E et al (2021) Evolutionary ecology of Agave: distribution patterns, phylogeny, and coevolution (an homage to Howard S. Gentry). Am J Bot 108:216–235. https://doi.org/10.1002/ajb2.1609

FAO (2020) Food and Agriculture Organization (2020) Statistical database, agriculture. faostat.fao.org

FCStone I (2018) Clima seco limita moagem de cana no Norte-Nordeste do Brasil, diz FCStone. NovaCana

Flores-Benítez S, Jiménez-Bremont JF, Rosales-Mendoza S et al (2007) Genetic transformation of Agave salmiana by Agrobacterium tumefaciens and particle bombardment. Plant Cell Tissue Organ Cult 91:215–224. https://doi.org/10.1007/s11240-007-9287-3

Gao J, Yang F, Zhang S et al (2014) Expression of a hevein-like gene in transgenic Agave hybrid No. 11648 enhances tolerance against zebra stripe disease. Plant Cell Tissue Organ Cult 119: 579–585. https://doi.org/10.1007/s11240-014-0557-6

Garcia-Moya E, Romero-Manzanares A, Nobel PS (2011) Highlights for Agave productivity. GCB Bioenergy 3:4–14. https://doi.org/10.1111/j.1757-1707.2010.01078.x

Gentry HS (1982) Agaves of continental North America. University of Arizona Press, Tucson

Giamalva MJ, Clarke SJ, Stein JM (1984) Sugarcane hybrids of biomass. Biomass 6:61–68. https://doi.org/10.1016/0144-4565(84)90008-8

Goldemberg J (2008) The Brazilian biofuels industry. Biotechnol Biofuels 1:6. https://doi.org/10.1186/1754-6834-1-6

Grassi MCB, Pereira GAG (2019) Energy-cane and RenovaBio: Brazilian vectors to boost the development of Biofuels. Ind Crop Prod 129:201–205. https://doi.org/10.1016/j.indcrop.2018.12.006

Gross S (2021) Australia takes on Mexico's tipple (just don't call it tequila). Bloomberg

Holtum JAM, Chambers D, Morgan T, Tan DKY (2011) Agave as a biofuel feedstock in Australia. GCB Bioenergy 3:58–67. https://doi.org/10.1111/j.1757-1707.2010.01083.x

Hunt ER, Zakir NJD, Nobel PS (1987) Water costs and water revenues for established and rain-induced roots of Agave deserti. Funct Ecol 1:125. https://doi.org/10.2307/2389715

IPCC (2014) Climate change 2014: synthesis report. Contribution of working groups I, II and III to the fifth assessment report of the intergovernmental panel on climate change. IPCC, Genebra

Isabel Enríquez-Salazar M, Veana F, Aguilar CN et al (2017) Microbial diversity and biochemical profile of aguamiel collected from Agave salmiana and A. atrovirens during different seasons of year. Food Sci Biotechnol 26:1003–1011. https://doi.org/10.1007/s10068-017-0141-z

Jean P (2015) Active ingredient with cutaneous application obtained from metschnikowia agaves and uses for improving the state of the skin

Jeong PE, Chul SW, Woo LD, et al (2020) Cosmetic composition comprising enzyme-treated extract of agave syrup as active ingredient

Jiménez-Hidalgo I, Virgen-Calleros G, Martínez-de la Vega O et al (2004) Identification and characterisation of bacteria causing soft-rot in Agave tequilana. Eur J Plant Pathol 110:317–331. https://doi.org/10.1023/B:EJPP.0000019791.81935.6d

Keeley JE, Rundel PW (2003) Evolution of CAM and C 4 carbon-concentrating mechanisms. Int J Plant Sci 164:S55–S77. https://doi.org/10.1086/374192

Kim M, Day DF (2011) Composition of sugar cane, energy cane, and sweet sorghum suitable for ethanol production at Louisiana sugar mills. J Ind Microbiol Biotechnol 38:803–807. https://doi.org/10.1007/s10295-010-0812-8

Klein-Marcuschamer D, Oleskowicz-Popiel P, Simmons BA, Blanch HW (2012) The challenge of enzyme cost in the production of lignocellulosic biofuels. Biotechnol Bioeng 109:1083–1087. https://doi.org/10.1002/bit.24370

Kogo BK, Kumar L, Koech R (2021) Climate change and variability in Kenya: a review of impacts on agriculture and food security. Environ Dev Sustain 23:23–43. https://doi.org/10.1007/s10668-020-00589-1

Kuo TM, Van Middlesworth JF, Wolf WJ (1988) Content of raffinose oligosaccharides and sucrose in various plant seeds. J Agric Food Chem 36:32–36. https://doi.org/10.1021/jf00079a008

Laborde Aguirre AE, Valencia Gallegos JA, Sergio HVJ, et al (2010) Proceso para preparar una mezcla termoplástica polimérica de fibras residuos de agave y aditivos oxo-biodegradativos para preparar artículos de plastico biodegradables. 37

Laborde Aguirre AE, Valencia Gallegos JA, Hernández Valdéz JS, et al (2013) Process for preparing a thermoplastic polymer mixture based on agave fibers and oxo-degradation additives for preparing biodegradable plastic articles. 15

LADB (1997) Mexico trade briefs: South African Tequila, China Chemical. Avocados, Sourcemex

Läderach P, Ramirez-Villegas J, Navarro-Racines C et al (2017) Climate change adaptation of coffee production in space and time. Clim Chang 141:47–62. https://doi.org/10.1007/s10584-016-1788-9

Lambers H, Chapin FS, Pons TL (2008) Plant physiological ecology. Springer, New York, NY

Landell MGDA, Scarpari MS, Xavier MA et al (2013) Residual biomass potential of commercial and pre-commercial sugarcane cultivars. Sci Agric 70:299–304. https://doi.org/10.1590/S0103-90162013000500003

Leal MRLV (2007) The potential of sugarcane as an energy source. Proc Int Soc Sugar Cane Technol 26:23

Leng G, Huang M (2017) Crop yield response to climate change varies with crop spatial distribution pattern. Sci Rep 7:1463. https://doi.org/10.1038/s41598-017-01599-2

Liu G-L, Fu G-Y, Chi Z, Chi Z-M (2014) Enhanced expression of the codon-optimized exo-inulinase gene from the yeast Meyerozyma guilliermondii in Saccharomyces sp. W0 and bioethanol production from inulin. Appl Microbiol Biotechnol 98:9129–9138. https://doi.org/10.1007/s00253-014-6079-7

Lledías F, Gutiérrez J, Martínez-Hernández A et al (2020) Mayahuelin, a Type I ribosome inactivating protein: characterization, evolution, and utilization in phylogenetic analyses of agave. Front Plant Sci 11:573. https://doi.org/10.3389/fpls.2020.00573

Long SP, Marshall-Colon A, Zhu X-G (2015a) Meeting the global food demand of the future by engineering crop photosynthesis and yield potential. Cell 161:56–66. https://doi.org/10.1016/j.cell.2015.03.019

Long SP, Karp A, Buckeridge MS et al (2015b) Feedstocks for biofuels and bioenergy. In: Bioenergy and sustainability: bridging the gaps. SCOPE, Paris, pp 302–347

Machado Filho G, Nascimento I, Sakai T et al (2018) Brix analysis and green corn cob productivy by nitrogen fertilization time. Appl Res Agrotechnol 11:33–41

Mancilla-Margalli NA, López MG (2006) Water-soluble carbohydrates and fructan structure patterns from agave and dasylirion species. J Agric Food Chem 54:7832–7839. https://doi.org/10.1021/jf060354v

Marin FR, Rattalino Edreira JI, Andrade JF, Grassini P (2021) Sugarcane yield and yield components as affected by harvest time. Sugar Tech 23:819–826. https://doi.org/10.1007/s12355-020-00945-5

Matsuoka S, Ferro J, Arruda P (2009) The Brazilian experience of sugarcane ethanol industry. Vitr Cell Dev Biol Plant 45:372–381. https://doi.org/10.1007/s11627-009-9220-z

Matsuoka S, Kennedy AJ, dos Santos EGD et al (2014) Energy cane: its concept, development, characteristics, and prospects. Adv Bot 2014:1–13. https://doi.org/10.1155/2014/597275

Medina JC (1954) O Sisal. Secretaria da Agricultura do Estado de São Paulo

Michel-Cuello C, Juárez-Flores BI, Aguirre-Rivera JR, Pinos-Rodríguez JM (2008) Quantitative characterization of nonstructural carbohydrates of mezcal Agave (Agave salmiana Otto ex Salm-Dick). J Agric Food Chem 56:5753–5757. https://doi.org/10.1021/jf800158p

Monja-Mio KM, Herrera-Alamillo MA, Sánchez-Teyer LF, Robert ML (2019) Breeding strategies to improve production of agave (Agave spp.). In: Advances in plant breeding strategies: industrial and food crops. Springer International Publishing, Cham, pp 319–362

Monja-Mio KM, Olvera-Casanova D, Herrera-Herrera G et al (2020) Improving of rooting and ex vitro acclimatization phase of Agave tequilana by temporary immersion system (BioMINT™). Vitr Cell Dev Biol Plant 56:662–669. https://doi.org/10.1007/s11627-020-10109-5

Moore FC, Baldos U, Hertel T, Diaz D (2017) New science of climate change impacts on agriculture implies higher social cost of carbon. Nat Commun 8:1607. https://doi.org/10.1038/s41467-017-01792-x

Morán JI, Alvarez VA, Cyras VP, Vázquez A (2008) Extraction of cellulose and preparation of nanocellulose from sisal fibers. Cellulose 15:149–159. https://doi.org/10.1007/s10570-007-9145-9

Morgan TJ, Turn SQ, Sun N, George A (2016) Fast pyrolysis of tropical biomass species and influence of water pretreatment on product distributions. PLoS One 11:e0151368. https://doi.org/10.1371/journal.pone.0151368

Mumm RH, Goldsmith PD, Rausch KD, Stein HH (2014) Land usage attributed to corn ethanol production in the United States: sensitivity to technological advances in corn grain yield, ethanol conversion, and co-product utilization. Biotechnol Biofuels 7:61. https://doi.org/10.1186/1754-6834-7-61

Nishizawa A, Yabuta Y, Shigeoka S (2008) Galactinol and raffinose constitute a novel function to protect plants from oxidative damage. Plant Physiol 147:1251–1263. https://doi.org/10.1104/pp.108.122465

Nobel PS (1988) Environmental biology of agaves and cacti. Cambridge University Press

Nobel PS (1994) Remarkable agaves and cacti. Oxford University Press

Nobel PS (2010) Desert wisdom/agaves and cacti: CO2, water, climate change. iUniverse, New York, NY

Nobel PS, Garcia-Moya E, Quero E (1992) High annual productivity of certain agaves and cacti under cultivation. Plant Cell Environ 15:329–335. https://doi.org/10.1111/j.1365-3040.1992.tb00981.x

Novak JM, Sigua GC, Ducey TF et al (2019) Designer biochars impact on corn grain yields, biomass production, and fertility properties of a highly-weathered ultisol. Environments 6:64. https://doi.org/10.3390/environments6060064

Novel PS, De Cortazar VG (1987) Interception of photosynthetically active radiation and predicted productivity for Agave Rosettes. Photosynthesis 21:261–272

OECD/FAO (2021) OECD-FAO agricultural outlook 2021–2030. OECD

Ortiz-Cano H, Hernandez-Herrera JA, Hansen NC et al (2020) Pre-columbian rock mulching as a strategy for modern agave cultivation in arid marginal lands. Front Agron 2. https://doi.org/10.3389/fagro.2020.00010

Owen NA, Griffiths H (2014) Marginal land bioethanol yield potential of four crassulacean acid metabolism candidates (Agave fourcroydes, Agave salmiana, Agave tequilana and Opuntia ficus-indica) in Australia. GCB Bioenergy 6:687–703. https://doi.org/10.1111/gcbb.12094

Owen NA, Fahy KF, Griffiths H (2016) Crassulacean acid metabolism (CAM) offers sustainable bioenergy production and resilience to climate change. GCB Bioenergy 8:737–749. https://doi.org/10.1111/gcbb.12272

Parascanu MM, Sanchez N, Sandoval-Salas F et al (2021) Environmental and economic analysis of bioethanol production from sugarcane molasses and agave juice. Environ Sci Pollut Res. https://doi.org/10.1007/s11356-021-15471-4

Pérez-López AV, Simpson J (2020) The sweet taste of adapting to the desert: fructan metabolism in agave species. Front Plant Sci 11. https://doi.org/10.3389/fpls.2020.00324

Pérez-Pimienta JA, López-Ortega MG, Sanchez A (2017) Recent developments in Agave performance as a drought-tolerant biofuel feedstock: agronomics, characterization, and biorefining. Biofuels Bioprod Biorefin 11:732–748. https://doi.org/10.1002/bbb.1776

Radding C (2012) The children of mayahuel: * agaves, human cultures, and desert landscapes in Northern Mexico. Environ Hist Durh N C 17:84–115. https://doi.org/10.1093/envhis/emr118

Ragauskas AJ, Beckham GT, Biddy MJ et al (2014) Lignin valorization: improving lignin processing in the biorefinery. Science (80-) 344:1246843. https://doi.org/10.1126/science.1246843

Raya FT, Marone MP, Carvalho LM et al (2021) Extreme physiology: biomass and transcriptional profiling of three abandoned Agave cultivars. Ind Crop Prod 172:114043. https://doi.org/10.1016/j.indcrop.2021.114043

Reboita MS, Gan MA, da Rocha RP, Ambrizzi T (2010) Regimes de precipitação na América do Sul: uma revisão bibliográfica. Rev Bras Meteorol 25:185–204. https://doi.org/10.1590/S0102-77862010000200004

Rezende GDSP, de Resende MD V, de Assis TF (2014) Eucalyptus breeding for clonal forestry, pp 393–424

Rodríguez-Garay B (2016) Somatic embryogenesis in Agave spp. In: Loyola-Vargas VM, Ochoa-Alejo N (eds) Somatic embryogenesis: fundamental aspects and applications. Springer International Publishing, Cham, pp 267–282

Rodríguez-Garay B, Lomelí-Sención JA, Tapia-Campos E et al (2009) Morphological and molecular diversity of Agave tequilana Weber var. Azul and Agave angustifolia Haw. var. Lineño. Ind Crops Prod 29:220–228. https://doi.org/10.1016/j.indcrop.2008.05.007

Sabea H (2008) Mastering the landscape? Sisal plantations, land, and labor in Tanga Region, 1893-1980s. Int J Afr Hist Stud 41:22

Salinas C, Handford M, Pauly M et al (2016) Structural modifications of fructans in Aloe barbadensis Miller (Aloe Vera) grown under water stress. PLoS One 11:e0159819. https://doi.org/10.1371/journal.pone.0159819

Silva ORRF, Beltrão NEDM (1999) O agronegócio do sisal no Brasil. embrapa, Brasília, DF

Silva JPR, Reboita MS, Escobar GCJ (2019) Caracterização da Zona de Convergência do Atlântico Sul em Campos Atmosféricos recentes. Rev Bras Climatol 25. https://doi.org/10.5380/abclima.v25i0.64101

Silva PD, Cruz R, Casal S (2021) Sugars and artificial sweeteners in soft drinks: a decade of evolution in Portugal. Food Control 120:107481. https://doi.org/10.1016/j.foodcont.2020.107481

Singh AK, Pérez-López AV, Simpson J, Castro-Camus E (2020) Three-dimensional water mapping of succulent Agave victoriae-reginae leaves by terahertz imaging. Sci Rep 10:1404. https://doi.org/10.1038/s41598-020-58277-z

Smith GF (2017) Producing organic alcohol and a tequila-like liquor from Agave americana L. (Asparagaceae subfam. Agavoideae/Agavaceae) at Graaff-Reinet in the Eastern Cape Province of South Africa: challenges to establish an industry based on a naturalised, alien succ. Bradleya 35:15–32. https://doi.org/10.25223/brad.n35.2017.a3

Somerville C, Youngs H, Taylor C et al (2010) Feedstocks for lignocellulosic biofuels. Science (80-) 329:790–792. https://doi.org/10.1126/science.1189268

Subedi R, Delwar Akbar D, Nanjappa Ashwath N et al (2017) Assessing the viability of growing agave tequilana as a biofuel feedstock in Queensland, Australia. Int J Energy Econ Policy 7: 172–180

Swaminathan MS, Kesavan PC (2012) Agricultural research in an era of climate change. Agric Res 1:3–11. https://doi.org/10.1007/s40003-011-0009-z

Tetreault D, McCulligh C, Lucio C (2021) Distilling agro-extractivism: Agave and tequila production in Mexico. J Agrar Chang 21:219–241. https://doi.org/10.1111/joac.12402

Tew TL, Cobill RM (2008) Genetic improvement of sugarcane (Saccharum spp.) as an energy crop. In: Genetic improvement of bioenergy crops. Springer, New York, NY, pp 273–294

Trebicki P (2020) Climate change and plant virus epidemiology. Virus Res 286:198059. https://doi.org/10.1016/j.virusres.2020.198059

UNESCO (2020) Botanical Garden (Orto Botanico), Padua. In: World Herit. List. https://whc.unesco.org/en/list/824/

Vaca-Navarro D, Gutiérrez-Vaca C, Rucoba-García A et al (2019) Estudio de viabilidad económica y comercial para el prototipo de una cosechadora troceadora de agave. Rev Tecnol en Marcha. https://doi.org/10.18845/tm.v32i7.4254

Valenzuela-Zapata AG, Nabhan GP (2004) ¡Tequila!: a natural and cultural history. University of Arizona Press, Tucson

Van den Ende W (2013) Multifunctional fructans and raffinose family oligosaccharides. Front Plant Sci 4:247. https://doi.org/10.3389/fpls.2013.00247

Vázquez-Martínez O, Núñez-Palenius HG, Balch EMP-M et al (2022) In vitro-propagation of Agave tequilana Weber cv. azul in a Temporary Immersion System. Phyton (B Aires) 91:83–96. https://doi.org/10.32604/phyton.2022.017281

Viator RP, Richard EP (2012) Sugar and energy cane date of planting effects on cane, sucrose, and fiber yields. Biomass Bioenergy 40:82–85. https://doi.org/10.1016/j.biombioe.2012.02.002

Vital A (2021) Falta de matéria-prima reduzirá produção de açúcar na safra 2021/22. J. da Cana

Waclawovsky AJ, Sato PM, Lembke CG et al (2010) Sugarcane for bioenergy production: an assessment of yield and regulation of sucrose content. Plant Biotechnol J 8:263–276. https://doi.org/10.1111/j.1467-7652.2009.00491.x

Winter K, Smith J (1996) Crassulacean acid metabolism: biochemistry, ecophysiology and evolution

Yan X, Corbin KR, Burton RA, Tan DKY (2020) Agave: a promising feedstock for biofuels in the water-energy-food-environment (WEFE) nexus. J Clean Prod 261:121283. https://doi.org/10.1016/j.jclepro.2020.121283

Yang L, Lu M, Carl S et al (2015a) Biomass characterization of Agave and Opuntia as potential biofuel feedstocks. Biomass Bioenergy 76:43–53. https://doi.org/10.1016/j.biombioe.2015.03.004

Yang X, Cushman JC, Borland AM et al (2015b) A roadmap for research on crassulacean acid metabolism (CAM) to enhance sustainable food and bioenergy production in a hotter, drier world. New Phytol 207:491–504. https://doi.org/10.1111/nph.13393

Chapter 19
The New Biorefineries: Integration with New Technologies for Carbon Capture and Utilization to Produce Bioethanol

Marilene Pavan

Abstract There is a pressing need to move from the traditional linear economy, where production and consumption patterns lead to large amounts of single-use waste, to a circular economy, prioritizing environmental, social, and economic welfare. Specifically, for the bioethanol industry, process integration with new technologies for carbon capture and utilization (CCU) represents a step towards an increase in carbon utilization efficiency and revenues while tackling significant greenhouse gases emissions. Microbial gas fermentation, demonstrated at a commercial scale, offers a feasible alternative to increasing bioethanol volume production via the utilization of low-value biogenic carbon waste without threatening food, land, and water. Locally, these biotechnology routes translate in job creation, advancement of rural areas, value creation through scientific developments, and a leadership position in an emergent climate economy. New scientific breakthroughs bring the potential to address challenges commonly associated with CCU such as the demand for high amounts of energy to transform the CO_2 molecule into high-density fuels. Finally, a solid and comprehensive policy framework to fund and credit CCU must be considered the main driver for the maturing and large-scale adoption and acceptance of these technologies.

19.1 Introduction

The recent IPCC report of 2021 reinforces that the continuing growth in CO_2 concentrations in the atmosphere is due to emissions from human activities (Sixth Assessment Report 2021). More specifically, the total anthropogenic greenhouse gas (GHG) emissions have continued to increase and CO_2 remains the primary anthropogenic GHG, accounting for 76% of emissions (AR5 Synthesis Report 2021). Furthermore, CO_2 from fossil fuel combustion and industrial processes contributed to about 78% of the total GHG emission increase from 1970 to 2010 (AR5 Synthesis

M. Pavan (✉)
LanzaTech Inc, Skokie, IL, USA
e-mail: marilene.pavan@lanzatench.com

C. R. Soccol et al. (eds.), *Liquid Biofuels: Bioethanol*, Biofuel and Biorefinery Technologies 12, https://doi.org/10.1007/978-3-031-01241-9_19

Report 2021). The positive net balance in emissions can be demonstrated by the fact that while the emissions from the energy, buildings, industry, transport, agriculture, and land usage account for the emission of 39 Gt CO_2/year, the land and ocean sinks account for the removal of 21 GtCO_2/year, leaving a current net atmospheric growth of approximately 18 GtCO_2/year (Hepburn et al. 2019). In addition, the fraction of emissions removed from the atmosphere by natural sinks decreases with higher CO_2 concentrations (Sixth Assessment Report 2021).

According to the Paris Agreement, the risks and impacts of climate change consequences could be contained under the circumstances of "holding the increase in the global average temperature to well below 2 °C above pre-industrial levels and pursuing efforts to limit the temperature increase to 1.5 °C above pre-industrial levels" (Nations U 2015). Therefore, it is essential to understand the remaining carbon budget and its relationship with increasing temperatures above these recommended limits. A total of 2390 GtCO_2 of anthropogenic CO_2 was emitted between 1850 and 2019 (Sixth Assessment Report 2021). The remaining carbon budgets (starting from 1 January 2020) for limiting warming to 1.5 °C, 1.7 °C, and 2.0 °C are estimated at 500 GtCO_2, 850 GtCO_2, and 1350 GtCO_2, respectively (Sixth Assessment Report 2021). If a yearly CO_2 emission between 18 and 42 GtCO_2 (Sixth Assessment Report 2021; Hepburn et al. 2019) is considered, the planet has an alarmingly low remaining budget before global surface air temperature increases beyond targeted limits (Sixth Assessment Report 2021; AR5 Synthesis Report 2021; Millar et al. 2017; Kriegler et al. 2018).

The severity of this panorama clearly indicates how important each carbon molecule we produce and consume should be considered. To reduce GHG emissions to acceptable levels, a suite of carbon CCU technologies must be considered, funded, developed, and commercially deployed, and no carbon waste should be tolerated. Only ambitious goals and deployments of net-zero and negative emissions technologies in a timely manner can deliver results that will keep us in the range of the Paris Agreement goals (Sixth Assessment Report 2021; AR5 Synthesis Report 2021). In fact, this portfolio approach is even more relevant given that different geographic locations and economic and cultural realities make it impossible to adopt a unique technology to solve all problems successfully (Zhike and Shasha 2021; Ajwani-Ramchandani et al. 2021). In addition, it seems unrealistic to replace all fossil-based value chains entirely with carbon-free alternatives, as many materials and chemical products will continue to depend on carbon, either because it provides the necessary energy density (fuels) or because, by definition, organic chemistry products contain carbon (chemical industry) (Gabrielli et al. 2020). In this sense, we prefer to adopt the term defossilization, as in (Gabrielli et al. 2020), instead of decarbonization of the production chain of materials, chemicals, and fuels.

One of the most successful molecules produced from renewable sources is ethanol. Bioethanol is a versatile molecule that can be used as fuel and as a building block for various chemicals and materials (Dagle et al. 2020; Posada et al. 2013), and, compared to gasoline, it can reduce GHG emissions by up to 83% (Pereira et al. 2019). Furthermore, it is readily available, can be safely transported and stored for long periods, and the chemistry required to convert this feedstock into several chemicals is already known (Dagle et al. 2020; Posada et al. 2013). Moreover, advances in bioethanol production efficiency and feedstock diversification may lead to an even wider geographic adoption and market penetration at competitive prices (Formann et al. 2020).

Though less pollutant (Pereira et al. 2019), there is still room for improvement in bioethanol production due to significant carbon losses during the production process. According to (de Souza Noel Simas Barbosa et al. 2017), in a regular Brazilian sugarcane mill for ethanol production, only 17% of the carbon, approximately, goes to the bioethanol molecule, while the rest mainly remains in the field (36%) or is exhausted via biomass combustion (32%) and sugarcane fermentation (9%). This under-utilized carbon could be captured and transformed into valuable bioproducts, avoiding mining more fossil carbon and contributing to energy security.

The gas fermentation biological pathway is one attractive, commercially proven way to maximize ethanol production via waste carbon utilization (Teixeira et al. 2018; Heffernan et al. 2020; Liew et al. 2016). Microorganisms, such as anaerobic acetogens, have the ability to fix gaseous oxides from a range of different feedstocks including agricultural waste, without the need for multiple pre-treatments and hydrolysis steps. These acetogens offer the potential to convert virtually all carbon available in the feedstock into valuable products giving gas fermentation technology a key advantage over sugar and cellulosic fermentation (Liew et al. 2016). Sources of carbon residues in a bioethanol plant can come from sugar fermentation, co-generation plant combustion, and plant straw and bagasse residues. Biomass can be gasified to syngas, a mixture of mainly CO, CO_2, and H_2, which, in the optimum composition, serve as a carbon source and energy for the conversion of the feedstock carbon in its entirety (Liew et al. 2016; Köpke and Simpson 2020; De Tissera et al. 2019). Additional reducing agents to increase process efficiency may come from the electrolysis of CO_2 to produce CO or water to produce H_2 via renewable energy. Water electrolysis to produce H_2, supplied with renewable energy, is a carbon-neutral, robust technology to produce a valuable energy carrier as H_2. Though a key energy vector, hydrogen still poses challenges for storage and transportation. Hence, it is attractive to use this molecule to store energy in biofuels, as bioethanol, which is easier to store, transport, and have already the necessary infrastructure for distribution and utilization.

Thus, this chapter reviews the microbial gas fermentation technology, the different technologies available for carbon capture, and their integration with enabling technologies as H_2 production fueled by renewable power, leading to an integrated, economically circular bioethanol refinery.

19.2 New Technologies for Carbon Utilization: Microbial Gas Fermentation

While the concept of a circular economy may not provide permanent CO_2 removal from the atmosphere, the production of recycled carbon-based fuels, chemicals, and materials makes sense to reduce emissions via waste CO_2 capture and utilization that displaces fossil fuel mining and usage (Hepburn et al. 2019). It also can integrate sectors like waste management, energy, and research and development, create new markets, jobs, and investments, and attribute value to carbon waste streams (Kircher 2021). Furthermore, the creation of valuable products from waste carbon streams brings the opportunity to potentially offset the cost of capture and extra energy generation that might be needed in the process (Hepburn et al. 2019). In this scenario, the CCU route to defossilization offers an exciting opportunity to substitute high-energy reduced fossil-based carbon feedstocks with residual carbon oxides, such as the oxidized low energy-carbon in the CO_2 molecule, provided renewable production of H_2 is supplied to reduce the CO_2. Cheap carbon feedstocks can be sourced from industrial off-gases, agricultural and municipal solid waste (MSW) gasification, or the pure residual CO_2 released by the sugar fermentation process (Gabrielli et al. 2020) (Fig. 19.1). Additionally, several pathways are proven technically feasible today for electrical reduction and carbon dioxide utilization (Hepburn et al. 2019; Köpke and Simpson 2020; Mikkelsen et al. 2010; Bushuyev et al. 2018; Liu et al. 2020). These pathways mainly differ in technology maturity level and potential for carbon removal (Hepburn et al. 2019).

Industrial gas fermentation of waste carbon oxides by carbon-fixing chemolithoautotrophic bacteria such as anaerobic acetogens, offers a commercially available solution to increase the productivity of sustainable fuels without competing with food or land (Köpke and Simpson 2020; Liew et al. 2016) while avoiding new carbon mining (Fig. 19.1). Key advantages of microbial gas fermentation, compared with first generation (1G) sugar fermentation and second generation (2G) cellulosic fermentation ethanol production, include the utilization of the biogenic carbon in its entirety via gaseous carbon capture and waste biomass gasification, the increase of chemical production without threats to food, land, and water, and the lower cost of the feedstock (Liu et al. 2020). Furthermore, compared with the Fischer-Tropsh process, gas fermentation offers the advantage of occurring at low temperature (37 °C) and atmospheric pressure, offering significant energy and cost savings (Liew et al. 2016). In addition, the high enzymatic specificity of biological conversions of anaerobic acetogens such as *Clostridium autoethanogenum* enables the utilization of a broad range of different gas mixtures while keeping the energy efficiency of around 80% provided H_2 is present (Hu et al. 2013). The microorganism provides resilience to tolerate gas streams with fluctuating gas compositions and contaminants (Köpke and Simpson 2020). Finally, higher product selectivity, use of diverse carbon oxide waste streams as feedstock (industrial off-gases, gasified agricultural and urban waste, among others), product flexibility, and high conversion efficiencies contribute to its economic viability.

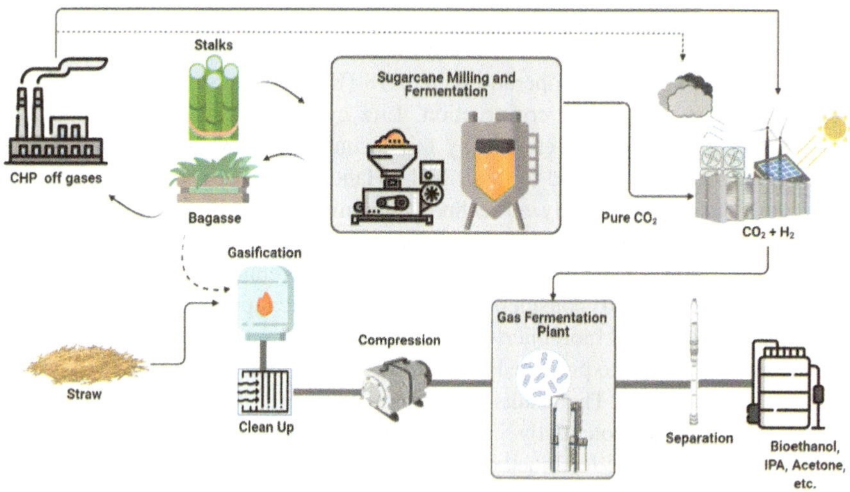

Fig. 19.1 Integrated Biorefinery. Sugar fermentation offers a pure CO_2 stream that can be captured without a purification step and, associated with carbon-free energy, can be fixed in high-energy fuels through microbial cell factories. Straw and bagasse residues (dashed line) can also be gasified, producing syngas to feed the gas fermentation plant. Still, CO_2 can also be captured directly from the atmosphere or the co-generation heat and power plant, providing extra feedstock to produce bioethanol. Optimal gas composition is achieved by CO_2 and/or H_2O electrolysis to produce CO and $H_{2,}$ respectively. Additionally, the configuration of a gas fermentation plant can be easily modified to accept different microorganisms producing different end-products, providing a high level of flexibility to a single facility (Köpke and Simpson 2020). This figure was designed via Biorender.com

Acetogens utilize the Wood-Ljungdahl pathway, one of the few carbon fixation pathways able to reduce CO_2 directly. Using gas as their only carbon and energy source, acetogens convert CO or $CO_2 + H_2$ into Acetyl-CoA (Liew et al. 2016; Ragsdale and Pierce 2008; Valgepea et al. 2018). In this metabolic process, electron donors might include CO (carbon monoxide), which can serve as both carbon and energy sources, and H_2, which can be produced via renewable powered water electrolysis, for example (Claassens et al. 2018). H_2 supplies the necessary reducing equivalents in the form of electrons to increase the carbon efficiency of the overall process when CO_2 is present (Köpke and Simpson 2020). Another approach is the electrochemical reduction of CO_2 (CO_2 electrolysis) to produce CO (Haas et al. 2018). However, though this is a simple reduction reaction, only requiring 2 protons and electrons, this technology is not as mature as water hydrolysis (Somoza-Tornos et al. 2021; Roh et al. 2020). This combination of hydrogen produced by electrolysis of water with renewable electricity and CO_2 captured from concentrated sources or the air is usually described in the literature as electrofuels (E-fuels) or Power-to-X (de Vasconcelos and Lavoie 2019; Fackler et al. 2021). These technologies can convert renewable electricity to chemicals and fuels that can be more easily stored and transported, involving minimum impact on GHG emissions.

The microbial production of ethanol on a large scale via carbon oxides fermentation is a technology that has already been commercialized by LanzaTech Inc., which the first plant started operations in 2018 (Köpke and Simpson 2020), with more plants currently under construction. Life cycle assessments of sustainable aviation fuel produced from ethanol by this technology showed up to 90% GHG emissions reduction compared to fossil fuels (Handler et al. 2016).

Recently, two studies analyzed techno-economic (Huang et al. 2020) and environmental (Lee et al. 2021) impacts of the bioprocess integration of a microbial gas fermentation facility (Daniell et al. 2012) to a corn biorefinery in the US. The first study (Huang et al. 2020) considers a hybrid-bioelectrical process to convert waste CO_2 streams within the biorefinery into ethanol. The framework considered combines water electrolysis to produce H_2, electrolysis to produce CO from CO_2, and gas fermentation to ethanol. The results show that with onsite CO_2 conversion, the total ethanol yield can be potentially improved by 45%, in agreement with previous carbon balance analyses for bioethanol production (de Souza Noel Simas Barbosa et al. 2017; Corrêa do Lago et al. 2012). Furthermore, the main parameters analyzed that would lead to the cost feasibility of the process are (1) high CO_2 electrolysis energy efficiency (above 70% theoretical), (2) high CO_2 electrolysis conversion efficiency (above 50% CO_2 input), and (3) low electricity cost (below 0.02/kWh). Moreover, additional results show that the feedstock cost is the main contributor in the 1G process, while utilities (due to the electrolysis steps) and CO_2 separation are the main cost drivers in the integrated workflow.

The second analysis (Lee et al. 2021) focuses on the integrated biorefinery's potential GHG emissions (carbon intensity, CI), as CCU technologies are usually energy-intensive. Accordingly, the results show that the sources of electricity and hydrogen are the critical drivers of cell-based CO_2-to-EtOH GHG emissions. While with wind electricity, the designed scenario shows near-zero CI ethanol, it can increase up to 531 gCO_2e/MJ if today's US Midwest electricity mix is used.

Both studies provide a comprehensive and necessary techno-economic and life-cycle analysis of an integrated 1G corn ethanol biorefinery with a gas fermentation facility supported by carbon capture processes and CO_2 and H_2 electrolysis. The analysis shows that the conversion of waste CO_2 into additional ethanol via renewable energy plus microbial gas fermentation would significantly increase biorefinery production, lead to a near-zero carbon emission ethanol, and has the potential to be cost-effective, increasing the biorefinery revenues.

The following sessions will explore the enabling technologies to commission highly efficient, integrated 1G and gas fermentation biorefineries.

19.3 Carbon Capture

Several technologies and materials are available to capture gaseous CO_2 from different sources, with different compositions, and with different concentrations (Nanda et al. 2016; Mondal et al. 2012; Wang and Song 2020). These include mainly physical and chemical absorption, adsorption, cryogenic, and membrane processes. Also, depending on the production process and ultimate CO_2 capture goal, these technologies might be applied to different configurations such as pre, post-combustion capture, oxyfuel combustion, or direct air capture (DAC) (Mondal et al. 2012; Wang and Song 2020).

The first part of this section will explore the different configurations that power, industrial, and biorefinery plants may adopt to capture CO_2 efficiently. The different materials and technologies that can be used for CO_2 separation and their applicability to different carbon capture frameworks are summarized in Table 19.1. Finally, we will also review the approaches for biogenic carbon capture in bioethanol refineries, which could lead to a decrease in the carbon intensity of their processes.

In general, the efficiency (and costs associated) of these technologies mainly depends on the CO_2 concentration in the source gas, as the CO_2 purification step is considered the costliest part of the process. Other parameters are also evaluated, such as the technology state-of-art, maturity level, and main advantages and limitations of each configuration.

19.3.1 Introduction to Carbon Capture Technologies

19.3.1.1 Pre-combustion

In the pre-combustion process, O_2 is purified from the air, and substoichiometric amounts are used to partially oxidate the fuel in a gasifier at elevated pressures (typically in the range of 30–70 atm) (Fig. 19.2). The partial oxidation of the fuel results in syngas (CO and H_2) (Mondal et al. 2012). Steam is then added to the syngas, leading to a Water Gas Shift (WGS) reaction that shifts CO and water into CO_2 and more H_2. The high concentrated (15–50%) CO_2 in this high-pressured gas mixture is then captured from the exhaust stream before reaching the combustion chamber (hence the process name), and the remaining H_2 is used as an input to a combined cycle, for example, to produce electricity (Mondal et al. 2012; Pre-Combustion Carbon Capture Research 2021). Adopting a pre-combustion configuration should consider the required addition of a chemical plant to the original plant, which translates into higher capital costs (Mondal et al. 2012).

Capturing CO_2 from a pre-combustion process brings some advantages. Due to the high concentration of CO_2 in the resulting gas mixture, low-cost materials, as physical solvents, can be used to capture this CO_2 with an efficiency of around 95% (Porter et al. 2017; Pre-Combustion CO2 Capture 2021; Bagnato and Sanna 2018; Theo et al. 2016), and lower energy levels will be required for its regeneration. Moreover, the overall process will be more affordable than post-combustion technologies, for example, where CO_2 is more diluted (5–15%) (Mondal et al. 2012; Pre-Combustion Carbon Capture Research 2021). Today, commercially available pre-combustion carbon capture technologies will cost around $60/tonne, considering

Table 19.1 Processes and materials used in CO_2 capture technologies. Based on Nanda et al. (2016), Mondal et al. (2012), Wang and Song (2020), Markewitz et al. (2012), Kearns et al. (2021), Samanta et al. (2011), Song et al. (2019))

Technology	Suitable for	Examples of process and materials	Pros	Cons
Chemical absorption	Pre-combustion, oxycombustion, post-combustion, DAC	Amine solutions such as monoethanol amine (MEA), diethanol amine (DEA), N-methyldiethanolamine (MDEA), and di-2-propanolamine (DIPA), alkali compounds	Mature and reliable technology, suitable for low-concentrated CO_2 streams	Regeneration cost of the solvent, solvent degradation, corrosion
Physical absorption	Pre-combustion, oxyfuel combustion	Selecsol, Rectisol, Purisol	Mature and reliable technology	Energy intensive processes
Adsorption	Pre-combustion, post-combustion, oxyfuel combustion, DAC	Solid adsorbents include activated carbon, zeolites, and molecular sieves and regenerative methods include pressure swing adsorption (PSA), temperature swing adsorption (TSA), and electrical swing adsorption (ESA).	No waste water, lower energy requirement (compared to cryogenic and chemical absorption)	Low selectivity and capacity, lower removal efficiency (compared to absorption and cryogenic)
Cryogenic	Pre-combustion, oxyfuel combustion	Compression and cooling of the gas mixtures in several stages to induce phase changes of CO_2 in flue gases	Produces liquid CO_2, no chemical absorbent is required, process occur at atmospheric pressure	Expensive, other gases in the mixture interfere with cooling and cause corrosion, feasible only under specific circumstances
Membranes	Pre-combustion, oxyfuel combustion	Gas separation membrane, gas absorption membrane	Lower energy requirements	Low removal efficiency (low capacity), its feasibility is decreased when concentration of CO_2 in feed stream is below 20%

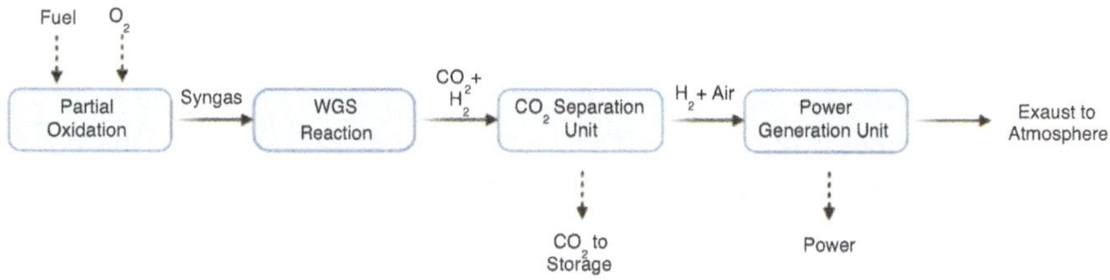

Fig. 19.2 Pre-combustion process. Adapted from (Mondal et al. 2012)

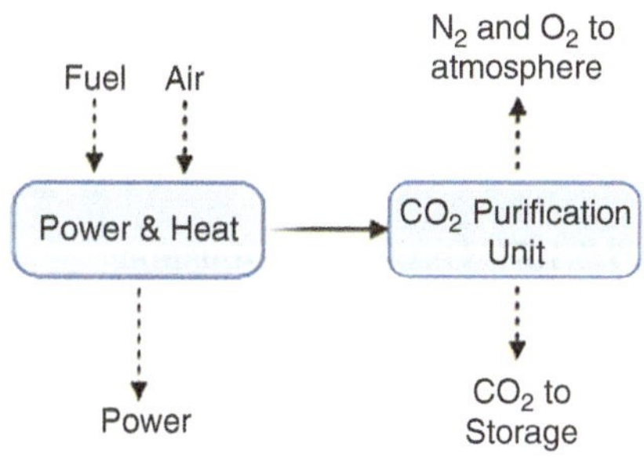

Fig. 19.3 Post-combustion CO_2 capture process

the capture of CO_2 generated by an integrated gasification combined cycle (IGCC) power plant (Pre-Combustion Carbon Capture Research 2021).

In a bioethanol refinery operating with a pre-combustion carbon capture process, waste biomass could be gasified to produce hydrogen for electricity and syngas for gas fermentation. A recent study (Agrawal and Rao 2018) shows that the capture of CO_2 produced from biomass gasification, using MEA based absorption, has the potential to capture approximately 90% of the original feedstock carbon, which could result in up to 111.5 kg of CO_2 captured per GJ of energy produced, representing an excellent opportunity for generating extra carbon feedstock while curbing large amounts of GHG emissions from bioethanol production.

19.3.1.2 Post-combustion

In post-combustion applications, the flue gas resulting from air-fired combustion passes through a washing column where the solvents absorb CO_2. The CO_2-rich solvent is then heated, releasing high-purity CO_2, then reused in the absorber (Bagnato and Sanna 2018) (Fig. 19.3). The technology is mature and commercially available. One of its advantages is that existing power and industrial plants can be retrofitted to include post-combustion carbon capture technology in their configuration with minimal structural changes.

CO_2 separation methods based on chemical solvents such as monoethanolamine (MEA) and diethanolamine (DEA) are the most mature and widely adapted for post-combustion capture (Bagnato and Sanna 2018; Energy Agency I Technology Perspectives Energy Special Report 2020). Using MEA, up to 98% CO_2 removal with a purity of over 99% has been achieved (Bagnato and Sanna 2018). However, problems with high rates of corrosion and degradation caused by amines need to be addressed.

Another significant obstacle for large-scale technology adoption is the power output reduction due to the energy-intensive CO_2 separation process (Mokhtar et al. 2012). This is due to the low concentration of CO_2 in the gas stream and the presence of larger quantities of impurities, meaning that a large volume of gas needs to be handled, resulting in large equipment sizes and high capital costs (Mondal et al. 2012). It is estimated, for example, that a 90% capture of CO_2 from post-combustion flue gas by amine absorption will consume about 30% of the power generated by a power plant, at the cost of 40–100$ per ton of CO_2 captured (Chao et al. 2021), which might be economically unfeasible for some scenarios.

19.3.1.3 Oxyfuel Combustion

Oxyfuel combustion is an approach to producing energy that uses O_2 purified from the air combined with fuel to produce the steam that moves the turbines and produces electricity. In this process, the fuel is combined with pure O_2 obtained from the air after removing N_2 and other elements via an air separation unit (ASU) (Fig. 19.4). The reaction of fuel and pure O_2 results in CO_2 and H_2O. In this case, without the presence of N_2 in the input gas, part of the CO_2-rich flue gas is recycled back to the boiler to keep temperatures under safe limits. After the water is separated from the flue gas mixture, highly pure CO_2 can then be captured through a cryogenic purification unit (CPU), for example (Perrin et al. 2014). In the oxy-combustion with carbon capture system, the energy penalty comes mainly from the ASU, which produces the high volumes of required O_2 at the expense of 190 kWh/tO_2 (Perrin et al. 2014; Bailera et al. 2017; Hu et al. 2013). In this specific case, thermal integration between the air separation unit and the steam cycle could alleviate the energy penalty of the integrated system (Perrin et al. 2014; Bailera et al. 2017).

If used, this CO_2 can be a valuable feedstock to produce hydrocarbons (as ethanol, for example) when H_2 is provided. The hydrogen can be produced via electrolysis, where electrical energy is stored in H_2, producing O_2 as a by-product. In

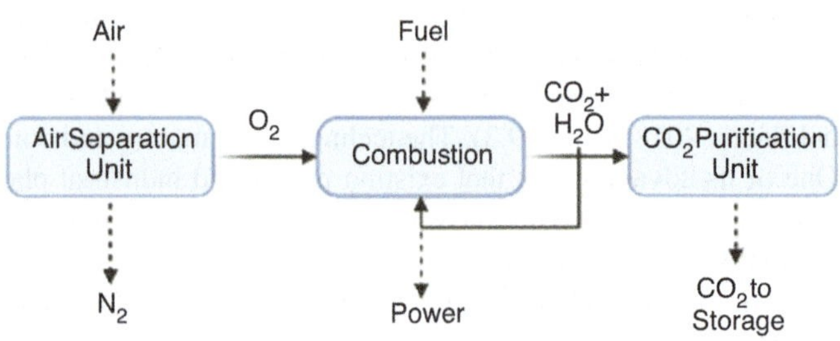

Fig. 19.4 CO_2 capture via oxy-combustion. Adapted from (Mondal et al. 2012; CO2 capture technologies: oxy combustion with CO2 capture - Global CCS Institute 2021).

this integrated framework, not only the H_2 could be used as a source of energy to reduce CO_2, but the by-produced O_2 could then replace the ASU (and its energy penalty) partially or entirely (depending on the electrolyzer throughput, oxy-combustion plant size, and oxy-combustion fuel and efficiency) (Bailera et al. 2017; Xiong et al. 2011; Ahn and Kim 2020). The integration of oxy-combustion with carbon capture for a biomass power plant has been modeled before (Bailera et al. 2017). In this case, due to biomass's lower LHV (low heat value) efficiency, the total substitution of the ASU would be prohibitive. According to (Bailera et al. 2017), a biomass power plant would require 160 alkaline electrolyzers to generate enough by-product O_2 to substitute the ASU entirely. Still, partial ASU energy penalty support should not be discarded.

19.3.1.4 Direct Air Capture (DAC)

Technologies that can contribute to carbon sequestration, as Direct Air Capture (DAC), can lead to relevant CO_2 mitigation provided the rate of DAC reaches the gigaton scale/year over the next decades (Beuttler et al. 2019). More specifically, DAC refers to a range of technological solutions that can extract CO_2 from ambient air—present today in the atmosphere at an average concentration of 405.5 parts per million—at virtually any location on the planet (Beuttler et al. 2019). This CO_2 can be used to produce valuable chemicals and fuels for sectors that are hard to decarbonize and electrify, such as aviation. In this scenario, DAC can contribute to closing the carbon cycle and reducing CO_2 emissions if low-carbon energy is employed to approach carbon neutrality over its entire life cycle (Deutz and Bardow 2021; Direct Air Capture – Analysis – IEA 2021).

The benefits of direct CO_2 capture include its limited land and water footprint (so it is not constrained due to resource limitations) and the possibility of locating plants close to the storage or utilization sites, eliminating the need for long-distance CO_2 transport. The choice of location should also consider the energy source needed to run the plant, determining whether the system is carbon-neutral, and the energy costs (Direct Air Capture – Analysis – IEA 2021). Although carbon removal technologies such as DAC are not an excuse for delayed action, they can be an important part of the suite of technology options used to achieve climate goals (Direct Air Capture – Analysis – IEA 2021).

There are two DAC technologies at a commercial scale today: absorption, which uses liquid solvents, and adsorption, using solid sorbents. DAC based on absorption typically uses aqueous hydroxy sorbents like alkali and alkali-earth hydroxides that, in contact with CO_2 (step 1: carbon capture), produce water and carbonate. Then, temperatures of 900 °C release the CO_2 (step 2: regeneration), so the pure CO_2 can be utilized as a feedstock, and the solvent can be reused (Deutz and Bardow 2021; Shayegh et al. 2021). By contrast, DAC based on adsorption can employ a range of solid sorbents, for example, alkali carbonates, amines supported on oxides, solid organic materials, and metal-organic frameworks (Deutz and Bardow 2021). In contrast to absorption, DAC by adsorption can operate at low regeneration temperatures (<100 °C), and it allows for more flexibility as its modular design can be scaled up to more easily (Deutz and Bardow 2021; Direct Air Capture – Analysis – IEA 2021; Shayegh et al. 2021).

Both government and private industries have been investing in DAC technologies. The Department of Energy of the US (DOE) recently announced an investment of $12 million in federal funding to support projects with the goals of creating tools that will increase the amount of CO_2 captured by DAC, decrease the cost of materials, and improve the energy efficiency of carbon removal operations (DOE Announces $12 Million for Direct Air Capture Technology | Department of Energy 2021). In the private sector, the leading players include Carbon Engineering, Global Thermostat, and ClimeWorks (Gambhir and Tavoni 2019). The latter is a Swiss company that launched in September 2021 the world's first large-scale direct air capture plant, in Iceland. According to ClimeWorks, the plant can capture 4000 tons of CO_2/year (Climeworks Latest Direct Air Capture Plant 2021) via modular CO_2 collectors that can be stacked and are powered solely by renewable energy or energy-from-waste. In the first DAC step, the carbon dioxide is captured on an adsorption/desorption process, at ambient conditions, on the surface of a highly selective filter material (alkaline-functionalized adsorbents) that sits inside the collectors. Then, after the filter material is full of carbon dioxide, the collector is closed. In this case, the CO_2 capture is combined with underground carbon storage provided by another company called Carbfix (We Turn CO2 into Stone 2021). In the second DAC step, desorption (regeneration) is performed through a temperature vacuum swing. The temperature is increased to ~100 °C, releasing high-purity, high-concentration carbon dioxide that can be collected (Beuttler et al. 2019; Deutz and Bardow 2021; How Direct Air Capture Helps Reverse Climate Change | Climeworks 2021). According to the company, the process results in a commercial carbon capture efficiency between 85.4% and 93.1% (Deutz and Bardow 2021) and the production of gaseous CO_2 at 1 bar with a purity level of >99.8%. Independent of the DAC technology, the regeneration process requires a significant amount of energy, and long-term energy requirement projections by ClimeWorks based on current technology assumptions for the DAC process are expected at around 2000 kWh per ton of CO_2 (400 kWh electrical and 1600 kWh thermal (Beuttler et al. 2019; Deutz and Bardow 2021; How Direct Air Capture Helps Reverse Climate Change | Climeworks 2021)). Therefore, the carbon footprint of captured CO_2 is mainly dependent on the electricity supply. Considering the ratio of avoided CO_2 emissions from DAC cradle-to-gate LCA analysis to CO_2 captured, it reaches almost 100% for wind power. In this case, GHG emissions rely mainly on constructing the DAC plant and the adsorbent production, which reduces the carbon capture efficiency by 0.6% and 2.4%, respectively (Deutz and Bardow 2021). Nevertheless, in general, the whole process would have a 90% efficiency.

The CO_2 in the atmosphere is much more dilute than industrial concentrated streams, contributing to the higher energy needs and costs for DAC relative to other CO_2 capture technologies and applications. Costs and energy needs may also vary according to the type of technology (solid or liquid) and whether the captured CO_2 will be geologically stored or used immediately at low pressure. CO_2 needs to be compressed at very high pressure to be injected into geological formations. This step increases both the capital and operating costs of the plant (Direct Air Capture – Analysis – IEA 2021). The costs of direct air capture have recently been assessed to be between \$500 and \$600 per tonne of CO_2 (Shayegh et al. 2021), with the potential to achieve \$200 per tonne of CO_2 in the following decades (Hepburn et al. 2019; Shayegh et al. 2021). In fact, according to more optimistic projections, costs per ton of CO_2 permanently removed from the atmosphere and safely stored via mineralization is calculated to come down to around \$100 per ton of CO_2 within a decade (Beuttler et al. 2019), which is considered the threshold for the DAC technology feasibility (Lackner and Azarabadi 2021; National Academies of Sciences E and M 2018), opening up early markets for DAC CO_2. It is important to note that independent DAC studies and projections show high uncertainties about cost estimation, though (Shayegh et al. 2021).

19.3.2 Biogenic Carbon Capture

Biogenic CO_2 emissions are related to those from the natural carbon cycle and those resulting from the combustion of biomass, biogas upgrading to methane, and industrial fermentation processes (Biogenic Carbon - Science and Climate 2021; Rodin et al. 2020). Biogenic carbon removal pathways and capture technologies are expected to play a key role in the transition to a net-zero energy system in which the amount of CO_2 released into the atmosphere is equivalent to the amount being removed. Options for biogenic CO_2 removal include nature-based solutions (e.g., afforestation, reforestation, restoration of coastal and marine habitats), measures to enhance naturally occurring processes (e.g., land management approaches to increase the carbon content in the soil, biochar) (Direct Air Capture – Analysis – IEA 2021), as well as the capture of waste CO_2 from agricultural and bioenergy processes (Naims 2016).

An important source of biogenic CO_2 is the bioethanol industry. Biomass is an attractive carbon feedstock compared to fossil fuels as it is considered a carbon-neutral energy source. This assumption comes from the fact that emissions from biomass energy conversion will be later offset by photosynthesis (Liu et al. 2016).

In a regular bioethanol plant, waste biogenic CO_2 originates mainly from three different sources: biomass combustion, waste biomass in the form of bagasse and straw, and CO_2 released from fermentation (de Souza Noel Simas Barbosa et al. 2017). Part of the feedstock carbon is also wasted through vinasse production (CH_4) (Formann et al. 2020). Though biomass is considered a carbon-neutral bioenergy resource, the waste CO_2 resulting from the bioethanol production process could be captured and utilized to produce even more ethanol, bringing extra revenues to the biorefinery without competing with land, forest, and food and reducing the overall carbon intensity of the process (Köpke and Simpson 2020; KircherManfred 2014). Furthermore, the higher availability of bioethanol for fuel and chemicals brings the potential to displace large amounts of fossil fuel-based products, which can be translated into extra revenues depending on local carbon offset policies.

19.3.2.1 CO_2 from Sugar Fermentation

The CO_2 stream from sugar fermentation is unique as it is virtually pure, so it does not require the costly CO_2 purification step commonly associated with carbon capture technologies, and its capture can be easily adapted to an existing refinery (Sanchez et al. 2018). Cost estimates for CO_2 capture and compression from fermentation are typically around \$30/tCO_2, among the lowest of all CO_2 point sources (Sanchez et al. 2018). In comparison, the ranges of estimates of capture and compression costs for large-scale industrial processes that emit dilute combustion gases such as steel/iron industries and coal-fired power plants are \$100 and 100–120 \$/tCO_2, respectively (Sanchez et al. 2018; Wheatcrd 2014). In Europe, the carbon capture cost of fermentation streams is estimated at 10 Euros/tCO_2 compared to 75 Euros/tCO_2 of aluminum production, where CO_2 is less concentrated and mixed with other gases (Naims 2016).

For each liter of ethanol produced, 0.76 kg of CO_2 is released to the atmosphere due to fermentation. According to (Sanchez et al. 2018), the 216 existing biorefineries in the US emit 45 MtCO_2 annually from fermentation, of which 60% could be captured and compressed for pipeline transport. The process of capturing this CO_2 is already implemented commercially in the US. For instance, the Illinois Industrial CCS project in Decatur, Illinois, captures 1 MtCO_2/y using dehydration and compression from a corn ethanol facility (Finley 2014). In Europe, the potential to capture fermentation CO_2 is estimated at 4.4–5.71 Mt. CO_2/year (Rodin et al. 2020). In 2019 sugarcane ethanol production in Brazil reached 36 billion liters, which might be translated to the emissions of 27.4 MtCO_2 in 1 year (Análise De Conjuntura Dos Biocombustíveis 2019).

19.3.2.2 Biomass Gasification

The syngas produced via gasification, the thermochemical conversion of solid or liquid organic materials, have a high heating power and can be used for energy generation (Molino et al. 2016). The gas can then be converted to bioethanol, a final product with a higher economic value than electricity (Corrêa do Lago et al. 2012). The biomass gasification process produces carbon-neutral heat, power, and hydrogen, important feedstock to produce advanced biofuels. Suppose this biomass comes from agricultural residues to produce renewable power. In that case, it allows for the carbon on biomass to be used in almost its entirety, and the extra sustainable energy production will not compete with land or food, put a higher value on and help manage the otherwise unvalued waste, and will contribute to displacing fossil fuels that significantly contribute to global warming.

Table 19.2 Main components of sugarcane bagasse and syngas derived from sugarcane bagasse gasification

Component % (% w/w, dry basis) of Brazilian Sugarcane	Range from Literature (based on (Canilha et al. 2012))
Cellulose	38.4–45.5
Hemicellulose	22.7–27.0
Lignin	19.1–32.4
Ash	1–2.8
Extractives	4.6–9.1
Component % (ash-free, sugarcane bagasse)	**Based on (Widjaya et al.** 2018; **Jordan and Akay** 2012)
C	46.62
H	6.45
O	45.66
N	1.21
S	0.03
Syngas Composition % (sugarcane bagasse, downdraft gasification)	**Based on (Sulaiman et al.** 2013)
CO	24.77
H_2	28.25
CO_2	44.89
CH_4	2.09

Gasification is an endothermic reaction, and it represents partial thermal oxidation of the carbon in the feedstock via a gasifying carrier, such as air, oxygen, steam, or carbon dioxide (Puig-Arnavat et al. 2010; Molino et al. 2016). The typical conversion efficiencies of lignocellulosic biomass are higher than 50% (Puig-Arnavat et al. 2010; Widjaya et al. 2018). The syngas produced is a gas mixture of carbon monoxide (CO), hydrogen (H_2), methane (CH_4), and carbon dioxide (CO_2), as well as light hydrocarbons, such as ethane and propane, and heavier hydrocarbons, such as tars. Contaminants such as sulphidric (H_2S), chloridric acid (HCl), and nitrogen (N_2) can also be present in the gas composition, and, hence, a cleaning step might be necessary. The presence of such contaminants depends on the biomass type treated and on the operational conditions of the gasification process. An exothermic reaction provides the energy necessary to carry out the gasification process by the oxidation of part of the biomass. This energy is used in the subsequent endothermic phases, such as biomass drying, feedstock pyrolysis (devolatilization), and reduction (Puig-Arnavat et al. 2010; Molino et al. 2016).

A pre-treatment step (size reduction, densification, torrefaction, among others) might also be necessary to suit the biomass feedstock to the gasification systems (Widjaya et al. 2018). The main components of sugarcane biomass and its gasification product are described in Table 19.2.

The waste biomass produced during agricultural processes is an important carbon source. Each 1000 kg of sugarcane stalks, for example, produces 65 kg of inexpensive carbon as crop residues (straw) which are usually left in the field to protect soil

(Corrêa do Lago et al. 2012). However, only 50% of the straw is necessary to protect the field. Therefore, the remaining 50% could be gasified, and the resulting carbon in the gas mixture could be converted to additional bioethanol (Liakakou et al. 2021). The gasification of biomass-based feedstock and consequent conversion to synthetic gas (syngas) is shown to be technically feasible through at least three different gasification methodologies (Liakakou et al. 2021).

In a regular sugarcane mill, heat and power are produced via sugarcane waste bagasse combustion. The annual Brazilian sugarcane production generates approximately 84 Mt. bagasse as agricultural residue (Corrêa do Lago et al. 2012), which is burned to produce heat and electricity. Part of this electricity supplies the biorefinery, and the excess is sold to the grid (Osaki and Seleghim 2017). However, the net efficiency for heat and electricity generation from biomass combustion is usually low, ranging from 20% to 40% (Caputo et al. 2005; Puig-Arnavat et al. 2010), while gasification heat conversion can reach 30-70% (Widjaya et al. 2018; Heidenreich and Foscolo 2015). Considering a gasification efficiency of 50% and an energy content of 16 MJ/kg for non-woody biomass (Widjaya et al. 2018), the potential for energy generation from residual Brazilian sugarcane production alone would be 1,344,000,000 GJ/year. The CO_2 released from the bagasse combustion process could also be captured and utilized. However, this post-combustion CO_2 stream is more diluted, and it contains several contaminants that might affect the CO_2 capture and separation performance, consequently making it cost-prohibitive today (Sanchez et al. 2018).

While syngas composition is not largely affected by the biomass composition, its LHV may vary depending on the moisture, gasification technology, and operating variables. Moreover, the amount of syngas may range in 1–3 Nm^3/kg on a dry basis, with an LHV spanning over 4–15 MJ/Nm^3 (Molino et al. 2016).

Gasifying technologies differ in terms of mode of contact between the feedstock and the gasifying agent, the mode and rate of heat transfer, and the residence time of the feed material into the reaction zone (Molino et al. 2016). The different gasification technologies are thoroughly reviewed in (Puig-Arnavat et al. 2010; Widjaya et al. 2018; Molino et al. 2016). Table 19.3 (based on (Puig-Arnavat et al. 2010; Molino et al. 2016)) brings the main characteristics and how these technologies compare specifically for biomass gasification.

Simulation results produced by (de Medeiros et al. 2017) have shown the feasibility of producing ethanol from sugarcane biomass residues via gasification and microbial gas fermentation. Achieving energy self-sufficiency is possible with an ethanol yield of 0.33 m^3 per metric ton of dry sugarcane bagasse, considering a production plant with an annual capacity of 71,000 m^3. Furthermore, the financial analysis predicted the MESP to be 706 US$/$m^3$ and, though this value is higher than the current market price of hydrous ethanol (currently at US$ 634.25, (Etanol - Centro de Estudos Avançados em Economia Aplicada - CEPEA-Esalq/USP 2021)), it demonstrates the potential for competitiveness in relation to other biomass conversion technologies.

Table 19.3 Technologies for biomass gasification

Technology	Main advantages	Main disadvantages	Cost	Biomass gasification requirements	Biomass gasification gas quality
Entrained flow reactor	High carbon conversion, there are no problems of scale-up	Large oxidant requirements, Requires the reduction of size and preparation supply	Moderate capital and maintenance costs	Requires specific features of the feed fuel in terms of particle size, moisture content, and a constant composition	Ensure a very low tar content with better quality syngas than a fluidized and fixed bed reactor
Fixed bed reactor (updraft)	Moderate to high carbon conversion efficiency, the high thermal efficiency can handle materials with high humidity, robust technology	Limited scale-up potential, high tar, and ash content	Limited capital and maintenance costs	Fluidized and fixed bed reactors are more flexible than an entrained flow reactor operating with larger particle sizes and a wider range of moisture contents; however, fixed bed reactor rarely is capable of using raw feedstock	Low syngas quality
Fluidized bed reactor (Bubbling fluidized bed)	High carbon conversion, can handle materials with different characteristics, suitable for highly reactive fuels such as biomass and municipal waste pre-treated	Pre-treatments need with heterogeneous materials, restrictions on the size	High investment costs and maintenance costs	The high mineral content of non-woody biomass can create problems in fluidized bed reactors	Syngas from a fluidized bed gasifier has better quality than the syngas produced by a fixed bed reactor
Rotary kiln	Good efficiency, there are problems of scale-up	Low capacity of heat exchange, low-efficiency heat	Reduced investment costs, high maintenance costs	Biomass feedstock with light or without pre-treatment can be supplied to the reactor	Syngas from a rotary kiln reactor has the worst quality
Plasma technology	Extremely short reaction times, content minimal polluting compounds in syngas,	Low efficiency, non-continuous process, the necessity of auxiliary fuel for obtaining a	High capital and maintenance costs	Biomass feedstock with light or without pre-treatment can be supplied to the reactor; however, the technology is	Provides the best syngas quality

(continued)

Table 19.3 (continued)

Technology	Main advantages	Main disadvantages	Cost	Biomass gasification requirements	Biomass gasification gas quality
	there are no problems of scale-up	homogeneous temperature inside the reactor		mainly focused on municipal waste, with little experience with biomass	

19.4 H_2 Production via Renewable Power

19.4.1 Importance of H_2 on Microbial Gas Fermentation

High carbon and energy efficiency is critical to driving an economically sustainable circular economy based on waste CO_2 utilization. H_2, with a low heating value of 119.9 MJ/kg, is a key energy vector for producing chemicals and fuels from CO_2 (Rego de Vasconcelos and Lavoie 2019). As discussed before, the supply of H_2 to acetogens, such as *Clostridium autoethanogenum*, would address energy limitations of anaerobic acetogens metabolism and, consequently, decrease the carbon loss during the microbial catalysis process (Köpke and Simpson 2020; Valgepea et al. 2018). In fact, Valgepea et al. (2018) demonstrated that supplying H_2 to CO-limited *Clostridium autoethanogenum* culture significantly impacts carbon distribution with a fourfold reduction in carbon loss as CO_2 (61% vs. 17%) and a proportional increase of flux to ethanol (15% vs. 61%). In addition, H_2 supplementation lowered the molar acetate/ethanol ratio by fivefold, providing important insights on gas composition for efficient microbial gas fermentation towards ethanol production.

These results show that H_2 availability indeed provided reducing power via H_2 oxidation and saved redox as cells reduced all the CO_2 supplied to formate directly using H_2 in the metabolic Wood–Ljungdahl pathway. In fact, stoichiometrically, when only CO is present in the gas stream, two thirds of the carbon are lost in form of CO_2 (Molitor et al. 2016; Drake et al. 2008). In contrast, at an H_2:CO ratio of 2 and H_2:CO_2 ratio of 3, theoretically all the carbon in the gases might be converted to ethanol, though at a higher thermodynamic cost (Köpke and Simpson 2020; Valgepea et al. 2018; Wilkins and Atiyeh 2011).

Anaerobic acetogens show high energetic efficiency for the substrate to product conversion, between 70% and 90%. In addition, bacterial growth also displays high electron consumption rate which can reach 100 μmol electrons per second per gram dry cell weight (Claassens et al. 2019), which makes the system robust to intermittent renewable electricity (Huang et al. 2020; Claassens et al. 2019; Haas et al. 2018) to produce the necessary reducing power. Therefore, economically viability of CO_2 utilization will heavily depend on the costs of the renewable energy sources.

19.4.2 H_2 Production

H_2 is an energy carrier not naturally available and hence needs to be synthesized. The most used technologies to produce H_2 today utilize fossil fuels as the primary source for its production. Concerns related to CO_2 emissions make obvious the need to look for renewable, carbon-neutral pathways to produce H_2, such as photovoltaic panels and wind turbines. Today, H_2 is mainly produced from fossil fuels, via steam reforming. Renewable solutions include biomass gasification, water electrolysis

Table 19.4 Characteristics of the main technologies to produce H_2 via water electrolysis

	Alkaline electrolysis	PEM	SOE
Maturity level	Mature technology, large scale	Medium Scale	Research stage/small scale
Electrolyte	Aqueous KOH solution	Proton exchange membranes	Zirconia-based materials
H_2 gas quality (%)	99.5–99.9	Greater than 99.99	–
Partial reaction (Cathode)	$2H_2O + 2e^- \rightarrow H_2 + 2OH^+$	$2H+ 2e^- \rightarrow H_2$	$H_2O + 2e^- \rightarrow H_2 + O^{2-}$
Partial reaction (Anode)	$2OH^- \rightarrow ½ O_2 + H_2O + 2e^-$	$H_2O \rightarrow ½ O_2 + 2H + 2e^-$	–
Energy consumption (stack—kWh/Nm^3)	4.2–4.8	4.4–5.0	3.0
Efficiency (stack—$\%_{LHV}$)	63–71	60–68	100
Efficiency (kWh of electricity/kg of H_2)	51	58	n/a
Lifetime (stack, operating hours)	80,000	40,000	n/a
H^2 Production per stack (Nm^3/h)	1400	400	<10
Investment costs (€/kW)	750–1500	1200–2100	>2000
Maintenance costs (% of investment costs per year)	2–3	2–5	n/a
Dynamic operation	Difficult to adapt to variable renewable energy sources	Provides a better adaptation with intermittent energy systems	–

("green hydrogen") and fossil-based steam reforming with carbon storage ("blue hydrogen").

Water electrolysis is considered one of the most promising technologies to produce carbon-neutral H_2 and, hence, to enable a sustainable energy system. Today, the main water electrolysis technologies are alkaline (ALK), Polymer Electrolyte Membrane (PEM), and solid oxide electrolysis (SOE) (Rego de Vasconcelos and Lavoie 2019; Buttler and Spliethoff 2018). Each technology's main characteristics are summarized in Table 19.4 (based on (Rego de Vasconcelos and Lavoie 2019; Buttler and Spliethoff 2018; Study on Development of Water Electrolysis in the EU Final Report 2014; IRENA 2018)).

ALK electrolyzers are a well-consolidated technology with a longer lifetime. In contrast, PEM electrolyzers can operate more flexibly than ALK technology, making them more suitable for intermittent energy systems, and they are rapidly entering commercial deployment. SOE technology holds the potential for greater efficiencies compared to ALK and PEM, but it is a less mature technology, only demonstrated at a small scale, and demands higher investment costs (Table 19.4). Nonetheless, the system provides the attractive possibility to produce syngas via the co-electrolysis of CO_2 and H_2O (Küngas 2020).

The high overall energy demand of water electrolysis reactions varies between 283.5 and 291.6 kJ/mol H_2, which is supplied by heat and electricity (Buttler and Spliethoff 2018). Hence, the technology is most likely to achieve cost-effectiveness—and consequent broader adoption and scalability—through high electrolyzer utilization rates, electrolyzer flexibility and efficiency, and low-cost renewable electricity (Gabrielli et al. 2020; IRENA 2018).

19.4.3 Renewable Energy Sources (RES)

The production of bioethanol today relies mainly on the heterotrophic microbial conversion of plant-based biomass via fermentation. Although a carbon-neutral approach, the overall solar-to-bioethanol energy conversion from plant biomass is very low, where the final product contains only 0.18–0.20% of the available solar energy (Conrado et al. 2013; Claassens et al. 2016). This is mainly due to the inherent inefficient carbon fixation via photosynthesis (with a maximum conversion efficiency of solar energy to the biomass of 4.6–6%) and important carbon losses during the conversion processes (de Souza Noel Simas Barbosa et al. 2017; Zhu et al. 2008). Hence, these characteristics combined represent challenges in scalability and underutilization of land resources, leading to competition with other cultures for land, water, and nutrients. Additional, more efficient solutions to address these limitations are needed to increase energy conversion and carbon utilization from plant-based biomass.

In contrast to photosynthesis, photovoltaic (PV) panels have an energy conversion efficiency of around 20% (Gürtürk et al. 2018; Most Efficient Solar Panels 2021; Leger et al. 2021). Furthermore, renewable energy costs are steadily declining year after year. Between 2010 and 2020 the cost of electricity from PV fell 85%; today, solar PV cost of electricity is US$ 0.057/kWh in the US (Renewable Power Generation Costs 2020). In fact, most existing US coal plants have currently higher costs than solar PV and onshore wind, due to the very competitive costs for those two technologies (Renewable Power Generation Costs 2020). The GHG emissions related to PV-based hydrogen production, around 2–3 kg CO_2 eq./kg H_2, are mainly due to emissions related to module manufacturing process (Bhandari et al. 2014).

With one of the lowest GHG emissions rate and competitive prices, wind-powered electricity is also an attractive energy source for H_2 production. The cost of electricity for new offshore wind projects is currently quoted at USD 0.084/kWh, while onshore wind projects it stands at USD 0.039/kWh (Renewable Power Generation Costs 2020). The Global Warming Potential (GWP) for wind-based electrolysis is very low at 0.97 kg CO_2 eq./kg H_2. These emissions come mainly from wind turbine production and operation and hydrogen compression and storage in a smaller scale. In contrast, hydrogen generation via steam methane reforming of natural gas varies from 8.9 to 12.9 kg CO_2 eq./kg H_2.

The accentuated decrease of renewable power generation costs in the last decade are driven mainly by technologies improvements, scalability, and developed supply chains (Renewable Power Generation Costs 2020). Finally, with remarkable learning rates, solar and wind-based technologies for energy generation have also provided invaluable insights on technology deployment and scalability (Renewable Power Generation Costs 2020).

19.5 Support to the Technological Deployment of Advanced CCU Technologies

19.5.1 Policies

The deployment of comprehensive policies and robust regulatory frameworks is paramount to support the appropriate scale-up of CO_2 capture and utilization technologies. This is especially important for technologies that generate no revenue or other obvious market benefits (A policy strategy for carbon capture and storage – analysis – IEA 2021; Andrés González-Garay et al. 2019). The incentives and instruments might include the creation of carbon credits (and trading systems) per ton of CO_2 avoided incentivizing verifiable CO_2 emissions reductions and removals from the atmosphere (Grassi and Pereira 2019; EU Emissions Trading System (EU ETS) 2021), carbon removal obligations (Bednar et al. 2021), the creation of penalties for emissions (Norway, for example, currently has a carbon tax of between USD 18/tCO_2 and USD 70/tCO_2), cap and trade systems, the cost-sharing between public and private sectors in the riskier technology early stages, and investment in operations and infrastructure to deploy new technologies at scale (Shayegh et al. 2021; A policy strategy for carbon capture and storage – analysis – IEA 2021; Org and Kennedy 2019).

One example of a tax credit is the 45Q—credit for carbon oxide sequestration (Carbon Capture Coalition 2021). Today, by this system, a DAC plant of a 1 MtCO_2 size, for example, would be eligible for the 45Q tax credit in the US, which would provide USD 35 per ton of CO_2 used in enhanced oil recovery and USD 50 per ton for CO_2 storage (Direct Air Capture – Analysis – IEA 2021). Another example is the California Low Carbon Fuel credit, which applies specifically to the production of low-carbon transportation fuels. These credits were being traded at around USD 180/t CO_2 in 2019 (Direct Air Capture – Analysis – IEA 2021).

In December 2017, the Brazilian government launched the National Biofuels Policy, named RenovaBio (ANP - Agencia Nacional do Petroleo RenovaBio 2020). This policy package combines the commitment assumed at COP21 with the need to implement mechanisms to increase the price stability of biofuels and thus their economic attractiveness (Grassi and Pereira 2019). The program is based on defined targets to decrease the carbon intensity (CI) of the fuel matrix. To achieve these targets, the fuel distributors will have to acquire bonds (equivalent to carbon credits) called CBIOS on the stock exchange market (Grassi and Pereira 2019). Each CBIO corresponds to one tonne of CO_2 no longer emitted and is generated from biofuel production. Though bioethanol production via microbial gas fermentation would be a perfect candidate to participate in this trading system, currently, the model to verify the environmental performance of biofuels producers does not include this technology. Hence, its expansion to new technologies will undoubtedly increase the offer of renewable energy in the country (Klein et al. 2019) and attract investments.

Furthermore, creating regional collaborative hubs formed by universities, public departments, and industries to analyze the different carbon capture and utilization pathways is needed to spark innovation and provide accurate information for decision-makers (Hepburn et al. 2019; Focus Areas – The REMADE Institute 2021; Programs - BioMADE 2021).

Moreover, costs tend to come down as companies learn by doing (Bui et al. 2018). This could be due to improvements in the production process, technological breakthroughs, supply chain cost reductions, R&D knowledge spillover, advances in automation, public policies, and social acceptance (Lackner and Azarabadi 2021; Bui et al. 2018). Complementary paths for cost reduction include corporate sector initiatives—as pledges to become carbon negative—and CO_2 purchasing contracts (Hepburn et al. 2019; A policy strategy for carbon capture and storage – analysis – IEA 2021; Bui et al. 2018).

19.5.2 Genetic Engineering of Non-model Organisms

To further strengthen the potential of acetogens to synthesize valuable bioproducts from waste carbon, advances in strain engineering of non-model microorganisms are essential. These anaerobic acetogens offer an advantage over model organisms such as *E. coli* and *S. cerevisiae* as many of them have evolved to utilize non-conventional feedstocks (as gaseous carbon-based compounds) to produce complex molecules (Riley and Guss 2021; Czajka et al. 2017). To fully explore their bioconversion potential, a robust genetic toolbox for efficient gene deletion and insertion, microbial transformation, and high-throughput genetic engineering and screening is needed to leverage their metabolic strength (Köpke and Simpson 2020; Liew et al. 2016; Fackler et al. 2021; Riley and Guss 2021; Czajka et al. 2017; Karim et al. 2020; Hillson et al. 2019). Additionally, private and public investment to support the scale-up of bioprocesses is necessary as large-scale infrastructure projects are capital intensive (Programs - BioMADE 2021; Global Status of CCS 2020).

19.6 Discussion

Though it is generally agreed that a linear profile of production and consumption is unsustainable and an important roadblock to achieving climate goals, it seems unreasonable to assume that less consumption and a carbon-free economy are solutions to these challenges. Conversely, the circular economy concept brings the potential to curb fossil fuel resources and mitigate GHG emissions via the capture and storage or utilization of waste carbons.

The deployment of efficient and economically viable technologies for CCU and CO_2 and H_2 electrolysis via renewable energy will lead to their broader adoption by heavy GHG emitters. At the same time, large volumes of carbon oxide emissions will be mitigated while providing cheap carbon feedstock for the production of carbon-based products, closing the carbon cycle, and promoting a circular bioeconomy operating in a net-zero-CO_2 emissions framework.

It is clear, for all scenarios and workflows considered in this chapter, that a solid policy framework that supports the development and scientific breakthrough is paramount to dilute the risks and costs associated with the deployment of new technologies and their scale-up, create new markets, promote education and acceptance, and, ultimately, lead to the achievement of the climate goals in a timely manner.

Among the different routes available, one should not consider a *winner* but rather a portfolio of alternatives that is best suitable to a specific geographical, cultural, economic, and infrastructural context. In addition, the diversification in energy production will contribute to a more robust resilience to face climate changes.

Biogenic carbon lost during bagasse combustion, farmed sugar fermentation, and agricultural waste biomass are important carbon residues in the bioethanol production chain. The process integration with innovative technologies using these residual fractions could significantly improve the energy output, avoid the competition by land and food, avoid the mining of below-ground carbon, and represent an additional revenue source for the biorefinery both in the form of higher ethanol yields and in carbon credits. In this chapter, emphasis is placed on a commercially available microbial gas fermentation technology. This successful microbial bioprocess employs an anaerobic acetogen (*Clostridium autoethanogenum*) that efficiently fixes residual carbon oxides to ethanol. It is worth mentioning that other microbial fermentation technologies are also under development to metabolize waste carbon. These include microorganisms such as methanotrophs, aerobic hydrogenotrophs, algae, and the anaerobic acetogens themselves to produce ethanol and other bioproducts at scale.

For the country, this integration might represent the development of rural areas, the adoption and development of cutting-edge technologies with high potential for economic revenue, increased biofuels, chemicals, and materials production and export, the decrease of dependence on imported fossil fuels, and a global leadership position on a new, unavoidable world economy based on climate, resilience, and justice.

Acknowledgements The author would like to thank Maryann Maas, Chad Haynes, Sarah Eisenlord, and Michael Köpke from LanzaTech for support and critical review of the manuscript. The author also wishes to thank Prof. Dr. Goncalo Amarante G. Pereira for the encouragement and guidance throughout this writing process.

Conflict of Interest Marilene Pavan is an employee of LanzaTech, which has a commercial interest in gas fermentation.

References

A policy strategy for carbon capture and storage – analysis – IEA. https://www.iea.org/reports/a-policy-strategy-for-carbon-capture-and-storage. Accessed 6 out 2021

Agrawal H, Rao AB (2018) Estimating the potential of pre-combustion CO2 capture in bio-hydrogen, Bio-Cng and bio-ethanol production plants. SSRN Electron J. https://doi.org/10.2139/SSRN.3365634

Ahn JH, Kim TS (2020) Effect of oxygen supply method on the performance of a micro gas turbine-based triple combined cycle with oxy-combustion carbon capture. Energy 211:119010. https://doi.org/10.1016/J.ENERGY.2020.119010

Ajwani-Ramchandani R, Figueira S, Torres de Oliveira R, Jha S (2021) Enhancing the circular and modified linear economy: the importance of blockchain for developing economies. Resour Conserv Recycl 168:105468. https://doi.org/10.1016/J.RESCONREC.2021.105468

Análise De Conjuntura Dos Biocombustíveis (2019)

Andrés González-Garay S, Frei M, Al-Qahtani A et al (2019) Plant-to-planet analysis of CO 2 -based methanol processes. Energy Environ Sci 12:3425–3436. https://doi.org/10.1039/C9EE01673B

ANP - Agencia Nacional do Petroleo RenovaBio. http://www.anp.gov.br/producao-de-biocombustiveis/renovabio. Accessed 31 ago 2020

AR5 Synthesis Report: Climate Change 2014 – IPCC. https://www.ipcc.ch/report/ar5/syr/. Accessed 5 out 2021

Bagnato G, Sanna A (2018) Membrane considerations and plant design for pre-combustion CO2 capture. Curr Trends Futur Dev Membr Carbon Dioxide Sep by Using Membr 415–435. https://doi.org/10.1016/B978-0-12-813645-4.00015-5

Bailera M, Kezibri N, Romeo LM et al (2017) Future applications of hydrogen production and CO2 utilization for energy storage: hybrid power to gas-oxycombustion power plants. Int J Hydrog Energy 42:13625–13632. https://doi.org/10.1016/J.IJHYDENE.2017.02.123

Bednar J, Obersteiner M, Baklanov A et al (2021) Operationalizing the net-negative carbon economy. Nature 596:1. https://doi.org/10.1038/S41586-021-03723-9

Beuttler C, Charles L, Wurzbacher J (2019) The role of direct air capture in mitigation of anthropogenic greenhouse gas emissions. Front Clim 0:10. https://doi.org/10.3389/FCLIM.2019.00010

Bhandari R, Trudewind CA, Zapp P (2014) Life cycle assessment of hydrogen production via electrolysis - a review. J Clean Prod 85:151–163

Biogenic Carbon - Science and Climate. https://climatechange.ucdavis.edu/climate-change-definitions/biogenic-carbon/. Accessed 5 out 2021

Bui M, Adjiman CS, Bardow A et al (2018) Carbon capture and storage (CCS): the way forward. Energy Environ Sci 11:1062. https://doi.org/10.1039/c7ee02342a

Bushuyev OS, De Luna P, Dinh CT et al (2018) What should we make with CO2 and how can we make it? Joule 2:825–832. https://doi.org/10.1016/J.JOULE.2017.09.003

Buttler A, Spliethoff H (2018) Current status of water electrolysis for energy storage, grid balancing and sector coupling via power-to-gas and power-to-liquids: a review. Renew Sust Energ Rev 82: 2440–2454. https://doi.org/10.1016/J.RSER.2017.09.003

Canilha L, Chandel AK, Dos Santos S, Milessi T et al (2012) Bioconversion of sugarcane biomass into ethanol: an overview about composition, pretreatment methods, detoxification of hydrolysates, enzymatic saccharification, and ethanol fermentation. J Biomed Biotechnol 2012. https://doi.org/10.1155/2012/989572

Caputo AC, Palumbo M, Pelagagge PM, Scacchia F (2005) Economics of biomass energy utilization in combustion and gasification plants: effects of logistic variables. Biomass Bioenergy 28:35–51. https://doi.org/10.1016/J.BIOMBIOE.2004.04.009

Carbon Capture Coalition. https://carboncapturecoalition.org/45q-legislation/. Accessed 6 out 2021

Chao C, Deng Y, Dewil R et al (2021) Post-combustion carbon capture. Renew Sust Energ Rev 138:110490. https://doi.org/10.1016/J.RSER.2020.110490

Claassens NJ, Sousa DZ, dos Santos VAPM et al (2016) Harnessing the power of microbial autotrophy. Nat Rev Microbiol 1411(14):692–706. https://doi.org/10.1038/nrmicro.2016.130

Claassens NJ, Sánchez-Andrea I, Sousa DZ, Bar-Even A (2018) Towards sustainable feedstocks: a guide to electron donors for microbial carbon fixation. Curr Opin Biotechnol 50:195–205. https://doi.org/10.1016/J.COPBIO.2018.01.019

Claassens NJ, Cotton CAR, Kopljar D, Bar-Even A (2019) Making quantitative sense of electromicrobial production. Nat Catal 2:437–447. https://doi.org/10.1038/s41929-019-0272-0

Climeworks Latest Direct Air Capture Plant. https://climeworks.com/orca. Accessed 8 set 2021

CO2 capture technologies: oxy combustion with CO2 capture - Global CCS Institute. https://www.globalccsinstitute.com/resources/publications-reports-research/co2-capture-technologies-oxy-combustion-with-co2-capture/. Accessed 5 out 2021

Conrado RJ, Haynes CA, Haendler BE, Toone EJ (2013) Electrofuels: a new paradigm for renewable fuels. Adv Biofuels Bioprod 9781461433484:1037–1064. https://doi.org/10.1007/978-1-4614-3348-4_38

Corrêa do Lago A, Bonomi A, Cavalett O et al (2012) Sugarcane as a carbon source: the Brazilian case. Biomass Bioenergy 46:5–12. https://doi.org/10.1016/j.biombioe.2012.09.007

Czajka J, Wang Q, Wang Y, Tang YJ (2017) Synthetic biology for manufacturing chemicals: constraints drive the use of non-conventional microbial platforms. Appl Microbiol Biotechnol 10120(101):7427–7434. https://doi.org/10.1007/S00253-017-8489-9

Dagle RA, Winkelman AD, Ramasamy KK et al (2020) Ethanol as a renewable building block for fuels and chemicals. Ind Eng Chem Res 59:4843–4853. https://doi.org/10.1021/ACS.IECR.9B05729

Daniell J, Köpke M, Simpson SD (2012) Commercial biomass syngas fermentation

de Medeiros EM, Posada JA, Noorman H et al (2017) Hydrous bioethanol production from sugarcane bagasse via energy self-sufficient gasification-fermentation hybrid route: simulation and financial analysis. J Clean Prod 168:1625–1635. https://doi.org/10.1016/J.JCLEPRO.2017.01.165

de Souza Noel Simas Barbosa L, Hytönen E, Vainikka P (2017) Carbon mass balance in sugarcane biorefineries in Brazil for evaluating carbon capture and utilization opportunities. Biomass Bioenergy 105:351–363. https://doi.org/10.1016/j.biombioe.2017.07.015

De Tissera S, Köpke M, Simpson SD et al (2019) Syngas biorefinery and syngas utilization. Adv Biochem Eng Biotechnol 166:247–280

de Vasconcelos BR, Lavoie JM (2019) Recent advances in power-to-X technology for the production of fuels and chemicals. Front Chem 7:392

Deutz S, Bardow A (2021) Life-cycle assessment of an industrial direct air capture process based on temperature–vacuum swing adsorption. Nat Energy 62(6):203–213. https://doi.org/10.1038/s41560-020-00771-9

Direct Air Capture – Analysis – IEA. https://www.iea.org/reports/direct-air-capture. Accessed 5 out 2021

DOE Announces $12 Million for Direct Air Capture Technology | Department of Energy. https://www.energy.gov/articles/doe-announces-12-million-direct-air-capture-technology. Accessed 8 set 2021

Drake HL, Gößner AS, Daniel SL (2008) Old acetogens, new light. Ann N Y Acad Sci 1125:100–128

Energy Agency I Technology Perspectives Energy Special Report on Carbon Capture Utilisation and Storage CCUS in clean energy transitions (2020). https://www.iea.org/reports/ccus-in-clean-energy-transitions

Etanol - Centro de Estudos Avançados em Economia Aplicada - CEPEA-Esalq/USP. https://www.cepea.esalq.usp.br/br/indicador/etanol.aspx. Accessed 31 out 2021

EU Emissions Trading System (EU ETS). https://ec.europa.eu/clima/eu-action/eu-emissions-trading-system-eu-ets_en. Accessed 6 out 2021

Fackler N, Heijstra BD, Rasor BJ et al (2021) Stepping on the gas to a circular economy: accelerating development of carbon-negative chemical production from gas fermentation. Annu Rev Chem Biomal Eng 12:439–470. https://doi.org/10.1146/annurev-chembioeng

Finley RJ (2014) An overview of the Illinois Basin – Decatur Project. Greenh Gases Sci Technol 4:571–579. https://doi.org/10.1002/GHG.1433

Focus Areas – The REMADE Institute. https://remadeinstitute.org/focus-areas. Accessed 7 out 2021

Formann S, Hahn A, Janke L et al (2020) Beyond sugar and ethanol production: value generation opportunities through sugarcane residues. Front Energy Res. https://doi.org/10.3389/fenrg.2020.579577

Gabrielli P, Gazzani M, Mazzotti M (2020) The role of carbon capture and utilization, carbon capture and storage, and biomass to enable a net-zero-CO2 emissions chemical industry. Ind Eng Chem Res 59:7033–7045. https://doi.org/10.1021/ACS.IECR.9B06579

Gambhir A, Tavoni M (2019) Direct air carbon capture and sequestration: how it works and how it could contribute to climate-change mitigation. One Earth 1:405–409. https://doi.org/10.1016/J.ONEEAR.2019.11.006

Global Status of CCS 2020

Grassi MCB, Pereira GAG (2019) Energy-cane and RenovaBio: Brazilian vectors to boost the development of Biofuels. Ind Crop Prod 129:201–205. https://doi.org/10.1016/j.indcrop.2018.12.006

Gürtürk M, Benli H, Ertürk NK (2018) Effects of different parameters on energy – Exergy and power conversion efficiency of PV modules. Renew Sust Energ Rev 92:426–439. https://doi.org/10.1016/J.RSER.2018.04.117

Haas T, Krause R, Weber R et al (2018) Technical photosynthesis involving CO2 electrolysis and fermentation. Nat Catal. https://doi.org/10.1038/s41929-017-0005-1

Handler RM, Shonnard DR, Griffing EM et al (2016) Life cycle assessments of ethanol production via gas fermentation: anticipated greenhouse gas emissions for cellulosic and waste gas feedstocks. Ind Eng Chem Res. https://doi.org/10.1021/acs.iecr.5b03215

Heffernan JK, Valgepea K, de Souza Pinto Lemgruber R et al (2020) Enhancing CO2-valorization using clostridium autoethanogenum for sustainable fuel and chemicals production. Front Bioeng Biotechnol. https://doi.org/10.3389/fbioe.2020.00204

Heidenreich S, Foscolo PU (2015) New concepts in biomass gasification. Prog Energy Combust Sci 46:72–95. https://doi.org/10.1016/J.PECS.2014.06.002

Hepburn C, Adlen E, Beddington J et al (2019) The technological and economic prospects for CO2 utilization and removal. Nature 575:87–97. https://doi.org/10.1038/s41586-019-1681-6

Hillson N, Caddick M, Cai Y et al (2019) Building a global alliance of biofoundries. Nat Commun 10:2040. https://doi.org/10.1038/s41467-019-10079-2

How Direct Air Capture Helps Reverse Climate Change | Climeworks. https://climeworks.com/co2-removal. Accessed 8 set 2021

Hu Y, Li X, Li H, Yan J (2013) Peak and off-peak operations of the air separation unit in oxy-coal combustion power generation systems. Appl Energy 112:747–754. https://doi.org/10.1016/J.APENERGY.2012.12.001

Huang Z, Grim G, Schaidle J, Tao L (2020) Using waste CO2 to increase ethanol production from corn ethanol biorefineries: techno-economic analysis. Appl Energy 280:115964. https://doi.org/10.1016/J.APENERGY.2020.115964

IRENA (2018) Hydrogen from renewable power: technology outlook for the energy transition

Jordan CA, Akay G (2012) Occurrence, composition and dew point of tars produced during gasification of fuel cane bagasse in a downdraft gasifier. Biomass Bioenergy 42:51–58. https://doi.org/10.1016/J.BIOMBIOE.2012.03.014

Karim AS, Liew F (Eric), Garg S, et al (2020) Modular cell-free expression plasmids to accelerate biological design in cells. Synth Biol 5:ysaa019. https://doi.org/10.1093/synbio/ysaa019

Kircher M (2021) Bioeconomy – present status and future needs of industrial value chains. New Biotechnol 60:96–104. https://doi.org/10.1016/J.NBT.2020.09.005

KircherManfred (2014) The emerging bioeconomy: industrial drivers, global impact, and international strategies. https://home.liebertpub.com/ind 10:11–18. https://doi.org/10.1089/IND.2014.1500

Klein BC, Chagas MF, Watanabe MDB et al (2019) Low carbon biofuels and the New Brazilian National Biofuel Policy (RenovaBio): a case study for sugarcane mills and integrated sugarcane-microalgae biorefineries. Renew Sust Energ Rev. https://doi.org/10.1016/j.rser.2019.109365

Köpke M, Simpson SD (2020) Pollution to products: recycling of 'above ground' carbon by gas fermentation. Curr Opin Biotechnol 65:180–189

Kriegler E, Luderer G, Bauer N et al (2018) Pathways limiting warming to 1.5°C: a tale of turning around in no time? Philos Trans A Math Phys Eng Sci 376(2119):20160457. https://doi.org/10.1098/rsta.2016.0457

Küngas R (2020) Review—electrochemical CO 2 reduction for CO production: comparison of low- and high-temperature electrolysis technologies. J Electrochem Soc. https://doi.org/10.1149/1945-7111/ab7099

Lackner KS, Azarabadi H (2021) Buying down the cost of direct air capture. Ind Eng Chem Res 60: 8196–8208. https://doi.org/10.1021/ACS.IECR.0C04839

Lee U, Hawkins TR, Yoo E et al (2021) Using waste CO2 from corn ethanol biorefineries for additional ethanol production: life-cycle analysis. Biofuels Bioprod Biorefin 15:468–480. https://doi.org/10.1002/BBB.2175

Leger D, Matassa S, Noor E et al (2021) Photovoltaic-driven microbial protein production can use land and sunlight more efficiently than conventional crops. Proc Natl Acad Sci U S A 118: e2015025118. https://doi.org/10.1073/PNAS.2015025118

Liakakou ET, Infantes A, Neumann A, Vreugdenhil BJ (2021) Connecting gasification with syngas fermentation: comparison of the performance of lignin and beech wood. Fuel 290. https://doi.org/10.1016/J.FUEL.2020.120054

Liew FM, Martin ME, Tappel RC et al (2016) Gas fermentation-a flexible platform for commercial scale production of low-carbon-fuels and chemicals from waste and renewable feedstocks. Front Microbiol

Liu W, Zhang Z, Xie X et al (2016) Analysis of the global warming potential of biogenic CO 2 emission in life cycle assessments OPEN. Nat Publ Gr. https://doi.org/10.1038/srep39857

Liu Z, Wang K, Chen Y et al (2020) Third-generation biorefineries as the means to produce fuels and chemicals from CO2. Nat Catal 33(3):274–288. https://doi.org/10.1038/s41929-019-0421-5

Markewitz P, Kuckshinrichs W, Leitner W et al (2012) Worldwide innovations in the development of carbon capture technologies and the utilization of CO2. Energy Environ Sci 5:7281–7305. https://doi.org/10.1039/C2EE03403D

Mikkelsen M, Mikkel Jørgensen C, Krebs F (2010) The teraton challenge. A review of fixation and transformation of carbon dioxide. Energy Environ Sci 3:43–81. https://doi.org/10.1039/B912904A

Millar RJ, Fuglestvedt JS, Friedlingstein P et al (2017) Emission budgets and pathways consistent with limiting warming to 1.5 °C. Nat Geosci 10(10):741–747. https://doi.org/10.1038/ngeo3031

Mokhtar M, Ali MT, Khalilpour R et al (2012) Solar-assisted post-combustion carbon capture feasibility study. Appl Energy 92:668–676. https://doi.org/10.1016/J.APENERGY.2011.07.032

Molino A, Chianese S, Musmarra D (2016) Biomass gasification technology: the state of the art overview. J Energy Chem 25:10–25. https://doi.org/10.1016/J.JECHEM.2015.11.005

Molitor B, Richter H, Martin ME et al (2016) Carbon recovery by fermentation of CO-rich off gases – turning steel mills into biorefineries. Bioresour Technol 215:386–396. https://doi.org/10.1016/J.BIORTECH.2016.03.094

Mondal MK, Balsora HK, Varshney P (2012) Progress and trends in CO2 capture/separation technologies: a review. Energy 46:431–441. https://doi.org/10.1016/J.ENERGY.2012.08.006

Most Efficient Solar Panels 2021 – clean energy reviews. https://www.cleanenergyreviews.info/blog/most-efficient-solar-panels. Accessed 3 nov 2021

Naims H (2016) Economics of carbon dioxide capture and utilization—a supply and demand perspective. Environ Sci Pollut Res 2322(23):22226–22241. https://doi.org/10.1007/S11356-016-6810-2

Nanda S, Reddy SN, Mitra SK, Kozinski JA (2016) The progressive routes for carbon capture and sequestration. Energy Sci Eng 4:99–122. https://doi.org/10.1002/ESE3.117

National Academies of Sciences E and M (2018) Negative emissions technologies and reliable sequestration: a research agenda. Negat Emiss Technol Reliab Sequestration. 10.17226/25259

Nations U (2015) What is the Paris Agreement? | UNFCCC. United Nations Climate Change

Org W, Kennedy KM (2019) Issue brief issue brief putting a price on carbon: evaluating a carbon price and complementary policies for a 1.5°C world 2|. https://www.wri.org/research/putting-price-carbon-evaluating-carbon-price-and-complementary-policies-15deg-world

Osaki MR, Seleghim P (2017) Bioethanol and power from integrated second generation biomass: a Monte Carlo simulation. Energy Convers Manag 141:274–284. https://doi.org/10.1016/J.ENCONMAN.2016.08.076

Pereira LG, Cavalett O, Bonomi A et al (2019) Comparison of biofuel life-cycle GHG emissions assessment tools: the case studies of ethanol produced from sugarcane, corn, and wheat. Renew Sustain Energy Rev

Perrin N, Paufique C, Leclerc M (2014) Latest performances and improvement perspective of oxycombustion for carbon capture on coal power plants. Energy Procedia 63:524–531. https://doi.org/10.1016/J.EGYPRO.2014.11.057

Porter RTJ, Fairweather M, Kolster C et al (2017) Cost and performance of some carbon capture technology options for producing different quality CO2 product streams. Int J Greenh Gas Control 57:185–195. https://doi.org/10.1016/J.IJGGC.2016.11.020

Posada JA, Patel AD, Roes A et al (2013) Potential of bioethanol as a chemical building block for biorefineries: preliminary sustainability assessment of 12 bioethanol-based products. Bioresour Technol 135:490–499. https://doi.org/10.1016/J.BIORTECH.2012.09.058

Pre-Combustion Carbon Capture Research | Department of Energy. https://www.energy.gov/fecm/science-innovation/carbon-capture-and-storage-research/carbon-capture-rd/pre-combustion-carbon. Accessed 5 out 2021

Pre-Combustion CO2 Capture | netl.doe.gov. https://netl.doe.gov/coal/carbon-capture/pre-combustion. Accessed 5 out 2021

Programs - BioMADE. https://biomade.org/programs/. Accessed 7 out 2021

Puig-Arnavat M, Bruno JC, Coronas A (2010) Review and analysis of biomass gasification models. Renew Sust Energ Rev 14:2841–2851. https://doi.org/10.1016/J.RSER.2010.07.030

Ragsdale SW, Pierce E (2008) Acetogenesis and the Wood–Ljungdahl pathway of CO2 fixation. Biochim Biophys Acta - Proteins Proteom 1784:1873–1898. https://doi.org/10.1016/J.BBAPAP.2008.08.012

Rego de Vasconcelos B, Lavoie J-M (2019) Recent advances in power-to-X technology for the production of fuels and chemicals. Front Chem 0:392. https://doi.org/10.3389/FCHEM.2019.00392

Renewable Power Generation Costs in 2020. https://www.irena.org/publications/2021/Jun/Renewable-Power-Costs-in-2020. Accessed 4 Nov 2021

Riley LA, Guss AM (2021) Approaches to genetic tool development for rapid domestication of non-model microorganisms. Biotechnol Biofuels 14:30

Rodin V, Lindorfer J, Böhm H, Vieira L (2020) Assessing the potential of carbon dioxide valorisation in Europe with focus on biogenic CO2. J CO2 Util 41:101219. https://doi.org/10.1016/J.JCOU.2020.101219

Roh K, Bardow A, Bongartz D et al (2020) Early-stage evaluation of emerging CO2 utilization technologies at low technology readiness levels. Green Chem 22:3842–3859. https://doi.org/10.1039/C9GC04440J

Samanta A, Zhao A, Shimizu GKH et al (2011) Post-combustion CO2 capture using solid sorbents: a review. Ind Eng Chem Res 51:1438–1463. https://doi.org/10.1021/IE200686Q

Sanchez DL, Johnson N, McCoy ST et al (2018) Near-term deployment of carbon capture and sequestration from biorefineries in the United States. Proc Natl Acad Sci U S A 115:4875–4880. https://doi.org/10.1073/PNAS.1719695115

Shayegh S, Bosetti V, Tavoni M (2021) Future prospects of direct air capture technologies: insights from an expert elicitation survey. Front Clim 0:46. https://doi.org/10.3389/FCLIM.2021.630893

Kearns D, Liu H, Consoli C (2021) Technology Readiness and Costs of CCS. Global CCS Institute. Available at: https://www.globalccsinstitute.com/wpcontent/uploads/2021/03/Technology-Readiness-and-Costs-for-CCS-2021-1.pdf. Accessed 1 Jun 2022

Sixth Assessment Report. https://www.ipcc.ch/report/ar6/wg1/. Accessed 21 set 2021

Somoza-Tornos A, Guerra OJ, Crow AM et al (2021) Process modeling, techno-economic assessment, and life cycle assessment of the electrochemical reduction of CO2: a review. iScience 24:102813. https://doi.org/10.1016/J.ISCI.2021.102813

Song C, Liu Q, Deng S et al (2019) Cryogenic-based CO2 capture technologies: State-of-the-art developments and current challenges. Renew Sust Energ Rev 101:265–278. https://doi.org/10.1016/J.RSER.2018.11.018

Study on Development of Water Electrolysis in the EU Final Report E4tech Sàrl with Element Energy Ltd for the Fuel Cells and Hydrogen Joint Undertaking (2014)

Sulaiman SA, Karim MF, Nazmi M et al (2013) On gasification of different tropical plant-based biomass materials. Asian J Sci Res 6:245–253. https://doi.org/10.3923/AJSR.2013.245.253

Teixeira LV, Moutinho LF, Romão-Dumaresq AS (2018) Gas fermentation of C1 feedstocks: commercialization status and future prospects. Biofuels Bioprod Biorefin 12:1103–1117. https://doi.org/10.1002/BBB.1912

Theo WL, Lim JS, Hashim H et al (2016) Review of pre-combustion capture and ionic liquid in carbon capture and storage. Appl Energy 183:1633–1663. https://doi.org/10.1016/J.APENERGY.2016.09.103

Valgepea K, de Souza Pinto Lemgruber R, Abdalla T et al (2018) H2 drives metabolic rearrangements in gas-fermenting Clostridium autoethanogenum. Biotechnol Biofuels 11:1–15. https://doi.org/10.1186/S13068-018-1052-9

Wang X, Song C (2020) Carbon capture from flue gas and the atmosphere: a perspective. Front Energy Res 0:265. https://doi.org/10.3389/FENRG.2020.560849

We Turn CO2 into Stone. https://www.carbfix.com/. Accessed 8 set 2021

Wheatcrd (2014) Cost of capturing CO 2 from industrial sources. Office of Fossil Energy

Widjaya ER, Chen G, Bowtell L, Hills C (2018) Gasification of non-woody biomass: a literature review. Renew Sust Energ Rev 89:184–193. https://doi.org/10.1016/J.RSER.2018.03.023

Wilkins MR, Atiyeh HK (2011) Microbial production of ethanol from carbon monoxide. Curr Opin Biotechnol 22:326–330. https://doi.org/10.1016/J.COPBIO.2011.03.005

Xiong J, Zhao H, Chen M, Zheng C (2011) Simulation study of an 800 MWe oxy-combustion pulverized-coal-fired power plant. Energy Fuels 25:2405–2415. https://doi.org/10.1021/EF200023K

Zhike L, Shasha L (2021) How financial development affects CO 2 emissions: a spatial econometric analysis. J Environ Manag 277:111397. https://doi.org/10.1016/J.JENVMAN.2020.111397

Zhu XG, Long SP, Ort DR (2008) What is the maximum efficiency with which photosynthesis can convert solar energy into biomass? Curr Opin Biotechnol 19:153–159. https://doi.org/10.1016/J.COPBIO.2008.02.004

Chapter 20
New Technologies for Bioethanol Production: Patents and Innovation

Susan Grace Karp, Ariane Fátima Murawski de Mello, Leonardo Wedderhoff Herrmann, Alexander da Silva Vale, Walter José Martinez-Burgos, Carolina Mene Savian, and Carlos Ricardo Soccol

Abstract The aim of this chapter is to assess the stage of technological development and innovation in bioethanol production. For this purpose, two patent searches were conducted, one focused on pretreatment and saccharification and another on fermentation processes. The status of innovation was projected from industrial data. In the field of pretreatment and saccharification, the number of published patent documents is still growing, and the leading patent applicants are the USA, China, Brazil, and India, which are worldwide major producers of ethanol. Among the technologies, chemical strategies are predominant (49%), and the enzymatic treatments are present in 26% of the evaluated documents. The use of green solvents appears in recent documents, mostly since 2015. Regarding fermentation processes, the main patent applicants were China and the USA with 37% and 33% of participation, respectively, and most of the documents were published in 2010. Among the novel technologies, special focus was given to the development of strains able to ferment pentoses and grow at high temperatures. The processes of first-generation bioethanol production have been extensively applied in industrial scale for the last five decades and represent consolidated technologies. Second-generation ethanol facilities, however, still face technological challenges that require new solutions.

20.1 Introduction

The energetic security, the reduction of fossil fuels utilization and the mitigation of climate change have encouraged different governmental initiatives to support biofuel and bioenergy production, such as the mandatory blending of gasoline with biofuel in Brazil, United States, and Canada (Aghbashlo et al. 2017, 2018; Mandegari et al.

S. G. Karp · A. F. Murawski de Mello · L. Wedderhoff Herrmann · A. da Silva Vale · W. J. Martinez-Burgos · C. Mene Savian · C. R. Soccol (✉)
Department of Bioprocess Engineering and Biotechnology, Federal University of Paraná, Curitiba, PR, Brazil
e-mail: soccol@ufpr.br

C. R. Soccol et al. (eds.), *Liquid Biofuels: Bioethanol*, Biofuel and Biorefinery Technologies 12, https://doi.org/10.1007/978-3-031-01241-9_20

2018). To meet the demand for these biofuels, platforms of first-generation ethanol, i.e., those that produce ethanol from the fermentation of traditional crops, were developed and consolidated using a wide range of food commodities, such as grains and sugarcane (Rodríguez Carpio et al. 2021). However, the use of those feedstocks can be a threat to food security (Aghbashlo et al. 2016). Therefore, the development of second-generation ethanol production technologies that use waste materials, especially lignocellulosic biomass, as feedstock, was stimulated, especially through the valorization of crop and plant residues (Mandegari et al. 2018). The use of waste biomass as renewable feedstock can also reduce environmental impacts and convert waste organic materials into high-value products (Hosseinpour et al. 2017; Aghbashlo et al. 2018).

The stage of technological development of bioethanol production can be assessed from the information available in patent documents. Although much information on scientific and technological development can be retrieved from scientific papers, a patent search can provide essential information related to the stage of technological maturity (time evolution of patent filings/publications), technology holders (assignees) and market interests (countries where the technology is protected). It is important to differentiate technological development from innovation. Innovation means diffusion, in other words, that the technology has reached considerable scale and has had an impact on society.

The processes of bioethanol production, particularly of first-generation ethanol, have been extensively applied in industrial scale for the last five decades, although there are important recent advances, especially related to the use of starchy raw materials. Second-generation ethanol plants started to appear in the last decade, and technological improvements are still necessary in the stages of pretreatment and fermentation.

In this chapter, a patent search was performed for two important technological areas in bioethanol production: pretreatment and saccharification processes and fermentation processes. Since the scope of technological solutions in these two areas is wide, special focus was given to advanced and sustainable technologies. In the case of pretreatment and saccharification, the use of green solvents and enzymatic processes was targeted. In the case of fermentation processes, the consumption of pentoses, the use of immobilized yeast, simultaneous saccharification and fermentation and consolidated bioprocessing were the focus of the search. To assess the status of innovations in bioethanol production, public information about industrial facilities was collected.

20.2 Pretreatment and Saccharification Processes

The production of ethanol through microbiological fermentation requires the availability of adequate substrates and nutrients to be assimilated by the fermenting microorganism. Usually, these substrates are mono- or disaccharides. This is a particular challenge in the production of second-generation bioethanol since the substrates are not readily available. The most common substrate for the production of second-generation bioethanol is lignocellulosic biomass. This biomass, however, is composed of several complex and long-chain sugars, especially cellulose, hemicellulose and lignin, which are not available for direct assimilation and use. Therefore, it must undergo pretreatment and saccharification processes, in order to release fermentable monosaccharides, such as glucose, fructose, xylose, and arabinose (Mosier et al. 2005; Rastogi and Shrivastava 2017).

Examples of lignocellulosic biomass include rice straw, wheat straw, sugarcane bagasse, corn stover, switchgrass, and other plant-based materials that remain as byproducts of other production chains. Each of them contains around 40–50% cellulose, 25–30% hemicellulose, 15–20% lignin, and small quantities of pectinase, proteins, and minerals (Saini et al. 2014; Tayyab et al. 2018). In more details, cellulose is a polymer made exclusively of glucose arranged into a crystalline and ordered structure, linked by β-1,4-glycosidic bonds. Hemicellulose, on the other hand, presents an irregular and amorphous structure with branches, including fructose, xylose, arabinose, mannose, and other monosaccharides among its monomers. The most difficult to degrade is lignin, formed by phenylpropane units linked through ether and carbon-carbon bonds (Tayyab et al. 2018; Rezania et al. 2020). In order to convert these three polymers into hexoses and pentoses, which are fermentable sugars, the most common strategies are physical, chemical, and biological breakage of the biomass, or even a combination of them.

In terms of technological development, a search among filed patents, which protect the recently developed products and processes around the world, is an interesting strategy to provide hints about how the biomass pretreatments are being performed nowadays. In this section, a brief description about each treatment process linked with the patents in the area is presented. An overview of the world leading countries in patent filings, common strategies, and other patent related information is also provided.

20.2.1 Chemical Pretreatments

The chemical pretreatment is one of the first processes designed for biomass, wherein a chemical compound is added to the material to disrupt the lignocellulosic scaffold, and it is mainly related to acid and alkaline hydrolysis. Acid treatments are especially effective in turning hemicellulose into a soluble form and removing it from the structure, leaving an amorphous cellulose and a disrupted matrix. The most common acids utilized are sulfuric, phosphoric, nitric, and acetic acids, and they are applied in diluted concentrations. However, the more diluted the acid, more heat is required to reach a sufficient fermentable sugar yield, with temperatures usually above 180 °C (Saha et al. 2005; Gonzales et al. 2016; Rezania et al. 2020). Acid techniques are classical, and their use for ethanol production dates back to 1990, such as in the patent documents US5125977A and WO1994008027A1, which described the hydrolysis of hemicellulose xylan and corn fibers, respectively, with sulfuric acid (Grohmann and Torget 1992; Grohmann et al. 1994). Recent patents are focused in testing alternative lignocellulosic biomass, such as coconut leaf (BR102019005092A2) (Albuquerque et al. 2020), algae (KR2096445B1) (Kim and Kim 2019), sawdust (IN201811040608A) (Sharma et al. 2018), and sisal (BR102017027916A2) (Xavier et al. 2017), all of them using sulfuric acid, with the addition of nitric or phosphoric acid.

The alkaline treatment, on the other hand, promotes an increased removal of lignin, breaking the alkyl-aryl linkages and resulting in an enhanced surface area full of pores. Sodium hydroxide, ammonium hydroxide, sulfite, and lime are the chemicals frequently used for this purpose. Most of the alkaline processes are conducted in ambient temperature, but for longer periods, even days, and are usually coupled with other treatments to reach better saccharification quality (Bali et al. 2015; Kim et al. 2016). The use of alkaline treatments was less frequent in patent documents as compared to acid treatments, even though promising results were reported. Recent approaches, for example, the one described in the patent document CN109652466A, utilized ammonium compounds in the disruption of corn stover, corn cob, bagasse, and wheat straw, with high efficiency (Wang et al. 2019). In another example, rice straw, bagasse, corncob, corn fibers, pineapple skin, and bamboo could be treated at ambient temperature with the use of sodium hydroxide and ammonium hydroxide (DE102017125090A1) (Kraikul et al. 2019). One interesting approach was the use of lignocellulosic kitchen waste, e.g., peels, bagasse, and napkins, for the production of ethanol, after treatment with alkali solution at 30–40 °C for 3 h (CN106011181A) (LTD 2016).

The chemical processes also include the use of organic solvents, ionic liquids, and other examples. The application of these solvents is associated with ecological advantages, mainly due to recovery and reuse of the component, as well as reduced harm compared to a concentrated acid or alkali. Organic solvents, such as acetone, ethanol, ethylene glycol, and methanol, are capable of separating lignin, hemicellulose, and cellulose into highly pure fractions. If the process occurs between 100 and 250 °C with organic or aqueous-organic solvent, the treatment is also called *organosolv*, and it presents high efficiency besides the increased recovery rate (Sun and Chen 2008; Shuai and Luterbacher 2016). The patent WO2017222084A1, for instance, described the use of ethanol, 1-butanol, 2-methyl-1-propanol, 2-butanol, and acetone as organic solvents for degrading herbal biomass (Koyama et al. 2016). Another patent reports the use of plant waste water sludge treated with benzene, toluene, n-hexane, dichloromethane, carbon tetrachloride, and carbon disulfide as non-polar organic solvents (CN106477831A) (Wu et al. 2016).

The ionic liquid pretreatment is considered one of the most sustainable and ecological methods, and the interest for its use is growing recently. As the name suggests, they are composed entirely of ions in the liquid state, having low vapor pressure and melting point, and high thermal stability. Imidazolium-based, pyridinium-based, pyrrolidinium-based, ammonium-based, phosphonium-based, and sulfonium-based solvents are alternatives of ionic liquids. Their strong hydrogen bond acceptors degrade the lignin while allowing enzymatic activity, however, they are difficult to recover and reuse (Espinoza-Acosta et al. 2014; Moniruzzaman and

Goto 2019). There is a considerable amount of patents for ionic liquid pretreatments, including the degradation of hemp hurds (WO2020214835A1) (Tulaphol and Sathitsuksanoh 2020), hardwood, softwood, grass (WO2016105538A1) (Shi et al. 2015), poplar, corn stover, and switchgrass (WO2015123627A1) (Schall and Farahani 2015). The denomination of green solvent and the ecological advantages of several ionic liquids are also mentioned in the patent document IN201721026359A (Kotia et al. 2017).

Recent works bring another kind of chemical compound, the eutectic solvent, as one of the most ecological treatments. Eutectic liquids are mixtures that present lower melting points compared to the isolated components and generate intermolecular hydrogen bonds. They show biocompatibility and high biodegradability, are cost effective, can be synthesized easily, and present no toxicity. Examples of eutectic solvents include mixtures of urea with sugars or organic acids, choline chloride with alcohols, quaternary ammonium salts, amides and carboxylic acids (Dai et al. 2013; Kumar et al. 2016; Rastogi and Shrivastava 2017). Two patents of the year 2019 were found utilizing this type of solvent to treat biomass, the first applied to algal material (WO2020053118A1) (Angelis et al. 2019), and the second applied to lignocellulosic biomass and kitchen waste (CN109942519A) (Hou et al. 2019).

20.2.2 Physical Pretreatments

Physical pretreatments, namely mechanical breakage, extrusion, and the use of electrolysis or radiation, consist in physical principles applied to the biomass to separate smaller pieces and/or disrupt chemical structures. The mechanical breakage, as simple as it may look, consists in the reduction of biomass size to particles, enhancing the surface area for posterior treatment. The use of milling or grinding is common when the biomass size is considerable, as it is in straw or bagasse, and particle sizes of less than 500 μm can be produced. However the energy consumption is high, and the liberation of monomers usually requires additional processes (Marrs et al. 2016; Tsapekos et al. 2017). Some patent documents, such as the patent number US2010203607A1, describe mechanical devices that enhance the shear rate to disrupt several lignocellulosic materials (Medoff et al. 2010).

Extrusion is a technique wherein the matter is submitted to high pressure or forced through a small hole or grid. The extreme shear rate and pressure causes the structure to disrupt, similarly to the mechanical breakage (Karunanithy and Muthukumarappan 2010; Zheng and Rehmann 2014). There are few patent documents that focus on the detailing of extrusion methods. The patent WO2016054132A1 described an extrusion treatment for the disruption of biomass derived from corn, wheat, rye, barley, milo, sorghum, and their combinations (Winsness et al. 2015).

Microwaves can also be used to vibrate the polar bonds until they break. They require less energy and do not produce inhibitors of enzymatic activity, degrading lignocellulose with easy operation steps. Other types of radiation, for instance, ultraviolet (UV) light, have similar approaches, being applied in the first steps of the pretreatment or before the enzymatic hydrolysis (Hu and Wen 2008; Ma et al. 2009). The patent document AU2017202376A1, for example, claimed the use of UV, X and gamma rays to irradiate several lignocellulosic materials from paper and textile industries (Medoff and Masterman 2017). Another patent document claimed the electron beam radiation to disrupt different kinds of biomass as well (US20130084613A1) (Medoff and Masterman 2012).

Another example of pretreatment is the steam explosion. The lignocellulosic matter is submitted to highly pressurized saturated steam for some minutes, followed by a sudden release in the pressure. This causes the cell wall to break and the hemicellulose and lignin to solubilize. This technique can also be classified as a physicochemical treatment, but for the purposes of this study, as the disruption is mainly caused by pressure difference, it was classified as a physical method for posterior analysis (Jacquet et al. 2015; Bonfiglio et al. 2019). Several methods use steam explosion to degrade wheat straw, corn stover, soybean straw (CN108998139A) (Hu et al. 2018), rice husk, grass, garden waste, wood waste (WO2019053750A1) (Ingolfsson et al. 2018), and corn stalk (CN108588166A) (Feng et al. 2018).

20.2.3 Physicochemical Pretreatments

A combination of physical forces and chemical compounds can also be applied to degrade biomass, especially when combining a solvent or reagent with high temperatures and pressures. One example is the hydrothermal treatment, which consists in maintaining the biomass at 160–240 °C in water. As it does not use chemical compounds, it is considered a green technology (Saha et al. 2013; Hashemi et al. 2019). The patent CN112064394A, for example, described the treatment of cellulose pulp with hydrothermal techniques, followed by the addition of ammonia and subsequent heating (Wang et al. 2020). Another document reported the utilization of hydrothermal strategies to break corncob, followed by acid treatment (CN109704917A) (Zhao et al. 2019).

Other physicochemical pretreatments comprise the simultaneous use of a chemical reagent, such as acids, alkalis or solvents, and variations of pressure, temperature, shear rate or radiation. The patent CN109704917A mentioned the use of weak acid gas explosion to treat crop straws with a rapid temperature increase (Lu 2019). The patent CN110791540A claimed a microwave alkali baking treatment, used to disrupt corn, rice, wheat, bean stalk, sorghum, alfalfa, and other lignocellulosic biomass (Meng et al. 2019). The patent WO2019094444A1 described the use of acid extrusion followed by an alkaline treatment to break corn syrup, molasses, silage, agricultural residues, corn stover, bagasse, sorghum, nuts, nut shells, coconut shells, among others (Tudman and Chesonis 2018).

20.2.4 Biological Pretreatments

Biological treatments involve the application of biomolecules and/or living organisms in the degradation of biomass. They are usually performed after other pretreatments, although these previous steps are not strictly necessary. The utilization of enzymes and microorganisms requires milder conditions of temperature, pH, and pressure as compared to physicochemical treatments, presenting lower energy cost. Additionally, such reaction conditions avoid the formation of toxic compounds (Sharma et al. 2017; Zabed et al. 2018).

Enzymes are protein-based biological catalysts that reduce the activation energy of a reaction providing a faster pathway. Each type of enzyme favors a different reaction, and several of them, such as cellulases, beta-glucosidases, xylanases, pectinases, laccases, and phytases can be applied for biomass degradation. Their use, individually or combined, liberates the specifically desired fermentable sugars, such as glucose, fructose, sucrose, xylose, arabinose, and other oligosaccharides. The efficiency of enzymatic hydrolysis is highly dependent on substrate's surface area (Zabed et al. 2018; Koupaie et al. 2019). There is a large number of patents filed in this area, reaching almost two times the number of documents related to acid, alkali, and other chemical treatments separately. The patent EP3725891A1, for instance, described the use of cellulases, hemicellulases, and beta-glucosidases in reutilization of lignocellulosic waste water (Adsul et al. 2020). The patent US2021087544A1 also depicted several combinations of cellulases and hemicellulases together with pectinases for saccharification (Alrik et al. 2020). The patent IN201911040938A claimed indigenous alpha-amylase and commercial glucoamylase for starchy substrates, especially from wheat (Kocher and Razdan 2019).

Besides enzymes, microorganisms can be used to degrade lignocellulosic compounds. They grow in the surface of the biomass and produce extracellular enzymes for the degradation of lignin, hemicellulose, and cellulose, with the aim of obtaining nutrients. The most frequently applied organisms are the white-rot, soft-rot, and brown-rot fungi, which are filamentous fungi of the phylum Basidiomycota. They naturally produce lignin-degrading enzymes, such as peroxidases and laccases, and are efficient in disrupting biomass for their growth. Other fungi, such as *Trichoderma reesei*, and some bacteria, such as *Clostridium* sp., can also produce the enzymatic cocktail for biomass pretreatment (Saha et al. 2016; Machado and Ferraz 2017). The patent WO2021029828A1 reported the production of cellulolytic enzymes from the filamentous fungus *Talaromyces pinophilus* to degrade herbaceous material, agricultural residue, forestry residue, waste paper, and pulp and paper mill residue (Geng et al. 2020). The patent WO2019083244(A2,A3) claimed the use of another fungus, *Penicillium* sp., to disrupt corn, sugarcane, beet, rice, reed, sunflower, switchgrass, rapeseed, wood, cotton, and fruit residues (Yoon 2018). The patent WO2011131667A1 claimed a genetically modified *Saccharomyces cerevisiae* to treat biomass, with the addition of xylose isomerase, xylose reductase, and xylitol dehydrogenase genes (Klaassen et al. 2011).

20.2.5 Patent Overview

An overview in the filed patent documents from 2010 to 2021 in the area of biomass pretreatment destined to bioethanol production can indicate how these technologies are being applied and in which direction the market is growing, since the patent system is used by companies, universities, and individuals to protect recently developed products and processes. A search in the Derwent Innovations Index database, from Web of Science and connected to the World Intellectual Property Organization, was performed, comprising patents filed worldwide. The time period was set between 2010 and 2021 with the search terms [(TS = (bioethanol OR fuel ethanol)) AND TS = ((physicochemical pretreatment) OR (acid pretreatment) OR (alkaline pretreatment) OR (enzymatic hydrolysis) OR (ionic solvents) OR (hydrothermal) OR (steam explosion) OR (eutectic solvents)) NOT TS = ((apparatus) OR (device))], totalizing 696 documents. The same search was performed in the Latipat database, which comprises the Latin American patents, with the proper translated terms, reaching 35 documents. All of the documents were manually filtered, and the patents not related to biomass pretreatment intended for bioethanol production were removed. A total of 279 documents was selected, wherein an analysis of publication year, International Patent Classification (IPC) codes utilized, countries and regions of origin, and type of applied pretreatment was conducted.

When analyzing the publication year, it is important to highlight that, due to the secrecy period of usually 18 months (Vandenberghe et al. 2020), patents filed in 2020 and 2021 may not appear. Figure 20.1 shows the profile of accumulated patent documents (Fig. 20.1a) and the most used IPC codes (Fig. 20.1b). It is possible to observe that the number of published patent documents is still growing along the years, although in a slower rate as compared to the beginning of the decade. This may indicate that the technology development is tending to maturity.

Evaluating the IPC codes utilized to identify and separate each patent, a total of 1734 codes were used, and the most frequent groups belong to the C12P, C12N, C07C, C12R, and D21C groups (Fig. 20.1b). The major group identified as the first C in the code represents the area of chemistry and metallurgy, and the D group represents textiles and paper industry, which are extremely pertinent to the concern of degrading cellulose, hemicellulose, and lignin. The C12P, C12N, and C12R codes relate to the utilization of enzymes and microorganisms, the first associated to the production of a chemical compound, the second to the fermentation and use per se, and the third related to the alcoholic fermentation. The frequent use of these codes indicates the extreme importance that the biological treatment acquired over the years, especially in terms of using enzymatic hydrolysis as the second treatment to obtain fermentable sugars. The C07C is the code for chemical compounds production, such as alcohol, as the treatments are destined to bioethanol production. The D21C is associated with the production of cellulose through the removal of hemicellulose, lignin, and other components of the plant biomass, which is used in the paper industry area, and is extremely relevant for biomass disruption.

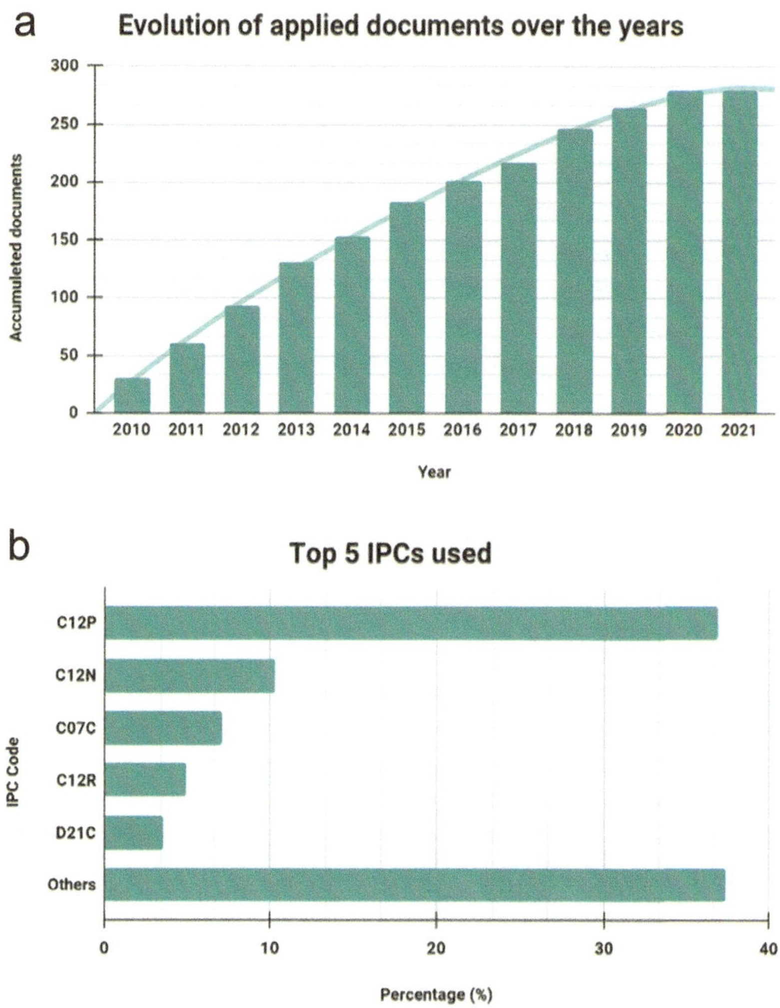

Fig. 20.1 Number of patents related to biomass pretreatment intended for ethanol production according to: (**a**) publication year (accumulated) and (**b**) attributed International Patent Classification code (IPC) code. Source: Derwent Innovations Index and Latipat databases (2021)

The profile of countries and regions of origin immediately reflects the world greatest producers of ethanol, with the USA, China, and Brazil ahead in the rank (Fig. 20.2a). The USA is the current first place in ethanol production among the countries, reaching more than 15 million gallons in 2019. There are around 200 ethanol-producing plants in the country, and the great majority of them uses corn as biomass (International BBI 2021). China, the second position, is recognized as an important producer of agricultural commodities, with a high score for ethanol production as well. Besides, the number of patent filings from China is very high for several areas beyond ethanol, and the investments in biomass pretreatment are

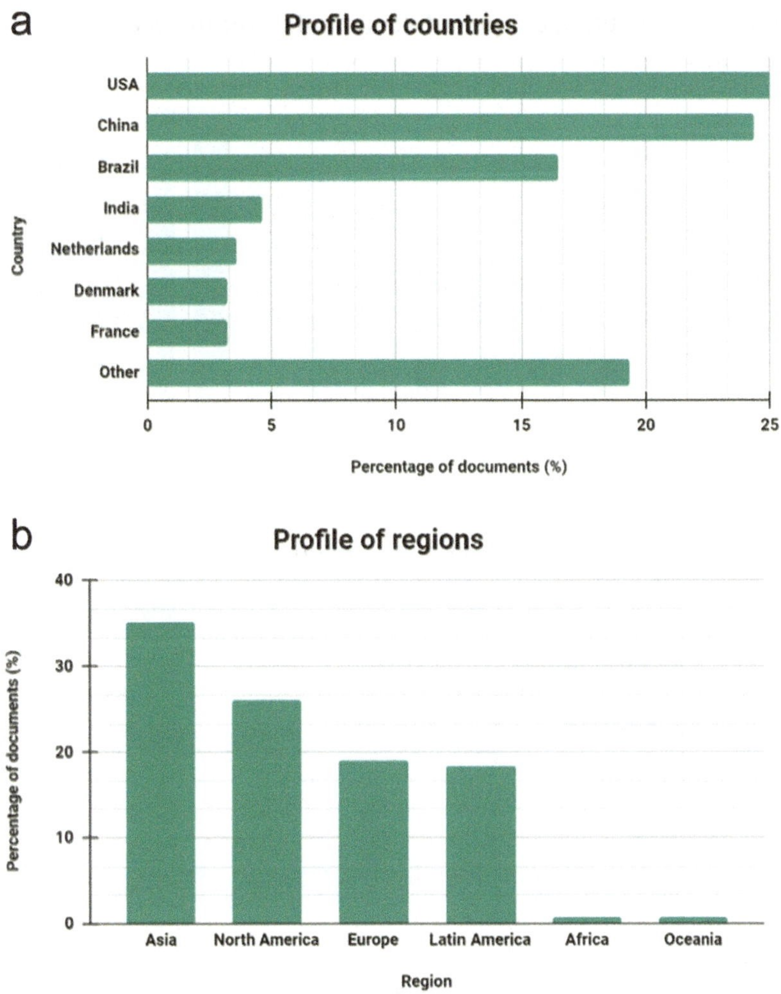

Fig. 20.2 Number of patents related to biomass pretreatment intended for ethanol production according to: (**a**) country of origin and (**b**) region of origin. Source: Derwent Innovations Index and Latipat databases (2021)

largely stimulated (2017). Brazil, located in third place, is the second worldwide producer of ethanol, mostly from sugarcane but also from corn in significant amount. Brazil was one of the first countries to stimulate ethanol production in the year of 1975, through the pioneer biofuel policy named *Pró-Álcool* (Alisson 2016). India is also a strong agro-industrial country, and is urging as a bioethanol producer as well (Renewable Fuels Association 2020). In the European countries, such as Netherlands, Denmark, and France, although the production of ethanol does not reach that of the USA or Brazil, there is a significant number of companies focused on the development of pretreatments, mainly enzymes. Much of the European substrates are starchy crops, such as wheat and maize, with the addition of sugar beet, a sucrose-rich material (EUBIA 2018). When the patents are distributed by region (Fig. 20.2b), it is possible to observe that Asia takes the first place. This is mainly caused by the fact that almost the only contributors of North America and Latin America are USA and Brazil respectively, while Asia counts with China, India, Japan, Indonesia, and South Korea. However, the distribution is even between these regions, together with Europe.

Another important aspect to evaluate is the distribution among the types of pretreatment (Fig. 20.3a). The enzymatic pretreatment is by far the most used strategy, around two times the second position. This strengthens the importance of this biological technique since it is applied generally after the chemical or physical breakage and right before the fermentation. Besides, the quantity of enzymes and the different effects they show in each lignocellulosic substrate was not so frequently tested as the solvents, acids, alkalis, and physical forces years before, presenting a potential to still grow in number of patents. Beyond the enzymatic hydrolysis, most of the significant strategies are from the chemical area (Fig. 20.3b), such as the use of acid, different chemicals, alkalis, organic solvents, and ionic liquids. The main type of treatment that includes the physical influence, classified as physicochemical, is the hydrothermal technique. This surely represents that the strategies are being directed to biological and chemical ways, possibly due to the increased interest in more sustainable and ecological methods nowadays.

20.3 Fermentation Processes

In order to assess the status of scientific and technological advances in fermentation processes for the production of ethanol, a search of patent documents was performed for the period 2010–2021. The search was carried out in the Derwent Innovations Index database using the following keywords in the topic [TS = ((ethanol OR bioethanol OR fuel alcohol) AND (pentose* OR xylose* OR arabinose* OR immobil* yeast* OR "simultaneous saccharification and fermentation" OR "consolidated bioprocessing") AND (sugarcane OR cane OR corn OR maize))]. A total of 592 patent documents were found. From these, 175 were selected, after a critical analysis of the title and summary to exclude documents that were not of interest to this topic. Among the patent documents, processes, utility models, and genetically modified microbial strains were found.

The main patent applicants in this area were China and the USA with 37% and 33% of the documents, respectively (Fig. 20.4a). One of the main technology holders was expected to be the USA, as the Country is the main ethanol producer in the world with approximately 53% of production (Renewable-Fuels Association 2021). Brazil, the second-largest ethanol producer, had only a 2% of participation in patent filings.

On the other hand, it was observed that the year in which most patent documents were published was 2010 with 44 patents (Fig. 20.4b). This could be explained by

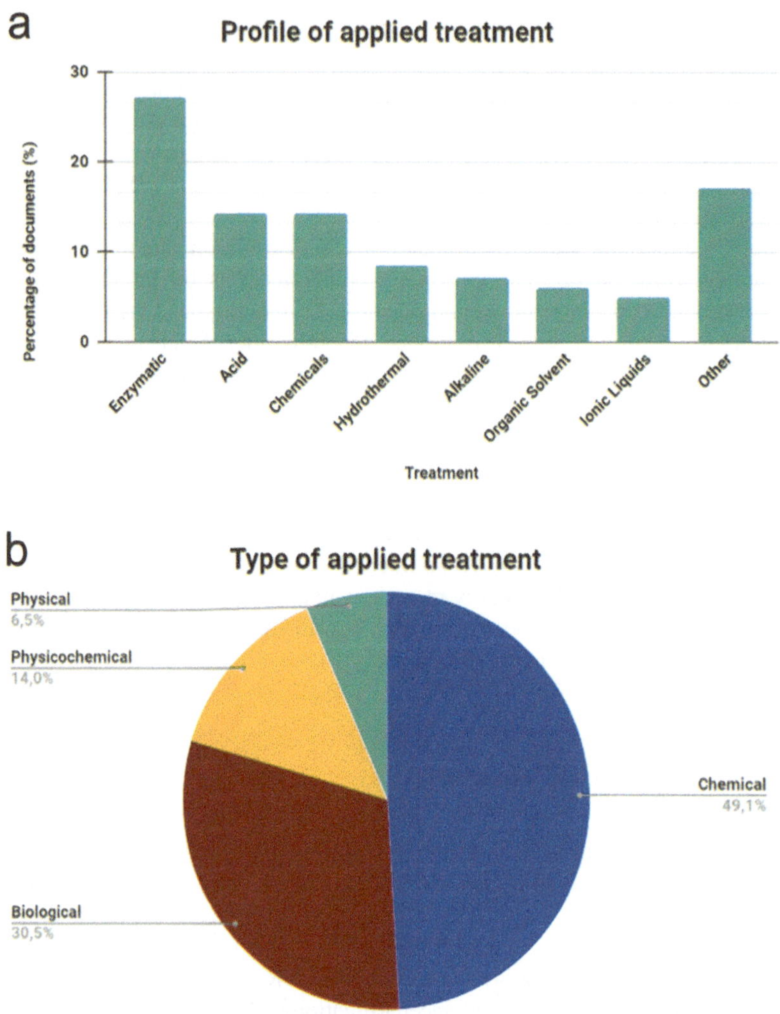

Fig. 20.3 Number of patents related to biomass pretreatment intended for ethanol production according to: (**a**) specific type of treatment and (**b**) percentage of general type of treatment. Source: Derwent Innovations Index and Latipat databases (2021)

energy crises and green movements in search of alternative fuels (Olivo et al. 2011). A similar situation was observed with the patent filings for the production of hydrogen from renewable sources (Olivo et al. 2011; Martinez-Burgos et al. 2021).

The top-two most recurrent IPC codes were C12P (Fermentation or enzyme-using processes to synthesize a desired chemical compound or composition or to separate optical isomers from a racemic mixture) and C12N (Microorganisms or enzymes; compositions thereof; propagating, preserving, or maintaining microorganisms;

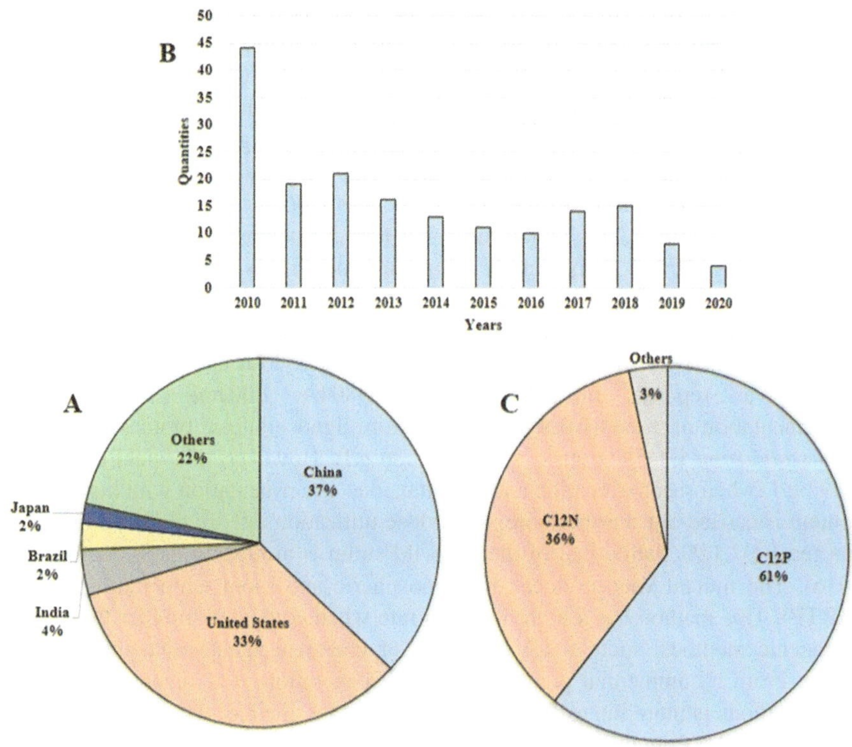

Fig. 20.4 (**a**) Distribution of patent documents on bioethanol production by countries of origin. (**b**) New patent documents on bioethanol production published per year. (**c**) Distribution of patent documents by International Patent Classification codes (IPC). Date of search: September 10th, 2021

mutation or genetic engineering; culture media) (World-Intellectual-Property-Organization 2021), with 60.5% and 36%, respectively (Fig. 20.4c).

The conversion of sugars from plant biomass into ethanol is complex and depends on the development of microorganisms with high fermentation capacity. Although *S. cerevisiae* is widely used to produce first-generation bioethanol, these yeasts generally have low fermentative efficiency to produce second-generation bioethanol. This is due to their inability to ferment pentoses, in addition to being inhibited by the compounds generated during the pre-treatment of lignocellulosic biomass and having low ethanol production at temperatures above 40 °C (Jansen et al. 2017). Fermentation at high temperatures is desirable for the production of bioethanol from lignocellulosic biomass, as this condition can reduce the consumption of water and energy used to cool the devices and thus reduce the production cost (Madeira-Jr and Gombert 2018). To solve these problems, some alternatives involving the genetic improvement of microorganisms are shown in Table 20.1.

The patent JP2013188156A suggests the use of the mutant yeast *Schizosaccharomyces japonicus* SS4-5. This strain is capable of fermenting hexoses from the saccharification of lignocellulosic biomass at 44 °C and in a range of pH 3 to 7. These fermentation parameters reduce the probability of contamination by bacteria and wild yeasts since the presence of these microorganisms during fermentation can cause significant losses. The high temperature also allows for simultaneous fermentation and distillation, reducing process time and increasing productivity. However, this mutant does not ferment pentoses, and it is necessary to carry out a second fermentation using microorganisms capable of producing ethanol from these sugars (Shindo 2012).

To solve this problem, an alternative was the development of the *Spathaspora passalidarum* U1-58 mutant capable of fermenting xylose and glucose simultaneously, as reported in patent CN105505804A. Ethanol production by co-fermentation of these hydrolyzed sugars from lignocellulosic biomass using the *S. passalidarum* U1-58 strain can reach 43.26 g/L. However, this yield can reach 49.92 g/L when saccharification and simultaneous co-fermentation with high solids content is carried out. Furthermore, the xylose utilization rate by the U1-58 mutant can reach 97.12%, while that of the the wild strain is around 57.37% (Yefu et al. 2016). The mutant was produced by atmospheric and room temperature plasma (ARTP). This method has a high mutation rate when compared to other traditional mutagenic methods, such as UV radiation or chemical agents (Ottenheim et al. 2018). From a commercial point of view, the generation of a mutant strain from these methods is more interesting, as they are not classified as genetically modified organisms (GMO). In this way, it is not necessary to pay attention to the legal norms (Twardowski and Małyska 2015). Although the generation of new microbial strains by random mutations has these advantages, these methods are often laborious and time-consuming (Yu et al. 2020).

Advances in "Omics" technologies, mainly Genomics, Transcriptomics, and Proteomics, opened new paths for the genetic improvement of microorganisms. For example, although *S. cerevisiae* does not metabolize pentoses like xylose, it can assimilate the xylulose isomer via the pentose phosphate pathway. Therefore, the patent WO2010070549A1 described the use of recombinant DNA technology to produce a strain of *S. cerevisiae* with the ability to convert xylose to xylulose. In fungi that metabolize xylose naturally, this process is carried out in two steps: (1) xylose is reduced to xylitol by the action of xylose reductase (EC 1.1.1.21), (2) xylitol is converted to xylulose by xylitol dehydrogenase (EC 1.1.1.9). In bacteria, this metabolism is simpler, as only the enzyme xylose isomerase (EC 5.3.1.5) is needed to convert xylose to xylulose. However, when this bacterial gene is expressed by *S. cerevisiae*, it usually results in an inactive enzyme and the exact reasons are not entirely clear. However, the patent document WO2010070549A1 presented a strain of *S. cerevisiae* that expresses the enzyme xylose isomerase derived from a species of the genus *Lactococcus* in its active form, allowing the yeast to produce ethanol from xylose (Rønnow et al. 2010).

The patent WO 2012067571A1 also disclosed a strain of *S. cerevisiae* genetically modified to ferment xylose. In this work, the authors opted for a different strategy.

Table 20.1 Patent documents describing the genetic improvement of microorganisms for bioethanol production, obtained in the Derwent Innovations Index database

Patent number (Publication)	Title	Technology	References
CN112375694-A	C6/C5 co-fermented *Saccharomyces cerevisiae* capable of relieving high xylose utilization and high robustness antagonism and application thereof	Development of a mutant *S. cerevisiae* strain with the ability to ferment pentoses and hexoses from rice straw hydrolysis	Xiaoming et al. (2020)
CN110591933-A	Engineering strain for efficiently producing ethanol and xylitol by fermenting xylose	*K. marxianus* with high fermentative capacity due to deletion of the "native" genes' xylose reductase and xylitol dehydrogenase and insertion of three exogenous genes: xylose reductase, galactose permease, and glucose/xylose symporter	Jia (2019)
CN105505804-A	Mutant strain capable of efficiently fermenting xylose and method of using mutant strain for fermentation to produce ethanol	*Spathaspora passalidarum* U1-58 mutant yeast with the ability to ferment glucose and xylose simultaneously	Yefu et al. (2016)
WO2015011572-A1	Genetically modified rumen microbes for production of alcohol and allied downstream products from lignocellulosic feedstock	Isolation of rumen bacteria from a group of ruminant animals; identify genes encoding cellulases and hemicellulases; cloning of these genes in the bacteria *Ruminoccocus albus*. To perform the hydrolysis of lignocellulosic biomass with these genetically modified microorganisms. The sugars obtained can be fermented into ethanol by commercial yeast	Mutalik (2013)
WO2014035458-A1	Expression of enzymes in yeast for lignocellulose derived oligomer CBP	Development of *S. cerevisiae* strains for the expression of 8 cellulolytic and hemicellulolytic enzymes for simultaneous hydrolysis and fermentation of lignocellulosic biomass	Mcbride et al. (2012)
WO2012067571-A1	New strains of *Saccharomyces cerevisiae*	*S. cerevisiae* with the ability to ferment xylose due to the insertion of three exogenous genes: xylose reductase, xylitol dehydrogenase, and xylulokinase	Albers et al. (2012)

(continued)

Table 20.1 (continued)

Patent number (Publication)	Title	Technology	References
JP2013188156-A	New yeast and method for producing ethanol using the same	A mutant strain of *S. cerevisiae* with the ability to ferment hexoses from saccharification of lignocellulosic biomass at 44 °C and pH 3–7	Shindo (2012)
WO2012138942-A1	Methods for the improvement of product yield and production in a microorganism through the addition of alternate electron acceptors	Development of new metabolic pathways to reduce or eliminate glycerol production and increase ethanol production	Argyros et al. (2011)
WO2012103385-A2	Biocatalysts synthesizing deregulated cellulases	A genetic modification was performed on the *Clostridium phytofermentans* bacteria that allows the microorganism to synthesize cellulases even in media with high glucose concentrations. The strains also showed a high ethanol production, with a theoretical yield of more than 90%	O'mullan et al. (2011)
WO2011011796-A2	Methods and compositions for improving sugar transport, mixed sugar fermentation, and production of biofuels	Development of a recombinant strain expressing a transmembrane protein responsible for sugar transport, increasing ethanol production	Glass et al. (2011)
WO2010070549-A1	Microorganism expressing xylose isomerase	*S. cerevisiae* that expresses the enzyme xylose isomerase in its active form, allowing the yeast to produce ethanol from xylose	Rønnow et al. (2010)
WO2012068310-A2	Compositions and methods for improved saccharification of genetically modified plant-derived biomass	Genetically modified microorganisms for heterologous expression of hydrolytic enzymes, improving the process of simultaneous saccharification and fermentation	Gray (2010)

Instead of expressing a single bacterial enzyme, they inserted the genes xylose reductase, xylitol dehydrogenase, and xylulokinase. One of the disadvantages of using this system is that high xylitol production can occur while reducing ethanol production, which is not desired.

However, the strain had a low xylitol production and the ethanol produced corresponded to 48% of the theoretical yield (Albers et al. 2012).

The patent CN110591933A presented a strain of *Kluyveromyces marxianus* with high efficiency of conversion of glucose and xylose obtained from a corn cob hydrolysate in ethanol and xylitol. Generally, wild *K. marxianus* yeasts have the ability to metabolize pentoses and usually show good fermentation performance at elevated temperatures. Therefore, the strategy used to improve these characteristics was the knockout of the "native" genes that encode the enzymes xylose reductase and xylitol dehydrogenase present in *K. marxianus* and the insertion of three exogenous genes: (1) xylose reductase from the fungus *Neurospora crassa*, (2) mutant gene N376F encoding a galactose permease and (3) glucose/xylose symporter (Jia 2019).

Other patent documents were mainly focused on immobilization systems for enzymes applied in the processes of saccharification of biomass or microorganisms used in fermentation, since these catalysts are very expensive and, therefore, their recovery and reuse is necessary (Almulaiky et al. 2021; Atiroğlu et al. 2021). Liuqin et al. (2018) in the patent document CN109988757A presented a semi-continuous method for the production of bioethanol using immobilized yeasts. The *S. cerevisiae* yeasts can be fixed in a spherical reticulated trellis of vegetable, animal, or synthetic fiber. According to the authors of the invention, the use of this type of inoculum can have greater efficiency and longer-lasting activity and shorter fermentation times between 6 and 8 h. In the patent document CN110117588A, Liuqin et al. (2018) presented a new fermentation bioreactor for the production of bioethanol using immobilized yeast. The bioreactor tank contains stratified grids that serve to fix the fibrous material where the yeasts are immobilized. The fibrous material is organized in the tank forming several layers, provided with liquid sprayers, which facilitate the contact between the yeasts and the must, which is recirculated. According to the inventors, the equipment is designed to improve fermentation efficiency in alcohol production processes. These innovations are mainly focused on having a fully active inocula and thus reducing the time of the adaptation phase in microbial growth.

Another way to optimize the production of bioethanol from lignocellulosic biomass is the simultaneous saccharification and fermentation of cellulose. Li et al. (2013), in the patent document US20140287473A1, presented a process where hydrolysis of cellulose in fermentable sugars takes place simultaneously to the transformation of sugars into ethanol. Initially, lignocellulosic biomass is mechanically and thermally pretreated. Subsequently, the fermentation broth is prepared using a filamentous fungus and yeasts or alcohol producing bacteria. The fungi secreted the enzymes that later carried out the saccharification of the polysaccharide and yeasts and bacteria fermented the sugars into ethanol. Duan et al. (2015) in the patent document WO2016044606A1 described a simultaneous saccharification and fermentation method to produce bioethanol from a granular starch paste in the presence of benzoate, this in concentrations between 0.05 and 0.3 g/L. The corn slurry is inoculated with *Aspergillus kawachi*, *Aspergillus niger*, and alcohol producing yeasts. These microorganisms produce alpha-amylase, glucoamylase, and ethanol, respectively. According to the authors, the presence of benzoate improves ethanol yield and fermentation time.

20.4 Innovations in Bioethanol Production

The processes of first-generation bioethanol production have been extensively applied in industrial scale for the last five decades and represent consolidated technologies. Cellulosic ethanol production requires biomass conditioning and pretreatment due to its complexity and its recalcitrant structure mainly composed by cellulose, hemicellulose, and lignin (Longati et al. 2018; Batista et al. 2019). Therefore, it demands a technological development that still limits the commercial scale production (Bomtempo and Soares 2016).

Despite the complexity, there are two Brazilian plants of second-generation ethanol: Raízen and Granbio, and both started operating in 2014. The Granbio plant (Bioflex® 1), located in the state of Alagoas, is capable of producing 60 million liters of cellulosic ethanol per year, using sugarcane residue, straw and bagasse as feedstock (GranBio 2021a). To produce this biofuel in an economically viable way, the platform uses a sugarcane biomass developed by the company (Cana-Vertix®), which was obtained from the crossing of ancestral and commercial hybrid species that generated a higher fiber content plant with longer lifespan, which can grow in degraded areas, allowing to explore and increase the productivity of such regions. Furthermore, the Cana-Vertix® needs less water and inputs to grow (Granbio 2021b). The integrated process for bioethanol production in Bioflex® 1 consists of two technologies: the GP+®, responsible for lignocellulose refining into fermentable sugars and co-generation of bioenergy; and the AVAP® technology, capable of producing pure lignocellulosic fermentable sugar and different biochemical products, such as butanol, butanediol, and lactic acid from different types of biomass, such as softwoods, corn residues, and tobacco stalks (Granbio 2019). The separation of cellulose, hemicellulose, and lignin is made by commodity SO_2 and ethanol, and chemicals are recycled in a close circuit (Granbio 2021c). For the pre-treatment, a hydrothermal-mechanical technology using steam (Proesa™ technology) is applied, developed by Beta Renewables (Granbio 2021d; Beta Renewables 2017; Neto et al. 2018), followed by enzymatic hydrolysis using a Novozymes blend and subsequential fermentation of the available sugars by a yeast strain developed by themselves, that can ferment C5 and C6 sugars (Beta Renewables 2017; Neto et al. 2018).

The Raízen plant started its production in November 2014, in São Paulo, and has the capacity to produce 40 million liters of ethanol per year. The plant is coupled to its first-generation structure and uses sugarcane bagasse left from the first-generation production as feedstock (Neto et al. 2018).

As summarized in Table 20.2, other cellulosic ethanol plants were implemented in a large-scale. Also in 2014, the Project LIBERTY plant, from POET-DSM Advanced Biofuels, started its production in Iowa (USA), with a capacity to produce 20 million gallons per year of ethanol from corn cobs, leaves, stalks, and husks (DSM 2014). The company developed its own enzyme mix that has a wide pH range and tolerates high temperatures for the hydrolysis step (DSM 2012, 2017). They also

Table 20.2 Commercial scale cellulosic ethanol producer companies and applied technologies

Company	Location	Feedstock	Gallons production per year ($\times 10^6$)	Feedstock pre-treatment	Hydrolysis	Fermentation	References
Granbio	Alagoas (Brazil)	Sugarcane residue, straw and bagasse	60	Proesa™ Steam explosion	Enzyme cocktail (Novozymes)	Celere-2L® recombinant yeast	GranBio (2021d), Beta Renewables (2017), Chandel et al. (2018), Neto et al. (2018), Gonzalez-Contreras et al. (2021)
Raízen	Piracicaba (Brazil)	Sugarcane bagasse	40	–	–	–	Neto et al. (2018)
Beta Renewables	Crescentino (Italy)	Wheat straw, rice straw and *Arundo donax*	50.7	Proesa™ Steam explosion	Novozymes blend	Modified yeast	ETIP Bioenergy (2021), Sanford et al. (2016), Gonzalez-Contreras et al. (2021)
POET-DSM	Iowa (USA)	Corn cobs, leaves, stalks and husks	20	Thermochemical (Steam explosion and two-stage dilute acid)	Enzyme blend	Modified yeast	DSM (2011, 2012, 2014, 2017)
Abengoa	Kansas (USA)	Stalks, leaves and cobs corn; wood waste, and non-food energy crops	25	Thermochemical (Steam explosion and single stage diluted acid)	Cellulases from genetically engineered microorganism	Modified yeast	Abengoa (2013), Chandel et al. (2018)
DuPont	Iosa (USA)	Corn leftover stalks and leaves	30	Thermochemical (Steam explosion and ammonium hydroxide-based)	–	Modified *Zymomonas mobilis* strain	AIChE ChEnected (2015), ETIP Bioenergy (2021), Lynd et al. (2017), Chandel et al. (2018)

Note: Beta Renewables and Abengoa are no longer operating

modified a yeast strain capable of converting pentoses (C5) e hexoses (C6) from lignocellulosic material (DSM 2011).

The DuPont 2G ethanol plant in Iowa produced 30 million gallons each year from corn leftover stalks and leaves. The process used modified *Zymomonas mobilis* strain to convert C5 and C6 sugar into ethanol (AIChE ChEnected 2015; ETIP Bioenergy 2021; Lynd et al. 2017). This plant is no longer operating, and this can be possibly attributed to the technoeconomic challenges associated to regulatory uncertainties.

Abengoa company also expanded its activities to the second-generation ethanol production field by constructing a 2G bioethanol plant in Kansas (USA). The biorefinery produced ethanol from agricultural wastes, such as stalks, leaves, corn cobs, wood waste, and non-food energy crops. The pre-treatment consisted in a thermochemical process, and hydrolysis and fermentation occured through the application of a genetically engineered microorganism, which produced cellulases, and a yeast (Abengoa 2013). Besides, the plant combined enzymatic hydrolysis and electric production (Abengoa 2013). This plant is no longer operating.

In Italy, Beta Renewables large-scale cellulosic ethanol plant used wheat straw, rice straw, and *Arundo donax* as feedstocks, and its production was based on the Proesa™ process, using a Novozymes blend in the hydrolysis step. The plant energy was generated using the lignin extracted from the biomass, and the excess electricity was sold to the local grid (ETIP Bioenergy 2021). This plant is no longer operating because of financial issues.

20.5 Conclusion

This chapter presented an overview on patent applications related to pretreatment, saccharification and fermentation processes for bioethanol production. In the field of pretreatment and saccharification, the number of published patent documents is still growing, and the leading countries are the USA, China, Brazil and India, which are worldwide major producers of ethanol. Among the technologies, chemical strategies are predominant (49%), and the enzymatic treatments are present in 26% of the evaluated documents. The use of green solvents appears in recent documents, mostly from 2015 on. Regarding fermentation processes, the main patent applicants were China and the USA with 37% and 33% of participation, respectively, and most of the documents were published in 2010. Among the technologies, special focus was given to the development of strains able to ferment pentoses and grow at high temperatures. The processes of first-generation bioethanol production have been extensively applied in industrial scale for the last five decades and represent consolidated technologies. Second-generation ethanol facilities, however, still face technological challenges that require new solutions.

References

Abengoa (2013) Annual report, industrial production abengoa abengoa in the USA

Adsul M, Sandhu SK, Singhania RR, Gupta RP (2020) Obtaining high titer of enzyme mixture comprising cellulases, hemicellulases and beta-glucosidases in reutilization of waste water, comprises e.g. preparing fermentation media, and inoculating media with *Penicillium funiculosum*. EP3725891(A1)

Aghbashlo M, Tabatabaei M, Karimi K (2016) Exergy-based sustainability assessment of ethanol production via *Mucor indicus* from fructose, glucose, sucrose, and molasses. Energy 98:240–252. https://doi.org/10.1016/j.energy.2016.01.029

Aghbashlo M, Tabatabaei M, Mohammadi P et al (2017) Neat diesel beats waste-oriented biodiesel from the exergoeconomic and exergoenvironmental point of views. Energy Convers Manag 148:1–15. https://doi.org/10.1016/j.enconman.2017.05.048

Aghbashlo M, Mandegari M, Tabatabaei M et al (2018) Exergy analysis of a lignocellulosic-based biorefinery annexed to a sugarcane mill for simultaneous lactic acid and electricity production. Energy 149:623–638

AIChE ChEnected (2015) Dupont's New Cellulosic Ethanol Plant Is Open for Business

Albers E, Olsson L, Koppram R (2012) New strains of *Saccharomyces cerevisiae*. WO2012067571-A1

Albuquerque ECMC, Silva JS, Barbosa KL, Rocha MSRS (2020) Produção de etanol de segunda geração a partir da bioconversão da folha do coqueiro como biomassa lignocelulósica. BR102019005092 (A2)

Alisson E (2016) Proálcool: uma das maiores realizações do Brasil baseadas em ciência e tecnologia | AGÊNCIA FAPESP. https://agencia.fapesp.br/proalcool-uma-das-maiores-realizacoes-do-brasil-baseadas-em-ciencia-e-tecnologia/24432/

Almulaiky YQ, Khalil NM, El-Shishtawy RM et al (2021) Hydroxyapatite-decorated ZrO2 for α-amylase immobilization: toward the enhancement of enzyme stability and reusability. Int J Biol Macromol 167:299–308. https://doi.org/10.1016/j.ijbiomac.2020.11.150

Alrik PL, Sagt CMJ, Schooneveld-Bergmans ME, Damveld RA (2020) New host cell useful in enzyme composition for saccharifying lignocellulosic material, comprises four different heterologous polynucleotides chosen from polynucleotides encoding celluloses, hemicelluloses and pectinases. US2021087544-A1

Angelis ARD, Castaldo F, Angelis ARD (2019) Extraction of bio-oil used for production of diesel fuel, involves preparing eutectic solvent, treating algal biomass with solvent, separating solid residue by filtration or centrifugation and drying. WO2020053118(A1)

Argyros A, Barrett T, Caiazza N et al (2011) Methods for the improvement of product yield and production in a microorganism through the addition of alternate electron acceptors. WO2012138942-A1

Atiroğlu V, Atiroğlu A, Özacar M (2021) Immobilization of α-amylase enzyme on a protein @metal–organic framework nanocomposite: A new strategy to develop the reusability and stability of the enzyme. Food Chem 349:129127. https://doi.org/10.1016/j.foodchem.2021.129127

Bali G, Meng X, Deneff JI et al (2015) The effect of alkaline pretreatment methods on cellulose structure and accessibility. ChemSusChem 8:275–279. https://doi.org/10.1002/CSSC.201402752

Batista G, Souza RBA, Pratto B et al (2019) Effect of severity factor on the hydrothermal pretreatment of sugarcane straw. Bioresour Technol 275:321–327. https://doi.org/10.1016/j.biortech.2018.12.073

Beta Renewables (2017) PROESA® TECHNOLOGY

Bomtempo JV, Soares G (2016) Opinion: why are the first commercial 2g ethanol plants almost experimental?

Bonfiglio F, Cagno M, Rey F et al (2019) Pretreatment of switchgrass by steam explosion in a semi-continuous pre-pilot reactor. Biomass Bioenergy 121:41–47. https://doi.org/10.1016/J.BIOMBIOE.2018.12.013

Chandel AK, Garlapati VK, Singh AK et al (2018) The path forward for lignocellulose biorefineries: bottlenecks, solutions, and perspective on commercialization. Bioresour Technol 264:370–381

Dai Y, van Spronsen J, Witkamp GJ et al (2013) Natural deep eutectic solvents as new potential media for green technology. Anal Chim Acta 766:61–68. https://doi.org/10.1016/J.ACA.2012.12.019

DSM (2011) DSM strengthens yeast technology leadership for 2G biofuels

DSM (2012) DSM enzymes for cellulosic ethanol qualified by DONG Energy

DSM (2014) First commercial-scale cellulosic ethanol plant in the U.S. opens for business

DSM (2017) POET-DSM plans on-site enzyme manufacturing facility at Project Liberty

Duan G, Gouthro M, Shetty JK et al (2015) Simultaneous saccharification and fermentation process in the presence of benzoate. WO2016044606A1

Espinoza-Acosta JL, Torres-Chávez PI, Carvajal-Millán E et al (2014) Ionic liquids and organic solvents for recovering lignin from lignocellulosic biomass. Bioresources 9:3660–3687. https://doi.org/10.15376/biores.9.2.3660-3687

ETIP Bioenergy (2021) Cellulosic Ethanol (CE)

EUBIA (2018) Bioethanol – European Biomass Industry Association

Feng G, Ying P, Fang Y et al (2018) Production of biobutanol and bioethanol involves using corn stalk as main raw material, treating straw, preparing fermentation medium, and producing alcohol. CN108588166-A

Geng A, Manglekar RR, Anli G (2020) *Talaromyces pinophilus* used for producing cellulolytic enzyme complex and in various other industries such as textiles, food stuffs, medicines and animal feeds, medium comprises inorganic salts and nutrients suitable for growth of strain. WO2021029828(A1)

Glass NL, Tian C, Beeson WT et al (2011) Methods and compositions for improving sugar transport, mixed sugar fermentation, and production of biofuels. WO2011011796-A2

Gonzales RR, Sivagurunathan P, Kim SH (2016) Effect of severity on dilute acid pretreatment of lignocellulosic biomass and the following hydrogen fermentation. Int J Hydrog Energy 41: 21678–21684. https://doi.org/10.1016/J.IJHYDENE.2016.06.198

Gonzalez-Contreras M, Lugo-Mendez H, Sales-Cruz M, Lopez-Arenas T (2021) Synthesis, design and evaluation of intensified lignocellulosic biorefineries – case study: ethanol production. Chem Eng Process Process Intensif 159:108220. https://doi.org/10.1016/j.cep.2020.108220

GranBio (2019) GranBio Acquires 100% Equity of the North American Biotechnology Firm American Process Inc. (API). http://www.granbio.com.br/press-releases/granbio-acquires-100-equity-of-the-north-american-biotechnology-firm-american-process-inc-api/

Granbio (2021a) Tecnologia de Biorefinaria GreenPower+®. http://www.granbio.com.br/conteudos/tecnologia-de-biorefinaria-greenpower/

Granbio (2021b) O que é Cana-Energia? http://www.granbio.com.br/conteudos/cana-energia/#:~:text=A%20cana%2Denergia%20%C3%A9%20uma,e%20gera%C3%A7%C3%A3o%20de%20energia%20renov%C3%A1vel

Granbio (2021c) AVAP® Biorefinery Technology. http://www.granbio.com.br/en/site/conteudos/avap-biorefinery-technology/#:~:text=The%20patented%20AVAP%C2%AE%20technology,also%20biofuels%2C%20including%20for%20aircraft

Granbio (2021d) PROESA. http://www.granbio.com.br/conteudos/proesa/

Gray K (2010) Compositions and methods for improved saccharification of genetically modified plant-derived biomass. WO2012068310-A2

Grohmann K, Torget RW (1992) Two-stage acid hydrolysis of xylan in biomass-first at low and then at high temp. with intermediate xylose recovery, provides high xylose yield for fermentation to ethanol. US5125977(A)

Grohmann K, Wyman CE, Hinman ND (1994) Increasing ethanol yield from corn fibre and related products-by treating with sulphuric acid and cellulase, to convert hemicellulose and cellulose to soluble sugars, then fermenting. WO1994008027(A1)

Hashemi SS, Karimi K, Mirmohamadsadeghi S (2019) Hydrothermal pretreatment of safflower straw to enhance biogas production. Energy 172:545–554. https://doi.org/10.1016/J.ENERGY.2019.01.149

Hosseinpour S, Aghbashlo M, Tabatabaei M, Mehrpooya M (2017) Estimation of biomass higher heating value (HHV) based on the proximate analysis by using iterative neural network-adapted partial least squares (INNPLS). Energy 138:473–479. https://doi.org/10.1016/j.energy.2017.07.075

Hou Q, Ju M, Zhen M et al (2019) Preparing 5-alkoxymethylfurfural from saccharide comprises e.g. using ionic liquid containing bromide ions or eutectic solvent containing bromide ions as reaction and organic solvent as extraction phase and converting saccharide material. CN109942519(A)

Hu Z, Wen Z (2008) Enhancing enzymatic digestibility of switchgrass by microwave-assisted alkali pretreatment. Biochem Eng J 38:369–378. https://doi.org/10.1016/J.BEJ.2007.08.001

Hu C, Shi Y, Sun R et al (2018) Preparing straw forming fuel comprises e.g. pulverizing and sieving straw, mixing sieved granules and water, performing steam explosion, collecting steam blasting materials, and mixing steam blasting materials and waste liquid. CN108998139(A)

Ingolfsson O, Matthiasson A, Ingolfsson S (2018) Waste conversion system for conversion of organic fractions of municipal and wider community waste to fuels for use in transportation, with all solid, has carbon dioxide inlet which is provided within lower portion of scrubbing unit. WO2019053750(A1)

International BBI (2021) Ethanol producer magazine – the latest news and data about ethanol production

Jacquet N, Maniet G, Vanderghem C et al (2015) Application of steam explosion as pretreatment on lignocellulosic material: a review. Ind Eng Chem Res 54:2593–2598. https://doi.org/10.1021/IE503151G

Jansen MLA, Bracher JM, Papapetridis I et al (2017) *Saccharomyces cerevisiae* strains for second-generation ethanol production: from academic exploration to industrial implementation. FEMS Yeast Res 17:1–20. https://doi.org/10.1093/femsyr/fox044

Jia Z (2019) Engineering strain for efficiently producing ethanol and xylitol by fermenting xylose. CN110591933-A

Karunanithy C, Muthukumarappan K (2010) Influence of extruder temperature and screw speed on pretreatment of corn stover while varying enzymes and their ratios. Appl Biochem Biotechnol 162:264–279. https://doi.org/10.1007/S12010-009-8757-Y

Kim JW, Kim JK (2019) Producing bio sugar used to produce bioethanol, comprises e.g. performing acid treatment of *Scenedesmus obliquus* using acid e.g. nitric acid, performing pressurized hot water extraction, and detoxifying by adding activated carbon. KR2096445-B1

Kim JS, Lee YY, Kim TH (2016) A review on alkaline pretreatment technology for bioconversion of lignocellulosic biomass. Bioresour Technol 199:42–48. https://doi.org/10.1016/J.BIORTECH.2015.08.085

Klaassen P, Gielesen BEM, Suylekom VGP, Heijne WHM (2011) New cell comprising xylose isomerase gene or xylose reductase and xylitol dehydrogenase and araA, araB and araD, genes, used for producing fermentation product from a sugar composition. WO2011131667(A1)

Kocher GS, Razdan N (2019) Producing bioethanol from damaged wheat by decontaminating damaged wheat samples, incubating samples with ethanol, milling wheat grains, solid state fermentation, enzymatic hydrolysis, producing reduced sugars and fermenting. IN201911040938-A

Kotia V, Noronha S, Patti A, Macfarlane D (2017) Biofuel production from biomass comprising e.g. cellulose, by synthesizing distillable green, pre-treating biomass, washing slurry/solids with methanol, separating solids and liquid fraction, and adding hydrolyzing enzymes and buffer. IN201721026359-A

Koupaie EH, Dahadha S, Lakeh AAB et al (2019) Enzymatic pretreatment of lignocellulosic biomass for enhanced biomethane production - a review. J Environ Manag 233:774–784. https://doi.org/10.1016/J.JENVMAN.2018.09.106

Koyama Y, Hirota K, Yamao S et al (2016) Manufacture of plant-derived biomass-derived product for producing bioethanol, involves treating plant-derived biomass, separating hemicellulose, mixing with solvent comprising organic solvent, or mixture of organic solvent and water. WO2017222084(A1)

Kraikul N, Techanan W, Viriya-Empikul N, Kraithong W (2019) Pretreatment of lignocellulosic biomass in production of e.g. fuel, involves contacting lignocellulosic biomass with alkaline solution, contacting with steam, and reducing pressure and temperature at preset temperature reduction rate. DE102017125090(A1)

Kumar AK, Parikh BS, Pravakar M (2016) Natural deep eutectic solvent mediated pretreatment of rice straw: bioanalytical characterization of lignin extract and enzymatic hydrolysis of pretreated biomass residue. Environ Sci Pollut Res 2310(23):9265–9275. https://doi.org/10.1007/S11356-015-4780-4

Li M, Mitchinson C, Steele L (2013) Methods and compositions for enhanced production of organic substances from fermenting microorganisms. US20140287473A1

Liuqin H, Qingguo L, Nan Z et al (2018) A kind of method that surface immobilized yeast semicontinuous fermentation prepares ethyl alcohol. CN109988757A

Longati AA, Lino ARA, Giordano RC et al (2018) Defining research and development process targets through retro-techno-economic analysis: The sugarcane biorefinery case. Bioresour Technol 263:1–9. https://doi.org/10.1016/j.biortech.2018.04.102

LTD SLYIPOC (2016) Preparation of fuel ethanol from kitchen garbage by crushing kitchen garbage, adding water, alkali treating, washing and filtering, obtaining alkali pretreated raw material, boiling under acidic condition and hydrolyzing. CN106011181(A,B)

Lu Y (2019) Preparation of high calorific value fuel used in biomass power plant, involves pretreating crop straw by weak acid gas explosion, separating cellulose, drying hemicellulose and lignin, and subjecting to semi-carbonization treatment. CN109704917(A)

Lynd LR, Liang X, Biddy MJ et al (2017) Cellulosic ethanol: status and innovation. Curr Opin Biotechnol 45:202–211. https://doi.org/10.1016/j.copbio.2017.03.008

Ma H, Liu WW, Chen X et al (2009) Enhanced enzymatic saccharification of rice straw by microwave pretreatment. Bioresour Technol 100:1279–1284. https://doi.org/10.1016/J.BIORTECH.2008.08.045

Machado A da S, Ferraz A (2017) Biological pretreatment of sugarcane bagasse with basidiomycetes producing varied patterns of biodegradation. Bioresour Technol 225:17–22. https://doi.org/10.1016/J.BIORTECH.2016.11.053

Madeira-Jr JV, Gombert AK (2018) Towards high-temperature fuel ethanol production using *Kluyveromyces marxianus*: on the search for plug-in strains for the Brazilian sugarcane-based biorefinery. Biomass Bioenergy 119:217–228. https://doi.org/10.1016/j.biombioe.2018.09.010

Mandegari M, Farzad S, Görgens JF (2018) A new insight into sugarcane biorefineries with fossil fuel co-combustion: techno-economic analysis and life cycle assessment. Energy Convers Manag 165:76–91. https://doi.org/10.1016/j.enconman.2018.03.057

Marrs G, Zamora-Cristales R, Sessions J (2016) Forest biomass feedstock cost sensitivity to grinding parameters for bio-jet fuel production. Renew Energy 99:1082–1091. https://doi.org/10.1016/J.RENENE.2016.07.071

Martinez-Burgos WJ, de Souza CE, Pedroni Medeiros AB et al (2021) Hydrogen: current advances and patented technologies of its renewable production. J Clean Prod 286:124970. https://doi.org/10.1016/j.jclepro.2020.124970

Mcbride JE, Wiswall E, Shikhare I et al (2012) Expression of enzymes in yeast for lignocellulose derived oligomer cbp. WO2014035458-A1

Medoff M, Masterman TC (2012) Making butanol involves irradiating lignocellulosic starting material in modified manufacturing facility to form low recalcitrance irradiated material; and converting the material to butanol by using lignocellulosic saccharification unit. US20130084613(A1)

Medoff M, Masterman T (2017) Making alcohol and carboxylic acid or its salt or ester by treating carbon-containing material with e.g. radiation, converting portion of treated material using microorganism, and capturing carbon dioxide generated and/or lignin liberated. AU2017202376(A1)

Medoff M, Masterman TC, Masterman T (2010) Preparing product e.g. fuel, ethanol involves utilizing existing making facility designed to produce lactose-based ethanol to produce product from non-lactose feedstock, while maintaining system to convert feedstock by using microorganism. US2010203607-A1; WO2010093835-A2; WO2010093835-A3

Meng B, Han S, Han M, Zheng W (2019) Fast pretreatment of biomass by microwave alkaline baking includes cutting biomass raw materials, crushing, spraying biomass with alkaline aqueous solution, standing and placing container in microwave environment for radiation treatment. CN110791540-A

Moniruzzaman M, Goto M (2019) Ionic liquid pretreatment of lignocellulosic biomass for enhanced enzymatic delignification. Adv Biochem Eng Biotechnol 168:61–77. https://doi.org/10.1007/10_2018_64

Mosier N, Wyman C, Dale B et al (2005) Features of promising technologies for pretreatment of lignocellulosic biomass. Bioresour Technol 96:673–686. https://doi.org/10.1016/J.BIORTECH.2004.06.025

Mutalik VS (2013) Genetically modified rumen microbes for production of alcohol and allied downstream products from lignocellulosic feedstock. WO2015011572-A1

Neto AC, Guimarães MJOC, Freire E (2018) Business models for commercial scale second-generation bioethanol production. J Clean Prod 184:168–178

O'Mullan P, Patel J, Schmalisch M et al (2011) Biocatalysts synthesizing deregulated cellulases. WO2012103385-A2

Olivo C, Lebedeva I, Chu C et al (2011) A patent analysis on advanced biohydrogen technology development and commercialisation : Scope and competitiveness. Int J Hydrog Energy 36: 14103–14110. https://doi.org/10.1016/j.ijhydene.2011.04.100

Ottenheim C, Nawrath M, Wu JC (2018) Microbial mutagenesis by atmospheric and room - temperature plasma (ARTP): the latest development. Bioresour Bioprocess. https://doi.org/10.1186/s40643-018-0200-1

Rastogi M, Shrivastava S (2017) Recent advances in second generation bioethanol production: an insight to pretreatment, saccharification and fermentation processes. Renew Sust Energ Rev 80: 330–340. https://doi.org/10.1016/j.rser.2017.05.225

Renewable Fuels Association (2020) Annual U.S. & World Fuel Ethanol Production

Renewable Fuels Association (2021) Annual Fuel Ethanol Production U.S. and World Ethanol Production

Rezania S, Oryani B, Cho J et al (2020) Different pretreatment technologies of lignocellulosic biomass for bioethanol production: an overview. Energy 199:117457. https://doi.org/10.1016/j.energy.2020.117457

Rodríguez Carpio R, de Carvalho MS, Elias AM et al (2021) Multi-objective optimization of a 1G-2G biorefinery: a tool towards economic and environmental viability. J Clean Prod 284: 125431. https://doi.org/10.1016/j.jclepro.2020.125431

Rønnow B, Andersen TH, Sibbesen O (2010) Microorganism expressing xylose isomerase. WO2010070549-A1

Saha BC, Iten LB, Cotta MA, Wu YV (2005) Dilute acid pretreatment, enzymatic saccharification, and fermentation of rice hulls to ethanol. Biotechnol Prog 21:816–822. https://doi.org/10.1021/BP049564N

Saha BC, Yoshida T, Cotta MA, Sonomoto K (2013) Hydrothermal pretreatment and enzymatic saccharification of corn stover for efficient ethanol production. Ind Crop Prod 44:367–372. https://doi.org/10.1016/J.INDCROP.2012.11.025

Saha BC, Qureshi N, Kennedy GJ, Cotta MA (2016) Biological pretreatment of corn stover with white-rot fungus for improved enzymatic hydrolysis. Int Biodeterior Biodegradation 109:29–35. https://doi.org/10.1016/J.IBIOD.2015.12.020

Saini JK, Saini R, Tewari L (2014) Lignocellulosic agriculture wastes as biomass feedstocks for second-generation bioethanol production: concepts and recent developments. 3 Biotech 5:337–353. https://doi.org/10.1007/S13205-014-0246-5

Sanford K, Chotani G, Danielson N, Zahn JA (2016) Scaling up of renewable chemicals. Curr Opin Biotechnol 38:112–122

Schall CA, Farahani SV (2015) Pretreating lignocellulosic biomass for conversion to sugar comprises contacting the biomass with an oxidizing agent to remove or decompose the lignin component, and contacting the produced lipoxygenase biomass with an ionic liquid. WO2015123627(A1)

Sharma HK, Xu C, Qin W (2017) Biological pretreatment of lignocellulosic biomass for biofuels and bioproducts: an overview. Waste Biomass Valori 102(10):235–251. https://doi.org/10.1007/S12649-017-0059-Y

Sharma S, Saini S, Jha S, Goyal P (2018) Producing bioethanol from lignocellulosic biomass waste, involves pulverizing biomass waste comprising cellulose, hemicellulose and/or lignin by sieving and drying, pretreating biomass waste with dilute acid, incinerating and hydrolyzing. IN201811040608-A

Shi J, Simmons BA, Singh S, Sun J (2015) Fermenting or saccharifying biomass involves contacting biomass containing polysaccharide and ionic liquid to form first solution, mixing first solution with carbon dioxide at lower pH, and adding obtained mixture with enzyme. WO2016105538-A1

Shindo A (2012) New yeast and method for producing ethanol using the same. JP2013188156-A

Shuai L, Luterbacher J (2016) Organic solvent effects in biomass conversion reactions. ChemSusChem 9:133–155. https://doi.org/10.1002/CSSC.201501148

Sun F, Chen H (2008) Organosolv pretreatment by crude glycerol from oleochemicals industry for enzymatic hydrolysis of wheat straw. Bioresour Technol 99:5474–5479. https://doi.org/10.1016/J.BIORTECH.2007.11.001

Tayyab M, Noman A, Islam W et al (2018) Bioethanol production from lignocellulosic biomass by environment-friendly pretreatment methods: A review. Appl EcolEnviron Res 16:225–249. https://doi.org/10.15666/aeer/1601_225249

Tsapekos P, Kougias PG, Egelund H et al (2017) Mechanical pretreatment at harvesting increases the bioenergy output from marginal land grasses. Renew Energy 111:914–921. https://doi.org/10.1016/J.RENENE.2017.04.061

Tudman S, Chesonis A (2018) Low-energy intensive method for producing cellulose from biomass by pretreating biomass with fibrillation, acid, elevated temperature and pressure through extruder, treating solids fraction to alkaline pH, and separating solubilized lignin. WO2019094444(A1)

Tulaphol S, Sathitsuksanoh N (2020) Production of levulinic acid for e.g. fuel, involves hydrolyzing cellulose present in cellulose-rich product to produce glucose-rich product, dehydrating glucose to produce 5-hydroxymethyl furfural, and hydrolyzing 5-hydroxymethyl furfural. WO2020214835-A1

Twardowski T, Małyska A (2015) Uninformed and disinformed society and the GMO market. Trends Biotechnol 33:1–3. https://doi.org/10.1016/j.tibtech.2014.11.006

Vandenberghe LP de S, Pandey A, Carvalho JC et al (2020) Solid-state fermentation technology and innovation for the production of agricultural and animal feed bioproducts. Syst Microbiol Biomanufact. https://doi.org/10.1007/s43393-020-00015-7

Wang W, Zhuang X, Wang Q et al (2019) Producing a zero-emission cellulose fuel ethanol involves collecting crop waste, crushing it to more than twenty mesh, washing and removing dust, separating it after solid-liquid separation, and drying it for use. CN109652466-A

Wang H, Sun R, Zhou J et al (2020) Pretreating reeds to obtain cellulose and modified lignin comprises e.g. adding reeds into water for hydrothermal pretreatment, carrying out solid-liquid separation, adding aqueous ammonia water and treating. CN112064394(A)

Winsness DJ, Cernohous JJ, Montgomery RW (2015) Processing biobased feedstock used in composition for forming article, involves extruding biobased feedstock to form extrudate, and separating portion of thermally labile component from extrudate to form treated biobased feedstock. WO2016054132(A1)

World-Intellectual-Property-Organization (2021) International patent classification

Wu S, Wu Y, Huang S (2016) Producing liquid fuels for sludge stage conversion, involves extracting light organic component in sludge directly as liquid fuel and adopting liquefaction method to heavy organic component in sludge. CN106477831(A)

Xavier FD, Conceição MM, Cavalcante GC, Silva RT (2017) Producing second generation ethanol and xylitol from sisal fiber and bagasse used in fuel industry, food industry, sweetening industry, and pharmaceutical industry, involves performing chemical pretreatment of biomass in reactor. BR102017027916-A2

Xiaoming B, Jianzhi Z, Fangqing W et al (2020) C6/C5 co-fermented *Saccharomyces cerevisiae* capable of relieving high xylose utilization and high robustness antagonism and application thereof. CN112375694-A

Yefu C, Dongguang X, Baozhong L, Xuewu G (2016) Mutant strain capable of efficiently fermenting xylose and method of using mutant strain for fermentation to produce ethanol. CN105505804-A

Yoon KP (2018) New fibrinolytic fungus i.e. *Penicillium* isolated from soil, useful for improving pretreatment and saccharification efficiency of biomass e.g. corn stover, corn cobs, sugarcane bagasse and rice straw for producing biofuels, or bioplastics. WO2019083244(A2,A3)

Yu Q, Li Y, Wu B (2020) Novel mutagenesis and screening technologies for food microorganisms: advances and prospects. Appl Microbiol Biotechnol 104:1517–1531

Zabed H, Sultana S, Sahu JN, Qi X (2018) An overview on the application of ligninolytic microorganisms and enzymes for pretreatment of lignocellulosic biomass. Recent Adv Biofuels Bioenergy Util 53–72. https://doi.org/10.1007/978-981-13-1307-3_3

Zhao C, Chu D, Xin Y (2019) Chemical conversion of corncob furfural residue to bioethanol comprises e.g. pulverizing, adding to hydrothermal reaction kettle, hydrolyzing, flowing into fixed bed reactor or slurry bed reactor, and adding supported metal catalyst. CN109704917(A)

Zheng J, Rehmann L (2014) Extrusion pretreatment of lignocellulosic biomass: a review. Int J Mol Sci 15:18967–18984. https://doi.org/10.3390/IJMS151018967

Correction to: How Would Solid Oxide Fuel Cells and Bioethanol Impact in Electric Mobility Transition?

Fábio Coutinho Antunes, Raissa Venâncio, Gustavo Doubek, and Hudson Zanin

Correction to: Chapter 17 in: C. R. Soccol et al. (eds.), *Liquid Biofuels: Bioethanol*, Biofuel and Biorefinery Technologies 12, https://doi.org/10.1007/978-3-031-01241-9_17

Owing to an oversight on the author's part, the original version of the book was inadvertently published with incorrect figure for Figure 17.5 which has been updated as shown below.

In addition, the following references have been added to the reference list and the corresponding citations have been included in the text:

Binelli ARR, Tasić MB, Filho RM (2016) Catalytic steam reforming of ethanol for hydrogen production: brief status. Chem Ind Chem Eng Q 22(4):327–332. https://doi.org/10.2298/CICEQ160216017B

Binelli ARR, Maciel Filho R, Jardini AL (2017) "Processo para Obtenção de Placas de Microcanais para Microreatores Químicos a Placas Assim Obtidas," BR10201203232

The chapter has been corrected and approved by the authors.

The updated version of this chapter can be found at
https://doi.org/10.1007/978-3-031-01241-9_17

C. R. Soccol et al. (eds.), *Liquid Biofuels: Bioethanol*, Biofuel and Biorefinery Technologies 12, https://doi.org/10.1007/978-3-031-01241-9_21

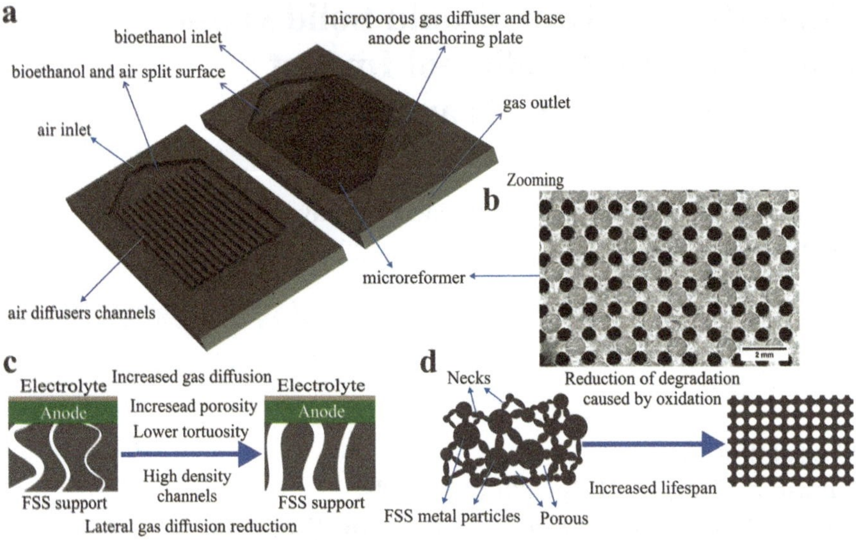

Fig. 17.5 (**a**) project developing for FSS micro-reformer support integrated to the ICs for cell anchoring, containing the organized microchannels produced by metal additive manufacturing; (**b**) micro-reformer volume element zooming (channels with ~200 μm of diameter); (**c**) engineering FSS support for enhancement performance obtained by Nielsen et al. (2018) and (**d**) the impacts that this proposal can bring prototyping micro-channels organized structure as micro-reformers. Adapted from Nielsen et al. (2018) and Zhou et al. (2021)

The manufacturer's authorised representative in the EU is Springer Nature Customer Service Centre GmbH, Europaplatz 3, 69115 Heidelberg, Germany. If you have any concerns regarding our products, please contact ProductSafety@springernature.com

Printed and bound by CPI Group (UK) Ltd, Croydon, CR0 4YY

28/07/2026

02179236-0001